Plant Cell Biology

Plant Cell Biology

Edited by **Agatha Wilson**

Callisto Reference
New York

Published by Callisto Reference,
106 Park Avenue, Suite 200,
New York, NY 10016, USA
www.callistoreference.com

Plant Cell Biology
Edited by Agatha Wilson

International Standard Book Number: 978-1-63239-679-2 (Hardback)

Printed in China.

Contents

Preface

Cell biology studies the structure and function of cells. It has many branches but plant cell biology is particularly important because plants are the only organisms that are capable of preparing their own food. Plant cells significantly differ from animal cells in a lot of aspects such as, they are larger in size; in addition to this plant cells contain chloroplast, cell walls, and vacuoles. Due to technological advancements in microscopy, the research in this field has been revolutionised. This book strives to provide a fair idea about this discipline and to help develop a better understanding of the latest advances in this field. While understanding the long-term perspectives of these topics, this book makes an effort in highlighting their impact as a modern tool for the growth of the field.

Various studies have approached the subject by analyzing it with a single perspective, but the present book provides diverse methodologies and techniques to address this field. This book contains theories and applications needed for understanding the subject from different perspectives. The aim is to keep the readers informed about the progress in the field; therefore, the contributions were carefully examined to compile novel researches by specialists from across the globe.

Indeed, the job of the editor is the most crucial and challenging in compiling all chapters into a single book. In the end, I would extend my sincere thanks to the chapter authors for their profound work. I am also thankful for the support provided by my family and colleagues during the compilation of this book.

Editor

1

Protein import into isolated pea root leucoplasts

*Chiung-Chih Chu and Hsou-min Li**

Institute of Molecular Biology, Academia Sinica, Taipei, Taiwan

Leucoplasts are important organelles for the synthesis and storage of starch, lipids and proteins. However, molecular mechanism of protein import into leucoplasts and how it differs from that of import into chloroplasts remain unknown. We used pea seedlings for both chloroplast and leucoplast isolations to compare within the same species. We further optimized the isolation and import conditions to improve import efficiency and to permit a quantitative comparison between the two plastid types. The authenticity of the import was verified using a mitochondrial precursor protein. Our results show that, when normalized to Toc75, most translocon proteins are less abundant in leucoplasts than in chloroplasts. A precursor shown to prefer the receptor Toc132 indeed had relatively more similar import efficiencies between chloroplasts and leucoplasts compared to precursors that prefer Toc159. Furthermore we found two precursors that exhibited very high import efficiency into leucoplasts. Their transit peptides may be candidates for delivering transgenic proteins into leucoplasts and for analyzing motifs important for leucoplast import.

Keywords: leucoplasts, plastid, root, protein import, translocon

Edited by:
Bo Liu,
University of California, Davis, USA

Reviewed by:
Kentaro Inoue,
University of California, Davis, USA
Danny Schnell,
University of Massachusetts, USA

****Correspondence:***
Hsou-min Li,
Institute of Molecular Biology,
Academia Sinica, Nankang,
Taipei 11529, Taiwan
mbhmli@gate.sinica.edu.tw

Introduction

Plastids are essential plant organelles responsible for functions ranging from photosynthesis and biosynthesis of amino acids and fatty acids to assimilation of nitrogen and sulfur (Leister, 2003; Sakamoto et al., 2008). All plastids are derived from embryonic proplastids and differentiate into different functional types in different tissues, for example chloroplasts in green tissues for photosynthesis, chromoplasts in petals and fruits for carotenoid pigment accumulation, and leucoplasts in non-green tissues for nutrient storage (Wise, 2006). To perform these specific functions, different plastid types require different proteins (Kleffmann et al., 2007; Brautigam and Weber, 2009; Barsan et al., 2012).

Although plastids have their own genome, most plastid proteins are encoded by the nuclear genome and synthesized in the cytosol as a larger precursor with an N-terminal extension called the transit peptide. Almost all of our current understanding of the molecular mechanism of plastid protein import is derived from studies with chloroplasts. Import across the chloroplast envelope membranes is mediated by the TOC and TIC (translocon at the outer/inner envelope membrane of chloroplasts) machinery. More than fifteen TOC and TIC components have been identified. Among them, Toc159 and Toc34 are receptors mediating the initial binding of precursor proteins to chloroplasts. Toc75 functions as the channel for protein translocation across the outer membrane. Major components of the TIC machinery includes Tic110 and the 1-MDa channel complex containing Tic20, Tic56, Tic100, and Tic214. Tic110 functions as the stromal site receptor for transit peptides and to act as a scaffold for tethering other translocon components located in the stroma. Hsp93, Hsp90C, and cpHsc70 are three chaperone motors important for the translocation of proteins into the stroma. Tic40 is a co-chaperone coordinating the actions of Tic110 and Hsp93 (for reviews see Li and Chiu, 2010; Shi and Theg, 2013; Paila et al., 2015).

The *Arabidopsis* Toc159 proteins are encoded by a four-gene family: Toc159, Toc132, Toc120, and Toc90. Toc159 and Toc132 are the major isoforms, with Toc90 and Toc120 as their functional redundant homologs, respectively. Toc159 preferentially binds transit peptides of photosynthetic proteins while Toc132 seems to prefer transit peptides of housekeeping proteins. Therefore, even though their substrate preference is not absolute, it is generally thought that Toc159 is the major receptor for photosynthetic proteins while Toc132 is important in mediating the import of non-photosynthetic housekeeping proteins into chloroplasts (Bauer et al., 2002; Ivanova et al., 2004; Kubis et al., 2004; Smith et al., 2004; Inoue et al., 2010; Bischof et al., 2011).

Leucoplasts are colorless, non-pigment-containing plastids. They are usually found in storage tissues and include the starch-storing amyloplasts, oil- and lipid-storing elaioplasts and protein-storing proteinoplasts (Wise, 2006). Most grain- and root-type food crops use leucoplasts to synthesize and store the nutrients that are used to feed the majority of the world population. However, despite the economic importance of leucoplasts, little is known about how proteins are imported into leucoplasts. It has been speculated that leucoplasts may use a similar basic import mechanism to that of chloroplasts (Jarvis and Lopez-Juez, 2013) but leucoplasts clearly have a preference for importing certain non-photosynthetic proteins (Wan et al., 1996; Yan et al., 2006; Primavesi et al., 2008). For example, only the transit peptide of the non-photosynthetic ferredoxin III and FtsZ precursors, but not that of the small subunit of RuBP carboxylase precursor (prRBCS), directs the import of GFP into leucoplasts in endosperms of transgenic wheat (Primavesi et al., 2008). Using leucoplasts isolated from caster seeds and chloroplasts isolated from pea leaves and the ratio of precursor to mature proteins after import as an assessment of import efficiency, it has been shown that prRBCS was imported preferentially into chloroplasts, while precursors to RuBP carboxylase activase (prRCA) and ferredoxin-NADP$^+$-oxidoreductase (prFNR) were imported equally well into both plastids (Wan et al., 1996). Using leucoplasts and chloroplasts isolated from pea roots and leaves, respectively, it has been shown that prRBCS could not be imported into leucoplasts at all while two other non-photosynthetic precursors could (Yan et al., 2006). The plastid selectivity is determined by the transit peptide of each precursor as swapping of transit peptides resulted in interchanged plastid preference (Wan et al., 1995, 1996; Yan et al., 2006). It has been suggested that the different substrate preferences of Toc159 and Toc132 may also account for the import preferences of the two plastids (Yan et al., 2006; Li and Teng, 2013). Indeed, in *Arabidopsis*, Toc159 is about 15-fold higher in leaves than in roots while Toc132 is only about fivefold higher in leaves (Ivanova et al., 2004). Furthermore, the *toc159* knockout mutant has severe import defects in mesophyll-cell chloroplasts (Bauer et al., 2000) but imports proteins normally into root plastids (Yu and Li, 2001). On the other hand, *toc132 toc120* double mutations cause structural abnormalities in root plastids (Kubis et al., 2004), and the *toc132* mutant is a root gravitropism mutant (Stanga et al., 2009), supporting that Toc159 is more important for leaf chloroplasts and Toc132 is more important for root leucoplasts.

As a first step in characterizing the molecular mechanism of protein import into leucoplasts, we optimized the *in vitro* protein import system of isolated leucoplasts to improve import efficiency and to permit a more quantitative comparison between leucoplasts and chloroplasts. We found that leucoplasts have a different stoichiometry of translocon components compared to chloroplasts. Precursors we tested fell into at least three different degrees of preference toward chloroplasts in our *in vitro* import system.

Materials and Methods

Plastid Isolation and Plastid Number Counting

Leucoplasts and chloroplasts were prepared from pea seedlings (*Pisum sativum* cv. Green Arrow, De Bruyn Seed Co., Zeeland, MI, USA) grown at 20°C on vermiculite for 4–5 days in the dark, and for 7 days under a 12-h photoperiod with a light intensity of approximately 150 μmol m^{-2} s^{-1}, respectively. Leucoplasts were prepared from pea roots as described (Bowsher et al., 1989) with modifications described below. Approximately 20~30 g of roots were collected and washed twice with 100 mL homogenization buffer (50 mM Tricine-KOH, pH 7.9, 330 mM sorbitol, 1 mM $MgCl_2$, 2 mM EDTA) in the presence of 1% BSA and reducing agents (2 mM ascorbic acid, 0.1 mM DTT, and 1.2 mM glutathione). The roots were cut into small pieces in another 100 mL of the same buffer, and were homogenized with a domestic blender by two 5-s pulses. The homogenate was filtered through four layers of Miracloth and centrifuged at 4,000 *g* for 3 min at 4°C. The leucoplast pellet was resuspended with 10 mL of homogenization buffer, underlaid with 10 mL 10% (v/v) Percoll™ in TS buffer (50 mM Tricine-KOH, pH 7.9, 330 mM sorbitol), and centrifuged with a swing-bucket rotor at 4,000 *g* for 5 min at 4°C. The pellet was first washed with 30 mL TS buffer and again with 30 mL of import buffer (50 mM HEPES-KOH, pH 8.0, 330 mM sorbitol), by centrifugation at 4,000 *g* at 4°C for 3 min. The final pellet was gently resuspended with 300 μL of import buffer.

Chloroplast isolation from pea leaves was conducted as described (Perry et al., 1991) except 2 mM ascorbic acid, 0.1 mM DTT, and 1.2 mM glutathione were added to the grinding buffer used for homogenization. Isolated chloroplasts were adjusted to 1 mg chlorophyll mL^{-1} in import buffer.

To count the plastid number and estimate the plastid size, fractions of isolated leucoplast and chloroplast suspensions were counted in the Multisizer 3 Coulter Counter (Beckman Coulter) using a 30 μm-aperture tube. The same plastid preparations were subjected to BCA assays (Thermo) to determine the protein concentration. It is noted here that the sensitivity of the BCA assay is protein dependent (Smith et al., 1985) and since leucoplasts and chloroplasts have different protein compositions, which may affect the results of the determination. However, the assay should still be able to provide an estimation of the amount of proteins in the two plastids.

Plasmid Construction and Translation of Precursors

The coding region of soybean mitochondrial alternative oxidase precursor (prAOX, accession number X68702) was amplified from a soybean leaf cDNA pool by the forward primer AOX-F1-*Xho*I (5′-gactcgagatgatgatgatgatgagccgc-3′) and the reverse primer AOX-R1-*Sal*I (5′-gcgtcgacttagtgataaccaattggagcagc-3′). The PCR products were digested with *Xho*I and *Sal*I and cloned into the *Xho*I/*Sal*I site of pSP72. The sequence of prAOX was confirmed by sequencing and the plasmid was named pSP72-prAOX. The expression of prAOX was under the control of the SP6 promoter. Plasmids encoding prRBCS, pea prTic40, prFd-protA$_{His}$ (Smith et al., 2004), prPDH E1α (At1g01090), and prCpn10-2 (At3g60210) have been described (Teng et al., 2012). [^{35}S]Met-labeled prPDH E1α was generated by *in vitro* transcription for synthesizing RNA followed by *in vitro* translation using the rabbit reticulocyte lysate system (Promega). All other precursors were synthesized using the TNT Coupled Wheat Germ Extract or Reticulocyte Lysate system (Promega).

Protein Import and Post-Import Analyses

To compare import efficiency on an equal protein basis, 18 μL [^{35}S]Met-labeled precursors were incubated with isolated plastids equivalent to 500 μg proteins in the presence of 3 mM ATP in import buffer in a final volume of 200 μL. To compare import efficiency on an equal plastid number basis, 113.67 μg leucoplast proteins and 500 μg chloroplast proteins were used instead. The import reactions were carried out at room temperature for 25 min and stopped by transferring to a new tube containing 1 mL cold import buffer. The plastids were pelleted at 3,000 g at 4°C for 3 min and resuspended in 200 μL import buffer. The leucoplast suspensions were underlaid with 1 mL 10% Percoll™ (v/v) in import buffer and the chloroplast suspensions were laid on top of a 40% Percoll™ (v/v) cushion and centrifuged in a swinging-bucket rotor at 2,900 *g* and 4°C for 6 min to isolate intact plastids. The plastids were washed once with import buffer. Thermolysin treatments of *in vitro* translation products and leucoplasts after import were performed as described (Perry et al., 1991), and intact leucoplasts were re-isolated as described above. Protein concentrations of the plastid samples were measured with the BCA kit (Thermo). Samples were analyzed by SDS-PAGE. Quantification of gel bands was performed using the Fuji FLA5000 PhosphorImager (Fujifilm).

Isolation of cDNA Encoding Partial Pea Toc132AG and Generation of Antibodies against Pea Toc132AG and OEP24

Because the cDNA sequence of pea *Toc132* was not available in the NCBI database, the *Toc132* cDNA sequences from two closely related species, *Medicago truncatula* and *Glycine max*, were aligned and two highly conserved regions in the acidic domain (A domain) and GTPase domain (G domain) were chosen to design primers (Supplementary Figure S1A) for cloning. First-stand cDNA was synthesized from

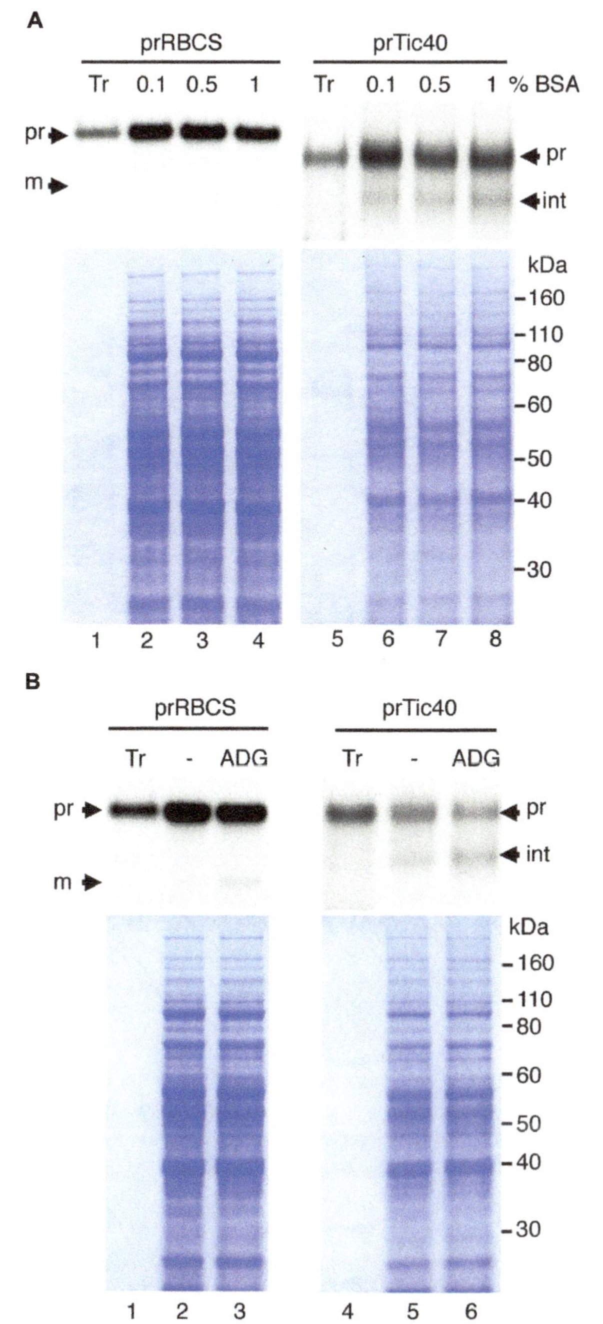

FIGURE 1 | Increasing BSA concentration and adding reducing agents to the homogenization buffer increases the import of prRBCS and prTic40 into leucoplasts. (A) Roots of 4-days-old pea seedlings grown in the dark were harvested and homogenized in homogenization buffer (see Materials and Methods) containing 0.1, 0.5, or 1% BSA. Isolated leucoplasts (500 μg protein) were then incubated with *in vitro*-translated [^{35}S]Met-prRBCS or [^{35}S]Met-prTic40 under import conditions for 25 min. After import, intact leucoplasts were re-isolated and analyzed by SDS-PAGE. The gels were stained with Coomassie blue and dried for fluorography. Twenty micrograms of proteins were loaded. Tr, 1% equivalent of the *in vitro*-translated proteins used in each import reaction. **(B)** Same as **(A)** except the homogenization buffer was supplemented with 1% BSA, and with or without 2 mM ascorbic acid, 0.1 mM DTT, and 1.2 mM glutathione (ADG). pr, precursor form; int, imported intermediate form; m, imported mature form.

TABLE 1 | The sizes and protein contents of the isolated plastids.

Plastid type	Plastid size (μm)	Protein content (pg/plastid)
Chloroplasts (7-days-old light-grown leaves)	3.24 ± 0.23	8.13 ± 1.42
Leucoplasts (4-days-old dark-grown roots)	1.81 ± 0.21	1.85 ± 0.63

total pea-root RNA and used as templates to amplify the partial coding region of pea Toc132 A and G domains with the degenerate forward primer psToc132A-*Nde*I-F2 (5′-gggcatatggawsttggagatgacaagatagagg-3′) and the reverse primer psToc132G-*Xho*I-R1 (5′-gggctcgagtgcacttgcacgatcaaagctaaa-3′). The sequence without the primer regions (Supplementary Figure S1B) has been submitted to the GenBank and received the accession number KT033462. The PCR products were cloned into the pCR™-Blunt II-TOPO® (Invitrogen) to generate pTOPO-psToc132AG. After confirming the sequence, the DNA fragment of psToc132AG was excised by *Nde*I and *Xho*I from pTOPO-psToc132AG. Because the psToc132AG cDNA fragment has an endogenous *Xho*I site, only the longer psToc132 DNA fragment, in which *Xho*I cut at the *Xho*I site on the vector backbone, was subcloned into the *Nde*I and *Xho*I site of pET22b (Invitrogen) to generate the plasmid pET22b-psToc132AG-His_6. The sequence was confirmed again and the pET22b-psToc132AG-His_6 plasmid was transformed into the *Escherichia coli* strain BL21 (DE3) for protein induction. Protein expression of psToc132AG-His_6 was induced by 1 mM IPTG at 37°C for 3 h, purified by TALON resins (CLONTECH Laboratories) with 50 mM imidazole, dialyzed with 50 mM Tris-HCl, pH 8.0, concentrated by Amicon Ultra-15 (Millipore) and used to raise the anti-Toc132 antibodies used in this study. During preparing the manuscript, another clone encoding a partial pea Toc132AG domains has been reported (Chang et al., 2014) and the amino acid sequence is different from the psToc132AG clone we obtained. Full-length pea OEP24 cDNA (EMBL accession number AJ001009) was amplified from a first-strand cDNA library of pea leaves and cloned into the *Nde*I/*Xho*I site of pET22b. OEP24 protein was then expressed and purified from *E. coli* and used for antibody production. Antibodies against other TOC and TIC proteins were prepared as described (Chou et al., 2003; Tu et al., 2004). Antibodies against cpHsc70 and Lhcb1 were purchased from Agrisera (AS08 348 and AS01 004, respectively), and the antibody against *Arabidopsis* plastid PGI was a gift of Dr. Jychian Chen.

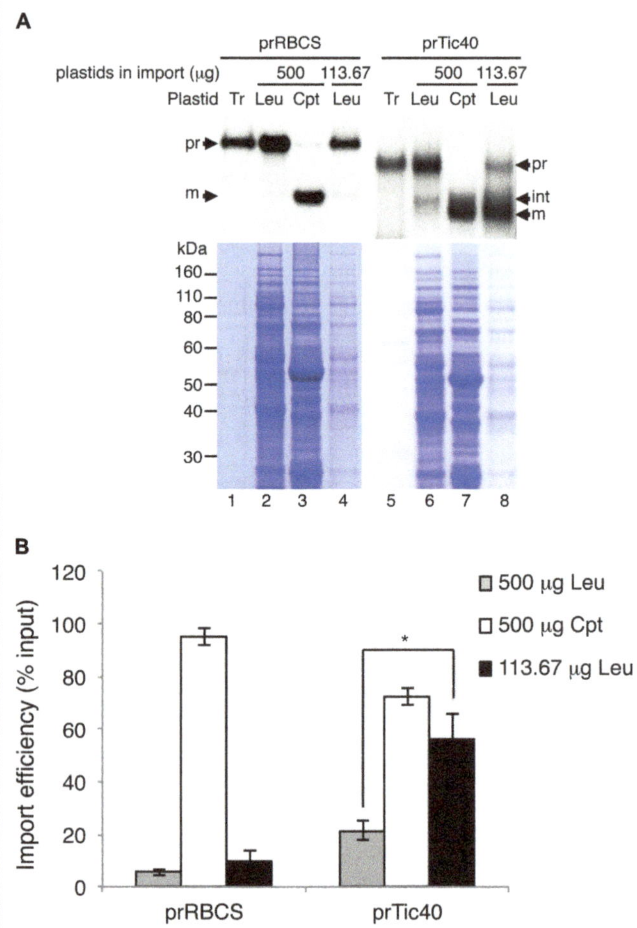

FIGURE 2 | Reducing the leucoplast concentration in import reactions increases import efficiency. (A) Isolated leucoplasts (500 or 113.67 μg of proteins) and chloroplasts (500 μg of proteins) were incubated under import conditions with 18 μL *in vitro*-translated [^{35}S]Met-prRBCS or [^{35}S]Met-prTic40 and 3 mM ATP in import buffer in a final volume of 200 μL for 25 min. After import, intact leucoplasts and chloroplasts were re-isolated and analyzed by SDS-PAGE. The gels were stained with Coomassie blue and dried for fluorography. 4% of the plastids in each import reaction were loaded. Tr, 1% equivalent of the *in vitro*-translated proteins used in each import reaction. Cpt, chloroplasts; Leu, leucoplasts; pr, precursor form; int, imported intermediate form; m, imported mature form. **(B)** Imported proteins in experiments as shown in **(A)** were quantified and the import efficiencies were calculated. Import efficiency was defined as % of [^{35}S]-labeled precursor proteins used that were found as imported mature or intermediate proteins in re-isolated plastids. The values have been corrected for the difference in number of methionine residues among the various forms. Data shown are mean ± SD of three independent experiments, $^*p < 0.05$ (Student's *t*-test). The *p*-value of 500 μg Cpt versus 113.67 μg Leu of prTic40 import is 0.095.

Results

Optimization of the Protocol for Leucoplast Isolation

Several reports have described protein import into isolated leucoplasts (Wan et al., 1996; Yan et al., 2006). We first followed an established protocol to isolate leucoplasts from roots of pea seedlings grown for 4–5 days in the dark (Bowsher et al., 1989). Leucoplasts isolated from pea roots have been studied extensively. They are on average 2.5 μm in diameter and contain starch grains, plastoglobuli, and also a few lamellae-type structures (Emes and England, 1986; Bowsher et al., 1989). When we tested the import of several precursor proteins from our collection, we found only prTic40 was imported and only to a small degree (data not shown). We therefore sought to improve the import efficiency of isolated leucoplasts. Because EDTA was reported to have a protective effect on recovering intact chloroplasts (Somerville

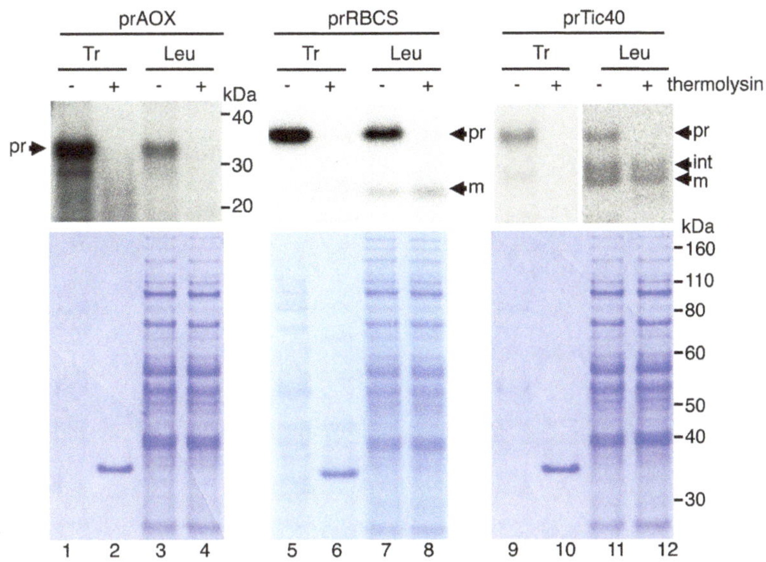

FIGURE 3 | The mitochondrial precursor protein prAOX was not imported into isolated leucoplasts. Isolated leucoplasts (113.67 μg) were incubated with *in vitro*-translated [^{35}S]Met-prAOX, [^{35}S]Met-prRBCS, or [^{35}S]Met-prTic40 under import conditions for 25 min. After import, the leucoplasts were treated with 0.2 mg/mL thermolysin for 30 min on ice, and intact leucoplasts were re-isolated and analyzed by SDS-PAGE. 4% of the plastids in each import reaction were loaded. The gels were stained with Coomassie blue and dried for fluorography. Tr, 1% equivalent of the *in vitro*-translated proteins used in each import reaction. Leu, leucoplasts; pr, precursor form; int, imported intermediate form; m, imported mature form.

and Ogren, 1981) and BSA may protect plastids during isolation by limiting the activity of proteases (Seigneurin-Berny et al., 2008), we increased the concentrations of EDTA and BSA in the homogenization buffer. As shown in **Figure 1A**, the amount of processed Tic40 slightly increased and precursors associated with the leucoplasts slightly decreased with higher concentrations of BSA (**Figure 1A**, compare lanes 6–8). Precursors of Tic40 are processed twice upon import into chloroplasts, generating an intermediate and a mature form (Li and Schnell, 2006; Tripp et al., 2007). The leucoplast imported prTic40 we observed was mostly the intermediate form (**Figure 1A**, int). We also noticed that poor import occasionally occurred concomitant with dark brown leucoplast pellets. We surmised that some oxidative stress might have occurred during the isolation. It has been shown that pea roots contain a higher phenol oxidase activity than do pea leaves (Henry et al., 1979) and it is also possible that isolated leucoplasts are more sensitive to oxidative damages than do isolated chloroplasts. We therefore added reducing agents (2 mM ascorbic acid, 0.1 mM DTT, and 1.2 mM glutathione, designated as ADG in **Figure 1B**) to the homogenization buffer (Schulz et al., 2004; Aronsson and Jarvis, 2011). Indeed after adding the reducing agents, we no longer observed dark-brown leucoplast pellets and the import efficiency of prTic40 further increased (**Figure 1B**, lanes 5–6). Under the modified conditions, a very small amount of import could even be detected for prRBCS (**Figure 1B**, lane 3).

Reducing the Number of Leucoplasts in Import Reactions Increases Import Efficiency

After optimizing the leucoplast isolation conditions, we next tried to compare the import behaviors of various precursors into leucoplasts and chloroplasts. An equal amount of total proteins (Yan et al., 2006) or an equal number of plastids (Wan et al., 1996) was used as the basis when comparing import efficiencies of different plastids. To determine the number of plastids, the isolated leucoplasts and chloroplasts were counted using the Multisizer 3 Coulter Counter (Beckman Coulter). We used chloroplasts isolated from 7-days-old seedlings for robust import and to minimize the age difference between the leucoplast and chloroplast samples. From three independent preparations of leucoplasts and chloroplasts, the average size of the isolated leucoplasts and chloroplasts was estimated to be 1.81 ± 0.21 μm and 3.24 ± 0.23 μm, respectively (**Table 1**). The same plastid preparations were then used for protein concentration determination and the average protein content per plastid was calculated to be 1.85 ± 0.63 and 8.13 ± 1.42 pg/plastid for leucoplasts and chloroplasts, respectively (**Table 1**). We then used protein content as an estimate of plastid numbers throughout our analyses. For example, 500 μg of plastid proteins would represent approximately 2.71×10^8 leucoplasts and 6.15×10^7 chloroplasts. For import into chloroplasts, we used 500 μg of chloroplasts (~6.15×10^7 plastids) in a 200 μL import reaction. We then either used 500 μg of leucoplasts for comparison on an equal protein basis or 113.67 μg of leucoplasts (~6.15×10^7 plastids) for comparison on an equal plastid number basis. After import, 4% of each import reaction was analyzed by SDS-PAGE. Interestingly, a lower concentration of leucoplasts (113.67 μg/reaction) in the import reactions resulted in a significant increase of prTic40 import efficiency (**Figure 2A**, compare lanes 6–8), compared to when 500 μg/reaction were used, which was the amount we used for **Figure 1**. Using this lower leucoplast input, the import efficiency of prTic40 was

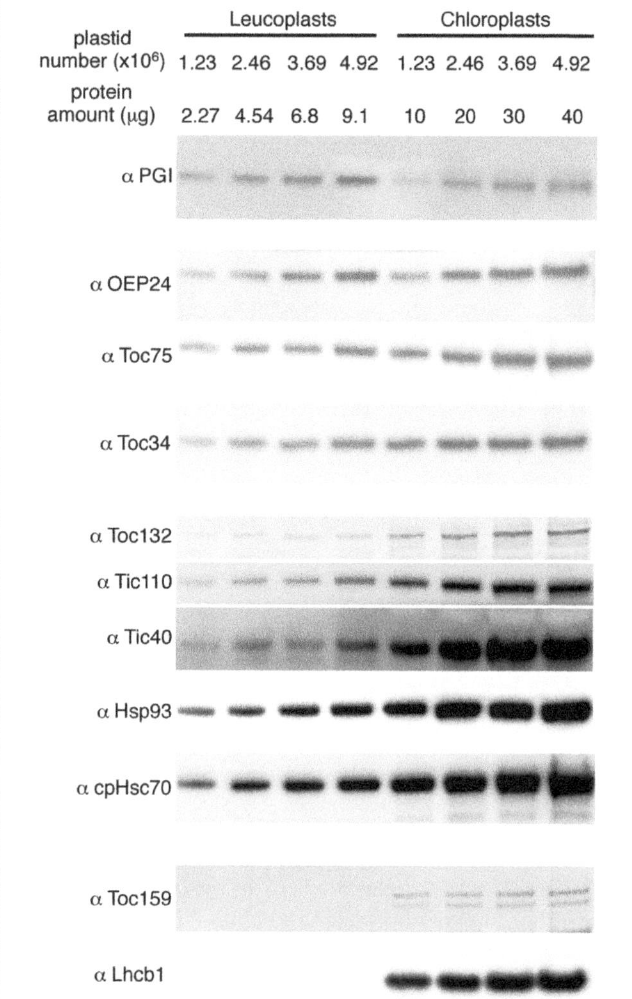

FIGURE 4 | Translocon protein abundance in leucoplasts and chloroplasts compared on an equal number of plastids basis. Various amounts of isolated leucoplasts and chloroplasts were analyzed by SDS-PAGE and immunoblotting with antibodies indicated at left. The amount of protein loaded in each lane and the equivalent number of plastids are indicated above the lanes.

almost equal to its efficiency into chloroplasts and most of the imported proteins were in the mature form (**Figure 2A**, lanes 7 and 8). The import efficiency of prRBCS into leucoplasts was also slightly increased when a lower amount of leucoplasts was used, but the efficiency was still much lower than its efficiency into chloroplasts (**Figures 2A**, lanes 2–4, **2B**). From these data, we decided to compare the import of different precursors into leucoplasts and chloroplasts on an equal plastid number basis in all subsequent analyses.

Mitochondrial Precursor prAOX is not Imported into Isolated Leucoplasts

The leucoplast preparations were reported to have about 1% contamination of mitochondria (Emes and England, 1986; Bowsher et al., 1989). Furthermore, it has been reported that chloroplast precursors can be imported into isolated pea mitochondria (Rudhe et al., 2002). To verify that the import we observed was not due to mis-sorting of plastid precursors to the mitochondria in the leucoplast preparation, we isolated cDNA clone encoding the soybean mitochondrial alternative oxidase precursor protein (prAOX), which is one of the most widely used precursor proteins for studying protein import into isolated plant mitochondria. The precursor form of prAOX is about 36 kDa. After being imported into isolated mitochondria, it is processed to a 32-kDa mature protein, which is protected exogenously added protease (Whelan et al., 1995; Rudhe et al., 2002; Lister et al., 2007). [^{35}S]Met-labeled prAOX was synthesized and used in the leucoplast import assays as prTic40 and prRBCS. If the import we observed was due to import of prRBCS and prTic40 into the mitochondria in the leucoplast preparation, then a mitochondrial precursor like prAOX should import equally well or even better. As shown in **Figure 3**, although some prAOX signals were detected after import, these signals were degraded after treating the leucoplasts with thermolysin. No imported mature protein was detected even after over-exposure of the fluorograph (**Figure 3**, lanes 3 and 4). In contrast, both prRBCS and prTic40 produced mature proteins after import, and these mature proteins were thermolysin resistant, indicating that they were inside the leucoplasts (**Figure 3**, lanes 7, 8, 11, and 12).

Most TOC/TIC Components are more Abundant in Chloroplasts than in Leucoplasts

To investigate the translocon composition of leucoplasts, representative translocon proteins were analyzed by immunoblots. We used phosphoglucoisomerase (PGI), a plastid enzyme in the starch biosynthesis pathway and expressed in both leaves and roots, as a non-translocon stromal reference protein. We also used OEP24, a non-translocon plastid outer membrane protein expressed in both leaves and roots (Pohlmeyer et al., 1998), for a general assessment of the amount of the envelope proteins. When compared on the basis of an equal number of plastids, the amount of PGI in chloroplasts was similar or slightly lower than that in leucoplasts (**Figure 4**). The amount of OEP24 and the translocon proteins Toc75 and Toc34 was about 1.5x to 2.5x higher in chloroplasts than in leucoplasts. In comparison, abundances of Toc132, Tic110, Hsp93, cpHsc70, and Tic40 were much higher in chloroplasts. Toc159 is known to be extremely sensitive to protease and is easily degraded during plastid isolation (Bolter et al., 1998). In chloroplasts, we detected, in addition to the full-length Toc159, at least one lower molecular weight band that is most likely a partially degraded Toc159 fragments. We could not detect any full-length Toc159 either in our leucoplast preparation (**Figure 4**) or when roots were directly ground up in SDS-PAGE sample buffer in liquid nitrogen (data not shown). Very low amounts of smaller fragments were detected in leucoplasts (data not shown) but our antibody was not specific enough to exclude cross-reactivity with other Toc159 members. Nonetheless, even if Toc159 is present in the leucoplasts of pea roots, it is much less abundant than in chloroplasts. In addition, no chlorophyll *a/b* binding protein of photosystem II (Lhcb1) could be detected, suggesting that our leucoplasts were not contaminated with chloroplasts.

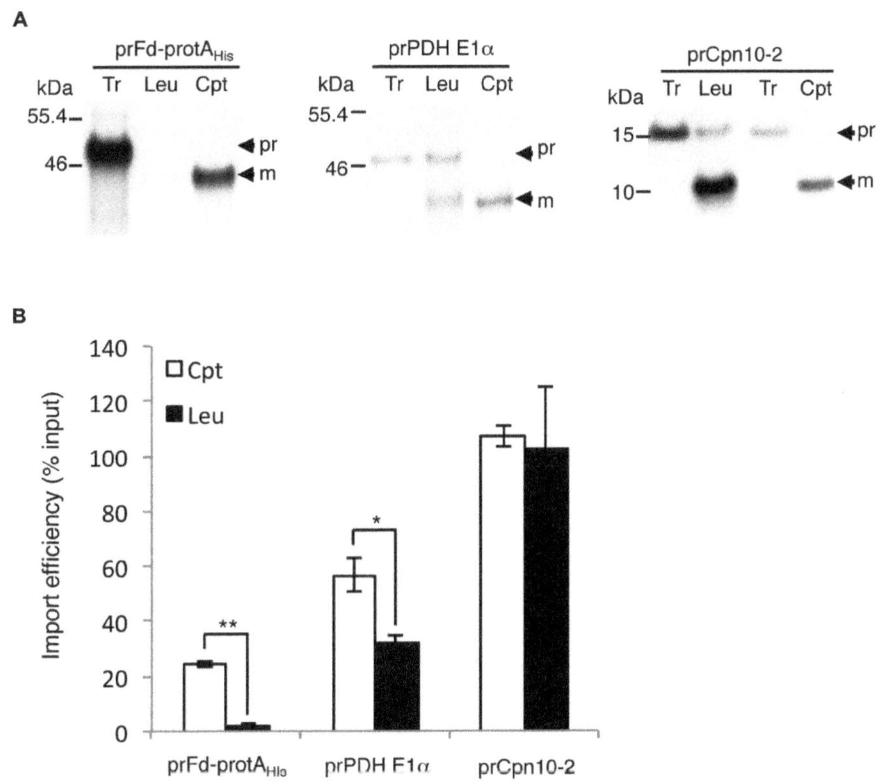

FIGURE 5 | Precursor proteins show different plastid preferences. (A) Isolated leucoplasts (113.67 μg protein) and chloroplasts (500 μg protein) were incubated with various *in vitro*-translated [^{35}S]Met-precursors under import conditions for 25 min. After import, intact leucoplasts and chloroplasts were re-isolated and analyzed by SDS-PAGE. The gels were stained with Coomassie blue and dried for fluorography. 4% of the plastids in each import reaction were loaded. Tr, 1% (for prFd-protA$_{His}$ and prPDH E1α) or 1.2% (for prCpn10-2) equivalent of the *in vitro*-translated proteins used in each import reaction. Cpt, chloroplasts; Leu, leucoplasts; pr, precursor form; m, imported mature form. **(B)** Imported mature proteins in experiments as shown in **(A)** were quantified and the import efficiencies were calculated. Import efficiency was defined as % of [^{35}S]-labeled precursor proteins used that were found as mature proteins in re-isolated plastids. The values have been corrected for the difference in number of methionine residues between the precursor and mature forms. Data shown are mean ± SD of three independent experiments, $^{*}p < 0.05$, $^{**}p < 0.001$ and the *p*-value of prCpn10-2 Cpt versus Leu is 0.67 (Student's *t*-test).

Import of prFd-protA$_{His}$, prPDH E1α, and prCpn10-2 into Leucoplasts

We had determined that pea root leucoplasts had a lower amount of Toc132 and almost no Toc159. If the interactions with these Toc159 family members contribute to plastid preference, a precursor that prefers Toc159 should import much better into chloroplasts, while a precursor that prefers Toc132 should exhibit a smaller difference. Our data with prRBCS indeed support this (**Figure 2B**). We further tested another two precursors that have been directly demonstrated to select between Toc159 and Toc132: the transit peptide of ferredoxin precursor (prFd) prefers Toc159 while the transit peptide of pyruvate dehydrogenase E1α subunit precursor (prPDH E1α) prefers Toc132 (Ivanova et al., 2004; Smith et al., 2004; Inoue et al., 2010). For prFd, we used the construct prFd-protA$_{His}$, which contains ferredoxin transit peptide fused to Staphylococcal protein A (protA; Smith et al., 2004). As shown in **Figure 5**, prFd-protA$_{His}$ imported well into chloroplasts but was nearly undetectable in leucoplasts, similar to the results for prRBCS (**Figure 2**). In comparison, although prPDH E1α also imported better into chloroplasts, its import was at least half as strong in leucoplasts in our *in vitro* system.

We further tested several additional precursors but did not find any precursors with higher import efficiency into leucoplasts than into chloroplasts in our system. However, we did find several precursors similar to prTic40 that exhibited no import bias between leucoplasts and chloroplasts in our *in vitro* system. One example was the precursor of chaperonin 10-2 (prCpn10-2), which imported very well into both chloroplasts and leucoplasts (**Figure 5B**). The import pathways prCpn10-2 and prTic40 used may be partially different from those of prRBCS and prPDH E1α.

Discussion

We performed initial quantifications of translocon components of pea root leucoplasts. Considering that the diameter of chloroplasts we used is about 1.7 times that of leucoplasts, the chloroplast surface area should be about three times that of leucoplasts. However, the amount of chloroplast Toc75 and Toc34 is only about twice that in leucoplasts. Therefore, the density of Toc75 and Toc34 is not necessarily higher in chloroplasts. If the non-translocon protein OEP24 is present in about the same density in chloroplasts and leucoplasts, then the density of Toc75 and Toc34 in the two plastids would also be similar. On the other hand, other essential translocon components, like Tic110, Hsp93 and cpHsc70, are clearly less abundant in leucoplasts. We did not analyze the amount of other essential components like Tic20 and Tic56 (Kikuchi et al., 2013; Köhler et al., 2015) due to unavailability of antibodies recognizing the pea orthologous well and their amounts in leucoplasts remain to be determined. However, based on our current data, it is likely that similar numbers of Toc75 and Toc34 subunits are coupled with lower numbers of other translocon components in leucoplasts. This composition may reflect that, under normal physiological conditions, the protein import demand of leucoplasts is not as high as young chloroplasts. Once precursors are bound to the TOC complex, leucoplasts can afford a lower import rate than chloroplasts.

It has been proposed that the precursor selectivity between leucoplasts and chloroplasts may also be due to the distinct substrate preference of Toc159 versus Toc132 (Yan et al., 2006; Li and Teng, 2013). We therefore tested the three precursors that have been directly shown to prefer one receptor to the other. In agreement with the much higher abundance of Toc159 in leaf chloroplasts, prRBCS and prFd-protA$_{His}$, which show a preference for Toc159, also imported much better into chloroplasts. Toc132 is still more abundant in chloroplasts but is present in leucoplasts at a reasonable amount. And indeed, prPDH E1α, which shows a preference for Toc132, had a leucoplast import efficiency about 50% that of chloroplasts in our current *in vitro* system. Unfortunately the transit peptide of prPDH E1α is the only transit peptide that has been shown biochemically to prefer Toc132 over Toc159. The importance of Toc132 in leucoplast import will need to be tested with more precursors with clear preferences for Toc132.

We did not find any precursors with higher import efficiency into leucoplasts than into chloroplasts in our system. It is possible that our leucoplast *in vitro* import system is still not as optimized as that of chloroplasts. However, compared to prRBCS/prFd and prPDH E1α, we found a third group of precursors that have similar import efficiencies into both plastids in our current *in vitro* system. Previous reports have indicated that prFNR, prRCA and the precursor of leucoplast pyruvate kinase also imported equally well into the two plastids (Wan et al., 1995, 1996). Although in those reports leucoplasts and chloroplasts were isolated from different species, it is likely that these precursors share some import characteristics with prTic40 and prCpn10-2 that we analyzed here. Unlike precursors that clearly prefer Toc159 or Toc132, the import of these precursors did not decline in leucoplasts. They may use some novel component present in similar amounts in the two plastids or they may simply bypass the need for the Toc159 family protein and use Toc34 and Toc75 directly. However, it is also possible that these precursors import into leucoplast through distinct pathways. It would be interesting to further study their import mechanism and identify the motifs in their transit peptides that confer efficient leucoplast import.

The transit peptide of prRBCS is the most widely used transit peptide for directing passenger proteins into plastids in transgenic plants, even when the passenger proteins were meant for non-green plastids like leucoplasts. For example, the prRBCS transit peptide was used to deliver the bacterial carotene desaturase into rice grain leucoplasts for the development of the Golden Rice (Ye et al., 2000; Paine et al., 2005). As shown here and other reports (Wan et al., 1996; Yan et al., 2006; Primavesi et al., 2008), the import efficiency of this transit peptide into leucoplasts is rather poor. The precursors that exhibited equally high efficiency into both chloroplasts and leucoplasts in our *in vitro* system, like prTic40 and prCpn10-2, offer promising potential that their transit peptides can confer better import of passenger proteins in transgenic plant leucoplasts than the prRBCS transit peptide. They could be valuable tools in manipulations of crops in which leucoplasts provide the major nutrient source.

Author Contributions

C-CC and H-mL designed the experiments and prepared the manuscript. C-CC performed the experiments.

Acknowledgments

We thank Dr. Tung-Yuan Ho for providing the Multisizer 3 for plastids counting, Ms. May-Jen Lu for technical assistance in using the counter, Dr. Danny Schnell for the prFd-protA$_{His}$ construct, and Dr. Jychian Chen for the PGI antibodies. We also thank Ms. Ming-Yan Kuo for the preparation of the anti-OEP24 antibody, Ms. Po-Ting Chan for cloning the psToc132AG fragment and Mr. Yang-Tsung Lin for preparation of the psToc132AG protein and anti-psToc132AG antibody. We acknowledge the assistance of the Institute of Molecular Biology Scientific English Editing Core in preparing this manuscript. This work was supported by grants to H-mL from Ministry of Science and Technology (MOST 103-2321-B-001-053) and Academia Sinica of Taiwan.

References

Aronsson, H., and Jarvis, R. P. (2011). Rapid isolation of *Arabidopsis* chloroplasts and their use for in vitro protein import assays. *Methods Mol. Biol.* 774, 281–305. doi: 10.1007/978-1-61779-234-2_17

Barsan, C., Zouine, M., Maza, E., Bian, W., Egea, I., Rossignol, M., et al. (2012). Proteomic analysis of chloroplast-to-chromoplast transition in tomato reveals metabolic shifts coupled with disrupted thylakoid biogenesis machinery and elevated energy-production components. *Plant Physiol.* 160, 708–725. doi: 10.1104/pp.112.203679

Bauer, J., Chen, K., Hiltbunner, A., Wehrli, E., Eugster, M., Schnell, D., et al. (2000). The major protein import receptor of plastids is essential for chloroplast biogenesis. *Nature* 403, 203–207. doi: 10.1038/35003214

Bauer, J., Hiltbrunner, A., Weibel, P., Vidi, P.-A., Alvarez-Huerta, M., Smith, M. D., et al. (2002). Essential role of the G-domain in targeting of the protein import receptor atToc159 to the chloroplast outer membrane. *J. Cell Biol.* 159, 845–854. doi: 10.1083/jcb.200208018

Bischof, S., Baerenfaller, K., Wildhaber, T., Troesch, R., Vidi, P. A., Roschitzki, B., et al. (2011). Plastid proteome assembly without Toc159: photosynthetic protein import and accumulation of N-acetylated plastid precursor proteins. *Plant Cell* 23, 3911–3928. doi: 10.1105/tpc.111.092882

Bolter, B., May, T., and Soll, J. (1998). A protein import receptor in pea chloroplasts, Toc86, is only a proteolytic fragment of a larger polypeptide. *FEBS Lett.* 441, 59–62.

Bowsher, C. G., Hucklesby, D. P., and Emes, M. J. (1989). Nitrite reduction and carbohydrate metabolism in plastids purified from roots of *Pisum sativum* L. *Planta* 177, 359–366. doi: 10.1007/BF00403594

Brautigam, A., and Weber, A. P. (2009). Proteomic analysis of the proplastid envelope membrane provides novel insights into small molecule and protein transport across proplastid membranes. *Mol. Plant* 2, 1247–1261. doi: 10.1093/mp/ssp070

Chang, W., Soll, J., and Bolter, B. (2014). A new member of the psToc159 family contributes to distinct protein targeting pathways in pea chloroplasts. *Front. Plant Sci.* 5:239. doi: 10.3389/fpls.2014.00239

Chou, M. L., Fitzpatrick, L. M., Tu, S. L., Budziszewski, G., Potter-Lewis, S., Akita, M., et al. (2003). Tic40, a membrane-anchored co-chaperone homologue in the chloroplast protein translocon. *EMBO J.* 22, 2970–2980. doi: 10.1093/emboj/cdg281

Emes, M. J., and England, S. (1986). Purification of plastids from higher plant roots. *Planta* 168, 161–166. doi: 10.1007/BF00402959

Henry, E. W., DeMorrow, J. M., and Richard, L. B. (1979). The effects of phenolic compounds on peroxidase and polyphenol oxidase in dwarf pea (*Pisum sativum* var. Little marvel) tissue. *Z. Pflanzenphysiol.* 92, 221–239. doi: 10.1016/S0044-328X(79)80004-5

Inoue, H., Rounds, C., and Schnell, D. J. (2010). The molecular basis for distinct pathways for protein import into *Arabidopsis* chloroplasts. *Plant Cell* 22, 1947–1960. doi: 10.1105/tpc.110.074328

Ivanova, Y., Smith, M. D., Chen, K., and Schnell, D. J. (2004). Members of the Toc159 import receptor family represent distinct pathways for protein targeting to plastids. *Mol. Biol. Cell* 15, 3379–3392. doi: 10.1091/mbc.E03-12-0923

Jarvis, P., and Lopez-Juez, E. (2013). Biogenesis and homeostasis of chloroplasts and other plastids. *Nat. Rev. Mol. Cell Biol.* 14, 787–802. doi: 10.1038/nrm3702

Kikuchi, S., Bedard, J., Hirano, M., Hirabayashi, Y., Oishi, M., Imai, M., et al. (2013). Uncovering the protein translocon at the chloroplast inner envelope membrane. *Science* 339, 571–574. doi: 10.1126/science.1229262

Kleffmann, T., von Zychlinski, A., Russenberger, D., Hirsch-Hoffmann, M., Gehrig, P., Gruissem, W., et al. (2007). Proteome dynamics during plastid differentiation in rice. *Plant Physiol.* 143, 912–923. doi: 10.1104/pp.106.090738

Köhler, D., Montandon, C., Hause, G., Majovsky, P., Kessler, F., Baginsky, S., et al. (2015). Characterization of chloroplast protein import without Tic56, a component of the 1-megadalton translocon at the inner envelope membrane of chloroplasts. *Plant Physiol.* 167, 972–990. doi: 10.1104/pp.114.255562

Kubis, S., Patel, R., Combe, J., Bedard, J., Kovacheva, S., Lilley, K., et al. (2004). Functional specialization amongst the *Arabidopsis* Toc159 family of chloroplast protein import receptors. *Plant Cell* 16, 2059–2077. doi: 10.1105/tpc.104.023309

Leister, D. (2003). Chloroplast research in the genomic age. *Trends Genet.* 19, 47–56. doi: 10.1016/S0168-9525(02)00003-3

Li, H.-m., and Chiu, C. C. (2010). Protein transport into chloroplasts. *Annu. Rev. Plant Biol.* 61, 157–180. doi: 10.1146/annurev-arplant-042809-112222

Li, H.-m., and Teng, Y. S. (2013). Transit peptide design and plastid import regulation. *Trends Plant Sci.* 18, 360–366. doi: 10.1016/j.tplants.2013.04.003

Li, M., and Schnell, D. J. (2006). Reconstitution of protein targeting to the inner envelope membrane of chloroplasts. *J. Cell Biol.* 175, 249–259. doi: 10.1083/jcb.200605162

Lister, R., Carrie, C., Duncan, O., Ho, L. H., Howell, K. A., Murcha, M. W., et al. (2007). Functional definition of outer membrane proteins involved in preprotein import into mitochondria. *Plant Cell* 19, 3739–3759. doi: 10.1105/tpc.107.050534

Paila, Y. D., Richardson, L. G., and Schnell, D. J. (2015). New insights into the mechanism of chloroplast protein import and its integration with protein quality control, organelle biogenesis and development. *J. Mol. Biol.* 427, 1038–1060. doi: 10.1016/j.jmb.2014.08.016

Paine, J. A., Shipton, C. A., Chaggar, S., Howells, R. M., Kennedy, M. J., Vernon, G., et al. (2005). Improving the nutritional value of Golden Rice through increased pro-vitamin A content. *Nat. Biotechnol.* 23, 482–487. doi: 10.1038/nbt1082

Perry, S. E., Buvinger, W. E., Bennett, J., and Keegstra, K. (1991). Synthetic analogues of a transit peptide inhibit binding or translocation of chloroplastic precursor proteins. *J. Biol. Chem.* 266, 11882–11889.

Pohlmeyer, K., Soll, J., Grimm, R., Hill, K., and Wagner, R. (1998). A high-conductance solute channel in the chloroplastic outer envelope from Pea. *Plant Cell* 10, 1207–1216. doi: 10.1105/tpc.10.7.1207

Primavesi, L. F., Wu, H., Mudd, E. A., Day, A., and Jones, H. D. (2008). Visualisation of plastids in endosperm, pollen and roots of transgenic wheat expressing modified GFP fused to transit peptides from wheat SSU RubisCO, rice FtsZ and maize ferredoxin III proteins. *Transgenic Res.* 17, 529–543. doi: 10.1007/s11248-007-9126-7

Rudhe, C., Chew, O., Whelan, J., and Glaser, E. (2002). A novel in vitro system for simultaneous import of precursor proteins into mitochondria and chloroplasts. *Plant J.* 30, 213–220. doi: 10.1046/j.1365-313X.2002.01280.x

Sakamoto, W., Miyagishima, S. Y., and Jarvis, P. (2008). Chloroplast biogenesis: control of plastid development, protein import, division and inheritance. *Arabidopsis Book* 6, e0110. doi: 10.1199/tab.0110

Schulz, A., Knoetzel, J., Scheller, H. V., and Mant, A. (2004). Uptake of a fluorescent dye as a swift and simple indicator of organelle intactness: import-competent chloroplasts from soil-grown *Arabidopsis*. *J. Histochem. Cytochem.* 52, 701–704. doi: 10.1177/002215540405200514

Seigneurin-Berny, D., Salvi, D., Joyard, J., and Rolland, N. (2008). Purification of intact chloroplasts from *Arabidopsis* and spinach leaves by isopycnic centrifugation. *Curr. Protoc. Cell Biol.* Chap. 3, Unit 3.30. doi: 10.1002/0471143030.cb0330s40

Shi, L. X., and Theg, S. M. (2013). The chloroplast protein import system: from algae to trees. *Biochim. Biophys. Acta* 1833, 314–331. doi: 10.1016/j.bbamcr.2012.10.002

Smith, M. D., Rounds, C. M., Wang, F., Chen, K., Afitlhile, M., and Schnell, D. J. (2004). atToc159 is a selective transit peptide receptor for the import of nucleus-encoded chloroplast proteins. *J. Cell Biol.* 165, 323–334. doi: 10.1083/jcb.200311074

Smith, P. K., Krohn, R. I., Hermanson, G. T., Mallia, A. K., Gartner, F. H., Provenzano, M. D., et al. (1985). Measurement of protein using bicinchoninic acid. *Anal. Biochem.* 150, 76–85. doi: 10.1016/0003-2697(85)90442-7

Somerville, C. R., and Ogren, W. L. (1981). Photorespiration-deficient mutants of *Arabidopsis thaliana* lacking mitochondrial serine transhydroxymethylase activity. *Plant Physiol.* 67, 666–671. doi: 10.1104/pp.67.4.666

Stanga, J. P., Boonsirichai, K., Sedbrook, J. C., Otegui, M. S., and Masson, P. H. (2009). A role for the TOC complex in *Arabidopsis* root gravitropism. *Plant Physiol.* 149, 1896–1905. doi: 10.1104/pp.109.135301

Teng, Y. S., Chan, P. T., and Li, H.-m. (2012). Differential age-dependent import regulation by signal peptides. *PLoS Biol.* 10:e1001416. doi: 10.1371/journal.pbio.1001416

Tripp, J., Inoue, K., Keegstra, K., and Froehlich, J. E. (2007). A novel serine/proline-rich domain in combination with a transmembrane domain is required for the insertion of AtTic40 into the inner envelope membrane of chloroplasts. *Plant J.* 52, 824–838. doi: 10.1111/j.1365-313X.2007.03279.x

Tu, S. L., Chen, L. J., Smith, M., Su, Y.-S., Schnell, D., and Li, H.-m. (2004). Import pathways of chloroplast interior proteins and the outer-membrane protein OEP14 converge at Toc75. *Plant Cell* 16, 2078–2088. doi: 10.1105/tpc.104.023952

Wan, J., Blakeley, S. D., Dennis, D. T., and Ko, K. (1995). Import characteristics of a leucoplast pyruvate kinase are influenced by a 19-amino-acid domain within the protein. *J. Biol. Chem.* 270, 16731–16739. doi: 10.1074/jbc.270.28.16731

Wan, J., Blakeley, S. D., Dennis, D. T., and Ko, K. (1996). Transit peptides play a major role in the preferential import of proteins into leucoplasts and chloroplasts. *J. Biol. Chem.* 271, 31227–31233. doi: 10.1074/jbc.271.49.31227

Whelan, J., Hugosson, M., Glaser, E., and Day, D. A. (1995). Studies on the import and processing of the alternative oxidase precursor by isolated soybean mitochondria. *Plant Mol. Biol.* 27, 769–778. doi: 10.1007/BF00020229

Wise, R. R. (2006). "The diversity of plastid form and function," in *The Structure and Function of Plastids*, eds R. R. Wise and J. K. Hoober (Dordrecbt: Springer Netherlands).

Yan, X., Khan, S., Hase, T., Emes, M. J., and Bowsher, C. G. (2006). Differential uptake of photosynthetic and non-photosynthetic proteins by pea root plastids. *FEBS Lett.* 580, 6509–6512. doi: 10.1016/j.febslet.2006.10.057

Ye, X., Al-Babili, S., Kloti, A., Zhang, J., Lucca, P., Beyer, P., et al. (2000). Engineering the provitamin A (beta-carotene) biosynthetic pathway into (carotenoid-free) rice endosperm. *Science* 287, 303–305. doi: 10.1126/science.287.5451.303

Yu, T.-S., and Li, H.-m. (2001). Chloroplast protein translocon components atToc159 and atToc33 are not essential for chloroplast biogenesis in guard cells and root cells. *Plant Physiol.* 127, 90–96. doi: 10.1104/pp.127.1.90

Conflict of Interest Statement: The authors declare that the research was conducted in the absence of any commercial or financial relationships that could be construed as a potential conflict of interest.

2

Cytomixis doesn't induce obvious changes in chromatin modifications and programmed cell death in tobacco male meiocytes

*Sergey Mursalimov[1], Natalya Permyakova[1], Elena Deineko[1], Andreas Houben[2] and Dmitri Demidov[2]**

[1] Institute of Cytology and Genetics, Siberian Branch, Russian Academy of Sciences, Novosibirsk, Russia, [2] Leibniz Institute of Plant Genetics and Crop Plant Research, Stadt Seeland, Germany

Cytomixis is a poorly studied process of nuclear migration between plant cells. It is so far unknown what drives cytomixis and what is the functional state of the chromatin migrating between cells. Using immunostaining, we have analyzed the distribution of posttranslational histone modifications (methylation, acetylation, and phosphorylation) that reflect the functional state of chromatin in the tobacco microsporocytes involved in cytomixis. We demonstrate that the chromatin in the cytomictic cells does not differ from the chromatin in intact microsporocytes according to all 14 analyzed histone modification types. We have also for the first time demonstrated that the migrating chromatin contains normal structures of the synaptonemal complex (SC) and lacks any signs of apoptosis. As has been shown, the chromatin migrating between cells in cytomixis is neither selectively heterochromatized nor degraded both before its migration to another cell and after it enters a recipient cell as micronuclei. We also showed that cytomictic chromatin contains marks typical for transcriptionally active chromatin as well as heterochromatin. Moreover, marks typical for chromosome condensation, SC formation and key proteins required for the formation of bivalents were also detected at migrated chromatin.

Edited by:
Anne-Catherine Schmit, Institut de Biologie Moléculaire des Plantes – CNRS UPR2357, France

Reviewed by:
Animesh Kumar Datta, University of Kalyani, India
Robyn Lynette Overall, University of Sydney, Australia

***Correspondence:**
Dmitri Demidov
demidov@ipk-gatersleben.de

Keywords: Cytomixis, histone modifications, euchromatin/heterochromatin, synaptonemal complex, TUNEL assay, *Nicotiana tabacum*

INTRODUCTION

Cytomixis is an enigmatic migration of nuclei between plant cells through intercellular channels of a special type (cytomictic channels), which are considerably larger than plasmodesmata. This phenomenon has been observed in 100s of higher plant species, mostly during microsporogenesis (for details, see the reviews by Lone and Lone, 2013 and Mursalimov et al., 2013b).

So far, the mechanisms underlying formation of cytomictic channels and migration of nuclei through them between plant microsporocytes have been studied in some detail. As has been shown, cytomixis is most frequently observed during meiotic prophase I (Negron-Ortiz, 2007; Liu et al., 2012; Barton et al., 2014). Cytomictic channels may be formed in the cell wall based on plasmodesmata or independently of them (Wang et al., 2002, 2004; Yu et al., 2004; Mursalimov et al., 2010, 2013a). Not just chromatin or chromosomes alone migrate through the cytomictic channels, rather the nucleus with all its components (chromatin, nuclear matrix, and nucleolus) enclosed within the intact nuclear membrane (Mursalimov and Deineko, 2011) can be transfered. In some cases, the nucleus may pass from a donor to a recipient cell as a whole to form binucleated microsporocytes (Mursalimov and Deineko, 2015); however, a more frequent situation is when the migrated fragments of the nucleus produce micronuclei in the cytoplasm of recipient cells (Mursalimov and Deineko, 2011, 2015; Barton et al., 2014).

It has been shown earlier that cytomixis may result in unreduced, polyploid, aneuploid, and sterile pollen (Falistocco et al., 1995; Ghaffari, 2006; Negron-Ortiz, 2007; Lavia et al., 2011; Pécrix et al., 2011; Mursalimov and Deineko, 2015). However, it is unclear what kind of chromatin migrates between cytomictic cells at the early meiotic stages, and what is the level of transcriptional activity and whether it has been damaged. To clarify the functional state of the migrating chromatin we analyzed the distribution of some histone modifications in cytomictic cells. The corresponding marks are highly conserved for all the eukaryotes and make it possible to detect euchromatic and heterochromatic regions, double-strand DNA breaks and the process of chromosome condensation and cohesion (Manzanero et al., 2000; Fuchs and Schubert, 2012; He et al., 2014).

In this work, we used immunostaining to analyze the distribution of major posttranslational histone modifications (methylation, acetylation, and phosphorylation) in tobacco cytomictic microsporocyte nuclei as well as in the micronuclei formed by cytomixis. Immunostaining and electron microscopy have allowed us to detect and examine the synaptonemal complex (SC) elements in the migrating chromatin and the cells involved in cytomixis have been analyzed for signs of DNA damage using internucleosomal DNA fragmentation and TUNEL assays.

MATERIALS AND METHODS

Plant Material

Nicotiana tabacum L. cv. ($2n = 4x = 48$) Petit Havana SR1 line was used in the work. The plants were grown in a greenhouse with a photoperiod of 16/8 h (day/night) at a temperature of 22/18°C (day/night).

Immunostaining of Squash Preparations

Anthers were fixed with freshly prepared 4% paraformaldehyde in Phosphate-buffered saline (PBS, pH 7.3) for 15–30 min on ice. Anthers were then washed in PBS 3 min × 15 min on ice, squashed into suspension in PBS and then microsporocyte cell walls were digested at 37°C for 30 min in a mixture of 1% Pectolyase (Sigma), 1% Cytohelicase (Sigma) 0.7% Cellulase R-10 (Duchefa), and 0.7% Cellulase (Calbiochem) dissolved in PBS. Cells were centrifuged at 7000 *g* for 6 min at 4°C, resuspended with PBS and centrifuged again at 7000 *g* for 6 min at 4°C. Then cells with a minimum volume of PBS (10–20 μl) were transferred to poly-L-lysine-slides and squashed between a glass slide and cover slip. After freezing in liquid nitrogen for 60 s, the cover slips were removed and the slides were transferred immediately into PBS.

To avoid non-specific antibody binding, slides were incubated for 1 h in PBS with 3% BSA and 0.1% Tween 20 at 37°C in a prewarmed moisture chamber, then primary antibodies were applied to the slides.

The primary antibodies used were mouse anti-histone H3S10ph (Abcam, cat. no. 14955); rabbit anti-H3K4me3 (Millipore, cat. no. 07-473); anti-H3K9me2 (Active Motif, cat. no. 39239); anti-H3K27me2 (EMD Millipore, cat. no. 07-452); anti-H3K4me2 (EMD Millipore, cat. no. ABE250); anti-H3T11ph (Upstate, cat. no. 07-492); anti-H2A.XS139ph (Cell Signaling, cat. no. 2577); anti-H3K9ac (Thermo Scientific, cat. no. MA5-11195); anti-H3K27me3 (EMD Millipore, cat. no. 07-449); anti-H3K27ac (Abcam, cat. no. 45173); anti-H3K14ac (Abcam, cat. no. 52946); anti-H3K18ac (Abcam, cat. no. 1191); anti-H2AT121ph (MyBioSource, cat. no. 004447); rat anti-H3Ser28ph (Sigma, cat. no. H9908); rabbit anti-Asy1 and guinea pig anti-Zyp1 from maize (kindly provided by Rachel Wang, Cande group, Berkeley). All the primary antibodies were diluted at 1:200 in PBS, 1% BSA, 0.1% Tween 20.

After overnight incubation at 4°C in a moisture chamber and washing three times for 15 min in PBS with 0.1% Tween 20, the slides were incubated for 1 h at 37°C in a prewarmed moisture chamber with Alexa Fluor 594- and DyLight 488-conjugated anti-rabbit IgG (Jackson ImmunoResearch, cat. no. 711-585-152 and 111-485-144); Alexa Fluor 488-conjugated anti-mouse IgG (Molecular probe, cat. no. A-11001); Alexa Fluor 594-conjugated anti-guinea pig IgG (Molecular probe, cat. no. A-11076); FITC-conjugated anti-rat IgG (Dianova, cat. No 112-095-003), diluted at 1:250 in PBS, 1% BSA, 0.1% Tween 20. Then the slides were washed three times for 15 min in PBS with 0.1% Tween 20, and mounted in ProLong Gold antifade (Life Technologies) containing 4′,6-diamidino-2-phenylindole (DAPI).

Fluorescent signals were observed using an epifluorescence microscope Olympus BX-61. Images were captured with an ORCA-ER CCD camera (Hamamatsu), analyzed, and processed with Adobe Photoshop CS4 software (Adobe Systems Incorporated). Single DAPI images were presented in gray as the best way to discriminate chromosome structures. In the merged images the DAPI staining is blue. From 100 to 200 nuclei were used for the immunostaining with each antibody.

Immunostaining of PEG-Embedded Sections

Anthers at different stages of development were fixed, embedded in polyethylene glycol 1500, cut in semithin sections with a thickness of 10 μm on a LEICA RM2265 microtome (LEICA, Wetzlar, Germany) and immunostained according to Houben et al. (2011). Mouse anti-H3S10p (Abcam, cat. no. 14955) was used as a primary antibody and Alexa Fluor 488-conjugated anti-mouse IgG (Molecular probe, cat. no. A-11001) as a secondary one. The callose wall was stained with aniline blue (0.5% in PBS). Note that tobacco cells after PEG-embedding have cytoplasmic autofluorescence (presented in green with callose wall). Image capture and processing were performed as described above.

Terminal Deoxynucleotidyl Transferase dUTP Nick End Labeling (TUNEL) Assay

Tobacco anthers were used like for immunostaining of chromosomes by squashing method. TUNEL assay was performed using DeadEnd Fluorometric TUNEL system (Promega, cat. no. G3250) according to the manufacturer's instruction. Meiotic and somatic cells after heat shock were used as positive controls according to (Fan and Xing, 2004). Tobacco anthers and leaves that had been frozen in liquid nitrogen were sealed in plastic bags to prevent water access and heated in a water bath at 42°C for 120 min.

Isolation of Nuclear DNA and DNA Laddering Analysis

Total DNA was isolated from anthers and leaves with CTAB (Sambrook et al., 1989). Then DNA was separated by electrophoresis in a 2% agarose gel. The gel was then stained with ethidium bromide for visualization. For positive control, somatic cells after heat shock were used (Fan and Xing, 2004).

Electron Microscopy

Sample preparation for transmission electron microscopy was as described previously (Mursalimov and Deineko, 2012). Probes were studied in Libra120 (Carl Zeiss, Germany) transmission electron microscope at an accelerating voltage of 80 kV.

RESULTS

In order to assess the functional state of the chromatin migrating between cytomictic cells, we used immunostaining to analyze the distribution of main posttranslational histone modifications in intact tobacco microsporocyte nuclei, nuclei in the process of migration, and micronuclei formed in recipient cells as a result of cytomixis. In tobacco microsporogenesis, cytomixis is most frequently observed during meiotic prophase I (Sidorchuk et al., 2007; Mursalimov and Deineko, 2012, 2015). Correspondingly, we have analyzed the histone modifications in microsporocyte nuclei starting from the leptotene to metaphase I, after which the cytomictic chromatin is indistinguishable in dividing cells. In order to demonstrate migration of chromatin between cytomictic cells, cytoplasm autofluorescence signals, denoting cell boundaries were added to merged images.

Migrating Chromatin is Transcriptionally Potent

The following types of histone modifications were selected as markers for transcriptionally potent chromatin: histone H3K4me2, H3K4me3, H3K9ac, H3K14ac, H3K18ac, and H3K27ac (Horst et al., 2009; Jang et al., 2011; Du et al., 2013).

Intensive and uniform immunosignals were obtained in nuclei of intact tobacco microsporocytes in the zygotene–pachytene stage using antibodies recognizing dimethylated and trimethylated histone H3 at lysine 4 (**Figures 1A,D**). The signals are similarly distributed in cytomictic nuclei (**Figures 1B,E**; *arrows*). The intensity and distribution of immunopatterns in nuclei migrating between cells (**Figures 1B,E**; *arrows*) are identical to those in intact cell nuclei (**Figures 1A,D**). The signal in recipient cells, which thus have additional chromatin, also does not differ from the norm (**Figures 1B,E**). No differences are observed between the part of migrating nuclei still remaining in the donor cell versus the part that already entered the recipient cell (**Figures 1B,E**; *arrows*). The level of histone H3 di- and trimethylation at lysine 4 also remains unchanged in the micronuclei formed in recipient cells after cytomixis (**Figures 1C,F**; *arrows*).

The profile of histone H3 acetylation at lysines 9, 14, 18, and 27 in tobacco microsporocytes is analogous to the H3K4 methylation profile (**Figure 2**). Intensive signals are uniformly distributed over the nuclei of intact cells (**Figures 2A,D,G,J**) as well as over the nuclei of cytomictic cells (**Figures 2B,E,H,K**). Chromatin in the process of migration to another cell (**Figures 2B,E,H,K**; *arrows*) does not display obvious differences in the distribution and intensity of the immunosignal as compared to the remaining part of the migrating nucleus still remaining in the donor cell. The micronuclei formed in recipient cells do not lose their normal acetylation level on completion of cytomixis (**Figures 2C,F,I,L**; *arrows*).

Thus, the histone H3 methylation and acetylation profiles in the chromatin migrating between cytomictic cells correspond to the profile of transcriptionally potent chromatin, and these markers do not disappear in the micronuclei formed in recipient cells after completion of cytomixis.

The Heterochromatization Level of Migrating Chromatin does not Change

The following histone modification types were selected as the markers for heterochromatin: H3K9me2, H3K27me2, and H3K27me3. Uniform immunosignals of medium intensity over the entire nuclei in tobacco microsporocytes were observed, in addition to individual bright signals, which most likely correspond to regions enriched in heterochromatin (**Figures 3A,D**).

The chromatin that migrates between cytomictic cells (**Figures 3B,E,G**; *arrows*) neither before nor after migration reveals any sign of additional heterochromatization. Moreover, the micronuclei formed after cytomixis are not hypermethylated at lysines 9 and 27 of histone H3 in the recipient cell (**Figures 3C,F,G**; *arrows*). Most likely, heterochromatic blocks are randomly distributed in the migrating nuclei. Several bright immunosignals are either observable in the migrating chromatin (**Figure 3B**, *yellow arrow;* **Figures 3E,G**; *white arrows*) or may be completely absent (**Figure 3B**, *white arrow*). Correspondingly, heterochromatin blocks may be present in different quantities in the micronuclei formed after cytomixis (**Figure 3C** *yellow arrow;* **Figure 3F**, *white arrow;* and **Figure 3G**, *yellow arrows*) or may be completely absent (**Figure 3C** *white arrow;* **Figure 3G**, *turquoise arrow*).

Thus, chromatin that leaves the cytomictic cell is neither previously heterochromatized (hypermethylated at lysines 9 and 27 of histone H3) nor inactivated during migration and upon entering the recipient cells as micronuclei.

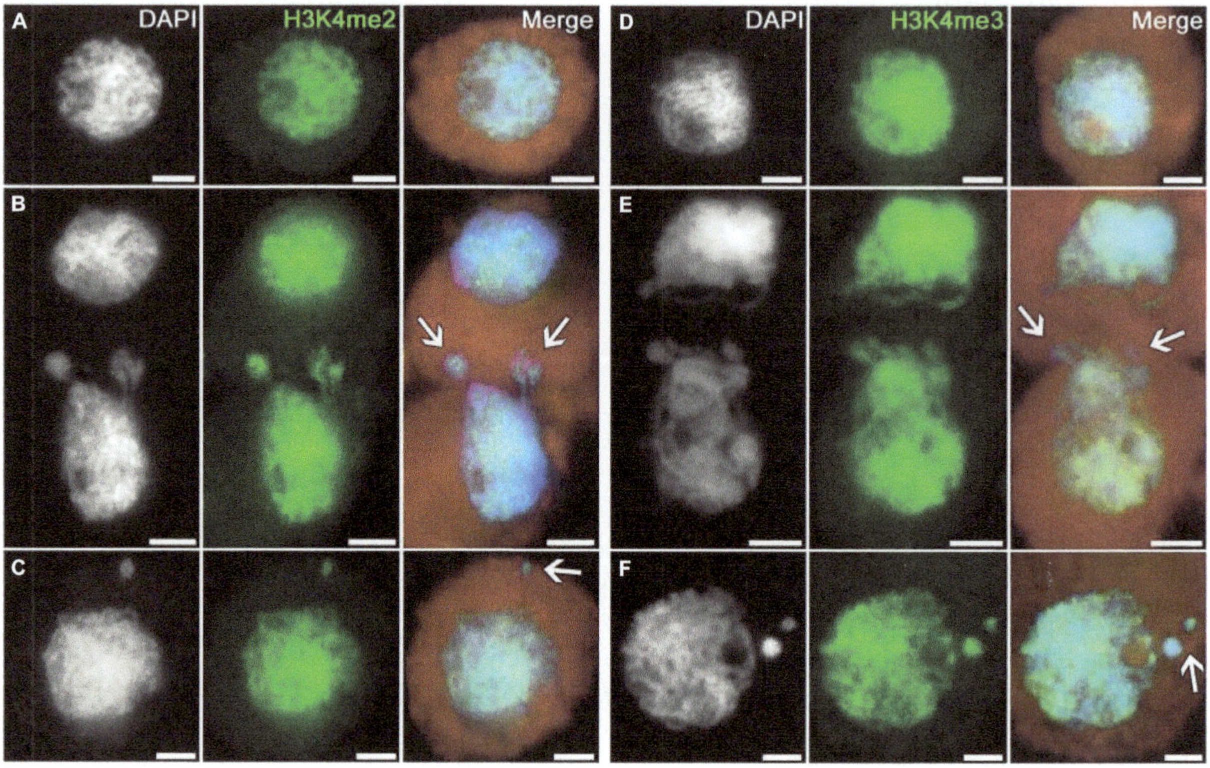

FIGURE 1 | Distributions of dimethylated and trimethylated histone H3K4 in the tobacco microsporocytes in meiotic prophase I by cytomixis. **(A)** H3K4me2 and **(D)** H3K4me3 in intact microsporocytes; **(B)** H3K4me2 and **(E)** H3K4me3 in cytomictic microsporocytes (*arrows* denote the migrating part of nucleus that entered the recipient cell); and **(C)** H3K4me2 and **(F)** H3K4me3 in microsporocytes after cytomixis (*arrows* denote micronuclei). Red, cytoplasm autofluorescence in merged images; bars, 5 μm.

Migrated Chromatin Contains Marks of Histone Phosphorylation that Correlate with Condensation and Separation of Chromosomes

Phosphorylation of histones is an important marker of chromosome condensation and segregation during cell division, repair of damaged DNA, and apoptosis (Thiriet and Hayes, 2005; Houben et al., 2007). In cytomictic cells, we studied the distributions of the following histone phosphorylation types: H3S10ph, H3S28ph, H3T11ph, H2AT121ph, and H2A.XS139ph.

The phosphorylation of histone H3 at serine 10 in tobacco microsporocytes is initiated in early meiotic prophase I, being detectable as intensive double signals putatively localized to the pericentromeric regions of chromosomes (**Figures 4A,D–J**). In cytomixis, these signals are distributed over the migrating nuclei in a random manner (**Figure 4A**): they are detectable in the migrating part of the nucleus (**Figure 4A**, *white arrow*) and appear in micronuclei (**Figure 4A**, *yellow arrow*) or may be absent in the migrating chromatin, as evidenced by micronuclei lacking the signal (**Figure 4A**, *turquoise arrow*).

By meiotic metaphase I, the signal indicating the phosphorylation of histone H3 at serine 10 spreads along the entire chromosome lengths (**Figure 4B**). At the same stage, the chromatin of both the recipient cell and cytomictic micronuclei after cytomixis undergo normal phosphorylation as in the intact nuclei (**Figure 4C**).

Thus, microsporocytes display normal phosphorylation of histone H3 at serine 10 dynamics in both the chromatin migrating between cells in cytomixis and the cytomictic micronuclei, matching the current meiotic stage and the degree of chromosome condensation.

Antibodies against histone H3S10ph modification was selected for comparing the cytomixis patterns in anther squash preparations and sections of the anthers embedded in PEG with retained native cell structure (**Figures 4D,E**). Examination of these sections demonstrates that the cytomixis pattern in the specimens embedded in PEG with preserved cell wall does not differ from that observed in squash preparations. Nuclei similarly migrate between cells in the zygotene–pachytene and form micronuclei displaying different quantities of immunosignals in recipient cells (**Figures 4D,E**, *arrows*).

Thus, we have demonstrated that the manipulations used for squash preparations do not influence the cytological pattern of cytomixis in tobacco microsporocytes, thereby proving that the immunostaining results thus obtained are adequate.

Double immunostaining of tobacco microsporocyte nuclei in the zygotene (bouquet stage) with anti-H3S10ph and H3K27me2 showed a clustering of immunosignals at the opposite poles of the nuclei (**Figure 4F**). At pachytene, this clustering disappears,

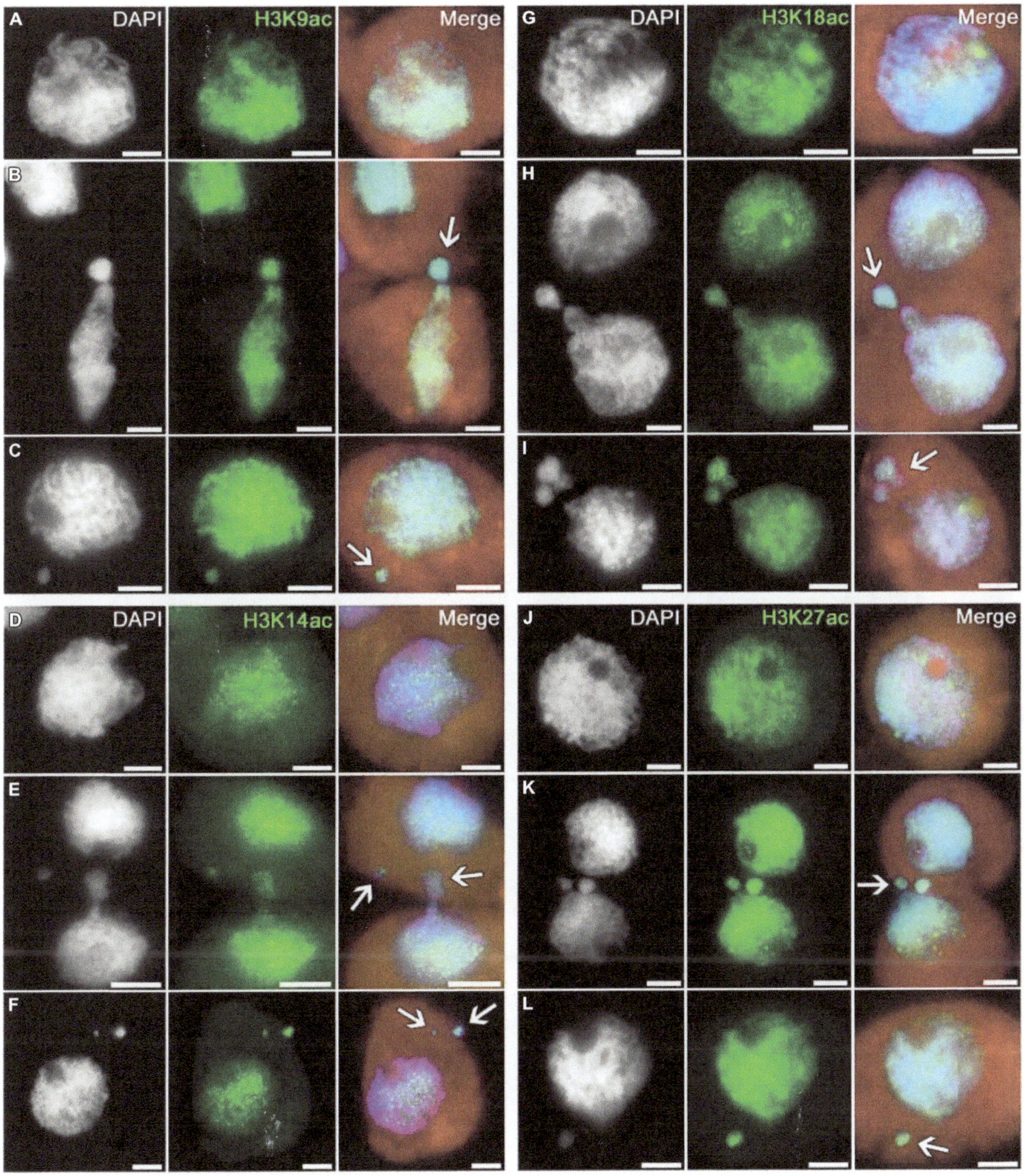

FIGURE 2 | Distributions of acetylated histone H3 in tobacco microsporocytes in meiotic prophase I by cytomixis. (A) H3K9ac, **(D)** H3K14ac, **(G)** H3K18ac, and **(J)** H3K27ac in intact microsporocytes; **(B)** H3K9ac, **(E)** H3K14ac, **(H)** H3K18ac, and **(K)** H3K27ac in cytomictic microsporocytes (*arrows* denote the migrating part of nucleus that entered the recipient cell); **(C)** H3K9ac, **(F)** H3K14ac, **(I)** H3K18ac, and **(L)** H3K27ac in microsporocytes after cytomixis (*arrows* denote micronuclei). Red, cytoplasm autofluorescence in merged images; bars, 5 μm.

and the signals spread over the nucleus but do not colocalize (**Figure 4G**). In case of cytomixis, the H3S10ph and H3K27me2 signals independently appear in the migrating part of the nucleus (**Figure 4H**, *arrow*). Cytomictic micronuclei may contain only H3S10ph signal (**Figure 4I**, *white arrow*), only H3K27me2 signal (**Figure 4I**, *yellow arrow*), or both signals at once (**Figure 4J**, *arrow*).

The immunosignals indicating phosphorylation of histone H3 at threonine 11 (a marker of chromatin condensation in cell division (Houben et al., 2005)) and serine 28 (coupled with sister

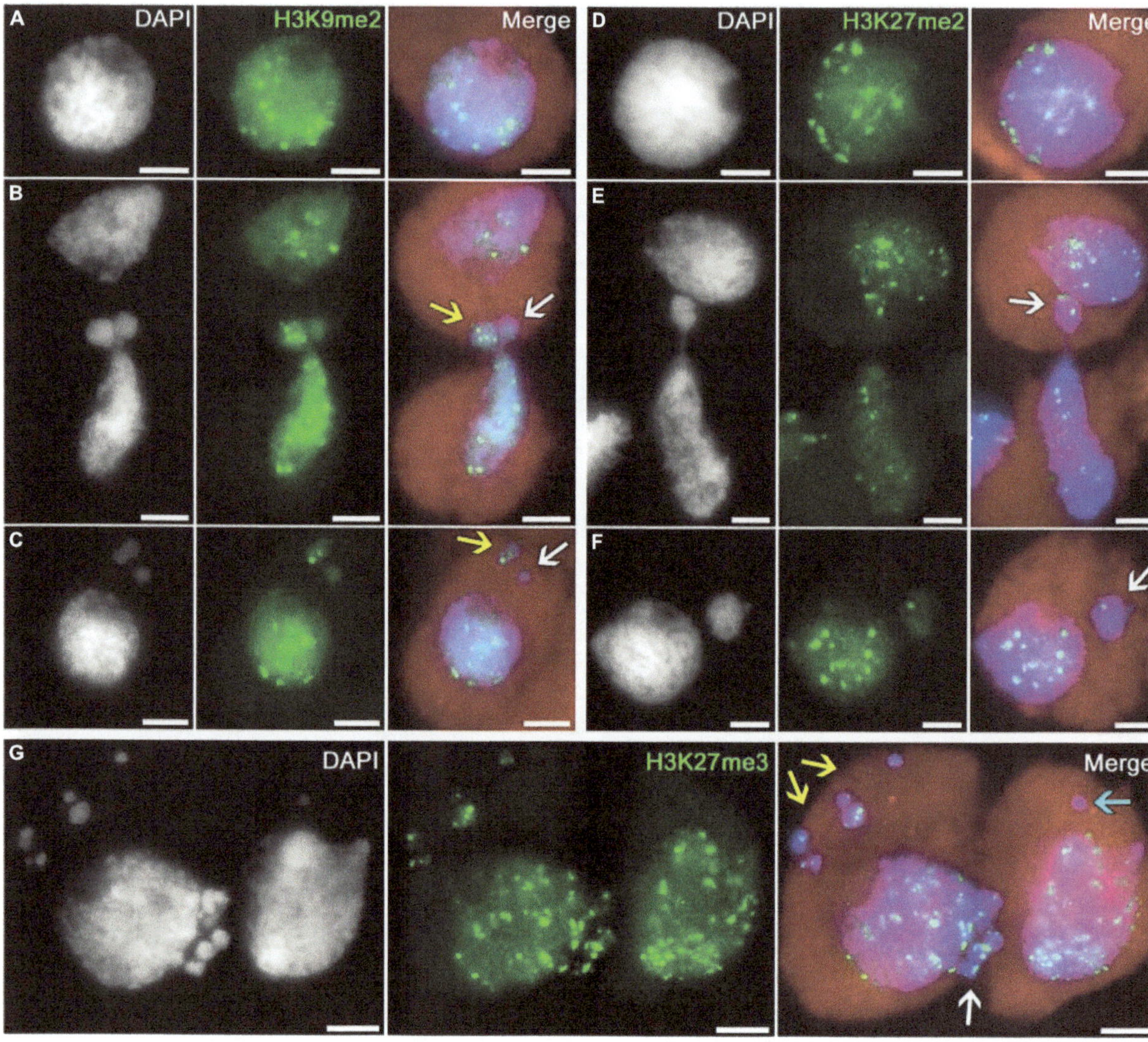

FIGURE 3 | Distributions of dimethylated histone H3K9 and di- and trimethylated H3K27 in tobacco microsporocytes in meiotic prophase I by cytomixis. (A) H3K9me2 and **(D)** H3K27me2 in intact microsporocytes; **(B)** H3K9me2 and **(E)** H3K27me2 in cytomictic microsporocytes (*arrows* denote the migrating part of nucleus that entered the recipient cell); **(C)** H3K9me2 and **(F)** H3K27me2 in microsporocytes after cytomixis (*arrows* denote micronuclei); and **(G)** H3K27me3 in cytomictic microsporocytes (*white arrow*, the migrating part of nucleus that entered the recipient cell; *yellow* and *turquoise arrows*, micronuclei). Red, cytoplasm autofluorescence in merged images; bars, 5 μm.

chromatid cohesion in cell division (Gernand et al., 2003)) and phosphorylation of histone H2A at threonine 121 (involved in Shugoshin binding, preserving cohesion of centromeric regions in cell division; Wang et al., 2012) during meiotic prophase I display the distributions similar to H3S10ph in intact tobacco microsporocyte nuclei, migrating nuclei in cytomixis, and the micronuclei formed after cytomixis (**Figures 5A–I**).

The migrating chromatin may contain different numbers of immunosignals (**Figures 5B,E,H;** *arrows*) whereas the micronuclei formed after cytomixis may either contain the signals (**Figures 5C,F,I;** *white arrows*) or lack them (**Figures 5F,I;** *yellow arrows*).

Thus, the phosphorylation patterns of histones H3 and H2A in cytomictic nuclei and micronuclei do not differ at all from the phosphorylation of these histones in intact microsporocyte nuclei, and match the normal degree of phosphorylation characteristic of the current meiotic stage.

The phosphorylation of histone H2A.X at serine 139 (γH2AX) is a marker of DNA double-strand breaks (Rogakou et al., 1999; Xiao et al., 2009). In the zygotene–pachytene, anti-H2A.XS139ph detects in tobacco microsporocytes numerous small signals dispersed over both the intact (**Figure 5J**) and migrating (**Figure 5K**) nuclei. Emergence of such double-strand DNA breaks is associated with recombination processes, taking place at this meiotic stage (Chicheportiche et al., 2007). There are no visible changes in the number of immunosignals in migrating chromatin (**Figure 5K**, *arrow*) or in cytomictic micronuclei (**Figure 5L**, *arrows*).

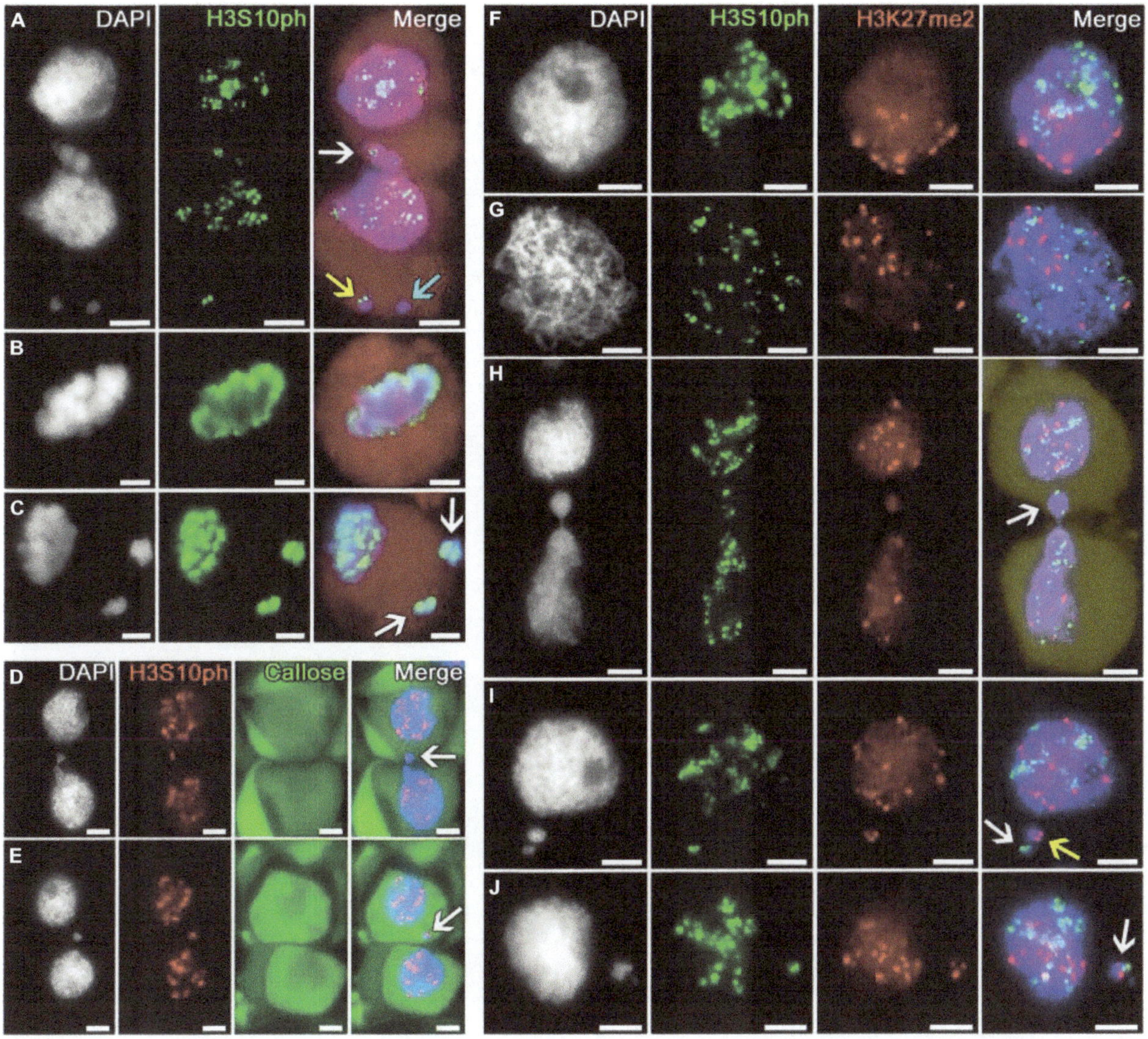

FIGURE 4 | Distribution of phosphorylated histone H3S10 in tobacco microsporocytes and its colocalization with dimethylated H3K27 by cytomixis. **(A)** H3S10ph in cytomictic microsporocytes (*white arrow* denotes the migrating part of nucleus that entered the recipient cell; *yellow* and *turquoise arrows*, micronuclei); H3S10ph in **(B)** an intact microsporocyte and **(C)** a microsporocyte after cytomixis in meiotic metaphase I (*arrows*, the chromatin of micronuclei; *red*, cell cytoplasm); **(D,E)** H3S10ph in cytomictic microsporocytes in sections of anthers embedded in PEG (*arrows*, the migrating part of nucleus that entered the recipient cell). The callose wall was stained with aniline blue (cytoplasmic autofluorescence presented in green); **(F)** double labeling of H3S10ph and H3K27me2 in intact microsporocytes in the zygotene and **(G)** pachytene; **(H)** double labeling of H3S10ph and H3K27me2 in cytomictic microsporocytes (*arrow* denotes the migrating part of nucleus that entered the recipient cell; *yellow*, cell cytoplasm); and **(I,J)** double labeling of H3S10ph and H3K27me2 in the microsporocytes after cytomixis (*arrows* denote micronuclei); bars, 5 μm.

Migrating Chromatin Contains Synaptonemal Complex

The SC is a chromosome structure formed during meiotic prophase I that provides a physical connection between homologous chromosomes. We have examined the presence of SC structures in cytomictic chromatin using electron microscopy and immunostaining for the SC proteins ZIP1 and ASY1.

The immunostaining for SC proteins in tobacco microsporocytes has demonstrated that the nuclei in zygotene–pachytene contain the ZIP1 and ASY1 proteins, which do not colocalize with one another (**Figure 6A**). Most likely, the transverse filament proteins ZIP1 appear after the axial SC elements, ASY1, disappear. The cytomictic cells display the immunosignals of both proteins (**Figures 6B,C**), which do not colocalize and gradually come and go during prophase, similar to the pattern characteristic of intact cells. Depending on the prophase stage, either ASY1 (**Figure 6B**, *arrows*) or ZIP1 (**Figure 6C**, *arrow*) may be prevalent in the migrating chromatin.

In the electron micrographs, SC is visible in the migrating nucleus before chromatin enters the cytomictic channel (**Figure 6D**, *black arrowheads*), while the chromatin

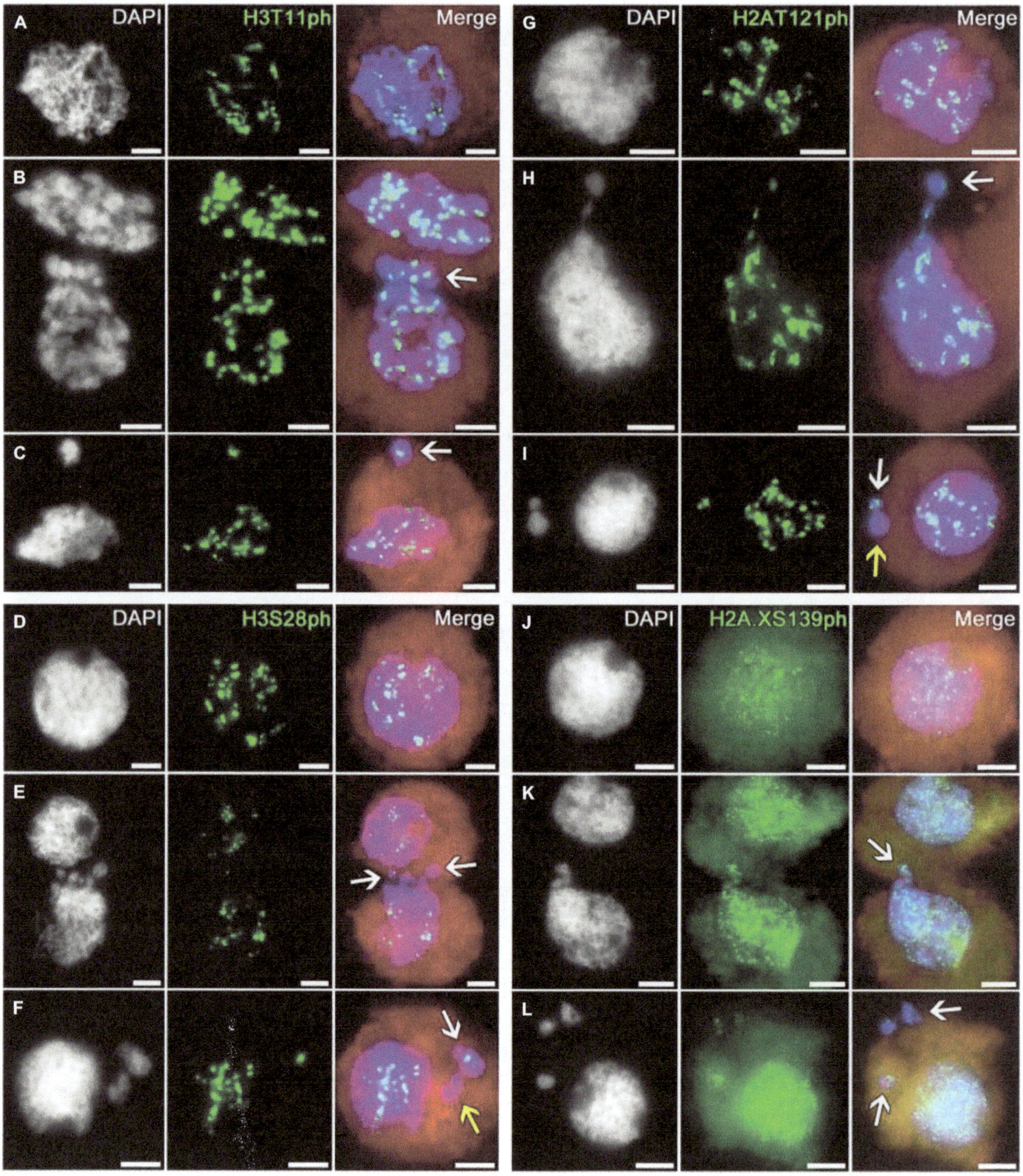

FIGURE 5 | Distributions of phosphorylated histones H3T11, H3S28, H2AT121, and H2A.XS139 in tobacco microsporocytes in meiotic prophase I by cytomixis. (A) H3T11ph, **(D)** H3S28ph, **(G)** H2AT121ph, and **(J)** H2A.XS139ph in intact microsporocytes; **(B)** H3T11ph, **(E)** H3S28ph, **(H)** H2AT121ph, and **(K)** H2A.XS139ph in cytomictic microsporocytes (*arrows* denote the migrating part of nucleus that entered the recipient cell); and **(C)** H3T11ph, **(F)** H3S28ph, **(I)** H2AT121ph, and **(L)** H2A.XS139ph in microsporocytes after cytomixis (*arrows* denote micronuclei). Red, cytoplasm autofluorescence in merged images; bars, 5 μm.

inside the channels becomes denser and looks as a black structure (**Figures 6D,E**, *yellow arrows*) with undetectable SC elements. However, the chromatin restores its structure immediately upon leaving the cytomictic channel and entering the recipient cell (**Figure 6E**, *black arrowhead*).

Thus, immunostaining and electron microscopy confirm the presence of SC-like structures in the migrating chromatin

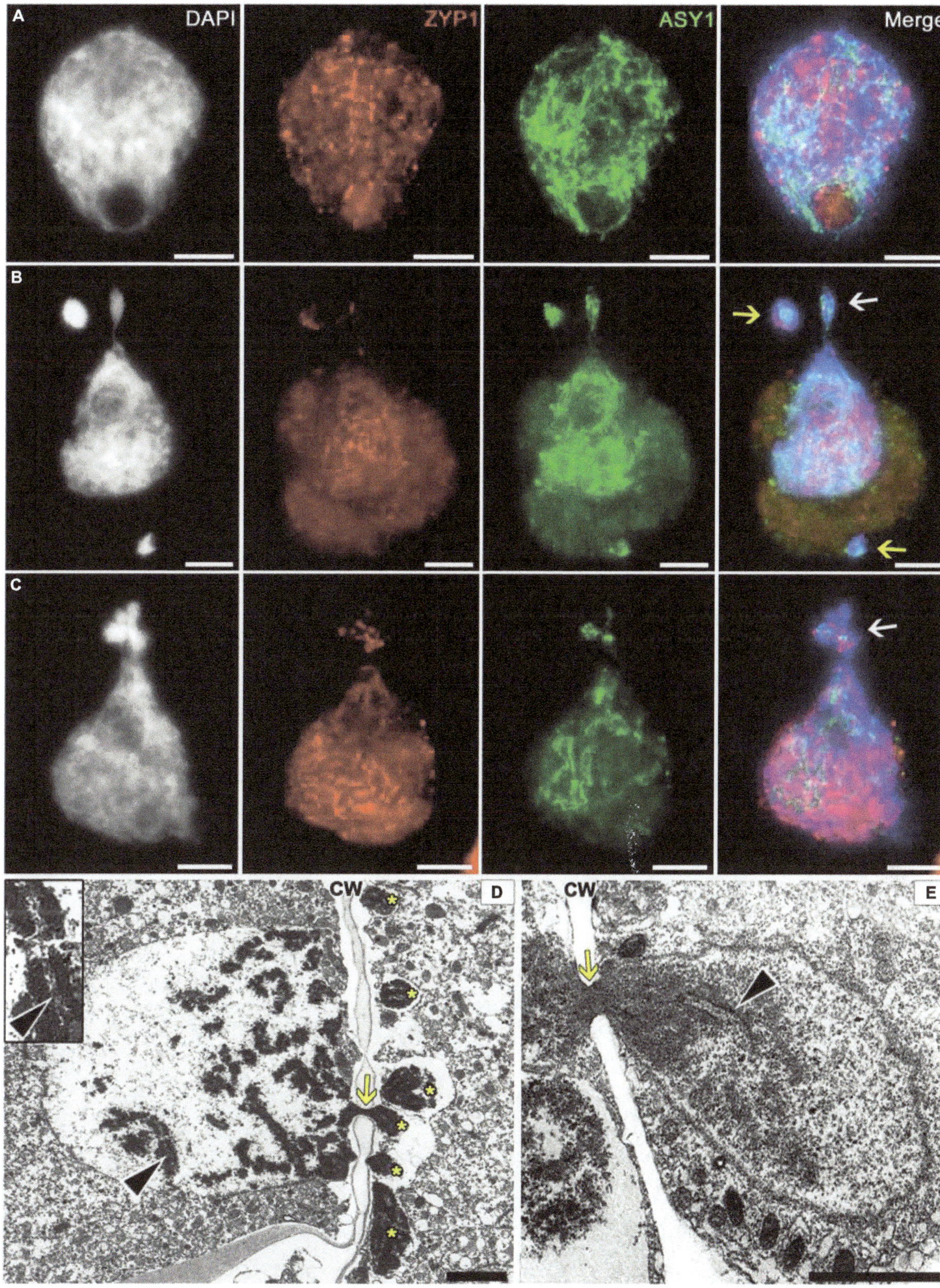

FIGURE 6 | Localization of the synaptonemal complex (SC) in tobacco cytomictic microsporocytes by (A–C) immunostaining and (D,E) electron microscopy. Distributions of the ZIP1 and ASY1 proteins in **(A)** intact microsporocytes; **(B,C)** cytomictic microsporocytes (*white arrows* denote the migrating part of nucleus and *yellow arrows,* micronuclei); **(D,E)** the SC (*black arrowheads*) in migrating nuclei (*asterisks,* the part of the nucleus that entered recipient cell; *yellow arrows,* cytomictic channel filled with chromatin; and *CW,* cell wall); bars, 5 μm in **(A–C)** and 2 μm in **(D,E)**.

and the micronuclei formed in the recipient cells after cytomixis.

The Cytomictic Cells Lack Any Signs of Apoptotic DNA Degradation

In order to test whether the processes of programmed cell death (PCD) are triggered in the cell after it has acquired "surplus" chromatin owing to cytomixis, we have searched for the signs of DNA damage characteristic of PCD. The signs suggesting internucleosomal DNA fragmentation were searched for by electrophoretic separation of the genomic DNA isolated from tobacco microsporocytes in the zygotene–pachytene (when the cytomictic rate is maximum). The TUNEL assay was used to search *in situ* for the signs of DNA damage in cytomictic tobacco microsporocytes.

TUNEL assay has shown that the cytomictic cells display no signs of DNA degradation both when the nuclei migrate through the cytomictic channel and on completion of migration and formation of micronuclei (**Figures 7A,B**; *arrows*). The nucleus of a recipient cell after the micronuclei are formed also displays no signs of DNA degradation (**Figure 7B**). TUNEL assay gave positive results only for the positive control, namely, the tobacco microsporocytes with cytomictic micronuclei (**Figure 7C**, *arrow*) and somatic anther cells (**Figure 7D**, *arrow*) heat-shocked to induce apoptosis-like changes. The test for the presence of internucleosomal DNA fragmentation was also positive only in the case of the positive (heat-shocked) control (**Figure 7E**, *lane 1*), while the genomic DNA isolated from intact tobacco microsporocytes (**Figure 7E**, *lane 2*) and leaf cells (**Figure 7E**, *lane 3*) displayed no signs of internucleosomal fragmentation. Since only up to 4% of cells exhibiting cytomixis in SR1 plants (Sidorchuk et al., 2007), a large amount (5 μg) of genomic DNA was assayed, but even in this case no DNA fragmentation was detected. Thus, any signs of the DNA damages characteristic of PCD are absent in the tobacco microsporocytes undergoing cytomixis.

DISCUSSION

Cytomixis was discovered over a century ago; however, the mechanism driving the migration of chromatin from one cell to another is still vague. On the one hand, it is believed that cytomixis is a pathological process associated with elimination of damaged chromatin/cells (Kravets, 2011; Barton et al., 2014).

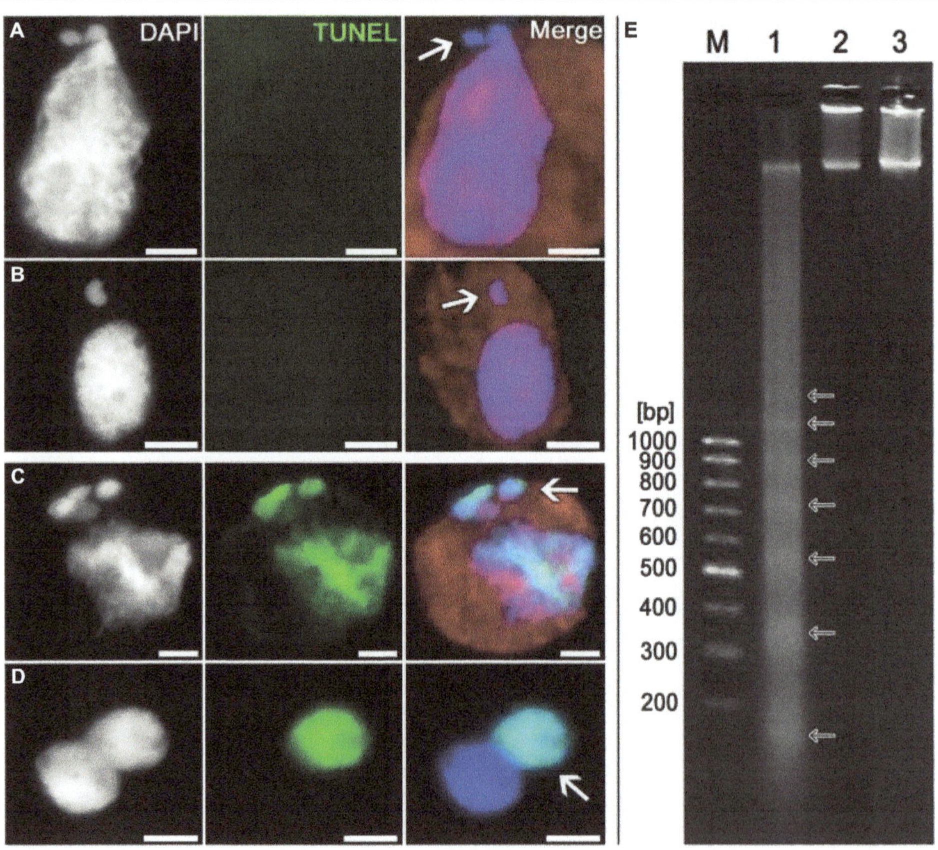

FIGURE 7 | Detection of DNA fragmentation in cytomictic cells by TUNEL assay and agarose gel electrophoresis. TUNEL assay data for the tobacco microsporocytes **(A)** by cytomixis (*arrow* denotes the migrating part of the nucleus) and **(B)** after cytomixis (*arrow* denotes the micronucleus); positive control after heat shock of **(C)** microsporocytes after cytomixis (*arrows* denote micronuclei) and **(D)** somatic cells (*arrow*, TUNEL-positive nucleus); bars, 5 μm. **(E)** DNA electrophoretic pattern (agarose gel): M, molecular weight marker; 1, positive control; 2, DNA of untreated microsporocytes; and 3, DNA of untreated leaf cells.

On the other hand, cytomixis is assumed to produce gametes with different chromosomes number (Falistocco et al., 1995; Ghaffari, 2006; Negron-Ortiz, 2007; Pécrix et al., 2011; Lavia et al., 2011). Yet none of these hypotheses have any direct experimental confirmation. First and foremost, this is associated with the absence of methods enabling the tracing of a microsporocyte from the moment when it acquired additional chromatin by cytomixis to the moment when it gives a tetrad of microspores and then gametes. Moreover, it is impossible to unambiguously identify migrated chromatin in microsporogenesis after meiotic metaphase I, and far less to prove its cytomictic origin, since no specific markers for cytomictic chromatin have been yet discovered.

Nonetheless, we have succeeded in demonstrating that the migrating chromatin in case of cytomixis in tobacco microsporogenesis lacks any signs of altered chromatin structure and that the cells involved in this process display no signs of PCD.

Cytomixis does not Change the State of Migrating Chromatin

If either migrating chromatin or whole cells were degraded in case of cytomixis, the markers characteristic of this process would be detectable. One of the most illustrative examples of such chromatin degradation is the meiotic silencing and fragmentation of the male zebra finch germline-restricted chromosome (Goday and Pigozzi, 2010; Schoenmakers et al., 2010). In this case, the chromatin is eliminated from the nucleus during male meiotic prophase I and undergoes characteristic changes in its histone modifications. The eliminated chromatin loses the di- and trimethylation of histone H3 at lysine 4 (markers of euchromatin), is subject to pronounced di- and trimethylation of histone H3 at lysine 9 (markers of heterochromatin) as well as to surplus phosphorylation of histone H2A.X at serine 139 (marker of DNA double-strand breaks), thus becoming TUNEL-positive (Goday and Pigozzi, 2010; Schoenmakers et al., 2010). Such changes are characteristic of the inactivated chromatin destined to be eliminated. However, despite that cytomixis in tobacco takes place at the same stage of male meiosis as chromosome elimination in the zebra finch, the analogous changes in tobacco cytomictic chromatin are undetectable.

Using six euchromatin markers - H3K4me2, H3K4me3, H3K9ac, H3K14ac, H3K18ac, and H3K27ac—we have demonstrated that the chromatin migrating between cytomictic cells does not change its state. These chromatin marks also retained in the micronuclei formed after cytomixis in recipient cells. These data are also confirmed by analysis of the distributions of heterochromatin markers in migrating chromatin, namely, H3K9me2, H3K27me2, and H3K27me3. Before migration to another cell, the chromatin is not selectively heterochromatized nor is it inactivated in the micronuclei after cytomixis. The constitutive heterochromatin blocks detectable in tobacco microsporocyte nuclei are not associated with cytomixis and are randomly included into the migrating chromatin; note that there are no obvious changes of immunosignals in the micronuclei after cytomixis.

Analysis of the key markers for DNA damage—H2A.XS139ph (γH2AX), TUNEL, and DNA internucleosomal fragmentation—also suggests the absence of changes in migrating chromatin. When entering another cell and upon cytomixis completion, the migrating chromatin displays no signs of DNA damage. Any signs of DNA damage are also absent in the nuclei of recipient cells.

In the case of the zebra finch, the germline-restricted chromosome fragments remain detectable in the cell cytoplasm as a densely packed chromatin body until the end of meiosis (Goday and Pigozzi, 2010; Schoenmakers et al., 2010). Micronucleation is also observed in plant microsporogenesis, for example, of *Avena sativa* and *Glycine max,* when part of chromosomes remains beyond the nucleus during division, further separating as affected small microsporocytes, and eventually giving small sterile pollen (Baptista-Giacomelli et al., 2000; Ortega and Dorrance, 2011). Despite careful analysis, we have not detected any small microsporocytes during microsporogenesis of tobacco plants (Mursalimov and Deineko, 2015).

Thus, our data do not support the hypothesis that the migrated chromatin is degraded in case of cytomixis, or the cells involved in this process die before the tetrad is formed.

Migrated Chromatin Contains Histone Marks Which Correlate with Chromosome Segregation

The phosphorylation of histone H3 at serines 10 and 28 and threonine 11, as well as of histone H2A at threonine 121, is critical for chromosome condensation and segregation in dividing cells (Manzanero et al., 2000; Houben et al., 2005, 2007; Kawashima et al., 2010).

We have shown that the phosphorylation of histones in the cytomictic microsporocytes does not differ from the normal process. The distribution of the above listed markers assumes correct chromatin condensation and segregation. Moreover, the case study of H3S10ph shows that the phosphorylation of histones also continues after migration in the micronuclei formed as a result of cytomixis. After completion of cytomixis, the micronuclei carrying the H3S10ph signal as individual loci are formed in tobacco microsporocytes in the pachytene; however, the chromatin in micronuclei by metaphase I is phosphorylated over the entire length, as during normal meiosis. This suggests that the recipient cell discerns no difference between the introduced and its own chromatin, so that the former is subject to further phosphorylation and receives the same stage-specific signals from the cytoplasm as the chromatin of the own recipient cell nuclei. Since the phosphorylation of histone H3 at serines 10, 28 as well as of histone H2A at threonine121 is coupled with active function of the centromeres and, this suggests a normal function of the latter in the migrating chromatin.

The presence of the phosphorylation signal of histone H2A.X at serine 139 in the cytomictic chromatin during zygotene–pachytene implies the presence of double-strand DNA breaks, which are associated with recombination, taking place at this meiotic stage (Chicheportiche et al., 2007). This suggests that recombination in the migrating chromatin is similar to the remaining chromatin of microsporocytes at this meiotic stage. However, recombination is feasible only provided that chromosomes are paired and bivalents are likely formed.

Electron microscopy examination and immunostaining of the SC proteins ZIP1 and ASY1 in tobacco microsporocytes confirm the presence of SC in the migrating chromatin. Note also that the ZIP1 and ASY1 proteins were only detected in microsporocyte nuclei in the zygotene–pachytene, and do not colocalize as in *Arabidopsis* (Higgins et al., 2005). Most likely, the filament proteins ZIP1 appear after the axial SC elements, ASY1, disappear, as is observed in the wheat (Khoo et al., 2012). Thus, the migrating chromatin shows histone marks for correct chromosome condensation and segregation according to the expected state at this meiotic stage, and furthermore contain key protein components of SC.

Signals of double labeling with antibodies against H3S10ph and H3K27me2 in tobacco microsporocytes has shown that they cluster at the opposite poles of the nucleus in the zygotene–pachytene (bouquet) stage, which is characteristic of centromeric and telomeric chromosome regions (Wen et al., 2012; Zhang et al., 2013). The H3S10ph signal is detected in the pericentromeric chromosome regions in mitosis and the second meiotic division (Manzanero et al., 2000; Houben et al., 2007); there are good grounds to suppose that phosphorylation of H3S10 in meiotic prophase I starts in the pericentromeric chromosome regions (Oliver et al., 2013). It is also known that the H3 dimethylated at lysine 27 can be detected in subtelomeric chromosome regions (Carchilan et al., 2007). The clustering of the H3K27me2 signal at the nuclear pole opposite to the centromeres at the bouquet stage suggests that this signal in tobacco microsporocyte nuclei marks the subtelomeric regions of some chromosomes. This assumption allows us to assess the integrity of the chromosomes migrating to other cells in case of cytomixis. In particular, the micronuclei that carry the "centromeric" signal and several "telomeric" ones are observed in the recipient cells after cytomixis, which is characteristic of a whole chromosome. The probability that the cytomictic chromatin containing centromeres and telomeres represents a normal bivalent remains to be tested. The fate of the cytomictic chromatin where only one of the signals is detected is questionable.

Thus, we have demonstrated that the chromatin migrating from one cytomictic cell to another shows no signs of chromatin alterations and DNA damage.

ACKNOWLEDGMENTS

We thank Kathrin Kumke (IPK, Gatersleben) for her help in the tissue section preparation. The work was supported by the DAAD (A/14/00237), the Russian Foundation for Basic Research (grant no. 14-04-00992 A) and the German Research Foundation, SFB 648 "Molecular mechanisms of information processing in plants." The electron microscopy was performed at the Joint Access Center for Microscopic Analysis of Biological Objects with the Siberian Branch of the Russian Academy of Sciences.

REFERENCES

Baptista-Giacomelli, F. R., Pagliarini, M. S., and Almeida, J. L. (2000). Elimination of micronuclei from microspores in a Brazilian oat (*Avena sativa* L.) variety. *Genet. Mol. Biol.* 23, 681–684. doi: 10.1590/S1415-47572000000300029

Barton, D. A., Cantrill, L. C., Law, A. M. K., Phillips, C. G., Sutton, B. G., and Overall, R. L. (2014). Chilling to zero degrees disrupts pollen formation but not meiotic microtubule arrays in *Triticum aestivum* L. *Plant. Cell Environ.* 37, 2781–2794. doi: 10.1111/pce.12358

Carchilan, M., Delgado, M., Ribeiro, T., Costa-Nunes, P., Caperta, A., Morais-Cecílio, L., et al. (2007). Transcriptionally active heterochromatin in rye B chromosomes. *Plant Cell* 19, 1738–1749. doi: 10.1105/tpc.106.046946

Chicheportiche, A., Bernardino-Sgherri, J., Massy, B., and Dutrillaux, B. (2007). Characterization of Spo11-dependent and independent phospho-H2AX foci during meiotic prophase I in the male mouse. *J. Cell Sci.* 120, 1733–1742. doi: 10.1242/jcs.004945

Du, Z., Li, H., Wei, Q., Zhao, X., Wang, C., Zhu, Q., et al. (2013). Genome-wide analysis of histone modifications: H3K4me2. H3K4me3, H3K9ac, and H3K27ac in *Oryza sativa* L. *J. Mol. Plant.* 6, 1463–1472. doi: 10.1093/mp/sst018

Falistocco, E., Tosti, N., and Falcinelli, M. (1995). Cytomixis in pollen mother cells of diploid Dactylis, one of the origins of 2n gametes. *J. Hered.* 86, 448–453.

Fan, T., and Xing, T. (2004). Heat shock induces programmed cell death in wheat leaves. *Biol. Plant.* 48, 389–394. doi: 10.1023/B:BIOP.0000041091.03068.ee

Fuchs, J., and Schubert, I. (2012)). Chromosomal distribution and functional interpretation of epigenetic histone marks. *Plant Cytogenet.* 4, 231–253. doi: 10.1007/978-0-387-70869-0_9

Gernand, D., Demidov, D., and Houben, A. (2003). The temporal and spatial pattern of histone H3 phosphorylation at serine 28 and serine 10 is similar in plants but differs between mono- and polycentric chromosomes. *Cytogenet. Genome. Res.* 101, 172–176. doi: 10.1159/000074175

Ghaffari, S. M. (2006). Occurrence of diploid and polyploid microspores in Sorghum bicolor (Poaceae) is the result of cytomixis. *Afr. J. Biotechnol.* 5, 1450–1453.

Goday, C., and Pigozzi, M. I. (2010). Heterochromatin and histone modifications in the germline-restricted chromosome of the zebra finch undergoing elimination during spermatogenesis. *Chromosoma* 119, 325–336. doi: 10.1007/s00412-010-0260-2

He, S., Yan, S., Wang, P., Zhu, W., Wang, X., Shen, Y., et al. (2014). Comparative analysis of genome-wide chromosomal histone modification patterns in maize cultivars and their wild relatives. *PLoS ONE* 9:e97364. doi: 10.1371/journal.pone.0097364

Higgins, J. D., Sanchez-Moran, E., Armstrong, S. J., Jones, G. H., and Franklin, F. C. H. (2005). The *Arabidopsis* synaptonemal complex protein ZYP1 is required for chromosome synapsis and normal fidelity of crossing over. *Genes Dev.* 19, 2488–2500. doi: 10.1101/gad.354705

Horst, I., Offermann, S., Dreesen, B., Niessen, M., and Peterhansel, C. (2009). Core promoter acetylation is not required for high transcription from the phosphoenolpyruvate carboxylase promoter in maize. *Epigenetics Chromatin* 2:17. doi: 10.1186/1756-8935-2-17

Houben, A., Demidov, D., Caperta, A. D., Karimi, R., Agueci, F., and Vlasenko, L. (2007). Phosphorylation of histone H3 in plants - A dynamic affair. *Biochim. Biophys. Acta Gene Struct. Expr.* 1769, 308–315. doi: 10.1016/j.bbaexp.2007.01.002

Houben, A., Demidov, D., Rutten, T., and Scheidtmann, K. H. (2005). Novel phosphorylation of histone H3 at threonine 11 that temporally correlates with condensation of mitotic and meiotic chromosomes in plant cells. *Cytogenet. Genome Res.* 109, 148–155. doi: 10.1159/000082394

Houben, A., Kumke, K., Nagaki, K., and Hause, G. (2011). CENH3 distribution and differential chromatin modifications during pollen development in rye (*Secale cereale* L.). *Chromosom. Res.* 19, 471–480. doi: 10.1007/s10577-011-9207-6

Jang, I. C., Chung, P. J., Hemmes, H., Jung, C., and Chua, N. H. (2011). Rapid and reversible light-mediated chromatin modifications of *Arabidopsis* phytochrome a locus. *Plant Cell* 23, 459–470. doi: 10.1105/tpc.110.080481

Kawashima, S. A., Yamagishi, Y., Honda, T., Ishiguro, K., and Watanabe, Y. (2010). Phosphorylation of H2A by Bub1 prevents chromosomal instability through localizing shugoshin. *Science* 327, 172–177. doi: 10.1126/science.1180189

Khoo, K. H., Able, A. J., and Able, J. A. (2012). The isolation and characterisation of the wheat molecular ZIPper I homologue. TaZYP1. *BMC Res.* 5:106. doi: 10.1186/1756-0500-5-106

Kravets, E. (2011). The role of cell selection for pollen grain fertility after treatment of barley sprouts (*Hordeum distichum* L.) with UV-B irradiation. *Acta Biol. Slov.* 54, 31–41.

Lavia, G. I., Ortiz, A. M., Robledo, G., Fernández, A., and Seijo, G. (2011). Origin of triploid *Arachis pintoi* (*Leguminosae*) by autopolyploidy evidenced by FISH and meiotic behaviour. *Ann. Bot.* 108, 103–111. doi: 10.1093/aob/mcr108

Liu, Y., Hui, R. K., Deng, R. N., Wang, J. J., Wang, M., and Li, Z. Y. (2012). Abnormal male meiosis explains pollen sterility in the polyploid medicinal plant *Pinellia ternata* (Araceae). *Genet. Mol. Res.* 11, 112–120. doi: 10.4238/2012.January.17.1

Lone, A., and Lone, S. (2013). Cytomixis – a well known but less understood phenomenon in plants. *Int. J. Recent Sci. Res.* 4, 347–352.

Manzanero, S., Arana, P., Puertas, M. J., and Houben, A. (2000). The chromosomal distribution of phosphorylated histone H3 differs between plants and animals at meiosis. *Chromosoma* 109, 308–317. doi: 10.1007/s004120000087

Mursalimov, S., Sidorchuk, Y., and Deineko, E. (2013a). The role of spherosome-like vesicles in formation of cytomictic channels between tobacco microsporocytes. *Biol. Plant.* 57, 291–297. doi: 10.1007/s10535-012-0276-y

Mursalimov, S. R., Sidorchuk, Y. V., and Deineko, E. V. (2013b). New insights into cytomixis: specific cellular features and prevalence in higher plants. *Planta* 238, 415–423. doi: 10.1007/s00425-013-1914-0

Mursalimov, S. R., Baiborodin, S. I., Sidorchuk, Y. V., Shumny, V. K., and Deineko, E. V. (2010). Characteristics of the cytomictic channel formation in *Nicotiana tabacum* L. pollen mother cells. *Tsitol. Genet.* 44, 19–24. doi: 10.3103/S0095452710010032

Mursalimov, S. R., and Deineko, E. V. (2011). An ultrastructural study of cytomixis in tobacco pollen mother cells. *Protoplasma* 248, 717–724. doi: 10.1007/s00709-010-0234-5

Mursalimov, S. R., and Deineko, E. V. (2012). An ultrastructural study of microsporogenesis in tobacco line SR1. *Biologia* 67, 369–376. doi: 10.2478/s11756-012-0005-1

Mursalimov, S. R., and Deineko, E. V. (2015). How cytomixis can form unreduced gametes in tobacco. *Plant Syst. Evol.* 301, 1293–1297. doi: 10.1007/s00606-014-1150-5

Negron-Ortiz, V. (2007). Chromosome numbers, nuclear DNA content, and polyploidy in Consolea (Cactaceae), an endemic cactus of the Caribbean Islands. *Am. J. Bot.* 94, 1360–1370. doi: 10.3732/ajb.94.8.1360

Oliver, C., Pradillo, M., Corredor, E., and Cuñado, N. (2013). The dynamics of histone H3 modifications is species-specific in plant meiosis. *Planta* 238, 23–33. doi: 10.1007/s00425-013-1885-1

Ortega, M. A., and Dorrance, A. E. (2011). Microsporogenesis of Rps8/rps8 heterozygous soybean lines. *Euphytica* 181, 77–88. doi: 10.1007/s10681-011-0422-1

Pécrix, Y., Rallo, G., Folzer, H., Cigna, M., Gudin, S., and Bris, M. L. (2011). Polyploidization mechanisms: temperature environment can induce diploid gamete formation in *Rosa* sp. *J. Exp. Bot.* 62, 3587–3597. doi: 10.1093/jxb/err052

Rogakou, E. P., Boon, C., Redon, C., and Bonner, W. M. (1999). Megabase chromatin domains involved in DNA double-strand breaks in vivo. *J. Cell Biol.* 146, 905–915. doi: 10.1083/jcb.146.5.905

Sambrook, J., Fritsch, E. F., and Maniatis, T. (1989). *Molecular Cloning: A Laboratory Manual.* Cold Spring Harbor, NY: Cold Spring Harbor Laboratory Press.

Schoenmakers, S., Wassenaar, E., Laven, J. S. E., Grootegoed, J. A., and Baarends, W. M. (2010). Meiotic silencing and fragmentation of the male germline restricted chromosome in zebra finch. *Chromosoma* 119, 311–324. doi: 10.1007/s00412-010-0258-9

Sidorchuk, Y. V., Deineko, E. V., and Shumny, V. K. (2007). Peculiarities of cytomixis in pollen mother cells of transgenic tobacco plants (*Nicotiana tabacum* L.) with mutant phenotype. *Cell Tissue Biol.* 1, 570–576. doi: 10.1134/S1990519X07060144

Thiriet, C., and Hayes, J. J. (2005). Chromatin in need of a fix: phosphorylation of H2AX connects chromatin to DNA repair. *Mol. Cell* 18, 617–622. doi: 10.1016/j.molcel.2005.05.008

Wang, M., Tang, D., Luo, Q., Jin, Y., Shen, Y., Wang, K., et al. (2012). BRK1, a Bub1-related kinase, is essential for generating proper tension between homologous kinetochores at metaphase I of rice meiosis. *Plant Cell* 24, 4961–4973. doi: 10.1105/tpc.112.105874

Wang, X. Y., Nie, X. W., Guo, G. Q., Pan, Y. F., and Zheng, G. C. (2002). Ultrastructural characterization of the cytoplasmic channel formation between pollen mother cells of David lily. *Caryologia* 55, 161–169. doi: 10.1080/00087114.2002.10589272

Wang, X. Y., Yu, C. H., Li, X., Wang, C. Y., and Zheng, G. C. (2004). Ultrastructural aspects and possible origin of cytoplasmic channels providing intercellular connection in vegetative tissues of anthers. *Russ. J. Plant Physiol.* 51, 110–120.

Wen, R., Moore, G., and Shaw, P. J. (2012). Centromeres cluster de novo at the beginning of meiosis in *Brachypodium distachyon*. *PLoS ONE* 7:e44681. doi: 10.1371/journal.pone.0044681

Xiao, A., Li, H., Shechter, D., Ahn, S. H., Fabrizio, L. A., Erdjument-Bromage, H., et al. (2009). WSTF regulates the H2A.X DNA damage response via a novel tyrosine kinase activity. *Nature* 457, 57–62. doi: 10.1038/nature 07668

Yu, C. H., Guo, G. Q., Nie, X. W., and Zheng, G. C. (2004). Cytochemical localization of pectinase activity in pollen mother cells of tobacco during meiotic prophase?and its relation to the formation of secondary plasmodesmata and cytoplasmic channels. *Acta Bot. Sin.* 46, 1443–1453.

Zhang, J., Pawlowski, W. P., and Han, F. (2013). Centromere pairing in early meiotic prophase requires active centromeres and precedes installation of the synaptonemal complex in maize. *Plant Cell* 25, 3900–3909. doi: 10.1105/tpc.113.117846

Conflict of Interest Statement: The authors declare that the research was conducted in the absence of any commercial or financial relationships that could be construed as a potential conflict of interest.

3

Sunflower centromeres consist of a centromere-specific LINE and a chromosome-specific tandem repeat

Kiyotaka Nagaki[1], Keisuke Tanaka[2], Naoki Yamaji[1], Hisato Kobayashi[2] and Minoru Murata[1]*

[1] Applied Genomics Unit, Institute of Plant Science and Resources, Okayama University, Kurashiki, Japan, [2] NODAI Genome Research Center, Tokyo University of Agriculture, Setagaya, Japan

The kinetochore is a protein complex including kinetochore-specific proteins that plays a role in chromatid segregation during mitosis and meiosis. The complex associates with centromeric DNA sequences that are usually species-specific. In plant species, tandem repeats including satellite DNA sequences and retrotransposons have been reported as centromeric DNA sequences. In this study on sunflowers, a cDNA-encoding centromere-specific histone H3 (CENH3) was isolated from a cDNA pool from a seedling, and an antibody was raised against a peptide synthesized from the deduced cDNA. The antibody specifically recognized the sunflower CENH3 (HaCENH3) and showed centromeric signals by immunostaining and immunohistochemical staining analysis. The antibody was also applied in chromatin immunoprecipitation (ChIP)-Seq to isolate centromeric DNA sequences and two different types of repetitive DNA sequences were identified. One was a long interspersed nuclear element (LINE)-like sequence, which showed centromere-specific signals on almost all chromosomes in sunflowers. This is the first report of a centromeric LINE sequence, suggesting possible centromere targeting ability. Another type of identified repetitive DNA was a tandem repeat sequence with a 187-bp unit that was found only on a pair of chromosomes. The HaCENH3 content of the tandem repeats was estimated to be much higher than that of the LINE, which implies centromere evolution from LINE-based centromeres to more stable tandem-repeat-based centromeres. In addition, the epigenetic status of the sunflower centromeres was investigated by immunohistochemical staining and ChIP, and it was found that centromeres were heterochromatic.

Edited by:
Inna Lermontova,
Leibniz Institute for Plant Genetics and Crop Plant Research, Germany

Reviewed by:
Andreas Houben,
Leibniz Institute for Plant Genetics and Crop Plant Research, Germany
Jiri Macas,
Biology Centre of the Academy of Sciences, Czech Republic

***Correspondence:**
Kiyotaka Nagaki
nagaki@rib.okayama-u.ac.jp

Keywords: centromere, sunflower (*Helianthus annuus*), centromeric histone H3, centromeric DNA, ChIP-seq

INTRODUCTION

Asterales is the most diverged order of dicots and includes 11 families and 27,000 species. Among the 11 families, Asteraceae is the largest and includes the sunflower and daisy. Sunflowers (*Helianthus annuus* L., $2n = 2x = 34$, genome size = 2.43 Gb/haploid) are one of the most important crops in Asterales because their seeds can be used for oil production (Bennett et al., 1982). Sunflowers have been genetically and cytogenetically investigated (Feng et al., 2013), and a genome sequencing project is now in progress (http://sunflowergenome.org/). For karyotypic analyses, repetitive DNA sequences including rDNA, sunflower-specific tandem repeats and bacterial artificial chromosome (BAC) clones have been used as probes, but none of these probes showed centromeric localization (Ceccarelli et al., 2007; Talia et al., 2010; Feng et al., 2013). In other words, at present, there is no genetic and cytogenetic marker for sunflower centromeres.

The kinetochore is a special protein complex formed on centromeric regions that ensures equal and accurate distribution of chromatids to daughter cells during mitosis and meiosis (Amor et al., 2004). Among the constitutive proteins of kinetochores, centromere-specific histone H3 (CENH3) acts as a base for assembling other kinetochore proteins, and its presence epigenetically determines the kinetochore position (Perpelescu and Fukagawa, 2011). The first identified CENH3 was CENP-A in humans (Earnshaw and Rothfield, 1985), and its orthologs have been isolated from 11 of 63 APG III orders, including Poales, Asparagales, Rosales, Fabales, Malpighiales, Malvales, Brassicales, Myrtales, Solanales, Asterales, and Apiales, in this decade (Talbert et al., 2002; Zhong et al., 2002; Nagaki et al., 2004, 2005, 2012b; Nagaki and Murata, 2005; Sanei et al., 2011; Wang et al., 2011; Neumann et al., 2012, 2015; Dunemann et al., 2014; Masonbrink et al., 2014; He et al., 2015; Maheshawari, 2015). However, no CENH3 has been isolated from Asterales species.

The CENH3s possess relatively conserved histone fold domains (HFDs) and highly variable N-terminal tails, and the HFD has been shown to be important for centromere localization (Vermaak et al., 2002; Black et al., 2004; Lermontova et al., 2006). The Loop 1 region on the HFD is longer than that of canonical histone H3, which is common among CENH3s. This region allows more compact packing of nucleosomes with CENH3 compared to those with canonical histone H3 (Black et al., 2004; Tek et al., 2010, 2011; Nagaki et al., 2012a; Maheshawari, 2015).

CENH3 is a component of core histones in the centromeric regions, and it directly binds to DNA. Additionally, CENH3 localizes to functional centromeres (Warburton et al., 1997; Nasuda et al., 2005; Han et al., 2006). Therefore, centromeric DNA sequences have been identified in plant species by chromatin immunoprecipitation (ChIP) using anti-CENH3 antibodies (Zhong et al., 2002; Nagaki et al., 2003, 2004, 2009, 2011, 2012a,b; Nagaki and Murata, 2005; Houben et al., 2007; Tek et al., 2010, 2011; Wang et al., 2011; Gong et al., 2012; Neumann et al., 2012; He et al., 2015). In most cases, species-specific microsatellites, minisatellites, macrosatellites, and retrotransposons have been identified as the centromeric sequences, and these sequences are located on all centromeric regions in the species (Zhong et al., 2002; Nagaki et al., 2003, 2009, 2011; Nagaki and Murata, 2005; Houben et al., 2007; Tek et al., 2010, 2011; Wang et al., 2011; Neumann et al., 2012; He et al., 2015).

Histone modifications are the key components in epigenetics, and they have been much investigated in this decade (Desvoyes et al., 2014; Sharma et al., 2015). In plant species, two different types of heterochromatin distribution are reported to be related to the genome size of the species (Houben et al., 2003). In plant species with a genome size smaller than 510 Mb, chromocenters appear on interphase nuclei, and heterochromatic modifications occur specifically on the chromocenters. *A. thaliana* is a good example for small-genome species, as it shows heterochromatic modifications on centromeric and pericentromeric regions not only at interphase but also at metaphase. Euchromatic modifications of the species were observed in the regions with no heterochromatic modifications (Houben et al., 2003; Jasencakova et al., 2003). In contrast, no chromocenters appear on interphase nuclei in plant species with genomes larger than 530 Mb. However, in these species, heterochromatic and euchromatic modifications are observed to disperse on interphase nuclei (Houben et al., 2003), and neither modification appears on centromeric and pericentromeric regions during metaphase.

The epigenetic modifications around centromeric regions have also been analyzed by ChIP using centromeric DNA sequences. In human cells, the centromere-specific histone H3 variants (CENP-A) coexist with heterochromatic modified histones at almost all stages during the cell cycle, but the histones are instantaneously modified with a euchromatic modification from anaphase to early G1 (Ohzeki et al., 2012). In rice, a CENH3-binding region in a centromere, Cen8, was revealed by ChIP to be heterochromatic (Nagaki et al., 2004). However, euchromatic markers were also detected in genic regions and 167-bp CentO variants in rice centromeres (Yan et al., 2006; Zhang et al., 2013).

In this study, we isolated a cDNA encoding CENH3 from sunflowers and raised a peptide antibody against CENH3. The antibody showed centromere-specific signals in immunostaining and immunohistochemical staining experiments. A ChIP-Seq experiment was also conducted to isolate DNA sequences that coexist with CENH3. Additionally, the epigenetic status of the centromeres was successfully revealed by immunohistochemical staining using anti-CENH3 and anti-modified histone antibodies as well as by ChIP-qPCR using these antibodies and isolated DNA sequences from ChIP-Seq.

MATERIALS AND METHODS

Plant Material

Sunflower seeds (*H. annuus* L., $2n = 2x = 34$, number 1802-065684) were obtained from a commercial source (LIC, Okayama, Japan).

Identification of a Sunflower Expressed Sequence Tag (EST) Encoding CENH3

An EST sequence encoding sunflower CENH3 (HaCENH3) was identified from the gene indices using the tblastn program (http://compbio.dfci.harvard.edu/tgi/) and the amino acid sequence of NtCENH3-1 (GenBank accession number: BAH03514, Nagaki et al., 2009) as a query.

RNA Isolation and PCR

Total RNA was isolated from a 3-day-old sunflower seedling using the RNeasy Plant Mini kit (Qiagen, Hilden, Germany). To determine the full-length cDNA sequence of a sunflower CENH3 gene, rapid amplification of cDNA ends (RACE) was conducted. For 3′RACE, the primer HaCENH3-3RACE (**Supplementary Table 1**) was designed from an EST encoding a putative sunflower CENH3 sequence found during the BLAST search; this primer was used with the SMARTer RACE cDNA Amplification Kit (Clontech, CA, USA). Another primer, HaCENH3-5RACE (**Supplementary Table 1**), was designed from the sequences determined by 3′RACE and was used to determine the 5′ end.

Sequencing and Sequence Analyses

The 3′- and 5′-RACE products were cloned into a pGEM-T easy vector (Promega, WI, USA) and sequenced from both ends using a BigDye Terminator v1.1 cycle sequencing kit and an ABI PRISM 3130xl genetic analyzer (Applied Biosystems, CA, USA). A putative amino acid sequence was deduced from the DNA sequences and used as a query sequence for a protein BLAST search on the NCBI website (http://blast.ncbi.nlm.nih.gov/Blast.cgi?CMD=Web&PAGE_TYPE=BlastHome). The deduced HaCENH3 amino acid sequence was aligned with orthologs identified by the protein BLAST search and canonical histone H3 of rice using the Clustal X software program (Thompson et al., 1997). Phylogenetic relationships among the CENH3s were analyzed by the neighbor-joining method (Saitou and Nei, 1987).

Immunostaining

Based on the deduced HaCENH3 amino acid sequence, a peptide corresponding to the N-terminus of HaCENH3 (H_2N-ARTKHPAKRSSGIPADGRSS-COOH) was synthesized and injected into two rabbits. The raised antisera were purified using an affinity Sepharose column consisting of the aforementioned peptide.

Immunostaining was conducted as previously described (Nagaki et al., 2012b). In brief, root tips of 3-day-old sunflowers were fixed in microtubule stabilizing buffer (50 mM PIPES, pH 6.9, 5 mM MgSO4, and 5 mM EGTA) containing 3% (w/v) paraformaldehyde and 0.2% (v/v) Triton X-100. The fixed tips were washed and digested with a mixture of 1% (w/v) cellulase Onozuka RS (Yakult Pharmaceutical Industry, Tokyo, Japan) and 0.5% (w/v) pectolyase Y-23 (Seishin Pharmaceuticals, Tokyo, Japan) and then compressed onto slides coated with poly-L-lysine (Matsunami, Osaka, Japan). A 1:100 dilution of the purified anti-HaCENH3 antibody and monoclonal anti-modified-histone antibodies produced in mice (anti-histone H3 dimethyl K4 (H3K4me2): MBL (Nagoya, Japan) MABI0303 and anti-histone H3 dimethyl K9 (H3K9me2): MBL MABI0317) were applied to the slides. To detect acetylations of histone H4, an anti-histone H4 acetyl (H4Ac) antibody raised in rabbits (Millipore, MA, USA: #06-598) was used. The antibodies were detected using 1:1000 dilutions of Alexa Fluor 555-labeled anti-rabbit antibodies (Molecular Probes, OR, USA) and Alexa Fluor 488-labeled anti-mouse antibodies (Molecular Probes), respectively. Chromosomes were counterstained with 0.1 μg/ml 4,6-diamino-2-phenylindole (DAPI). Immunosignals and stained chromosomes were captured using a chilled charge-coupled device (CCD) camera, AxioCam HR (Carl Zeiss, Oberkochen, Germany), and images were pseudo-colored and processed using AxioVision software (Carl Zeiss).

Immunohistochemical Staining

Immunohistochemical staining was conducted as described with minor modifications (Yamaji and Ma, 2007; Nagaki et al., 2012b). Three-day-old sunflower roots were fixed as described above for immunostaining, and the fixed roots were sectioned at 100-μm thickness using a microslicer (LinearSlicer PRO10; Dosaka EM). These sections were transferred onto slides and then macerated. For three-color detection, 1:100 dilutions of the purified anti-HaCENH3 rabbit antibody and anti-α-tubulin mouse antibody (Sigma, MO, USA: T6199) were applied to the slides. For four-color detection, 1:100 dilutions of the purified anti-HaCENH3 rabbit antibody, anti-α-tubulin rat antibody (Abcam, Cambridge, UK: ab64332) and monoclonal anti-H3K9me2 mouse antibody (MBL MABI0317) were applied to the slides. After washing in PBS, the primary antibodies were detected using 1:1000 diluted secondary antibodies, the Alexa Fluor 555-labeled anti-rabbit antibodies and the Alexa Fluor 488-labeled anti-mouse antibodies for the three-color detection and Alexa Fluor 647-labeled anti-rabbit antibodies (Molecular Probes), Alexa Fluor 488-labeled anti-rat antibodies (Molecular Probes), and Alexa Fluor 546-labeled anti-mouse antibodies (Molecular Probes) for the four-color detection. Then, nuclei and chromosomes were counterstained with DAPI. Immunosignals and stained chromosomes were observed with a laser-scanning confocal microscope (LSM700; Carl Zeiss). The obtained data were analyzed using AxioVision software.

Chromatin Immunoprecipitation (ChIP)

ChIP was performed as previously described (Nagaki et al., 2012b) with minor modifications using the anti-HaCENH3 antibody and the anti-modified histone antibody (anti-H3K9me2: MBL #MABI0317, anti-H3K4me2: MBL #MABI0303 and anti-H4Ac: Millipore #06-598). Nuclei were isolated from the leaves of 1-month-old sunflowers and then digested with micrococcal nuclease (Sigma) to produce chromatin. Following overnight incubation of the chromatin with the antibodies at 4°C, the antibodies were captured using Dynabeads Protein G (Invitrogen, CA, USA). For mock experiments, a normal rabbit serum was used instead of the antibodies. DNA was purified from the chromatin with the captured antibodies by phenol/chloroform extraction followed by ethanol precipitation.

ChIP-Seq and Repeatexplorer Analysis

ChIP-Seq was conducted using precipitated DNA from the input and HaCENH3 fractions in the ChIP. Libraries were constructed using the NEBNext ChIP-Seq Library Prep Reagent Set for Illumina (New England Biolabs, MA, USA), and the libraries were read by MiSeq (Illumina, CA, USA) with the paired-end 2 × 300 bp protocol. Conversion of raw base-call data to sequence data in the fastq format, identification of reads derived from each sample by index sequences, and adapter trimming were performed using MiSeq reporter 2.3.32. The sequence data were analyzed by a similarity-based clustering program, RepeatExplorer (http://www.repeatexplorer.org) (Novák et al., 2010) with default parameters.

qPCR

qPCR was conducted using SYBR Premix Ex Taq II (Tli RNaseH Plus) (Takara, Shiga, Japan) and primers for qPCR (**Supplementary Table 1**) with a StepOne instrument (Applied Biosystems). The primers were designed based on the sequences in the clusters of the RepeatExplorer analysis. The precipitated DNA in the ChIP experiment was used as a template, and the mock was used as a negative control. Relative enrichment (RE) was calculated by the following formula: RE = amount of the sequence in the antibody fraction/amount of the sequence in the mock. The qPCR results were assessed by Student's t-test.

Fluorescence *in Situ* Hybridization (FISH)

Probes were amplified using primer sets designed based on the sequences in clusters of the RepeatExplorer analysis (**Supplementary Table 1**) using sunflower genomic DNA as a template. The amplified DNA was cloned into pGEM T-easy, and sequences were confirmed using a BigDye Terminator v1.1 cycle sequencing kit and an ABI PRISM 3130xl genetic analyzer. To detect nucleolar organizing regions, an 18S-5.8S-28SrDNA clone from wheat (pTa71) was used (Gerlach and Bedbrook, 1979). To characterize sunflower chromosomes, a reported 386-bp tandem repetitive sequence, HAG004N15 (Ceccarelli et al., 2007), was amplified using specific primers (**Supplementary Table 1**) and cloned. The sequence was confirmed as described above.

FISH analysis of mitotic chromosomes was performed as previously described (Nagaki et al., 2011). Chromosomes were prepared from the root tips of 3-day-old sunflowers. The plasmid DNA was labeled by nick translation using a DIG-Nick Translation Mix (Roche, Basel, Switzerland) or a Biotin-Nick Translation Mix (Roche). The digoxigenin- and biotin-labeled probes were visualized using rhodamine-conjugated anti-digoxigenin antibody (Roche) and Alexa Fluor 488-conjugated streptavidin (Molecular Probes), respectively.

RESULTS

Isolation and Sequence Analyses of CENH3 in Sunflowers

To identify the HaCENH3 gene, a BLAST search was conducted in a sunflower EST database of the gene index project using the amino acid sequence of NtCENH3-1 as a query. One EST group (TC48348) containing a 675-bp sequence showed 68% identity to the query sequence. However, because of low quality in the first 150 bp of the EST, no start codon was found in the sequence. To obtain a full-length cDNA of the gene, 5′- and 3′-RACE experiments were performed using sunflower seedling cDNA as a template. As a result, a putative full-length cDNA containing a 429-bp ORF that encodes 143 amino acids (GenBank accession number LC075743) was obtained.

The amino acid sequence deduced from the ORF showed similarity to the sequences of some other plants, with CENH3 from *Daucus muricatus* showing the highest similarity (70%). The amino acid sequence of HaCENH3 was aligned with those of CENH3s from other plant species and rice canonical histone H3 (**Supplementary Image 1**). The alignment indicated that the N-terminal amino acid at position 14–44 of HaCENH3 did not show any similarity to those of the other CENH3s or the canonical histone H3, and it also indicated that HaCENH3 possessed a longer loop 1 domain than canonical histone H3. The longer loop 1 is a feature of CENH3s.

In a phylogenetic analysis using the alignment, HaCENH3 was classified into a dicot clade (**Figure 1**). As expected, HaCENH3 was found to be most closely related to the CENH3s of *Daucus* species, but the sequence was placed outside of a clade containing the *Daucus* CENH3s.

Centromere Localization of HaCENH3

An anti-HaCENH3 antibody was raised against a synthetic peptide comprising N-terminal amino acid residues 2–21 of the deduced HaCENH3 amino acid sequence (**Supplementary Image 1**). To confirm its specificity to the centromeres, immunostaining using the antibody was conducted, and centromere-specific immunosignals appeared on all sunflower chromosomes (**Figure 2**). Additionally, in immunohistochemical staining, microtubule signals on all chromosomes were associated with all of the anti-HaCENH3 immunosignals at metaphase (**Supplementary Movie 1**).

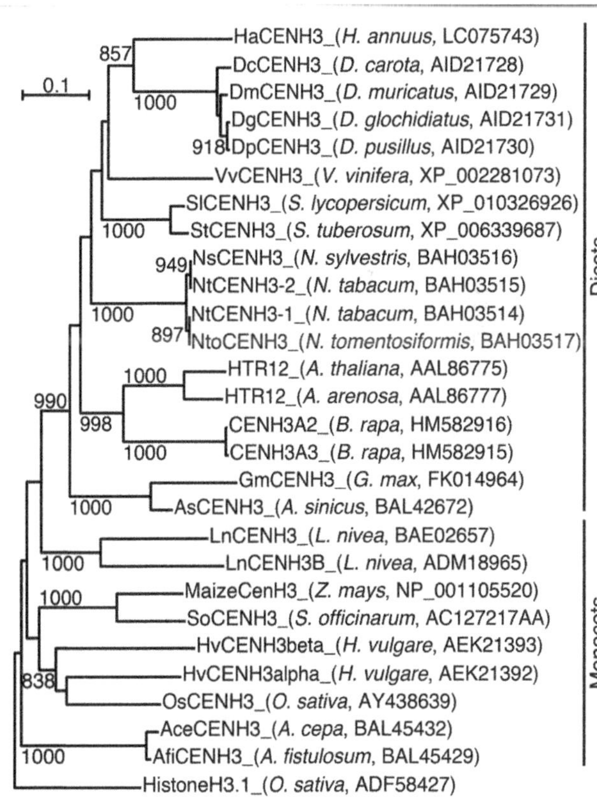

FIGURE 1 | Phylogenetic tree based on the amino acid sequences of plant CENH3. The species name and GenBank accession number are indicated in parentheses. Rice canonical histone H3 was used as an outgroup. Bootstrap values greater than 800 in 1000 tests are indicated on the branches.

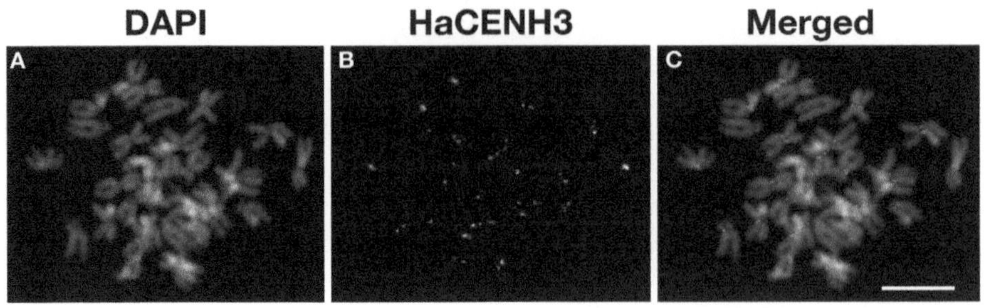

FIGURE 2 | Immunostaining of sunflower metaphase chromosomes using an anti-HaCENH3 antibody. (A) DAPI-stained chromosomes. **(B)** Immunosignals of an anti-HaCENH3 antibody. **(C)** Merged image of **(A,B)**. Scale bar, 10 μm.

DNA Sequences that Interact with Centromeric Nucleosomes

To investigate DNA sequences that coexist with HaCENH3, a ChIP-Seq experiment was conducted using the anti-HaCENH3 antibody and chromatin extracted from sunflower leaves. DNA fragments from the input and the HaCENH3 fraction in the ChIP were sequenced using MiSeq with the paired-end 2 × 300 bp protocol and deposited in DDBJ (Accession number: DRA003719). After the initial quality checks, 1,795,002 paired reads from the input and 2,012,000 paired reads from the HaCENH3 fraction were analyzed using the RepeatExplorer program. By this analysis, a total of 372 clusters containing at least 0.01% of the used sequences were generated (**Supplementary Data Sheet 1**). Then, enrichment ratios (ERs) were calculated by the following formula: ER = HaCENH3 ChIP reads/input reads in the each cluster, and the clusters were sorted by ER (**Supplementary Table 2**). Out of the 372 clusters formed, 41 clusters showed an ER higher than 2.0.

To confirm the ChIP-Seq results, ChIP-qPCR was conducted using the DNA sequences in five clusters (HaCENH3CL1, 20, 22, 124, and 289) selected from the 41 clusters having an ER higher than 2.0 (**Figure 3**). In the ChIP-qPCR, a sunflower ubiquitin gene (GenBank accession number: X14333) was used as a negative (non-centromeric) control. The DNA sequences from all five clusters were significantly increased (27-fold for HaCENH3CL1, 154-fold for HaCL289, 289-fold for HaCENH3CL22, 344-fold for HaCENH3CL22, and 370-fold for HaCENH3CL124) compared with the negative control in the CENH3 fractions ($P < 0.01$ via Student's t-test, $n = 4$), whereas a non-centromeric control sequence, HAG004N15, did not increase ($P = 0.14$ via Student's t-test, $n = 4$). Since the ER in the ChIP-Seq and the RE in the ChIP-qPCR showed a high correlation coefficient ($r = 0.78$) according to the Pearson product-moment correlation, the ChIP-qPCR data support the ChIP-Seq results. The significant increase in HaCENH3CL1 showing the minimum ER (2.0) among the five clusters in the ChIP-qPCR suggested that sequences in the 41 clusters having an ER higher than 2.0 in the ChIP-Seq coexisted with HaCENH3 in the sunflower genome.

The HaCENH3CL124 had the highest ER (52.0) and contained 187-bp tandem repeat sequences (**Supplementary Table 2**). The consensus DNA sequence of the tandem repeat was relatively AT-rich (58.3%), and no sequence similarity to DNA sequences in the GenBank database was found.

Of the 41 clusters, the program suggested that 25 clusters contained long interspersed nuclear element (LINE)-like sequences, and the sequences involve an ORF encoding an endonuclease (ENDO) and reverse transcriptase (RT) complex of LINE (**Supplementary Table 2**). These sequences show approximately 60% similarity to each other. Almost all RepeatExplorer cluster graphs of the LINE showed typical patterns of retrotransposons (line shape) rather than typical patterns of tandem repeats (star or ring shape) (**Supplementary Image 2**). As exceptional cases, two clusters, HaCENH3CL78 and HaCENH3CL115, involved 12 and four ENDO/RT-related reads, and showed star shape. Since RepeatExplorer splits some satellite repeats with long monomers, e.g., rDNA, into multiple clusters, connections of the LINE-like sequences clusters were investigated (**Supplementary Table 3** and **Supplementary Image 3**). In the investigation, rDNA was used as a positive control of a satellite repeat with long monomers, and 17 clusters related to rDNA were found in the investigation. Ends of the rDNA clusters showed similarity to other clusters. On the other hand, almost all of the LINE clusters did not show frequent similarity hits observed among the rDNA clusters. As exceptional cases, two clusters, HaCENH3CL78 and HaCENH3CL115, showed frequent similarity hits to HaCENH3CL8, suggesting these are not LINE-like elements. Twenty-two of the remaining 23 LINE-like clusters showed ERs higher than 10.0. Additionally, junction of the elements and insertion sites were surveyed (**Supplementary Image 4**). If clusters include ends of mobile elements, junctions should be visible as sites with heterologous sequences at ends of contigs in the cluster. For example, a clear junction was observed in an end of HaCENH3CL189 showing a typical cluster graph of LTR-retrotransposon, and similar junctions were also observed in some of the LINE-like clusters. However, sunflower DNA sequences that were similar to the LINE-like sequences were not found in the GenBank DNA database. These results suggested that most centromeric regions of sunflowers are not involved in the current sunflower genome sequencing project because the project has not involved repetitive DNA sequence-rich regions (http://sunflowergenome.org/). An additional 12 of 41 clusters

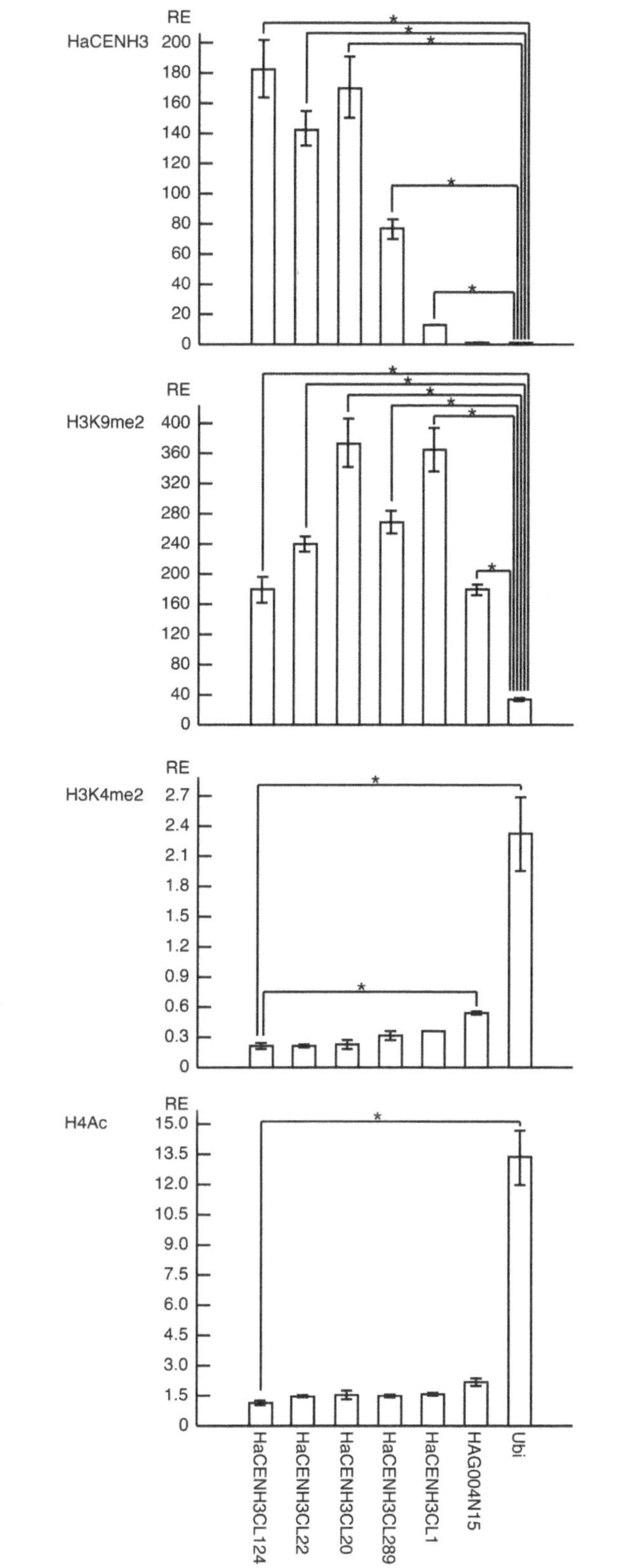

FIGURE 3 | ChIP-qPCR analysis of centromeric DNA sequences of sunflowers. The columns and error bars represent the average relative enrichments (REs) and the standard errors from four independent ChIP

(Continued)

FIGURE 3 | Continued
reactions, respectively. The coding region of the sunflower ubiquitin gene (X14333) was used as a non-centromeric (negative) control in the anti-HaCENH3 ChIP and as a negative control in the anti-H3K9me2 ChIP. Since HaCENH3CL124 showed the lowest REs among the sequences in the anti-H3K4me2 and H4Ac ChIP, HaCENH3CL124 was used as a negative control. The statistical significance of differences between the negative controls and other sequences was determined using Student's *t*-test (*$P < 0.01$).

contained transposable element-like sequences, whereas DNA sequences in the remaining four clusters showed no homology to those registered in the GenBank DNA database.

To confirm the centromeric localization of the DNA sequences immunoprecipitated with anti-HaCENH3 in the ChIP-Seq and ChIP-qPCR, these sequences were used as probes in FISH analysis (**Figure 4**). A probe containing the HaCENH3CL124 sequence (pHaCENH3CL124-1, GenBank accession number: LC075744) showed centromeric signals on a pair of chromosomes with a secondary constriction (**Figures 4A–D**). To identify the chromosomes showing the HaCENH3CL124 signal, another FISH experiment in which the repeat and a reported FISH marker, HAG004N15, were used as probes was conducted; it revealed that the chromosome in question was chromosome 8 in Ceccarelli's report (Ceccarelli et al., 2007) (**Figures 4E–M**). A cloned probe containing the LINE-like sequence from HaCENH3CL20 (pHaCENH3CL20-1, GenBank accession number: LC075745) showed centromeric signals on almost all of the chromosomes (**Figures 4N–Q**). Since the LINE-like sequence showed centromeric signals, we named it HaCEN-LINE. In FISH using HaCENH3CL20-PCR products as probes, much stronger signals on all the centromeric and pericentromeric regions were found, but faint signals were also observed on the arm regions (**Figures 4R–U**). These results imply that the PCR products contain not only some HaCEN-LINE variants but also some non-centromeric LINE variants; centromeric variants are on all of the centromeres. Two Ty3-related DNA sequences from HaCENH3CL1 (pHaCENH3CL1-1, GenBank accession number: LC075747) and HaCENH3CL189 (pHaCENH3CL189-1, GenBank accession number: LC075746), when used as FISH probes, showed dispersed patterns with some strong spots on centromeres (**Supplementary Image 5**). These partial centromere localizations coincided with the partial enrichment of the sequences in the ChIP-Seq and ChIP-qPCR (**Figure 3** and **Supplementary Table 2**), suggesting that these localize on both centromeric and non-centromeric arm regions.

Epigenetic Status of Sunflower Centromeres

Post translational histone modifications of sunflower centromeres were investigated by immunostaining with four different antibodies against H3K4me2, H3K9me2, H3K9Ac, and H4Ac (**Figures 5**, **6** and **Supplementary Image 6**). In the interphase cells, HaCENH3 signals scattered as dots on nuclei, and polar organization, Rabl orientation, and chromocenters were not observed (**Figures 5A–E**, **6A–E** and **Supplementary Images 6A–E**). H3K4me2, one of the

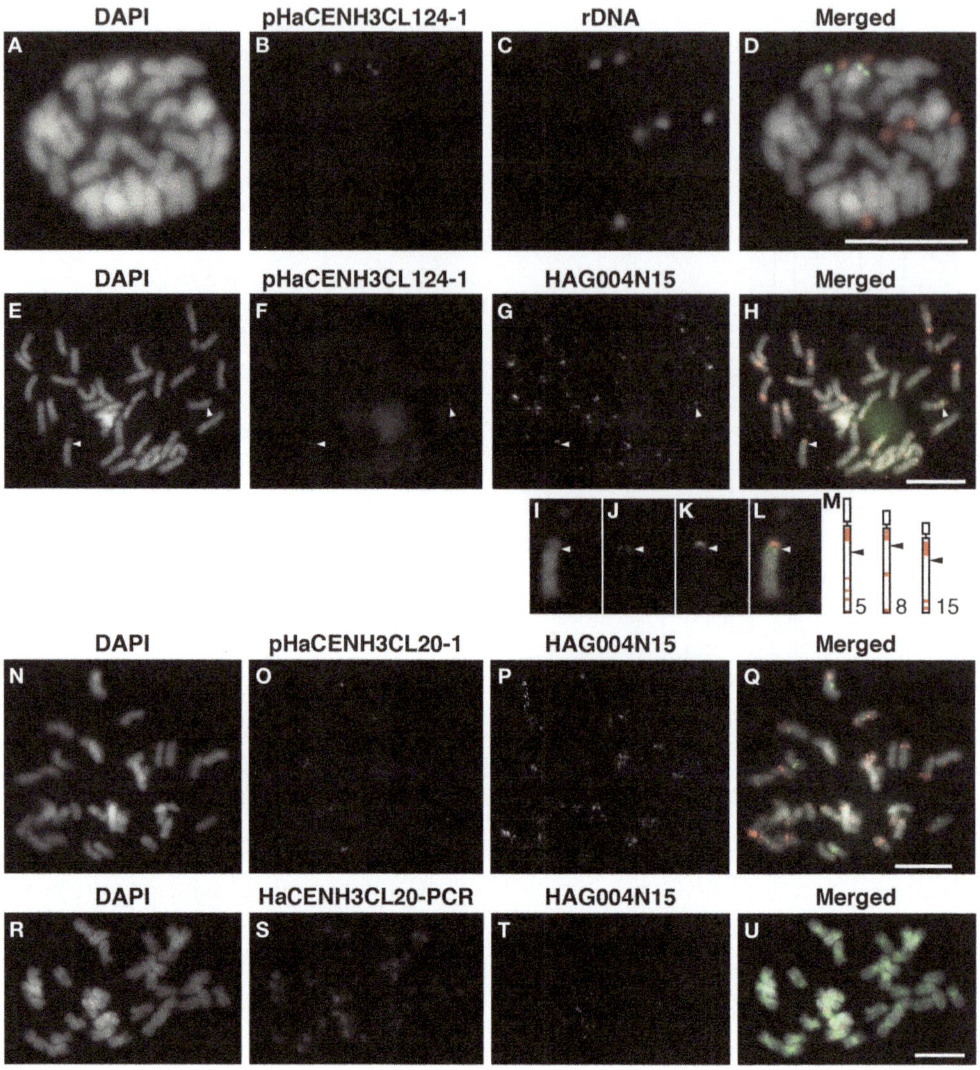

FIGURE 4 | FISH using the enriched sequence from ChIP-Seq. **(A,E,I,N,R)** DAPI-stained sunflower chromosomes. **(B,F,J)** FISH signals of pHaCENH3CL124-1. **(C)** FISH signals of rDNA. **(G,K,P,T)** FISH signals of HAG004N15. **(O)** FISH signals of pHaCENH3CL20-1. **(S)** FISH signals of the PCR products of HaCENH3CL20. **(D)** A merged image of **(A–C)**. **(H)** A merged image of **(E–G)**. **(L)** A merged image of **(I–K)**. **(Q)** A merged image of **(N–P)**. **(U)** A merged image of **(R–T)**. **(I–L)** Enlarged images of **(E–H)**. **(M)** Karyograms from Ceccarelli's report (2007). White arrowheads in **(E–H)** and black arrowheads in **(M)** indicate centromeres on the HaCENH3CL124-positive chromosomes and the reported chromosomes, respectively. Scale bar, 10 μm.

representative euchromatic modifications, was detected on almost all regions of nuclei with many small dot signals, but the dot signals did not overlap with the HaCENH3 signals (**Figures 5A–E**). Similarly, the H3K4me2 signals did not co-localize with the HaCENH3 signals at prophase or metaphase (**Figures 5F–O**). Another euchromatic modification, H3K9Ac, showed a similar tendency at interphase and prophase (**Supplementary Images 6A–E**), but almost all H3K9Ac signals disappeared at metaphase (**Supplementary Images 6F–J**). Additionally, another euchromatic modification, H4Ac, showed stronger signals on the arms of metaphase chromosomes, but the signals on the centromeric and pericentromeric regions were weaker than those on the arms (**Supplementary Images 6K–M**).

Immunosignals from H3K9me2 that represented heterochromatin status were distributed on an entire region of nuclei with dots in the interphase cells, but the dot sizes were larger than these of H3K4me2 (**Figures 6A–E**). Although the larger dot signals did not overlap with the HaCENH3 signals, the HaCENH3 signals were colocalized with faint H3K9me2 signals. In the prophase and metaphase cells, the H3K9me2 signals appeared mainly on the chromosome arms, whereas centromeres and pericentric regions of chromosomes showed faint signals for histone modification (**Figures 6F–O**). Co-localization of HaCENH3 and H3K9me2 was confirmed by immunohistochemical staining using a laser-scanning confocal microscope (**Figures 6P–W**). On metaphase chromosomes, the HaCENH3 signals were colocalized with weak H3K9me2 signals,

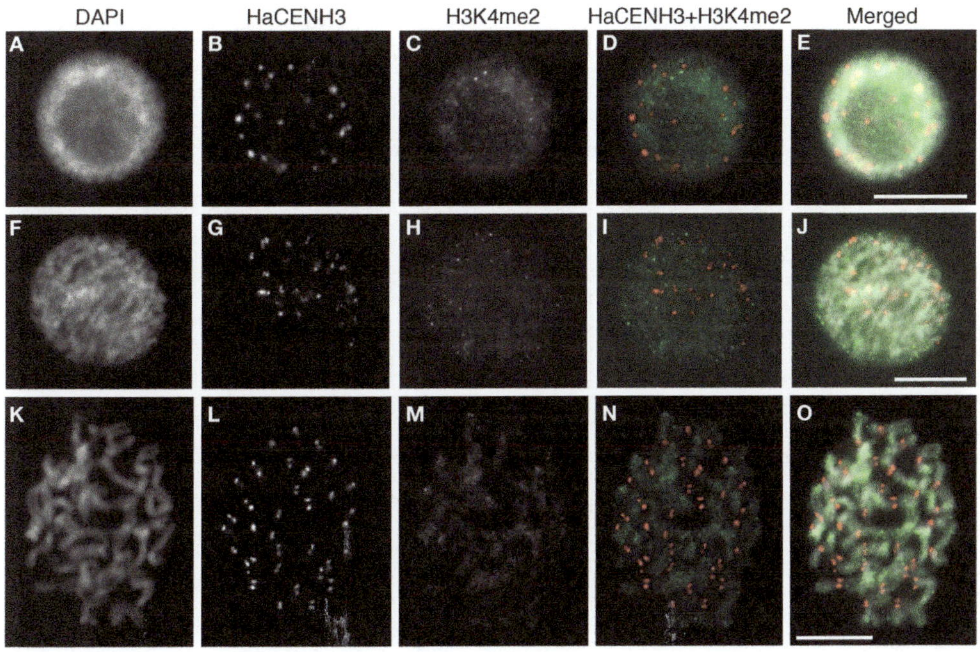

FIGURE 5 | Immunostaining using the anti-HaCENH3 and anti-H3K4me2 antibodies. (A–E) An interphase nucleus. **(F–J)** Prophase chromosomes. **(K–O)** Metaphase chromosomes. Scale bar, 10 μm.

and both arms showed strong H3K9me2 (**Figures 6U,V**). In a horizontal-scanning of centromeres on a pair of chromatids (**Figures 6U,W**), a DAPI-staining peak appeared at the middle of the paired chromatids, and HaCENH3 signals were detected on both sides of the paired chromatids. On the other hand, H3K9me2 showed no peaks in the scan, and weak H3K9me2 signals appeared constantly on the centromeric region.

Histone modifications on centromeric DNA sequences were also investigated by ChIP-qPCR (**Figure 3**). In the analyses, the ubiquitin gene was used as a negative control for a heterochromatin marker, H3K9me2, and as a positive control for two euchromatin markers, H3K4me2 and H4Ac. In ChIP-qPCR with the anti-H3K9me2 antibody, all five HaCENH3-positive sequences and a non-centromeric repetitive sequence (HAG004N15) significantly accumulated compared with the negative control ($P < 0.01$ via Student's t-test, $n = 4$, **Figure 3**). For modification of H3K4me2, HaCENH3CL124 showed the lowest RE (0.21), and the REs of the positive control (ubiquitin) and HAG004N15 were 2.32 and 0.54, respectively (**Figure 3**). These two sequences significantly accumulated compared with HaCENH3CL124 ($P < 0.01$ via Student's t-test, $n = 4$). Similarly, HaCENH3CL124 showed the lowest RE (1.14) for H4Ac, and only the positive control (ubiquitin, RE = 13.36) increased significantly ($P = 0.003$ via Student's t-test, $n = 4$) compared with HaCENH3CL124 (**Figure 3**). Since leaf cells usually do not divide, these data suggest that all HaCENH3-positive sequences were involved with heterochromatin at interphase.

DISCUSSION

In this study, a CENH3-encoding cDNA was identified in sunflower. Based on the amino acid sequence deduced from the cDNA sequence, a peptide corresponding to the N-terminal 20 amino acids was synthesized. An antibody against the synthesized peptide recognized centromeres on all sunflower chromosomes. Using this antibody, ChIP-Seq analysis was applied and succeeded in isolating centromeric DNA sequences from sunflowers. Additionally, the epigenetic status at the centromeric DNA sequences was investigated by ChIP-qPCR with antibodies against modified histones.

Usually, centromeric DNA sequences in plant species consist of species-specific tandem repeats and retrotransposons, and these sequences are located on all centromeric regions (Zhong et al., 2002; Nagaki et al., 2003, 2009, 2011; Nagaki and Murata, 2005; Houben et al., 2007; Tek et al., 2010, 2011; Wang et al., 2011; Neumann et al., 2012; He et al., 2015). In this study of sunflower, we identified two types of centromeric DNA sequences (**Supplementary Table 2** and **Figure 4**). One was a LINE-like sequence, HaCEN-LINE, which showed centromeric signals on all of the chromosomes (**Figures 4N–U**). Although some tandem repeats evolved from retrotransposons were reported in potato (Gong et al., 2012), no centromere-specific LINE-like elements have been reported. The cluster graphs of HaCEN-LINEs in RepeatExplorer showed line shape (**Supplementary Image 2**), implying their retroelement form rather than modified tandem repeat form. Therefore, this is

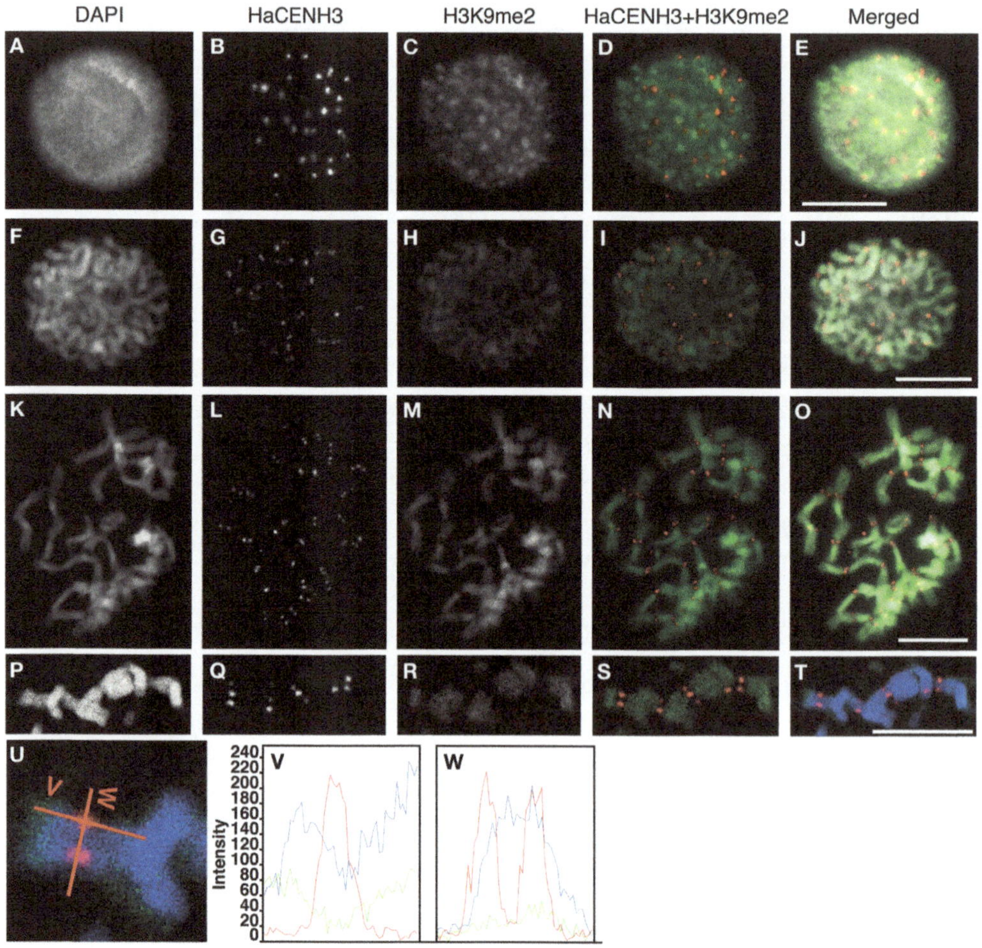

FIGURE 6 | Immunostaining and immunohistochemical staining using the anti-HaCENH3 and anti-H3K9me2 antibodies. (A–O) Immunostaining images. **(A–E)** An interphase nucleus. **(F–J)** Prophase chromosomes. **(K–O)** Metaphase chromosomes. **(P–T)** Immunohistochemical staining images of metaphase chromosomes. Scale bar, 10 μm. **(U)** Scanning of metaphase chromosomes. Scanned positions are indicated as red lines. **(V)** A vertical scan of the chromosome. **(W)** A horizontal scan of the chromosome.

the first report describing a centromere-specific LINE. Although the centromere-targeting mechanisms of transposable elements remain unknown, the present study indicated that the LINE-like sequence can also target centromeric regions in the same manner as retrotransposons. Another centromeric DNA sequence in sunflowers was a 187-bp tandem repeat, which was located on a single pair of chromosomes (**Figure 4**). Such chromosome-specific centromeric DNA sequences have been reported in chickens (Shang et al., 2010), tobacco (Nagaki et al., 2012a), and potatoes (Gong et al., 2012); however, they are not very common. The enrichment of the tandem repeats in HaCENH3 ChIP was much higher than that of the LINE (**Figure 3**), suggesting that the tandem repeat may be more useful for building more stable centromeres than the LINE. As discussed previously for other species (Nagaki et al., 2004, 2012a; Shang et al., 2010; Gong et al., 2012), sunflower centromeres may also be undergoing centromeric DNA evolution to equalize centromeric DNA sequences among chromosomes; the tandem repeats may form stabilized centromeres on the all sunflower chromosomes after this evolutionary event.

With the exception of phosphorylation, other histone modifications on the centromeric regions are not well characterized in plants (Desvoyes et al., 2014; Sharma et al., 2015). According to Houben et al. (2003), plant species with a genome smaller than 510 Mb formed chromocenters at interphase, and H3K9me2 preferentially occurred on the chromocenters. In *Arabidopsis*, H3K9me2 signals appeared on all centromeric and pericentromeric regions of metaphase chromosomes. Euchromatin-specific histone modifications, such as H3K4me2 and H4Ac, were observed in an inverse pattern compared to that of the H3K9me2 (Houben et al., 2003; Jasencakova et al., 2003). In plant species with genomes larger than 530 Mb, such as barley, no chromocenters were observed, and dispersed H3K9me2 signals were observed at interphase; their low level of modification appears on centromeric and pericentromeric regions of metaphase chromosomes (Fuchs et al., 2006). Occasionally, H4Ac were detected at centromeric and pericentromeric regions of metaphase chromosomes in barley and field beans (Jasencakova et al., 2000, 2001). In this study, sunflowers showed dispersed modification patterns for H3K4me2, H3K9Ac and H3K9me2 at interphase and a lower level of H3K9me2 modification on metaphase centromeres (**Figures 5, 6**). However, the irregular modification patterns observed at metaphase in barley and field bean were not detected at metaphase in sunflower. The genome size of sunflower was estimated to be 2.43 Gb (Bennett et al., 1982), and the observed distribution patters of the modified histones in this study (**Figures 5, 6** and **Supplementary Image 6**) coincided with those in plants with large genomes (Houben et al., 2003).

To determine histone modification events on each different centromeric DNA sequence, ChIP analysis has higher resolution and quantitative capacity than immunostaining. In rice, ChIP revealed that centromeric DNA sequences excepting the genic regions on chromosome 8 were heterochromatic (Nagaki et al., 2004; Yan et al., 2006). Subsequently, different histone modifications were observed in CentO sequences (Zhang et al., 2013). CentO sequences with regular size units (155 bp) were enriched in OsCENH3 ChIP, whereas 167-bp CentO variants were enriched in a euchromatic modification, H3K4me2, ChIP rather than OsCENH3. In the present study with sunflowers, two centromeric DNA sequences showed heterochromatic status (**Figure 3**), suggesting that the status of plant centromeres at interphase was heterochromatic. Additionally, immunohistochemical staining revealed the existence of H3K9me2 on the centromeres at metaphase (**Figure 6**). Furthermore, immunostaining showed that the level of H3K9Ac in the interphase nuclei was higher than that on metaphase chromosomes (**Supplementary Image 6**). Although we could not quantify the H3K9Ac level on centromeres, the increased level of H3K9Ac in the interphase nuclei implies that it increased on centromeres as well. In the case of human cells, the combination of rigid cell cycle control and ChIP made it possible to detect H3K9Ac within a short range of time during the cell cycle (Ohzeki et al., 2012). To investigate H3K9Ac and other modifications on plant centromeres, a more accurate quantification system, such as that utilized in human cells, is required in the future.

ACKNOWLEDGMENTS

This research was supported by the Cooperative Research Grant of the Genome Research for BioResource, NODAI Genome Research Center, Tokyo University of Agriculture.

SUPPLEMENTARY MATERIAL

The Supplementary Material for this article can be found online at: http://journal.frontiersin.org/article/10.3389/fpls.2015.00912

Supplementary Image 1 | Amino acid sequence alignment of plant CENH3 and a canonical histone H3. Species name and GenBank accession numbers are indicated in parentheses. A red box indicates the amino acid residues used for raising a peptide antibody against HaCENH3. A magenta box indicates the position of Loop 1.

Supplementary Image 2 | RepeatExplorer cluster graphs of the enriched sequence from ChIP-Seq. RepeateExplorer cluster graphs were indicated with their cluster name. Repeat types are indicated by magenta (tandem repeat), orange (LINE), green (LTR-retrotransposon), and black (others). 5SrDNA and HAG004N15 are examples of tandem repeats.

Supplementary Image 3 | Connection of RepeatExplorer cluster graphs of rDNA and HaCEN-LINE. Connection of RepeateExplorer cluster graphs were indicated with their cluster name.

Supplementary Image 4 | Junction of HaCEN-LINE and its insertion sites. Junction of HaCEN-LINE and its insertion sites were indicated with their cluster name.

Supplementary Image 5 | FISH using the accumulated sequence from ChIP-Seq. (A,D) DAPI-stained sunflower chromosomes. **(B)** FISH signals of pHaCENH3CL1-1. **(E)** FISH signals of pHaCENH3CL189-1. **(C)** A merged image of **(A,B)**. **(F)** A merged image of **(D,E)**. Scale bar, 10 μm.

Supplementary Image 6 | Immunostaining using the anti-HaCENH3, anti-H3K9Ac, and anti-H4Ac antibodies. (A–E) An interphase nucleus (top) and prophase chromosomes (bottom). **(F–M)** Metaphase chromosomes. **(A,F,K)** DAPI staining. **(B,G)** Immunosignals of anti-HaCENH3. **(C,H)** Immunosignals of anti-H3K9Ac. **(L)** Immunosignals of anti-H4Ac. **(D)** A merged image of **(B,C)**. **(E)** A merged image of **(A–C)**. **(I)** A merged image of **(G,H)**. **(J)** A merged image of **(F–H)**. **(M)** A merged image of **(K,L)**. Scale bar, 10 μm.

Supplementary Movie 1 | A video including immunohistochemical images. HaCENH3 (red), α-tubulin (green), and DAPI (gray).

Supplementary Data Sheet 1 | A FASTA file containing HaCENH3CLs.

Supplementary Table 1 | Primers used in this study.

Supplementary Table 2 | Clusters possessing an enrichment ratio (ER) higher than 2.0 in the RepeatExplorer analysis.

Supplementary Table 3 | Similarity hits on rDNA and HaLINE clusters.

REFERENCES

Amor, D. J., Kalitsis, P., Sumer, H., and Choo, K. H. (2004). Building the centromere: from foundation proteins to 3D organization. *Trends Cell Biol.* 14, 359–368. doi: 10.1016/j.tcb.2004.05.009

Bennett, M. D., Smith, J. B., and Heslopharrison, J. S. (1982). Nuclear-DNA amounts in angiosperms. *Proc. R. Soc. Lond. B Biol. Sci.* 216, 179–199. doi: 10.1098/rspb.1982.0069

Black, B. E., Foltz, D. R., Chakravarthy, S., Luger, K., Woods, V. L. Jr., and Cleveland, D. W. (2004). Structural determinants for generating centromeric chromatin. *Nature* 430, 578–582. doi: 10.1038/nature02766

Ceccarelli, M., Sarri, V., Natali, L., Giordani, T., Cavallini, A., Zuccolo, A., et al. (2007). Characterization of the chromosome complement of *Helianthus annuus* by *in situ* hybridization of a tandemly repeated DNA sequence. *Genome* 50, 429–434. doi: 10.1139/G07-019

Desvoyes, B., Fernández-Marcos, M., Sequeira-Mendes, J., Otero, S., Vergara, Z., and Gutierrez, C. (2014). Looking at plant cell cycle from the chromatin window. *Front. Plant Sci.* 5:369. doi: 10.3389/fpls.2014.00369

Dunemann, F., Schrader, O., Budahn, H., and Houben, A. (2014). Characterization of centromeric histone H3 (CENH3) variants in cultivated and wild carrots (*Daucus* sp.). *PLoS ONE* 9:e98504. doi: 10.1371/journal.pone.0098504

Earnshaw, W. C., and Rothfield, N. (1985). Identification of a family of human centromere proteins using autoimmune sera from patients with scleroderma. *Chromosoma* 91, 313–321. doi: 10.1007/BF00328227

Feng, J., Liu, Z., Cai, X., and Jan, C. C. (2013). Toward a molecular cytogenetic map for cultivated sunflower (*Helianthus annuus* L.) by landed BAC/BIBAC clones. *G3 (Bethesda)* 3, 31–40. doi: 10.1534/g3.112.004846

Fuchs, J., Demidov, D., Houben, A., and Schubert, I. (2006). Chromosomal histone modification patterns–from conservation to diversity. *Trends Plant Sci.* 11, 199–208. doi: 10.1016/j.tplants.2006.02.008

Gerlach, W. L., and Bedbrook, J. R. (1979). Cloning and characterization of ribosomal RNA genes from wheat and barley. *Nucleic Acids Res.* 7, 1869–1885. doi: 10.1093/nar/7.7.1869

Gong, Z., Wu, Y., Koblízková, A., Torres, G. A., Wang, K., Iovene, M., et al. (2012). Repeatless and repeat-based centromeres in potato: implications for centromere evolution. *Plant Cell* 24, 3559–3574. doi: 10.1105/tpc.112.100511

Han, F., Lamb, J. C., and Birchler, J. A. (2006). High frequency of centromere inactivation resulting in stable dicentric chromosomes of maize. *Proc. Natl. Acad. Sci. U.S.A.* 103, 3238–3243. doi: 10.1073/pnas.0509650103

He, Q., Cai, Z., Hu, T., Liu, H., Bao, C., Mao, W., et al. (2015). Repetitive sequence analysis and karyotyping reveals centromere-associated DNA sequences in radish (*Raphanus sativus* L.). *BMC Plant Biol.* 15:105. doi: 10.1186/s12870-015-0480-y

Houben, A., Demidov, D., Gernand, D., Meister, A., Leach, C. R., and Schubert, I. (2003). Methylation of histone H3 in euchromatin of plant chromosomes depends on basic nuclear DNA content. *Plant J.* 33, 967–973. doi: 10.1046/j.1365-313X.2003.01681.x

Houben, A., Schroeder-Reiter, E., Nagaki, K., Nasuda, S., Wanner, G., Murata, M., et al. (2007). CENH3 interacts with the centromeric retrotransposon cereba and GC-rich satellites and locates to centromeric substructures in barley. *Chromosoma* 116, 275–283. doi: 10.1007/s00412-007-0102-z

Jasencakova, Z., Meister, A., and Schubert, I. (2001). Chromatin organization and its relation to replication and histone acetylation during the cell cycle in barley. *Chromosoma* 110, 83–92. doi: 10.1007/s004120100132

Jasencakova, Z., Meister, A., Walter, J., Turner, B. M., and Schubert, I. (2000). Histone H4 acetylation of euchromatin and heterochromatin is cell cycle dependent and correlated with replication rather than with transcription. *Plant Cell* 12, 2087–2100. doi: 10.1105/tpc.12.11.2087

Jasencakova, Z., Soppe, W. J. J., Meister, A., Gernand, D., Turner, B. M., and Schubert, I. (2003). Histone modifications in Arabidopsis - high methylation of H3 lysine 9 is dispensable for constitutive heterochromatin. *Plant J.* 33, 471–480. doi: 10.1046/j.1365-313X.2003.01638.x

Lermontova, I., Schubert, V., Fuchs, J., Klatte, S., Macas, J., and Schubert, I. (2006). Loading of *Arabidopsis* centromeric histone CENH3 occurs mainly during G2 and requires the presence of the histone fold domain. *Plant Cell* 18, 2443–2451. doi: 10.1105/tpc.106.043174

Maheshawari, S. (2015). Naturally occurring differences in CENH3 affect chromosome segregation in Zygotic Mitosis of hybrid. *PLoS Genet.* 11:e1004970. doi: 10.1371/journal.pgen.1004970

Masonbrink, R. E., Gallagher, J. P., Jareczek, J. J., Renny-Byfield, S., Grover, C. E., Gong, L., et al. (2014). CenH3 evolution in diploids and polyploids of three angiosperm genera. *BMC Plant Biol.* 14:383. doi: 10.1186/s12870-014-0383-3

Nagaki, K., Cheng, Z., Ouyang, S., Talbert, P. B., Kim, M., Jones, K. M., et al. (2004). Sequencing of a rice centromere uncovers active genes. *Nat. Genet.* 36, 138–145. doi: 10.1038/ng1289

Nagaki, K., Kashihara, K., and Murata, M. (2005). Visualization of diffuse centromeres with centromere-specific histone H3 in the holocentric plant *Luzula nivea*. *Plant Cell* 17, 1886–1893. doi: 10.1105/tpc.105.032961

Nagaki, K., Kashihara, K., and Murata, M. (2009). A centromeric DNA sequence colocalized with a centromere-specific histone H3 in tobacco. *Chromosoma* 118, 249–257. doi: 10.1007/s00412-008-0193-1

Nagaki, K., and Murata, M. (2005). Characterization of CENH3 and centromere-associated DNA sequences in sugarcane. *Chromosome Res.* 13, 195–203. doi: 10.1007/s10577-005-0847-2

Nagaki, K., Shibata, F., Kanatani, A., Kashihara, K., and Murata, M. (2012a). Isolation of centromeric-tandem repetitive DNA sequences by chromatin affinity purification using a HaloTag7-fused centromere-specific histone H3 in tobacco. *Plant Cell Rep.* 31, 771–779. doi: 10.1007/s00299-011-1198-4

Nagaki, K., Shibata, F., Suzuki, G., Kanatani, A., Ozaki, S., Hironaka, A., et al. (2011). Coexistence of NtCENH3 and two retrotransposons in tobacco centromeres. *Chromosome Res.* 19, 591–605. doi: 10.1007/s10577-011-9219-2

Nagaki, K., Talbert, P. B., Zhong, C. X., Dawe, R. K., Henikoff, S., and Jiang, J. M. (2003). Chromatin immunoprecipitation reveals that the 180-bp satellite repeat is the key functional DNA element of *Arabidopsis thaliana* centromeres. *Genetics* 163, 1221–1225.

Nagaki, K., Yamamoto, M., Yamaji, N., Mukai, Y., and Murata, M. (2012b). Chromosome dynamics visualized with an anti-centromeric histone H3 antibody in Allium. *PLoS ONE* 7:e51315. doi: 10.1371/journal.pone.0051315

Nasuda, S., Hudakova, S., Schubert, I., Houben, A., and Endo, T. R. (2005). Stable barley chromosomes without centromeric repeats. *Proc. Natl. Acad. Sci. U.S.A.* 102, 9842–9847. doi: 10.1073/pnas.0504235102

Neumann, P., Navrátilová, A., Schroeder-Reiter, E., Koblížková, A., Steinbauerová, V., Chocholová, E., et al. (2012). Stretching the rules: monocentric chromosomes with multiple centromere domains. *PLoS Genet.* 8:e1002777. doi: 10.1371/journal.pgen.1002777

Neumann, P., Pavlíková, Z., Koblížková, A., Fuková, I., Jedličková, V., Novák, P., et al. (2015). Centromeres off the hook: massive changes in centromere size and structure following duplication of CenH3 gene in Fabeae species. *Mol. Biol. Evol.* 32, 1862–1879. doi: 10.1093/molbev/msv070

Novák, P., Neumann, P., and Macas, J. (2010). Graph-based clustering and characterization of repetitive sequences in next-generation sequencing data. *BMC Bioinformatics* 11:378. doi: 10.1186/1471-2105-11-378

Ohzeki, J., Bergmann, J. H., Kouprina, N., Noskov, V. N., Nakano, M., Kimura, H., et al. (2012). Breaking the HAC Barrier: histone H3K9 acetyl/methyl balance regulates CENP-A assembly. *EMBO J.* 31, 2391–2402. doi: 10.1038/emboj.2012.82

Perpelescu, M., and Fukagawa, T. (2011). The ABCs of CENPs. *Chromosoma* 120, 425–446. doi: 10.1007/s00412-011-0330-0

Saitou, N., and Nei, M. (1987). The neighbor-joining method: a new method for reconstructing phylogenic trees. *Mol. Biol. Evol.* 4, 406–425.

Sanei, M., Pickering, R., Kumke, K., Nasuda, S., and Houben, A. (2011). Loss of centromeric histone H3 (CENH3) from centromeres precedes uniparental chromosome elimination in interspecific barley hybrids. *Proc. Natl. Acad. Sci. U.S.A.* 108, E498–E505. doi: 10.1073/pnas.1103190108

Shang, W. H., Hori, T., Toyoda, A., Kato, J., Popendorf, K., Sakakibara, Y., et al. (2010). Chickens possess centromeres with both extended tandem repeats and short non-tandem-repetitive sequences. *Genome Res.* 20, 1219–1228. doi: 10.1101/gr.106245.110

Sharma, S. K., Yamamoto, M., and Mukai, Y. (2015). Immuno-cytogenetic manifestation of epigenetic chromatin modification marks in plants. *Planta* 241, 291–301. doi: 10.1007/s00425-014-2233-9

Talbert, P. B., Masuelli, R., Tyagi, A. P., Comai, L., and Henikoff, S. (2002). Centromeric localization and adaptive evolution of an *Arabidopsis* histone H3 variant. *Plant Cell* 14, 1053–1066. doi: 10.1105/tpc.010425

Talia, P., Greizerstein, E., Quijano, C. D., Peluffo, L., Fernandez, L., Fernández, P., et al. (2010). Cytological characterization of sunflower by *in situ* hybridization using homologous rDNA sequences and a BAC clone containing highly represented repetitive retrotransposon-like sequences. *Genome* 53, 172–179. doi: 10.1139/G09-097

Tek, A. L., Kashihara, K., Murata, M., and Nagaki, K. (2010). Functional centromeres in soybean include two distinct tandem repeats and a retrotransposon. *Chromosome Res.* 18, 337–347. doi: 10.1007/s10577-010-9119-x

Tek, A. L., Kashihara, K., Murata, M., and Nagaki, K. (2011). Functional centromeres in Astragalus sinicus include a compact centromere-specific histone H3 and a 20-bp tandem repeat. *Chromosome Res.* 19, 969–978. doi: 10.1007/s10577-011-9247-y

Thompson, J. D., Gibson, T. J., Plewniak, F., Jeanmougin, F., and Higgins, D. G. (1997). The CLUSTAL_X windows interface: flexible strategies for multiple sequence alignment aided by quality analysis tools. *Nucleic Acids Res.* 25, 4876–4882. doi: 10.1093/nar/25.24.4876

Vermaak, D., Hayden, H. S., and Henikoff, S. (2002). Centromere targeting element within the histone fold domain of Cid. *Mol. Cell. Biol.* 22, 7553–7561. doi: 10.1128/MCB.22.21.7553-7561.2002

Wang, G., He, Q., Cheng, Z., Talbert, P. B., and Jin, W. (2011). Characterization of CENH3 proteins and centromere-associated DNA sequences in diploid and allotetraploid Brassica species. *Chromosoma* 120, 353–365. doi: 10.1007/s00412-011-0315-z

Warburton, P. E., Cooke, C. A., Bourassa, S., Vafa, O., Sullivan, B. A., Stetten, G., et al. (1997). Immunolocalization of CENP-A suggests a distinct nucleosome structure at the inner kinetochore plate of active centromeres. *Curr. Biol.* 7, 901–904. doi: 10.1016/S0960-9822(06)00382-4

Yamaji, N., and Ma, J. F. (2007). Spatial distribution and temporal variation of the rice silicon transporter Lsi1. *Plant Physiol.* 143, 1306–1313. doi: 10.1104/pp.106.093005

Yan, H., Ito, H., Nobuta, K., Ouyang, S., Jin, W., Tian, S., et al. (2006). Genomic and genetic characterization of rice Cen3 reveals extensive transcription and evolutionary implications of a complex centromere. *Plant Cell* 18, 2123–2133. doi: 10.1105/tpc.106.043794

Zhang, T., Talbert, P. B., Zhang, W., Wu, Y., Yang, Z., Henikoff, J. G., et al. (2013). The CentO satellite confers translational and rotational phasing on cenH3 nucleosomes in rice centromeres. *Proc. Natl. Acad. Sci. U.S.A.* 110, E4875–E4883. doi: 10.1073/pnas.1319548110

Zhong, C. X., Marshall, J. B., Topp, C., Mroczek, R., Kato, A., Nagaki, K., et al. (2002). Centromeric retroelements and satellites interact with maize kinetochore protein CENH3. *Plant Cell* 14, 2825–2836. doi: 10.1105/tpc.006106

Conflict of Interest Statement: The authors declare that the research was conducted in the absence of any commercial or financial relationships that could be construed as a potential conflict of interest.

4

Down-regulation of multiple CDK inhibitor *ICK/KRP* genes promotes cell proliferation, callus induction and plant regeneration in *Arabidopsis*

Yan Cheng[1], Han Liu[1], Ling Cao[1,2], Sheng Wang[2], Yongpeng Li[1], Yuanyuan Zhang[1], Wei Jiang[1], Yongming Zhou[1]* and Hong Wang[2]*

[1] National Key Laboratory of Crop Genetic Improvement, College of Plant Science and Technology, Huazhong Agricultural University, Wuhan, China, [2] Department of Biochemistry, University of Saskatchewan, Saskatoon, SK, Canada

Edited by:
Bo Liu,
University of California, Davis, USA

Reviewed by:
Ming Yang,
Oklahoma State University, USA
Sherryl Bisgrove,
Simon Fraser University, Canada

****Correspondence:***
Hong Wang,
Department of Biochemistry,
University of Saskatchewan,
Saskatoon, SK S7N 5E5, Canada
hong.wang@usask.ca;
Yongming Zhou,
National Key Laboratory of Crop
Genetic Improvement, College
of Plant Science and Technology,
Huazhong Agricultural University,
Wuhan 430070, China
ymzhou@mail.hzau.edu.cn

The ICK/KRP cyclin-dependent kinase (CDK) inhibitors are important plant cell cycle regulators sharing only limited similarity with the metazoan CIP/KIP family of CDK inhibitors. Information is still limited regarding the specific functions of different *ICK/KRP* genes *in planta*. We have shown previously that down-regulation of multiple CDK inhibitor *ICK/KRP* genes up-regulates the E2F pathway and increases cell proliferation, and organ and seed sizes in *Arabidopsis*. In this study, we observed that the quintuple *ick1/2/5/6/7* mutant had more cells in the cortical layer of the root apical meristem (RAM) than the wild type (Wt) while its RAM length was similar to that of the Wt, suggesting a faster cell cycle rate in the quintuple mutant. We further investigated the effects of down-regulating *ICK* genes on tissue culture responses. The cotyledon explants of *ick1/2/5/6/7* could form callus efficiently in the absence of cytokinin and also required a lower concentration of 2,4-D for callus induction compared to the Wt plants, suggesting increased competence for callus induction in the mutant. In addition, the quintuple *ick* mutant showed enhanced abilities to regenerate shoots and roots, suggesting that increased competence to enter the cell cycle in the quintuple mutant might make it possible for more cells to become proliferative and be utilized to form shoots or roots. These findings indicate that CDK activity is a major factor underlying callus induction and increased cell proliferation can enhance *in vitro* organogenesis.

Keywords: *Arabidopsis thanliana*, cell cycle, cyclin-dependent kinase, CDK inhibitor, *ICK/KRP*, cell proliferation, callus induction, plant regeneration

Introduction

Cell division is fundamental to plant growth, development, and reproduction. In eukaryotes, cyclin-dependent kinases (CDKs) control cell division cycle, and their activities are in turn modulated by different factors (Morgan, 1997). Among them, CDK inhibitors are crucial negative regulators which inhibit CDK activity through direct protein binding. CDK inhibitor genes were initially identified in mammalian and yeast (Sherr and Roberts, 1995). The first plant CDK inhibitor gene *ICK1* (Inhibitor of CDK) was discovered in *Arabidopsis* (Wang et al., 1997) and there are seven *ICK* genes (also refered to as *KRPs*) in *Arabidopsis* (De Veylder et al., 2001). To date, *ICK/KRP* genes have been identified from different plant species such as tobacco (Jasinski et al., 2003), maize

(Coelho et al., 2005), rice (Barrôco et al., 2006), tomato (Bisbis et al., 2006), apple (Malladi and Johnson, 2011), and avocado (Sabag et al., 2013).

Tissue culture and plant regeneration from explants have many different applications. When proper stimuli are given, somatic plant cells may form adventitious embryos, root, or shoots (De Klerk et al., 1997). Plant regeneration usually takes one of the two pathways: somatic embryogenesis and organogenesis (Davey and Anthony, 2010). The plant regeneration process can be divided conceptually into the following three phases: (1) dedifferentiation, during which the cells acquire competence to respond to the induction stimuli; (2) induction, during which the competent cells are induced to enter particular morphologic pathways; and (3) realization, during which the calli undergo morphological differentiation and development (see review by Duclercq et al., 2011). The organogenesis pathway is more often the route encountered in micropropagation, haploid production and plant transformation (Duclercq et al., 2011). During *in vitro* organogenesis, callus induction is followed by shoot and root regeneration. It has been demonstrated in a wide range of plant species that generally a high cytokinin (CK) to auxin ratio induces shoot organogenesis, whereas a low ratio results in root development (Sangwan and Sangwan-Norreel, 1990; De Klerk et al., 1997; Davey and Anthony, 2010). In addition to the exogenous plant hormones, other conditions such as nutrient concentrations, sugar sources, and induction time on culture media can also affect the frequency of plant regeneration (Christianson and Warnick, 1983).

In recent years, considerable progress has been made in understanding the developmental events during *de novo* organogenesis and its underlying molecular mechanisms. Certain genes involved in shoot and root organogenesis processes have been identified (see reviews: Duclercq et al., 2011; Motte et al., 2014). Most of these genes are involved in auxin and CK pathways or shoot meristem maintenance. In addition, *ESR1* and *ESR2* encoding transcriptional factors of the *AP2/ERF* family, are identified as enhancers of shoot regeneration in *Arabidopsis* (Banno et al., 2001; Ikeda et al., 2006), and *CUP-SHAPED COTYLEDON1 (CUC1)*, *CLAVATA3/EMBRYO SURROUNDING REGION-RELATED PEPTIDE (CLE2)* and *GCN5-related N-acetyltransferase 1 (GNAT1)* were identified as *ESR1* up-regulated genes (Matsuo et al., 2009), while *CUC1*, *Cyclin D1;1* and *ARABIDOPSIS PHOSPHOTRANSMITTER 6 (AHP6)* were identified as *ESR2* up-regulated genes (Ikeda et al., 2006), which might also be involved in shoot regeneration.

Cell division is a prerequisite to both callus induction and shoot/root regeneration during organogenesis. Although there is considerable amount of knowledge on the functions of cell cycle regulators in the cell cycle, relative little is known about their involvement *in vitro* plant regeneration. Several previous studies showed that cell cycle regulators can affect callus induction. Overexpression of a D-type cyclin has been shown to increase callus induction frequency and callus growth rate in *Arabidopsis* (Riou-Khamlichi et al., 1999; Cockcroft et al., 2000). In rice, inducible expression of a rice CDK-activating kinase also increases callus induction of tobacco leaf explants (Yamaguchi et al., 2003).

In a previous study, we have reported the effects of down-regulating ICK/KRP CDK inhibitors on plant growth and development using a series of *ick* mutants (Cheng et al., 2013a). The multiple *ick* mutants particularly the quintuple mutant had increased CDK activity, up-regulated E2F-RB pathway and enhanced cell proliferation. In this study, we investigated the effects of *ICK* down-regulation on root cell proliferation and further on callus induction and plant regeneration.

Materials and Methods

Plant Materials and Growth Conditions

The T-DNA insertion lines for the five *Arabidopsis ICK/KRP* genes used in this study have been described previously (Cheng et al., 2013a). For plants in soil, *Arabidopsis* plants were grown at 21°C under 16 h/8 h day/night photoperiod in a plant growth room. For seedlings in plates, seeds were sterilized as described (Valvekens et al., 1988) and sowed on ½-strength solid MS medium (1/2 MS, 1% sucrose, 0.7% agar, pH 5.8). The plates were placed vertically in the tissue culture room with the temperature of 22°C and photoperiod of 16 h/8 h day/night.

Root Length and Root Meristem Size Analysis

The seeds of wild type (Wt) and *ick1/2/5/6/7* quintuple mutant were sterilized and sowed on square plates containing ½-strength solid MS medium. The plates were placed vertically in the plant tissue culture room. Everyday from 2 to 6 days after germination (DAG), 16–25 seedlings from each line were removed from the plates for root length and meristem size analyses. The root length was measured with a ruler. To measure root meristem size, the root was removed and immersed in Hoyer's solution chloral hydrate/water/glycerol (3:0.8:0.4) on a glass slide. After 30 min treatment, the slide was covered with a coverslip. The root meristem was observed with DIC (differential interference contrast) under a microscope (Nikon ECLIPSE 80i). The root meristem size was represented by the number of meristematic cortex cells, which was counted as described (Casamitjana-Martíneznez et al., 2003; Ioio et al., 2007). The length of cortex cells in the mature zone was determined at 6 DAG. For each root, an image was taken and cell length measured for 6–10 cortex cells along the mature zone using Image J (http://rsb.info.nih.gov/ij).

Callus Induction and Growth Analyses

Sterilized Wt and mutant *Arabidopsis* seeds were sowed on square plates containing solid 1/2 MS medium. Seven days after sowing, cotyledons were cut into explants of approximately 4 mm × 4 mm in size. The explants were placed onto 1/2 MS medium containing 0.2 mg/ml 2,4-D (2,4-dichlorophenoxy-acetic acid) or containing both 0.2 mg/ml 2,4-D and 0.2 mg/ml 6-BA (6-benzylaminopurine) solidified with 0.7% agarose. Ten plates for each treatment (about 15 explants in each plate) were used for callus induction and growth analyses. Every the seventh day, the explants were examined to obtain the frequency of explants with callus. Then, the explants were transferred to a fresh plate containing the same medium. The weight of the plate before and after the transfer was measured and the average weight of calli was obtained [=(weight of plate after transfer - weight of plate before transfer)/number of the calli].

To determine the minimal 2,4-D concentration for callus induction, root segments, and excised cotyledons were incubated on $^{1}/_{2}$-strength solid MS medium supplemented with different concentrations of 2,4-D. (for the first batch: 0 mg/L, 0.1mg/L, 0.15 mg/L, and 0.2 mg/L; for the second batch: 0.005 mg/L, 0.01 mg/L, 0.02mg/L, 0.03 mg/L, 0.04 mg/L, 0.05 mg/L). Callus induction frequency was obtained after 10 and 20 days of culture for root explants and after 20 days for cotyledon explants. Fresh weigh was obtained after 20 days of culture. For each treatment, 5–6 plates with 32 root or cotyledon explants in each plate were used.

Root and Shoot Regeneration Analysis

Sterilized Wt and mutant seeds were sowed on square plates containing $^{1}/_{2}$-strength MS solid medium, with the plates placed vertically in the tissue culture room. Seven days after sowing, excised roots were cut into 3–5 mm segments and transferred onto callus induction medium (CIM) containing Gamborg's B5 salt and vitamins (Gamborg et al., 1968), 2% sucrose, 0.5 g/L MES, 0.48 mg/L 2,4-D, 0.043 mg/L kinetin (KT), and 0.7% agarose. After 7 days of culture on CIM, the root explants with callus were transferred either onto shoot induction medium (SIM) containing Gamborg's B5 and vitamins, 2% sucrose, 0.5 g/L MES, 1 mg/L isopentenyladenine (2-ip), 0.15 mg/L indole-3-acetic acid (IAA), and 0.7% agarose for regenerating shoots, or onto root induction medium (RIM) containing Gamborg's B5 and vitamins, 2% sucrose, 0.5 g/L MES, 0.87 mg/L IAA and 0.7% agarose for regenerating roots (Yasutani et al., 1994). For shoot regeneration analysis, each line had 9–10 SIM plates with each plate having ~50 explants. After 30 days of culture, the calli were surveyed to obtain the frequency of shoot regeneration for each plate.

To determine shoot and root regeneration in different *ick* mutant lines, calli were first induced from root explants as described above. The explants with callus were then transferred onto SIM for shoot induction or RIM for root induction, with 9–10 plates for each treatment. For shoot induction each plate had ~40 cultured root explants (root explants with callus), while for root induction each plate about 50 cultured root explants transferred from the CIM. After 20 days of culture on SIM or RIM, the number of calli with regenerating shoots was counted and shoot regeneration frequency obtained for each plate. For root regeneration, we determined the frequency of explants with root induction (=number of calli with regenerating roots/total number of calli). Since the number of roots on each callus varied greatly, we grouped the explants into four categories: (1) no root, (2) 1–5 roots, (3) 6–10 roots and (4) more than 10 roots. The percentages of the four categories were calculated for each line.

RNA Extractions and Real-time PCR

Arabidopsis total RNA was isolated using TRIzol reagent (Invitrogen) according to manufacturer's instructions. First-strand cDNA synthesis and real-time PCR analysis were performed as described previously (Zhang et al., 2015). The primers for real-time PCR are listed in Supplementary Table S1.

Statistical Analysis

The Mann–Whitney two-tailed *U* test was used for analyzing the differences in callus induction rates from cotyledon explants and root induction rates from root explants. Fisher's least significant difference (LSD) method was used for multiple comparisons of shoot regeneration rates of the Wt and mutants. For the other differences between the mutant and Wt, The Student's *t*-test (*T*-test) method was used and the analysis was performed.

Results

Cell Division was Accelerated in Root Cortex Meristematic Cells of *ick1/2/5/6/7* Mutant

In our previous study, we established a series of T-DNA insertion lines in which one to five *ICK* genes were knocked out, and observed phenotypical changes in triple, quadruple and quintuple mutants. Notably the quintuple mutant *ick1 ick2 ick5 ick6 ick7* (referred to *as ick1/2/5/6/7* for shoot) has larger leaves, petals, and seeds than the Wt (Cheng et al., 2013a), suggesting that cell proliferation is promoted in the *ick1/2/5/6/7* quintuple mutant, as a result of down-regulating *ICK* genes.

To determine more specifically how cell proliferation is affected in the mutant, we examined cell production in the root. First, we analyzed the root growth of the *ick1/2/5/6/7* mutant and Wt plants from 2 to 6 days after seed plating. As shown in **Figure 1A**, the primary root length for *ick1/2/5/6/7* quintuple mutant was very similar to that of the Wt. We then investigated the cortex cells in the mature zone of the roots at 6-day stage after germination (DAG). The length of cortex cells in different positions along each root was measured, and the total cortex cell number estimated based on the root length and average cortex cell length. The results showed that the average length of the cortex cells in the mature zone of *ick1/2/5/6/7* quintuple mutant was reduced compared with that of Wt (137.7 ± 3.6 um compared to 168.9 ± 4.1 um in Wt). Since the mutant and the Wt lines had a similar root length (36.3 ± 3.9 mm and 36.3 ± 3.2 mm at 6 DAG, respectively), the total cell number in a cortex cell file along the mature zone of *ick1/2/5/6/7* quintuple mutant (264.0 ± 8.8) was significantly higher than that of the Wt (229.1 ± 8.2; **Figures 1B,C**; **Figure 1B** shows the relative ratio of the mutant to Wt).

To determine whether the quintuple mutant has a larger root meristem, we performed a time-course analysis on root meristem size following Ioio's method (Ioio et al., 2007). In this assay, the root-meristem size was expressed as the number of cortex cells in a file extending from the quiescent center (QC) to the first elongated cell (**Figure 1D**). We found that the roots of *ick1/2/5/6//7* quintuple mutant and Wt had a similar final root meristem size. However, the *ick1/2/5/6/7* quintuple mutant reached the final size 4 DAG, while the Wt reached this final size 5 DAG (**Figure 1E**), suggesting an accelerated rate of cell division and reduced cell elongation in the mutant. These results imply that more cells in the cortex of the quintuple mutant are likely due to a faster cell production rate, instead of a larger root meristem.

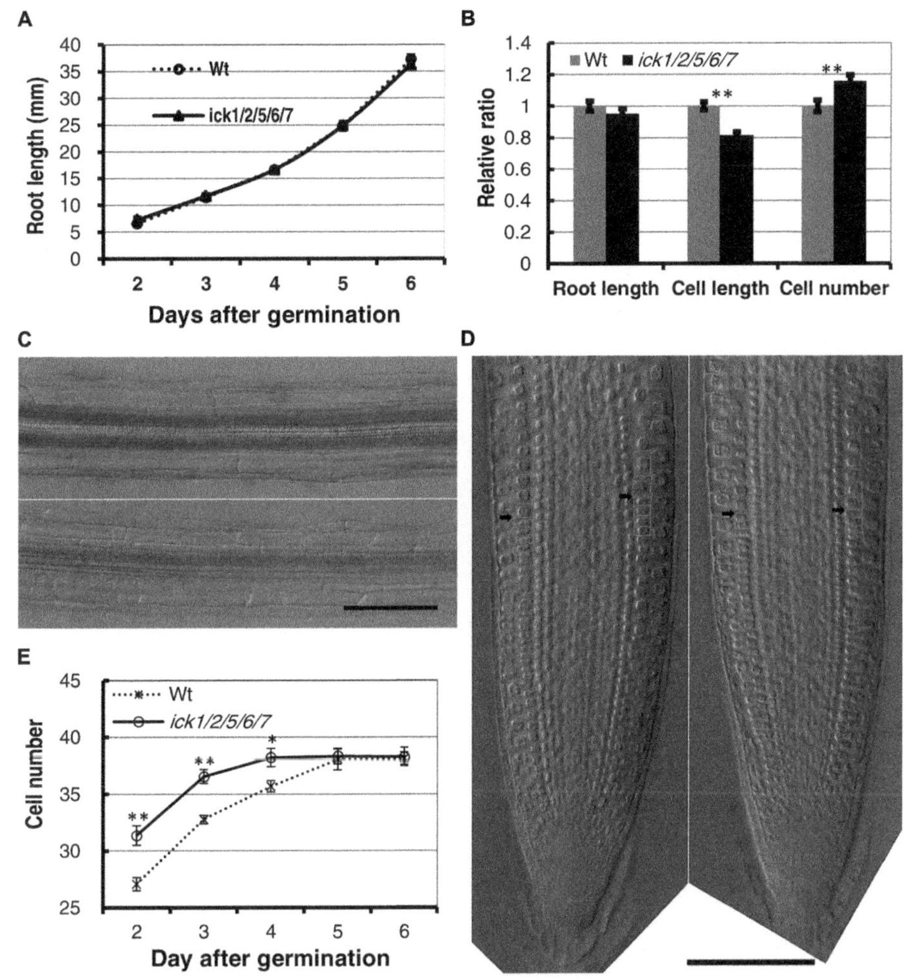

FIGURE 1 | Cell length, cell number, and root apical meristem size of wild type (Wt) and *ick1/2/5/6/7* mutant *Arabidopsis* plants. (A) Primary root length of the Wt and *ick1/2/5/6/7* quintuple mutant at 2–6 days after germination (DAG). Each datum point represents the average of 16–25 seedlings. **(B)** Relative root length, cortex cell length and cortex cell number of the mature zone of the Wt and *ick1/2/5/6/7* mutant plants at 6 DAG (relative ratio = value of mutant/value of Wt). For root length and cell length analyses, 20–22 seedlings were used. For cell length analysis, at least six cells in the mature zone were measured for each root tip. **(C)** Differential interference contrast (DIC) images showing the cortex cells of the Wt (upper) and *ick1/2/5/6/7* quintuple mutant (lower) roots. Two cortex cells of each line are marked in red. **(D)** Longitudinal view of root tips of 6-day-old Wt and *ick1/2/5/6/7* quintuple (right) plants. The root tips were placed on a glass slide, covered with a coverslip, and viewed under a DIC microscope. The cortex cells around the transition zone are marked in red, and black arrowheads indicate the boundary between meristem and elongation zone. **(E)** Meristem size of Wt and *ick1/2/5/6/7* quintuple mutant plants at 2-6 DAG. The root meristem is expressed as the number of cortex cells in a file extending from the quiescent center (QC) to the first elongated cell. The Student's *t*-test was used for analyzing the difference between the Wt and mutant [error bar = SE (standard error); $*P < 0.05$, $**P < 0.01$]. Each datum point represents the average of 16–25 seedlings. Bars = 100 um.

Down-regulation of Five *ICKs* Increased Callus Induction

To further understand the impact of *ICK* down-regulation, we examined tissue culture responses since cell proliferation is critical for callus and plant regeneration. Cotyledon explants of both Wt and *ick1/2/5/6/7* mutant produced calli efficiently on 1/2 MS medium containing both 0.2 mg/ml 2,4-D and 0.2 mg/ml 6-BA. On 1/2 MS medium containing 0.2 mg/ml 2,4-D, however, there was a higher frequency of callus induction for the mutant explants (**Figures 2A,B**). For instance, 98.4% of the *ick1/2/5/6/7* mutant explants produced calli, compared 69.1% for the Wt (**Figure 2C**). In addition, the calli of quintuple mutants were much larger with lightly greenish color, while those of the Wt yellower and smaller (**Figure 2B**). These results indicate that down-regulation of the five *ICK* genes enhances callus formation and reduces CK requirement for callus induction.

To determine callus growth rate, the explants were transferred to fresh callus induction plates every week, and the callus growth was obtained by weighing the plate immediately after the transfer and on the seventh day of culture. As shown in **Figure 2D** and Supplementary Figure S1, the calli of *ick1/2/5/6/7* grew faster than those of Wt in the presence of 6-BA and 2,4-D or 2,4-D only. Those results indicate that down-regulation of the five *ICK* genes also enhances callus growth.

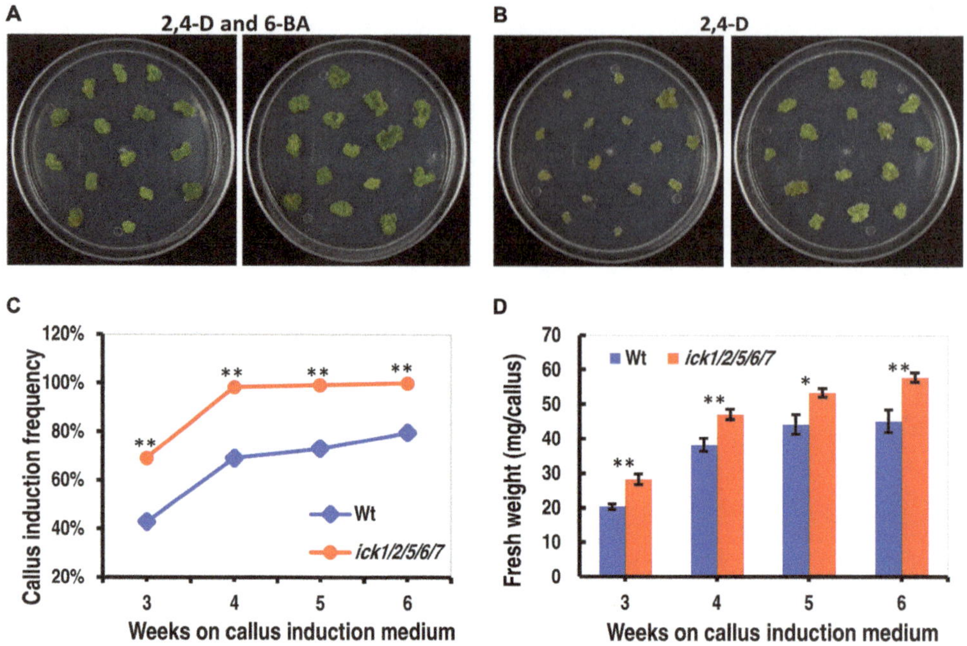

FIGURE 2 | Callus induction from cotyledon explants and callus growth of Wt and *ick1/2/5/6/7* plants. (A,B) Representative plates showing callus induction from cotyledon explants of Wt (left) and *ick1/2/5/6/7* mutant (right) in the presence of 2,4-D and 6-BA **(A)** or 2,4-D alone **(B)**. Similar results were obtained in three independent experiments. In each experiment, 10 plates were used for each treatment, with each plate having around 15 explants. The data from one experiment are shown here. **(C)** Callus induction frequency of the Wt and *ick1/2/5/6/7* mutant explants on culture medium containing 2,4-D alone. The difference in the induction rate between the Wt and mutant at each time point was compared by Mann–Whitney *U* test. **(D)** Fresh weight of calli from the Wt and *ick1/2/5/6/7* mutant cotyledon explants after 3–7 weeks of culture on culture medium containing both 2,4-D and 6-BA. The differences between the Wt and mutant were analyzed by Student's *t*-test (error bar = SE; $^{*}P < 0.05$, $^{**}P < 0.01$).

Auxin Dependency for Callus Induction was Decreased in the *ick1/2/5/6/7* Mutant

To further confirm that *ICK* down-regulation reduces auxin requirement for callus induction, we determined the minimal 2,4-D concentration for callus induction from root explants of both Wt and mutant plants. In this assay, the root segments (about 5 mm in length) were incubated on the 1/2 MS medium supplemented with different concentrations of 2,4-D. We first used 2,4-D concentrations of 0, 0.05, 0.1, 0.15, and 0.2 mg/L. Neither the Wt nor the quintuple mutant showed callus induction on 1/2 MS medium without 2,4-D after 20 days of culture; whereas, on the culture plates containing 0.05, 0.1, 0.15, or 0.2 mg/L 2,4-D, almost all segments of both lines generated calli (Supplementary Figures S2A,B). We then used a series of lower 2,4-D concentrations of 0.005, 0.01, 0.02, 0.03, 0.04, and 0.05 mg/L. The root explants of both lines produced no callus at 0 and 0.005 mg/L 2,4-D, and almost all root explants of both lines produced calli at 0.02, 0.03, 0.04, and 0.05 mg/L 2,4-D after 20 days of culture (**Figure 3A**). At 0.01 mg/ml 2,4-D, the callus induction frequency of *ick1/2/5/6/7* mutant (61.1%) was significantly higher than that of the Wt (19.9%; **Figure 3B**). This observation suggests that the *ick1/2/5/6/7* mutant needs a lower concentration of auxin for callus induction compared to Wt. Moreover, for the 2,4-D concentrations at which the callus induction frequencies of the two lines were comparable, the fresh weights of calli from the mutant root explants were significantly higher than those of Wt (**Figure 3C** and Supplementary Figure S2C), consistent with observation made with the cotyledon explants.

We also determined the minimal 2,4-D concentration required for callus induction from cotyledonary petiole of both lines. When the 2,4-D concentration was higher than 0.05 mg/ml, almost all of the cotyledonary petiole explants of *ick1/2/5/6/7* produced calli (Supplementary Figures S3A,B). When the 2,4-D concentration was lower than 0.1 mg/ml, the callus induction frequency of the *ick1/2/5/6/7* mutant was significantly higher than that of the Wt (**Figures 4A,B** and Supplementary Figures S3A,B). At 0.005 mg/ml 2,4-D, 33% of *ick1/2/5/6/7* cotyledonary petioles showed callus induction, while no cotyledonary petioles of the Wt did (**Figures 4A,B**). Also, at various concentrations of 2,4-D the calli of the *ick1/2/5/6/7* mutant were significantly larger than those of Wt (**Figure 4C** and Supplementary Figure S3C), further confirming that the callus of *ick1/2/5/6/7* mutant was growing faster than the Wt callus.

Shoot and Root Regeneration was Enhanced in *ick1/2/5/6/7* Mutant

In addition to callus induction, shoot and root regeneration is another important aspects of plant tissue culture. Thus, we investigated the ability of the root explants to regenerate

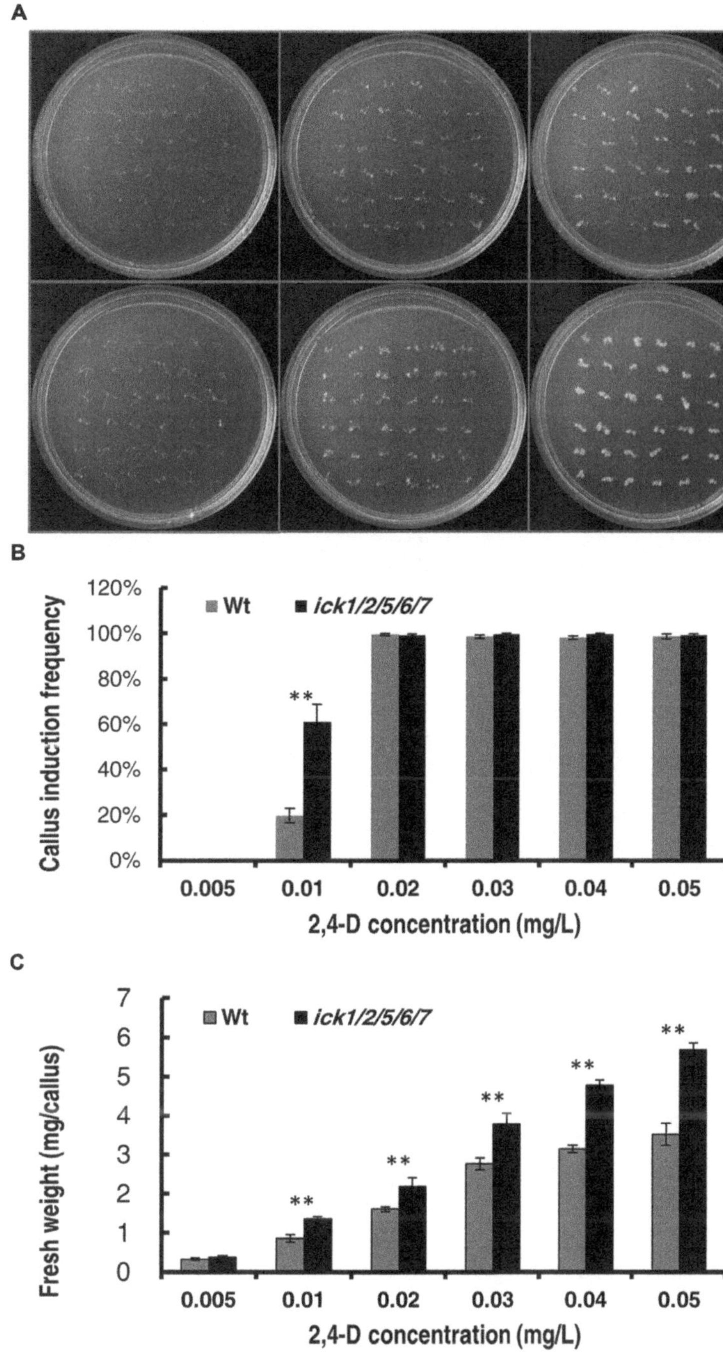

FIGURE 3 | Callus induction from root explants of the Wt and *ick1/2/5/6/7* mutant at different 2,4-D concentrations. Root segments were cultured on 1/2 MS medium containing the indicated concentrations of 2,4-D for 20 days. **(A)** Representative plates showing the root explants of Wt (the upper row) and *ick1/2/5/6/7* mutant (the lower row) on 1/2 MS medium containing 0.05, 0.01, and 0.05 mg/L (the first–third column, respectively) 2,4-D. **(B)** Callus induction rate of root explants of Wt and *ick1/2/5/6/7* mutant after 20 days of culture. **(C)** Fresh weight of root explants with callus of Wt and *ick1/2/5/6/7* after 20 days of culture. The bars show the mean values of 4–5 plates. Student's *t*-test was used for analyzing the differences in induction rate and fresh weight between the Wt and mutant (error bar = SE; $^{**}P < 0.01$).

shoots. Root explants (about 5 mm segments) were first cultured on CIM. After 7 days, explants with callus were transferred to SIM. As shown in Supplementary Figure S4, the shoot regeneration frequency of *ick1/2/5/6/7* quintuple mutant was significantly higher than that of Wt after 30 days of culture. Also, each of the explants of *ick1/2/5/6/7* quintuple mutant on average regenerated more roots than that of Wt. Those results indicate that the *ick1/2/5/6/7* quintuple

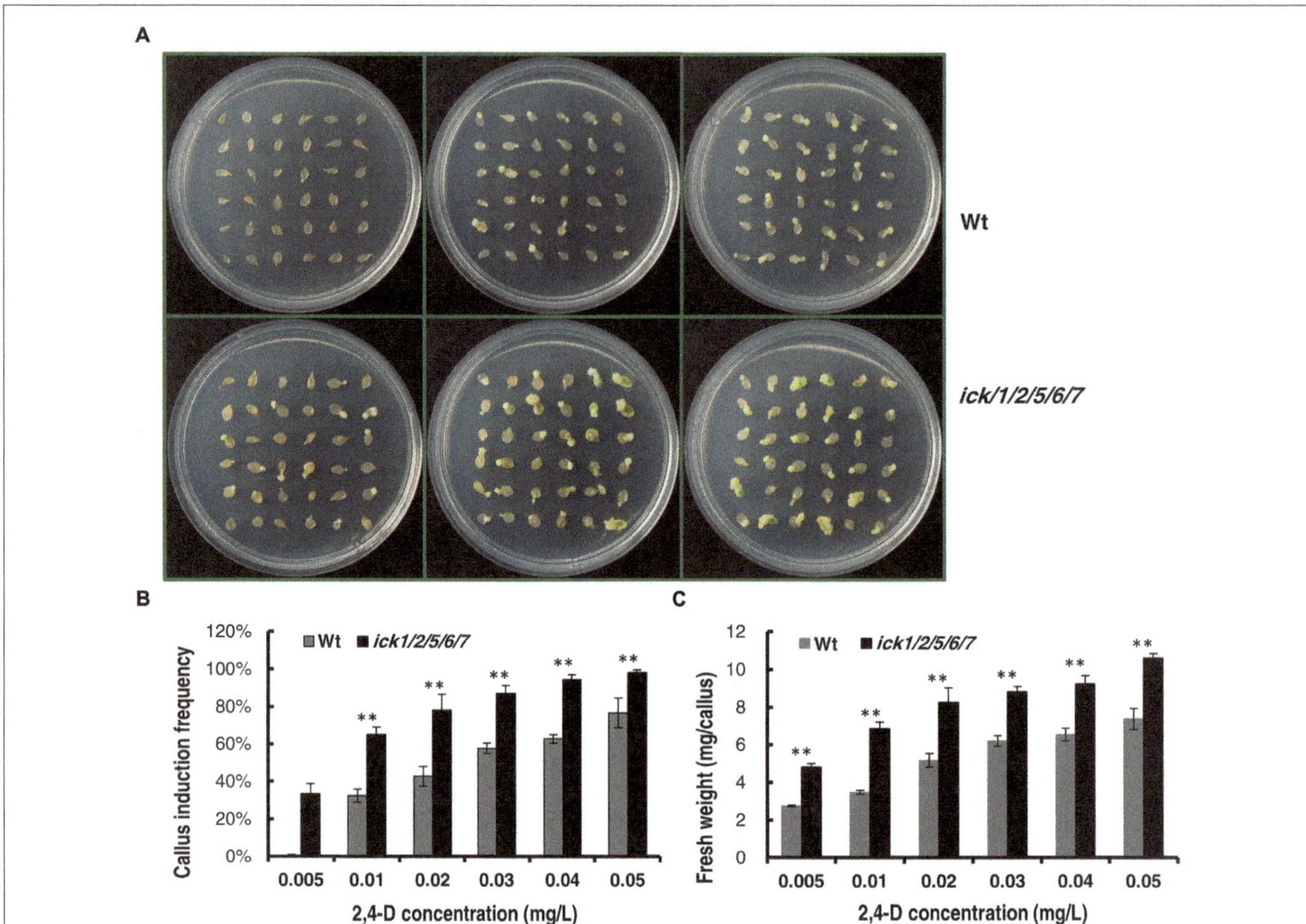

FIGURE 4 | Callus induction from cotyledon explants of the Wt and *ick1/2/5/6/7* mutant at different 2,4-D concentrations. The cotyledons were excised form 7-day-old seedlings and cultured on 1/2 MS medium for 20 days. **(A)** Representative plates showing the cotyledon explants of Wt (the upper row) and *ick1/2/5/6/7* mutant (the lower row) on 1/2 MS medium containing 0.005 mg/L, 0.02g/L, 0.04 mg/L (the first–third column) 2,4-D. **(B)** Callus induction rate of the cotyledon explants of *ick1/2/5/6/7* and Wt on the indicated media. **(C)** Fresh weight of the cotyledon explants with callus of *ick1/2/5/6/7* and Wt on the indicated media. The bars show the mean induction rates of 4–5 plates. Student's *t*-test was used for analyzing the differences in induction rate and fresh weight between the Wt and mutant (error bar = SE; $^{**}P < 0.01$).

mutant has a stronger ability to regenerate shoots and roots.

Disruption of *ICK* Genes Additively Promoted Shoot and Root Regeneration

Our previous results on a series T-DNA mutants showed that the effects from down-regulating *ICK* genes become more evident as more *ICK* genes are disrupted (Cheng et al., 2013a). Therefore, we selected a series of single, double, triple, quadruple, and quintuple mutants to determine whether such an additive effect of multiple loci also exists for shoot and root regeneration. For this analysis, root explants with callus were transferred onto SIM or RIM (RIM) for shoot or root regeneration. After culturing on SIM for 20 days, the single *ick1* mutant had a similar frequency of calli with regenerating shoots to the Wt, and there was a trend of increasing regeneration frequency from the single to the quintuple mutant (**Figure 5A**). Although the differences among Wt, *ick1/2*, *ick1/2/7*, *ick1/2/6/7* did not reach a significant level, the quintuple *ick1/2/5/6/7* mutant with the highest frequency showed significant differences from the other lines (**Figure 5B**).

Calli were also cultured on RIM for 20 days, although most calli of all the lines produced roots, the calli of *ick1/2/6/7* and *ick1/2/5/6/7* mutants had visibly more roots than Wt and *ick1*, *ick1/2* and *ick1/2/7* mutants (Supplementary Figure S5A). We first compared the root regeneration rate of the six lines. The root regeneration frequencies of *ick1/2/6/7* and *ick1/2/5/6/7* were significantly higher than those of other lines (Supplementary Figure S5B). To better characterize the number of roots per callus, the calli were grouped into four categories, with 0, 1–5, 6–10, and more than 10 roots per callus, respectively. The calli of *ick1/2/6/7* and *ick1/2/5/6/7* mutants had significant more roots than the calli of other lines (Supplementary Figure S5C). Those results indicate that down-regulation of *ICK* genes additively enhances both the abilities for shoot and root regeneration.

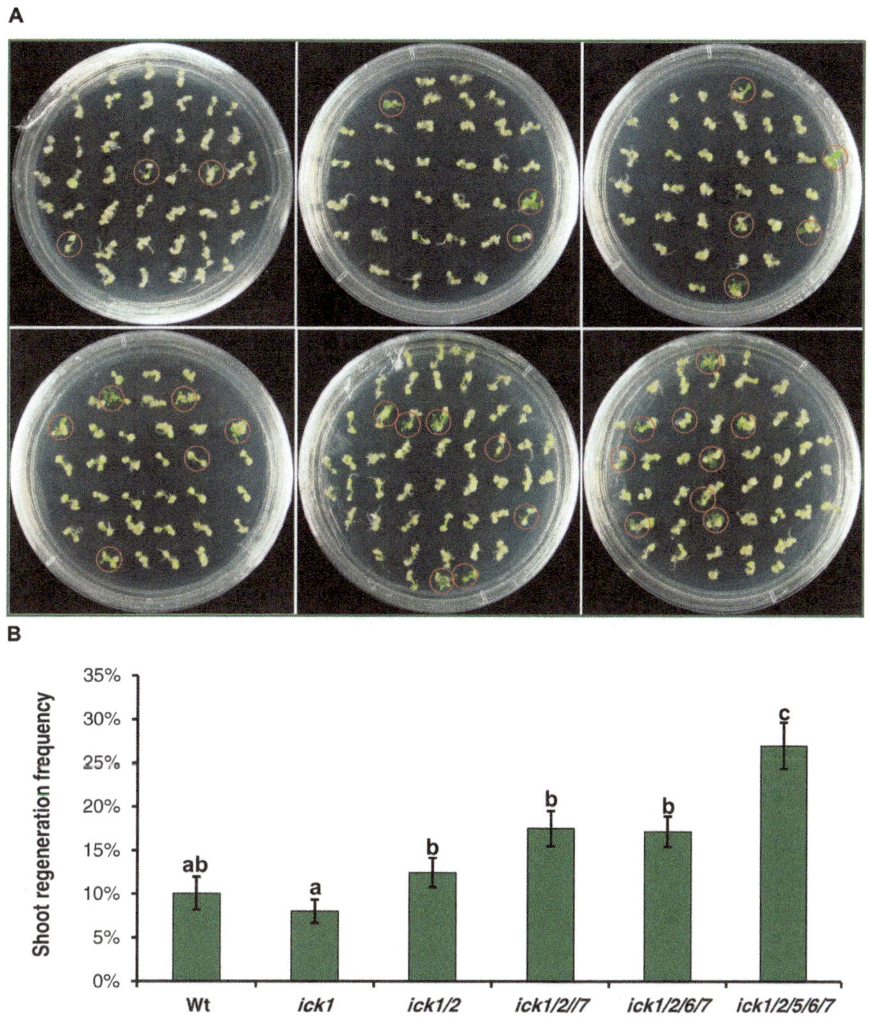

FIGURE 5 | Shoot regeneration from root-derived calli of Wt and various *ick* mutant plants. Root explants were cultured on callus induction medium (CIM) first. After 7 days, root explants with callus were transferred onto shoot induction medium (SIM) and cultured for 30 days. For each line, 9–10 plates were used with each plate having about 40 explants. **(A)** Representative plates showing shoot regeneration on SIM. The plates in the first row are: Wt (left), *ick1* (middle), *ick1/2* (right), and in the second row, *ick1/2/7* (left), *ick1/2/6/7* (middle), and *ick1/2/5/6/7* (right). The red circles mark the calli with regenerating shoots. **(B)** Frequency of shoot regeneration in Wt and *ick* mutants (Mean ± SE). The significant differences among different lines were analyzed by Fisher's least significant difference (LSD) method, and are indicated by different lowercase letters ($P < 0.05$).

E2F-dependent Genes and Shoot Regeneration Related Genes were Mostly Up-regulated During Shoot Regeneration of *ick1/2/5/6/7*

In our previous studies, we have demonstrated that the E2F-dependent genes are up-regulated in the *ick1/2/5/6/7* seedlings (Cheng et al., 2013a). The expression levels of the same group of E2F-dependent genes that function in cell cycle, DNA synthesis, chromatin structure, metabolism, plant development, cell structure, and light signaling/photosynthesis (Ramirez-Parra et al., 2003; de Jager et al., 2009) were analyzed in the SIM incubated callus of *ick1/2/5/6/7* mutant and Wt. Of the 24 genes analyzed, 20 had a higher level of expression in the mutant, with 17 of them having a relative fold expression higher than 1.19 (equal to the Log2 value of 0.25). The four down-regulated genes are *MCM3*, *HTH/EDA17*, *KICP-02*, and *CCA1* (**Figure 6A**). This result suggests that E2F-pathway was also enhanced in the callus of *ick1/2/5/6/7* mutant as observed in the seedlings (Cheng et al., 2013a).

It has been demonstrated that a number of genes were up-regulated during *de novo* shoot regeneration (Duclercq et al., 2011; Motte et al., 2014). To investigate whether the increased shoot regeneration ability of *ick1/2/5/6/7* mutant is accompanied by the up-regulation of these *de novo* regeneration related genes, a group of 12 genes were selected and their expression levels analyzed in the calli cultured on SIM for shoot regeneration. Of the 12 genes, *ESR1* and *ESR2* are known as the enhancers of shoot regeneration (Banno et al., 2001; Matsuo et al., 2009), *CUC1*, *CLE2*, and *GNAT1* are induced by ESR1 (Matsuo et al., 2009), and *WUS*, *STM*, *PIN1*, *REV*, *FIL*, *ATML1*, and *CLV3* are up-regulated

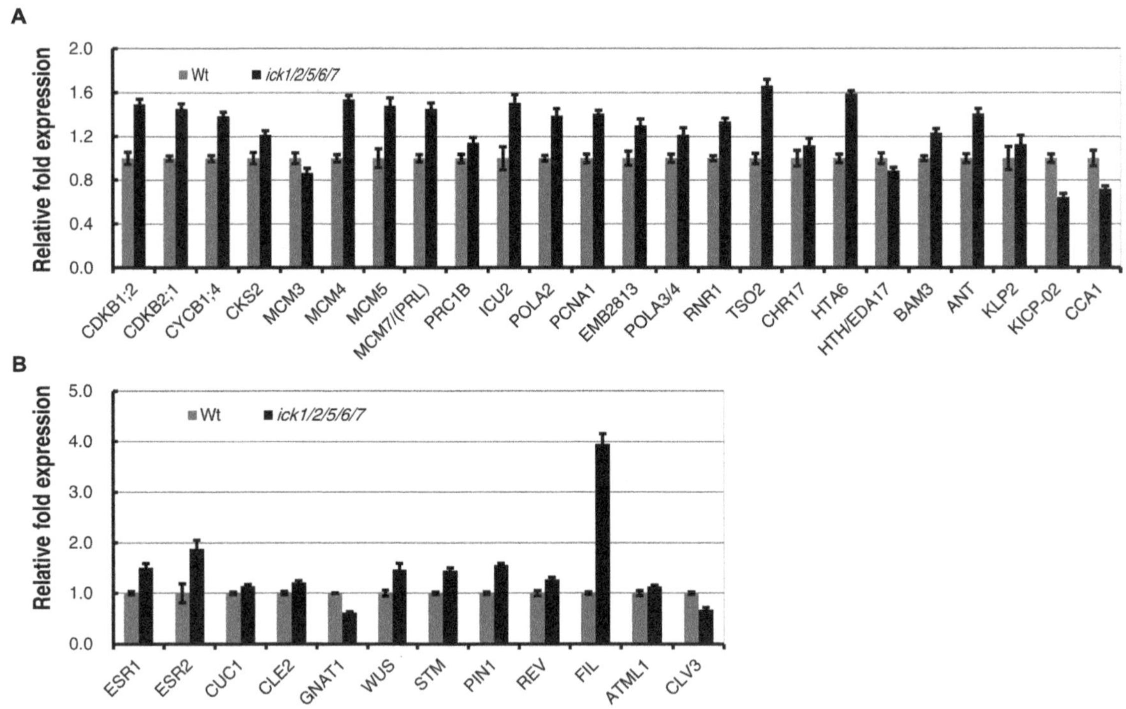

FIGURE 6 | Expression of E2F-dependent and shoot regeneration related genes during *in vitro* shoot regeneration of the Wt and *ick1/2/5/6/7* mutant. Root explants were cultured on CIM for 7 days and then transferred onto SIM. Total RNA was extracted after 7 days of culture on SIM. Each value on the vertical axes indicates a relative fold expression level calculated by scaling to the *Ubq10c* transcript level. Data present the average of four biological repeats and error bars indicated standard errors. Each biological repeat has four technical replicates. **(A)** E2F-dependent genes. **(B)** Shoot regeneration related genes.

during *de novo* shoot formation (Gordon et al., 2007; Duclercq et al., 2011). Among these 12 genes, 10 were up-regulated in the calli of *ick1/2/5/6/7* mutant, while only 2 (*GNAT1* and *CLV3*) were down-regulated (**Figure 6B**). Interestingly, the *FIL* was highly up-regulated in the callus of *ick1/2/5/6/7* with about fourfold of expression relative to that in the Wt calli.

Discussion

Down-regulation of Five CDK Inhibitor Genes Promotes Cell Proliferation in Roots

Root growth is determined by the balance between cell division and cell elongation (Beemster and Baskin, 1998). The defined cortical layer development in *Arabidopsis* provides a good tool to investigate the cell proliferation rate (Ivanov and Dubrovsky, 1997). Several studies using this approach have revealed that a reduced root apical meristem (RAM) size is responsible for the observed inhibition of primary root growth under different conditions (Westet al., 2004; Ubeda-Tomás et al., 2009; González-García et al., 2011). In this study, we observed that the same number of cortex meristematic cells in *ick1/2/5/6/7* quintuple mutant generated more but smaller cells in the mature zone compared to the Wt, indicating an accelerated rate of cell division in the mutant. Previously, we have shown that CDK activity (most likely CDKA) is enhanced in the *ick* mutants (Cheng et al., 2013a), which is likely to be responsible for the increased rate of cell division in roots.

CDK Kinase Activity is a Major Factor Underlying Callus Induction

In various plant species, callus induction depends on exogenous application of both auxin and CK. The factors underlying callus induction/repression and the genes involved in these processes have been reviewed recently by Ikeuchi et al. (2013). It has been observed that the transgenic *Arabidopsis* overexpressing a putative CK receptor *CKI1* could produce callus efficiently in the absence of CK (Kakimoto, 1996), indicating the significance of CK in callus induction. Also, leaf explants of transgenic *Arabidopsis* constitutively expressing a cell cycle regulator gene *CYCD3* could also produce calli in the absence of exogenous CK (Riou-Khamlichi et al., 1999). It has been shown that CK promotes G1/S and G2/M transitions through regulating CDK activities (Sieberer et al., 2003; Del Pozo et al., 2005). Furthermore, overexpression of rice *R2*, a CDK-activating kinase, also results in CK-independent callus induction in tobacco (Yamaguchi et al., 2003). In this study, we showed that callus could be induced in the *ick1/2/5/6/7* quintuple mutant in the absence of CK. The increased CDK activity in the *ick1/2/5/6/7* mutant must have lowered the threshold requirement for CK as well as auxin for cells in the explants to enter and progress through the cell cycle. In addition, antisense expression of *Nicto;CYCA3;2* in tobacco has been observed to impairs callus formation (Yu et al., 2003). Together, these results suggest that regulation of CDK activity is a key determinant of callus induction.

It is well known that auxin along with auxin signaling modules is required for callus formation (Fan et al., 2012; Perianez-Rodriguez et al., 2014). Part of the auxin requirement may be for up-regulating certain cell cycle genes. It has been shown that *Arabidopsis* CDKA is induced by auxin (Hemerly et al., 1993). Our results showing that knockdown of *ICK* genes also lowers the threshold requirement for auxin in terms of callus induction suggests that the effect of auxin on callus induction is at least partially through CDK.

It has been reported that transgenic *Arabidopsis* plants overexpressing two TFs genes, *HB52* and *CRF3*, exhibit spontaneous callus formation without exogenous phytohormone in some organs (Xu et al., 2012). It is not known whether CDK activity is enhanced in those transgenic plants. However, genome-wide transcriptome profiling during callus initiation has revealed the up-regulation of many cell-cycle related genes (Xu et al., 2012). Thus, it is possible that *HB52* and *CRF3* promote callus formation in the absence of exogenous hormones through these cell cycle genes.

Increased Cell Proliferation Enhances *In Vitro* Organogenesis

In the classical scheme for plant regeneration, the explants undergo dedifferentiation to obtain pluripotency during callus induction. The calli are then transferred to SIM or RIM to induce shoots and roots, respectively. Sometimes, shoots and roots can be induced at the same time. During callus induction, founder meristem cells arise in the pericycle of root explants (Atta et al., 2009). Depending on the subsequent culture conditions, the cell fate of organ primordia is determined to be either shoot or root identity (Christianson and Warnick, 1983; Atta et al., 2009). During *in vitro* organogenesis, auxin is the main phytohormone for root organogenesis, while CK promotes shoot organogenesis (Skoog and Miller, 1957; Duclercq et al., 2011; da Rocha Correa et al., 2012). Thus, the auxin-CK crosstalk is important for organ formation and identity determination (Gordon et al., 2009; Besnard et al., 2011; Cheng et al., 2013b; Zhao et al., 2013). Motte et al. (2014) reviewed the mutants with altered regeneration phenotypes, and noticed that most of the genes are related to meristem maintenance, and auxin and CK signaling.

Our findings that the quintuple *ick* mutant with increased CDK activity showed increased shoot/root organogenesis from both the root and cotyledonary explants indicate enhanced competence for cell proliferation can also promote organogenesis. We speculate that under the auxin and CK conditions favoring organ formation, increased competence to enter the cell cycle in the quintuple mutant makes it possible for more cells to become proliferative and capable of forming shoots or roots, while under conditions favoring unorganized growth increased competence of cell proliferation makes it easier to induce callus formation. Consistent with this suggestion, application of cell cycle inhibitors during SIM incubation was shown to significantly impair organogenesis in *Arabidopsis* (Che et al., 2007). In addition, *Arabidopsis ESR1* and *ESR2*, belonging to the *AP2/EAR* family transcription factors, both could enhance shoot regeneration (Banno et al., 2001; Ikeda et al., 2006). Overexpression of *ESR2* has been shown to up-regulate cell cycle genes (Ikeda et al., 2006), suggesting that ESRs may promote shoot regeneration by up-regulating cell cycle machinery and cell proliferation.

We further observed that E2F-dependent genes and shoot regeneration related genes generally showed higher levels of expression in the *ick1/2/5/6/7* during shoot regeneration compared to Wt plants (**Figures 6A,B**) suggesting that enhanced shoot regeneration is accompanied by the up-regulation of a consort of genes involved in the regeneration process. Interestingly, among them, *FIL* was highly up-regulated in *ick1/2/5/6/7* (**Figure 6B**). Although the regulation of *FIL* expression in the quintuple mutant during shoot regeneration is unknown, it is interesting to note that *FIL* gene encodes a *YABBY* (*YAB*) family putative transcription factor that has been implicated in specifying abaxial cell identities and thus being involved in development of leaves and floral organs, and in meristem activity (Sawa et al., 1999; Lugassi et al., 2010).

Conclusion

In this study, we have demonstrated that down-regulation of CDK inhibitor genes results in enhanced shoot/root regeneration. To date, while efficient plant regeneration system has been established in a range of plant species, other plant species remain recalcitrant. Since regulation of the cell cycle by CDK is conserved through plants and all eukaryotes, it is tempting to speculate that callus induction can be enhanced through modulating CDK activity in other plants as well. *In vitro* plant regeneration has been optimized most empirically by testing a variety of hormonal and culture conditions. The realization that CDK regulation plays a key role provides molecular means to enhance plant regeneration and possibly plant transformation for applications in different plants, particularly crop species.

Acknowledgments

We thank the *Arabidopsis* Biological Resource Center and Nottingham *Arabidopsis* Stock Centre for the T-DNA insertion mutants. HW is grateful for financial support (Discovery grant) from the Natural Sciences and Engineering Research Council of Canada, YZ is grateful for financial support from the Ministry of Science and Technology of China (grants 2006CB101604 and 2011AA10A104) and the National Natural Science Foundation of China (30471097), and YC is grateful for financial support from China Postdoctoral Science Foundation (2014M562036).

References

Atta, R., Laurens, L., Boucheron-Dubuisson, E., Guivarc'h, A., Carnero, E., Giraudat-Pautot, V., et al. (2009). Pluripotency of *Arabidopsis xylem* pericycle underlies shoot regeneration from root and hypocotyl explants grown in vitro. *Plant J.* 57, 626–644. doi: 10.1111/j.1365-313X.2008.03715.x

Banno, H., Ikeda, Y., Niu, Q. W., and Chua, N. H. (2001). Overexpression of *Arabidopsis* ESR1 induces initiation of shoot regeneration. *Plant Cell* 13, 2609–2618. doi: 10.1105/tpc.13.12.2609

Barrôco, R. M., Peres, A., Droual, A. M., De Veylder, L., De Wolf, J., Mironov, V., et al. (2006). The cyclin-dependent kinase inhibitor Orysa. KRP1 plays an important role in seed development of rice. *Plant Physiol.* 142, 1053–1064.

Beemster, G. T., and Baskin, T. I. (1998). Analysis of cell division and elongation underlying the developmental acceleration of root growth in *Arabidopsis thaliana*. *Plant Physiol.* 116, 1515–1526. doi: 10.1104/pp.116.4.1515

Besnard, F., Vernoux, T., and Hamant, O. (2011). Organogenesis from stem cells in planta: multiple feedback loops integrating molecular and mechanical signals. *Cell. Mol. Life Sci.* 68, 2885–2906. doi: 10.1007/s00018-011-0732-4

Bisbis, B., Delmas, F., Joubès, J., Sicard, A., Hernould, M., Inzé, D., et al. (2006). Cyclin-dependent kinase (CDK) inhibitors regulate the CDK-cyclin complex activities in endoreduplicating cells of developing tomato fruit. *J. Biol. Chem.* 281, 7374–7383. doi: 10.1074/jbc.M506587200

Casamitjana-Martínez, E., Hofhuis, H. F., Xu, J., Liu, C. M., Heidstra, R., and Scheres, B. (2003). Root-specific CLE19 overexpression and the sol1/2 suppressors implicate a CLV-like pathway in the control of *Arabidopsis* root meristem maintenance. *Curr. Biol.* 13, 1435–1441. doi: 10.1016/S0960-9822(03)00533-5

Che, P., Lall, S., and Howell, S. H. (2007). Developmental steps in acquiring competence for shoot development in *Arabidopsis* tissue culture. *Planta* 226, 1183–1194. doi: 10.1007/s00425-007-0565-4

Cheng, Y., Cao, L., Wang, S., Li, Y., Shi, X., Liu, H., et al. (2013a). Downregulation of multiple CDK inhibitor ICK/KRP genes upregulates the E2F pathway and increases cell proliferation, and organ and seed sizes in *Arabidopsis*. *Plant J.* 75, 642–655. doi: 10.1111/tpj.12228

Cheng, Z. J., Wang, L., Sun, W., Zhang, Y., Zhou, C., Su, Y. H., et al. (2013b). Pattern of auxin and cytokinin responses for shoot meristem induction results from the regulation of cytokinin biosynthesis by AUXIN RESPONSE FACTOR3. *Plant Physiol.* 161, 240–251. doi: 10.1104/pp.112.203166

Christianson, M., and Warnick, D. (1983). Competence and determination in the process of in vitro shoot organogenesis. *Dev. Biol.* 95, 288–293. doi: 10.1016/0012-1606(83)90029-5

Cockcroft, C. E., den Boer, B. G., Healy, J. S., and Murray, J. A. (2000). Cyclin D control of growth rate in plants. *Nature* 405, 575–579. doi: 10.1038/35014621

Coelho, C. M., Dante, R. A., Sabelli, P. A., Sun, Y., Dilkes, B. P., Gordon-Kamm, W. J., et al. (2005). Cyclin-dependent kinase inhibitors in maize endosperm and their potential role in endoreduplication. *Plant Physiol.* 138, 2323–2336. doi: 10.1104/pp.105.063917

da Rocha Correa, L., Troleis, J., Mastroberti, A., Mariath, J., and Fett-Neto, A. (2012). Distinct modes of adventitious rooting in *Arabidopsis thaliana*. *Plant Biol.* 14, 100–109. doi: 10.1111/j.1438-8677.2011.00468.x

Davey, M. R., and Anthony, P. (2010). *Plant Cell Culture: Essential Methods*. Hoboken, NJ: John Wiley & Sons.

de Jager, S. M., Scofield, S., Huntley, R. P., Robinson, A. S., den Boer, B. G., and Murray, J. A. (2009). Dissecting regulatory pathways of G1/S control in *Arabidopsis*: common and distinct targets of CYCD3; 1. E2Fa and E2Fc. *Plant Mol. Biol.* 71, 345–365. doi: 10.1007/s11103-009-9527-5

De Klerk, G. J., Arnholdt-Schmitt, B., Lieberei, R., and Neumann, K. H. (1997). Regeneration of roots, shoots and embryos: physiological, biochemical and molecular aspects. *Biol. Plant.* 39, 53–66. doi: 10.1023/A:1000369309303

Del Pozo, J. C., Lopez-Matas, M., Ramirez-Parra, E., and Gutierrez, C. (2005). Hormonal control of the plant cell cycle. *Physiol. Plant.* 123, 173–183. doi: 10.1111/j.1399-3054.2004.00420.x

De Veylder, L., Beeckman, T., Beemster, G. T., Krols, L., Terras, F., Landrieu, I., et al. (2001). Functional analysis of cyclin-dependent kinase inhibitors of *Arabidopsis*. *Plant Cell* 13, 1653–1668. doi: 10.2307/3871392

Duclercq, J., Sangwan-Norreel, B., Catterou, M., and Sangwan, R. S. (2011). De novo shoot organogenesis: from art to science. *Trends Plant Sci.* 16, 597–606. doi: 10.1016/j.tplants.2011.08.004

Fan, M., Xu, C., Xu, K., and Hu, Y. (2012). Lateral organ boundaries domain transcription factors direct callus formation in *Arabidopsis* regeneration. *Cell Res.* 22, 1169–1180. doi: 10.1038/cr.2012.63

Gamborg, O. L. C., Miller, R. A., and Ojima, K. (1968). Nutrient requirements of suspension cultures of soybean root cells. *Exp. Cell Res.* 50, 151–158. doi: 10.1016/0014-4827(68)90403-5

González-García, M. P., Vilarrasa-Blasi, J., Zhiponova, M., Divol, F., Mora-García, S., Russinova, E., et al. (2011). Brassinosteroids control meristem size by promoting cell cycle progression in *Arabidopsis* roots. *Development* 138, 849–859. doi: 10.1242/dev.057331

Gordon, S. P., Chickarmane, V. S., Ohno, C., and Meyerowitz, E. M. (2009). Multiple feedback loops through cytokinin signaling control stem cell number within the *Arabidopsis* shoot meristem. *Proc. Natl. Acad. Sci. U.S.A.* 106, 16529–16534. doi: 10.1073/pnas.0908122106

Gordon, S. P., Heisler, M. G., Reddy, G. V., Ohno, C., Das, P., and Meyerowitz, E. M. (2007). Pattern formation during de novo assembly of the *Arabidopsis* shoot meristem. *Development* 134, 3539–3548. doi: 10.1242/dev.010298

Hemerly, A. S., Ferreira, P., de Almeida Engler, J., Van Montagu, M., Engler, G., and Inzé, D. (1993). cdc2a expression in *Arabidopsis* is linked with competence for cell division. *Plant Cell* 5, 1711–1723. doi: 10.1105/tpc.5.12.1711

Ikeda, Y., Banno, H., Niu, Q. W., Howell, S. H., and Chua, N. H. (2006). The ENHANCER OF SHOOT REGENERATION 2 gene in *Arabidopsis* regulates CUP-SHAPED COTYLEDON 1 at the transcriptional level and controls cotyledon development. *Plant Cell Physiol.* 47, 1443–1456. doi: 10.1093/pcp/pcl023

Ikeuchi, M., Sugimoto, K., and Iwase, A. (2013). Plant callus: mechanisms of induction and repression. *Plant Cell* 25, 3159–3173. doi: 10.1105/tpc.113.116053

Ioio, R. D., Linhares, F. S., Scacchi, E., Casamitjana-Martinez, E., Heidstra, R., Costantino, P., et al. (2007). Cytokinins determine *Arabidopsis* root-meristem size by controlling cell differentiation. *Curr. Biol.* 17, 678–682. doi: 10.1016/j.cub.2007.02.047

Ivanov, V. B., and Dubrovsky, J. G. (1997). Estimation of the cell-cycle duration in the root apical meristem: a model of linkage between cell-cycle duration, rate of cell production, and rate of root growth. *Int. J. Plant Sci.* 158, 757–763. doi: 10.1086/297487

Jasinski, S., Leite, C. S., Domenichini, S., Stevens, R., Raynaud, C., Perennes, C., et al. (2003). NtKIS2, a novel tobacco cyclin-dependent kinase inhibitor is differentially expressed during the cell cycle and plant development. *Plant Physiol. Biochem.* 41, 667–676. doi: 10.1016/S0981-9428(03)00082-2

Kakimoto, T. (1996). CKI1, a histidine kinase homolog implicated in cytokinin signal transduction. *Science* 274, 982–985. doi: 10.1126/science.274.5289.982

Lugassi, N., Nakayama, N., Bochnik, R., and Zik, M. (2010). A novel allele of FILAMENTOUS FLOWER reveals new insights on the link between inflorescence and floral meristem organization and flower morphogenesis. *BMC Plant Biol.* 10:131. doi: 10.1186/1471-2229-10-131

Malladi, A., and Johnson, L. K. (2011). Expression profiling of cell cycle genes reveals key facilitators of cell production during carpel development, fruit set, and fruit growth in apple (Malus × domestica Borkh). *J. Exp. Bot.* 62, 205–219. doi: 10.1093/jxb/erq258

Matsuo, N., Mase, H., Makino, M., Takahashi, H., and Banno, H. (2009). Identification of ENHANCER OF SHOOT REGENERATION 1-upregulated genes during in vitro shoot regeneration. *Plant Biotechnol.* 26, 385–393. doi: 10.5511/plantbiotechnology.26.385

Morgan, D. O. (1997). Cyclin-dependent kinases: engines, clocks, and microprocessors. *Annu. Rev. Cell Dev. Biol.* 13, 261–291. doi: 10.1146/annurev.cellbio.13.1.261

Motte, H., Vereecke, D., Geelen, D., and Werbrouck, S. (2014). The molecular path to in vitro shoot regeneration. *Biotechnol. Adv.* 32, 107–121. doi: 10.1016/j.biotechadv.2013.12.002

Perianez-Rodriguez, J., Manzano, C., and Moreno-Risueno, M. A. (2014). Post-embryonic organogenesis and plant regeneration from tissues: two sides of the same coin? *Front. Plant Sci.* 5:219. doi: 10.3389/fpls.2014.00219

Ramirez-Parra, E., Fründt, C., and Gutierrez, C. (2003). A genome-wide identification of E2F-regulated genes in *Arabidopsis*. *Plant J.* 33, 801–811. doi: 10.1046/j.1365-313X.2003.01662.x

Riou-Khamlichi, C., Huntley, R., Jacqmard, A., and Murray, J. A. (1999). Cytokinin activation of *Arabidopsis* cell division through a D-type cyclin. *Science* 283, 1541–1544. doi: 10.1126/science.283.5407.1541

Sabag, M., Ari, G. B., Zviran, T., Biton, I., Goren, M., Dahan, Y., et al. (2013). PaKRP, a cyclin-dependent kinase inhibitor from avocado, may facilitate exit from the cell cycle during fruit growth. *Plant Sci.* 213, 18–29. doi: 10.1016/j.plantsci.2013.08.007

Sangwan, R., and Sangwan-Norreel, B. (1990). "Genetic transformation and plant improvement," in *The Impact of Biotechnology on Agriculture*, eds A. Altman and P. M. Hasegawa (Berlin: Springer), 299–337.

Sawa, S., Watanabe, K., Goto, K., Kanaya, E., Morita, E. H., and Okada, K. (1999). FILAMENTOUS FLOWER, a meristem and organ identity gene of *Arabidopsis*, encodes a protein with a zinc finger and HMG-related domains. *Genes Dev.* 13, 1079–1088. doi: 10.1101/gad.13.9.1079

Sherr, C. J., and Roberts, J. M. (1995). Inhibitors of mammalian G1 cyclin-dependent kinases. *Genes Dev.* 9, 1149–1163. doi: 10.1101/gad.9.10.1149

Sieberer, T., Hauser, M. T., Seifert, G. J., and Luschnig, C. (2003). PROPORZ1, a putative *Arabidopsis* transcriptional adaptor protein, mediates auxin and cytokinin signals in the control of cell proliferation. *Curr. Biol.* 13, 837–842. doi: 10.1016/S0960-9822(03)00327-0

Skoog, F., and Miller, C. (1957). Chemical regularion of growth and organ formation in plant fissue cultured. *In vitro. Symp. Soc. Exp. Biol.* 11, 118–131. doi: 10.1038/jid.2014.273

Ubeda-Tomás, S., Federici, F., Casimiro, I., Beemster, G. T., Bhalerao, R., Swarup, R., et al. (2009). Gibberellin signaling in the endodermis controls *Arabidopsis* root meristem size. *Curr. Biol.* 19, 1194–1199. doi: 10.1016/j.cub.2009.06.023

Valvekens, D., Van Montagu, M., and Van Lijsebettens, M. (1988). *Agrobacterium tumefaciens*-mediated transformation of *Arabidopsis thaliana* root explants by using kanamycin selection. *Proc. Natl. Acad. Sci. U.S.A.* 85, 5536–5540. doi: 10.1073/pnas.85.15.5536

Wang, H., Zhou, Y., Gilmer, S., Whitwill, S., and Fowke, L. C. (1997). A plant cyclindependent kinase inhibitor gene. *Nature* 386, 451–452. doi: 10.1038/386451a0

West, G., Inzé, D., and Beemster, G. T. (2004). Cell cycle modulation in the response of the primary root of *Arabidopsis* to salt stress. *Plant Physiol.* 135, 1050–1058. doi: 10.1104/pp.104.040022

Xu, K., Liu, J., Fan, M., Xin, W., Hu, Y., and Xu, C. (2012). A genome-wide transcriptome profiling reveals the early molecular events during callus initiation in *Arabidopsis* multiple organs. *Genomics* 100, 116–124. doi: 10.1016/j.ygeno.2012.05.013

Yamaguchi, M., Kato, H., Yoshida, S., Yamamura, S., Uchimiya, H., and Umeda, M. (2003). Control of in vitro organogenesis by cyclin-dependent kinase activities in plants. *Proc. Natl. Acad. Sci. U.S.A.* 100, 8019–8023. doi: 10.1073/pnas.1332637100

Yasutani, I., Ozawa, S., Nishida, T., Sugiyama, M., and Komamine, A. (1994). Isolation of temperature-sensitive mutants of *Arabidopsis thaliana* that are defective in the redifferentiation of shoots. *Plant Physiol.* 105, 815–822.

Yu, Y., Steinmetz, A., Meyer, D., Brown, S., and Shen, W. H. (2003). The tobacco A-type cyclin. Nicta; CYCA3; 2, at the nexus of cell division and differentiation. *Plant Cell* 15, 2763–2777.

Zhang, Y., Li, B., Huai, D., Zhou, Y., and Kliebenstein, D. J. (2015). The conserved transcription factors, MYB115 and MYB118, control expression of the newly evolved benzoyloxy glucosinolate pathway in *Arabidopsis thaliana*. *Front. Plant Sci.* 6:343. doi: 10.3389/fpls.2015.00343

Zhao, X. Y., Su, Y. H., Zhang, C. L., Wang, L., Li, X. G., and Zhang, X. S. (2013). Differences in capacities of in vitro organ regeneration between two *Arabidopsis* ecotypes Wassilewskija and Columbia. *Plant Cell Tiss. Org.* 112, 65–74. doi: 10.1007/s11240-012-0216-8

Conflict of Interest Statement: The authors declare that the research was conducted in the absence of any commercial or financial relationships that could be construed as a potential conflict of interest.

5

The *Arabidopsis* COX11 Homolog is Essential for Cytochrome *c* Oxidase Activity

Ivan Radin[1], Natanael Mansilla[2], Gerhard Rödel[1]* and Iris Steinebrunner[3]

[1] Institute for Genetics, Department of Biology, Technische Universität Dresden, Dresden, Germany, [2] Instituto de Agrobiotecnología del Litoral-Consejo Nacional de Investigaciones Científicas y Técnicas, Universidad Nacional del Litoral, Santa Fe, Argentina, [3] Department of Biology, Technische Universität Dresden, Dresden, Germany

Edited by:
Marinus Pilon,
Colorado State University, USA

Reviewed by:
Olena Vatamaniuk,
Cornell University, USA
Manuel González-Guerrero,
Universidad Politécnica de Madrid, Spain
Lola Peñarrubia,
Universitat de Valencia, Spain

***Correspondence:**
Gerhard Rödel
gerhard.roedel@tu-dresden.de

Members of the ubiquitous COX11 (cytochrome *c* oxidase 11) protein family are involved in copper delivery to the COX complex. In this work, we characterize the *Arabidopsis thaliana* COX11 homolog (encoded by locus At1g02410). Western blot analyses and confocal microscopy identified *Arabidopsis* COX11 as an integral mitochondrial protein. Despite sharing high sequence and structural similarities, the *Arabidopsis* COX11 is not able to functionally replace the *Saccharomyces cerevisiae* COX11 homolog. Nevertheless, further analysis confirmed the hypothesis that *Arabidopsis* COX11 is essential for COX activity. Disturbance of *COX11* expression through knockdown (KD) or overexpression (OE) affected COX activity. In KD lines, the activity was reduced by ~50%, resulting in root growth inhibition, smaller rosettes and leaf curling. In OE lines, the reduction was less pronounced (~80% of the wild type), still resulting in root growth inhibition. Additionally, pollen germination was impaired in *COX11* KD and OE plants. This effect on pollen germination can only partially be attributed to COX deficiency and may indicate a possible auxiliary role of COX11 in ROS metabolism. In agreement with its role in energy production, the *COX11* promoter is highly active in cells and tissues with high-energy demand for example shoot and root meristems, or vascular tissues of source and sink organs. In *COX11* KD lines, the expression of the plasma-membrane copper transporter *COPT2* and of several copper chaperones was altered, indicative of a retrograde signaling pathway pertinent to copper homeostasis. Based on our data, we postulate that COX11 is a mitochondrial chaperone, which plays an important role for plant growth and pollen germination as an essential COX complex assembly factor.

Keywords: COX11 (cytochrome *c* oxidase 11), copper chaperone, mitochondria, COX complex, root development, pollen development, *Arabidopsis thaliana*

INTRODUCTION

The cytochrome *c* oxidase (COX) is a crucial component of the mitochondrial respiratory chain, which is of utmost importance for providing cellular energy. Dysfunction of COX inevitably brings processes that depend on respiration-based energy supply to a sudden halt: respective *Saccharomyces (S.) cerevisiae* mutants stop growing on non-fermentable carbon sources (Merz and Westermann, 2009) as do COX-deficient *Chlamydomonas reinhardtii* mutants under heterotrophic conditions (Colin et al., 1995). In *Arabidopsis thaliana*, compromised COX function prevents development beyond the embryo-stage (Attallah et al., 2011; Steinebrunner et al., 2011; Welchen et al., 2012; Dahan et al., 2014).

The multi-subunit COX harbors two copper (Cu) centers, which are essential for the electron transfer through the complex to molecular oxygen (reviewed in Khalimonchuk and Rödel, 2005). One of the copper centers, Cu_A, lies in the subunit COX2, while the other one, Cu_B, is formed in the subunit COX1. The Cu ions for Cu_A and Cu_B are provided by the two copper chaperones SCO1 (for synthesis of cytochrome *c* oxidase 1) (Schulze and Rödel, 1988; Lode et al., 2000; Nittis et al., 2001) and COX11 (Banting and Glerum, 2006), respectively. Analysis of *Rhodobacter (R.) sphaeroides* mutants, which lack the *COX11* gene, confirms that the encoded protein is required for the insertion of Cu into the Cu_B site (Hiser et al., 2000). Yeast COX11 is N-terminally anchored in the inner mitochondrial membrane with the majority of the protein protruding into the intermembrane space (IMS) (Carr et al., 2005; Khalimonchuk et al., 2005). *In vitro* studies with SCO1 and COX11 from *S. cerevisiae* indicate that the two proteins most likely receive the Cu ions from the small soluble copper chaperone COX17 located in the IMS (Horng et al., 2004). While evidence was presented that the Cu transfer from COX11 to COX1 may occur co-translationally (Khalimonchuk et al., 2005), the exact molecular mechanism of the transfer is unknown. However, the Cu-binding mode has been well studied: COX11 forms a homodimer that binds two Cu ions, each of which is ligated by three cysteine residues (Carr et al., 2002). Based on the structure of the soluble *Sinorhizobium meliloti* COX11, it was suggested that two of these cysteines are derived from the conserved Cu-binding motif CxC of one subunit, while the third cysteine is provided by the respective Cu-binding motif of the second subunit (Banci et al., 2004). The exact role of a third conserved cysteine outside of the Cu-binding motif that is essential for COX11 function (Banting and Glerum, 2006) is less clear. Data obtained with COX11 from *R. sphaeroides* suggest that it might serve as a transient Cu-binding site during Cu insertion into COX1 (Thompson et al., 2010). *S. cerevisiae cox11* null mutants are respiratory deficient (Tzagoloff et al., 1990; Carr et al., 2002; Banting and Glerum, 2006), in agreement with the essential contribution of COX11 for COX assembly.

COX11 is a ubiquitous gene found in several bacterial genomes and every eukaryotic genome publicly available. However, no data on COX11 function exist for plants. Here, we close this gap by presenting an extensive characterization of the COX11 homolog in *Arabidopsis*, which includes expression pattern, subcellular localization, knockdown as well as overexpression effects. We show that COX11 is essential for COX activity and that it affects vital physiological processes such as pollen germination, development, and growth.

MATERIALS AND METHODS

Plant Material and Culture Conditions

A. thaliana Columbia (Col) 0 was used as the WT and background for all mutants generated. Plants with the silenced *COX11* gene were generated using an amiRNA designed with the online tool WMD3 (Web MicroRNA Designer; wmd3.weigelworld.org) (Ossowski et al., 2008). Overlap extension PCR (Pogulis et al., 1996) with specific primers (Supplementary Table 1) was used to create the amiRNA precursor for the selected target as described in WMD3 (Supplementary Figure 1). The vector pNB47 (Bologna et al., 2009) was kindly provided by Javier Palatnik (Institute of Molecular and Cellular Biology, Rosario, Argentina). This binary vector provides the precursor sequence for the miR319a as the template for overlap extension PCR and allows amiRNA expression in the plant under the CaMV *35S* promoter. The PCR product was cloned into the pNB47 using the *Pst*I and *Bam*HI restriction sites, which replaced the miR319a precursor sequence. *Arabidopsis* plants were transformed by the floral dip procedure (Clough and Bent, 1998). Selected plants were confirmed by PCR using specific primers (Supplementary Table 1).

Two independent T2-generation lines, KD1 and KD2, were obtained. The KD2 line was used in the T2 generation only, because the KD effect was lost in the T3 generation (Supplementary Figure 2A). From the KD1 line, two separate seed batches (KD1-1 and KD1-2; progeny of two individual T2 generation plants) were harvested and used in the T3 generation unless stated otherwise. The T-DNA insertion lines for the locus At1g02410 [Supplementary Figure 3; SALK_105793, SALK_003445C (Alonso et al., 2003), SAIL_603_G12, SAIL_861_D09, SAIL_683_B03 (McElver et al., 2001)] were provided by The Nottingham *Arabidopsis* Stock Centre and *Arabidopsis* Biological Resource Center. Presence and positions of the T-DNA insertion were confirmed by PCR with T-DNA and genomic DNA specific primers (listed in Supplementary Table 1).

All other constructs were cloned applying the Gateway cloning technology (LifeTechnologies, USA) according to the manufacturer's instructions. Primers used for cloning are listed in the Supplementary Table 1. The plasmid pENRT223-*COX11* (Yamada et al., 2003), or clone G61127, obtained from *Arabidopsis* Biological Resource Center, was used as a template or starting plasmid for all constructs. For localizations experiments, the *COX11* cDNA was inserted into the pGWB553 destination vector (Nakagawa et al., 2007) to generate the *35S:COX11-mRFP* (monomeric red fluorescent protein) construct. To generate the *35S:mRFP* construct, the *mRFP* cDNA was inserted into the pGWB502 destination vector (Nakagawa et al., 2007). For generation of the GUS lines, the putative *COX11* promoter region (−419 to +366 bp with the transcription start site being +1) was recombined into the pMDC163 plasmid (Curtis and Grossniklaus, 2003). For *COX11* overexpression lines (OE), the *COX11* cDNA was cloned into the pGWB514 and pGWB502 (Nakagawa et al., 2007) destination vectors to obtain the *35S:COX11-3HA* (OE1) and *35S:COX11* (OE2) constructs, respectively. For both OE lines, T3-generation homozygous lines were selected and used for all experiments, except for the experiments shown in **Figure 3**, for which hemizygous T3-generation *35S:*COX11-*3HA* plants were used. Constructs were generally transformed into Col-0 background via the floral dip method (Clough and Bent, 1998). Only the two mRFP lines

35S:COX11-mRFP and *35S:mRFP* were also transformed into the plant background expressing green fluorescent protein targeted to mitochondria (mt-GFP) (Nelson et al., 2007). The genomic T-DNA insertion for all constructs, except for *COX11:GUS*, were confirmed by PCR with primers specific for individual constructs (listed in Supplementary Table 1). The *COX11:GUS* insertion was confirmed by direct GUS staining of seedlings.

Plants were grown on sterile selective or non-selective Murashige and Skoog (MS) + sucrose (1 or 2%) plates. For selection, 30 or 50 $\mu g\ mL^{-1}$ of kanamycin and/or 20 $\mu g\ mL^{-1}$ of hygromycin were added. Alternatively, plants were cultured on soil [Einheitserde, type P, Pätzer, Sinntal-Jossa, Germany; mixed with sand 4:1, fertilized by watering with 0.1% (v/v) Wuxal Basis, Aglukon].

Plants were cultured in growth chamber with light intensity of 150 $\mu mol\ m^{-2}\ s^{-1}$, relative humidity of 35% and day/night temperatures of 24/21°C, respectively. Two types of day/night cycles were used: long day (16-h day) and short day (10-h day).

Yeast Material and Culture Conditions

S. cerevisiae wild-type (WT) strain BY4741 [Accession no. Y00000; *MATa, his3Δ1, leu2Δ0, met15Δ0, ura3Δ0, (rho⁺)] and Δcox11* deletion strain [Accession no. Y06479; *MATa, his3Δ1, leu2Δ0, met15Δ0, ura3Δ0, cox11::kanMX4, (rho⁺)*] were obtained from EUROSCARF (Frankfurt, Germany). For complementation assays, the full-length yeast and *Arabidopsis COX11* sequences were inserted by Gateway cloning into yeast expression vectors (Alberti et al., 2007) [pAG415ADH-ccdB-3HA (for moderate overexpression, single copy vector with *ADH* promoter), pAG425ADH-ccdB-3HA (for strong overexpression, high copy vector with *ADH* promoter) and pAG415COX11-ccdB-3HA (for native expression, single copy vector in which the *ADH* promoter was replaced by 1500 bp of the *COX11* promoter region)]. Two chimeric proteins (CHYM-1 and -2) were generated by overlap extension PCR (Pogulis et al., 1996) and inserted in the same expression vectors. Cloning primers are listed in Supplementary Table 1. Yeast cells were transformed as described in Gietz and Schiestl (2007).

Transformed yeast strains were cultured on minimal media [0.5% (w/v) ammonium sulfate, 0.19% (w/v) yeast nitrogen bases, 2% (w/v) glucose, 2.5% (w/v) agar, and selection amino acids]. For complementation assays, strains were spotted on either YPD [1% (w/v) yeast extract, 2% (w/v) peptone, 2% (w/v) glucose, 2% (w/v) agar] or YPEG [1% (w/v) yeast extract, 2% (w/v) peptone, 2% (w/v) glycerol, 3% (v/v) ethanol, 2% (w/v) agar] media. Copper was added to the YPEG media at the concentration 0.5 mM, which according to the literature is sub-toxic (Liang and Zhou, 2007). For spotting, yeast strains were inoculated into liquid minimal media and upon reaching the stationary phase, the cultures were sequentially diluted from 1.25×10^6 to 1.25×10^3 cells/ml in four steps (The assumption was that the OD_{600} of 0.1 equals 1×10^6 cells/ml). From these dilution series, 8 µl of each dilution were spotted on the appropriate solid media plate. Plates were incubated at 30°C for 2 days (YPD media), 4 days [WT and *Δ+Sc COX11* (= *Δcox*11 strain transformed with the expression vector containing the yeast *COX11* gene) strains on YPEG] and 20 days (other strains on YPEG).

Bioinformatic Analysis

For single and multiple protein sequence alignments, the EMBOSS Needle and Cluster Omega software (The European Bioinformatics Institute) (McWilliam et al., 2013), respectively, were used. For the prediction of cellular targeting and targeting signal cleavage sites, the programs TargetP (Emanuelsson et al., 2000) and MitoProt II (Claros and Vincens, 1996) were used. Transmembrane domains were predicted with the TMHMM2.0 (Krogh et al., 2001). The prediction of the proteins' secondary structures was performed with PSIPRED (McGuffin et al., 2000). *Arabidopsis* sequence information was obtained from The *Arabidopsis* Information Resource (Lamesch et al., 2012), while all other sequences were retrieved from the GeneBank (Benson et al., 2013).

Confocal Microscopy

Confocal imaging was performed on a Zeiss LSM780 upright microscope of the Light Microscopy Facility, a core facility of the BIOTEC/CRTD at Technische Universität Dresden. Colocalization images were made with the C-Apochromat 63×/1.20 W korr M27 objective. GFP and mRFP were simultaneously excited with 488 and 561-nm lasers, respectively and their fluorescences simultaneously detected in the range of 470 and 552 nm for GFP, and 575 and 650 nm for mRFP. Image analysis was performed with the Zeiss Zen2012 software (also used for acquisition) and the colocalization plugin JACoP (Bolte and Cordelières, 2006) of the Fiji software (Schindelin et al., 2012). For colocalization studies, 14-day-old seedlings (cultured on selective MS + 2% sucrose plates under long-day conditions) were imaged with water as imaging medium.

Protein and Mitochondria Isolation

For total protein isolation, *Arabidopsis* leaves from plants cultured on soil under short-day condition were mixed with 2 mL of QB buffer [100 mM potassium phosphate buffer pH 7.8, 1 mM EDTA, 1% (v/v) Triton X-100, 10% (v/v) glycerol, 1 mM DTT] (Ni et al., 1996) per 1 g of tissue and ground with a mortar and pestle. Samples were centrifuged at 14,000 g for 15 min at 4°C. The supernatant was saved and stored at −80°C.

Crude yeast mitochondria preparations used in the Supplementary Figure 6 were isolated as previously described (Gey et al., 2014). For mitochondria isolation, yeast strains were cultured in minimal media.

Crude *Arabidopsis* mitochondria preparations used in the Supplementary Figure 9 were isolated from etiolated seedlings as previously described (Steinebrunner et al., 2011). Crude mitochondria for the COX activity assay and the BN-PAGE (**Figure 6**), and pure mitochondria for the sodium carbonate extraction procedure (**Figure 3**) were isolated from leaves of plants cultured in short day as described in Keech et al. (2005). For the isolation of pure mitochondria, one modification was made: Cysteine was replaced in the grinding buffer with 0.005% (v/v) β-mercaptoethanol. Protein concentrations were determined with the DC protein assay kit from Bio-Rad (USA).

SDS-PAGE and Western Blot Analysis

For protein separation, the gradient (4–12%) NuPAGE electrophoresis system with the NuPAGE MOPS buffer from LifeTechnologies (USA) was used. After separation, proteins were transferred onto a polyvinylidene difluoride (PVDF) membrane (Immobilion-P, Millipore, USA) with a semi-dry blotter from Peqlab (Germany). For Western blot detection, ECL prime Western blotting detection reagents from GE Healthcare (UK) were used. Most primary antibodies were purchased from Agrisera, Sweden [cytosolic fructose-1,6-bisphosphatase (cFBP, 1:5000); cytochrome *c* oxidase subunit 2 (COX2, 1:2000); glycine decarboxylase complex subunit H (GDCH, 1:5000); rubisco large subunit (RbCL, 1:10,000); voltage-dependent anion-selective channel 1 (VDAC1, 1:10,000)]. Anti-HA and anti-GFP (both 1:1000) were obtained from Roche (Switzerland), while anti-mRFP (1:5000) was purchased from Rockland (USA). Antibodies against yeast phosphoglycerate kinase 1 (PGK1, 1:1000) and porin 1 (POR1, 1:1000) were obtained from LifeTechnologies (USA). Dilutions of the antibodies are indicated in parentheses. Anti-mouse and anti-rabbit horseradish peroxidase (HRP)-coupled secondary antibodies (1:10,000) were obtained from GE healthcare (UK). Between detections, the membranes were stripped by one 30-min incubation in stripping buffer [62.5 mM Tris-Cl pH 6.7, 2% SDS (w/v) and freshly added 100 mM β-mercaptoethanol] at 55°C.

Sodium Carbonate Extraction

Sodium carbonate extraction was performed as previously described (Fujiki et al., 1982), with some modifications. Pure mitochondria (100 μg each) were spun down with 15,000 g for 15 min at 4°C, resuspended in 500 μL of 0.2 M Na_2CO_3 and kept on ice for 1 h. The suspension was centrifuged, 220,000 g (max) for 1 h at 4°C. The pellet was resuspended in 10 mM Tris-Cl pH 7.6, while the supernatant was precipitated with TCA [trichloroacetic acid, final concentration of 10% (v/v)] over night at −25°C. Proteins were pelleted by centrifugation at 20,000 g for 1 h at 4°C. The pellet was rinsed with ice-cold 80% (v/v) acetone, spun down, dried, and finally resuspended in 10 mM Tris-Cl pH 7.6.

COX Complex Activity Measurement

COX complex activity was determined by monitoring the oxidation of reduced cytochrome *c* by crude mitochondria as previously described (Wigge and Gardeström, 1987). Because the crude mitochondria fractions had varying degrees of purity, the citrate synthase activity was also determined for each fraction and used for normalization of COX activity as introduced in Steinebrunner et al. (2014).

Blue-native PAGE and in-gel COX Activity Staining

Protein complexes present in crude mitochondria fractions were separated on a gradient blue-native (BN)-PAGE gel (3–13%, with a 3% stacking gel) as previously described (Schägger and Jagow, 1991; Steinebrunner et al., 2014). The in-gel COX activity staining was performed as described in Steinebrunner et al. (2014).

RNA Isolation and Quantitative Real-time RT-PCR (qPCR)

Total RNA was isolated from 50 to 75 mg of 14-day-old seedlings (cultured under long-day conditions on MS + 1% sucrose plates) with the RNeasy plant mini kit (Qiagen, Germany). Genomic DNA contaminations were removed with the on-column RNase-free DNase set (Qiagen, Germany). RNA quantity and quality were assessed using the Nanodrop ND-1000 (Peqlab, Germany) and BioAnalyser 2100 (Agilent, USA) device, respectively. The RNA integrity numbers (RIN) for most samples were in the range of 7.5–8.5. Total RNA (500 ng) was reverse transcribed with oligo$(dT)_{18}$ primers and the RevertAid first strand cDNA synthesis kit (Thermo Fisher Scientific, USA). Of all samples, control reverse transcription reactions without enzyme were set up to check for genomic DNA contaminations.

qPCR was performed with the 2× DyNAmo ColorFlash SYBR Green qPCR Master Mix from Thermo Fisher Scientific (USA). A 50-fold dilution of cDNA was used as a template. The efficiency and optimal concentrations of all primer pairs were experimentally determined (Supplementary Table 2). The following cycling conditions were selected for all reactions: 1× [95°C/8 min], 40× [95°C/10 s; 60°C/20 s; 72°C/15 s, plate read], 1× [95°C/10 s], melting curve from 65 to 95°C with a heating ramp of 0.5°C/5 s. The qPCR reactions were always run in triplicates in a C1000 thermo-cycler and CFX96 real-time system (Bio-Rad, USA). White non-skirted 96-well "low profile" PCR plates (Thermo Fisher Scientific, USA) were used and sealed with absolute qPCR film (Thermo Fisher Scientific, USA). The data was statistically analyzed with the Bio-Rad CFX Manager 3.1 software. Transcript levels were normalized with both *actin 2* (*ACT2*) and *protein phosphatase 2A subunit A3* (*PP2AA3*).

Root Growth Assay

For root length measurements, seedlings were cultured on MS plates + 1% (w/v) sucrose without antibiotics. In the heterozygous KD lines, WT seedlings, which appeared in the expected Mendelian ratios, could be easily distinguished from KD seedlings by their twice as long roots and were removed from the analysis. All seedlings were cultured in parallel. The mutant lines, however, were divided into three groups (KD1-1 and KD1-2/KD2/OE1 and OE2) and cultured on separate plates, but always with the WT on one-half of the same plate. The root lengths were normalized to WT from the same plate. For treatments, the chemicals were added to the medium in the desired concentration before plate pouring. Plates were kept in the vertical position under long-day conditions for 12 days. Each plate was photographed with the Olympus digital camera C-5000 Zoom, and root lengths were determined with the Root Detection 0.1.2 software (http://labutils.de/).

Pollen Viability Assay

Flowers were harvested from 10-week-old *Arabidopsis* plants grown under long-day conditions. Seeds were air-dried for 2 h and then slowly rehydrated in a humid chamber for 45 min at 30°C. Flowers were gently dipped into working solution [17% (w/v) sucrose with 2 μg mL^{-1} of FDA (fluorescein diacetate) (Heslop-Harrison and Heslop-Harrison, 1970)] to release the pollen. After 15 min, the pollen grains were observed in the same media under a fluorescence microscope (Axio Observer Z1 inverted microscope with AxioCamMR3, from Zeiss) with EC Plan-Neofluar 10×/0.30 Ph1 objective (Zeiss) and 38 HE Green Fluorescence Reflector (Zeiss). For image acquisition and analysis, AxioVision Rev 4.8 software was used. Fluorescent pollen grains were counted as viable.

Pollen Germination Assay

Flowers were prepared as described for the pollen viability assay above. Pollen was germinated in duplicates or triplicates with the sitting-drop method (Bou Daher et al., 2009). In the appropriate experiments, KCN was added directly to the germination media.

GUS Staining

GUS activity staining was performed as previously described (Wolf et al., 2007). For the assay, *COX11:GUS* plants were cultured either on MS + 2% sucrose plates or on soil under long-day conditions. For determination of GUS activity in pollen, x-gluc (5-bromo-4-chloro-3-indolyl-b-D-glucuronide) was added to a final concentration of 1 mM either directly to germination media (Bou Daher et al., 2009) or to 18% (w/v) sucrose for pollen rehydration. Pollen was germinated or rehydrated with the sitting-drop method for 4 h at 30°C and subsequently transferred to 37°C for over-night incubation.

Copper Content Measurement

The copper content was determined in leaves of 9-week-old plants cultured under short-day conditions by Inductively Coupled Plasma Optical Emission Spectrometry (ICP-OES). For ICP-OES analysis, samples were prepared as previously described (Cohu and Pilon, 2007; Tan et al., 2010).

RESULTS

Identification of a COX11 Family Member in *Arabidopsis*

To identify members of the COX11 family, the *Arabidopsis* Information Research (TAIR) genome database was searched for genes with sequence similarity to *COX11* genes from other species. A single candidate locus (At1g02410) was identified. The At1g02410 locus is transcribed and predicted to code for a 287-amino acid protein. The amino acid sequence of the *Arabidopsis* COX11 homolog (further referred to as *Arabidopsis* COX11) shares 32% identity and 45% similarity to the COX11 protein of *S. cerevisiae* (Carr et al., 2005; Banting and Glerum, 2006) (**Figure 1A**).

The yeast COX11 protein was shown to carry an N-terminal mitochondrial targeting sequence. In agreement with this, the *Arabidopsis* COX11 homolog is predicted to be targeted to mitochondria (0.711 and 0.93 probability according to the programs TargetP and MitoProtII, respectively), and to possess a cleavable N-terminal mitochondrial targeting signal (**Figure 1A**). Both, the yeast and the *Arabidopsis* COX11 protein are predicted to harbor a single transmembrane domain of 23 amino acids close to their N-terminal ends (**Figure 1A**). Interestingly, predicted secondary structures of the two proteins, particularly regarding the numbers and positions of β-sheets in C-terminal parts, also show high similarity (**Figure 1A**). The *Arabidopsis* COX11 protein contains six cysteines, with the N-terminal three being part of the putative cleavable mitochondrial targeting signal and hence probably not present in the mature protein. However, the remaining three cysteines are conserved among COX11 family members (Supplementary Figure 4) and have been shown to be essential for the function of yeast COX11 (Banting and Glerum, 2006). Cysteines C_{219} and C_{221} (*Arabidopsis* numbering) are part of the conserved copper-binding motif CFCF.

Arabidopsis COX11 cannot Functionally Complement a *COX11* Deletion Strain of *Saccharomyces Cerevisiae*

We tested whether the *Arabidopsis* homolog can functionally replace its *S. cerevisiae* counterpart and rescue the respiratory-deficient phenotype of yeast *Δcox11* strain (**Figure 1B**). We cloned the full-length yeast and *Arabidopsis COX11* coding sequences, as well as the sequences encoding two chimeric proteins, generated from parts of both homologs into yeast expression vectors. In CHYM-1, the C-terminal region was derived from yeast COX11 (containing the Cu-binding motif), while the transmembrane anchor and the N-terminal region including the mitochondrial targeting sequence stemmed from *Arabidopsis* COX11. In the other chimera (CHYM-2), the origin of the sequences was vice versa. The exchange site was between V_{112} and P_{113} (*Arabidopsis* numbering, **Figure 1A**).

Yeast Δ*cox11* strains that moderately overexpressed the respective constructs, were tested for respiratory competence (**Figure 1B**). The control (full-length yeast *COX11*) was able to restore respiratory competence to WT levels, while all other constructs failed to do so. The only exception was noted upon transformation with the *CHYM-1* construct that resulted in a weak and sporadic growth after long incubation (more than 20 days). Parallel spotting on fermentable glucose media (YPD) demonstrated viability of all yeast strains. The same results were obtained with native expression or strong overexpression of the various constructs (Supplementary Figure 5). All four constructs, irrespective of the origin of their targeting signal (yeast or *Arabidopsis*), were almost completely targeted to mitochondria (Supplementary Figure 6). Hence, incorrect intracellular targeting of the proteins can be excluded as the reason for the observed failure to complement the Δ*cox11* strain. In summary, despite their high similarity, the *Arabidopsis* COX11 protein cannot functionally replace its yeast counterpart.

Arabidopsis COX11 is Localized to Mitochondria

In order to examine the cellular localization of *Arabidopsis* COX11, its ORF was fused at its 3′ end to the sequence encoding mRFP and transformed into an *Arabidopsis* line expressing mitochondria-targeted GFP (mt-GFP) (Nelson et al., 2007). As a control, mRFP alone was expressed in the same genetic background. As expected, mRFP was detected in the cytoplasm of the control lines (**Figure 2A**). Upon fusion to COX11,

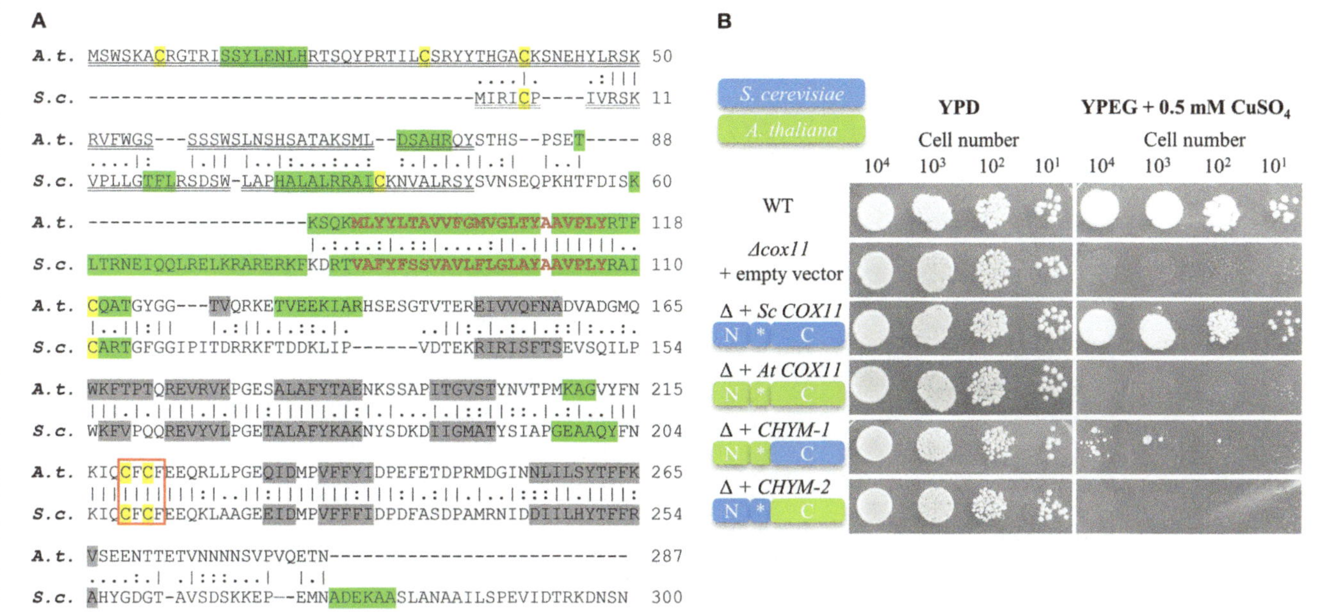

FIGURE 1 | The *Arabidopsis* COX11 homolog shares structural and sequence similarities with yeast COX11, but cannot functionally complement it. (A) The *A. thaliana* (A. t.) and *S. cerevisiae* (S. c.) protein sequence were retrieved from the GenBank database and aligned. The vertical lines (|) indicate identical residues between the two sequences, while the colons (:) and periods (.) represent conservative substitutions with strongly and weakly similar properties, respectively. Cysteines are labeled yellow. The putative copper-binding motifs are boxed. The predicted cleavable N-terminal targeting signals are double underlined. The predicted transmembrane domains are shown in bold red letters, while the predicted α-helices and β sheets are highlighted in green and gray, respectively. **(B)** Results from a respiratory competence test of yeast *Δcox11* stains moderately overexpressing either full-length *S. cerevisiae COX11*, *Arabidopsis COX11*, or sequences encoding the chimeric proteins CHYM-1 or -2. Schematic diagrams of the proteins depict their composition (N, N-terminal region; C, C-terminal region; *, transmembrane domain) and the origin of the respective regions (*S. cerevisiae*, blue; *A. thaliana*, green). Serial dilutions of yeast strains were spotted on fermentable glucose media (YPD) or respiratory media (YPEG, ethanol + glycerol media supplemented with 0.5 mM $CuSO_4$).

however, mRFP co-localized with mt-GFP-labeled mitochondria (**Figure 2B**), demonstrating that COX11 contains sequence information for mitochondrial targeting.

Imaging plants expressing only mt-GFP or mRFP with the same settings as used for the colocalization images ruled out a possible bleed-through between the GFP and mRFP channels (Supplementary Figure 7). No overlap between GFP and mRFP signals was detected. In addition, colocalization of COX11-mRFP and mt-GFP was evaluated by scatter-plotting of the two channel images (Supplementary Figure 8) and calculation of the colocalization coefficients (Supplementary Table 3). Western blot analyses of cell fractions (Supplementary Figure 9) corroborated the mitochondrial localization of COX11-mRFP.

In order to determine, whether *Arabidopsis* COX11 is a soluble or an integral membrane protein, pure mitochondria from a WT line expressing 3HA-tagged *COX11* were treated with sodium carbonate. Upon centrifugation, the two fractions containing the integral membrane and the soluble proteins were subjected to a Western blot analysis with HA-antibody. **Figure 3** shows that a protein of approximately 30 kDa corresponding to the mature COX11-3HA protein was mainly present in the insoluble pellet fraction, in agreement with the prediction of an integral membrane protein. Another prominent protein in the pellet fraction of about 60 kDa suggests COX11 dimer formation, as was shown for yeast COX11 (Carr et al., 2002). In addition, high molecular-weight aggregates were observed. The persistence of the COX11 dimers on SDS-PAGE gels is quite intriguing (**Figure 3** and Supplementary Figures 6, 9). Possibly the dimer stability results from the strong binding of binuclear Cu(I) cluster, which bridges the COX11 monomers (see Introduction). As outlined in a recent review (Palm-Espling et al., 2012), copper clusters often exhibit an extraordinary stability against chemical and thermal unfolding.

The small amount of COX11-3HA that was detected in the soluble fraction cannot be accounted for by cross-contamination of the two fractions, as the membrane marker COX2 (cytochrome *c* oxidase subunit 2) and the soluble marker GDCH (glycine decarboxylase complex subunit H) were exclusively present in the pellet and supernatant fractions, respectively. It remains unclear whether the soluble portion of COX11 reflects the physiological situation, or results from the artificial high-level expression of *COX11-3HA*.

The *COX11* Promoter is Active in Tissues with High Metabolic and/or Division Rates

In order to determine the tissue expression pattern of *COX11*, genomic DNA comprising about 400 bp upstream and downstream of the putative transcription start site (as annotated in the TAIR data base; Lamesch et al., 2012) was fused with the reporter gene β-*glucuronidase (GUS)* and transformed into the

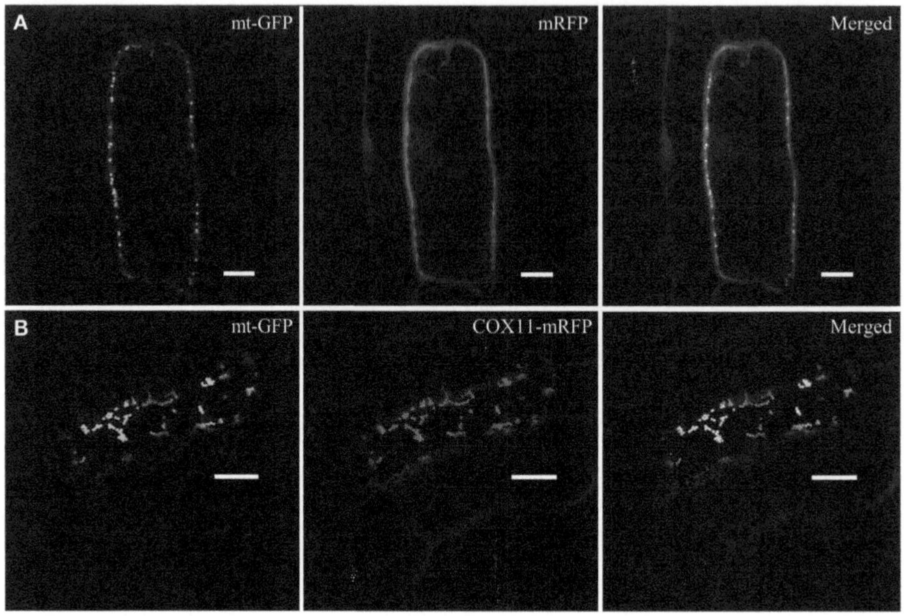

FIGURE 2 | COX11-mRFP is localized to mitochondria. LSM images of root cells from *Arabidopsis* lines with GFP-labeled mitochondria, expressing either *mRFP* alone **(A)** or *COX11-mRFP* **(B)**, are shown. The fluorescence of mRFP and GFP targeted to mitochondria (mt-GFP) was falsely colored magenta and green, respectively. Scale bars correspond to 10 μm each.

WT background. The downstream region contained almost the entire 5′ UTR including the first intron. **Figure 4** summarizes the results of the GUS staining.

After overnight imbibition, a strong promoter activity was observed in the whole embryo (**Figure 4A**). With the development of the embryo, the *COX11* promoter activity disappeared in most tissues or was below the detection limit of the GUS assay (**Figures 4B,F**). The GUS activity remained high in meristematic tissues of the shoot and root, as well as in the vascular tissues of source and sink organs (**Figures 4C,F,H**). In contrast, no *COX11* promoter activity was detected in the vascular tissue of transport organs such as the hypocotyl and shoot. These observations agree with reported *COX11* mRNA levels from publicly available microarray data compiled by Genevestigator. For example, *COX11* transcript levels in leaf rosettes are on average approximately two-fold higher than in hypocotyls.

Young leaves near the shoot meristem (**Figure 4F**) were also stained, indicating strong *COX11* promoter activity. With ongoing maturation, the staining became weaker, except for the vascular tissue and guard cells (**Figure 4G**). In fully developed leaves (**Figure 4C**), GUS activity was limited to the vascular tissue close to the leaf periphery. In flowers (**Figure 4D**), only the stamen vascular tissue and the stigma showed *COX11* promoter activity. Unlike the mature pollen (**Figure 4E**), the germinated pollen grain exhibited a strong GUS activity (**Figure 4I**). This was also observed in pollen grains rehydrated in 18% (w/v) sucrose (**Figure 4J**). In siliques (**Figure 4K**), the GUS activity was restricted to the abscission zone.

In summary, the *COX11* promoter was mainly active in tissues with high division and/or high metabolic rates. In this regard it resembles the expression patterns of other genes related to COX function, such as COX assembly factor *HCC1* (for homolog of copper chaperone SCO1) (Steinebrunner et al., 2011), the COX subunit *COX5b-1* (Welchen et al., 2004) and the cytochrome *c* encoding genes *CYCT-1* and *CYCT-2* (Welchen and Gonzalez, 2005).

Knockdown and Overexpression of *Arabidopsis COX11*

To further investigate the functions of *Arabidopsis COX11*, we aimed to analyze plant lines with no or with elevated expression of *COX11* [knock-out mutants (KO) and overexpressors (OE), respectively]. Five publicly available T-DNA insertion lines within the *COX11* locus (Supplementary Figure 3) were tested for *COX11* KO. For two of them, Sail_683_B03 and Salk_105793, the presence of the T-DNA insertion could not be confirmed. For the three other lines (Sail_603_G12, Sail_861_D09 and Salk_003445), the T-DNA insertions, all within the *COX11* promoter region, were detected by PCR and homozygous plant lines were selected. For the Sail_861_D09 line, which carries the T-DNA insertion closest to the putative transcription start site, the *COX11* mRNA level was exemplarily determined by qPCR. Unexpectedly, the level was 1.9-fold higher than that of the WT, making this line a weak OE. The *COX11* mRNA levels of the other two lines were not determined, as the sites of the T-DNA insertions in the *COX11* promoter are farther upstream compared with that of the Sail_861_D09 line. Therefore, it can be expected that their impact on *COX11* expression is less pronounced.

To compensate for unavailable KO plant lines, two knock-down lines (KD1 and KD2) were generated by employing

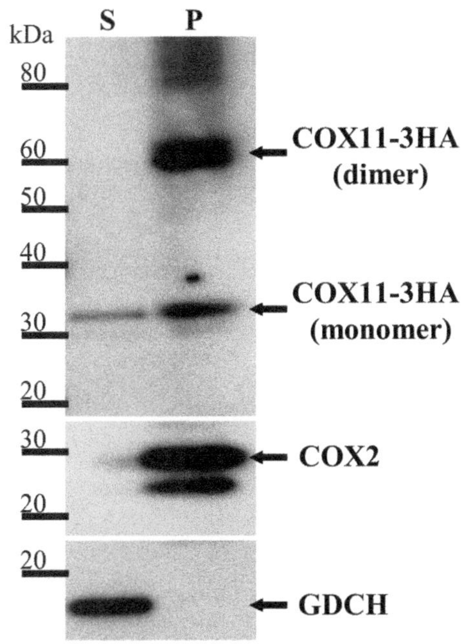

FIGURE 3 | COX11-3HA is an integral membrane protein. Pure mitochondria were isolated from plants expressing *COX11-3HA* and subjected to Na_2CO_3 treatment to separate soluble (S) from membrane and membrane associated proteins, which remained in the pellet (P). The S and P fractions (from 100 µg of mitochondria) were analyzed by Western blotting, using antibodies against HA (from human influenza hemagglutinin) to detect COX11-3HA of the theoretical molecular weight of 30 kDa after cleavage of the predicted targeting signal. The purity of the fractions was assessed by detection of the marker COX2 (cytochrome *c* oxidase subunit 2, theoretical molecular weight of 30 kDa) for integral and GDCH (glycine decarboxylation complex subunit H, theoretical molecular weight of 16 kDa) for soluble proteins. All proteins were detected on the same membrane, which was stripped between each detection.

the artificial microRNA (amiRNA) technology (see Materials and Methods). Homozygosity of the amiRNA-encoding T-DNA insertion appears to be lethal, as no respective mutant lines could be identified. Therefore, the seedlings had to be cultured on kanamycin to select for heterozygous KD plants. Because these kanamycin-selected seedlings were directly used for qPCR analyses, the influence of kanamycin on the transcript levels of the investigated genes was determined. With the exception of *COPT2* (copper transporter 2) (Supplementary Figure 10), only minor changes were observed. The *COX11* mRNA level was slightly increased (1.19-fold).

For KD1, seeds were harvested from two individual plants and the respective progeny, KD1-1 and KD1-2, was independently analyzed. In all three lines (KD1-1, KD1-2, and KD2), *COX11* mRNA levels were reduced to 30-40% compared with the WT (**Figure 5**).

Overexpression was obtained by placing the *COX11* cDNA under the control of the *35S* promoter in the WT background. Two lines, OE1 expressing *COX11* fused to the 3HA-tag sequence, and OE2 expressing untagged *COX11*, were propagated to homozygosity. The two lines showed a six- and four-fold overexpression of *COX11*, respectively, compared with the WT (**Figure 5**).

COX11 Knockdown and Overexpression Affects COX Activity

The main function of COX11 proteins in other organisms is the insertion of copper into the COX complex during its assembly. Therefore, we investigated the effect of *COX11* KD and OE on COX complex activity (**Figure 6**). Crude mitochondria were prepared from WT, KD and OE plants and assayed for COX activity by measuring the oxidation rate of reduced cytochrome *c*. Citrate synthase activity was used to normalize for the concentration of mitochondrial protein as described previously (Steinebrunner et al., 2014).

The KD lines showed a significant reduction in COX complex activity compared with the WT, about 45% for KD1-1 and KD1-2 and about 55% for KD2 (**Figure 6A**). Surprisingly, OE lines also exhibited a reduced COX activity (~80% of the WT) (**Figure 6A**). To confirm the specificity of the COX activity assay, reactions were compared in the presence or absence of the specific COX inhibitor KCN. In the absence of KCN, cytochrome *c* was continuously oxidized. In presence of KCN, this reaction was completely blocked (Supplementary Figure 11A). This result demonstrates that the COX complex activity was responsible for the oxidation of reduced cytochrome *c*.

To corroborate these data, COX complex activity was analyzed by a second method. Crude mitochondria fractions were separated by blue native (BN)-PAGE and subsequently subjected to in-gel COX activity staining (**Figure 6B**). Brown precipitate marks COX activity, which was the strongest in the size range of 250–400 kDa, corresponding to the COX monomer (230 kDa; Klodmann et al., 2011) and possibly a dimer. Based on the staining intensity, the OE lines and the WT had similar levels of activity, while the KD mutants showed much less COX activity. Especially for the KD1-1 and KD1-2 plants, the activity was so low that it was barely detectable. No brown staining was detected in the presence of KCN, verifying the specificity of the in-gel COX activity staining (Supplementary Figure 11B).

In the upper half of the BN gel, three additional COX activity bands were observed mainly in the WT and OE samples. These bands probably represent respiratory supercomplexes that include the COX complex (Eubel et al., 2004).

To test whether the reduced COX activity associated with *COX11* KD was accompanied by an altered abundance of COX subunits, we analyzed total protein extracts in a Western blot with COX2 antibody (**Figure 6C**). Densitometric analysis of COX2 abundance, after normalization with GDCH, revealed that COX2 steady-state level was reduced in KD plants, while in the OE lines it was the same or slightly higher compared with the WT. The reduction in KD plants is likely to result from degradation of unassembled respiratory complex subunits as has been reported in yeast and humans (Nakai et al., 1994; Banting and Glerum, 2006; Kovárová et al., 2012). In contrast, at the mRNA level, COX complex subunits were upregulated in both KD and OE plants, as shown exemplarily for *COX5b-1* (**Figure 6D**).

Previously, it was demonstrated, that in response to respiratory deficiency the gene encoding alternative oxidase 1a

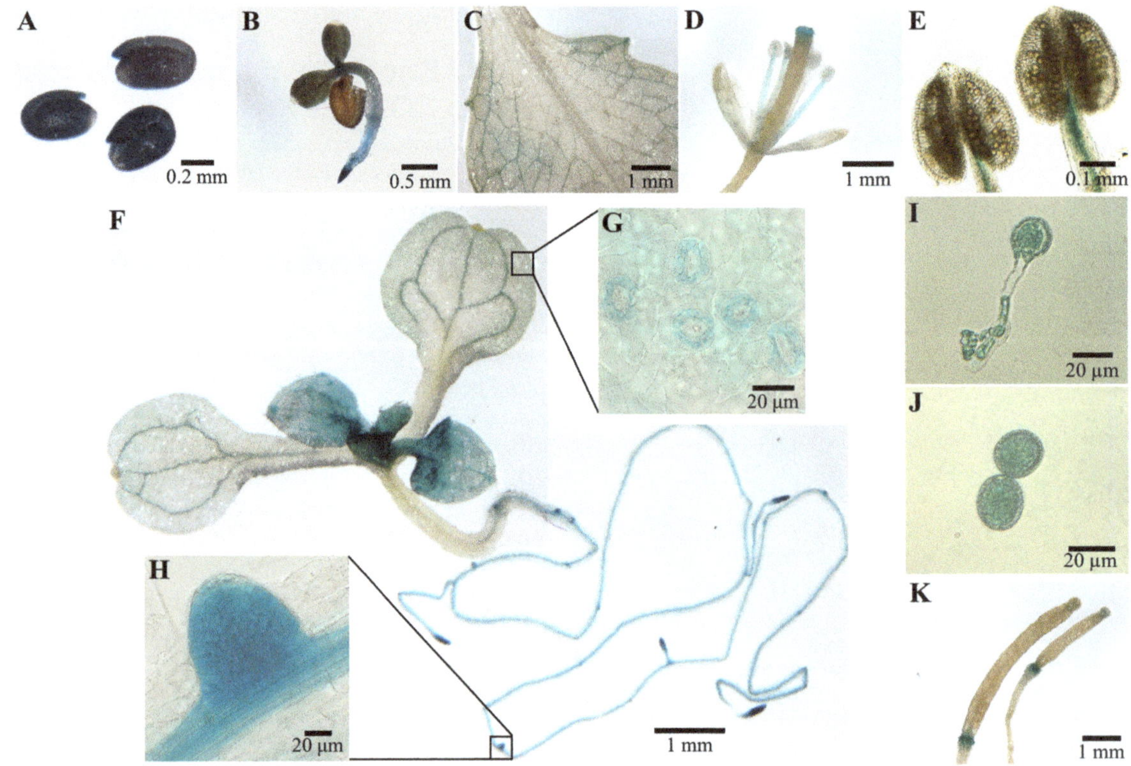

FIGURE 4 | *COX11:GUS* promoter analysis. Plants from three *COX11:GUS* lines (T2 generation) were assayed for GUS activity at various developmental stages by bright-field microscopy. Representative staining patterns are shown. **(A)** Embryos after overnight imbibition and removal of the seed coats. **(B)** A 3-day-old seedling. **(C)** Mature leaf. **(D)** *Arabidopsis* flower. **(E)** Anthers with mature pollen. **(F)** Two images of different parts of a 12-day-old seedling, stitched together. The insert panels show a section of its cotyledon **(G)** and a lateral root primordium **(H)** at a higher magnification. **(I)** Germinated pollen. **(J)** Pollen grain rehydrated in 18% sucrose. **(K)** Siliques at different developmental stages.

(AOX1a) was significantly upregulated (Yuan and Liu, 2012; Yang et al., 2014). In order to test whether this holds also true for the *COX11* KD lines, we determined their *AOX1a* mRNA levels (**Figure 6D**). We confirmed a negative correlation with the COX activity data, as the KD1-1 and KD1-2 lines exhibited a significantly higher increase of *AOX1a* mRNA than the KD2 line (3–4 vs. 1.6-fold). A similar slight increase of 1.4 was also observed in the OE lines.

Impaired Root Growth in *COX11* Knock-down and Overexpression Mutants

In order to investigate a possible effect of *COX11* KD and OE on plant development, we studied the phenotype of respective mutants (**Figure 7**). The rosettes of 7-week-old KD2 and OE mutants did not differ from the WT in size and morphology, whereas those of the KD1-1 and KD1-2 mutants were smaller with slightly curled leaves (**Figure 7A**). Apparently, the slightly higher COX activity rates in KD2 mutants compared with the KD1 knockdowns are sufficient to maintain WT leaf growth. Next, we analyzed roots, which are mainly dependent on respiration for energy supply. The KD lines displayed a pronounced difference in root length compared with the WT (**Figures 7B,C**): after 12 days of growth, the roots were about 50% shorter than in the WT. In the OE mutants, a reduction of root length was less obvious. Thus, the root lengths correlated positively with the COX complex activity of the plants.

Further confirmation that the observed reduction in root length is related to diminished COX complex activity came from experiments employing the specific inhibition of COX by KCN. WT seedlings were cultured on plates supplemented with different concentrations of KCN. As shown in **Figures 7D,E**, root growth was inhibited in a dosage-dependent manner, demonstrating that COX deficiency leads to reduction in root length. KCN treatment phenocopied the *COX11* mutants, arguing in favor of the KD and OE root phenotype resulting from impaired COX activity.

Environmental Copper Affects Root Growth of *COX11* Mutants

With the assumption that *Arabidopsis* COX11 is, like other homologs, a copper chaperone, the effect of Cu on root growth was analyzed. WT and *COX11* mutant seedlings were grown together on regular MS plates and on plates supplemented with $CuSO_4$ or with the copper chelator BCS (bathocuproinedisulfonic acid). After 12 days, root lengths were evaluated (**Figure 8**). For more clarity of the figure, the results for

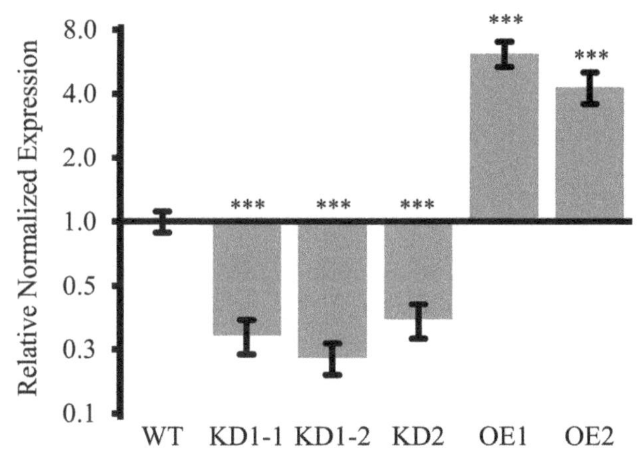

FIGURE 5 | *COX11* mRNA levels in KD and OE lines. *COX11* transcript levels were determined in KD and OE lines by qPCR. RNA was isolated from 14-day-old seedlings cultured on selective or non-selective MS plates. Mean values of mRNA levels in *COX11* mutants were normalized to WT and plotted on logarithmic scale (base two). Values and statistical significance (compared with the WT, $^{***}P < 0.001$), were calculated with the CFX manager software. Error bars represent ± standard deviation (SD). Individual values are listed in the Supplementary Table 5.

the KD2 line are depicted separately in Supplementary Figure 12. Under copper excess, the roots of KD1 plants grew less compared with the WT (**Figure 8A**). In contrast, the roots of OE grew better compared with the WT (**Figure 8B**). When seedlings were cultured in the presence of BCS, only the roots of KD1-2 plants grew shorter compared with the WT.

The data are supportive, but not conclusive of COX11 acting as a copper chaperone. Possibly, in the KD lines the reduced amounts of COX11 were unable to efficiently chelate excess copper in the IMS, which in turn exerted its toxic effect and led to reduced root growth. In the OE lines the opposite might have occurred, because the larger amounts of COX11 were able to bind excess copper.

COX11 Knockdown Elicits Expression Changes of Genes Involved in Copper Metabolism

Next, we were interested whether the disturbance of COX11 abundance, through KD or OE, affected the regulation of the other mitochondrial copper chaperones HCC1 and COX17-1, both of which have been implicated in COX assembly (Attallah et al., 2007, 2011; Steinebrunner et al., 2011, 2014). This was tested by determining their transcript levels in 14-day-old seedlings by qPCR analysis. Both genes were upregulated about two-fold in the KD lines, but remained almost unaffected or only slightly affected in the OE lines compared with the WT (**Figure 9**). Possibly this increase, as observed for the subunit *COX5b-1* (**Figure 6D**), reflects an attempt to counteract the COX deficiency of *COX11* mutants. This result could also indicate a presence of retrograde signaling pathways, which convey mitochondrial dysfunction of COX assembly and Cu metabolism.

Considering this retrograde signaling from mitochondria to the nucleus, we were interested to see whether other genes involved in copper metabolism were also affected in the *COX11* mutant lines. We analyzed the expression levels of *COPT2* (plasma-membrane copper transporter 2), *CSD1* (copper-zinc superoxide dismutase) and *ZIP2* (plasma-membrane metal-ion transporter), all known to be regulated by Cu. We noted a two-fold upregulation of the *COPT2* mRNA level in the KD seedlings, while no difference was observed between the OE and the WT seedlings (**Figure 9**). For *CSD1* and *ZIP2*, transcript levels were upregulated 2–2.5-fold and downregulated 1.5–2-fold in all lines, respectively. Apparently, the KD of *COX11* had a broad impact on the expression of genes involved in cellular copper homeostasis.

This result prompted us to determine the Cu content in mature leaves of KD mutants. However, as shown in the Supplementary Table 4, it was unaltered in the KD1-1, KD1-2 and OE mutants and even reduced in the KD2 line compared with the WT. An explanation could be that increased COPT2-mediated Cu uptake was limited to seedlings, where we observed the upregulation, or to other tissues than leaves. Indeed, roots for example have 2.4-fold higher *COPT2* mRNA levels than leaves (average microarray data from Genevestigator).

COX11 KD and OE Leads to Reduced Pollen Germination

It was previously reported that knockdown of rice *COX11* by RNA interference, specifically induced in the tapetum, was linked to defective pollen maturation (Luo et al., 2013). The authors proposed that COX11 has a secondary role in controlling the timing of the oxidative burst by reactive oxygen species (ROS) scavenging. This oxidative signal is as a key signal for the degeneration of the anther tapetum and pollen maturation. In order to test whether the same holds true for *Arabidopsis*, pollen viability of *COX11* mutants was checked by staining with fluorescein diacetate (FDA). Contrary to the situation in rice, more than 95% of the pollen grains in all lines proved to be viable (**Figure 10A**). However, we cannot rule out the possibility that our approach to generate a *35S*-promoter-driven KD of *COX11* was not effective enough to interfere with pollen viability.

As we observed a substantial *COX11* promoter activity in germinating pollen (**Figure 4I**), in the next step we analyzed the *in vitro* germination rates of KD and OE pollen. Pollen grains with extended tubes (**Figure 10B**, #1) were counted as germinated, while pollen grains without any or only small protrusions as non-germinated (**Figure 10B**, #2). Pollen germination rates were significantly reduced by about 20 and 30%, respectively, in the OE and KD lines compared with the WT, except for the KD2 line (**Figure 10C**). We speculate that this outlier was due to a loss of the KD effect in the pollen grains, as we observed that *COX11* transcripts had returned to WT levels in the next generation (Supplementary Figure 2A). We also attempted to phenocopy the pollen germination reduction in KD and OE lines by germination of WT pollen in the presence of KCN. This specific COX complex inhibitor led to small reduction of germination rates (Supplementary Figure 13). Addition of 100 μM KCN caused a reduction of about 30% in the germination rate of WT pollen. In summary, COX11 did not impair the pollen viability, but affected pollen germination.

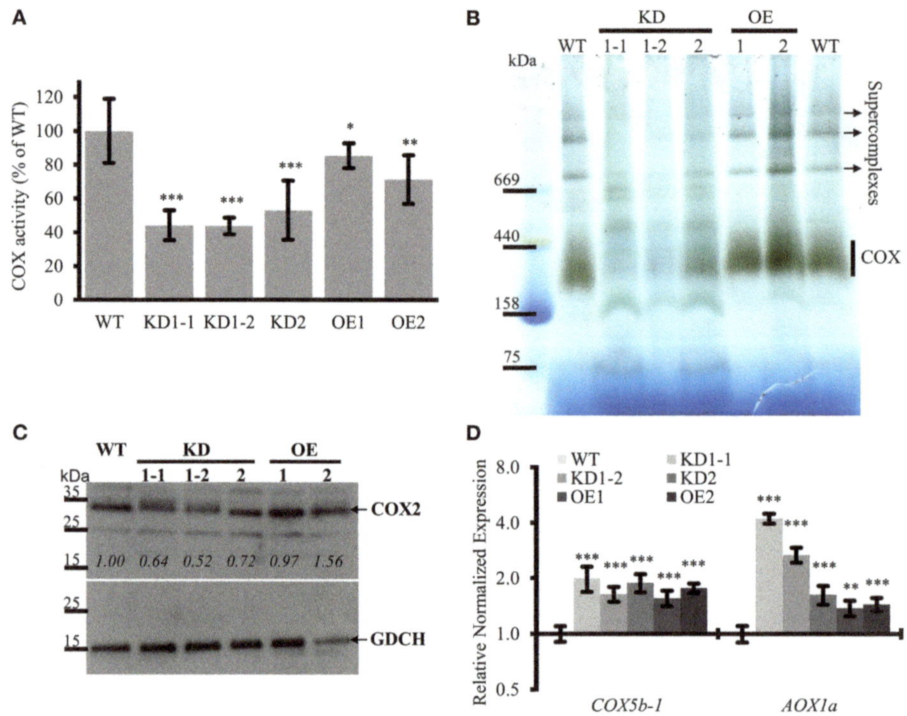

FIGURE 6 | COX complex activities are reduced in COX11 KD and OE mutants. (A) COX activities in KD and OE mutants and the WT were measured in three independent mitochondria isolations (each in technical triplicate) and normalized to the WT (=100%). Asterisks indicate statistical significance calculated with the Student's *t*-test (*$P < 0.05$; **$P < 0.01$; ***$P < 0.001$). Error bars represent ± SD. **(B)** Mitochondrial complexes from COX11 KD, OE and WT plants were separated by gradient BN-PAGE and subsequently stained in gel for COX complex activity visible as a brown precipitate. For WT, 100 µg of crude mitochondria was loaded, while for the mutants amounts were adjusted (based on citrate synthase activity) to compensate for contaminations. **(C)** Total leaf proteins from WT, KD, and OE plants (6 weeks old) were analyzed by Western blotting, using antibodies against COX2 (theoretical molecular weight of 30 kDa). The membrane was stripped and subsequently incubated with antibodies against GDCH (theoretical molecular weight of 16 kDa) as a loading control. In this experiment T4-generation KD1-1 plants were used (Supplementary Figure 2B). Values in italics indicate estimated COX2 abundance after densitometry analysis and normalization to GDCH amounts. **(D)** qPCR analysis of indicated transcript levels in WT and COX11 mutants. RNA was isolated from 14-day-old seedlings cultured on selective (KD mutants) or non-selective MS (WT) plates. Mean values of mRNA levels in COX11 mutants were normalized to WT and plotted on a logarithmic scale (base two). Values and statistical significance (compared with the WT, **$P < 0.01$, ***$P < 0.001$) were calculated with the CFX manager software. Error bars represent ± SD. All individual values are listed in Supplementary Table 5.

DISCUSSION

Several members of the COX11 protein family were shown to play a key role in Cu delivery to the COX subunit 1 during COX assembly (Hiser et al., 2000; Carr et al., 2002). In this work, we aimed to investigate if the *Arabidopsis* homolog exerts the same function.

Surprisingly, despite the high sequence and structural similarities *Arabidopsis* COX11 could not complement the respiratory deficiency of yeast *Δcox11* strain (**Figure 1B**). However, one has to keep in mind that COX assembly is a very complex process that requires strict spatial and temporal coordination of many proteins, including COX11 (reviewed in Soto et al., 2012). It is possible that specific protein-protein interactions have evolved differently in yeast and *Arabidopsis* mitochondria, preventing the functionality of *Arabidopsis* COX11 in yeast. The observed sporadic growth of yeast cells expressing *CHYM-1* (**Figure 1B**) might be caused by secondary mutations in some COX11 partner protein(s), allowing the functionality of CHYM-1. Additionally, it cannot be rigorously excluded that this growth is an artifact, caused by prolonged cultivation. Carr et al. (2005) reported similar results for complementation of yeast *Δcox11* strain with human COX11 (both full-length and chimeras). Obviously, it is not possible to conclude the function of homologous proteins solely based on sequence similarity.

Subsequently, we were able to show that *Arabidopsis* COX11 protein is localized to mitochondria (**Figure 2**), where it is anchored in the inner mitochondrial membrane (**Figure 3**) like the yeast COX11 (Carr et al., 2005; Khalimonchuk et al., 2005). Furthermore, we found that the COX activity was reduced in *COX11* KD and OE plant lines (**Figure 6A**). This result agrees with a function of *Arabidopsis* COX11 in the assembly the COX complex, as shown for other organisms (Hiser et al., 2000; Carr et al., 2005; Banting and Glerum, 2006). The question arises why overexpression of *COX11* also lowers COX activity. A high concentration of the COX11 protein could lead to its inactivation due to the disturbance of protein stoichiometry between COX11 and its auxiliary factor(s). For example, it was recently reported that in yeast, COX19 is required to maintain COX11 in an

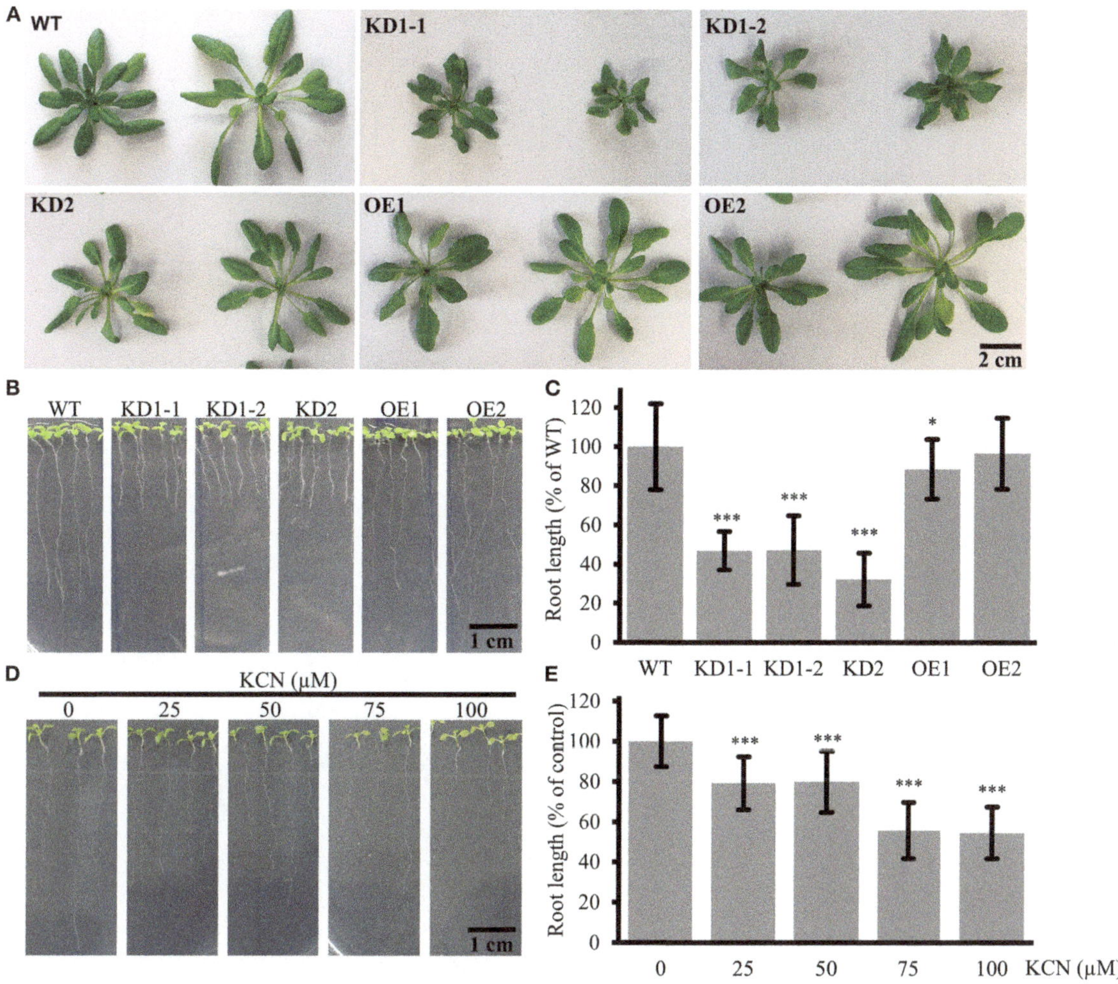

FIGURE 7 | Leaf and root phenotype of *COX11* KD and OE mutants. (A) Rosettes from 7-week-old plants cultured under short-day conditions. In this experiment T4-generation KD1-1 plants were used (Supplementary Figure 2B). The scale bar applies to all panels. **(B)** Roots of 12-day-old seedlings. The scale bar applies to all panels **(C)** Root lengths from three independent experiments were measured and normalized to the WT (=100%). Error bars represent ± SD for $n = 30$ (WT, KD1-1, KD1-2, OE1, OE2), 16 (KD2). Asterisks indicate statistical significance calculated with the Student's *t*-test ($^{*}P < 0.05$; $^{***}P < 0.001$) **(D)** WT roots grown in the presence or absence of KCN. The scale bar applies to all panels. **(E)** Root lengths from two independent experiments were measured and normalized to the untreated control (=100%). Error bars represent ± SD for $n = 40$ (0, 50, 100 µM KCN), 30 (25, 75 µM KCN). Asterisks indicate statistical significance calculated with the Student's *t*-test ($^{***}P < 0.001$). Absolute and normalized root lengths for **(C,E)** are given in Supplementary Table 5.

active state (Bode et al., 2015). A higher concentration of COX11 could result in a depletion of active COX19. As an alternative explanation, one could envisage the titration of essential factors such as Cu (reviewed in Prelich, 2012). The elevated amount of COX11 possibly competes with HCC1 for Cu loading from COX17, disturbing Cu transfer to COX2, subsequently leading to a reduced COX activity (**Figure 6A**). These hypotheses explain the phenotypes of OE plants cultured under normal conditions; however, they do not explain why OE plants, when exposed to excess copper grow better compared with the WT (**Figure 8B**). Possibly, the surplus COX11 in OE plants promotes root growth by sequestering excess Cu, thus alleviating Cu toxicity. This together with the results for KD plants (**Figure 8A**), where the reduced amount of COX11 probably fails to sequester excess Cu, is in favor of COX11 being a copper binding protein. We conclude that the *Arabidopsis* COX11 homolog is essential for assembly of the COX complex, possibly by inserting Cu into the COX1 subunit, as shown for other organisms (Hiser et al., 2000; Banting and Glerum, 2006; Thompson et al., 2010).

Support for a role of COX11 in energy metabolism is provided by its expression pattern (**Figure 4**). The *COX11* promoter was predominantly active in tissues with a high-energy demand, e.g., in meristems to sustain high division rates or in vascular tissues of leaves and roots to maintain phloem loading and unloading, respectively (Taiz and Zeiger, 2010). The promoter activity was also high in imbibed embryos, which need to repair and differentiate their mitochondria for ATP production prior to germination (reviewed in Weitbrecht et al., 2011). All these tissues depend on the continuous formation of respiratory chain

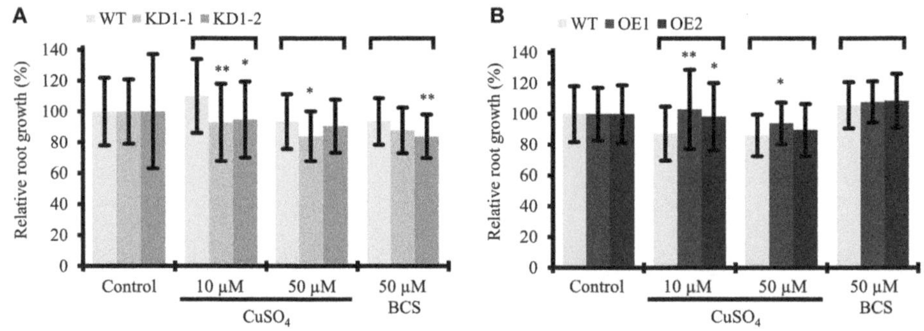

FIGURE 8 | Influence of Cu treatment on root growth of *COX11* mutants. Relative root growth of 12-day-old WT and KD **(A)** as well as WT and OE **(B)** seedlings cultured on plates with copper excess or deficiency (BCS). Root length is expressed relative to WT or mutant seedlings on control plates. Roots were measured in three independent experiments. Error bars represent ± SD for $n = \sim 40$ (WT), ~30 (KD1-1, KD1-2), and ~35 (OE1, OE2). Asterisks indicate statistical significance calculated with the Student's *t*-test (mutant line compared with WT, $^{*}P < 0.05$; $^{**}P < 0.01$). Individual values are listed in the Supplementary Table 5. BCS, bathocuproinedisulfonic acid.

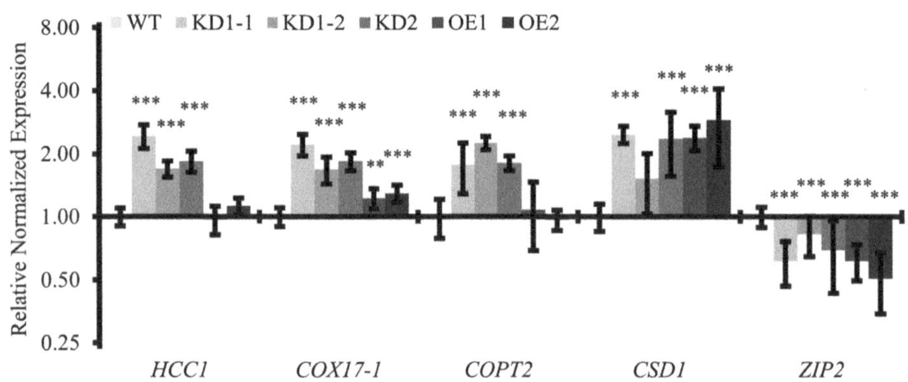

FIGURE 9 | Quantitative analysis of transcripts associated with copper metabolism. qPCR analysis of indicated transcripts in WT and *COX11* mutants. RNA was isolated from 14-day-old seedlings cultured on selective (KD) or non-selective MS (WT, OE) plates. Mean values of mRNA levels in *COX11* mutants were normalized to WT and plotted on a logarithmic scale (base two). Values and statistical significance (compared with the WT, $^{**}P < 0.01$; $^{***}P < 0.001$) were calculated with the CFX manager software. Error bars represent ± SD. Individual values are listed in the Supplementary Table 5.

complexes and hence require sufficient amounts of assembly factors like COX11.

In the course of our study, we observed an upregulation of several genes both in *COX11* KD and OE plants (**Figures 6D, 9**). The higher expression levels of COX subunits and assembly factors (*COX5b-1, HCC1,* and *COX17-1*) in the KD lines may reflect a cellular reaction to compensate for the reduced COX complex activity, by elevating the formation of COX subunits and the factors necessary for their assembly. Probably due to a less pronounced COX deficiency in OE lines, only the level of *COX5b-1* mRNA was significantly elevated. The upregulation of the expression of the *AOX1a* gene, both in KD and OE plants, may indicate a further counter-reaction of the cells against COX deficiency. Remarkably, in the KD lines the expression of the *COPT2* gene was also upregulated, which can possibly be regarded as an attempt of the cells to increase the cellular concentration of Cu to allow COX complex assembly. If indeed, the KD mutants were Cu deficient, their *CSD1* levels should be lower and *ZIP2* levels higher (Wintz et al., 2003) compared with the WT. However, both genes were regulated in the exact opposite way (**Figure 9**). Therefore, it seems unlikely that increased *COPT2* amounts in KD plants are related to Cu deficiency. Instead, the high number of stress-and ROS-responsive regulatory elements in the promoter region of *COPT2* (Peñarrubia et al., 2015) suggests that *COPT2* might be upregulated by stress conditions due to impaired COX function. In agreement with this, *CSD1*, a known ROS detoxifier, is upregulated in KD mutants (**Figure 9**).

Both the knockdown and overexpression of *COX11* interferes with the protein's function and has substantial phenotypic consequences. The most striking effect was the root growth inhibition in *COX11* KD plants (**Figures 7B,C**). The question arises whether energy deficiency due to reduced COX activity or other effects of COX11 disturbance are the underlying reason(s) for reduced root growth. The shorter-root phenotype positively correlated with a diminished COX activity (**Figure 6A**). Importantly, the same phenotype could be mimicked by

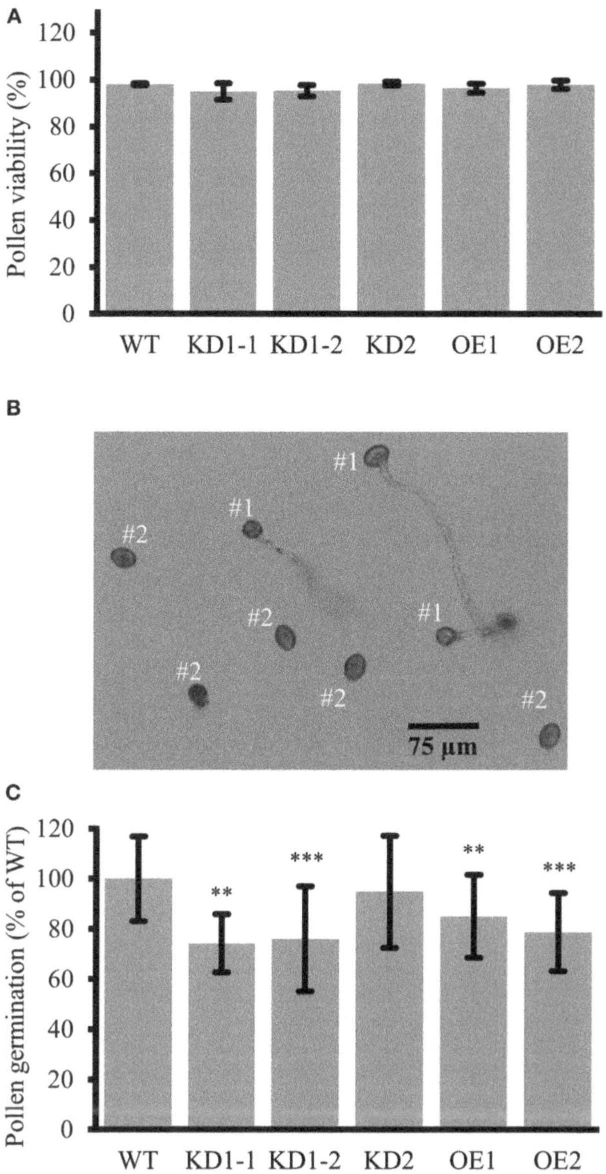

FIGURE 10 | Reduced pollen germination in *COX11* KD and OE mutants. (A) Pollen viability in *COX11* mutants compared with the WT. Error bars represent ± SD of means from four experiments. Between 400 and 600 pollen grains per plant line were evaluated **(B)** Definition of germinated (#1) vs. non-germinated (#2) pollen grains used for determining germination rates. **(C)** Pollen germination rates of KD and OE mutants were normalized to the WT (=100%). Error bars represent ± SD of means from 10 experiments. For each experiment, germination of 100–300 pollen grains was evaluated in duplicates or triplicates. Asterisks indicate statistical significance calculated with the Student's *t*-test (**$P < 0.01$; ***$P < 0.001$). Exact values are given in Supplementary Table 5.

treatment of WT seedlings with KCN (**Figures 7D,E**), which specifically inhibits COX. Therefore, it seems probable that the energy deficit caused by the lower COX activity results in root growth inhibition in *COX11* KD and OE plants. Similar root shortening was observed in other *Arabidopsis* mutants with impaired energy production due to deficient respiratory chain complexes (Yuan and Liu, 2012; Huang et al., 2013; Yang et al., 2014).

A more subtle, but still noticeable phenotypic change in KD mutants was curling of the leaf blades (**Figure 7A**). It was more pronounced in the KD1 lines, in which *COX11* expression was more strongly suppressed than in the KD2 line (**Figure 5**). This leaf phenotype was found to be associated with mitochondrial dysfunction, and can be explained by the high demand of energy required for morphogenetic processes, such as cell division (Van Aken et al., 2007; Gibala et al., 2009).

Previously, it was suggested that COX11 is involved in pollen maturation, because silencing of *COX11* in rice produced non-viable pollen (Luo et al., 2013). Furthermore, near-complete COX-deficient *Arabidopsis* mutants were found to produce non-viable pollen as well (Dahan et al., 2014). However, pollen viability was not affected in the *COX11* mutant lines of our study (**Figure 10A**), possibly because the knockdown efficiency was not strong enough. Instead, we observed that pollen germination was reduced, in both the KD and OE lines (**Figure 10C**). Again it is likely that this effect is at least partly due to the reduced COX activity, as KCN-induced COX deficiency also affected pollen germination (Supplementary Figure 13). However, this effect was significantly less pronounced than in case of root growth (**Figure 7E**). Additionally, the KD and OE pollen germination rates did not correlate well with the respective COX activities (**Figures 6A, 10C**). In a previous report, it was shown that petunia pollen germination does not rely on respiration and that aerobic glycolysis can provide sufficient energy for this process (Gass et al., 2005). Therefore, *Arabidopsis* COX11 may contribute to pollen germination not only via its function in energy supply, but also through unrelated function(s). A speculative fitting role could be in oxidative signaling. As shown by Speranza et al. (2012), a tightly regulated ROS appearance is essential for pollen activation and pollen tube emergence. *Arabidopsis* COX11 could participate in the ROS metabolism regulation as suggested for the COX11 homologs in rice (Luo et al., 2013) and yeast (Banting and Glerum, 2006; Khalimonchuk et al., 2007). Consequently, its knockdown or overexpression could disturb the ROS signaling and thus germination.

In summary, the characterization of *COX11* KD and OE mutants clearly documents the essential role of *Arabidopsis* COX11 for COX activity, and that this function strongly affects plant development and performance. Additionally, our data hint at an additional function of *Arabidopsis* COX11 in ROS signaling.

AUTHOR CONTRIBUTIONS

IR, GR, and IS conceived and designed the research. IR performed the research. NM designed and generated the KD lines. IR analyzed the data, with IS co-analyzing root growth and imaging data. IR, GR, and IS wrote the article.

FUNDING

This work was supported by a PhD fellowship to IR from the Dresden International Graduate School for Biomedicine and Bioengineering (DIGS-BB), which is funded by the DFG

(German Research Foundation). The grant STE 1455/5-1 from DFG awarded to IS supported a research stay of IR at the Universidad Nacional del Litoral in Santa Fe, Argentina, which included training in methods employed in this work. The publication fee for this article was covered by the DFG grant RO 1299/9-3.

ACKNOWLEDGMENTS

We thank Lucila Garcia and Daniel Gonzalez for providing us with the *35S:GUS* plant line; Tsuyoshi Nakagawa and Michael Curtis for supplying us with the pGWB and pMDC vectors, respectively; Simon Alberti for supplying us with yeast expression vectors; EUROSCARF for yeast strains; ABRC and NASC for distributing the *COX11* cDNA clone, mt-gk and T-DNA insertion lines; Joseph Ecker for cDNA clone and SALK lines, and Syngenta (SAIL lines) for generating them. We are also grateful to Andrea Voß (Leibniz-Institut für Festkörper- und Werkstoffforschung Dresden) for performing the ICP-OES analysis, Uta Gey for establishing the BN-PAGE protocol and the Light Microscopy Facility of the BIOTEC/CRTD at Technische Universität Dresden for their help with our confocal microscopy experiments.

REFERENCES

Alberti, S., Gitler, A. D., and Lindquist, S. (2007). A suite of gateway cloning vectors for high-throughput genetic analysis in *Saccharomyces cerevisiae*. *Yeast* 24, 913–919. doi: 10.1002/yea.1502

Alonso, J., Stepanova, A., Leisse, T., Kim, C., Chen, H., Shinn, P., et al. (2003). Genome-wide insertional mutagenesis of *Arabidopsis thaliana*. *Science* 301, 653–657. doi: 10.1126/science.1086391

Attallah, C. V., Welchen, E., and Gonzalez, D. H. (2007). The promoters of *Arabidopsis thaliana* genes *AtCOX17–1* and *–2*, encoding a copper chaperone involved in cytochrome *c* oxidase biogenesis, are preferentially active in roots and anthers and induced by biotic and abiotic stress. *Physiol. Plant.* 129, 123–134. doi: 10.1111/j.1399-3054.2006.00776.x

Attallah, C. V., Welchen, E., Martin, A. P., Spinelli, S. V., Bonnard, G., Palatnik, J. F., et al. (2011). Plants contain two SCO proteins that are differentially involved in cytochrome *c* oxidase function and copper and redox homeostasis. *J. Exp. Bot.* 62, 4281–4294. doi: 10.1093/jxb/err138

Banci, L., Bertini, I., Cantini, F., Ciofi-Baffoni, S., Gonnelli, L., and Mangani, S. (2004). Solution structure of Cox11, a novel type of beta-immunoglobulin-like fold involved in Cu_B site formation of cytochrome *c* oxidase. *J. Biol. Chem.* 279, 34833–34839. doi: 10.1074/jbc.M403655200

Banting, G. S., and Glerum, D. M. (2006). Mutational analysis of the *Saccharomyces cerevisiae* cytochrome *c* oxidase assembly protein Cox11p. *Eukaryot. Cell* 5, 568–578. doi: 10.1128/EC.5.3.568-578.2006

Benson, D. A., Cavanaugh, M., Clark, K., Karsch-Mizrachi, I., Lipman, D. J., Ostell, J., et al. (2013). GenBank. *Nucleic Acids Res.* 41, D36–D42. doi: 10.1093/nar/gks1195

Bode, M., Woellhaf, M. W., Bohnert, M., van der Laan, M., Sommer, F., Jung, M., et al. (2015). Redox-regulated dynamic interplay between Cox19 and the copper-binding protein Cox11 in the intermembrane space of mitochondria facilitates biogenesis of cytochrome *c* oxidase. *Mol. Biol. Cell.* 26, 2385–2401. doi: 10.1091/mbc.e14-11-1526

Bologna, N. G., Mateos, J. L., Bresso, E. G., and Palatnik, J. F. (2009). A loop-to-base processing mechanism underlies the biogenesis of plant microRNAs miR319 and miR159. *EMBO J.* 28, 3646–3656. doi: 10.1038/emboj.2009.292

Bolte, S., and Cordelières, F. P. (2006). A guided tour into subcellular colocalization analysis in light microscopy. *J. Microsc.* 224, 213–232. doi: 10.1111/j.1365-2818.2006.01706.x

Bou Daher, F., Chebli, Y., and Geitmann, A. (2009). Optimization of conditions for germination of cold-stored *Arabidopsis thaliana* pollen. *Plant Cell Rep.* 28, 347–357. doi: 10.1007/s00299-008-0647-1

Carr, H. S., George, G. N., and Winge, D. R. (2002). Yeast Cox11, a protein essential for cytochrome *c* oxidase assembly, is a Cu(I)-binding protein. *J. Biol. Chem.* 277, 31237–31242. doi: 10.1074/jbc.M204854200

Carr, H. S., Maxfield, A. B., Horng, Y. C. C., and Winge, D. R. (2005). Functional analysis of the domains in Cox11. *J. Biol. Chem.* 280, 22664–22669. doi: 10.1074/jbc.M414077200

Claros, G. M., and Vincens, P. (1996). Computational method to predict mitochondrially imported proteins and their targeting sequences. *Eur. J. Biochem.* 241, 779–786. doi: 10.1111/j.1432-1033.1996.00779.x

Clough, S. J., and Bent, A. F. (1998). Floral dip: a simplified method for *Agrobacterium*-mediated transformation of *Arabidopsis thaliana*. *Plant J.* 16, 735–743. doi: 10.1046/j.1365-313x.1998.00343.x

Cohu, C. M., and Pilon, M. (2007). Regulation of superoxide dismutase expression by copper availability. *Physiol. Plant.* 129, 747–755. doi: 10.1111/j.1399-3054.2007.00879.x

Colin, M., Dorthu, M. P., Duby, F., Remacle, C., Dinant, M., Wolwertz, R. M., et al. (1995). Mutations affecting the mitochondrial genes encoding the cytochrome oxidase subunit I and apocytochrome b of *Chlamydomonas reinhardtii*. *Mol. Gen. Genet.* 249, 179–184. doi: 10.1007/BF00290364

Curtis, M. D., and Grossniklaus, U. (2003). A gateway cloning vector set for high-throughput functional analysis of genes in planta. *Plant Physiol.* 133, 462–469. doi: 10.1104/pp.103.027979

Dahan, J., Tcherkez, G., Macherel, D., Benamar, A., Belcram, K., Quadrado, M., et al. (2014). Disruption of the *CYTOCHROME C OXIDASE DEFICIENT1* gene leads to cytochrome *c* oxidase depletion and reorchestrated respiratory metabolism in *Arabidopsis*. *Plant Physiol.* 166, 1788–1802. doi: 10.1104/pp.114.248526

Emanuelsson, O., Nielsen, H., Brunak, S., and von Heijne G. (2000). Predicting subcellular localization of proteins based on their N-terminal amino acid sequence. *J. Mol. Biol.* 300, 1005–1016. doi: 10.1006/jmbi.2000.3903

Eubel, H., Heinemeyer, J., and Braun, H.-P. (2004). Identification and characterization of respirasomes in potato mitochondria. *Plant Physiol.* 134, 1450–1459. doi: 10.1104/pp.103.038018

Fujiki, Y., Fowler, S., Shio, H., Hubbard, A. L., and Lazarow, P. B. (1982). Polypeptide and phospholipid composition of the membrane of rat liver peroxisomes: comparison with endoplasmic reticulum and mitochondrial membranes. *J. Cell Biol.* 93, 103–110. doi: 10.1083/jcb.93.1.103

Gass, N., Glagotskaia, T., Mellema, S., Stuurman, J., Barone, M., Mandel, T., et al. (2005). Pyruvate decarboxylase provides growing pollen tubes with a competitive advantage in petunia. *Plant Cell* 17, 2355–2368. doi: 10.1105/tpc.105.033290

Gey, U., Czupalla, C., Hoflack, B., Krause, U., and Rödel, G. (2014). Proteomic analysis reveals a novel function of the kinase Sat4p in *Saccharomyces cerevisiae* mitochondria. *PLoS ONE* 9:e103956. doi: 10.1371/journal.pone.0103956

Gibala, M., Kicia, M., Sakamoto, W., Gola, E. M., Kubrakiewicz, J., Smakowska, E., et al. (2009). The lack of mitochondrial AtFtsH4 protease alters *Arabidopsis* leaf morphology at the late stage of rosette development under short day photoperiod. *Plant J.* 59, 685–699. doi: 10.1111/j.1365-313X.2009.03907.x

Gietz, D., and Schiestl, R. (2007). High-efficiency yeast transformation using the LiAc/SS carrier DNA/PEG method. *Nat. Protoc.* 2, 31–34. doi: 10.1038/nprot.2007.13

Heslop-Harrison, J., and Heslop-Harrison, Y. (1970). Evaluation of pollen viability by enzymatically induced fluorescence; intracellular hydrolysis of fluorescein diacetate. *Stain Technol.* 45, 115–120. doi: 10.3109/10520297009085351

Hiser, L., Valentin, M. D., Hamer, A. G., and Hosler, J. P. (2000). Cox11p is required for stable formation of the CuB and magnesium centcers of cytochrome *c* oxidase. *J. Biol. Chem.* 275, 619–623. doi: 10.1074/jbc.275.1.619

Horng, Y. C., Cobine, P. A., Maxfield, A. B., Carr, H. S., and Winge, D. R. (2004). Specific copper transfer from the Cox17 metallochaperone to both Sco1 and Cox11 in the assembly of yeast cytochrome *c* oxidase. *J. Biol. Chem.* 279, 35334–35340. doi: 10.1074/jbc.M404747200

Huang, S., Taylor, N. L., Ströher, E., Fenske, R., and Millar, A. H. (2013). Succinate dehydrogenase assembly factor 2 is needed for assembly and activity of mitochondrial complex II and for normal root elongation in *Arabidopsis*. *Plant J.* 73, 429–441. doi: 10.1111/tpj.12041

Keech, O., Dizengremel, P., and Gardeström, P. (2005). Preparation of leaf mitochondria from *Arabidopsis thaliana*. *Physiol. Plant.* 124, 403–409. doi: 10.1111/j.1399-3054.2005.00521.x

Khalimonchuk, O., Bird, A., and Winge, D. R. (2007). Evidence for a pro-oxidant intermediate in the assembly of cytochrome oxidase. *J. Biol. Chem.* 282, 17442–17449. doi: 10.1074/jbc.M702379200

Khalimonchuk, O., Ostermann, K., and Rödel, G. (2005). Evidence for the association of yeast mitochondrial ribosomes with Cox11p, a protein required for the Cu(B) site formation of cytochrome *c* oxidase. *Curr. Genet.* 47, 223–233. doi: 10.1007/s00294-005-0569-1

Khalimonchuk, O., and Rödel, G. (2005). Biogenesis of cytochrome *c* oxidase. *Mitochondrion* 5, 363–388. doi: 10.1016/j.mito.2005.08.002

Klodmann, J., Senkler, M., Rode, C., and Braun, H. P. P. (2011). Defining the protein complex proteome of plant mitochondria. *Plant Physiol.* 157, 587–598. doi: 10.1104/pp.111.182352

Kovářová, N., Čížková Vrbacká, A., Pecina, P., Stránecký, V., Pronicka, E., Kmoch, S., et al. (2012). Adaptation of respiratory chain biogenesis to cytochrome *c* oxidase deficiency caused by *SURF1* gene mutations. *Biochim. Biophys. Acta* 1822, 1114–1124. doi: 10.1016/j.bbadis.2012.03.007

Krogh, A., Larsson, B., von Heijne, G., and Sonnhammer, E. L. L. (2001). Predicting transmembrane protein topology with a hidden Markov model: application to complete genomes. *J. Mol. Biol.* 305, 567–580. doi: 10.1006/jmbi.2000.4315

Lamesch, P., Berardini, T. Z., Li, D., Swarbreck, D., Wilks, C., Sasidharan, R., et al. (2012). The *Arabidopsis* Information Resource (TAIR): improved gene annotation and new tools. *Nucleic Acids Res.* 40, D1202–D1210. doi: 10.1093/nar/gkr1090

Liang, Q., and Zhou, B. (2007). Copper and manganese induce yeast apoptosis via different pathways. *Mol. Biol. Cell* 18, 4741–4749. doi: 10.1091/mbc.E07-05-0431

Lode, A., Kuschel, M., Paret, C., and Rödel, G. (2000). Mitochondrial copper metabolism in yeast: interaction between Sco1p and Cox2p. *FEBS Lett.* 485, 19–24. doi: 10.1016/S0014-5793(00)02176-1

Luo, D., Xu, H., Liu, Z., Guo, J., Li, H., Chen, L., et al. (2013). A detrimental mitochondrial-nuclear interaction causes cytoplasmic male sterility in rice. *Nat. Genet.* 45, 573–577. doi: 10.1038/ng.2570

McElver, J., Tzafrir, I., Aux, G., Rogers, R., Ashby, C., Smith, K., et al. (2001). Insertional mutagenesis of genes required for seed development in *Arabidopsis thaliana*. *Genetics* 159, 1751–1763.

McGuffin, L. J., Bryson, K., and Jones, D. T. (2000). The PSIPRED protein structure prediction server. *Bioinformatics* 15, 404–405. doi: 10.1093/bioinformatics/16.4.404

McWilliam, H., Li, W., Uludag, M., Squizzato, S., Park, Y. M., Buso, N., et al. (2013). Analysis tool web services from the EMBL-EBI. *Nucleic Acids Res.* 41, W597–W600. doi: 10.1093/nar/gkt376

Merz, S., and Westermann, B. (2009). Genome-wide deletion mutant analysis reveals genes required for respiratory growth, mitochondrial genome maintenance and mitochondrial protein synthesis in *Saccharomyces cerevisiae*. *Genome Biol.* 10:R95. doi: 10.1186/gb-2009-10-9-r95

Nakagawa, T., Suzuki, T., Murata, S., Nakamura, S., Hino, T., Maeo, K., et al. (2007). Improved Gateway binary vectors: high-performance vectors for creation of fusion constructs in transgenic analysis of plants. *Biosci. Biotechnol. Biochem.* 71, 2095–2100. doi: 10.1271/bbb.70216

Nakai, T., Mera, Y., Yasuhara, T., and Ohashi, A. (1994). Divalent metal ion-dependent mitochondrial degradation of unassembled subunits 2 and 3 of cytochrome *c* oxidase. *J. Biochem.* 116, 752–758.

Nelson, B. K., Cai, X., and Nebenführ, A. (2007). A multicolored set of *in vivo* organelle markers for co-localization studies in *Arabidopsis* and other plants. *Plant J.* 51, 1126–1136. doi: 10.1111/j.1365-313X.2007.03212.x

Ni, M., Dehesh, K., Tepperman, J. M., and Quail, P. H. (1996). GT-2: *in vivo* transcriptional activation activity and definition of novel twin DNA binding domains with reciprocal target sequence selectivity. *Plant Cell* 8, 1041–1059. doi: 10.1105/tpc.8.6.1041

Nittis, T., George, G. N., and Winge, D. R. (2001). Yeast Sco1, a protein essential for cytochrome *c* oxidase function is a Cu (I)-binding protein. *J. Biol. Chem.* 276, 42520–42526. doi: 10.1074/jbc.M107077200

Ossowski, S., Schwab, R., and Weigel, D. (2008). Gene silencing in plants using artificial microRNAs and other small RNAs. *Plant J.* 53, 674–690. doi: 10.1111/j.1365-313X.2007.03328.x

Palm-Espling, M. E., Niemiec, M. S., and Wittung-Stafshede, P. (2012). Role of metal in folding and stability of copper proteins *in vitro*. *Biochim. Biophys. Acta* 1823, 1594–1603. doi: 10.1016/j.bbamcr.2012.01.013

Peñarrubia, L., Romero, P., Carrió-Seguí, A., Andrés-Bordería, A., Moreno, J., and Sanz, A. (2015). Temporal aspects of copper homeostasis and its crosstalk with hormones. *Front. Plant Sci.* 6:255. doi: 10.3389/fpls.2015.00255

Pogulis, R., Vallejo, A., and Larry, P. (1996). "*In vitro* recombination and mutagenesis by overlap extension PCR," in *In Vitro Mutagenesis Protocols*, ed M. Trower (Totowa, NJ: Humana Press Inc.), 167–176.

Prelich, G. (2012). Gene overexpression: uses, mechanisms, and interpretation. *Genetics* 190, 841–854. doi: 10.1534/genetics.111.136911

Schägger, H., and von Jagow, G. (1991). Blue native electrophoresis for isolation of membrane protein complexes in enzymatically active form. *Anal. Biochem.* 199, 223–231. doi: 10.1016/0003-2697(91)90094-A

Schindelin, J., Arganda-Carreras, I., Frise, E., Kaynig, V., Longair, M., Pietzsch, T., et al. (2012). Fiji: an open-source platform for biological-image analysis. *Nat. Methods* 9, 676–682. doi: 10.1038/nmeth.2019

Schulze, M., and Rödel, G. (1988). *SCO1*, a yeast nuclear gene essential for accumulation of mitochondrial cytochrome *c* oxidase subunit II. *Mol. Gen. Genet.* 211, 492–498. doi: 10.1007/BF00425706

Soto, I. C., Fontanesi, F., Liu, J., and Barrientos, A. (2012). Biogenesis and assembly of eukaryotic cytochrome *c* oxidase catalytic core. *Biochim. Biophys. Acta* 1817, 883–897. doi: 10.1016/j.bbabio.2011.09.005

Speranza, A., Crinelli, R., Scoccianti, V., and Geitmann, A. (2012). Reactive oxygen species are involved in pollen tube initiation in kiwifruit. *Plant Biol. (Stuttg.)* 14, 64–76. doi: 10.1111/j.1438-8677.2011.00479.x

Steinebrunner, I., Gey, U., Andres, M., Garcia, L., and Gonzalez, D. H. (2014). Divergent functions of the *Arabidopsis* mitochondrial SCO proteins: HCC1 is essential for COX activity while HCC2 is involved in the UV-B stress response. *Front. Plant Sci.* 5:87. doi: 10.3389/fpls.2014.00087

Steinebrunner, I., Landschreiber, M., Krause-Buchholz, U., Teichmann, J., and Rödel, G. (2011). HCC1, the *Arabidopsis* homologue of the yeast mitochondrial copper chaperone SCO1, is essential for embryonic development. *J. Exp. Bot.* 62, 319–330. doi: 10.1093/jxb/erq269

Taiz, L., and Zeiger, E. (2010). *Plant physiology*. Sunderland: Sinauer Associates.

Tan, Y. F., O'Toole, N., Taylor, N. L., and Millar, A. H. (2010). Divalent metal ions in plant mitochondria and their role in interactions with proteins and oxidative stress-induced damage to respiratory function. *Plant Physiol.* 152, 747–761. doi: 10.1104/pp.109.147942

Thompson, A. K., Smith, D., Gray, J., Carr, H. S., Liu, A., Winge, D. R., et al. (2010). Mutagenic analysis of Cox11 of *Rhodobacter sphaeroides*: insights into the assembly of Cu(B) of cytochrome *c* oxidase. *Biochemistry* 49, 5651–5661. doi: 10.1021/bi1003876

Tzagoloff, A., Capitanio, N., Nobrega, M. P., and Gatti, D. (1990). Cytochrome oxidase assembly in yeast requires the product of COX11, a homolog of the *P. denitrificans* protein encoded by ORF3. *EMBO J.* 9, 2759–2764.

Van Aken, O., Pecenková, T., van de Cotte, B., De Rycke, R., Eeckhout, D., Fromm, H., et al. (2007). Mitochondrial type I prohibitins of *Arabidopsis thaliana* are required for supporting proficient meristem development. *Plant J.* 52, 850–864. doi: 10.1111/j.1365-313X.2007.03276.x

Weitbrecht, K., Müller, K., and Leubner-Metzger, G. (2011). First off the mark: early seed germination. *J. Exp. Bot.* 62, 3289–3309. doi: 10.1093/jxb/err030

Welchen, E., Chan, R. L., and Gonzalez, D. H. (2004). The promoter of the *Arabidopsis* nuclear gene *COX5b-1*, encoding subunit 5b of the mitochondrial cytochrome *c* oxidase, directs tissue-specific expression by a combination of positive and negative regulatory elements. *J. Exp. Bot.* 55, 1997–2004. doi: 10.1093/jxb/erh223

Welchen, E., and Gonzalez, D. H. (2005). Differential expression of the *Arabidopsis* cytochrome *c* genes *Cytc-1* and *Cytc-2.* evidence for the involvement of TCP-domain protein-binding elements in anther- and meristem-specific expression of the *Cytc-1* gene. *Plant Physiol.* 139, 88–100. doi: 10.1104/pp.105.065920

Welchen, E., Hildebrandt, T. M., Lewejohann, D., Gonzalez, D. H., and Braun, H. P. (2012). Lack of cytochrome *c* in *Arabidopsis* decreases stability of Complex IV and modifies redox metabolism without affecting Complexes I and III. *Biochim. Biophys. Acta* 1817, 990–1001. doi: 10.1016/j.bbabio.2012.04.008

Wigge, B., and Gardeström, P. (1987). "The effects of different ionic-conditions on the activity of cytochrome *c*-oxidase in purified plant mitochondria," in *Plant Mitochondria,* eds A. L. Moore and R. B. Beechey (New York, NY: Springer Science+Business Media), 127–130.

Wintz, H., Fox, T., Wu, Y.-Y., Feng, V., Chen, W., Chang, H.-S., et al. (2003). Expression profiles of *Arabidopsis thaliana* in mineral deficiencies reveal novel transporters involved in metal homeostasis. *J. Biol. Chem.* 278, 47644–47653. doi: 10.1074/jbc.M309338200

Wolf, C., Hennig, M., Romanovicz, D., and Steinebrunner, I. (2007). Developmental defects and seedling lethality in apyrase *AtAPY1* and *AtAPY2* double knockout mutants. *Plant Mol. Biol.* 64, 657–672. doi: 10.1007/s11103-007-9184-5

Yamada, K., Lim, J., Dale, J. M., Chen, H., Shinn, P., Palm, C. J., et al. (2003). Empirical analysis of transcriptional activity in the *Arabidopsis genome. Science* 302, 842–846. doi: 10.1126/science.1088305

Yang, L., Zhang, J., He, J., Qin, Y., Hua, D., Duan, Y., et al. (2014). ABA-Mediated ROS in mitochondria regulate root meristem activity by controlling *PLETHORA* expression in *Arabidopsis. PLoS Genet.* 10:e1004791. doi: 10.1371/journal.pgen.1004791

Yuan, H., and Liu, D. (2012). Functional disruption of the pentatricopeptide protein SLG1 affects mitochondrial RNA editing, plant development, and responses to abiotic stresses in *Arabidopsis. Plant J.* 70, 432–444. doi: 10.1111/j.1365-313X.2011.04883.x

Conflict of Interest Statement: The authors declare that the research was conducted in the absence of any commercial or financial relationships that could be construed as a potential conflict of interest.

6

Behavior of Leaf Meristems and Their Modification

Yasunori Ichihashi[1]* *and Hirokazu Tsukaya*[2,3]

[1] *RIKEN Center for Sustainable Resource Science, Yokohama, Japan,* [2] *Department of Biological Sciences, Graduate School of Science, The University of Tokyo, Tokyo, Japan,* [3] *Bio-Next Project, Okazaki Institute for Integrative Bioscience, National Institutes of Natural Sciences, Okazaki, Japan*

A major source of diversity in flowering plant form is the extensive variability of leaf shape and size. Leaf formation is initiated by recruitment of a handful of cells flanking the shoot apical meristem (SAM) to develop into a complex three-dimensional structure. Leaf organogenesis depends on activities of several distinct meristems that are established and spatiotemporally differentiated after the initiation of leaf primordia. Here, we review recent findings in the gene regulatory networks that orchestrate leaf meristem activities in a model plant *Arabidopsis thaliana*. We then discuss recent key studies investigating the natural variation in leaf morphology to understand how the gene regulatory networks modulate leaf meristems to yield a substantial diversity of leaf forms during the course of evolution.

Keywords: *Arabidopsis thaliana*, gene regulatory network, leaf development, leaf meristems, natural variation

Edited by:
Simon Gilroy,
University of Wisconsin - Madison, USA

Reviewed by:
Javier Palatnik,
Instituto de Biologia Molecular y Celular de Rosario, Argentina
Michael Lenhard,
Universität Potsdam, Germany

***Correspondence:**
Yasunori Ichihashi
yasunori.ichihashi@riken.jp

INTRODUCTION

A leaf is a flat lateral organ of the stem, and grows along three-dimensional axes: proximal-distal, medial-lateral, and adaxial-abaxial axes (Steeves and Sussex, 1989). In general, the leaf consists of a leaf blade and a leaf petiole, and the leaf is composed of different cell types including epidermal cells, palisade cells, spongy mesophyll cells, and xylem/phloem cells (Esau, 1977). Because cell proliferation and cell differentiation occur concurrently during leaf development (Donnelly et al., 1999), a single leaf maintains cells in different developmental stages such as mitotic cells, differentiating cells, and endoreduplicating cells. Therefore, an elaborated spatiotemporal regulation of organ and cellular morphologies should underpin the leaf formation.

Plants acquire the bulk of their energy from light capture by leaves, and for this reason the leaf is specialized for photosynthesis, respiration, and photoperception. Leaf shape has direct consequences on the efficiency of light capture, photosynthetic carbon fixation, and gas exchange (Nicotra et al., 2011; Chitwood et al., 2012). As a result, leaf morphology must be optimized in response to variations in environmental conditions. In addition, plant leaves are equipped with an array of structural, chemical and protein-based defenses against herbivores and pathogens, which often target leaves (Agrios, 2005). These multiple functions are accomplished by the heterologous organ and cellular morphologies in a single leaf.

Due to their sessile lifestyle, plants exhibit a variety of morphological and physiological leaf traits that have allowed adaptation to different natural habitats. Indeed, leaf structural traits such as shape, size, and venation pattern, and physiological traits such as photosynthetic mechanisms are diversified in angiosperms (Flood et al., 2011; Sack et al., 2012; Tsukaya, 2014a). There are pressing research questions regarding leaf formation and variation: How do cellular activities cause leaf formation? What are the gene regulatory networks controlling leaf development? How were the gene regulatory networks altered during evolution? In this review, we highlight recent findings on the activities of leaf meristem and their gene regulatory networks in the model plant

Arabidopsis thaliana. Knowledge gained from studies of *A. thaliana* has facilitated evolutionary developmental studies of leaf morphology, and we discuss recent investigations of the natural variation in leaf morphology.

LEAF MERISTEMS

What Are Leaf Meristems?

Plants have three major organs: leaves, stems, and roots. Stems and roots are directly derived from the shoot apical meristem (SAM) and the root apical meristem (RAM), respectively. The SAM and RAM maintain stem cells and exhibit indeterminate growth, which is an open-ended growth plan. On the other hand, leaves exhibit determinate growth, which is growth with a finite period of development. Meristems are defined, in a broad sense, as proliferating tissues regardless of presence of self-renewing stem cells, although the meristems have been controversial concept: Most molecular developmental biologists narrowly adopt a definition of meristems as proliferating tissues that maintain self-renewing stem cells, while meristematic tissues in leaves or stems are excluded by this definition at present because no stem cells have been observed in these tissues. (Esau, 1977; Tsukaya, 2014b). Notably WOX genes, which are key for sustaining stem cells both in SAM and RAM, are also important for the meristemactic activities in leaf primordia (Nardmann and Werr, 2013). The proliferative activity in leaf primordia is much stronger than that in the SAM, but cells that make a leaf come from restricted area of the primordium in angiosperms. Cell differentiation occurred subsequent to cell division makes the proliferative region in leaf separated spatially from SAM (**Figure 1A**). Therefore, leaf meristems producing leaf mesophyll cells as well as initial cells of stomata and veins are classified into intercalary meristems, that are meristematic tissues reside in a differentiating organ. Botanist Katherine Esau described in her textbook Plant Anatomy that a series of organogenesis steps in the leaf primordium depends on several distinct meristematic tissues including the plate meristem and the marginal meristem (Esau, 1977). The plate meristem consists of parallel layers of cells dividing anticlinally to play a major role in leaf growth. The marginal meristem, which is located at the edge of the leaf between the adaxial and abaxial surfaces, contributes to the establishment of tissue layers within the leaf.

These leaf meristems coordinate the orientation of the cell division plate, produce the main compartments of the leaf (the leaf blade and leaf petiole), and determine the proximal-distal gradient where the switch from cell proliferation to cell differentiation occurs in developing leaves (Donnelly et al., 1999; Ichihashi et al., 2011). This boundary is known as the cell cycle arrest front, and was assumed to progress with basipetal polarity, from the tip to the base of the leaf blade (Donnelly et al., 1999; Nath et al., 2003; White, 2006). However, recent detailed characterizations of the kinetics of cell division during leaf development have indicated that the cell-cycle arrest front does not progress gradually, but rather remains at an almost constant position, and is abolished abruptly (Kazama et al., 2010; Andriankaja et al., 2012; **Figure 1A**). Like the SAM or RAM, the leaf meristematic region maintains a constant size, but unlike the

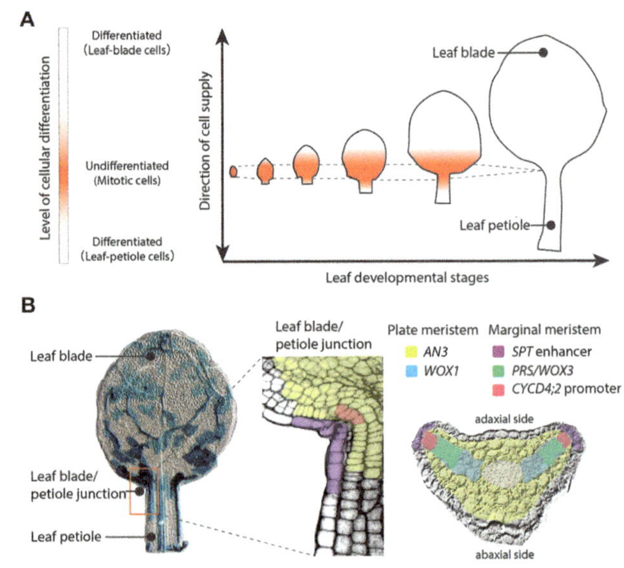

FIGURE 1 | Leaf meristems of *Arabidopsis thaliana*. (A) Leaf developmental stages showing the proliferative region (red). This meristematic region localizes at the leaf blade/petiole junction and produces both leaf-blade and leaf-petiole cells in a bidirectional manner. The region maintains a constant size over a limited time period. **(B)** Left image: A leaf primordium at 7 days after sowing with cell lineages indicated by blue staining (sectors were induced at 4 days after sowing). The middle and right images indicate the spatial differentiation of leaf meristems. The plate meristem is marked by *AN3* and *WOX1* gene expression domains and the marginal meristem is marked by the *SPT* enhancer, *PRS/WOX3*, and the promoter of *CYCD4;2*.

apical meristems, cell division ceases in the leaf after a certain time period. This leaf meristematic region is localized at the junction between the leaf blade and leaf petiole, and produces both the leaf-blade and leaf-petiole cells via cell divisions in a bidirectional manner (Ichihashi et al., 2011). In addition, leaf meristematic activity differs between tissue layers, and cell divisions directly related to the formation of veins and stomata occur throughout the period of leaf development (Donnelly et al., 1999; White, 2006; Ichihashi et al., 2011). Taken together, these studies reveal that leaves maintain their own meristems, and that the tightly controlled activity of these meristems directs the complex process of leaf tissue development.

Regulatory Mechanisms of Leaf Meristems

Molecular markers have been used to identify distinct regions within the leaf proliferative region of *A. thaliana*. (**Figure 1B**). The *ANGUSTIFOLIA3* (*AN3*) gene promoter is active in mesophyll cells just above the leaf blade/petiole junction within the leaf proliferating region (Horiguchi et al., 2005; Ichihashi et al., 2011; Kawade et al., 2013). *AN3* encodes a putative transcriptional coactivator homologous to human synovial sarcoma translocation protein, and is a positive regulator of cell proliferation in the leaf blade and leaf petiole (Kim and Kende, 2004; Horiguchi et al., 2005; Ichihashi et al., 2011). *AN3* transcripts accumulate only in mesophyll cells, but the AN3 protein moves across different leaf layers to coordinate proliferation between clonally independent leaf cells (Kawade et al., 2013). Although the exact spatiotemporal distribution of AN3 protein has to be characterized, *AN3* could mark the position of the plate meristem in leaf primordia. On the other hand, an enhancer trap line with T-DNA insertion in the 5′ region of *SPATULA* (*SPT*), along with other studies of the *SPT* promoter, show that *SPT* is expressed at the margin of the proliferative region in leaf primordia (Groszmann et al., 2010; Ichihashi et al., 2010). The promoter of a D-type cyclin gene, *CYCD4;2*, is active in a small number of cells directly adjacent to the marginal cells of the leaf primordia. Although the actual expression pattern of *CYCD4;2* is not identical to that observed in *pCYCD4;2* promoter studies (Kono et al., 2007), a specific *cis*-element seems to allow expression in this small cell population. *SPT* limits the size of the leaf proliferative region independently of *AN3* activity (Ichihashi et al., 2010), and overexpression of *CYCD4;2* promotes cell proliferation in leaves (Kono et al., 2007). Therefore, it appears that the promoter activities of *SPT* and *CYCD4;2* mark the position of the marginal meristem in leaf primordia. In addition, two WUSCHEL-RELATED HOMEOBOX (WOX) genes, *PRESSED FLOWER* (*PRS*)/*WOX3* and *WOX1*, which encode homeobox transcription factors, also redundantly promote leaf blade outgrowth (Nakata et al., 2012). *PRS/WOX3* is expressed mainly near the leaf margin (Nardmann et al., 2004) and may play a role in marginal meristem activity. *WOX1* is expressed within the two middle mesophyll layers located exactly between the adaxial and abaxial sides of the leaf blade (Nakata et al., 2012), and might be involved in plate meristem activity. Thus, leaf development depends on multiple leaf meristem activities with local controls of gene expression.

In addition to the local regulation factors, organ-level regulation of leaf meristem activity has also been identified in *A. thaliana*. *KLUH* (*KLU*) encodes the cytochrome P450 enzyme CYP78A5, which promotes organ growth, including growth of leaves, in a non-cell autonomous manner (Anastasiou et al., 2007; Adamski et al., 2009; Eriksson et al., 2010). This suggests that *KLU* is involved in generating a mobile growth factor. Computer simulation predicts that the *KLU*-dependent mobile growth factor might have less permeability or be regulated at the physical/biochemical level (Kazama et al., 2010). Computational modeling and time-lapse clonal analyses suggest that growth orientations are specified by a tissue polarity system that changes during leaf development, and that a basic pattern of growth rates across the leaf is established from an early developmental stage (Kuchen et al., 2012). Thus, organ-level regulation coordinates growth patterns at the cellular level to form leaf shape. Taken together, the studies discussed above indicate that leaf meristem activities are tightly controlled by both the local regulation systems in the plate and marginal meristem, and by mobile growth factor and tissue polarity information that functions at the organ level.

GENE REGULATORY NETWORKS OF LEAF MERISTEMS

Genes Functioning in Cell Proliferation

A number of genes responsible for cell proliferation in leaf primordia have been identified in studies of *A. thaliana* mutants (Gonzalez et al., 2012; Kalve et al., 2014; **Figure 2A**). As previously mentioned, *AN3* functions at the plate meristem to produce cells of both the leaf blade and the leaf petiole (Kim and Kende, 2004; Horiguchi et al., 2005; Ichihashi et al., 2011; Kawade et al., 2013). AN3 shows protein-protein interaction with GROWTHREGULATING FACTOR5 (GRF5) to promote cell proliferation (Horiguchi et al., 2005). *AN3* is also known as *GRF-INTERACTING FACTOR1* (*GIF1*), and other members of the *GIF* family, *GIF2* and *GIF3*, also promote cell proliferation in a redundant fashion (Lee et al., 2009). AN3 binds to the SWITCH/SUCROSE NONFERMENTING (SWI/SNF) chromatin remodeling complexes to regulate transcription during leaf development (Vercruyssen et al., 2014). *AN3* is also involved in the establishment of leaf identity in cotyledons via the repression of root fate during embryogenesis (Kanei et al., 2012). On the other hands *KLU* is expressed in the basal region of leaf primordia and generates a mobile growth factor (Anastasiou et al., 2007). *PRS/WOX3* and *WOX1* are also classified as activators of cell proliferation (Nakata et al., 2012). The auxin inducible gene *AUXIN-REGULATED GENE INVOLVED IN ORGAN SIZE* (*ARGOS*) increases the expression level of the D-type cyclin *CYCD3;1* gene through the regulation of the *AINTEGUMENTA* genes (Krizek, 1999; Mizukami and Fischer, 2000; Hu et al., 2003; Nole-Wilson et al., 2005). APC10 and CDC27a are subunits of the anaphase-promoting complex/cyclosome (APC/C), which functions at the G2 to M

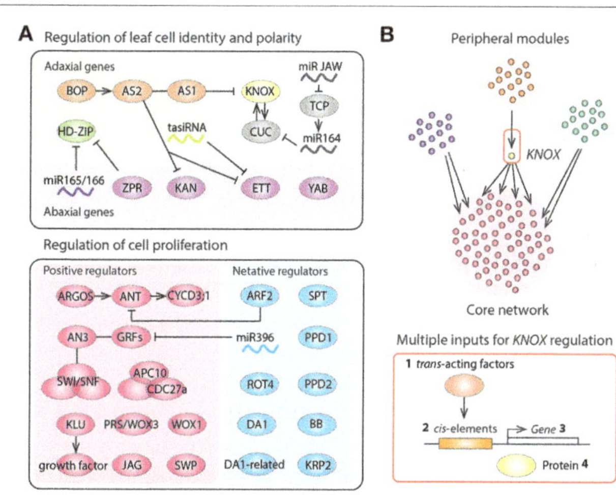

FIGURE 2 | Gene regulatory networks of leaf development.
(A) Regulators of leaf structural identification and leaf cell proliferation in *Arabidopsis thaliana*. Arrows, T bars, and lines indicate positive regulation, negative regulation, and protein-protein interactions, respectively.
(B) Schematic diagram representing the gene regulatory networks controlling tomato leaf development, which consists of several peripheral gene network modules and a core network having highly interconnected genes. *KNOX* appears as a bottleneck in the network, suggesting that *KNOX* was an evolutionary hot spot that was repeatedly recruited for generating natural variation in leaf shape. *KNOX* regulation occurs at multiple levels including (1) modulation of trans-acting factors regulating *KNOX* (Ichihashi et al., 2014), (2) promoter changes at *KNOX* (Piazza et al., 2010), (3) changes in *KNOX* expression patterns (Bharathan et al., 2002), and (4) changes in effective KNOX protein concentration (Kimura et al., 2008).

transition of the cell cycle and is also reported to regulate leaf cell proliferation (Rojas et al., 2009; Eloy et al., 2011). In addition, the C2H2 zinc finger protein JAGGED (JAG) and a subunit of the Mediator complex STRUWWELPETER (SWP) are also constituent factors that positively control cell proliferation in leaves (Autran et al., 2002; Ohno et al., 2004). All of these genes function mainly in the control of lateral organ growth and not in the SAM and RAM. Therefore, a specialized set of genes is utilized to maintain leaf meristem activities.

Negative regulators of cell proliferation are important for conferring determinate growth in the leaves. As previously mentioned, *SPT* is expressed in the marginal meristem of leaf primordia to restrict the size of the leaf proliferative region (Ichihashi et al., 2010). Given that *SPT* may also help to restrict the size of the RAM (Makkena and Lamb, 2013), *SPT* might play a general role in the control of meristematic sizes in roots and leaves. MicroRNA396 represses cell proliferation through regulation of the GRF family (Jones-Rhoades and Bartel, 2004; Liu et al., 2009; Rodriguez et al., 2010). The AUXIN RESPONSE FACTOR2 (ARF2) is a repressor of auxin signaling that represses *ANT* gene expression to inhibit cell proliferation (Horiguchi et al., 2006; Schruff et al., 2006; Lim et al., 2010). The short polypeptide ROTUNDIFOLIA4 (ROT4) also functions to repress cell proliferation, especially along the proximal-distal axis (Narita et al., 2004; Ikeuchi et al., 2011; Guo et al., 2015). Therefore, a microRNA, auxin and a short peptide, all of which can potentially move across cell layers, underlie the delicate mechanisms needed to shut down leaf meristem activities. In addition, the RING-finger protein BIG BROTHER (BB); two putative ubiquitin receptors, DA and DA-RELATED (DAR); two TIFY-type transcription factors, PEAPOD1 and PEAPOD2; and the cyclin-dependent kinase inhibitor gene KIP-RELATED PROTEIN2 (KRP2) are also known to negatively regulate leaf cell proliferation (De Veylder et al., 2001; Disch et al., 2006; White, 2006; Ferjani et al., 2007; Li et al., 2008).

Genes Functioning in Leaf Cell Identity and Polarity

Several genes playing a role in the coordination of the above cell proliferation regulators have been identified in *A. thaliana* (**Figure 2A**). The *BLADE-ON-PETIOLE* (*BOP*) genes and their direct target *ASYMMETRIC LEAVES2* (*AS2*), which interacts with *AS1*, are involved in the recruitment of the leaf founder cells from the SAM and in the establishment of all three-dimensional axes of the leaf (Semiarti et al., 2001; Iwakawa et al., 2002, 2007; Ha et al., 2003, 2004, 2007, 2010; Xu et al., 2003; Hepworth et al., 2005; Norberg et al., 2005; Zgurski et al., 2005; Fu et al., 2007; Ikezaki et al., 2010; Jun et al., 2010; Ichihashi et al., 2011; Kojima et al., 2011; Ishibashi et al., 2012; Chen et al., 2013). *BOP* and *AS1/2* repress the expression of class I *KNOTTED-like homeobox* (*KNOX*) genes, which help maintain the indeterminate growth of the SAM (Semiarti et al., 2001; Byrne et al., 2002; Ha et al., 2003; Lin et al., 2003; Phelps-Durr et al., 2005; Guo et al., 2008). Interestingly, the chromatin-remodeling protein HIRA and the Polycomb-repressive complex2 interact with AS1/2 to bring about *KNOX* gene silencing (Phelps-Durr et al., 2005; Ueno et al., 2007; Guo et al., 2008; Lodha et al., 2013). In addition, the microRNA JAW regulates the class II TEOSINTE BRANCHED1/CYCLOIDEA/PCF (TCP) genes, which are heterochronic regulators of the leaf maturation schedule and determine the developmental window for organogenesis (Nath et al., 2003; Palatnik et al., 2003; Efroni et al., 2008). Like AN3, the TCPs also interact with SWI/SNF chromatin remodeling complex components (Efroni et al., 2013), suggesting that dynamic reorganization of chromatin architecture might play an important role throughout leaf development.

An array of genes responsible for the regulation of the adaxial-abaxial axis has been identified. These genes competitively regulate of adaxial and abaxial identity, required for flat outgrowth of the lamina. The regulators of adaxial identity are *BOP*, *AS1/2*, a family of class III HOMEODOMAIN LEUCINE ZIPPER (HD-ZIP) transcription factors, and a trans-acting small-interfering RNA (*tasiRNA*) (McConnell et al., 2001; Garcia et al., 2006). The regulators of abaxial identity are microRNA165/166 and *LITTLE ZIPPER* (*ZPR*), which both repress *HD-ZIP* (Mallory et al., 2004; Wenkel et al., 2007); *KANADI* (*KAN*), which is repressed by *AS2* (Wu et al., 2008); *ETTIN/ARF3*, which is repressed by *AS2* and tasiRNA (Garcia et al., 2006; Iwasaki et al., 2013; Takahashi et al., 2013). YABBY (YAB) genes also interact with the abaxial identity systems, but they are essential in switching from the SAM program to the leaf-specific program (Sawa et al., 1999; Siegfried et al., 1999; Sarojam et al., 2010). Moreover, various metabolites, ribosomal proteins, and plastid signals have been reported to affect the establishment of the leaf adaxial-abaxial axis (Pinon et al., 2008; Yao et al., 2008; Horiguchi et al., 2011; Toyokura et al., 2011, 2015; Tameshige et al., 2013). Thus, housekeeping genes working in basic metabolic and cellular functions might play specific roles in leaf development (Tsukaya et al., 2013).

In addition to three-dimensional axes of whole leaf, additional growth axes are formed to develop leaf serrations in case of *Arabidopsis thaliana*. Auxin maxima along leaf margins are required for the outgrowth of serrations, and automatically formed through the activity of auxin efflux carrier *PIN-FORMED1* (Kawamura et al., 2010; Bilsborough et al., 2011). CUP-SHAPED COTYLEDON (CUC) genes, which are generally required for boundary formation between two organs, are also key players in the serration formation (Nikovics et al., 2006). *CUC2* is essential for robustly positioning and indenting individual serrations (Bilsborough et al., 2011). *TCP* fine-tunes the expressional patterns of *CUC* through the regulation of microRNA164 to shape the serrations (Nikovics et al., 2006; Koyama et al., 2007, 2010; Kawamura et al., 2010).

LEAF MORPHOLOGICAL EVOLUTION

Despite of the diversity in leaf morphology, the molecular mechanisms that give rise to developmental variation are incompletely understood. Evolutionary developmental biology (evo-devo) studies of plants and animals have revealed the importance of gene regulation in determining developmental variation (Blein et al., 2008; Kimura et al., 2008; Rebeiz et al., 2009; Yamaguchi et al., 2010; Loehlin and Werren, 2012). This suggests that the rewiring of developmental gene regulatory networks is a crucial causal factor driving morphological evolution (Peter and Davidson, 2011). Current evo-devo studies of plant morphology suggest that *KNOX* expression was recruited repeatedly to generate natural variation in leaf shape in several plant lineages (Bharathan et al., 2002; Kimura et al., 2008; Hay and Tsiantis, 2010; Piazza et al., 2010; Nakayama et al., 2012, 2014). Changes in the activity of other homeobox genes REDUCED COMPLEXITY also cause the repeated evolutionary modification of leaf morphology in Brassicaceae (Sicard et al., 2014; Vlad et al., 2014), suggesting that the regulation of homeobox genes was recruited repeatedly to influence leaf diversity, similar to homeobox genes contributing to animal body plan evolution (Pick and Heffer, 2012).

As we have seen, a number of leaf development genes have been identified, but the exact interactions between these genes has not been fully elucidated. The development of new genomic tools has enabled the generation of large datasets, which can be used to determine exactly how developmental gene modules are organized into a network hierarchy (Fischer and Smith, 2012). Genome-wide gene expression analyses have been performed for several plant species, to reveal the dynamic changes in gene expression during leaf development (Beemster et al., 2005; Li et al., 2010; Andriankaja et al., 2012; Ichihashi et al., 2014; Palmer et al., 2015). Ichihashi et al. (2014) used cross-species, tissue-specific, and large-scale RNA-seq data to reveal the gene regulatory networks controlling leaf development in the domesticated tomato and its wild relatives. Comparisons of the gene networks among species showed that a module regulating KNOX at the protein level had significant differences across species in a manner correlating with leaf complexity (Kimura et al., 2008; Macalister et al., 2012; Ichihashi et al., 2014). Interestingly, *KNOX* serves as a bridge connecting a peripheral gene network module to the core network that includes leaf cell proliferation regulators (Ichihashi et al., 2014). Therefore, it appears that *KNOX* is repeatedly co-opted to generate plant morphological diversity by virtue of its bottleneck location in the gene regulatory network (**Figure 2B**).

CONCLUSION

The leaf is a complex three-dimensional photochemical reactor whose form and function are determined by gene regulatory networks. Recent technical advances are being used to unravel the mysteries of the molecular mechanisms behind leaf development and evolution. For example, gene expression studies using tissue-specific promoters will further reveal the detailed functions of leaf meristems. The key leaf development genes *KNOX*, *TCP*, and *AN3* are involved in epigenetic regulation (Phelps-Durr et al., 2005; Ueno et al., 2007; Guo et al., 2008; Efroni et al., 2013; Lodha et al., 2013; Vercruyssen et al., 2014). Therefore, next-generation sequencing will be useful for characterizing the genome-wide changes in chromatin structure, DNA methylation, and histone modifications during leaf development and between plant species. Novel leaf structures such as the pitcher leaves of carnivorous plants and the unifacial leaves of monocots, are generated through tissue-specific changes in cell division (Yamaguchi et al., 2010; Fukushima et al., 2015). Future studies of leaf meristem activity in determining leaf shape will undoubtedly provide greater insights into the molecular mechanisms behind the substantial diversity of leaf forms in nature.

ACKNOWLEDGMENTS

We apologize to colleagues whose work could not be highlighted because of space limitations. This work was partially supported by the RIKEN Special Postdoctoral Researchers Program; Grant-in-Aid for Young Scientists (B; grant no. 15K18589) from the Ministry of Education, Culture, Sports, Science and Technology, Japan to YI; and by the Japan Society for the Promotion of Science (Grants-in-Aid for Creative Scientific Research and Scientific Research A), the Ministry of Education, Culture, Sports, Science and Technology, Japan (Scientific Research on Priority Areas and Scientific Research on Innovative Areas) to HT.

REFERENCES

Adamski, N. M., Anastasiou, E., Eriksson, S., O'Neill, C. M., and Lenhard, M. (2009). Local maternal control of seed size by KLUH/CYP78A5-dependent growth signaling. *Proc. Natl. Acad. Sci. U.S.A.* 106, 20115–20120. doi: 10.1073/pnas.0907024106

Agrios, G. N. (2005). *Plant Pathology*. Burlington, VT: Elsevier Academic Press.

Anastasiou, E., Kenz, S., Gerstung, M., MacLean, D., Timmer, J., Fleck, C., et al. (2007). Control of plant organ size by KLUH/CYP78A5-dependent intercellular signaling. *Dev. Cell* 13, 843–856. doi: 10.1016/j.devcel.2007.10.001

Andriankaja, M., Dhondt, S., De Bodt, S., Vanhaeren, H., Coppens, F., De Milde, L., et al. (2012). Exit from proliferation during leaf development in *Arabidopsis thaliana*: a not-so-gradual process. *Dev. Cell* 22, 64–78. doi: 10.1016/j.devcel.2011.11.011

Autran, D., Jonak, C., Belcram, K., Beemster, G. T., Kronenberger, J., Grandjean, O., et al. (2002). Cell numbers and leaf development in Arabidopsis: a functional analysis of the STRUWWELPETER gene. *EMBO J.* 21, 6036–6049. doi: 10.1093/emboj/cdf614

Beemster, G. T., De Veylder, L., Vercruysse, S., West, G., Rombaut, D., Van Hummelen, P., et al. (2005). Genome-wide analysis of gene expression profiles associated with cell cycle transitions in growing organs of Arabidopsis. *Plant Physiol.* 138, 734–743. doi: 10.1104/pp.104.053884

Bharathan, G., Goliber, T. E., Moore, C., Kessler, S., Pham, T., and Sinha, N. R. (2002). Homologies in leaf form inferred from KNOXI gene expression during development. *Science* 296, 1858–1860. doi: 10.1126/science.1070343

Bilsborough, G. D., Runions, A., Barkoulas, M., Jenkins, H. W., Hasson, A., Galinha, C., et al. (2011). Model for the regulation of *Arabidopsis thaliana* leaf margin development. *Proc. Natl. Acad. Sci. U.S.A.* 108, 3424–3429. doi: 10.1073/pnas.1015162108

Blein, T., Pulido, A., Vialette-Guiraud, A., Nikovics, K., Morin, H., Hay, A., et al. (2008). A conserved molecular framework for compound leaf development. *Science* 322, 1835–1839. doi: 10.1126/science.1166168

Byrne, M. E., Simorowski, J., and Martienssen, R. A. (2002). ASYMMETRIC LEAVES1 reveals knox gene redundancy in Arabidopsis. *Development* 129, 1957–1965.

Chen, X., Wang, H., Li, J., Huang, H., and Xu, L. (2013). Quantitative control of ASYMMETRIC LEAVES2 expression is critical for leaf axial patterning in Arabidopsis. *J. Exp. Bot.* 64, 4895–4905. doi: 10.1093/jxb/ert278

Chitwood, D. H., Headland, L. R., Filiault, D. L., Kumar, R., Jiménez-Gómez, J. M., Schrager, A. V., et al. (2012). Native environment modulates leaf size and response to simulated foliar shade across wild tomato species. *PLoS ONE* 7:e29570. doi: 10.1371/journal.pone.0029570

De Veylder, L., Beeckman, T., Beemster, G. T., Krols, L., Terras, F., Landrieu, I., et al. (2001). Functional analysis of cyclin-dependent kinase inhibitors of Arabidopsis. *Plant Cell* 13, 1653–1668. doi: 10.1105/tpc.13.7.1653

Disch, S., Anastasiou, E., Sharma, V. K., Laux, T., Fletcher, J. C., and Lenhard, M. (2006). The E3 ubiquitin ligase BIG BROTHER controls arabidopsis organ size in a dosage-dependent manner. *Curr. Biol.* 16, 272–279. doi: 10.1016/j.cub.2005.12.026

Donnelly, P. M., Bonetta, D., Tsukaya, H., Dengler, R. E., and Dengler, N. G. (1999). Cell cycling and cell enlargement in developing leaves of Arabidopsis. *Dev. Biol.* 215, 407–419. doi: 10.1006/dbio.1999.9443

Efroni, I., Blum, E., Goldshmidt, A., and Eshed, Y. (2008). A protracted and dynamic maturation schedule underlies Arabidopsis leaf development. *Plant Cell* 20, 2293–2306. doi: 10.1105/tpc.107.057521

Efroni, I., Han, S. K., Kim, H. J., Wu, M. F., Steiner, E., Birnbaum, K. D., et al. (2013). Regulation of leaf maturation by chromatin-mediated modulation of cytokinin responses. *Dev. Cell* 24, 438–445. doi: 10.1016/j.devcel.2013.01.019

Eloy, N. B., de Freitas Lima, M., Van Damme, D., Vanhaeren, H., Gonzalez, N., De Milde, L., et al. (2011). The APC/C subunit 10 plays an essential role in cell proliferation during leaf development. *Plant J.* 68, 351–363. doi: 10.1111/j.1365-313X.2011.04691.x

Eriksson, S., Stransfeld, L., Adamski, N. M., Breuninger, H., and Lenhard, M. (2010). KLUH/CYP78A5-dependent growth signaling coordinates floral organ growth in Arabidopsis. *Curr. Biol.* 20, 527–532. doi: 10.1016/j.cub.2010.01.039

Esau, K. (1977). *Anatomy of Seed Plants.* New York, NY: Wiley.

Ferjani, A., Horiguchi, G., Yano, S., and Tsukaya, H. (2007). Analysis of leaf development in fugu mutants of Arabidopsis reveals three compensation modes that modulate cell expansion in determinate organs. *Plant Physiol.* 144, 988–999. doi: 10.1104/pp.107.099325

Fischer, A. H., and Smith, J. (2012). Evo-devo in the era of gene regulatory networks. *Integr. Comp. Biol.* 52, 842–849. doi: 10.1093/icb/ics112

Flood, P. J., Harbinson, J., and Aarts, M. G. (2011). Natural genetic variation in plant photosynthesis. *Trends Plant Sci.* 16, 327–335. doi: 10.1016/j.tplants.2011.02.005

Fu, Y., Xu, L., Xu, B., Yang, L., Ling, Q., Wang, H., et al. (2007). Genetic interactions between leaf polarity-controlling genes and ASYMMETRIC LEAVES1 and 2 in Arabidopsis leaf patterning. *Plant Cell Physiol.* 48, 724–735. doi: 10.1093/pcp/pcm040

Fukushima, K., Fujita, H., Yamaguchi, T., Kawaguchi, M., Tsukaya, H., and Hasebe, M. (2015). Oriented cell division shapes carnivorous pitcher leaves of Sarracenia purpurea. *Nat. Commun.* 6, 6450. doi: 10.1038/ncomms7450

Garcia, D., Collier, S. A., Byrne, M. E., and Martienssen, R. A. (2006). Specification of leaf polarity in Arabidopsis via the trans-acting siRNA pathway. *Curr. Biol.* 16, 933–938. doi: 10.1016/j.cub.2006.03.064

Gonzalez, N., Vanhaeren, H., and Inzé, D. (2012). Leaf size control: complex coordination of cell division and expansion. *Trends Plant Sci.* 17, 332–340. doi: 10.1016/j.tplants.2012.02.003

Groszmann, M., Bylstra, Y., Lampugnani, E. R., and Smyth, D. R. (2010). Regulation of tissue-specific expression of SPATULA, a bHLH gene involved in carpel development, seedling germination, and lateral organ growth in Arabidopsis. *J. Exp. Bot.* 61, 1495–1508. doi: 10.1093/jxb/erq015

Guo, M., Thomas, J., Collins, G., and Timmermans, M. C. (2008). Direct repression of KNOX loci by the ASYMMETRIC LEAVES1 complex of Arabidopsis. *Plant Cell* 20, 48–58. doi: 10.1105/tpc.107.056127

Guo, P., Yoshimura, A., Ishikawa, N., Yamaguchi, T., Guo, Y., and Tsukaya, H. (2015). Comparative analysis of the RTFL peptide family on the control of plant organogenesis. *J. Plant Res.* 128, 497–510. doi: 10.1007/s10265-015-0703-1

Ha, C. M., Jun, J. H., and Fletcher, J. C. (2010). Control of Arabidopsis leaf morphogenesis through regulation of the YABBY and KNOX families of transcription factors. *Genetics* 186, 197–206. doi: 10.1534/genetics.110.118703

Ha, C. M., Jun, J. H., Nam, H. G., and Fletcher, J. C. (2004). BLADE-ON-PETIOLE1 encodes a BTB/POZ domain protein required for leaf morphogenesis in *Arabidopsis thaliana. Plant Cell Physiol.* 45, 1361–1370. doi: 10.1093/pcp/pch201

Ha, C. M., Jun, J. H., Nam, H. G., and Fletcher, J. C. (2007). BLADE-ON-PETIOLE 1 and 2 control Arabidopsis lateral organ fate through regulation of LOB domain and adaxial-abaxial polarity genes. *Plant Cell* 19, 1809–1825. doi: 10.1105/tpc.107.051938

Ha, C. M., Kim, G. T., Kim, B. C., Jun, J. H., Soh, M. S., Ueno, Y., et al. (2003). The BLADE-ON-PETIOLE 1 gene controls leaf pattern formation through the modulation of meristematic activity in Arabidopsis. *Development* 130, 161–172. doi: 10.1242/dev.00196

Hay, A., and Tsiantis, M. (2010). KNOX genes: versatile regulators of plant development and diversity. *Development* 137, 3153–3165. doi: 10.1242/dev.030049

Hepworth, S. R., Zhang, Y., McKim, S., Li, X., and Haughn, G. W. (2005). BLADE-ON-PETIOLE-dependent signaling controls leaf and floral patterning in Arabidopsis. *Plant Cell* 17, 1434–1448. doi: 10.1105/tpc.104.030536

Horiguchi, G., Fujikura, U., Ferjani, A., Ishikawa, N., and Tsukaya, H. (2006). Large-scale histological analysis of leaf mutants using two simple leaf observation methods: identification of novel genetic pathways governing the size and shape of leaves. *Plant J.* 48, 638–644. doi: 10.1111/j.1365-313X.2006.02896.x

Horiguchi, G., Kim, G. T., and Tsukaya, H. (2005). The transcription factor AtGRF5 and the transcription coactivator AN3 regulate cell proliferation in leaf primordia of *Arabidopsis thaliana. Plant J.* 43, 68–78. doi: 10.1111/j.1365-313X.2005.02429.x

Horiguchi, G., Mollá-Morales, A., Pérez-Pérez, J. M., Kojima, K., Robles, P., Ponce, M. R., et al. (2011). Differential contributions of ribosomal protein genes to *Arabidopsis thaliana* leaf development. *Plant J.* 65, 724–736. doi: 10.1111/j.1365-313X.2010.04457.x

Hu, Y., Xie, Q., and Chua, N. H. (2003). The Arabidopsis auxin-inducible gene ARGOS controls lateral organ size. *Plant Cell* 15, 1951–1961. doi: 10.1105/tpc.013557

Ichihashi, Y., Aguilar-Martínez, J. A., Farhi, M., Chitwood, D. H., Kumar, R., Millon, L. V., et al. (2014). Evolutionary developmental transcriptomics reveals a gene network module regulating interspecific diversity in plant leaf shape. *Proc. Natl. Acad. Sci. U.S.A.* 111, E2616–E2621. doi: 10.1073/pnas.1402835111

Ichihashi, Y., Horiguchi, G., Gleissberg, S., and Tsukaya, H. (2010). The bHLH transcription factor SPATULA controls final leaf size in *Arabidopsis thaliana. Plant Cell Physiol.* 51, 252–261. doi: 10.1093/pcp/pcp184

Ichihashi, Y., Kawade, K., Usami, T., Horiguchi, G., Takahashi, T., and Tsukaya, H. (2011). Key proliferative activity in the junction between the leaf blade and leaf petiole of Arabidopsis. *Plant Physiol.* 157, 1151–1162. doi: 10.1104/pp.111.185066

Ikeuchi, M., Yamaguchi, T., Kazama, T., Ito, T., Horiguchi, G., and Tsukaya, H. (2011). ROTUNDIFOLIA4 regulates cell proliferation along the body axis in Arabidopsis shoot. *Plant Cell Physiol.* 52, 59–69. doi: 10.1093/pcp/pcq138

Ikezaki, M., Kojima, M., Sakakibara, H., Kojima, S., Ueno, Y., Machida, C., et al. (2010). Genetic networks regulated by ASYMMETRIC LEAVES1 (AS1) and AS2 in leaf development in *Arabidopsis thaliana*: KNOX genes control five morphological events. *Plant J.* 61, 70–82. doi: 10.1111/j.1365-313X.2009.04033.x

Ishibashi, N., Kanamaru, K., Ueno, Y., Kojima, S., Kobayashi, T., Machida, C., et al. (2012). ASYMMETRIC-LEAVES2 and an ortholog of eukaryotic NudC domain proteins repress expression of AUXIN-RESPONSE-FACTOR and class 1 KNOX homeobox genes for development of flat symmetric leaves in Arabidopsis. *Biol. Open* 1, 197–207. doi: 10.1242/bio.2012406

Iwakawa, H., Iwasaki, M., Kojima, S., Ueno, Y., Soma, T., Tanaka, H., et al. (2007). Expression of the ASYMMETRIC LEAVES2 gene in the adaxial domain of Arabidopsis leaves represses cell proliferation in this domain and is critical for the development of properly expanded leaves. *Plant J.* 51, 173–184. doi: 10.1111/j.1365-313X.2007.03132.x

Iwakawa, H., Ueno, Y., Semiarti, E., Onouchi, H., Kojima, S., Tsukaya, H., et al. (2002). The ASYMMETRIC LEAVES2 gene of *Arabidopsis thaliana*, required for formation of a symmetric flat leaf lamina, encodes a member of a novel family of proteins characterized by cysteine repeats and a leucine zipper. *Plant Cell Physiol.* 43, 467–478. doi: 10.1093/pcp/pcf077

Iwasaki, M., Takahashi, H., Iwakawa, H., Nakagawa, A., Ishikawa, T., Tanaka, H., et al. (2013). Dual regulation of ETTIN (ARF3) gene expression by AS1-AS2, which maintains the DNA methylation level, is involved in stabilization of leaf adaxial-abaxial partitioning in Arabidopsis. *Development* 140, 1958–1969. doi: 10.1242/dev.085365

Jones-Rhoades, M. W., and Bartel, D. P. (2004). Computational identification of plant microRNAs and their targets, including a stress-induced miRNA. *Mol. Cell* 14, 787–799. doi: 10.1016/j.molcel.2004.05.027

Jun, J. H., Ha, C. M., and Fletcher, J. C. (2010). BLADE-ON-PETIOLE1 coordinates organ determinacy and axial polarity in Arabidopsis by directly activating ASYMMETRIC LEAVES2. *Plant Cell* 22, 62–76. doi: 10.1105/tpc.109.070763

Kalve, S., De Vos, D., and Beemster, G. T. (2014). Leaf development: a cellular perspective. *Front. Plant Sci.* 5:362. doi: 10.3389/fpls.2014.00362

Kanei, M., Horiguchi, G., and Tsukaya, H. (2012). Stable establishment of cotyledon identity during embryogenesis in Arabidopsis by ANGUSTIFOLIA3 and HANABA TARANU. *Development* 139, 2436–2446. doi: 10.1242/dev.081547

Kawade, K., Horiguchi, G., Usami, T., Hirai, M. Y., and Tsukaya, H. (2013). ANGUSTIFOLIA3 signaling coordinates proliferation between clonally distinct cells in leaves. *Curr. Biol.* 23, 788–792. doi: 10.1016/j.cub.2013.03.044

Kawamura, E., Horiguchi, G., and Tsukaya, H. (2010). Mechanisms of leaf tooth formation in Arabidopsis. *Plant J.* 62, 429–441. doi: 10.1111/j.1365-313X.2010.04156.x

Kazama, T., Ichihashi, Y., Murata, S., and Tsukaya, H. (2010). The mechanism of cell cycle arrest front progression explained by a KLUH/CYP78A5-dependent mobile growth factor in developing leaves of *Arabidopsis thaliana*. *Plant Cell Physiol.* 51, 1046–1054. doi: 10.1093/pcp/pcq051

Kim, J. H., and Kende, H. (2004). A transcriptional coactivator, AtGIF1, is involved in regulating leaf growth and morphology in Arabidopsis. *Proc. Natl. Acad. Sci. U.S.A.* 101, 13374–13379. doi: 10.1073/pnas.0405450101

Kimura, S., Koenig, D., Kang, J., Yoong, F. Y., and Sinha, N. (2008). Natural variation in leaf morphology results from mutation of a novel KNOX gene. *Curr. Biol.* 18, 672–677. doi: 10.1016/j.cub.2008.04.008

Kojima, S., Iwasaki, M., Takahashi, H., Imai, T., Matsumura, Y., Fleury, D., et al. (2011). Asymmetric leaves2 and Elongator, a histone acetyltransferase complex, mediate the establishment of polarity in leaves of *Arabidopsis thaliana*. *Plant Cell Physiol.* 52, 1259–1273. doi: 10.1093/pcp/pcr083

Kono, A., Umeda-Hara, C., Adachi, S., Nagata, N., Konomi, M., Nakagawa, T., et al. (2007). The Arabidopsis D-type cyclin CYCD4 controls cell division in the stomatal lineage of the hypocotyl epidermis. *Plant Cell* 19, 1265–1277. doi: 10.1105/tpc.106.046763

Koyama, T., Furutani, M., Tasaka, M., and Ohme-Takagi, M. (2007). TCP transcription factors control the morphology of shoot lateral organs via negative regulation of the expression of boundary-specific genes in Arabidopsis. *Plant Cell* 19, 473–484. doi: 10.1105/tpc.106.044792

Koyama, T., Mitsuda, N., Seki, M., Shinozaki, K., and Ohme-Takagi, M. (2010). TCP transcription factors regulate the activities of ASYMMETRIC LEAVES1 and miR164, as well as the auxin response, during differentiation of leaves in Arabidopsis. *Plant Cell* 22, 3574–3588. doi: 10.1105/tpc.110.075598

Krizek, B. A. (1999). Ectopic expression of AINTEGUMENTA in Arabidopsis plants results in increased growth of floral organs. *Dev. Genet.* 25, 224–236.

Kuchen, E. E., Fox, S., de Reuille, P. B., Kennaway, R., Bensmihen, S., Avondo, J., et al. (2012). Generation of leaf shape through early patterns of growth and tissue polarity. *Science* 335, 1092–1096. doi: 10.1126/science.1214678

Lee, B. H., Ko, J. H., Lee, S., Lee, Y., Pak, J. H., and Kim, J. H. (2009). The Arabidopsis GRF-INTERACTING FACTOR gene family performs an overlapping function in determining organ size as well as multiple developmental properties. *Plant Physiol.* 151, 655–668. doi: 10.1104/pp.109.141838

Li, P., Ponnala, L., Gandotra, N., Wang, L., Si, Y., Tausta, S. L., et al. (2010). The developmental dynamics of the maize leaf transcriptome. *Nat. Genet.* 42, 1060–1067. doi: 10.1038/ng.703

Li, Y., Zheng, L., Corke, F., Smith, C., and Bevan, M. W. (2008). Control of final seed and organ size by the DA1 gene family in *Arabidopsis thaliana*. *Genes Dev.* 22, 1331–1336. doi: 10.1101/gad.463608

Lim, P. O., Lee, I. C., Kim, J., Kim, H. J., Ryu, J. S., Woo, H. R., et al. (2010). Auxin response factor 2 (ARF2) plays a major role in regulating auxin-mediated leaf longevity. *J. Exp. Bot.* 61, 1419–1430. doi: 10.1093/jxb/erq010

Lin, W. C., Shuai, B., and Springer, P. S. (2003). The Arabidopsis LATERAL ORGAN BOUNDARIES-domain gene ASYMMETRIC LEAVES2 functions in the repression of KNOX gene expression and in adaxial-abaxial patterning. *Plant Cell* 15, 2241–2252. doi: 10.1105/tpc.014969

Liu, D., Song, Y., Chen, Z., and Yu, D. (2009). Ectopic expression of miR396 suppresses GRF target gene expression and alters leaf growth in Arabidopsis. *Physiol. Plant.* 136, 223–236. doi: 10.1111/j.1399-3054.2009.01229.x

Lodha, M., Marco, C. F., and Timmermans, M. C. (2013). The ASYMMETRIC LEAVES complex maintains repression of KNOX homeobox genes via direct recruitment of Polycomb-repressive complex2. *Genes Dev.* 27, 596–601. doi: 10.1101/gad.211425.112

Loehlin, D. W., and Werren, J. H. (2012). Evolution of shape by multiple regulatory changes to a growth gene. *Science* 335, 943–947. doi: 10.1126/science.1215193

MacAlister, C. A., Park, S. J., Jiang, K., Marcel, F., Bendahmane, A., Izkovich, Y., et al. (2012). Synchronization of the flowering transition by the tomato TERMINATING FLOWER gene. *Nat. Genet.* 44, 1393–1398. doi: 10.1038/ng.2465

Makkena, S., and Lamb, R. S. (2013). The bHLH transcription factor SPATULA regulates root growth by controlling the size of the root meristem. *BMC Plant Biol.* 13:1. doi: 10.1186/1471-2229-13-1

Mallory, A. C., Reinhart, B. J., Jones-Rhoades, M. W., Tang, G., Zamore, P. D., Barton, M. K., et al. (2004). MicroRNA control of PHABULOSA in leaf development: importance of pairing to the microRNA 5′ region. *EMBO J.* 23, 3356–3364. doi: 10.1038/sj.emboj.7600340

McConnell, J. R., Emery, J., Eshed, Y., Bao, N., Bowman, J., and Barton, M. K. (2001). Role of PHABULOSA and PHAVOLUTA in determining radial patterning in shoots. *Nature* 411, 709–713. doi: 10.1038/35079635

Mizukami, Y., and Fischer, R. L. (2000). Plant organ size control: AINTEGUMENTA regulates growth and cell numbers during organogenesis. *Proc. Natl. Acad. Sci. U.S.A.* 97, 942–947. doi: 10.1073/pnas.97.2.942

Nakata, M., Matsumoto, N., Tsugeki, R., Rikirsch, E., Laux, T., and Okada, K. (2012). Roles of the middle domain-specific WUSCHEL-RELATED HOMEOBOX genes in early development of leaves in Arabidopsis. *Plant Cell* 24, 519–535. doi: 10.1105/tpc.111.092858

Nakayama, H., Nakayama, N., Seiki, S., Kojima, M., Sakakibara, H., Sinha, N., et al. (2014). Regulation of the KNOX-GA gene module induces heterophyllic alteration in North American Lake Cress. *Plant Cell.* 26, 4733–4748. doi: 10.1105/tpc.114.130229

Nakayama, H., Yamaguchi, T., and Tsukaya, H. (2012). Acquisition and diversification of cladodes: leaf-like organs in the genus Asparagus. *Plant Cell* 24, 929–940. doi: 10.1105/tpc.111.092924

Nardmann, J., Ji, J., Werr, W., and Scanlon, M. J. (2004). The maize duplicate genes narrow sheath1 and narrow sheath2 encode a conserved homeobox gene function in a lateral domain of shoot apical meristems. *Development* 131, 2827–2839. doi: 10.1242/dev.01164

Nardmann, J., and Werr, W. (2013). Symplesiomorphies in the WUSCHEL clade suggest that the last common ancestor of seed plants contained at least four independent stem cell niches. *New Phytol.* 199, 1081–1092. doi: 10.1111/nph.12343

Narita, N. N., Moore, S., Horiguchi, G., Kubo, M., Demura, T., Fukuda, H., et al. (2004). Overexpression of a novel small peptide ROTUNDIFOLIA4 decreases cell proliferation and alters leaf shape in *Arabidopsis thaliana*. *Plant J.* 38, 699–713. doi: 10.1111/j.1365-313X.2004.02078.x

Nath, U., Crawford, B. C., Carpenter, R., and Coen, E. (2003). Genetic control of surface curvature. *Science* 299, 1404–1407. doi: 10.1126/science.1079354

Nicotra, A., Leigh, A., Boyce, K. C., Jones, C. S., Niklas, K. J., Royer, D. L., et al. (2011). The evolution and functional significance of leaf shape in the angiosperms. *Funct. Plant Biol.* 38, 535–552. doi: 10.1071/FP11057

Nikovics, K., Blein, T., Peaucelle, A., Ishida, T., Morin, H., Aida, M., et al. (2006). The balance between the MIR164A and CUC2 genes controls leaf margin serration in Arabidopsis. *Plant Cell* 18, 2929–2945. doi: 10.1105/tpc.106.045617

Nole-Wilson, S., Tranby, T. L., and Krizek, B. A. (2005). AINTEGUMENTA-like (AIL) genes are expressed in young tissues and may specify meristematic or division-competent states. *Plant Mol. Biol.* 57, 613–628. doi: 10.1007/s11103-005-0955-6

Norberg, M., Holmlund, M., and Nilsson, O. (2005). The BLADE ON PETIOLE genes act redundantly to control the growth and development of lateral organs. *Development* 132, 2203–2213. doi: 10.1242/dev.01815

Ohno, C. K., Reddy, G. V., Heisler, M. G., and Meyerowitz, E. M. (2004). The Arabidopsis JAGGED gene encodes a zinc finger protein that promotes leaf tissue development. *Development* 131, 1111–1122. doi: 10.1242/dev.00991

Palatnik, J. F., Allen, E., Wu, X., Schommer, C., Schwab, R., Carrington, J. C., et al. (2003). Control of leaf morphogenesis by microRNAs. *Nature* 425, 257–263. doi: 10.1038/nature01958

Palmer, N. A., Donze-Reiner, T., Horvath, D., Heng-Moss, T., Waters, B., Tobias, C., et al. (2015). Switchgrass (*Panicum virgatum* L) flag leaf transcriptomes reveal molecular signatures of leaf development, senescence, and mineral dynamics. *Funct. Integr. Genomics* 15, 1–16. doi: 10.1007/s10142-014-0393-0

Peter, I. S., and Davidson, E. H. (2011). Evolution of gene regulatory networks controlling body plan development. *Cell* 144, 970–985. doi: 10.1016/j.cell.2011.02.017

Phelps-Durr, T. L., Thomas, J., Vahab, P., and Timmermans, M. C. (2005). Maize rough sheath2 and its Arabidopsis orthologue ASYMMETRIC LEAVES1 interact with HIRA, a predicted histone chaperone, to maintain knox gene silencing and determinacy during organogenesis. *Plant Cell* 17, 2886–2898. doi: 10.1105/tpc.105.035477

Piazza, P., Bailey, C. D., Cartolano, M., Krieger, J., Cao, J., Ossowski, S., et al. (2010). *Arabidopsis thaliana* leaf form evolved via loss of KNOX expression in leaves in association with a selective sweep. *Curr. Biol.* 20, 2223–2228. doi: 10.1016/j.cub.2010.11.037

Pick, L., and Heffer, A. (2012). Hox gene evolution: multiple mechanisms contributing to evolutionary novelties. *Ann. N.Y. Acad. Sci.* 1256, 15–32. doi: 10.1111/j.1749-6632.2011.06385.x

Pinon, V., Etchells, J. P., Rossignol, P., Collier, S. A., Arroyo, J. M., Martienssen, R. A., et al. (2008). Three PIGGYBACK genes that specifically influence leaf patterning encode ribosomal proteins. *Development* 135, 1315–1324. doi: 10.1242/dev.016469

Rebeiz, M., Pool, J. E., Kassner, V. A., Aquadro, C. F., and Carroll, S. B. (2009). Stepwise modification of a modular enhancer underlies adaptation in a Drosophila population. *Science* 326, 1663–1667. doi: 10.1126/science.1178357

Rodriguez, R. E., Mecchia, M. A., Debernardi, J. M., Schommer, C., Weigel, D., and Palatnik, J. F. (2010). Control of cell proliferation in *Arabidopsis thaliana* by microRNA miR396. *Development* 137, 103–112. doi: 10.1242/dev.043067

Rojas, C. A., Eloy, N. B., Lima Mde, F., Rodrigues, R. L., Franco, L. O., Himanen, K., et al. (2009). Overexpression of the Arabidopsis anaphase promoting complex subunit CDC27a increases growth rate and organ size. *Plant Mol. Biol.* 71, 307–318. doi: 10.1007/s11103-009-9525-7

Sack, L., Scoffoni, C., McKown, A. D., Frole, K., Rawls, M., Havran, J. C., et al. (2012). Developmentally based scaling of leaf venation architecture explains global ecological patterns. *Nat. Commun.* 3, 837. doi: 10.1038/ncomms1835

Sarojam, R., Sappl, P. G., Goldshmidt, A., Efroni, I., Floyd, S. K., Eshed, Y., et al. (2010). Differentiating Arabidopsis shoots from leaves by combined YABBY activities. *Plant Cell* 22, 2113–2130. doi: 10.1105/tpc.110.075853

Sawa, S., Watanabe, K., Goto, K., Liu, Y. G., Shibata, D., Kanaya, E., et al. (1999). FILAMENTOUS FLOWER, a meristem and organ identity gene of Arabidopsis, encodes a protein with a zinc finger and HMG-related domains. *Genes Dev.* 13, 1079–1088. doi: 10.1101/gad.13.9.1079

Schruff, M. C., Spielman, M., Tiwari, S., Adams, S., Fenby, N., and Scott, R. J. (2006). The AUXIN RESPONSE FACTOR 2 gene of Arabidopsis links auxin signalling, cell division, and the size of seeds and other organs. *Development* 133, 251–261. doi: 10.1242/dev.02194

Semiarti, E., Ueno, Y., Tsukaya, H., Iwakawa, H., Machida, C., and Machida, Y. (2001). The ASYMMETRIC LEAVES2 gene of *Arabidopsis thaliana* regulates formation of a symmetric lamina, establishment of venation and repression of meristem-related homeobox genes in leaves. *Development* 128, 1771–1783.

Sicard, A., Thamm, A., Marona, C., Lee, Y. W., Wahl, V., Stinchcombe, J. R., et al. (2014). Repeated evolutionary changes of leaf morphology caused by mutations to a homeobox gene. *Curr. Biol.* 24, 1880–1886. doi: 10.1016/j.cub.2014.06.061

Siegfried, K. R., Eshed, Y., Baum, S. F., Otsuga, D., Drews, G. N., and Bowman, J. L. (1999). Members of the YABBY gene family specify abaxial cell fate in Arabidopsis. *Development* 126, 4117–4128.

Steeves, T. A., and Sussex, I. M. (1989). *Patterns in Plant Development.* Cambridge: Cambridge University Press.

Takahashi, H., Iwakawa, H., Ishibashi, N., Kojima, S., Matsumura, Y., Prananingrum, P., et al. (2013). Meta-analyses of microarrays of Arabidopsis asymmetric leaves1 (as1), as2 and their modifying mutants reveal a critical role for the ETT pathway in stabilization of adaxial-abaxial patterning and cell division during leaf development. *Plant Cell Physiol.* 54, 418–431. doi: 10.1093/pcp/pct027

Tameshige, T., Fujita, H., Watanabe, K., Toyokura, K., Kondo, M., Tatematsu, K., et al. (2013). Pattern dynamics in adaxial-abaxial specific gene expression are modulated by a plastid retrograde signal during *Arabidopsis thaliana* leaf development. *PLoS Genet.* 9:e1003655. doi: 10.1371/journal.pgen.1003655

Toyokura, K., Watanabe, K., Oiwaka, A., Kusano, M., Tameshige, T., Tatematsu, K., et al. (2011). Succinic semialdehyde dehydrogenase is involved in the robust patterning of Arabidopsis leaves along the adaxial-abaxial axis. *Plant Cell Physiol.* 52, 1340–1353. doi: 10.1093/pcp/pcr079

Toyokura, K., Yamaguchi, K., Shigenobu, S., Fukaki, H., Tatematsu, K., and Okada, K. (2015). Mutations in plastidial 5-aminolevulinic acid biosynthesis genes suppress a pleiotropic defect in shoot development of a mitochondrial gaba shunt mutant in Arabidopsis. *Plant Cell Physiol.* 56, 1229–1238. doi: 10.1093/pcp/pcv050

Tsukaya, H. (2014a). Comparative leaf development in angiosperms. *Curr. Opin. Plant Biol.* 17, 103–109. doi: 10.1016/j.pbi.2013.11.012

Tsukaya, H. (2014b). "Meristems," in *Atlas of Plant Cell Structure*, eds T. Noguchi, S. Kawano, H. Tsukaya, S. Matsunaga, A. Sakai, I. Karahara, and Y. Hayashi (Tokyo: Springer), 187–202.

Tsukaya, H., Byrne, M. E., Horiguchi, G., Sugiyama, M., Van Lijsebettens, M., and Lenhard, M. (2013). How do 'housekeeping' genes control organogenesis?-Unexpected new findings on the role of housekeeping genes in cell and organ differentiation. *J. Plant Res.* 126, 3–15. doi: 10.1007/s10265-012-0518-2

Ueno, Y., Ishikawa, T., Watanabe, K., Terakura, S., Iwakawa, H., Okada, K., et al. (2007). Histone deacetylases and ASYMMETRIC LEAVES2 are involved in the establishment of polarity in leaves of Arabidopsis. *Plant Cell* 19, 445–457. doi: 10.1105/tpc.106.042325

Vercruyssen, L., Verkest, A., Gonzalez, N., Heyndrickx, K. S., Eeckhout, D., Han, S. K., et al. (2014). ANGUSTIFOLIA3 binds to SWI/SNF chromatin remodeling complexes to regulate transcription during Arabidopsis leaf development. *Plant Cell* 26, 210–229. doi: 10.1105/tpc.113.115907

Vlad, D., Kierzkowski, D., Rast, M. I., Vuolo, F., Dello Ioio, R., Galinha, C., et al. (2014). Leaf shape evolution through duplication, regulatory diversification, and loss of a homeobox gene. *Science* 343, 780–783. doi: 10.1126/science.1248384

Wenkel, S., Emery, J., Hou, B. H., Evans, M. M., and Barton, M. K. (2007). A feedback regulatory module formed by LITTLE ZIPPER and HD-ZIPIII genes. *Plant Cell* 19, 3379–3390. doi: 10.1105/tpc.107.055772

White, D. W. (2006). PEAPOD regulates lamina size and curvature in Arabidopsis. *Proc. Natl. Acad. Sci. U.S.A.* 103, 13238–13243. doi: 10.1073/pnas.0604349103

Wu, G., Lin, W. C., Huang, T., Poethig, R. S., Springer, P. S., and Kerstetter, R. A. (2008). KANADI1 regulates adaxial-abaxial polarity in Arabidopsis by directly repressing the transcription of ASYMMETRIC LEAVES2. *Proc. Natl. Acad. Sci. U.S.A.* 105, 16392–16397. doi: 10.1073/pnas.0803997105

Xu, L., Xu, Y., Dong, A., Sun, Y., Pi, L., Xu, Y., et al. (2003). Novel as1 and as2 defects in leaf adaxial-abaxial polarity reveal the requirement for ASYMMETRIC LEAVES1 and 2 and ERECTA functions in specifying leaf adaxial identity. *Development* 130, 4097–4107. doi: 10.1242/dev.00622

Yamaguchi, T., Yano, S., and Tsukaya, H. (2010). Genetic framework for flattened leaf blade formation in unifacial leaves of Juncus prismatocarpus. *Plant Cell* 22, 2141–2155. doi: 10.1105/tpc.110.076927

Yao, Y., Ling, Q., Wang, H., and Huang, H. (2008). Ribosomal proteins promote leaf adaxial identity. *Development* 135, 1325–1334. doi: 10.1242/dev.017913

Zgurski, J. M., Sharma, R., Bolokoski, D. A., and Schultz, E. A. (2005). Asymmetric auxin response precedes asymmetric growth and differentiation of asymmetric leaf1 and asymmetric leaf2 Arabidopsis leaves. *Plant Cell* 17, 77–91. doi: 10.1105/tpc.104.026898

Conflict of Interest Statement: The authors declare that the research was conducted in the absence of any commercial or financial relationships that could be construed as a potential conflict of interest.

7

Arabidopsis LEAFY COTYLEDON1 controls cell fate determination during post-embryonic development

Mingkun Huang[1,2], Yilong Hu[1,2], Xu Liu[1], Yuge Li[1] and Xingliang Hou[1]*

[1] Key Laboratory of South China Agricultural Plant Molecular Analysis and Genetic Improvement, South China Botanical Garden, Chinese Academy of Sciences, Guangzhou, China, [2] University of the Chinese Academy of Sciences, Beijing, China

Arabidopsis LEAFY COTYLEDON1 (LEC1) transcription factor is a master regulator that shapes plant embryo development and post-embryonic seedling establishment. Loss-of-function of *LEC1* alters the cotyledon identity, causing the formation of ectopic trichomes, which does not occur in wild-type seedlings, implying that LEC1 might regulate embryonic cell fate determination during post-embryonic development. To test this hypothesis, we compared the expression of trichome development-related genes between the wild-type and the *lec1* mutant. We observed that transcripts of GLABROUS1 (*GL1*), *GL2*, and *GL3*, genes encoding the positive regulators in trichome development, were significantly upregulated, while the TRICHOMELESS1 (*TCL2*), ENHANCER OF TRY AND CPC1 (*ETC1*), and *ETC2* genes, encoding the negative regulators in trichome development, were downregulated in the *lec1* mutant. Furthermore, overexpression of *LEC1* activated the expressions of *TCL2*, CAPPICE (*CPC*), and *ETC1*, resulting in production of cotyledonary leaves with no or fewer trichomes during vegetative development. In addition, we demonstrated that LEC1 interacts with TCL2 in yeast and *in vitro*. A genetic experiment showed that loss-of-function of *GL2* rescued the ectopic trichome formation in the *lec1* mutant. These findings strongly support that LEC1 regulates trichome development, providing direct evidence for the role of LEC1 in cell fate determination during post-embryonic development.

Edited by:
Anja Geitmann,
Université de Montréal, Canada

Reviewed by:
Sergey Morozov,
Moscow State University, Russia
Biswapriya Biswavas Misra,
University of Florida, USA

****Correspondence:***
Xingliang Hou
houxl@scib.ac.cn

Keywords: *Arabidopsis*, LEAFY COTYLEDON1, trichome, cell fate determination, post-embryonic development

INTRODUCTION

In higher plants, embryogenesis generally terminates with a dormancy period for future sporophyte growth. Once conditions are favorable, seeds germinate, and undergo post-embryonic development, during which cells acquire specific fates. Extensive studies have shown that the specific cell fate determination during post-embryonic development is precisely controlled by multiple transcription factors (Peris et al., 2010; Perianez-Rodriguez et al., 2014). The plant Nuclear Factor Y (NF-Y) transcription factor LEAFY COTYLEDON1 (LEC1), a master regulator controlling embryogenesis (Meinke, 1992; West et al., 1994; Harada, 2001; Lee et al., 2003), was recently revealed to play a potential role in regulating post-embryonic development (Junker and Baumlein, 2012). *LEC1* is expressed in both developing embryo and post-embryonic seedlings, but not in the vegetative true leaves (Kwong et al., 2003; Warpeha et al., 2007; Le et al., 2010). Loss-of-function of *LEC1* causes a pleiotropic phenotype, including embryo desiccation intolerance and defects in the accumulation of seed storage compounds in developing seeds, as well as cotyledon with trichomes, premature development of vascular tissue and mesophyll cells, and short hypocotyls in early seedlings (Meinke, 1992; Meinke et al., 1994; West et al., 1994; Brocard-Gifford et al., 2003). By contrast, ectopic expression *LEC1* allows seedlings to remain in the embryonic state with an unexpanded cotyledon after germination and is sufficient to induce conversion of true leaves into embryonic structures that lack trichomes (Lotan et al., 1998; Junker et al., 2012). These findings support the multifunctional roles of LEC1 in embryonic and post-embryonic development. Trichomes originate from the epidermal cell layer whose nuclei have undergone multiple rounds of endoreduplication, which then enlarge and expand away from the surface (Hulskamp et al., 1994; Pattanaik et al., 2014). The regulation of trichome formation has been well documented in many studies. The R2R3 MYB transcription factor GLABROUS1 (GL1) was the first identified positive regulator involved in trichome development (Marks and Feldmann, 1989; Oppenheimer et al., 1991). GL1 interacts with the bHLH transcription factor GLABRA3 (GL3), its close homolog ENHANCER OF GLABRA3 (EGL3), and the WD40-repeat factor TRANSPARENT TESTA GLABRA1 (TTG1), resulting in the formation of the MYB-bHLH-WD-repeat trichome-positive transcription complex (Zhao et al., 2008). This protein complex subsequently induces trichome formation by activating the expression of the downstream gene *GLABRA2 (GL2)*, encoding a homeodomain-leucine zipper (HD-Zip) transcription factor (Rerie et al., 1994; Morohashi et al., 2007). Contrastingly, a small single-repeat R3 MYB protein family, including TRICHOMELESS1 (TCL1), TCL2, TRIPTYCHON (TRY), CAPPICE (CPC), ENHANCER OF TRY AND CPC1 (ETC1), and ETC2, plays redundant roles in the negative regulation of trichome development. These proteins compete with GL1 for binding to GL3/EGL3, resulting in the formation of an inactive protein complex that fails to activate *GL2* expression and thus represses trichome formation (Wang and Chen, 2014; Zhou et al., 2014).

Trichome formation is considered a specific characteristic that distinguishes true leaves from cotyledons in *Arabidopsis* (Chandler, 2008). A hypothesis for the ectopic trichome formation on the cotyledons of the *lec1* mutant is that the loss-of-function of *LEC1* might result in the disturbance of embryonic cell fate determination, allowing the precocious post-germination events in which cotyledons are partially converted into true leaves (West et al., 1994; Chandler, 2008). However, the mechanism by which LEC1 functions in cell fate determination during post-embryonic development remains elusive.

Given that LEC1 mutation causes remarkable ectopic trichome growth on cotyledons, the trichome-related genes might serve as good candidates to study the function of LEC1 in the post-embryonic development phase. Here, we demonstrated that LEC1 represses trichome cell differentiation during post-embryonic development by regulating trichome-related genes. In addition, LEC1 was shown to interact with TCL2, a repressor of trichome development, which indicated that LEC1 might couple with other transcription factors to co-regulate trichome formation. A genetic experiment showed that loss-of-function of *GL2* rescued the ectopic trichome development phenotype of the *lec1* mutant. These results strongly supported the view that *LEC1* functions in cell fate determination during the post-embryonic phase, providing new insights into *LEC1*'s role beyond the embryogenesis.

MATERIALS AND METHODS

Plant Materials and Growth Conditions

All *Arabidopsis* plants used in this study are in the Col genetic background. The mutant *gl2* was described previously (Wang et al., 2010). The mutant *lec1-4* (Salk_095699) was obtained from the *Arabidopsis* Biological Resource Center (http://www.arabidopsis.org). The estradiol-inducible *pER10:LEC1-MYC* transgenic line was obtained by kanamycin selection after transformation (Mu et al., 2008). The double mutant was generated by genetic crossing. After surface sterilization, seeds were sown on the half-strength Murashige and Skoog medium containing 0.8% agar and incubated at 4°C in the dark for 3 days. The seedlings were then transferred to the growth chamber at 22°C under long day conditions (16 h light/8 h dark). For estradiol treatment, 10 μM estradiol was added to the half-strength Murashige and Skoog medium plate and dimethyl sulfoxide, the solvent for estradiol, served as the mock treatment. After treatment, these seedlings were harvested for further analyses.

Microscopy

The formation of trichomes on the cotyledons and leaves was observed and photographed using a LEICA M165C stereoscope (Leica, Wetzlar, Germany).

Statistical Analysis

Five-day-old seedlings were observed to calculate frequency of the appearance of ectopic trichomes on cotyledons. At least 15 Col and 50 *lec1-4* seedlings were used in each experiment. Data represent mean ± SD of three independent repeats. To calculate the frequency of different trichome types on *lec1-4* cotyledons, at least 50 trichomes were analyzed in each experiment. Data represent the mean ± SD of seven independent repeats.

Gene Expression Analysis

Total RNA was extracted from 5-day-old *Arabidopsis* seedlings using the Plant RNA Kit (OMEGA, Atlanta, GA, USA), following the manufacturer's instructions. One microgram of total RNA was used for the reverse transcription reaction. The first-strand cDNA was synthesized using the M-MLV reverse transcriptase (Promega, Madison, WI, USA), according to the manufacturer's instructions. Quantitative real-time PCR was performed using the LightCycler480 system (Roche, Basel, Switzerland) in a total volume of 10 μl, with 0.25 μl of each primer (10 μM), 1 μl of cDNA product and 5 μl of SYBR Premix ExTaq (Takara, Tokyo, Japan). The PCR program included an initial denaturation step at

94°C for 1 min, followed by 40 cycles of 10 s at 94°C and 1 min at 60°C. Each sample was quantified at least in triplicate and normalized using the *TUB2* gene as the internal control. Primers used in this study are listed in Supplementary Table S1.

Yeast Two-Hybrid Assay

The full-length coding sequences of *LEC1* and *TCL2* were amplified and cloned into vectors pGBKT7 and pGADT7 (Clontech, Palo Alto, CA, USA), respectively. Primers are listed in Supplementary Table S1. Yeast two-hybrid assays were performed following the manufacturer's instructions of the Yeastmaker Yeast Transformation System 2 (Clontech). In brief, a single colony of yeast *AH109* was incubated at 30°C overnight (OD_{600} > 1.0). The cells were harvested by centrifugation and then resuspended in 25 mL ddH_2O. After re-harvesting the cells by centrifugation, a 1.5-ml sterile 1× Tris-EDTA/LiAc solution was added to prepare yeast-competent cells. For the yeast two-hybrid assay, the bait (0.5 μg) and/or prey (0.5 μg) plasmids with 0.1 mg of carrier DNA were co-transformed into the yeast-competent cells using Polyethylene glycol/LiAc solution. After incubation at 30°C for 30 min, 70 μL dimethyl sulfoxide was added and incubation continued at 42°C for 15 min. The cells were centrifuged and washed using Tris-EDTA solution. After transformation, the yeast cells were grown on SD/-Trp/-Leu/-His/-Ade medium for the interaction test.

Pull-Down Assay

To produce glutathione-S-transferase (GST)-TCL1, GST-TCL2, GST-CPC, GST-ETC1, His-LEC1, and His-GFP recombinant proteins, the full-length coding sequences of the tested genes were cloned into vector pGEX-4T-1 (Pharmacia, Piscataway, NJ, USA) or pQE30 (QIAGEN, Dusseldorf, Germany), respectively. Primers are listed in Supplementary Table S1. These constructs were transformed into the *Escherichia coli* Rosetta strain DE3 (Novagen, Billerica, MA, USA). One hundred milliliters of (OD_{600} ≈ 0.5) Rosetta strains harboring various vectors were incubated at 16°C for 16 h with 0.1 mM isopropyl β-D-1-thiogalactopyranoside. The soluble GST fusion proteins were purified using glutathione sepharose beads (Amersham Biosciences, Piscataway, NJ, USA), while His fusion proteins were purified using Ni- nitrilotriacetic acid agarose beads (QIAGEN), according to the manufacturer's instructions, respectively. For pull-down assays, 2 μg of His-LEC1 or His-GFP were incubated with immobilized GST or GST fusion proteins in the binding buffer (50 mM Tris-HCl, pH 8.0, 100 mM NaCl, and 1 mM EDTA) at 4°C for 4 h. After washing with binding buffer three times, proteins retained on the beads were subsequently resolved by sodium dodecyl sulfate polyacrylamide electrophoresis and detected with an anti-His antibody (GBI, Beijing, China).

RESULTS

Ectopic Trichome Formation on the Cotyledons of the *lec1* Mutant

Extensive studies have shown that LEC1 serves as a key regulator in embryo development, which is consistent with its main expression in the developing seed (Kwong et al., 2003). However, recent observations suggested a potential role of LEC1 in post-embryonic cell differentiation, including hypocotyl elongation and the formation of vascular tissue, mesophyll cells and trichomes (Warpeha et al., 2007; Junker and Baumlein, 2012; Junker et al., 2012; Supplementary Figure S1A). Serving as a general model system in the study of cell fate determination in *Arabidopsis* (Yang and Ye, 2013), we focused on trichome formation to reveal the function of LEC1 in post-embryonic development. A desiccation-tolerant leaky allele *lec1-4* whose mature seeds germinate normally (unpublished data) was used in this study. Consistent with previous reports, the wild-type cotyledons were glabrous, while ~85% of *lec1* cotyledons produced a varied number of ectopic trichomes (**Figure 1A**; Supplementary Table S2). Moreover, unlike the mostly three-branched trichomes developed on true leaves, the trichomes on *lec1* cotyledons were mainly unbranched (~44%) or two-branched (~55%), and rarely three-branched (less than 1%; **Figure 1B**; Supplementary Table S3). These observations implied incomplete conversion of the *lec1* cotyledons into true leaves. In addition, there was no significant difference observed in trichome numbers or the morphology of rosette leaves between the wild-type and *lec1*, which is consistent with the observation that no *LEC1* transcripts accumulated in the vegetative tissues (Supplementary Figures S1A,B).

Misregulation of Trichome-Related Genes in the *lec1* Mutant

Trichome formation on leaves has been well studied during the past two decades and a number of trichome-related genes were identified, which could be divided into two groups: trichome-positive and -negative genes (Yang and Ye, 2013). To examine whether LEC1's involvement in the inhibition of trichome formation on cotyledons occurs by mediating the regulation of trichome-related genes, five trichome-positive genes, and six trichome-negative genes were selected for quantitative real-time PCR analysis in wild-type and *lec1* seedlings. Interestingly, among the trichome-positive genes, the expressions of *GL1*, *GL2*, and *GL3* were significantly upregulated in the *lec1* mutant compared with the wild-type, while those of *EGL3* and *TTG1* showed no change (**Figure 2A**). Moreover, the expressions of the trichome-negative genes *TCL2*, *ETC1*, and *ETC2* were significantly downregulated in the *lec1* mutant compared with the wild-type, while those of *TRY* and *CPC* showed little change. The *TCL1* transcript was hardly detected in either the wild-type or *lec1* seedlings (**Figure 2B**). These results indicated that the ectopic trichome formation might be caused by misregulation of the trichome-related genes in the *lec1* mutant.

Overexpression of *LEC1* Activates Several Trichome-Negative Transcription Factors

Given that loss-of-function of *LEC1* resulted in misregulation of trichome-related genes, we wondered whether *LEC1* overexpression could activate or repress these genes. *LEC1* overexpression, driven by the 35S promoter, caused complex

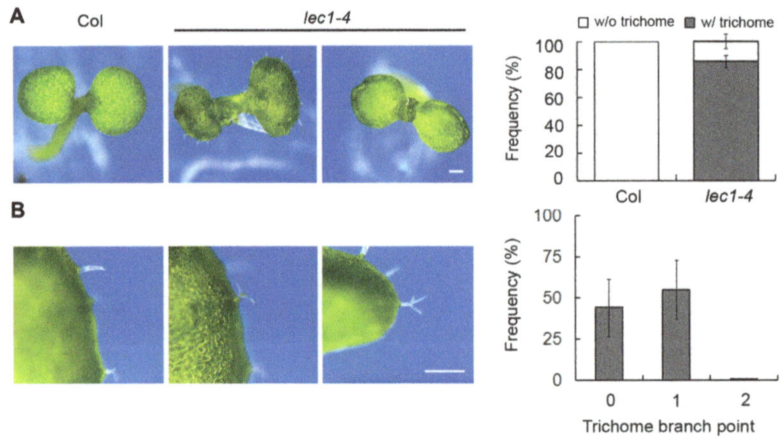

FIGURE 1 | Ectopic trichome formation on the *lec1-4* mutant cotyledons. (A) Comparison of 5-day-old Col wild-type and *lec1-4* mutant cotyledons. Bar = 0.5 mm. The right panel indicates the frequency of seedlings with (w/) or without (w/o) ectopic trichomes on their cotyledons. **(B)** Light micrographs of three types of trichome branch on *lec1-4* mutant cotyledons. Bar = 0.5 mm. The right panel indicates the frequency of different types of trichomes. 0 indicates no branch on a trichome; 1 indicates that the trichome has two branches; 2 indicates a trichome with three branches.

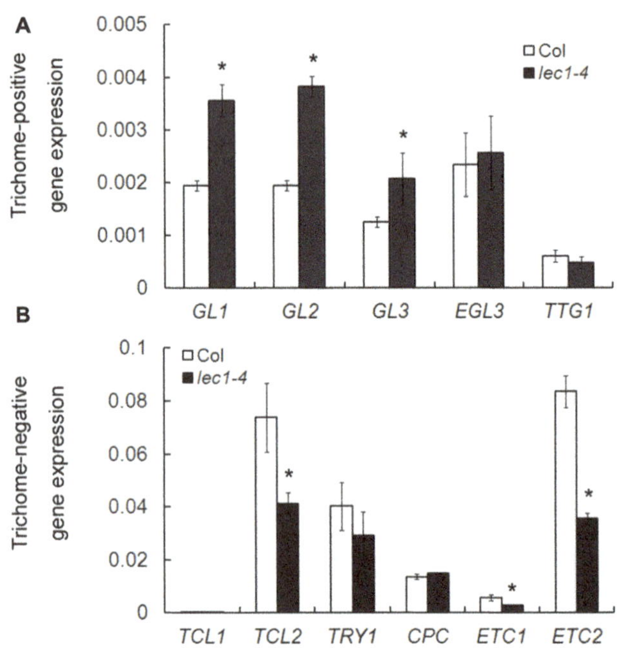

FIGURE 2 | Misregulation of trichome-related genes in the *lec1-4* mutant seedlings. (A) The expression analysis of trichome-positive genes in 5-day-old Col wild-type and *lec1-4* mutant seedlings. **(B)** The expression analysis of trichome-negative genes in 5-day-old Col wild-type and *lec1-4* mutant seedlings. *TUB2* was used as an internal control. Asterisks indicate significant differences between Col and *lec1-4* mutant ($p < 0.05$, by Student's *t*-test).

phenotypes and embryonic lethality (Lotan et al., 1998). Thus, instead, we employed an inducible transgenic plant harboring *pER10:LEC1-MYC*, in which *LEC1* expression could be significantly induced by estradiol (Mu et al., 2008; Supplementary Figure S2). Unexpectedly, the trichome-positive genes *GL1*, *GL2*, *GL3*, *EGL3*, and *TTG1* were not regulated by induced LEC1 (data not shown). Nevertheless, overexpression of *LEC1* activated the expressions of *TCL2*, *CPC*, and *ETC1*, but had no effect on those of *TCL1*, *TRY*, and *ETC2* (**Figure 3A**), which revealed the role of LEC1 in the regulation of trichome-negative genes. A previous study reported that *TCL2*, *CPC*, and *ETC1* overexpression lines all presented a glabrous leaves phenotype (Wang and Chen, 2014); therefore, we asked whether ectopic *LEC1* expression affects trichome formation on true leaves. The investigation demonstrated that cotyledons of *pER10:LEC1-MYC* under either mock or estradiol treatment displayed a similar glabrous phenotype to the wild-type (data not shown). Strikingly, under long-term induction by estradiol (14 days), the *pER10:LEC1-MYC* seedlings developed fewer trichomes and cotyledon-like rosette leaves (**Figure 3B**). Together with a previous report that the single-repeat R3 MYB quintuple mutant *try cpc etc1 etc3 tcl1* produced ectopic trichomes on cotyledons similar to the *lec1* mutant (Wang et al., 2008), our observations supported the hypothesis that LEC1 suppresses trichome formation via activation of *TCL2*, *CPC*, and *ETC1* expression.

LEC1 Physically Interacts with TCL2 *In Vitro* and in Yeast

LEAFY COTYLEDON1 belongs to the NF-Y family, whose members interact with other transcription factors to co-regulate target gene expression (Yamamoto et al., 2009). To investigate whether LEC1 couples with cofactors to mediate trichome formation, we performed a yeast two-hybrid screening assay. Interestingly, the trichome transcriptional repressor TCL2 was observed to interact with LEC1 in yeast (**Figure 4A**). We then performed a pull-down assay to confirm the interaction between LEC1 and TCL2 using purified GST- and His-tagged proteins (Supplementary Figure S3). As expected, GST-TCL2 interacted with His-LEC1, but not with the His-GFP protein, while GST

FIGURE 3 | Induced *LEC1* overexpression suppresses trichome formation. (A) The expression analysis of trichome-negative genes in 5-day-old *pER10:LEC1-MYC* transgenic seedlings with 10 μM estradiol or mock treatment. *TUB2* was used as an internal control. Asterisks indicate significant differences between estradiol and mock ($p < 0.05$, by Student's *t*-test). **(B)** Induced *LEC1* overexpression promotes conversion of true leaves into embryonic structures. Fourteen-day-old *pER10:LEC1-MYC* transgenic seedlings with 10 μM estradiol or mock treatment were used for phenotype observation. Red arrowheads indicate rosette leaves; white arrowheads indicate cotyledons. Bar = 1 mm.

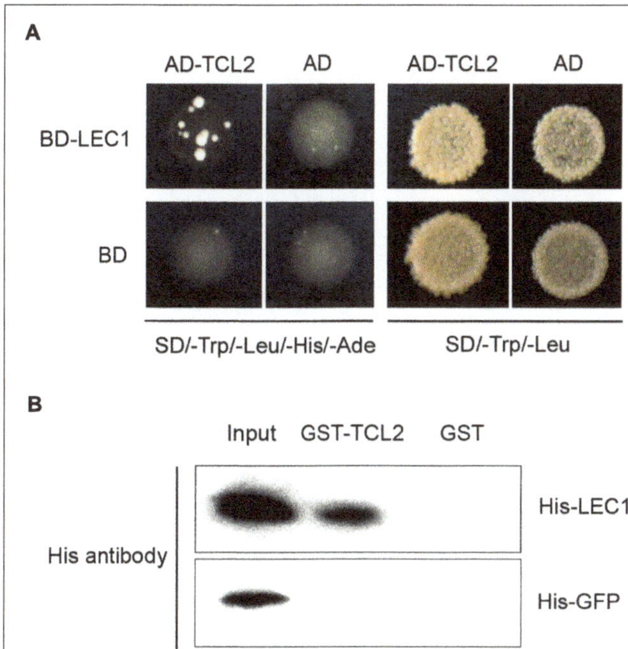

FIGURE 4 | TCL2 physically interacts with LEC1 in yeast and *in vitro*. (A) Yeast two-hybrid assays showing that TCL2 interacts with LEC1. Transformed yeast cells were grown both on SD/-Trp/-Leu/-His/-Ade and SD/-Trp/-Leu medium. **(B)** A pull-down assay showing the direct interaction between His-LEC1 and GST-TCL2 fusion proteins *in vitro*. His-LEC1 or His-GFP protein was incubated with immobilized GST or GST-TCL2 proteins and the immunoprecipitated fractions were detected by an anti-His antibody.

alone did not precipitate either of the His-tagged proteins (**Figure 4B**). TCL2 and its homologs TCL1, TRY, CPC, ETC1, and ETC2 exert a master role in repressing trichome formation during the whole life of plants; therefore, we further tested the binding between LEC1 and these homologous proteins. The pull-down results showed that LEC1 also interacted with TCL1, CPC, and ETC1 *in vitro* (Supplementary Figure S4); however, their bindings were not observed in the yeast two-hybrid assay, probably because of weak interactions in yeast (data not shown). These results indicated that, besides regulating the expression of the single-repeat R3 MYB genes, LEC1 might mediate trichome formation via direct protein interactions with these repressors, a result that requires further investigation.

GL2 Genetically Acts Downstream of LEC1

GLABRA2 functions as the key transcription activator in trichome development, whose expression is activated by the trichome-positive MYB-bHLH-WD and repressed by the trichome-negative single-repeat R3 MYB-bHLH-WD transcription complexes (Zhao et al., 2008). Loss-of-function of *GL2* completely abolished trichome formation on the first pair of true leaves in early seedlings (**Figure 5**). The elevated *GL2* expression in the *lec1* (**Figure 2A**) implied an epistatic effect of *GL2* on *LEC1* in trichome formation on cotyledons. To test this hypothesis, we created a *gl2 lec1* double mutant and found that similar to *gl2*, the *gl2 lec1* double mutant displayed glabrous true leaves (**Figure 5**). In contrast, the *gl2* mutation repressed the ectopic trichome formation on cotyledons of *lec1* seedlings remarkably (**Figure 5**), confirming the primary role of GL2 in trichome formation, regardless of the developmental stage. These results strongly supported the hypothesis that *GL2* is required for *LEC1*-mediated repression of trichome formation on cotyledons.

DISCUSSION

Post-embryonic development is a vital transition state from embryonic status to vegetative growth, during which seedlings sense various environmental cues, such as light and temperature, to enable young sporophyte survival. At this stage, cell fate determination occurs to promote post-embryonic organogenesis that arises from the primary meristems, during which the visible features, such as cotyledons extension, hypocotyl elongation, and the emergence of true leaves with trichomes, can be observed in seedlings. Studies have shown that multiple endogenous factors, including phytohormones and transcription regulators, are involved in post-embryonic seedling development (Perianez-Rodriguez et al., 2014). LEC1, the master regulator in embryogenesis, was recently reported to be involved in light-mediated post-embryonic hypocotyl elongation (Junker and

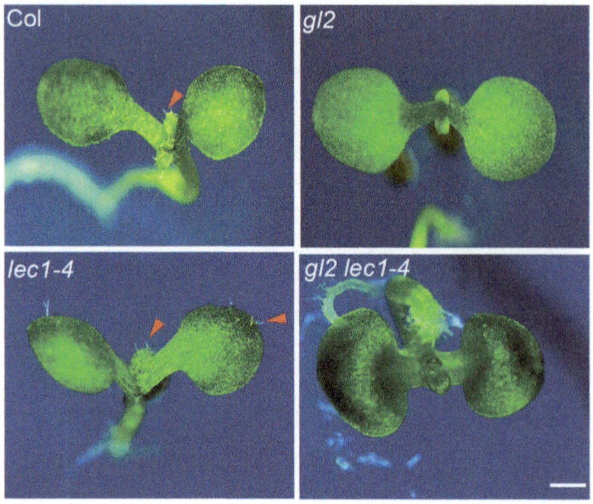

FIGURE 5 | Loss-of-function of *GL2* rescues the phenotype of ectopic trichome formation on *lec1* cotyledons. Ten-day-old Col, *gl2*, *lec1*, and *gl2 lec1* seedlings were used for the phenotype observation. A red arrowhead indicates the trichomes. Bar = 1 mm.

Baumlein, 2012). In this study, we provide molecular evidence that *LEC1* and its genetic downstream partner *GL2* are required for the repression of trichome formation on embryo-derived cotyledons, the characteristic that distinguishes them from true leaves. Mutation of *LEC1* resulted in the misregulation of several trichome-related genes and thus triggered trichome formation on cotyledons. By contrast, ectopic LEC1 activates the expression of the single-repeat *R3 MYB* genes and induces embryonic leaves with few or no trichomes. Furthermore, the results indicated that LEC1 might regulate trichome formation by interacting with other transcription factors (e.g., TCL2 and its homologs). Taken together, our findings support the hypothesis that LEC1 plays an essential role in cell fate determination during the post-embryonic development.

The fact that *LEC1* is expressed in both developing seeds and post-embryonic seedlings, combined with the pleiotropic phenotype of the *lec1* mutant, supports the multifunctional roles of *LEC1* in plant development (Junker and Baumlein, 2012; Supplementary Figure S1A); however, less is known about the downstream proteins of the LEC1-mediated regulatory process. A recent report indicated that LEC1 might regulate directly a number of genes involved in light, phytohormone signaling, and embryogenesis, including several bHLH and MYB transcription factors (Junker et al., 2012). However, using a chromatin immunoprecipitation assay, we observed that LEC1 does not target the selected trichome-related genes directly (data not shown), indicating that these genes might not act immediately downstream of LEC1. Instead, we observed that LEC1 interacts with the trichome development-related transcription repressors TCL1, TCL2, CPC, and ETC1 *in vitro*. *TCL1* was reported to be highly expressed in developing seeds and its protein targets the *GL1* promoter, while *TCL2* is highly expressed in cotyledons, rather than in developing seeds, implying distinct roles of these two homologous genes in trichome regulation (Wang et al., 2007; Gan et al., 2011). The interactions between LEC1 and these repressors provide a possible mechanism by which LEC1 regulates embryonic cell fate determination spatio-temporally, via coupling with different transcription factors. Nevertheless, whether LEC1 and TCL2 together determine the post-embryonic trichome initiation requires further investigation.

Phytohormones serve as crucial endogenous integrators to mediate plant development by controlling precisely exact cell division and differentiation (Lumba et al., 2010; Vanstraelen and Benkova, 2012). Gibberellin (GA) has been reported to act as a positive endogenous factor to promote trichome formation, because the GA biosynthesis defective mutant *ga1* displays fewer trichomes on its leaves (Perazza et al., 1998; Qi et al., 2014). LEC2 (Stone et al., 2001) and FUSCA3 (FUS3, Keith et al., 1994), two B3 domain-containing transcription factors, also defined as the *LEC* genes, promote *GL1* expression via direct activation of the GA biosynthesis gene, *ATGA3ox2*. When crossed to *ga1*, the ectopic trichome formation on *lec2* and *fus3* cotyledons was completely suppressed (Curaba et al., 2004). However, the ectopic trichomes were not abolished in *lec1 ga1* double mutant, which suggested that LEC1-mediated trichome formation is independent of GA (Curaba et al., 2004). Given that *LEC2* and *FUS3* act as potential downstream candidates of LEC1 (Junker et al., 2010), GA's effect on the function of LEC factors might be under the control of a complicated regulatory network. Abscisic acid is also involved in LEC1-mediated trichome formation. A previous study reported that the conversion of leaves into cotyledon-like structures is enhanced in the *LEC1* overexpression line when coupled with abscisic acid treatment (Junker et al., 2012). In addition, a number of abscisic acid-responsive genes were shown to be the direct targets of LEC1 (Junker et al., 2012). Nevertheless, the role of phytohormones in LEC1-mediated cell fate determination remains elusive and requires further investigation.

Accurate control of *LEC1* expression is critical for plants to regulate cell fate determination during development. Epigenetic regulation plays key roles in the exclusion of *LEC1* in the vegetative growth stage. PICKLE, a CHD3 chromatin-remodeling factor, represses *LEC1* transcriptional activation in vegetative tissue by modifying histone methylation. In the *pickle* mutant, the seedling root produced an embryonic structure, termed the "PICKLE root", caused by the ectopic expression of *LEC1* (Ogas et al., 1997, 1999; Li et al., 2005). In addition, the polycomb group proteins CLF and SWN were demonstrated to couple with PICKLE to repress the conversion from vegetative tissue to an embryonic structure (Chanvivattana et al., 2004; Aichinger et al., 2009). Histone deacetylases (e.g., HDA6 and HDA19) also contribute to the repression of embryonic properties by regulating several embryogenesis regulators, including *LEC1* (Tanaka et al., 2008). In addition to these epigenetic factors, some transcription factors, such as VP1/ABI3-LIKE and MYB115/118, were also reported as negative regulators of *LEC1* (Suzuki et al., 2007; Wang et al., 2009; Jia et al., 2013). Embryonic development originates from the zygote after fertilization in higher plants (Lau et al., 2012). High expression during embryogenesis indicates that LEC1 is released from the suppressing environment of the vegetative tissue, thus allowing it to controls embryonic and post-embryonic cell fate determination.

CONCLUSION

In this study, we revealed that LEC1 interacts with other transcription factors and regulates trichome-related genes expression to repress the trichome formation on cotyledons. These findings provide evidence that supports the hypothesis that LEC1, probably by integrating phytohormone signals and epigenetic regulation, mediates cell fate determination during the post-embryonic development phase, providing new insights into the role of LEC1 beyond embryogenesis.

ACCESSION NUMBERS

Sequence data from this article can be found in the *Arabidopsis* Genome Initiative database under the following accession numbers: *LEC1* (AT1G21970), *GL1* (AT3G27920), *GL2* (AT1G79840), *GL3* (AT5G41315), *EGL3* (AT1G63650), *TCL1* (AT2G30432), *TCL2* (AT2G30424), *TRY* (AT5G53200), *CPC* (AT2G46410), *ETC1* (AT1G01380), *ETC2* (AT2G30420), and *TUB2* (AT5G62690).

AUTHOR CONTRIBUTIONS

MH and XH designed the research. MH, YH, XL, and YL performed experiments. MH, YH, and XH analyzed data. MH and XH wrote the paper.

FUNDING

This work was supported by grants from the National Science Foundation of China (No. 31370342, 31301055, and 31300239) and "Hundred Talents" program of the Chinese Academy of Sciences.

ACKNOWLEDGMENT

We thank Shucai Wang for providing *gl2* mutant seeds, the *Arabidopsis* Biological Resource Centre for *lec1-4* seeds, and Jianru Zuo for *pER10:LEC1-MYC* plasmid.

REFERENCES

Aichinger, E., Villar, C. B. R., Farrona, S., Reyes, J. C., Hennig, L., and Kohler, C. (2009). CHD3 proteins and polycomb group proteins antagonistically determine cell identity in *Arabidopsis*. *PLoS Genet.* 5:605. doi: 10.1371/journal.pgen.1000605

Brocard-Gifford, I. M., Lynch, T. J., and Finkelstein, R. R. (2003). Regulatory networks in seeds integrating developmental, abscisic acid, sugar, and light signaling. *Plant Physiol.* 131, 78–92. doi: 10.1104/pp.011916

Chandler, J. W. (2008). Cotyledon organogenesis. *J. Exp. Bot.* 59, 2917–2931. doi: 10.1093/Jxb/Ern167

Chanvivattana, Y., Bishopp, A., Schubert, D., Stock, C., Moon, Y. H., Sung, Z. R., et al. (2004). Interaction of polycomb-group proteins controlling flowering in *Arabidopsis*. *Development* 131, 5263–5276. doi: 10.1242/Dev.01400

Curaba, J., Moritz, T., Blervaque, R., Parcy, F., Raz, V., Herzog, M., et al. (2004). AtGA3ox2, a key gene responsible for bioactive gibberellin biosynthesis, is regulated during embryogenesis by LEAFY COTYLEDON2 and FUSCA3 in *Arabidopsis*. *Plant Physiol.* 136, 3660–3669. doi: 10.1104/pp.104.047266

Gan, L. J., Xia, K., Chen, J. G., and Wang, S. C. (2011). Functional characterization of TRICHOMELESS2, a new single-repeat R3 MYB transcription factor in the regulation of trichome patterning in *Arabidopsis*. *BMC Plant Biol.* 11:176. doi: 10.1186/1471-2229-11-176

Harada, J. J. (2001). Role of *Arabidopsis* LEAFY COTYLEDON genes in seed development. *J. Plant Physiol.* 158, 405–409. doi: 10.1078/0176-1617-00351

Hulskamp, M., Misera, S., and Jurgens, G. (1994). Genetic dissection of trichome cell-development in *Arabidopsis*. *Cell* 76, 555–566. doi: 10.1016/0092-8674(94)90118-X

Jia, H. Y., Mccarty, D. R., and Suzuki, M. (2013). Distinct roles of LAFL network genes in promoting the embryonic seedling fate in the absence of VAL repression. *Plant Physiol.* 163, 1293–1305. doi: 10.1104/pp.113.220988

Junker, A., and Baumlein, H. (2012). Multifunctionality of the LEC1 transcription factor during plant development. *Plant Signal. Behav.* 7, 1718–1720. doi: 10.4161/psb.22365

Junker, A., Hartmann, A., Schreiber, F., and Baumlein, H. (2010). An engineer's view on regulation of seed development. *Trends Plant Sci.* 15, 303–307. doi: 10.1016/j.tplants.2010.03.005

Junker, A., Monke, G., Rutten, T., Keilwagen, J., Seifert, M., Thi, T. M., et al. (2012). Elongation-related functions of LEAFY COTYLEDON1 during the development of *Arabidopsis thaliana*. *Plant J.* 71, 427–442. doi: 10.1111/j.1365-313X.2012.04999.x

Keith, K., Kraml, M., Dengler, N. G., and Mccourt, P. (1994). fusca3 – a heterochronic mutation affecting late embryo development in *Arabidopsis*. *Plant Cell* 6, 589–600. doi: 10.1105/Tpc.6.5.589

Kwong, R. W., Bui, A. Q., Lee, H., Kwong, L. W., Fischer, R. L., Goldberg, R. B., et al. (2003). LEAFY COTYLEDON1-LIKE defines a class of regulators essential for embryo development. *Plant Cell* 15, 5–18. doi: 10.1105/tpc.006973

Lau, S., Slane, D., Herud, O., Kong, J. X., and Jurgens, G. (2012). Early embryogenesis in flowering plants: setting up the basic body pattern. *Annu. Rev. Plant Biol.* 63, 483–506. doi: 10.1146/annurev-arplant-042811-105507

Le, B. H., Cheng, C., Bui, A. Q., Wagmaister, J. A., Henry, K. F., Pelletier, J., et al. (2010). Global analysis of gene activity during *Arabidopsis* seed development and identification of seed-specific transcription factors. *Proc. Natl. Acad. Sci. U.S.A.* 107, 8063–8070. doi: 10.1073/pnas.1003530107

Lee, H. S., Fischer, R. L., Goldberg, R. B., and Harada, J. J. (2003). *Arabidopsis* LEAFY COTYLEDON1 represents a functionally specialized subunit of the CCAAT binding transcription factor. *Proc. Natl. Acad. Sci. U.S.A.* 100, 2152–2156. doi: 10.1073/pnas.0437909100

Li, H. C., Chuang, K., Henderson, J. T., Rider, S. D., Bai, Y., Zhang, H., et al. (2005). PICKLE acts during germination to repress expression of embryonic traits. *Plant J.* 44, 1010–1022. doi: 10.1111/j.1365-313X.2005.02602.x

Lotan, T., Ohto, M., Yee, K. M., West, M. A., Lo, R., Kwong, R. W., et al. (1998). *Arabidopsis* LEAFY COTYLEDON1 is sufficient to induce embryo development in vegetative cells. *Cell* 93, 1195–1205. doi: 10.1016/S0092-8674(00)81463-4

Lumba, S., Cutler, S., and Mccourt, P. (2010). Plant nuclear hormone receptors: a role for small molecules in protein-protein interactions. *Annu. Rev. Cell Dev. Biol.* 26, 445–469. doi: 10.1146/annurev-cellbio-100109-103956

Marks, M. D., and Feldmann, K. A. (1989). Trichome development in *Arabidopsis-thaliana*.1. T-DNA tagging of the Glabrous1 gene. *Plant Cell* 1, 1043–1050. doi: 10.1105/tpc.1.11.1043

Meinke, D. W. (1992). A homoeotic mutant of *Arabidopsis thaliana* with leafy cotyledons. *Science* 258, 1647–1650. doi: 10.1126/science.258.5088.1647

Meinke, D. W., Franzmann, L. H., Nickle, T. C., and Yeung, E. C. (1994). Leafy cotyledon mutants of *Arabidopsis*. *Plant Cell* 6, 1049–1064. doi: 10.1105/tpc.6.8.1049

Morohashi, K., Zhao, M. Z., Yang, M. L., Read, B., Lloyd, A., Lamb, R., et al. (2007). Participation of the *Arabidopsis* bHLH factor GL3 in trichome initiation regulatory events. *Plant Physiol.* 145, 736–746. doi: 10.1104/pp.107.104521

Mu, J., Tan, H., Zheng, Q., Fu, F., Liang, Y., Zhang, J., et al. (2008). LEAFY COTYLEDON1 is a key regulator of fatty acid biosynthesis in *Arabidopsis*. *Plant Physiol.* 148, 1042–1054. doi: 10.1104/pp.108.126342

Ogas, J., Cheng, J. C., Sung, Z. R., and Somerville, C. (1997). Cellular differentiation regulated by gibberellin in the *Arabidopsis thaliana* pickle mutant. *Science* 277, 91–94. doi: 10.1126/science.277.5322.91

Ogas, J., Kaufmann, S., Henderson, J., and Somerville, C. (1999). PICKLE is a CHD3 chromatin-remodeling factor that regulates the transition from embryonic to vegetative development in *Arabidopsis*. *Proc. Natl. Acad. Sci. U.S.A* 96, 13839–13844. doi: 10.1073/pnas.96.24.13839

Oppenheimer, D. G., Herman, P. L., Sivakumaran, S., Esch, J., and Marks, M. D. (1991). A myb gene required for leaf trichome differentiation in *Arabidopsis* is expressed in stipules. *Cell* 67, 483–493. doi: 10.1016/0092-8674(91)90523-2

Pattanaik, S., Patra, B., Singh, S. K., and Yuan, L. (2014). An overview of the gene regulatory network controlling trichorne development in the model plant, *Arabidopsis*. *Front. Plant Sci.* 5:259. doi: 10.3389/Fpls.2014.00259

Perazza, D., Vachon, G., and Herzog, M. (1998). Gibberellins promote trichome formation by up-regulating GLABROUS1 in *Arabidopsis*. *Plant Physiol.* 117, 375–383. doi: 10.1104/Pp.117.2.375

Perianez-Rodriguez, J., Manzano, C., and Moreno-Risueno, M. A. (2014). Post-embryonic organogenesis and plant regeneration from tissues: two sides of the same coin? *Front. Plant Sci.* 5:219. doi: 10.3389/Fpls.2014.00219

Peris, C. I. L., Rademacher, E. H., and Weijers, D. (2010). Green beginnings - pattern formation in the early plant embryo. *Plant Dev.* 91, 1–27. doi: 10.1016/S0070-2153(10)91001-6

Qi, T. C., Huang, H., Wu, D. W., Yan, J. B., Qi, Y. J., Song, S. S., et al. (2014). *Arabidopsis* DELLA and JAZ proteins bind the WD-Repeat/bHLH/MYB complex to modulate gibberellin and jasmonate signaling synergy. *Plant Cell* 26, 1118–1133. doi: 10.1105/tpc.113.121731

Rerie, W. G., Feldmann, K. A., and Marks, M. D. (1994). The Glabra2 gene encodes a homeo domain protein required for normal trichome development in *Arabidopsis*. *Genes Dev.* 8, 1388–1399. doi: 10.1101/gad.8.12.1388

Stone, S. L., Kwong, L. W., Yee, K. M., Pelletier, J., Lepiniec, L., Fischer, R. L., et al. (2001). LEAFY COTYLEDON2 encodes a B3 domain transcription factor that induces embryo development. *Proc. Natl. Acad. Sci. U.S.A.* 98, 11806–11811. doi: 10.1073/pnas.201413498

Suzuki, M., Wang, H. H. Y., and Mccarty, D. R. (2007). Repression of the LEAFY COTYLEDON 1/B3 regulatory network in plant embryo development by VP1/ABSCISIC ACID INSENSITIVE 3-LIKE B3 genes. *Plant Physiol.* 143, 902–911. doi: 10.1104/pp.106.092320

Tanaka, M., Kikuchi, A., and Kamada, H. (2008). The *Arabidopsis* histone deacetylases HDA6 and HDA19 contribute to the repression of embryonic properties after germination. *Plant Physiol.* 146, 149–161. doi: 10.1104/pp.107.111674

Vanstraelen, M., and Benkova, E. (2012). Hormonal interactions in the regulation of plant development. *Annu. Rev. Cell Dev. Biol.* 28, 463–487. doi: 10.1146/annurev-cellbio-101011-155741

Wang, S. C., Barron, C., Schiefelbein, J., and Chen, J. G. (2010). Distinct relationships between GLABRA2 and single-repeat R3 MYB transcription factors in the regulation of trichome and root hair patterning in *Arabidopsis*. *New Phytol.* 185, 387–400. doi: 10.1111/j.1469-8137.2009.03067.x

Wang, S. C., and Chen, J. G. (2014). Regulation of cell fate determination by single-repeat R3 MYB transcription factors in *Arabidopsis*. *Front. Plant Sci.* 5:133. doi: 10.3389/Fpls.2014.00133

Wang, S. C., Hubbard, L., Chang, Y., Guo, J. J., Schiefelbein, J., and Chen, J. G. (2008). Comprehensive analysis of single-repeat R3 MYB proteins in epidermal cell patterning and their transcriptional regulation in *Arabidopsis*. *BMC Plant Biol.* 8:81. doi: 10.1186/1471-2229-8-81

Wang, S. C., Kwak, S. H., Zeng, Q. N., Ellis, B. E., Chen, X. Y., Schiefelbein, J., et al. (2007). TRICHOMELESS1 regulates trichome patterning by suppressing GLABRA1 in *Arabidopsis*. *Development* 134, 3873–3882. doi: 10.1242/Dev.009597

Wang, X. C., Niu, Q. W., Teng, C., Li, C., Mu, J. Y., Chua, N. H., et al. (2009). Overexpression of PGA37/MYB118 and MYB115 promotes vegetative-to-embryonic transition in *Arabidopsis*. *Cell Res.* 19, 224–235. doi: 10.1038/Cr.2008.276

Warpeha, K. M., Upadhyay, S., Yeh, J., Adamiak, J., Hawkins, S. I., Lapik, Y. R., et al. (2007). The GCR1, GPA1, PRN1, NF-Y signal chain mediates both blue light and abscisic acid responses in *Arabidopsis*. *Plant Physiol.* 143, 1590–1600. doi: 10.1104/pp.106.089904

West, M., Yee, K. M., Danao, J., Zimmerman, J. L., Fischer, R. L., Goldberg, R. B., et al. (1994). LEAFY COTYLEDON1 is an essential regulator of late embryogenesis and cotyledon identity in *Arabidopsis*. *Plant Cell* 6, 1731–1745. doi: 10.1105/tpc.6.12.1731

Yamamoto, A., Kagaya, Y., Toyoshima, R., Kagaya, M., Takeda, S., and Hattori, T. (2009). *Arabidopsis* NF-YB subunits LEC1 and LEC1-LIKE activate transcription by interacting with seed-specific ABRE-binding factors. *Plant J.* 58, 843–856. doi: 10.1111/j.1365-313X.2009.03817.x

Yang, C. X., and Ye, Z. B. (2013). Trichomes as models for studying plant cell differentiation. *Cell. Mol. Life Sci.* 70, 1937–1948. doi: 10.1007/s00018-012-1147-6

Zhao, M., Morohashi, K., Hatlestad, G., Grotewold, E., and Lloyd, A. (2008). The TTG1-bHLH-MYB complex controls trichome cell fate and patterning through direct targeting of regulatory loci. *Development* 135, 1991–1999. doi: 10.1242/Dev.016873

Zhou, L. M., Zheng, K. J., Wang, X. Y., Tian, H. N., Wang, X. L., and Wang, S. C. (2014). Control of trichome formation in *Arabidopsis* by poplar single-repeat R3 MYB transcription factors. *Front. Plant Sci.* 5:262. doi: 10.3389/Fpls.2014.00262

Conflict of Interest Statement: The authors declare that the research was conducted in the absence of any commercial or financial relationships that could be construed as a potential conflict of interest.

8

Atypical centromeres in plants—what they *can* tell us

*Maria Cuacos, F. Chris H. Franklin and Stefan Heckmann**

School of Biosciences, University of Birmingham, Birmingham, UK

The centromere, visible as the primary constriction of condensed metaphase chromosomes, is a defined chromosomal locus essential for genome stability. It mediates transient assembly of a multi-protein complex, the kinetochore, which enables interaction with spindle fibers and thus faithful segregation of the genetic information during nuclear divisions. Centromeric DNA varies in extent and sequence composition among organisms, but a common feature of almost all active eukaryotic centromeres is the presence of the centromeric histone H3 variant cenH3 (a.k.a. CENP-A). These typical centromere features apply to most studied species. However, a number of species display "atypical" centromeres, such as holocentromeres (centromere extension along almost the entire chromatid length) or neocentromeres (ectopic centromere activity). In this review, we provide an overview of different atypical centromere types found in plants including holocentromeres, *de novo* formed centromeres and terminal neocentromeres as well as di-, tri- and metapolycentromeres (more than one centromere per chromosomes). We discuss their specific and common features and compare them to centromere types found in other eukaryotic species. We also highlight new insights into centromere biology gained in plants with atypical centromeres such as distinct mechanisms to define a holocentromere, specific adaptations in species with holocentromeres during meiosis or various scenarios leading to neocentromere formation.

Edited by:
Inna Lermontova,
Institute of Plant Genetics and Crop Plant Research, Germany

Reviewed by:
Kiyotaka Nagaki,
Okayama University, Japan
Paul Talbert,
Howard Hughes Medical Institute, USA

***Correspondence:**
Stefan Heckmann
s.heckmann@bham.ac.uk

Keywords: centromere, kinetochore, holocentric chromosomes, neocentromeres, cenH3, plants, mitosis, meiosis

CENTROMERE TYPES

Centromeres are chromosomal loci where kinetochores assemble. Kinetochore proteins mediate cell cycle regulation, sister chromatid cohesion, spindle microtubule attachment and chromosome movements (Lermontova et al., 2014). These functions are essential for genome stability by mediating faithful mitotic and meiotic chromosome segregation. Any failure leads to chromosome missegregation and ultimately genome instability.

Kinetochore establishment and centromere maintenance in active eukaryotic centromeres generally depends on the presence of the centromeric histone H3 variant cenH3 (also called CENP-A in mammals; De Rop et al., 2012). Although essential for genome integrity, contrary to expectation centromeric cenH3 localization is not specified by centromere specific DNA sequence(s) except in budding yeast (Clarke and Carbon, 1985). It is rather determined epigenetically. Centromere loci and centromeric DNAs are highly diverse varying dramatically in size and sequence composition between species. Centromeres can range in size from the 125 bp "point" centromeres in budding yeast up to mega bp-sized "regional" centromeres in humans and plants. In the most extreme case, poly- or holocentromeres can even extend over the entire chromosome length.

Although centromeric DNAs are not conserved often plant centromeres contain distinct satellite DNA sequences and families of long terminal repeat (LTR) retrotransposons (Houben and Schubert, 2003; Neumann et al., 2011). However, these repeats are neither necessary nor sufficient for centromere activity since gain of new centromeric activity over unique DNA sequences can occur (Nasuda et al., 2005; Han et al., 2006).

Centromeric DNAs are one of the fastest evolving sequences in eukaryotic genomes (Bensasson, 2011; Melters et al., 2013). It is interesting that such an essential and functionally conserved chromosomal locus has so rapidly evolved with regards to its structure, extension and DNA sequence composition. For instance, (i) the sequence composition and centromere extension vary dramatically between closely-related species, e.g., *Solanum* (Zhang et al., 2014) or *Oryza* species (Yi et al., 2013), and even between centromeres within one species, e.g., *Pisum sativum* (Macas et al., 2007), (ii) different centromere types, e.g., mono- and holocentromeres, have evolved between different insect lineages (Drinnenberg et al., 2014) and even between closely-related dodder species (Pazy and Plitmann, 1994, 1995), or (iii) albeit functionally similar, unconventional centromeres, e.g., cenH3- and CENP-C-independent insect holocentromeres (Drinnenberg et al., 2014) or "meta-polycentric" centromeres in *Pisum* or *Lathyrus* (Neumann et al., 2012, 2015), and unconventional kinetochores, e.g., kinetoplastid kinetochores devoid of any conventional components (Akiyoshi and Gull, 2014), have evolved.

Studies on "atypical" plant centromeres such as neocentromeres or holocentromeres have contributed to our general knowledge of the structure, regulation and function of centromeres. In this review, we focus on such unusual centromere types in plants, highlight recent discoveries and discuss their implications.

HOLOCENTROMERES

Most studied organisms possess one size-restricted centromere (monocentromere) per chromosome (**Figure 1A**). However, in various species so-called holocentromeres ("holo-" from Greek: entire) initially described by Schrader (1935), characterized by an almost chromosome-wide extension occur (**Figure 1B**). They are also called diffuse centromeres or polycentromeres—for the rest of the review we will use the terms holocentromere or holocentric chromosome. Holocentromeres evolved by convergent evolution in diverse eukaryotic lineages including green algae, invertebrates, and plants (Melters et al., 2012). Around 800 species as diverse as nematodes, spiders, and sedges are reported to possess holocentromeres.

CHROMOSOME CLASSIFICATION: MONOCENTROMERE VS. HOLOCENTROMERE

If a given chromosome is comparatively large, classical cytogenetic techniques are applicable for the classification as holo- or monocentric. In the case of a holocentromere, a mitotic metaphase chromosome lacks a primary constriction and during anaphase

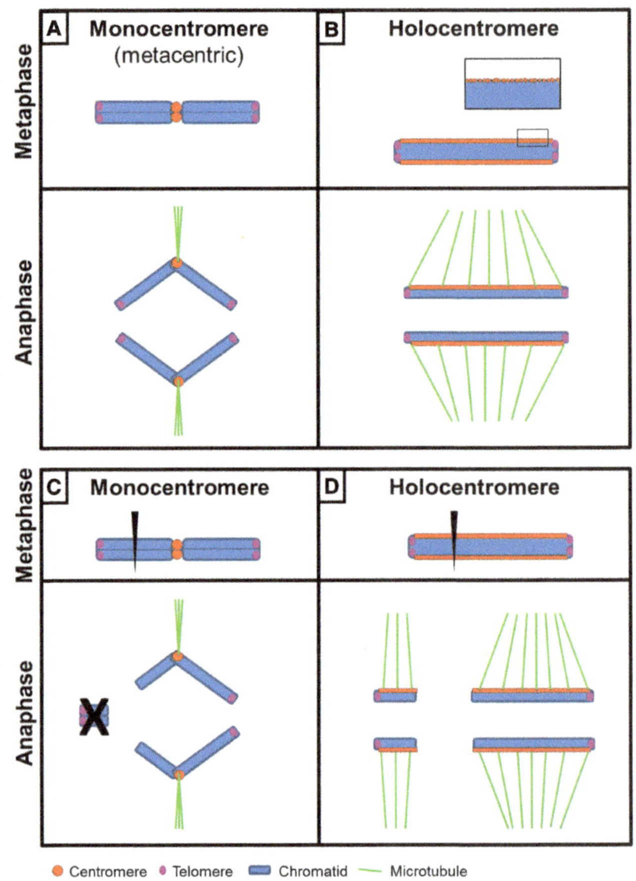

FIGURE 1 | Structure and behavior of a monocentric and a holocentric chromosome. (A) A metacentric chromosome shows a primary constriction during metaphase. During anaphase chromatids move as V-shaped structures due to microtubule attachment to the size-restricted centromere. **(B)** A holocentric chromosome shows an almost chromosome-wide centromere extension and no primary constriction during metaphase. Sister chromatids are not discernible. During anaphase spindle microtubule attachment to the holocentromere results in chromatids moving as linear bars parallel to the spindle. Inset, various centromeric subdomains fuse to one functional composite linear holocentromere at metaphase. **(C)** Breakage of a monocentric chromosome results in loss of the acentric chromosome fragment during anaphase, whereas **(D)** after chromosome breakage of a holocentric chromosome both fragments retain kinetic activity due to the almost chromosome-wide centromere extension and thus can be transmitted. Note absence of telomeric repeats at broken chromosome ends. In case of holocentric chromosomes of *Luzula elegans*, rapid telomere-mediated "chromosome healing" occurs (Jankowska et al., 2015).

chromatids move as linear bars parallel to the spindle due to almost chromosome-wide spindle attachment (holokinetic behavior; **Figure 1B**). If a chromosome is monocentric, it bears a primary constriction and the chromatids either move as a v-shaped structure (metacentric), as a linear bar perpendicular to the spindle pole (acrocentric) or as a configuration in between both extreme cases due to microtubule attachment to the defined size-restricted centromere (**Figure 1A**).

A more direct approach to classify the chromosome structure is through immunolocalization of centromere components. Although many components are functionally conserved, they are often divergent in sequence composition even between closely-related species. Thus, centromere-related antibodies are not universally available. The discovery of phosphorylation of threonine 120 of histone H2A (H2AThr120ph) as universal mark for active centromeres in plant species with mono- and holocentromeres will allow the classification of (comparatively large) chromosomes to be refined (Demidov et al., 2014).

HOLOCENTROMERES IN PLANTS

In flowering plants, holocentromeres are found among the monocots *Cyperaceae* (sedges), *Juncaceae* (rushes; Malheiros et al., 1947; Hakansson, 1958) and *Chionographis* (string flowers; Tanaka, 1977) as well as in dicots such as *Cuscuta* subgenus *Cuscuta* (dodders; Pazy and Plitmann, 1995), *Drosera* (sundews; Sheikh et al., 1995), or in the nutmeg tree *Myristica fragrans* (Flach, 1966). There is evidence for 228 plant species with holocentromeres (Melters et al., 2012). This number is likely an underestimate since for instance many species possess comparatively small chromosomes and thus chromosome classification is cytologically challenging. In addition, all *Juncaceae* and *Cyperaceae* are predicted to be holocentric (the genus *Carex* within *Cyperaceae* consists of around 2000 species, Reznicek, 1990), which actually suggests an even higher number of species with holocentromeres within only these two families. However, there are contradictory reports such as in the genus *Drosera*, where species are reported to possess holocentromeres (Kondo and Lavarack, 1984; Sheikh and Kondo, 1995; Sheikh et al., 1995) and monocentromeres (Shirakawa et al., 2011; Demidov et al., 2014).

KARYOTYPE EVOLUTION IN SPECIES WITH HOLOCENTROMERES

A chromosome-wide centromere organization allows, in theory, rapid karyotype evolution. Unlike monocentric chromosomes, basically each part of a given holocentric chromosome has centromere activity and thus theoretically a high probability of being transmitted during nuclear divisions after chromosome breakage (**Figures 1C,D**). This karyotype flexibility conferred by a holocentromere is reflected in (i) the extremely wide and almost continuous chromosome number found among related holocentric species, e.g., *Carex* 2n = 12–124, *Eleocharis* 2n = 6–196, or *Juncus* 2n = 18–170 (Bureš et al., 2013), (ii) interspecies chromosome number variation, e.g., *Eleocharis kamtschatica* with 2n = 41–47 (Yano and Hoshino, 2006), or (iii) the negative correlation between chromosome number and chromosome size in *Luzula* (e.g., Nordenskiold, 1951; Jankowska et al., 2015).

Typically double strand breaks (DSBs) in DNA are resolved by non-homologous end-joining or homologous recombination (Knoll et al., 2014). However, telomerase can add telomeric repeats at break sites leading to "chromosome healing" (Tsujimoto, 1993; Tsujimoto et al., 1999; Nelson et al., 2011). Monocentric chromosome healing at DSB sites results in deletion or loss of the distal acentric chromosome fragment and thus needs to be tightly regulated. Unlike monocentric chromosomes, stable transmission of (artificially induced) holocentric chromosome fragments during mitosis and meiosis suggests that chromosome healing is a more or less common phenomenon in holocentric chromosome species (e.g., Hakansson, 1954; Nordenskiold, 1964). Recently, Jankowska et al. (2015) showed that irradiation of holocentric *Luzula elegans* chromosomes results in a range of heteromorphic derived karyotypes. Independent of their size all chromosomes/fragments showed centromere activity and gradual telomere-mediated "chromosome healing." Newly formed telomere repeats were cytologically detectable 21 days after irradiation in ~50% of cases, increasing to >95% after 3 months. In the progenies of the irradiated plants all the chromosomes/fragments possessed telomeric repeats. This rapid and efficient *de novo* telomere formation is likely conferred by a telomerase-mediated healing process and important for fragment stabilization/karyotype fixation.

The combination of holokinetic chromosomes and rapid telomere formation at DSBs allows stable transmission of chromosome fragments and thus rapid karyotype evolution. Additionally, polyploidy and proliferation/removal of high copy sequences are involved in rapid genome evolution (e.g., Kuta et al., 2004; Zedek et al., 2010; Bozek et al., 2012). However, how holocentric chromosome species can deal with this cytological "chromosomal chaos" is largely unknown. In holocentric *Lepidoptera* with a high intraspecific cytogenetic variation, a high degree of synteny at fine scales is found, suggesting an adaptive mechanism (d'Alencon et al., 2010). Similar studies in plants are missing.

STRUCTURE AND BEHAVIOR OF HOLOCENTRIC CHROMOSOMES

In plant holocentromeres, cenH3 is found along mitotic chromosomes representing active centromeres as in species with monocentromeres (Nagaki et al., 2005). In some *Luzula* and *Rhynchospora* species with comparatively large chromosomes cenH3-positive chromosome regions form a groove-like structure except at chromosome ends during mitotic metaphase (Nagaki et al., 2005; Heckmann et al., 2011; Cabral et al., 2014; Wanner et al., 2015). It seems likely that a centromeric groove is a structural adaptation of relatively large holocentric chromosomes or a distinct evolutionary accommodation within certain genera. Ultrastructural analysis of mitotic *L. elegans* chromosomes showed that cenH3 containing chromatin is found at the periphery of each individual chromatid and that microtubules attach to cenH3- and not H2AThr120ph-chromatin during mitosis (Wanner et al., 2015). H2AThr120ph is enriched in the centromeric groove and completely absent along the axis where chromatids are in close contact, suggesting that H2AThr120ph is not involved in holocentric sister chromatid cohesion (Wanner et al., 2015). No differentiation between holocentric chromatids is found microscopically in *L. elegans* probably owing to almost chromosome-wide centromeric cohesion (Heckmann et al., 2011; Wanner et al., 2015). Mitotic sister chromatids are only discernible after staining sister-chromatid exchanges (Heckmann and Houben, 2013). Microtubule attachment regions are concentrated on the pericentromeric rims, possibly increasing attachment stability during separation of sister chromatids. Bundles of 2–4 individual microtubules are distributed along the entire centromere length with a mean distance between individual bundles of 300–400 nm during mitosis and 350–500 nm during meiosis (Heckmann et al., 2014a; Wanner et al., 2015). In *C. elegans* with much smaller holocentromeres the number of microtubules attachments is ~85 genome-wide or ~15 per chromosome (O'Toole et al., 2003). In summary, H2AThr120ph and cenH3 are found within the longitudinal centromeric groove of (large) holocentric plant metaphase chromosomes and microtubule attachment is enriched along the groove rim.

The almost chromosome-wide holocentromere architecture is also reflected in the distribution of epigenetic marks. The cell cycle-dependent phosphorylation of serine 10 or serine 28 of H3 typically enriched in pericentromere regions of monocentric plant chromosomes (Houben et al., 1999; Gernand et al., 2003) occurs in *Luzula* and *Rhynchospora* uniformly along the chromosomes (Gernand et al., 2003; Nagaki et al., 2005; Guerra et al., 2006). Similarly typical eu- and heterochromatin epigenetic marks or early/late DNA replicating chromatin domains are detected uniformly along *L. elegans* mitotic chromosomes at normal resolution (Heckmann et al., 2013). However, using super-high-resolution light microscopy interspersed units of various chromatin types were distinguished. Intermingling of different chromatin domains throughout the *L. elegans* genome is correlated with the distribution of highly repetitive DNA and likely reflects interplay between scattered chromosome-wide centromere organization and overall genome organization (Heckmann et al., 2013).

A scattered polycentric centromere arrangement is microscopically reflected in cenH3 dynamics during the cell cycle. During interphase cenH3 is found dispersed, in prophase as small foci along chromosomes and during metaphase as composite linear axial line along each sister chromatid (Buchwitz et al., 1999; Nagaki et al., 2005; Heckmann et al., 2011). In *C. elegans* a polycentric chromosome arrangement is revealed at fine scale resolution (Gassmann et al., 2012; Steiner and Henikoff, 2014). About 700 individual centromeric sites -single cenH3 nucleosomes flanked by well-spaced canonical nucleosomes- are preferentially found at dispersed sites of permissive chromatin (Steiner and Henikoff, 2015). CenH3 is also found with low density in roughly 2900 broad chromosome domains of low transcriptional activity and low nucleosome turnover that put together represent roughly half the genome (Gassmann et al., 2012; Steiner and Henikoff, 2014). Thus, *C. elegans* holocentromeres are polycentromeres consisting of individual point centromeres as the basic units of assembly.

A holo-/polycentric chromosome, composed of various centromeric subdomains, should have a high risk of misorientation during anaphase due to potential merotelic spindle attachment to individual subdomains. In *C. elegans*, chromokinesin KLP-19 counteracts persistent merotelic attachments (Powers et al., 2004). However, it is unclear how holocentric plants circumvent this holocentromere-associated challenge. In dicentric chromosomes, when a critical distance between two active centromeres is reached, the chromosome can break during anaphase due to not forming one functional centromere at metaphase (see below). In case of a holocentric chromosome, the distance between individual centromere subunits must be likewise restricted. In *C. elegans*, with comparatively small holocentric chromosomes, the genomic cenH3 distribution indicates a distance between individual centromere subunits of maximally 1.9 Mb ranging from 290 bp to 1.9 Mb with a median of 83 kb (Steiner and Henikoff, 2014). Notably, during divisions probably not all centromeric regions (only ~15%) are kinetically active in *C. elegans*, thus the maximum distance might be even higher. Similar studies in plants with (larger) holocentric chromosomes are, to date, lacking. Thus, it is unclear what the maximum functional inter-subunit distance tolerated between individual centromere units of a given holocentric plant chromosome is and additionally, it is unclear whether as in *C. elegans*, only a subset of these centromeric domains are kinetically active.

HOLOCENTROMERE IDENTITY

Recently, holocentromere-enriched satellite DNA sequences and retrotransposons preferentially bound by cenH3 were found in *R. pubera* (Marques et al., 2015) similar to most plant monocentromeres (Houben and Schubert, 2003; Neumann et al., 2011). Thus, also in species with holocentric chromosomes centromere-specific repetitive DNAs can occur. Stretched *Pisum* chromosomes show multiple centromeres consisting of satellite DNAs (Neumann et al., 2012, 2015). This "meta-polycentricity" may be an evolutionary link toward the development of holocentromeres in species such as *Rhynchospora*.

Unlike *Rhynchospora*, in *L. elegans* neither typical centromere-associated retrotransposons nor any holocentromere-associated satellite DNAs are found (Heckmann et al., 2013). Thus, cenH3 may be associated with a centromere-specific chromatin status rather than with specific centromeric DNA sequences. In *L. nivea* the 178-bp tandem repeat sequence LCS1 (Haizel et al., 2005) which shares some similarity with the centromeric tandem repeat RCS2 of rice (Dong et al., 1998; Nonomura and Kurata, 2001) has been described. Whether LCS1 plays a centromeric role is uncertain.

In *C. elegans* a short DNA motif is enriched at individual centromeric sites, however, it is likely not a direct target for cenH3 (Steiner and Henikoff, 2014). Accordingly, basically any DNA sequence can acquire centromere activity and extrachromosomal arrays are even segregated after few cycles in *C. elegans* (Stinchcomb et al., 1985; Yuen et al., 2011). Thus, in *C. elegans* centromeric nucleosomes are inherited epigenetically rather than being DNA sequence-dependent.

Limited available data suggest that there are different ways of defining holocentromeres with regards to centromeric sequences. In *R. pubera* there are centromere-specific satellite DNAs and retrotransposons, whereas in *L. elegans* and *C. elegans* no centromere-specific sequences are found. Thus, possibly different evolutionary scenarios with regards to centromere-specific DNA sequences led to the formation of holocentromeres. Further studies will clarify, e.g., how in *L. elegans* individual centromeric subunits are defined or whether between closely-related species with holo- and/or monocentromeres such as *Cuscuta* or *Drosera* different or similar centromeric DNA sequences occur.

MEIOTIC ADAPTATIONS OF HOLOCENTRIC CHROMOSOMES

Sexual reproduction is characterized by the process of meiosis, during which two consecutive rounds of chromosome segregation follow one single round of DNA replication generating haploid gametes. Cytologically, during the first meiotic division homologous chromosomes (homologs) are separated and during the second meiotic division chromatids disjoin. To allow faithful transmission typically homologs pair and perform reciprocal genetic exchange, termed crossover, physically linking homologs and thus ensuring balanced chromosome segregation during meiosis I. Proper chromosome segregation further depends on mono-orientation of fused sister kinetochores during meiosis I and on bi-orientation of sister kinetochores during meiosis II. In monocentromere species this is realized by a two-step loss of cohesion, i.e., along chromosome arms during meiosis I and at sister centromeres during meiosis II.

The two-step loss of cohesion is hampered in a holocentromere due to the lack of defined broad chromosomal centromere and arm domains allowing their spatial distinction. As adaptation, species with holocentromeres evolved different strategies to conduct faithful meiotic chromosome segregation, such as "chromosome remodeling" in *C. elegans* (Schvarzstein et al., 2010) or "functional monocentricity" in *Heteroptera* (Viera et al., 2009). Another alternative meiotic process is found in plants with holocentromeres characterized by separation of sister chromatids already during meiosis I (Cabral et al., 2014; Heckmann et al., 2014a). *Luzula* and *Rhynchospora* display a functional holocentromere throughout meiosis. Prophase I events are cytologically similar to those found in species with monocentric chromosomes including meiotic DSB induction and progression even in achiasmatic chromosomes of *R. tenuis*. At metaphase I, contrary to a monopolar orientation of sister monocentromeres, sister holocentromeres are unfused and interact, individually bi-orientated, with the meiotic spindle, resulting in a separation of sister chromatids already during meiosis I. The homologous non-sister chromatids are kept usually terminally linked by chromatin threads until anaphase II when they are separated ensuring haploidization. These chromatin threads are heterochromatic, enriched in satellite DNAs in *L. elegans* and possibly formed in a crossover-independent manner as their occurrence in achiasmatic chromosomes of *R. tenuis* suggests. Notably, monocentric chromosomes can also associate via chromatin threads (e.g., crane flies or Drosophila, LaFountain et al., 2002; Hughes et al., 2009), suggesting that terminal heterochromatin can be "sticky" enabling an achiasmatic association of homologs and the link of homologous non-sister chromatids in *Luzula* or *Rhynchospora* species. Whether an achiasmatic cohesin-mediated or another unknown mechanism is involved is unclear. Alternatively, these chromatin threads may be formed in a crossover-dependent manner in such a way that only distinct terminal chiasmata persist, being released later than those in interstitial regions. Although *R. tenuis* chromosomes are apparently achiasmatic, DSB and synaptonemal complex formation -prerequisites for crossover formation- suggest that distinct terminal crossovers mediating chromatin thread formation might occur in *R. tenuis*. Additionally, these threads are most likely not caused by catenated late replicating DNA, as homologs and non-sister chromatids are connected as, e.g., in *L. elegans*. Chromatin threads are not described for species with holocentromeres using a different meiotic mode. Therefore, chromatin threads are likely conducive for an inverted meiotic chromatid segregation process in species with holocentromeres (Heckmann et al., 2014b). Interestingly, reminiscent of the situation in monocentric species where centromeres are typically recombination cold spots (Yelina et al., 2012), in species with holocentromeres crossovers mainly occur in non-centromeric (i.e., terminal) chromosome regions. Thus, centromeric regions are crossover cold spots in both mono- and holocentromere species.

EVOLUTION OF HOLOCENTROMERES

An intriguing question is whether a holocentric or a monocentric chromosome structure appeared first during evolution. Nagaki et al. (2005) proposed for *Luzula* that a 90°direction turn of a monocentromere in an ancestral *Luzula* species together with subsequent centromere extension could be the basis of holocentricity. Neumann et al. (2012) proposed that spreading of centromere-competent satellite(s) was the cause of a transition from a monocentric to a polycentric chromosome structure in *L. nivea*. Villasante et al. (2007) proposed the "telomere to centromere" model that predicts an origin of holocentromeres from monocentromeres. The "centromere drive" hypothesis suggests a transition from mono- to holocentric chromosomes in order to suppress centromere drive (Malik and Henikoff, 2002). Melters et al. (2012) proposed that monocentricity was the ancestral chromosome configuration and that holocentricity evolved multiple times independently. Currently it is generally accepted that an independent transition from mono- into holocentromeres occurred in total on at least 13 occasions in eukaryotic lineages with the exception of vertebrates (four times in plants and nine times in animals; Melters et al., 2012).

In four holocentric insect lineages cenH3 and CENP-C were independently lost while other outer and inner kinetochore components such as NDC80 or MIS12 remain (Drinnenberg et al., 2014). In *Luzula*, *Rhynchospora*, and *C. elegans* this transition occurred without a loss of cenH3 (Buchwitz et al., 1999; Nagaki et al., 2005; Cabral et al., 2014). Two cenH3s are even found in *C. elegans*, Hcp-3 and Cpar-1 (Monen et al., 2005), and in *L. nivea*, LnCENH3-A and LnCENH3-B (Moraes et al., 2011). CenH3 is essential for mitosis but dispensable for male meiosis in *C. elegans* (Monen et al., 2005). Interestingly, the weakly expressed isoform, Cpar-1, specifically localizes to meiotic chromosomes and is cleaved in its N-terminal tail by separase at anaphase I (Monen et al., 2015). Whether these dynamics reflect a meiotic adaptation compensating for a holocentric chromosome structure during meiosis is unknown. In *L. nivea* two cenH3s are transcribed and at least LnCENH3-B is found in centromeric nucleosomes (Moraes et al., 2011). Whether both cenH3s are essential for centromere activity, are functionally diverged or redundant, and whether this duplication is related to holocentromere occurrence, is at present unclear.

Apparently, a transition from mono- into holocentromeres can evolve differently with regards to cenH3: either entire cenH3 loss in insects, "partial functional loss" being dispensable for male meiosis in *C. elegans*, or complete mitotic and meiotic retention of cenH3 in holocentric plants. Drinnenberg et al. (2014) proposed that an event early in the evolution of insects, e.g., a lineage-specific evolution of a centromeric protein similar to *Umbrea* in flies (Ross et al., 2013), enabled a cenH3 loss in holocentric insects. Unlike holocentric insects, in *L. nivea* and *C. elegans* holocentromeres coincide with two cenH3 variants. Also highly dynamic centromere architectures found in some *Fabeae* species positively correlate with the presence of two active cenH3 variants (Neumann et al., 2015). Possibly two cenH3 variants, albeit also found in monocentric species (e.g., Sanei et al., 2011), either enable or are the consequence of structural centromere changes. Thus, it is tempting to speculate that either any variation in the level of cenH3, loss or duplication, might enable centromere plasticity and thus structural centromere evolution or that structural changes in centromere architecture may render centromere-dependency on cenH3. In a nutshell, transitions from mono- into holocentromeres are likely based on distinct evolutionary scenarios rather than on one common pivotal event.

In *Cuscuta*, plants of the subgenera *Monogyna* and *Grammica* are reported to have monocentromeres while members of the subgenus *Cuscuta* are reported to have holocentromeres (Pazy and Plitmann, 1994, 1995). Similarly, some *Drosera* species have monocentromeres (Shirakawa et al., 2011; Demidov et al., 2014) while others are reported to have holocentromeres (Kondo and Lavarack, 1984; Sheikh and Kondo, 1995; Sheikh et al., 1995). These species potentially offer a great opportunity to gain further insight into the evolution of differing centromere types between closely-related species and of the mechanisms involved. Additionally, occurrence of holo- and monocentromeres between closely related *Cuscuta* or *Drosera* species might offer a possibility to generate hybrids between species with holo- and with monocentromeres. This would be an attractive model system to study two different centromere types within a hybrid.

Another striking question is why holocentricity has arisen multiple times during evolution in diverse eukaryotic lineages but not in all? One common explanation is their advantage in relation to DSBs when compared to monocentric chromosomes. However, widespread occurrence of monocentromeres and the fact that holocentromeres are not described (so far) in vertebrates suggest that potential advantages conferred by a holocentromere are counteracted by certain disadvantages; possibly the potential of merotelic chromosome attachments during nuclear divisions or the faithful segregation of meiotic chromosomes. This may explain why species with holocentromeres evolved diverse meiotic chromosome segregation processes and why manifold mechanisms for sexual and/or asexual reproduction are found in species with holocentromeres (e.g., in aphids, Manicardi et al., 2015).

DI-, TRI- AND META-POLYCENTRIC CHROMOSOMES

Dicentric chromosomes (chromosomes with two active centromeres) are typically unstable (**Figure 2A**). They form anaphase bridges which lead to chromosome breakage, as already observed by Barbara McClintock (1939) in maize. Dicentric chromosomes are reported in plant and non-plant species such as *Drosophila*, yeast and human (Stimpson et al., 2012). There are even reports about tricentric (with three centromeres; Zhang et al., 2010) and meta-polycentric (with up to five centromeres) chromosomes (Neumann et al., 2012, 2015). Typically, di- or tricentric chromosomes arise as a consequence of profound genome rearrangements (Stimpson et al., 2012) although naturally occurring di- and meta-polycentric chromosomes do exist.

Stabilization of di- and tricentric chromosomes can occur via different mechanisms. One mechanism is the epigenetic inactivation of one of the centromeres leading to a functional monocentric chromosome (**Figure 2B**). If two or three centromeres are different in size, the small centromere(s) is/are inactivated (Han et al., 2009a; Zhang et al., 2010). An inactivated centromere can be reactivated when detached from the active centromere, demonstrating that it retains centromeric capability (Schubert et al., 1995; Han et al., 2009a). Stabilization can also occur if one centromere exhibits functional dominance over the other(s). A tricentric wheat chromosome (Zhang et al., 2010) was found with one large and two small centromeres that due to being in close proximity, function as one unit. All three centromeres contain centromeric sequences, cenH3 (the small centromeres containing 30% the amount of the large centromere), H3S10ph and bind spindle microtubules. This tricentric chromosome exhibits features of dicentrics: the two smaller centromeres can be inactivated, positively correlating with increased amounts of H3K27me2/3 or, when the two smaller centromeres are active, chromosome breakage occurs. However, in 70% of progenies the intact tricentric chromosomes was transmitted, possibly due to dominant pulling forces of the large over the small centromeres during anaphase (Zhang et al., 2010).

Stabilization is further dependent on proximity between the two centromeres: if they are "close enough," both active centromeres can behave as a functional unit and orientate to the same pole (**Figure 2C**), whereas when a critical distance is reached between the two active centromeres the chromosome can break due to merotelic spindle attachments. The critical distance is estimated to be around 10 Mb in a human X chromosome (Sullivan and Willard, 1998) and up to 20 Mb in an engineered human dicentric chromosome (Higgins et al., 2005). A naturally occurring stable "dicentric" chromosome is found in rice (Wang et al., 2013). Two cenH3-binding domains composed of typical centromeric repeats are separated by ~400 kb 5S rDNA sequences that do not associate with cenH3. Also canonical centromeres contain blocks of H3- instead of cenH3-containing chromatin of considerable size, e.g., rice Cen8 (Nagaki et al., 2004; Yan et al., 2008), potato Cen9 and Cen11 (Gong et al., 2012) or "dicentric" maize chromosome 5 with an estimated "gap"

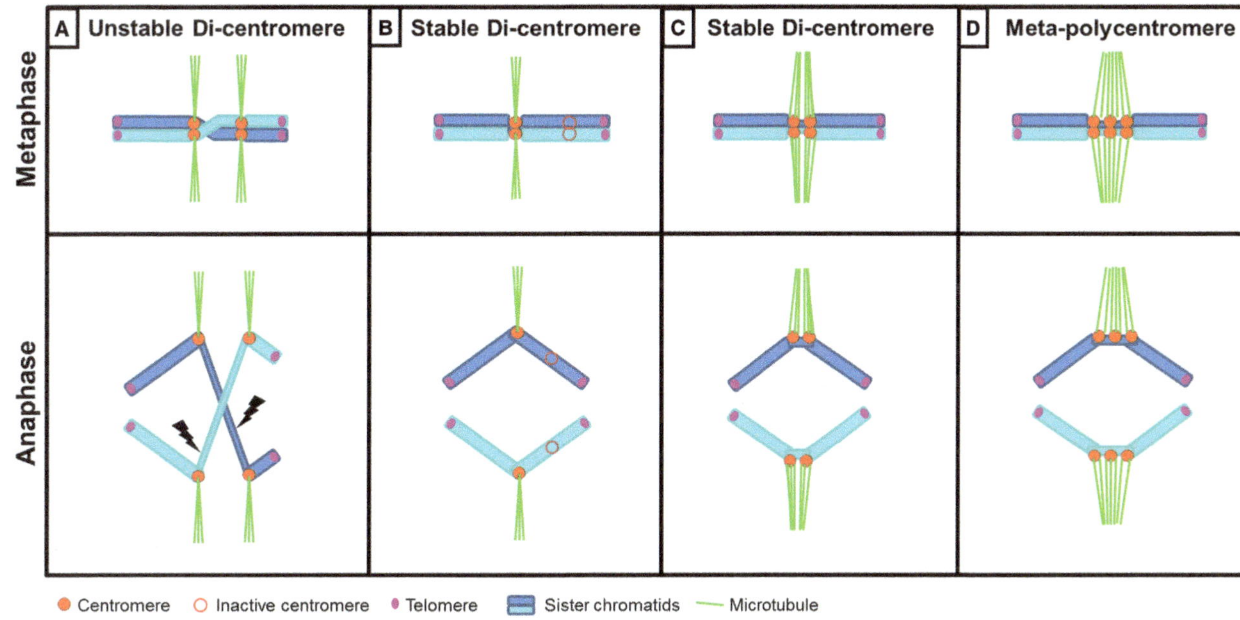

FIGURE 2 | Schematic representation of chromosomes with multiple centromeres. (A) A chromosome with two active centromeres (di-centromere or dicentric chromosome) is typically unstable. A twist between sister chromatids within the region between both centromeres leads to merotelic spindle attachment to two kinetochores on the same chromatid resulting in an anaphase bridge and subsequent chromosome breakage. **(B,C)** Stabilization of a dicentric chromosome can occur by **(B)** inactivation of one centromere, so that the chromosome behaves as monocentric or **(C)** when the close proximity between two active centromeres enables them to behave as one functional unit. **(D)** Meta-polycentric chromosome with three functional centromeres within one constriction.

of 2.8 Mb (Wolfgruber et al., 2009). In the most extreme case, meta-polycentric chromosomes in pea and closely related *Lathyrus* species contain three to five functional centromeres within stretched primary constrictions (Neumann et al., 2012, 2015; **Figure 2D**). In these meta-polycentric chromosomes of pea up to several megabases are estimated to lie between cenH3-containing domains (Neumann et al., 2012). Thus, it seems likely that as long as the distance between multiple centromeres per chromosome is limited, these centromeres can function together, similar to the situation found in holocentric chromosomes.

NEOCENTROMERES

A neocentromere is a chromosomal locus outside the endogenous centromere that acquires kinetic activity. They are described in various organisms including plants, yeast, flies, chicken and humans (Guerra et al., 2010; Burrack and Berman, 2012; Scott and Sullivan, 2014). The term "neocentromere" has been traditionally used to define two different phenomena: (i) a *de novo* centromere formation occurring after chromosome breakage or endogenous centromere inactivation, which allows transmission of the (re-arranged) chromosome (**Figure 3**), and (ii) the kinetic motility of terminal or subterminal heterochromatin, which is pulled to the cell poles during meiosis in plants (**Figure 4**).

DE NOVO CENTROMERES

In monocentric chromosome species occasionally an acentric chromosome fragment is stably transmitted during mitosis and meiosis due to neocentromere formation (**Figures 3A,B**).

In non-plant species, human neocentromeres are the best described to date. They usually appear in rearranged chromosomes associated with developmental delays or cancer and are typically isolated from clinical samples (Burrack and Berman, 2012). The first human neocentromere was reported by Voullaire et al. (1993) in chromosome 10. To date, neocentromeres are found in all human chromosomes except chromosome 19 (Marshall et al., 2008; Liehr et al., 2010; Klein et al., 2012), typically in euchromatic regions devoid of alpha-satellite DNAs characteristic of human centromeres (Alonso et al., 2010). These initial findings suggested that (alpha) satellite DNAs and/or heterochromatin are not necessary for (human) centromere activity. However, like endogenous centromeres, human neocentromeres form a primary constriction and contain all tested centromere-associated proteins except CENP-B whose localization is sequence-specific and requires a CENP-B box found in endogenous centromeres (Saffery et al., 2000).

In plants, *de novo* centromere formation has been documented in barley and maize. They appear in rearranged acentric chromosomes devoid of typical centromeric DNA sequences but contain centromeric proteins including cenH3 (Nasuda et al., 2005; Topp et al., 2009; Fu et al., 2013; Zhang et al., 2013; Liu et al., 2015).

In a barley telocentric chromosome derived from a 7HS isochromosome a neocentromere occurred close to the endogenous centromere (Nasuda et al., 2005). In maize, two *de novo* centromeres were reported in chromosome 3, one in the short arm (3S, Topp et al., 2009) and another one in a derivative of the long arm called Dp3a (Fu et al., 2013). In maize 3S cenH3 amounts varied between lines and low amounts of cenH3

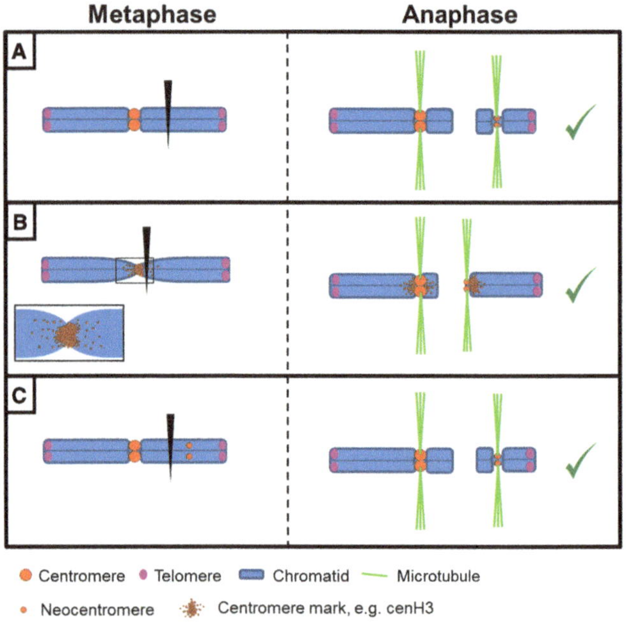

FIGURE 3 | Formation and behavior of *de novo* centromeres. (A) Following chromosome breakage, an acentric fragment can form a neocentromere allowing its proper transmission. **(B)** Chromosome breakage close to the endogenous centromere may lead to neocentromere formation due to presence or spreading of centromeric marks (e.g., cenH3) to pericentromeric regions. **(C)** Neocentromere formation in an intact chromosome leads to a dicentric chromosome structure. If this results in chromosome breakage, a centromere- and a neocentromere-containing fragment will result.

positively correlated with low transmission rates. CenH3 levels increased over generations and the neocentromere became more stable, thus accumulation of cenH3 over time likely stabilizes the neocentromere (Topp et al., 2009). The neocentromere in Dp3a formed within protein-coding genes (Fu et al., 2013), similar to, e.g., rice centromere 8 which contains actively transcribed genes (Nagaki et al., 2004; Yan et al., 2008). Although Dp3a was occasionally transmitted during meiosis, it got frequently lost during somatic divisions suggesting that the neocentromere was unstable. The cenH3-binding region was ~350 kb, considerable smaller than the megabase-sized cenH3 binding domains of canonical maize centromeres (Wolfgruber et al., 2009). Possibly the amount of cenH3 within the 350 kb is not sufficient for proper centromere activity and larger cenH3 amounts, potentially acquired over successive generations (Topp et al., 2009), are required for stabilization of neocentromeres. Neocentromeres are also found in supernumerary maize B chromosomes (Zhang et al., 2013; Liu et al., 2015) indicating that they are not exclusive to A chromosomes.

ORIGIN OF *DE NOVO* CENTROMERES

Why these neocentromeres appear and why specifically in a given chromosome region is an intriguing question. *De novo* neocentromeres might represent "latent" centromeres (Voullaire et al., 1993; Choo, 1997), locations of ancestral centromeres following centromere repositioning events (Rocchi et al., 2012). Alternatively, spreading of centromeric chromatin "marks" to adjacent chromosomal regions may explain neocentromeres arising close to an endogenous centromere (Maggert and Karpen, 2001; Nasuda et al., 2005; Shang et al., 2013; Thakur and Sanyal, 2013; Zhang et al., 2013; **Figure 3B**). In chicken, for instance, cenH3 is found in pericentromeric regions and that could trigger neocentromere formation when the endogenous centromere is removed (Shang et al., 2013).

For neocentromeres located far away from the canonical centromere, an interesting point arises: are they a *consequence*, or conversely a *cause*, of chromosome breakage? Generally a neocentromere is proposed to "rescue" acentric chromosome fragments allowing their transmission during cell divisions (hence, they would be a *consequence*). However, neocentromere formation could also be the *cause*. If a new region acquires the ability to assemble an active kinetochore, this could lead to a dicentric structure and subsequent chromosome breakage (**Figure 3C**). Possibly, both options exist: whereas experimentally centromere excision results in neocentromere formation (Ishii et al., 2008; Ketel et al., 2009; Shang et al., 2013; Thakur and Sanyal, 2013), artificially tethering of centromeric components to non-centromeric loci or overexpression of cenH3 can lead to ectopic kinetochore formation and chromosome instability (Heun et al., 2006; Barnhart et al., 2011; Mendiburo et al., 2011; Olszak et al., 2011; Teo et al., 2013). CenH3 is also found in non-centromeric chromatin in, e.g., human and chicken cells (Shang et al., 2013; Bodor et al., 2014) but whether these ectopic cenH3 domains can nucleate under certain conditions a *de novo* centromere is unclear. It has been proposed that small neocentromeres could be nucleated with relatively high frequency. While presence of the endogenous larger centromere prevents them from becoming active, in fragments detached from the main centromere this activation could occur (Liu et al., 2015).

TERMINAL NEOCENTROMERES IN PLANTS

Terminal neocentromeres were first described by Kattermann (1939) in rye who called them "T-chromosomes" for "terminal chromosomes." He described an activity in chromosomes from inbred lines resembling terminal centromeres. Prakken and Müntzing (1942) and Rhoades and Vilkomerson (1942) described in rye and maize respectively, that chromosome ends were *attracted* to the poles. The term "neo-centric" was used for the first time by Rhoades (1942). Later reports described similar chromosome activities, primarily in grasses but also in lilies and a moss (Dawe and Hiatt, 2004; **Figure 4A**).

Maize terminal neocentromeres appear in heterochromatic DNA domains known as knobs (Rhoades and Vilkomerson, 1942). Under standard conditions, knobs are inert. However, presence of a chromosome 10 carrying a distinct large knob called abnormal 10 (Ab10) renders them active as neocentromeres during meiosis (Rhoades and Vilkomerson, 1942). Ab10-dependent terminal neocentromere activity causes the preferential transmission - meiotic drive- of knobbed chromosomes to the egg cell during female meiosis (Rhoades, 1942). Terminal neocentromeres are

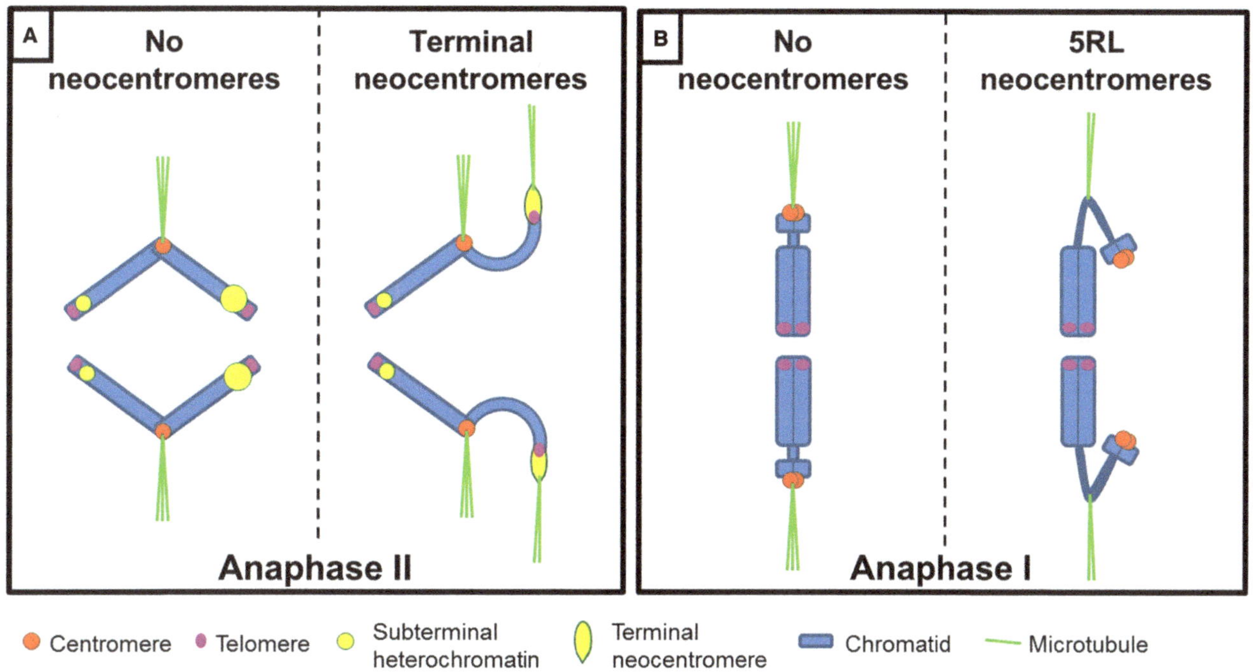

FIGURE 4 | Schematic representation of meiosis-specific neocentromeres in plants. (A) During anaphase II, terminal neocentromeres are visible as heterochromatic stretches directed toward the cell poles ahead of the centromere. Telomeric regions are not stretched to the poles. Large heterochromatic regions (represented as larger yellow circles) are more prone to form neocentromeres than small ones. **(B)** Neocentromere in rye 5RL arises at an interstitial heterochromatic constriction and can substitute for the centromere during anaphase I.

genetically regulated by a locus called *Smd1* (suppressor of meiotic drive 1; Dawe and Cande, 1996) and further unknown genes (Hiatt et al., 2002; Hiatt and Dawe, 2003). Smd1 regulates the activation of terminal neocentromeres and their subsequent preferential segregation (Dawe and Cande, 1996). Interestingly, terminal neocentromeres are necessary but not sufficient for the preferential transmission of Ab10 and knobbed chromosomes since deletion of *smd1* led to neocentromere activity without meiotic drive (Dawe and Cande, 1996). Moreover, the two major components of maize knobs, i.e., a 180-bp satellite repeat and a 350-bp tandemly repeated sequence (TR-1; Peacock et al., 1981; Ananiev et al., 1998; Gonzalez-Sanchez et al., 2007) are independently regulated in such a way that both repeats "compete" genetically to be preferentially transmitted (Hiatt et al., 2002; Kanizay et al., 2013). Thus, maize neocentromeres might constitute a mechanism by which knobs and associated loci are preferentially transmitted in a process resembling the behavior of "selfish" B chromosomes (Houben et al., 2014).

Maize terminal neocentromeres lack, based on immunological analysis, cenH3 and CENP-C (Dawe et al., 1999; Dawe and Hiatt, 2004) but interact with spindle microtubules albeit in a lateral way and not the canonical end-on fashion (Yu et al., 1997). The poleward motility of chromosome arms is not affected by the microtubule-stabilizing drug taxol suggesting that a meiosis specific microtubule-based motor protein could be responsible for their kinetic activity (Hiatt et al., 2002). Identification of such protein remains elusive.

Rye terminal neocentromeres were initially observed in inbred populations (Kattermann, 1939; Prakken and Müntzing, 1942) and later also in open pollinated rye varieties (Kavander and Viinikka, 1987; Manzanero and Puertas, 2003). During meiosis, all rye chromosomes can show terminal neocentromeres that interact with microtubules (Puertas et al., 2005) but they preferentially occur in chromosomes with large C-banding positive heterochromatic blocks (Manzanero and Puertas, 2003; **Figure 4A**). The repetitive sequences pSc200 and pSc250 (Vershinin et al., 1995), but not centromeric or telomeric sequences, are stretched at the neocentromeres (Manzanero and Puertas, 2003). Terminal heterochromatin domains of rye B chromosomes show no neocentromeric activity (Manzanero and Puertas, 2003). Thus, only a subset of terminal heterochromatin can acquire kinetic activity. Similar to maize, rye terminal neocentromeres are genetically regulated. Hayward (1962) reported that rye neocentromeres are controlled by a polygenic system and Puertas et al. (2005) proposed a model based on two *trans*-acting genes. Contrary to maize, no examples of meiotic drive are reported for rye terminal neocentromeres.

Terminal neocentromeres are active together with the endogenous centromere stretching the chromosome arms to the same, or the opposite pole of the centromere (Manzanero and Puertas, 2003). Thus, chromosomes behave as di- or polycentric during meiotic divisions but chromosome breakage does not occur and the endogenous kinetochore leads the chromosome movement to the poles (Yu et al., 1997; Manzanero and Puertas, 2003).

TABLE 1 | Comparison of the three types of neocentromeres described with regards to their main features.

Feature	*De novo* centromere	Terminal neocentromere	5RL neocentromere
Replaces the centromere	Yes	No	Occasionally
Sister-chromatid cohesion	Yes	No	Yes
Visible constriction	Yes	No	Yes
Genomic location	Eu- and heterochromatin	Heterochromatin	Heterochromatin
Centromeric proteins	Yes	No	No
Interaction with spindle microtubules	Yes (end-on)	Yes (lateral in maize, end-on in rye)	Yes (end-on)
Species	Human, chicken, yeast, Drosophila, barley, maize	Plants (e.g., maize, rye)	Rye, wheat-rye addition lines

Initial observations in maize suggested that terminal neocentromeres could enable migration of acentric fragments to the cell poles (Rhoades and Dempsey, 1966; Dawe and Cande, 1996). However, it was later demonstrated that only when an acentric fragment was brought to the equator plate (possibly by interaction with other bivalents) could it then migrate to the pole during anaphase I (Yu et al., 1997). Similarly, rye terminal neocentromeres fail to move acentric fragments unless they are physically linked (even by chromatin threads) to a centromere (Puertas et al., 2005). Therefore, a *cis*-acting centromere is essential for terminal neocentromere activity.

ORIGIN OF TERMINAL NEOCENTROMERES

Why would the chromosome ends acquire kinetic activity during meiosis? Frequently terminal neocentromeres appear in situations of genetic instability, such as inbred populations, interspecific hybrids, heat stress, X-ray irradiated plants or presence of abnormal chromosomes (Dawe and Hiatt, 2004). However, they also appear in normal cultivars of diploid rye. Neocentromeres in fission yeast (Ishii et al., 2008), *Drosophila* (Platero et al., 1999) and human (Marshall et al., 2008) preferentially occur in subtelomeric regions. This could indicate that subtelomeres are preferred domains to acquire centromeric activity or, in line with the "centromeres from telomeres" model (Villasante et al., 2007), that subtelomeric regions might have retained some of the ancestral features which (especially under unusual genomic conditions) allow them to recover centromeric activity. Recently, an interesting functional exchange of roles between centromeres and telomeres during yeast meiosis was found (Fennell et al., 2015): in mutants that fail to form a meiotic bouquet (attachment of telomeres to the nuclear envelope during meiotic prophase, essential for proper meiotic progression in some species) centromeres can functionally replace telomeres and (at least partially) rescue the mutant phenotype. This indicates that centromeres and telomeres are not completely independent functional entities.

Kinetochore-independent chromosome movements similar to terminal neocentromeres are also found in other organisms. For example during mitotic metaphase in the African blood lily *Haemanthus katherinae* a poleward force acts on chromosome arms facilitating movement of acentric chromosome fragments (Khodjakov et al., 1996), or in the germ line of *Parascaris*, large amounts of satellite DNAs (heterochromatin) are enriched at its chromosome termini functionally acting as centromeres but getting eliminated rapidly in somatic tissues (Pimpinelli and Goday, 1989). Whether similar mechanisms are involved in the regulation of terminal neocentromere activity is unclear.

OTHER TYPES OF NEOCENTROMERES

Key differences between *de novo* and terminal neocentromeres are that the latter do not mediate sister chromatid cohesion nor lead chromosome movement in the absence of an active endogenous centromere (**Table 1**). An intermediate situation is found in the long arm of chromosome 5 of rye (5RL; **Figure 4B**). A neocentromere in 5RL is found that is active only during meiosis, can coincide with the centromere(s) and is associated with heterochromatin (Manzanero et al., 2000, 2002), similar to terminal neocentromeres. However, unlike terminal neocentromeres, it mediates sister chromatid cohesion at anaphase I and leads chromosome movement when the endogenous centromere is inactive, similar to the situation found in *de novo* centromeres (Manzanero et al., 2000, 2002; Cuacos et al., 2011).

This neocentromere arises in an interstitial constriction in haploid rye, wheat-rye and wheat-Triticale hybrids and wheat-rye addition lines involving chromosome 5R, both in mono(telo)somic and di(telo)somic conditions (Schlegel, 1987; Manzanero et al., 2000, 2002; Cuacos et al., 2011). When the 5RL neocentromere is active, the constriction is cytologically stretched up to several times the chromosome length, reaching the cell poles before the onset of anaphase I (Manzanero et al., 2000, 2002; Cuacos et al., 2011). Interestingly, it can lead chromosome movement together with the centromere or alone (Manzanero et al., 2000, 2002; Cuacos et al., 2011) suggesting that this neocentromere, unlike terminal neocentromeres, does not require a *cis*-acting centromere to be active (**Figure 4B**).

The heterochromatic 5RL interstitial constriction lacks typical centromeric and telomeric sequences but contains the repetitive sequences pSc119.2 (Bedbrook et al., 1980; McIntyre et al., 1990) and UCM600 (Manzanero et al., 2002; Gonzalez-Garcia et al., 2011; Cuacos et al., 2011). Proteins accumulate at the constriction and a thin bundle of microtubules is end-on attached (Manzanero et al., 2002; Cuacos et al., 2011), showing that interaction between spindle microtubules and the constriction occurs. Notably, kinetic activity is cenH3- and CENP-C-independent (Cuacos, 2013) similar to maize terminal neocentromeres.

Varying frequencies of active rye 5RL neocentromeres in wheat-rye addition lines (Manzanero et al., 2000, 2002) suggested the influence of an environmental factor in its activity. Indeed, treating plants with an organophosphate pesticide significantly increased a basal frequency of ~10% neocentromeres up to ~50% (Cuacos et al., 2011). Concomitantly, an alteration of the meiotic spindle was found possibly facilitating the interaction of the spindle microtubules with the constriction (Cuacos et al., 2011). However, why this region acquires under certain conditions such ability is not well understood.

In *Bromus marginatus* and *B. pseudolaevipes* hybrids (Walters, 1952), *Aegilops markgrafii* (Schubert, 2011), and wheat chromosomes in the progeny from Triticale × tritordeum hybrids (Carvalho et al., 2008) a stretched constriction during meiosis, similar to the one found in rye 5RL, was described. In *A. markgrafii* this constriction is not related to neocentromeric activity since cenH3 and microtubules are exclusively found at the endogenous centromere (Schubert, 2011). Whether constrictions in *Bromus* or wheat acquire kinetic activity is not documented.

Heterochromatin within the 5RL neocentromere could explain its ability to maintain sister chromatid cohesion, as heterochromatin within pericentromeric regions does in native centromeres (Bernard et al., 2001; Nonaka et al., 2002). Moreover, a heterochromatic environment is likely conducive for neocentromere formation, as fission yeast (Ishii et al., 2008) and *Drosophila* (Platero et al., 1999; Olszak et al., 2011) neocentromeres are commonly found in heterochromatic regions. Also human neocentromeres arising at euchromatic domains recruit heterochromatic proteins (Saffery et al., 2000) and plant terminal neocentromeres are heterochromatin-associated. However, the latter do not maintain sister chromatid cohesion suggesting that heterochromatin is necessary but not sufficient to maintain sister chromatid cohesion.

NEO-CENTROMERE FORMATION AND EVOLUTION

Different models have been proposed to explain the origin of the centromere. Villasante et al. (2007) proposed that centromeres were derived from telomeres, in such a way that subtelomeric regions were the first proto-centromeres recognized by spindle microtubules. This model is, e.g., supported by telomere-like sequences found within centromeric regions of several species (Villasante et al., 2007 and references therein). The model assumes that centromeres were, from their origin, associated with repetitive DNA elements.

Another model proposes the opposite: centromeres originally formed on single-copy non-repetitive DNA loci and subsequently acquired highly repeated sequences (Gong et al., 2012; Rocchi et al., 2012). In support of this model, repeat-free centromeres have been found. For example in rice, the centromere of chromosome 8 contains low amounts of satellite DNAs (Nagaki et al., 2004) and some species lack the centromeric satellite CentO (Lee et al., 2005). Also, in potato five centromeres contain unique- or low-copy repetitive DNA (Gong et al., 2012). In non-plant species, repeat-less centromeres are found in chicken (Shang et al., 2010) and the genus Equus (Wade et al., 2009). This model is further supported by the fact that centromere repositioning has occurred in mammals and plants, i.e., a new centromere formed in a different chromosome region while the "old" one became inactive (Han et al., 2009b; Rocchi et al., 2012). Consistently, shattered remnants of centromeric satellites are found throughout plant and animal genomes (Rocchi et al., 2012; Lysak, 2014).

In potato, together with repeat-free centromeres there are centromeres containing megabases of repetitive satellite DNA with an extremely large monomer size up to 5 kb (Gong et al., 2012) which are not, or only rarely found in closely related species (Gong et al., 2012; Wang et al., 2014b; Zhang et al., 2014) suggesting a relatively recent species-specific amplification. The coexistence of centromeres with and without repetitive DNA suggests that accumulation of repeats could be a sudden, rather than progressive, process. These sequences could be incorporated from another centromere or constitute new repeats, for which retrotransposons might play a relevant role (Gong et al., 2012; Gao et al., 2015). Later accumulation or expansion of repeats could occur by several mechanisms such as replication slippage, rolling circle replication followed by reinsertion or transposition. Moreover, maize centromeric satellites are highly diverged from ancient maize centromere satellites and chromosome-specific satellites are not found, suggesting existence of a homogenization mechanism (Bilinski et al., 2015). Finally, maize chromosomes transferred to a species with larger genome such as oat experience a rapid expansion of centromere size (cenH3-binding domain; Wang et al., 2014a) agreeing with previous observations that correlated centromere and genome size in grasses (Zhang and Dawe, 2012). Altogether, this demonstrates that neo-centromere evolution can be highly dynamic.

CONCLUDING REMARKS

Despite striking progress made in plant centromere biology, many questions remain to be answered. How could such an essential chromosomal locus evolve structurally so diverse, from a point centromere up to a holocentromere? The more data on centromeres becomes available, the more challenging it is to define the centromere. The centromere could be best described as a "conserved function, but distinct structure, organization and features between organisms." In future, a better understanding in different organisms and in different centromere types of the precise nature of epigenetic mechanisms specifying centromere location, formation and maintenance as well as of the unsolved connection between epigenetics and genetics behind centromere biology is needed.

We anticipate that further studies of (plant) species with atypical centromeres will broaden our mainly monocentric chromosome-biased knowledge on centromere definition and will help in establishing more accurate models of centromere biology and evolution. Future research on plant centromere biology will also help plant genetic engineering (e.g., artificial chromosome or double haploid plant production) and thus ultimately help plant breeding.

ACKNOWLEDGMENT

This work was supported by the BBSRC (grant numbers BB/M004902/1 and BB/J007633/1).

REFERENCES

Akiyoshi, B., and Gull, K. (2014). Discovery of unconventional kinetochores in kinetoplastids. *Cell* 156, 1247–1258. doi: 10.1016/j.cell.2014.01.049

Alonso, A., Hasson, D., Cheung, F., and Warburton, P. E. (2010). A paucity of heterochromatin at functional human neocentromeres. *Epigenetics Chromatin* 3, 6. doi: 10.1186/1756-8935-3-6

Ananiev, E. V., Phillips, R. L., and Rines, H. W. (1998). A knob-associated tandem repeat in maize capable of forming fold-back DNA segments: are chromosome knobs megatransposons? *Proc. Natl. Acad. Sci. U.S.A.* 95, 10785–10790. doi: 10.1073/pnas.95.18.10785

Barnhart, M. C., Kuich, P. H., Stellfox, M. E., Ward, J. A., Bassett, E. A., Black, B. E., et al. (2011). HJURP is a CENP-A chromatin assembly factor sufficient to form a functional *de novo* kinetochore. *J. Cell Biol.* 194, 229–243. doi: 10.1083/jcb.201012017

Bedbrook, J. R., Jones, J., O'dell, M., Thompson, R. D., and Flavell, R. B. (1980). A molecular description of telometic heterochromatin in secale species. *Cell* 19, 545–560. doi: 10.1016/0092-8674(80)90529-2

Bensasson, D. (2011). Evidence for a high mutation rate at rapidly evolving yeast centromeres. *BMC Evol. Biol.* 11:211. doi: 10.1186/1471-2148-11-211

Bernard, P., Maure, J. F., Partridge, J. F., Genier, S., Javerzat, J. P., and Allshire, R. C. (2001). Requirement of heterochromatin for cohesion at centromeres. *Science* 294, 2539–2542. doi: 10.1126/science.1064027

Bilinski, P., Distor, K., Gutierrez-Lopez, J., Mendoza, G. M., Shi, J., Dawe, R. K., et al. (2015). Diversity and evolution of centromere repeats in the maize genome. *Chromosoma* 124, 57–65. doi: 10.1007/s00412-014-0483-8

Bodor, D. L., Mata, J. F., Sergeev, M., David, A. F., Salimian, K. J., Panchenko, T., et al. (2014). The quantitative architecture of centromeric chromatin. *Elife* 3, e02137. doi: 10.7554/eLife.02137

Bozek, M., Leitch, A. R., Leitch, I. J., Záveská Drábková, L., and Kuta, E. (2012). Chromosome and genome size variation in *Luzula* (*Juncaceae*), a genus with holocentric chromosomes. *Bot. J. Linn. Soc.* 170, 529–541. doi: 10.1111/j.1095-8339.2012.01314.x

Buchwitz, B. J., Ahmad, K., Moore, L. L., Roth, M. B., and Henikoff, S. (1999). A histone-H3-like protein in *C. elegans*. *Nature* 401, 547–548. doi: 10.1038/44062

Bureš, P., Zedek, F., and Marková, M. (2013). "Holocentric Chromosomes," in *Plant Genome Diversity*, Vol. 2, eds J. Greilhuber, J. Dolezel, and J. F. Wendel (Vienna: Springer), 187–208.

Burrack, L. S., and Berman, J. (2012). Neocentromeres and epigenetically inherited features of centromeres. *Chromosome Res.* 20, 607–619. doi: 10.1007/s10577-012-9296-x

Cabral, G., Marques, A., Schubert, V., Pedrosa-Harand, A., and Schlögelhofer, P. (2014). Chiasmatic and achiasmatic inverted meiosis of plants with holocentric chromosomes. *Nat. Commun.* 5, 5070. doi: 10.1038/ncomms6070

Carvalho, A., Guedes-Pinto, H., Heslop-Harrison, J. S., and Lima-Brito, J. (2008). Wheat neocentromeres found in F1 Triticale × Tritordeum hybrids (AABBRHch) After 5-Azacytidine Treatment. *Plant Mol. Biol. Rep.* 26, 46–52. doi: 10.1007/s11105-008-0021-z

Choo, K. H. (1997). Centromere DNA dynamics: latent centromeres and neocentromere formation. *Am. J. Hum. Genet.* 61, 1225–1233. doi: 10.1086/301657

Clarke, L., and Carbon, J. (1985). The structure and function of yeast centromeres. *Annu. Rev. Genet.* 19, 29–55. doi: 10.1146/annurev.ge.19.120185.000333

Cuacos, M. (2013). *Analysis of a Secondary Constriction in Chromosome 5R of Rye and its Neocentromeric Activity*. Ph.D. thesis, Universidad Complutense de Madrid, Madrid.

Cuacos, M., Gonzalez-Garcia, M., Gonzalez-Sanchez, M., Puertas, M. J., and Vega, J. M. (2011). Activation of rye 5RL neocentromere by an organophosphate pesticide. *Cytogenet. Genome Res.* 134, 151–162. doi: 10.1159/000325744

d'Alencon, E., Sezutsu, H., Legeai, F., Permal, E., Bernard-Samain, S., Gimenez, S., et al. (2010). Extensive synteny conservation of holocentric chromosomes in *Lepidoptera* despite high rates of local genome rearrangements. *Proc. Natl Acad. Sci. U.S.A.* 107, 7680–7685. doi: 10.1073/pnas.0910413107

Dawe, R. K., and Cande, W. Z. (1996). Induction of centromeric activity in maize by suppressor of meiotic drive 1. *Proc. Natl. Acad. Sci. U.S.A.* 93, 8512–8517. doi: 10.1073/pnas.93.16.8512

Dawe, R. K., and Hiatt, E. N. (2004). Plant neocentromeres: fast, focused, and driven. *Chromosome Res.* 12, 655–669. doi: 10.1023/B:CHRO.0000036607.74671.db

Dawe, R. K., Reed, L. M., Yu, H. G., Muszynski, M. G., and Hiatt, E. N. (1999). A maize homolog of mammalian CENPC is a constitutive component of the inner kinetochore. *Plant Cell* 11, 1227–1238. doi: 10.1105/tpc.11.7.1227

De Rop, V., Padeganeh, A., and Maddox, P. (2012). CENP-A: the key player behind centromere identity, propagation, and kinetochore assembly. *Chromosoma* 121, 527–538. doi: 10.1007/s00412-012-0386-5

Demidov, D., Schubert, V., Kumke, K., Weiss, O., Karimi-Ashtiyani, R., Buttlar, J., et al. (2014). Anti-phosphorylated histone H2AThr120: a universal microscopic marker for centromeric chromatin of mono- and holocentric plant species. *Cytogenet. Genome Res.* 143, 150–156. doi: 10.1159/000360018

Dong, F. G., Miller, J. T., Jackson, S. A., Wang, G. L., Ronald, P. C., and Jiang, J. M. (1998). Rice (*Oryza sativa*) centromeric regions consist of complex DNA. *Proc. Natl Acad. Sci. U.S.A.* 95, 8135–8140. doi: 10.1073/pnas.95.14.8135

Drinnenberg, I. A., Deyoung, D., Henikoff, S., and Malik, H. S. (2014). Recurrent loss of CenH3 is associated with independent transitions to holocentricity in insects. *eLife* 3, e03676. doi: 10.7554/eLife.03676

Fennell, A., Fernandez-Alvarez, A., Tomita, K., and Cooper, J. P. (2015). Telomeres and centromeres have interchangeable roles in promoting meiotic spindle formation. *J. Cell Biol.* 208, 415–428. doi: 10.1083/jcb.201409058

Flach, M. (1966). Diffuse centromeres in a dicotyledonous plant. *Nature* 209, 1369–1370. doi: 10.1038/2091369b0

Fu, S., Lv, Z., Gao, Z., Wu, H., Pang, J., Zhang, B., et al. (2013). *De novo* centromere formation on a chromosome fragment in maize. *Proc. Natl. Acad. Sci. U.S.A.* 110, 6033–6036. doi: 10.1073/pnas.1303944110

Gao, D., Jiang, N., Wing, R. A., Jiang, J., and Jackson, S. A. (2015). Transposons play an important role in the evolution and diversification of centromeres among closely related species. *Front. Plant Sci.* 6:216. doi: 10.3389/fpls.2015.00216

Gassmann, R., Rechtsteiner, A., Yuen, K. W., Muroyama, A., Egelhofer, T., Gaydos, L., et al. (2012). An inverse relationship to germline transcription defines centromeric chromatin in *C. elegans*. *Nature* 484, 534–537. doi: 10.1038/nature10973

Gernand, D., Demidov, D., and Houben, A. (2003). The temporal and spatial pattern of histone H3 phosphorylation at serine 28 and serine 10 is similar in plants but differs between mono- and polycentric chromosomes. *Cytogenet. Genome Res.* 101, 172–176. doi: 10.1159/000074175

Gong, Z., Wu, Y., Koblizkova, A., Torres, G. A., Wang, K., Iovene, M., et al. (2012). Repeatless and repeat-based centromeres in potato: implications for centromere evolution. *Plant Cell* 24, 3559–3574. doi: 10.1105/tpc.112.100511

Gonzalez-Garcia, M., Cuacos, M., Gonzalez-Sanchez, M., Puertas, M. J., and Vega, J. M. (2011). Painting the rye genome with genome-specific sequences. *Genome* 54, 555–564. doi: 10.1139/g11-003

Gonzalez-Sanchez, M., Gonzalez-Garcia, M., Vega, J. M., Rosato, M., Cuacos, M., and Puertas, M. J. (2007). Meiotic loss of the B chromosomes of maize is influenced by the B univalent co-orientation and the TR-1 knob constitution of the A chromosomes. *Cytogenet. Genome Res.* 119, 282–290. doi: 10.1159/000112075

Guerra, M., Brasileiro-Vidal, A. C., Arana, P., and Puertas, M. J. (2006). Mitotic microtubule development and histone H3 phosphorylation in the holocentric chromosomes of *Rhynchospora tenuis* (*Cyperaceae*). *Genetica* 126, 33–41. doi: 10.1007/s10709-005-1430-7

Guerra, M., Cabral, G., Cuacos, M., Gonzalez-Garcia, M., Gonzalez-Sanchez, M., Vega, J., et al. (2010). Neocentrics and holokinetics (holocentrics): chromosomes out of the centromeric rules. *Cytogenet. Genome Res.* 129, 82–96. doi: 10.1159/000314289

Haizel, T., Lim, Y. K., Leitch, A. R., and Moore, G. (2005). Molecular analysis of holocentric centromeres of *Luzula* species. *Cytogenet. Genome Res.* 109, 134–143. doi: 10.1159/000082392

Hakansson, A. (1954). Meiosis and pollen mitosis in x-rayed and untreated spikelets of *Eleocharis palustris*. *Hereditas* 15, 325–345. doi: 10.1111/j.1601-5223.1954.tb02976.x

Hakansson, A. (1958). Holocentric chromosomes in *Eleocharis*. *Hereditas* 44, 531–540. doi: 10.1111/j.1601-5223.1958.tb03498.x

Han, F., Gao, Z., and Birchler, J. A. (2009a). Reactivation of an inactive centromere reveals epigenetic and structural components for centromere specification in maize. *Plant Cell* 21, 1929–1939. doi: 10.1105/tpc.109.066662

Han, Y., Zhang, Z., Liu, C., Liu, J., Huang, S., Jiang, J., et al. (2009b). Centromere repositioning in cucurbit species: implication of the genomic impact from centromere activation and inactivation. *Proc. Natl. Acad. Sci. U.S.A.* 106, 14937–14941. doi: 10.1073/pnas.0904833106

Han, F., Lamb, J. C., and Birchler, J. A. (2006). High frequency of centromere inactivation resulting in stable dicentric chromosomes of maize. *Proc. Natl. Acad. Sci. U.S.A.* 103, 3238–3243. doi: 10.1073/pnas.0509650103

Hayward, M. D. (1962). Genetic control of neocentric activity in rye. *Heredity* 17, 439–441. doi: 10.1038/hdy.1962.43

Heckmann, S., and Houben, A. (2013). "Holokinetic centromeres," in *Plant Centromere Biology*, eds J. Jiang and J. A. Birchler (Oxford: Wiley-Blackwell), 83–94. doi: 10.1002/9781118525715.ch7

Heckmann, S., Jankowska, M., Schubert, V., Kumke, K., Ma, W., and Houben, A. (2014a). Alternative meiotic chromatid segregation in the holocentric plant *Luzula elegans*. *Nat. Commun.* 5:4979. doi: 10.1038/ncomms5979

Heckmann, S., Schubert, V., and Houben, A. (2014b). Holocentric plant meiosis: first sisters, then homologues. *Cell Cycle* 13, 3623–3624. doi: 10.4161/15384101.2014.986628

Heckmann, S., Macas, J., Kumke, K., Fuchs, J., Schubert, V., Ma, L., et al. (2013). The holocentric species *Luzula elegans* shows interplay between centromere and large-scale genome organization. *Plant J.* 73, 555–565. doi: 10.1111/tpj.12054

Heckmann, S., Schroeder-Reiter, E., Kumke, K., Ma, L., Nagaki, K., Murata, M., et al. (2011). Holocentric chromosomes of *Luzula elegans* are characterized by a longitudinal centromere groove, chromosome bending, and a terminal nucleolus organizer region. *Cytogenet. Genome Res.* 134, 220–228. doi: 10.1159/000327713

Heun, P., Erhardt, S., Blower, M. D., Weiss, S., Skora, A. D., and Karpen, G. H. (2006). Mislocalization of the Drosophila centromere-specific histone CID promotes formation of functional ectopic kinetochores. *Dev. Cell* 10, 303–315. doi: 10.1016/j.devcel.2006.01.014

Hiatt, E. N., and Dawe, R. K. (2003). Four loci on abnormal chromosome 10 contribute to meiotic drive in maize. *Genetics* 164, 699–709.

Hiatt, E. N., Kentner, E. K., and Dawe, R. K. (2002). Independently regulated neocentromere activity of two classes of tandem repeat arrays. *Plant Cell* 14, 407–420. doi: 10.1105/tpc.010373

Higgins, A. W., Gustashaw, K. M., and Willard, H. F. (2005). Engineered human dicentric chromosomes show centromere plasticity. *Chromosome Res.* 13, 745–762. doi: 10.1007/s10577-005-1009-2

Houben, A., Banaei-Moghaddam, A. M., Klemme, S., and Timmis, J. N. (2014). Evolution and biology of supernumerary B chromosomes. *Cell. Mol. Life Sci.* 71, 467–478. doi: 10.1007/s00018-013-1437-7

Houben, A., and Schubert, I. (2003). DNA and proteins of plant centromeres. *Curr. Opin. Plant Biol.* 6, 554–560. doi: 10.1016/j.pbi.2003.09.007

Houben, A., Wako, T., Furushima-Shimogawara, R., Presting, G., Künzel, G., Schubert, I., et al. (1999). The cell cycle dependent phosphorylation of histone H3 is correlated with the condensation of plant mitotic chromosomes. *Plant J.* 18, 675–679. doi: 10.1046/j.1365-313x.1999.00496.x

Hughes, S. E., Gilliland, W. D., Cotitta, J. L., Takeo, S., Collins, K. A., and Hawley, R. S. (2009). Heterochromatic threads connect oscillating chromosomes during prometaphase i in drosophila oocytes. *PLoS Genet.* 5:e1000348. doi: 10.1371/journal.pgen.1000348

Ishii, K., Ogiyama, Y., Chikashige, Y., Soejima, S., Masuda, F., Kakuma, T., et al. (2008). Heterochromatin integrity affects chromosome reorganization after centromere dysfunction. *Science* 321, 1088–1091. doi: 10.1126/science.1158699

Jankowska, M., Fuchs, J., Klocke, E., Fojtová, M., Polanská, P., Fajkus, J., et al. (2015). Holokinetic centromeres and efficient telomere healing enable rapid karyotype evolution. *Chromosoma* doi: 10.1007/s00412-015-0524-y [Epub ahead of print].

Kanizay, L. B., Albert, P. S., Birchler, J. A., and Dawe, R. K. (2013). Intragenomic conflict between the two major knob repeats of maize. *Genetics* 194, 81–89. doi: 10.1534/genetics.112.148882

Kattermann, G. (1939). Ein neuer karyotyp bei roggen. *Chromosoma* 1, 284–299. doi: 10.1007/BF01271637

Kavander, T., and Viinikka, Y. (1987). Neocentric activity in open-pollinated cultivars of rye. *Hereditas* 107, 229–233. doi: 10.1111/j.1601-5223.1987.tb00289.x

Ketel, C., Wang, H. S., Mcclellan, M., Bouchonville, K., Selmecki, A., Lahav, T., et al. (2009). Neocentromeres form efficiently at multiple possible loci in *Candida albicans*. *PLoS Genet.* 5:e1000400. doi: 10.1371/journal.pgen.1000400

Khodjakov, A., Cole, R. W., Bajer, A. S., and Rieder, C. L. (1996). The force for poleward chromosome motion in *Haemanthus* cells acts along the length of the chromosome during metaphase but only at the kinetochore during anaphase. *J. Cell Biol.* 132, 1093–1104. doi: 10.1083/jcb.132.6.1093

Klein, E., Rocchi, M., Ovens-Raeder, A., Kosyakova, N., Weise, A., Ziegler, M., et al. (2012). Five novel locations of Neocentromeres in human: 18q22.1, Xq27.1 approximately 27.2, Acro p13, Acro p12, and heterochromatin of unknown origin. *Cytogenet. Genome Res.* 136, 163–166. doi: 10.1159/000336648

Knoll, A., Fauser, F., and Puchta, H. (2014). DNA recombination in somatic plant cells: mechanisms and evolutionary consequences. *Chromosome Res.* 22, 191–201. doi: 10.1007/s10577-014-9415-y

Kondo, K., and Lavarack, P. S. (1984). Study of some Australian species of *Drosera* L. (Droseraceae). *Bot. J. Linn. Soc.* 88, 317–333. doi: 10.1111/j.1095-8339.1984.tb01579.x

Kuta, E., Bohanec, B., Dubas, E., Vizintin, L., and Przywara, L. (2004). Chromosome and nuclear DNA study on *Luzula*—a genus with holokinetic chromosomes. *Genome* 47, 246–256. doi: 10.1139/g03-121

LaFountain, J. R., Cole, R. W., and Rieder, C. L. (2002). Partner telomeres during anaphase in crane-fly spermatocytes are connected by an elastic tether that exerts a backward force and resists poleward motion. *J. Cell Sci.* 115, 1541–1549.

Lee, H. R., Zhang, W., Langdon, T., Jin, W., Yan, H., Cheng, Z., et al. (2005). Chromatin immunoprecipitation cloning reveals rapid evolutionary patterns of centromeric DNA in *Oryza* species. *Proc. Natl. Acad. Sci. U.S.A.* 102, 11793–11798. doi: 10.1073/pnas.0503863102

Lermontova, I., Sandmann, M., and Demidov, D. (2014). Centromeres and kinetochores of Brassicaceae. *Chromosome Res.* 22, 135–152. doi: 10.1007/s10577-014-9422-z

Liehr, T., Kosyakova, N., Weise, A., Ziegler, M., and Raabe-Meyer, G. (2010). First case of a neocentromere formation in an otherwise normal chromosome 7. *Cytogenet. Genome Res.* 128, 189–191. doi: 10.1159/000271471

Liu, Y., Su, H., Pang, J., Gao, Z., Wang, X. J., Birchler, J. A., et al. (2015). Sequential *de novo* centromere formation and inactivation on a chromosomal fragment in maize. *Proc. Natl. Acad. Sci. U.S.A.* 112, E1263–E1271. doi: 10.1073/pnas.1418248112

Lysak, M. A. (2014). Live and let die: centromere loss during evolution of plant chromosomes. *New Phytol.* 203, 1082–1089. doi: 10.1111/nph.12885

Macas, J., Neumann, P., and Navratilova, A. (2007). Repetitive DNA in the pea (*Pisum sativum* L.) genome: comprehensive characterization using 454 sequencing and comparison to soybean and *Medicago truncatula*. *BMC Genom.* 8:427. doi: 10.1186/1471-2164-8-427

Maggert, K. A., and Karpen, G. H. (2001). The activation of a neocentromere in Drosophila requires proximity to an endogenous centromere. *Genetics* 158, 1615–1628.

Malheiros, N., De Castro, K., and Camara, A. (1947). Chromosomas sem centromero localizado O caso da *Luzula* purpurea Link. *Agronomia Lusitana* 9, 51–74.

Malik, H. S., and Henikoff, S. (2002). Conflict begets complexity: the evolution of centromeres. *Curr. Opin. Genet. Dev.* 12, 711–718. doi: 10.1016/S0959-437X(02)00351-9

Manicardi, G. C., Mandrioli, M., and Blackman, R. L. (2015). The cytogenetic architecture of the aphid genome. *Biol. Rev. Camb. Philos. Soc.* 90, 112–125. doi: 10.1111/brv.12096

Manzanero, S., and Puertas, M. J. (2003). Rye terminal neocentromeres: characterisation of the underlying DNA and chromatin structure. *Chromosoma* 111, 408–415. doi: 10.1007/s00412-002-0224-2

Manzanero, S., Puertas, M. J., Jimenez, G., and Vega, J. M. (2000). Neocentric activity of rye 5RL chromosome in wheat. *Chromosome Res.* 8, 543–554. doi: 10.1023/A:1009275807397

Manzanero, S., Vega, J. M., Houben, A., and Puertas, M. J. (2002). Characterization of the constriction with neocentric activity of 5RL chromosome in wheat. *Chromosoma* 111, 228–235. doi: 10.1007/s00412-002-0211-7

Marques, A., Ribeiro, T., Neumann, P., Macas, J., Novak, P., Schubert, V., et al. (2015). Holocentromeres in *Rhynchospora* are associated with genome-wide centromere-specific repeat arrays interspersed amongst euchromatin. *Proc. Natl. Acad. Sci. U.S.A.* (in press).

Marshall, O. J., Chueh, A. C., Wong, L. H., and Choo, K. H. (2008). Neocentromeres: new insights into centromere structure, disease development, and karyotype evolution. *Am. J. Hum. Genet.* 82, 261–282. doi: 10.1016/j.ajhg.2007.11.009

McClintock, B. (1939). The behavior in successive nuclear divisions of a chromosome broken at meiosis. *Proc. Natl. Acad. Sci. U.S.A.* 25, 405–416. doi: 10.1073/pnas.25.8.405

McIntyre, C. L., Pereira, S., Moran, L. B., and Appels, R. (1990). New *Secale cereale* (rye) DNA derivatives for the detection of rye chromosome segments in wheat. *Genome* 33, 635–640. doi: 10.1139/g90-094

Melters, D. P., Bradnam, K. R., Young, H. A., Telis, N., May, M. R., Ruby, J. G., et al. (2013). Comparative analysis of tandem repeats from hundreds of species reveals unique insights into centromere evolution. *Genome Biol.* 14, R10–R10. doi: 10.1186/gb-2013-14-1-r10

Melters, D. P., Paliulis, L. V., Korf, I. F., and Chan, S. W. (2012). Holocentric chromosomes: convergent evolution, meiotic adaptations, and genomic analysis. *Chromosome Res.* 20, 579–593. doi: 10.1007/s10577-012-9292-1

Mendiburo, M. J., Padeken, J., Fulop, S., Schepers, A., and Heun, P. (2011). Drosophila CENH3 is sufficient for centromere formation. *Science* 334, 686–690. doi: 10.1126/science.1206880

Monen, J., Hattersley, N., Muroyama, A., Stevens, D., Oegema, K., and Desai, A. (2015). Separase cleaves the N-Tail of the CENP-A related protein CPAR-1 at the meiosis I metaphase-anaphase transition in *C. elegans*. *PLoS ONE* 10:e0125382. doi: 10.1371/journal.pone.0125382

Monen, J., Maddox, P. S., Hyndman, F., Oegema, K., and Desai, A. (2005). Differential role of CENP-A in the segregation of holocentric *C. elegans* chromosomes during meiosis and mitosis. *Nat. Cell Biol.* 7, 1248–1255. doi: 10.1038/ncb1331

Moraes, I. C., Lermontova, I., and Schubert, I. (2011). Recognition of *A. thaliana* centromeres by heterologous CENH3 requires high similarity to the endogenous protein. *Plant Mol. Biol.* 75, 253–261. doi: 10.1007/s11103-010-9723-3

Nagaki, K., Cheng, Z., Ouyang, S., Talbert, P. B., Kim, M., Jones, K. M., et al. (2004). Sequencing of a rice centromere uncovers active genes. *Nat. Genet.* 36, 138–145. doi: 10.1038/ng1289

Nagaki, K., Kashihara, K., and Murata, M. (2005). Visualization of diffuse centromeres with centromere-specific histone H3 in the holocentric plant *Luzula nivea*. *Plant Cell* 17, 1886–1893. doi: 10.1105/tpc.105.032961

Nasuda, S., Hudakova, S., Schubert, I., Houben, A., and Endo, T. R. (2005). Stable barley chromosomes without centromeric repeats. *Proc. Natl. Acad. Sci. U.S.A.* 102, 9842–9847. doi: 10.1073/pnas.0504235102

Nelson, A. D., Lamb, J. C., Kobrossly, P. S., and Shippen, D. E. (2011). Parameters affecting telomere-mediated chromosomal truncation in *Arabidopsis*. *Plant Cell* 23, 2263–2272. doi: 10.1105/tpc.111.086017

Neumann, P., Navratilova, A., Koblizkova, A., Kejnovsky, E., Hribova, E., Hobza, R., et al. (2011). Plant centromeric retrotransposons: a structural and cytogenetic perspective. *Mob. DNA* 2, 4. doi: 10.1186/1759-8753-2-4

Neumann, P., Navratilova, A., Schroeder-Reiter, E., Koblizkova, A., Steinbauerova, V., Chocholova, E., et al. (2012). Stretching the rules: monocentric chromosomes with multiple centromere domains. *PLoS Genet.* 8:e1002777. doi: 10.1371/journal.pgen.1002777

Neumann, P., Pavlíková, Z., Koblížková, A., Fuková, I., Jedličková, V., Novák, P., et al. (2015). Centromeres off the hook: massive changes in centromere size and structure following duplication of CenH3 gene in fabeae species. *Mol. Biol. Evol.* 32, 1862–1879. doi: 10.1093/molbev/msv070

Nonaka, N., Kitajima, T., Yokobayashi, S., Xiao, G., Yamamoto, M., Grewal, S. I., et al. (2002). Recruitment of cohesin to heterochromatic regions by Swi6/HP1 in fission yeast. *Nat. Cell Biol.* 4, 89–93. doi: 10.1038/ncb739

Nonomura, K. I., and Kurata, N. (2001). The centromere composition of multiple repetitive sequences on rice chromosome 5. *Chromosoma* 110, 284–291. doi: 10.1007/s004120100148

Nordenskiold, H. (1951). Cyto-taxonomical studies in the genus *Luzula*.1. somatic chromosomes and chromosome numbers. *Hereditas* 37, 325–355. doi: 10.1111/j.1601-5223.1951.tb02898.x

Nordenskiold, H. (1964). Effect of X-Irradiation on Diploid + Polyploid *Luzula*. *Hereditas* 51, 344. doi: 10.1111/j.1601-5223.1964.tb01940.x

O'Toole, E. T., Mcdonald, K. L., Mäntler, J., Mcintosh, J. R., Hyman, A. A., and Müller-Reichert, T. (2003). Morphologically distinct microtubule ends in the mitotic centrosome of *Caenorhabditis elegans*. *J. Cell Biol.* 163, 451–456. doi: 10.1083/jcb.200304035

Olszak, A. M., Van Essen, D., Pereira, A. J., Diehl, S., Manke, T., Maiato, H., et al. (2011). Heterochromatin boundaries are hotspots for *de novo* kinetochore formation. *Nat. Cell Biol.* 13, 799–808. doi: 10.1038/ncb2272

Pazy, B., and Plitmann, U. (1994). Holocentric chromosome behavior in *Cuscuta* (*Cuscutaceae*). *Plant Syst. Evol.* 191, 105–109. doi: 10.1007/BF00985345

Pazy, B., and Plitmann, U. (1995). Chromosome divergence in the genus *Cuscuta* and its systematic implications. *Caryologia* 48, 173–180. doi: 10.1080/00087114.1995.10797327

Peacock, W. J., Dennis, E. S., Rhoades, M. M., and Pryor, A. J. (1981). Highly repeated DNA sequence limited to knob heterochromatin in maize. *Proc. Natl. Acad. Sci. U.S.A.* 78, 4490–4494. doi: 10.1073/pnas.78.7.4490

Pimpinelli, S., and Goday, C. (1989). Unusual kinetochores and chromatin diminution in *Parascaris*. *Trends Genet.* 5, 310–315. doi: 10.1016/0168-9525(89)90114-5

Platero, J. S., Ahmad, K., and Henikoff, S. (1999). A distal heterochromatic block displays centromeric activity when detached from a natural centromere. *Mol. Cell* 4, 995–1004. doi: 10.1016/S1097-2765(00)80228-2

Powers, J., Rose, D. J., Saunders, A., Dunkelbarger, S., Strome, S., and Saxton, W. M. (2004). Loss of KLP-19 polar ejection force causes misorientation and missegregation of holocentric chromosomes. *J. Cell Biol.* 166, 991–1001. doi: 10.1083/jcb.200403036

Prakken, R., and Müntzing, A. (1942). A meiotic peculiarity in rye, simulating a terminal centromere. *Hereditas* 28, 441–482. doi: 10.1111/j.1601-5223.1942.tb03291.x

Puertas, M. J., Garcia-Chico, R., Sotillo, E., Gonzalez-Sanchez, M., and Manzanero, S. (2005). Movement ability of rye terminal neocentromeres. *Cytogenet. Genome Res.* 109, 120–127. doi: 10.1159/000082390

Reznicek, A. A. (1990). Evolution in sedges (*Carex, Cyperaceae*). *Can J. Bot.* 68, 1409–1432. doi: 10.1139/b90-180

Rhoades, M. M. (1942). Preferential segregation in maize. *Genetics* 27, 395–407.

Rhoades, M. M., and Dempsey, E. (1966). The effect of abnormal chromosome 10 on preferential segregation and crossing over in maize. *Genetics* 53, 989–1020.

Rhoades, M. M., and Vilkomerson, H. (1942). On the anaphase movement of chromosomes. *Proc. Natl. Acad. Sci. U.S.A.* 28, 433–436. doi: 10.1073/pnas.28.10.433

Rocchi, M., Archidiacono, N., Schempp, W., Capozzi, O., and Stanyon, R. (2012). Centromere repositioning in mammals. *Heredity (Edinb)* 108, 59–67. doi: 10.1038/hdy.2011.101

Ross, B. D., Rosin, L., Thomae, A. W., Hiatt, M. A., Vermaak, D., De La Cruz, A. F. A., et al. (2013). Stepwise evolution of essential centromere function in a *Drosophila neogene*. *Science* 340, 1211–1214. doi: 10.1126/science.1234393

Saffery, R., Irvine, D. V., Griffiths, B., Kalitsis, P., Wordeman, L., and Choo, K. H. (2000). Human centromeres and neocentromeres show identical distribution patterns of >20 functionally important kinetochore-associated proteins. *Hum. Mol. Genet.* 9, 175–185. doi: 10.1093/hmg/9.2.175

Sanei, M., Pickering, R., Kumke, K., Nasuda, S., and Houben, A. (2011). Loss of centromeric histone H3 (CENH3) from centromeres precedes uniparental chromosome elimination in interspecific barley hybrids. *Proc. Natl. Acad. Sci. U.S.A.* 108, E498–E505. doi: 10.1073/pnas.1103190108

Schlegel, R. (1987). Neocentric activity in chromosome 5R of rye revealed by haploidy. *Hereditas* 107, 1–6. doi: 10.1111/j.1601-5223.1987.tb00264.x

Schrader, F. (1935). Notes an the mitotic behavior of long chromosomes. *Cytologia* 6, 422–430. doi: 10.1508/cytologia.6.422

Schubert, I., Rieger, R., and Fuchs, J. (1995). Alteration of basic chromosome number by fusion-fission cycles. *Genome* 38, 1289–1292. doi: 10.1139/g95-170

Schubert, V. (2011). No neocentric activity on *Aegilops markgrafii* chromosome E. *Cytogenet. Genome Res.* 132, 100–104. doi: 10.1159/000318996

Schvarzstein, M., Wignall, S. M., and Villeneuve, A. M. (2010). Coordinating cohesion, co-orientation, and congression during meiosis: lessons from holocentric chromosomes. *Genes Dev.* 24, 219–228. doi: 10.1101/gad.1863610

Scott, K. C., and Sullivan, B. A. (2014). Neocentromeres: a place for everything and everything in its place. *Trends Genet.* 30, 66–74. doi: 10.1016/j.tig.2013.11.003

Shang, W. H., Hori, T., Martins, N. M., Toyoda, A., Misu, S., Monma, N., et al. (2013). Chromosome engineering allows the efficient isolation of vertebrate neocentromeres. *Dev. Cell* 24, 635-648. doi: 10.1016/j.devcel.2013.02.009

Shang, W. H., Hori, T., Toyoda, A., Kato, J., Popendorf, K., Sakakibara, Y., et al. (2010). Chickens possess centromeres with both extended tandem repeats and short non-tandem-repetitive sequences. *Genome Res.* 20, 1219–1228. doi: 10.1101/gr.106245.110

Sheikh, S. A., and Kondo, K. (1995). Differential staining with orcein, giemsa, cma, and dapi for comparative chromosome study of 12 species of australian *Drosera* (*Droseraceae*). *Am. J. Bot.* 82, 1278–1286. doi: 10.2307/2446251

Sheikh, S. A., Kondo, K., and Hoshi, Y. (1995). Study of diffused centromeric nature of *Drosera chromosomes*. *Cytologia (Tokyo)* 60, 43–47. doi: 10.1508/cytologia.60.43

Shirakawa, J., Nagano, K., and Hoshi, Y. (2011). A chromosome study of two centromere differentiating *Drosera* species, *D. arcturi* and *D. regia*. *Caryologia* 64, 453–463.

Steiner, F. A., and Henikoff, S. (2014). Holocentromeres are dispersed point centromeres localized at transcription factor hotspots. *Elife* 3, e02025. doi: 10.7554/eLife.02025

Steiner, F. A., and Henikoff, S. (2015). Diversity in the organization of centromeric chromatin. *Curr. Opin. Genet. Dev.* 31, 28–35. doi: 10.1016/j.gde.2015.03.010

Stimpson, K. M., Matheny, J. E., and Sullivan, B. A. (2012). Dicentric chromosomes: unique models to study centromere function and inactivation. *Chromosome Res.* 20, 595–605. doi: 10.1007/s10577-012-9302-3

Stinchcomb, D. T., Shaw, J. E., Carr, S. H., and Hirsh, D. (1985). Extrachromosomal DNA transformation of *Caenorhabditis elegans*. *Mol. Cell. Biol.* 5, 3484–3496. doi: 10.1128/MCB.5.12.3484

Sullivan, B. A., and Willard, H. F. (1998). Stable dicentric × chromosomes with two functional centromeres. *Nat. Genet.* 20, 227–228. doi: 10.1038/3024

Tanaka, N. (1977). Chromosome studies in *Chionographis* (*Liliaceae*). 1. holokinetic nature of chromosomes in *Chionographis*-japonica-maxim. *Cytologia* 42, 753–763. doi: 10.1508/cytologia.42.753

Teo, C. H., Lermontova, I., Houben, A., Mette, M. F., and Schubert, I. (2013). *De novo* generation of plant centromeres at tandem repeats. *Chromosoma* 122, 233–241. doi: 10.1007/s00412-013-0406-0

Thakur, J., and Sanyal, K. (2013). Efficient neocentromere formation is suppressed by gene conversion to maintain centromere function at native physical chromosomal loci in *Candida albicans*. *Genome Res.* 23, 638–652. doi: 10.1101/gr.141614.112

Topp, C. N., Okagaki, R. J., Melo, J. R., Kynast, R. G., Phillips, R. L., and Dawe, R. K. (2009). Identification of a maize neocentromere in an oat-maize addition line. *Cytogenet. Genome Res.* 124, 228–238. doi: 10.1159/000218128

Tsujimoto, H. (1993). Molecular cytological evidence for gradual telomere synthesis at the broken chromosome ends in wheat. *J. Plant Res.* 106, 239–244. doi: 10.1007/BF02344591

Tsujimoto, H., Usami, N., Hasegawa, K., Yamada, T., Nagaki, K., and Sasakuma, T. (1999). *De novo* synthesis of telomere sequences at the healed breakpoints of wheat deletion chromosomes. *Mol. Gen. Genet.* 262, 851–856. doi: 10.1007/s004380051150

Vershinin, A. V., Schwarzacher, T., and Heslop-Harrison, J. S. (1995). The large-scale genomic organization of repetitive DNA families at the telomeres of rye chromosomes. *Plant Cell* 7, 1823–1833. doi: 10.1105/tpc.7.11.1823

Viera, A., Page, J., and Rufas, J. S. (2009). Inverted meiosis: the true bugs as a model to study. *Genome Dyn* 5, 137–156. doi: 10.1159/000166639

Villasante, A., Abad, J. P., and Mendez-Lago, M. (2007). Centromeres were derived from telomeres during the evolution of the eukaryotic chromosome. *Proc. Natl. Acad. Sci. U.S.A.* 104, 10542–10547. doi: 10.1073/pnas.0703808104

Voullaire, L. E., Slater, H. R., Petrovic, V., and Choo, K. H. (1993). A functional marker centromere with no detectable alpha-satellite, satellite III, or CENP-B protein: activation of a latent centromere? *Am. J. Hum. Genet.* 52, 1153–1163.

Wade, C. M., Giulotto, E., Sigurdsson, S., Zoli, M., Gnerre, S., Imsland, F., et al. (2009). Genome sequence, comparative analysis, and population genetics of the domestic horse. *Science* 326, 865–867. doi: 10.1126/science.1178158

Walters, M. S. (1952). Spontaneous chromosome breakage and atypical chromosome movement in meiosis of the hybrid *Bromus marginatus* × *B. Pseudolaevipes*. *Genetics* 37, 8–25.

Wang, G., Li, H., Cheng, Z., and Jin, W. (2013). A novel translocation event leads to a recombinant stable chromosome with interrupted centromeric domains in rice. *Chromosoma* 122, 295–303. doi: 10.1007/s00412-013-0413-1

Wang, K., Wu, Y., Zhang, W., Dawe, R. K., and Jiang, J. (2014a). Maize centromeres expand and adopt a uniform size in the genetic background of oat. *Genome Res.* 24, 107–116. doi: 10.1101/gr.160887.113

Wang, L., Zeng, Z., Zhang, W., and Jiang, J. (2014b). Three potato centromeres are associated with distinct haplotypes with or without megabase-sized satellite repeat arrays. *Genetics* 196, 397–401. doi: 10.1534/genetics.113.160135

Wanner, G., Schroeder-Reiter, E., Ma, W., Houben, A., and Schubert, V. (2015). The ultrastructure of mono- and holocentric plant centromeres: an immunological investigation by structured illumination microscopy and scanning electron microscopy. *Chromosoma* doi: 10.1007/s00412-015-0521-1 [Epub ahead of print].

Wolfgruber, T. K., Sharma, A., Schneider, K. L., Albert, P. S., Koo, D. H., Shi, J., et al. (2009). Maize centromere structure and evolution: sequence analysis of centromeres 2 and 5 reveals dynamic Loci shaped primarily by retrotransposons. *PLoS Genet.* 5:e1000743. doi: 10.1371/journal.pgen.1000743

Yan, H., Talbert, P. B., Lee, H. R., Jett, J., Henikoff, S., Chen, F., et al. (2008). Intergenic locations of rice centromeric chromatin. *PLoS Biol.* 6:e286. doi: 10.1371/journal.pbio.0060286

Yano, O., and Hoshino, A. T. (2006). Cytological studies of aneuploidy in *Eleocharis kamtschatica* (*Cyperaceae*). *Cytologia* 71, 141–147. doi: 10.1508/cytologia.71.141

Yelina, N. E., Choi, K., Chelysheva, L., Macaulay, M., De Snoo, B., Wijnker, E., et al. (2012). Epigenetic remodeling of meiotic crossover frequency in *Arabidopsis thaliana* DNA methyltransferase mutants. *PLoS Genet.* 8:e1002844. doi: 10.1371/journal.pgen.1002844

Yi, C., Zhang, W., Dai, X., Li, X., Gong, Z., Zhou, Y., et al. (2013). Identification and diversity of functional centromere satellites in the wild rice species *Oryza brachyantha*. *Chromosome Res.* 21, 725–737. doi: 10.1007/s10577-013-9374-8

Yu, H. G., Hiatt, E. N., Chan, A., Sweeney, M., and Dawe, R. K. (1997). Neocentromere-mediated chromosome movement in maize. *J. Cell Biol.* 139, 831–840. doi: 10.1083/jcb.139.4.831

Yuen, K. W. Y., Nabeshima, K., Oegema, K., and Desai, A. (2011). Rapid *de novo* centromere formation occurs independently of heterochromatin protein 1 in *C. elegans* Embryos. *Curr. Biol.* 21, 1800–1807. doi: 10.1016/j.cub.2011.09.016

Zedek, F., Smerda, J., Smarda, P., and Bures, P. (2010). Correlated evolution of LTR retrotransposons and genome size in the genus *Eleocharis*. *BMC Plant Biol.* 10:265. doi: 10.1186/1471-2229-10-265

Zhang, B., Lv, Z., Pang, J., Liu, Y., Guo, X., Fu, S., et al. (2013). Formation of a functional maize centromere after loss of centromeric sequences and gain of ectopic sequences. *Plant Cell* 25, 1979–1989. doi: 10.1105/tpc.113.110015

Zhang, H., and Dawe, R. K. (2012). Total centromere size and genome size are strongly correlated in ten grass species. *Chromosome Res.* 20, 403–412. doi: 10.1007/s10577-012-9284-1

Zhang, H., Koblizkova, A., Wang, K., Gong, Z., Oliveira, L., Torres, G. A., et al. (2014). Boom-bust turnovers of megabase-sized centromeric DNA in *Solanum* species: rapid evolution of DNA sequences associated with centromeres. *Plant Cell* 26, 1436–1447. doi: 10.1105/tpc.114.123877

Zhang, W., Friebe, B., Gill, B. S., and Jiang, J. (2010). Centromere inactivation and epigenetic modifications of a plant chromosome with three functional centromeres. *Chromosoma* 119, 553–563. doi: 10.1007/s00412-010-0278-5

Conflict of Interest Statement: The authors declare that the research was conducted in the absence of any commercial or financial relationships that could be construed as a potential conflict of interest.

9

Isolation of Plant Photosystem II Complexes by Fractional Solubilization

*Patrycja Haniewicz[1], Davide Floris[2], Domenica Farci[2], Joanna Kirkpatrick[3], Maria C. Loi[2], Claudia Büchel[4], Matthias Bochtler[1,5] and Dario Piano[1,2]**

[1] Laboratory of Structural Biology, Department of Molecular Biology, International Institute of Molecular and Cell Biology, Warsaw, Poland, [2] Laboratory of Photosynthesis and Photobiology, Department of Life and Environmental Sciences, University of Cagliari, Cagliari, Italy, [3] Proteomics Core Facility, European Molecular Biology Laboratory, Heidelberg, Germany, [4] Laboratory of Plant Cell Physiology, Institute of Molecular Biosciences, Goethe-University Frankfurt, Frankfurt am Main, Germany, [5] Department of Bioinformatics, Institute of Biochemistry and Biophysics, Warsaw, Poland

Photosystem II (PSII) occurs in different forms and supercomplexes in thylakoid membranes. Using a transplastomic strain of *Nicotiana tabacum* histidine tagged on the subunit PsbE, we have previously shown that a mild extraction protocol with β-dodecylmaltoside enriches PSII characteristic of lamellae and grana margins. Here, we characterize residual granal PSII that is not extracted by this first solubilization step. Using affinity purification, we demonstrate that this PSII fraction consists of PSII-LHCII mega- and supercomplexes, PSII dimers, and PSII monomers, which were separated by gel filtration and functionally characterized. Our findings represent an alternative demonstration of different PSII populations in thylakoid membranes, and they make it possible to prepare PSII-LHCII supercomplexes in high yield.

Keywords: photosystem II, PSII-LHCII supercomplex, PSII-LHCII megacomplex, thylakoid membranes, *Nicotiana tabacum*, oligomeric state

Edited by:
Roman Sobotka,
Czech Academy of Sciences,
Czech Republic

Reviewed by:
Josef Komenda,
Institute of Microbiology of the ASCR,
Czech Republic
Bettina Ughy,
Biological Research Centre of the
Hungarian Academy of Sciences,
Hungary

Correspondence:
Dario Piano
dario.piano@unica.it

INTRODUCTION

Oxygenic photosynthesis is one of the key processes sustaining the life on our planet by providing the biosphere with oxygen and sugars. Photosystem II (PSII) is a membrane protein complex that plays an essential role in oxygenic photosynthesis. Sunlight drives the splitting of water into oxygen, electrons, and protons (Cardona et al., 2012), but it also causes photodamage, which is minimized by photoprotection in conditions of excess light and repaired by PSII turnover in all conditions. The distribution of PSII in different complexes within the membranes reflects PSII assembly and repair (Pokorska et al., 2009), as well as optimization for efficient usage of light while avoiding or limiting damage (Boekema et al., 1995; Dekker and Boekema, 2005; Takahashi et al., 2009; Watanabe et al., 2009; Pagliano et al., 2011). In lamellae and the marginal grana, PSII is assembled *de novo* or repaired. Exhausted PSII complexes formed in the grana cores migrate to grana margins and lamellae, while being replaced by new and fully functional complexes (Edelman and Mattoo, 2008; Nixon et al., 2010; Puthiyaveetil et al., 2014; Tomizioli et al., 2014). With respect to PSII, lamellae and grana margins may thus be considered as the "nursery", while the grana cores act as the "photochemical plant" of thylakoids.

Five main forms of PSII are thought to occur *in vivo*: the monomeric (PSIIm or C), the dimeric (PSIId or C_2), the incomplete form free of the antenna component CP43 (RC-CP47) and finally the two PSII complexes, consisting of several combinations of C_2 with the trimeric Light Harvesting Complex II (LHCII), which may bind C_2 strongly (S) or mildly (M) via the so-called minor antenna complexes (CP24, CP26, and CP29). Their assembly leads to higher photosynthetic units called PSII-LHCII supercomplex (PSIIsc or C_2S_2) and megacomplex (PSIImc or $C_2S_2M_2$; Boekema et al., 1999; Eshaghi et al., 1999; de Bianchi et al., 2008; Caffarri et al., 2009). The five PSII types are localized in different regions of the thylakoid membranes (Danielsson et al., 2006). In particular PSIIm and RC-CP47 are typically localized in the lamellae and peripheral part of the grana, being the main constituents of the "nursery" region, whereas PSIId, C_2S_2, $C_2S_2M_2$ are localized in the grana cores, being the main constituents of the functional side of the thylakoid membranes.

We have recently reported the purification of an inactive form of PSIIm that contains the subunit PsbS and appears as one of the main PSII forms associated with the lamellae region of the thylakoid membranes (Haniewicz et al., 2013). In this work, we now describe the fraction that is not solubilized by the mild-extraction protocol. This fraction, that requires harsher extraction conditions, was characterized in order to reveal whether it may contain the grana complexes opening the way for their isolation and characterization. The material solubilized in the second step is characterized by Blue native polyacrylamide gel electrophoresis (BN-PAGE) and mass spectrometry (MS), which together provide information on native molecular mass and subunit composition. After solubilization, granal thylakoids were also subjected to Immobilized-Metal Affinity Chromatography (IMAC) and subsequent Size Exclusion Chromatography (SEC). In the latter step, PSII complexes, supercomplexes, and megacomplexes could be separated, further characterized and compared using several techniques. This procedure allows the chromatography isolation of preparative amounts of highly pure C, C_2, C_2S_2 particles and represents a step further to their structural and functional characterization. The described procedure also provides a direct biochemical probe of PSII organization and distribution in thylakoid membranes.

MATERIALS AND METHODS

Growth of Tobacco Plants

Plant material was obtained from a transplastomic strain of *Nicotiana tabacum* that carries a hexa-histidine tag at the 5′ end of the gene coding for the PsbE subunit (Fey et al., 2008). The plants were grown for 10–12 weeks with a 50% relative humidity at a constant temperature of 25°C and under a light regime of 12 h/day, with a light intensity of 150–200 μmol photons/(s m^2).

Thylakoid Preparation

Thylakoid membranes were purified as reported previously by Fey et al. (2008), but in the last centrifugation step they were resuspended in 20 mM MES–NaOH, pH 6.5; 100 mM NaCl; 5 mM $MgCl_2$; 10 mM $NaHCO_3$; 12.5% (v/v) glycerol.

Thylakoids Solubilization and PSII Core Complex Purification by Affinity Chromatography

Thylakoid fractions were obtained routinely and with high reproducibility. Lamellar fractions were obtained by using the procedure reported in Haniewicz et al. (2013). Minimal modifications were introduced with the aim to ensure a complete removal of the peripheral grana (and lamellae) from the grana cores. Briefly, thylakoids membranes where solubilized for 30 min at 4°C and the final chlorophyll concentration was kept in a range between 2 and 3 mg/ml (SL) without major changes. After solubilization, the supernatant was separated from the unsolubilized fraction by spinning at 45000 × *g*. The unsolubilized granal fraction was homogenized and subsequently solubilized for 10 min at 4°C at a chlorophyll concentration of 1 mg/ml (SG), as reported in Fey et al. (2008) for the isolation of PSII complexes with LHC polypeptides bound. In both cases solubilization was carried out using 20 mM β-dodecylmaltoside (β-DDM).

Photosystem II samples were prepared routinely using Ni affinity chromatography. PSII was isolated following the procedure reported in Piano et al. (2010) with the only difference that the washing buffer was free of glycerol and betaine (20 mM MES–NaOH, pH 6.5; 100 mM NaCl; 10 mM $NaHCO_3$; 15 mM imidazole) and PSII cores were eluted using 40 mM MES–NaOH, pH 6.5; 20 mM NaCl; 5 mM $MgCl_2$; 1 mM $CaCl_2$; 10 mM $NaHCO_3$; 400 mM imidazole. The washing and the elution buffers contained 0.01% instead of 0.03% (w/v) β-DDM.

Size Exclusion Chromatography

The eluted fractions from Ni-NTA chromatography were pooled and concentrated using Vivaspin 20 ultrafiltration membranes with 100 kDa cutoff until a final volume of 200 μl. The protein sample was loaded on a home-made column of 80 ml bed volume Superose 6 resin (GE Healthcare) with a diameter of 10 mm leading to highly reproducible separations of specific protein complexes. Protein separation and column pre-equilibration were performed in gel filtration buffer (40 mM MES–NaOH, pH 6.5; 20 mM NaCl; 5 mM $MgCl_2$; 1 mM $CaCl_2$; 10 mM $NaHCO_3$; 0.01% (w/v) β-DDM).

Absorption Spectroscopy, Chlorophyll Determination, and Yield Calculation

The protein content in thylakoids and purified complexes was calculated referring to the Chl *a* and Chl *b* concentrations from three independent measurements. The analysis was done photometrically in 80% (v/v) acetone using a Pharmacia Biotech Ultrospec 4000 spectrophotometer and Chl concentrations were calculated according to Porra et al. (1989). Yield calculation in thylakoids and PSII complexes was expressed in mg or % of chlorophylls. Measurements were performed on samples from 10 independent purifications starting from different thylakoid stocks. For each of these 10 purifications was calculated the amount of thylakoids and the amounts of the specific PSII complexes isolated (C-PsbS, C, C_2, C_2S_2) expressing them in mg or % of Chls with respect to the initial thylakoid amount. Finally, the amounts for each of the 5 classes (thylakoids, C-PsbS, C, C_2, C_2S_2) were averaged respect to the 10 independent purifications (**Table 4**). The values were expressed as means ± standard deviations.

Polyacrylamide Gel Electrophoresis

Blue native polyacrylamide gel electrophoresis was routinely used as a reference to cross check the correct SG solubilization, the correct SEC profiles and to select the specific pools to be used for further structural and functional tests. According to Schägger and von Jagow (1991), native electrophoresis was performed using 3–12% (w/v) continuous gradient gels. PSII complexes and thylakoid samples at 0.2 mg Chl/ml were mixed with 0.25 volumes of Coomassie Blue Solution (5% (v/v) serva Blue G, 750 mM aminocaproic acid, 35% (w/v) sucrose). The electrophoresis was carried out at 205 V for 5 h for PSII complexes, while for thylakoids it consisted in a run at 60 V for 12 h. In both cases the run was performed at 4°C. After the electrophoretic run, the gels were stained with Coomassie brilliant blue G250.

Mass Spectrometry

The BN-PAGE gel bands from the SG samples or from the SEC fractions were excised and analyzed. Samples were processed as described in Farci et al. (2014). Protein groups were assigned to bands, and qualitative estimates of protein abundance were based on an "index" obtained by dividing unweighted spectral counts (spectral count) by protein mass (kDa). Proteins were either excluded (index below 0.25), or divided into groups with indices between 0.25 and 0.50 (marked with + in **Table 3**), 0.50 and 0.75 (marked with ++ in **Table 3**) and indices higher than 0.75 (marked with +++ in **Table 3**). Higher indices indicate greater abundance of the protein in the samples.

Oxygen Evolution

Oxygen evolution rates under light saturation were measured using a Clark-type electrode (Hansatech, England) at 20°C. In the reaction mixture the samples were added to gel filtration buffer enriched with freshly prepared electron acceptors (1 mM 2,6-dichloro-*p*-benzoquinone and 1 mM ferricyanide). The measurements were carried out at a Chls concentration of 100 μg/ml for thylakoids and 50 μg/ml for PSII samples. Activity was tested with three independent measurements on the same preparation and the values were expressed as means ± standard deviations.

Electron Microscopy

Size Exclusion Chromatography fractions of different PSII complexes from three independent purifications were checked by Transmission Electron Microscopy (TEM). Samples were diluted in gel filtration buffer and applied on glow-discharged carbon

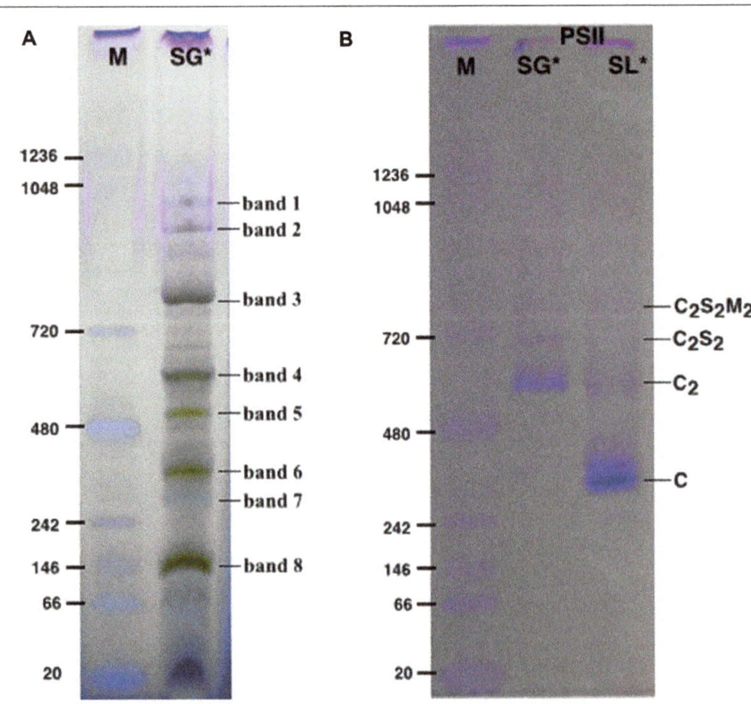

FIGURE 1 | The solubilized grana cores (SG), when resolved by Blue native polyacrylamide gel electrophoresis (BN-PAGE; A) separated into a pattern of bands equivalent to specific thylakoid complexes. The bands were attributed to a given complex on the basis of their mass spectrometry (MS) analysis (see **Table 1**). The PSII pools purified by NiNTA affinity chromatography are resolved by BN-PAGE **(B)**. The lane SG is the PSII pool purified from the solubilized grana fraction, whereas lane SL represents the pool of PSII purified from solubilized lamellae. In lanes M the molecular marker (M) was loaded. $C_2S_2M_2$, C_2S_2, C_2 and C are the PSII-LHCII megacomplexes, PSII-LHCII supercomplexes, PSII dimers and PSII monomers, respectively. *The BN-PAGE used must be considered reliable in the mass range between 66 and 480 kDa according to the molecular marker (M). Above this range, the heaviest bands appear significantly under estimated in weight.

TABLE 1 | Mass spectrometry (MS) analysis performed on the bands of the SG samples resolved by BN-PAGE (see Figure 1A).

Band	PSI (LHCI)	PSII (LHCII)	Cyt b_6f complex	ATP synthase	Complexes
1	PsaB, D, E, F, K CAB4, 7, 40	PsbA, B, C, D, E, O, Q CP29	–	–	PSI and PSII higher complexes
2	PsaA, B, D, F, K CAB40	PsbA, B, C, D, Q	–	–	
3	PsaA, B, D, F, L,H,K CAB4,16, 21, 40, lhca	PsbA, B, C, D, E, O CAB7, 36, 40, CP24, 26, 29	–	atpA, B	PSII megacomplex ($C_2S_2M_2$), PSI-LHCII ATPase
4	PsaA, B, D, E, F, L, CAB4, 25, lhca	PsbA, B, C, D, E, O, S CAB7, 13, 36, 40, CP29, lhcb	–	atpA, B, C, E, F atpα, γ	PSII supercomplex (C_2S_2), PSI-LHCII ATPase
5	PsaB CAB16, 21, 40	PsbA, B, C, D, E, O, S CAB7, 13, 36, 50, CP26, 29	–	–	PSII dimer (C_2) PSI
6	CAB21, 40	PsbA, B, C, D, E, O, Q CAB7, 16, 36, 50, CP26, 29	Cyt. f subunit subunit IV	–	PSII monomer (C), cytb_6f complex
7	CAB40	PsbA, B, C, D CAB7, 36, CP24, 26	–	–	PSII incomplete monomer
8	CAB21, 40	PsbC, D, O, P, Q, S CAB7, 36, 50, CP24, 26, 29	Cyt. f subunit	atpβ	Free subunits

For a more detailed analysis see Supplementary Table S1.

coated copper grids (400 mesh) followed by negative staining using filtered 2% uranyl acetate. Electron microscopy was performed in a CM12 electron microscope (Philips, Eindhoven, Netherlands) operated at 80 kV. Images were recorded under low dose conditions (total dose ~25e-/Å^2) with a ES500W camera (Gatan, Pleasanton, CA, USA) at a magnification of 110 kx.

RESULTS

Composition of SL and SG Fractions

In order to characterize PSII species associated to granal thylakoids, we solubilized membranes in two steps. In the first step, we used a previously described mild and long extraction procedure that selectively solubilizes only the peripheral part of the thylakoids (lamellae and grana margins; SL) leading to the chromatography isolation of monomeric PSII that binds the subunit PsbS (see protocol B samples in Table 1 of Haniewicz et al., 2013). The fraction of thylakoids not solubilized during the preparation of SL was pelleted by centrifugation, resuspended and finally subjected to a second solubilization, leading to samples with a PSII content representative for grana cores (SG) that were then subjected to BN-PAGE.

SG samples migrated on BN-PAGE resolving in bands with apparent masses from 1050 to 60 kDa (**Figure 1A**). The content in thylakoid complexes of each band was identified by MS. According to this analysis, PSII was mainly present in the $C_2S_2M_2$, C_2S_2, C_2 and C forms (**Table 1**, Supplementary Table S1). When incubated on ice for more than 6 h, the SL samples, but not the SG, were characterized by a tendency to precipitate, indicating an insufficient solubilization, which led to difficulties in their characterization by BN-PAGE (data not shown). Taken together, these data suggested large differences in the properties of lamellae and grana cores of the thylakoid membranes.

Oxygen Evolving Activity of SL and SG Fractions

The oxygen evolution capacity of SL and SG samples was tested. The SL samples had a minimal activity of 19 μmol O_2/mg Chl h, while the SG samples evolved 10-fold more oxygen (186 μmolO_2/mg Chl h), suggesting the presence of abundant and functional PSII (**Table 2**). These findings are consistent with the accepted view of the lamellae as the assembly and/or repair region of thylakoids in which the PSII particles are mainly inactive, and of the grana in which PSII is in an optimal chemical-physical environment that keeps it very active (Aro et al., 2005).

TABLE 2 | Rates of oxygen evolution of solubilized lamellar (SL) thylakoids, solubilized granal (SG) thylakoids and of PSII-LHCII megacomplexes (C2S2M2), PSII-LHCII supercomplexes (C_2S_2), PSII dimers (C2), and PSII monomers (C).

Thylakoids*		PSII purified samples*			
SL	SG	PSII monomers (C)	PSII dimers(C_2)	PSII-LHCII supercomplexes (C_2S_2)	PSII-LHCII megacomplexes ($C_2S_2M_2$)
19 ± 2	186 ± 5	960 ± 5	1360 ± 12	1030 ± 10	not determined

*For each sample, the activity was tested with three independent measurements and the values obtained were expressed in μmol O2/mg Chl h as means ± standard deviations. The $C_2S_2M_2$ complexes were not characterized because of their low amounts and their C_2S_2 impurities. The activity of the $C_2S_2M_2$ complexes was not determined because of their low amounts and their C_2S_2 impurities. *Values represent means ± standard deviations of three independent measurements.*

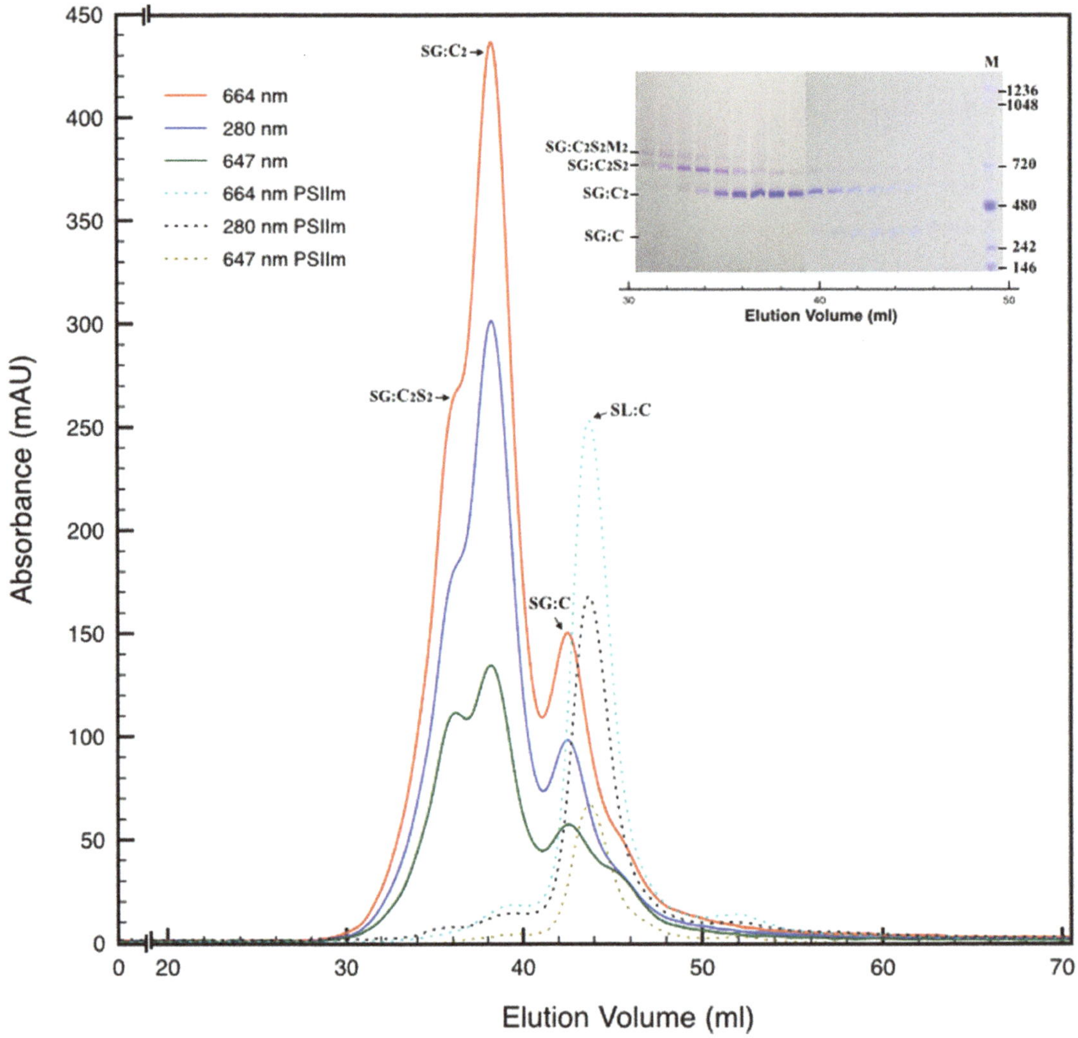

FIGURE 2 | Size exclusion chromatography of the PSII pool isolated from SG thylakoids by affinity chromatography (solid lines). All the measurements were recorded at three different wavelengths: 280 nm (proteins), 664 nm (chlorophyll *a*), 647 nm (chlorophyll *b*). In the inset are shown the elution fractions analyzed by BN-PAGE, confirming the partial separation of several oligomeric states of PSII. M is the molecular marker. The mostly monomeric PSII pool isolated from SL thylakoids by affinity chromatography (dotted lines) was used as a mass reference. In the chromatograms and in the inset SG:$C_2S_2M_2$, SG:C_2S_2, SG:C_2, and SG:C are the PSII-LHCII megacomplexes, PSII-LHCII supercomplexes, PSII dimers, and PSII monomers of grana origin (SG), respectively; SL:C are PSII monomer from lamellae (SL).

Oligomeric States of PSII Complexes from the SG Fraction

The SG and SL samples were subjected to further PSII separation steps after solubilization. The histidine tag on PsbE of the transplastomic plants made it possible to purify PSII by Ni-NTA (Fey et al., 2008). After purification, the oligomeric profile of the obtained SG-PSII was assessed by BN-PAGE and the composition compared with the already characterized SL-PSII (Haniewicz et al., 2013), showing a significant difference between the oligomeric patterns of the two PSII samples (**Figure 1B**). As clearly shown in **Figure 1B**, mainly C and sometimes incomplete PSII forms such as RC-CP47 can be obtained from SL, while SG is a mixture of C_2 and C_2S_2 with the frequent presence of small $C_2S_2M_2$ amounts.

The Different SG-PSII Species can be Separated by Size Exclusion Chromatography

Next, we attempted to separate the different PSII components obtained by the Ni-NTA chromatography from the SG-PSII samples by means of SEC. SG-PSII samples resolved into a characteristic and reproducible profile consisting of a shoulder and two peaks (**Figure 2**). In contrast, the same procedure for the SL-PSII samples led to a single peak corresponding to monomeric PSII confirming the BN-PAGE analysis on the NiNTA pool (**Figures 2** and **1B**) and previous results (Haniewicz et al., 2013). The collected fractions from the SEC analysis on the SG-PSII samples were checked by BN-PAGE and thus divided into four pools corresponding to samples enriched in

TABLE 3 | Mass spectrometry analysis performed on the bands of the SEC fractions resolved by BN-PAGE.

		Name	Accession Number	Mass (kDa)	Qualitative (unweighted subunit presence)				Quantitative* (weighted subunit presence)			
					C	C_2	C_2S_2	$C_2S_2M_2$	C	C_2	C_2S_2	$C_2S_2M_2$
PSII-LHCII (super/megacomplexes)	PSII monomers and dimers	PSB_A (D1)	PSBA_TOBAC	39	+	+	+	+	+++	+++	+++	+++
		PSB_B (CP47)	PSBB_TOBAC	56	+	+	+	+	+++	+++	+++	+++
		PCB_C (CP43)	PSBC_TOBAC	52	+	+	+	+	+++	+++	+++	+++
		PSB_D (D2)	PSBD_TOBAC	40	+	+	+	+	+++	+++	+++	+++
		PSB_E (cytb559)	PSBE_TOBAC	9	+	+	+	+	+++	+++	+++	+
		PSB_H	PSBH_TOBAC	8	+	+	+	+	+++	+++	+++	+++
		PSB_L	PBL_TOBAC	4	−	+	+	−	−	+++	+++	−
		PSB_O (33kDa)	Q84QE8	35	+	+	+	+	−	++	+++	+
		PSB_O (33kDa)	PSBO_TOBAC	35	−	+	+	+	−	++	+++	+
		PSB_R	PSBR_TOBAC	14	−	+	+	+	−	+	+	−
	CAB and LhcII proteins	Lhcb1 (CB24)	CB24_TOBAC	28	+	+	+	+	−	−	+++	+++
		Lhcb1 (CB27)	CB27_TOBAC	28	−	+	+	+	−	−	+++	+++
		Lhcb1 (CB22)	CB22_TOBAC	28	−	−	+	+	−	−	+++	++
		Lhcb1 (CB25)	CB25_TOBAC	28	−	−	+	−	−	−	+++	−
		Lhcb2 (CB23)	CB23_TOBAC (+1)	29	−	+	+	+	−	−	+++	++
		Lhcb3	A0A076L1Y1_TOBAC (+1)	29	−	−	+	+	−	−	−	−
		Lhcb4 (CP29)	Q0PWS7_TOBAC	31	+	+	+	+	−	−	+++	+++
		Lhcb5 (CP26)	Q0PWS5_TOBAC	30	−	+	+	+	−	−	+++	+++
		Lhcb6 (CP24)	Q0PWS6_TOBAC	27	−	+	+	+	−	−	+	+
Plastidial ATP synthase		ATP_A (α-subunit)	ATPA_TOBAC	55	−	+	−	−	−	−	−	−
		ATP_B (β-subunit)	ATPB_TOBAC (+1)	54	−	+	−	−	−	−	−	−
−		37kDa inner membrane polypeptide	Q40501_TOBAC	38	−	+	+	+	−	−	−	−
Vacuolar H^+-ATPase		subunit B	Q9M5Z8_TOBAC	54	−	−	+	+	−	−	−	−

*In the table are shown the protein composition and the relative genes for each of the PSII types separated (**Figure 2** inset). $C_2S_2M_2$, C_2S_2, C_2, C are the PSII-LHCII megacomplexes, PSII-LHCII supercomplexes, PSII dimers, and PSII monomers, respectively. *Unweighted spectrum count/MASS (kDa) = i; + = $0.25 \leq i \leq 0.50$; ++ = $0.50 \leq i \leq 0.75$; +++ = $i > 0.75$ (for details see Materials and Methods).*

a specific complex primarily identified by its apparent size on BN-PAGE (**Figure 2** inset). The identity of each complex was finally confirmed by MS analysis (**Table 3**, Supplementary Table S2). Furthermore, the C_2S_2 and C_2 complexes were also analyzed by TEM (**Figure 3**), demonstrating the sizes reported for C_2 and C_2S_2 (Aro et al., 2005; Nield and Barber, 2006), the latter most easily discernible from the side views (circled in **Figure 3A**). Unfortunately, $C_2S_2M_2$ complexes could not be analyzed by TEM, because they were obtained in low amounts and eluted at the beginning of the C_2S_2 peak in SEC.

The Isolated PSII Types Evolve Oxygen at Very Different Rates

The different PSII complexes obtained by SEC were tested for their oxygen evolving capacity. The pool of $C_2S_2M_2$ was excluded from this analysis since, as mentioned above, it was not perfectly separated from the C_2S_2. Each one of the others pools, represented by a dominant PSII type, were found to be fully functional and characterized by oxygen evolution rates comparable with other reports in literature (Kihara et al., 2014; **Table 2**). As oxygen evolution per Chl is reported, and because C_2S_2 binds about three times more Chls per reaction center than cores (around 210/reaction center as opposed to around 70; Ferreira et al., 2004; Liu et al., 2004; Loll et al., 2005; Barros et al., 2009), the C_2S_2 complexes are characterized by the highest activity.

Mass Spectrometry Analysis on the Four PSII Types Showed Significant Differences in Subunit Composition

The BN-PAGE bands obtained from the SEC fractions (**Figure 2** inset) were excised and directly analyzed by MS to characterize subunit content (**Table 3**, Supplementary Table S2). The main core subunits D1, D2, CP43, CP47, PsbE, PsbH, PsbL, PsbO, and PsbR were found in this analysis. The subunit PsbL, an important dimerization factor (Suorsa et al., 2004), was absent from the C and $C_2S_2M_2$ types. PsbR was identified only in the C_2 and C_2S_2 types, while PsbF was not found. The partial or complete absence of these three small peptides is most likely due to their small masses, which limit their successful MS identification (Shi et al., 2012). The PsbO subunit was found in both isoforms. The C_2 and C forms appeared to contain lower amounts of this protein, even though they showed a fair water splitting capacity (**Table 2**). PsbP and PsbQ, the other main water splitting subunits, were not identified in the analyzed complexes. Specific subunits were characteristic for a given oligomeric state. As expected, it was found that the subunits CP24, CP26, and CP29 were present just in traces or absent in the C and C_2, while they were found to be characteristic for C_2S_2 and $C_2S_2M_2$. Also the proteins Lhcb1, Lhcb2 and Lhcb3 were found associated with the C_2S_2 and $C_2S_2M_2$ complexes. In particular Lhcb3 was found to be present only in traces and Lhcb1 was present in four different isoforms. One of these, CB25, was specific for the C_2S_2 form.

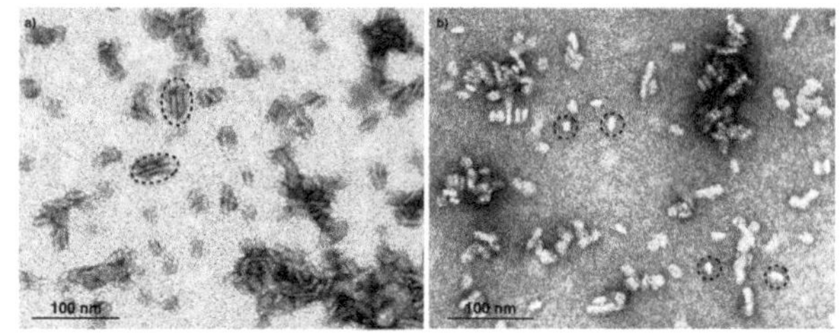

FIGURE 3 | Electron micrographs of negatively stained C_2S_2 complexes (A) and C_2 (B) separated by size exclusion chromatography. The suitable fractions in terms of purity were chosen on the basis of their band pattern on BN-PAGE. Dotted ellipses **(A)** and dotted circles **(B)** surround specific examples of C_2S_2 complex side views and C_2 particles, respectively.

DISCUSSION

The Oxygen Evolving Properties Reflect the PSII Cycle

The distribution of PSII in thylakoid membranes reflects the PSII cycle (Tomizioli et al., 2014). Grana cores sustain water splitting at the highest rates. Lamellae and grana margins are mainly required to recycle and replace PSII, which at these stages is still exposed to light, but requires protection. In agreement with this general picture, PSII isolated from SL samples is inactive, monomeric and retains PsbS. The presence of PsbS and low PSII activity in SL samples fit with the low activity of lamellar PSII reported earlier and with the role of PsbS in non-photochemical quenching (Allen, 2002; Munekage et al., 2002; Gerotto et al., 2015). PsbS may keep PSII in an inactive and protected state for assembly or repair. PSII from the SG samples is a mix of four different types that are very active in oxygen evolution, also in agreement with the current model for the PSII cycle.

PsbP and PsbQ

All active PSII complexes from grana lacked the PsbP and PsbQ subunits, which were found earlier in the SL fractions (Haniewicz et al., 2013). As PsbP and PsbQ are thought to enhance the PSII activity and the Oxygen Evolving Complex (OEC) stability (Bricker and Frankel, 2011), this finding is surprising, but it is not unprecedented. PsbP and PsbQ were also absent after a very mild membrane solubilization in a study on D1 subunit processing (Figure 5B from Che et al., 2013). However, the two subunits were present in several earlier PSII preparations, at least judging from SDS-PAGE gel migration patterns (Boekema et al., 2000; Nield et al., 2000; Caffarri et al., 2009). Mummadisetti et al. (2014) also detected PsbP and PsbQ by Western blotting, but, unfortunately, they did not determine the oligomeric PSII state by BN-PAGE and/or SEC. Boekema et al. (1998) detected PsbP in PSII-LHCII supercomplexes, but other studies did not corroborate this finding (Caffarri et al., 2009; Barera et al., 2012). Thus, PsbP and PsbQ may only be present in some assembly stages of PSII complexes or, more likely, in the very active forms of PSII they may be loosely bound due to their high sensitivity to damage.

Lhcb3 and CP24

Low levels of Lhcb3 and CP24 were detected for both $C_2S_2M_2$ and C_2S_2 complexes. The latter has not been reported to contain these two subunits. However, special hemi-types, intermediate between the C_2S_2 and the $C_2S_2M_2$ types, which retain both extra subunits, have been described and could fit well with the co-presence in our C_2S_2 samples. In Caffarri et al. (2009) is reported a so-called C_2SM complex containing only one S trimer and one M trimer that are bridged through CP29, CP24, and CP26 to the dimeric core. This complex has almost the same mass (1040 kDa) as the C_2S_2 complex (1100 kDa), and therefore co-migrates in

TABLE 4 | Purification yields for the isolated PSII types.

	Thylakoid membranes	Purification				
		Mild solubilization (SL)	Harsh solubilization (SG)			
		C-PsbS	C	C_2	C_2S_2	$C_2S_2M_2$
Mass (mg)	34.97 ± 3.8	0.95 ± 0.11	0.19 ± 0.01	9.55 ± 0.81	1.91 ± 0.18	Not determined
Yield (%)	100	2.73 ± 0.31	0.55 ± 0.04	27.30 ± 2.33	5.45 ± 0.51	Not determined

*$C_2S_2M_2$ yield was not determined because of their low amounts and their C_2S_2 impurities. *Values are expressed in mg or % of chlorophylls and represent means ± standard deviations of 10 independent purifications (see details in Materials and Methods).*

BN-PAGE (and also co-sediments in a sucrose gradient; Caffarri et al., 2009). Another hemi-type, called $C_2S_2^+$, is reported to have similar characteristics (Pietrzykowska et al., 2014). It is likely that these incomplete forms are assembly intermediates on the way to $C_2S_2M_2$, or result from its disintegration, either in membranes or during solubilization or purification.

Isolation of Supercomplexes by Coupling Differential Solubilization and Chromatography

The organization of the thylakoids and in particular the distribution of pigment-protein complexes within are very well characterized (Danielsson et al., 2006). This and other works all demonstrate substantial heterogeneity of the PSII species in the thylakoid membranes (Boekema et al., 1995; Dekker and Boekema, 2005; Danielsson et al., 2006; Takahashi et al., 2009; Watanabe et al., 2009; Pagliano et al., 2011). Here, we present a differential solubilization of the thylakoid membranes, which leads to the isolation of samples with different oligomeric patterns of PSII and other thylakoid complexes (**Figure 1A**). Conceptually, our method to access to the grana cores fraction, in which is found highly active PSII, is similar to the method described by Berthold et al. (1981), the so-called BBY preparation, which also removes lamellae and peripheral granal regions. In contrast to the BBY method, the new method has the advantage to avoid the breakdown of granal PSII core complexes (C and RC-CP47 are found at most in small quantities).

CONCLUSION

It has to be remarked that the mild nature of the solubilization protocol permits a full separation between PSII monomers and dimers, and makes it possible to prepare C_2S_2 in larger quantities without density gradients on sucrose or Percoll (Ghanotakis et al., 1987; Caffarri et al., 2004), (**Table 4**).

AUTHOR CONTRIBUTIONS

DP conceived the study, participated in its design and coordination, carried out the membranes preparation, participated in the biochemical studies, and drafted the manuscript. PH participated in the design of the study, participated the biochemical studies, and participated in the membranes preparation. D. Floris carried out the membranes preparation, participated in the biochemical studies. D. Farci carried out the membranes preparation, participated in the biochemical studies. JK carried out the MS analysis. ML participated in the preparation of membranes. CB contributed to the electron microscopy analysis and helped to draft the manuscript. MB participated in the design of the study and helped to draft the manuscript.

ACKNOWLEDGMENTS

This work was carried out with the support of the program Homing Plus (Foundation for Polish Science) grant No: 2012-6/10 co-financed by the European Union under the European Regional Development Funds; the program PRELUDIUM (National Science Centre) grant number DEC-2012/05/05/N/NZ1/01922; the Marie Curie program "European Reintegration Grant" (PERG05-GA-2009-247789); the program "FSE SARDEGNA 2007-2013, Legge Regionale 7 agosto 2007, n. 7, Promozione della ricerca scientifica e dell'innovazione tecnologica in Sardegna".

REFERENCES

Allen, J. (2002). Photosynthesis of ATP—electrons, proton pumps, rotors, and poise. *Cell* 110, 273–276.

Aro, E. M., Suorsa, M., Rokka, A., Allahverdiyeva, Y., Paakkarinen, V., Saleem, A., et al. (2005). Dynamics of photosystem II: a proteomic approach to thylakoid protein complexes. *J. Exp. Bot.* 56, 347–356. doi: 10.1093/jxb/eri041

Barera, S., Pagliano, C., Pape, T., Saracco, G., and Barber, J. (2012). Characterization of PSII-LHCII supercomplexes isolated from pea thylakoid membrane by one-step treatment with α- and β-dodecyl-D-maltoside. *Philos. Trans. R. Soc. Lond. B Biol. Sci.* 367, 3389–3399. doi: 10.1098/rstb.2012.0056

Barros, T., Royant, A., Standfuss, J., Dreuw, A., and Kühlbrandt, W. (2009). Crystal structure of plant light-harvesting complex shows the active, energy-transmitting state. *EMBO J.* 28, 298–306. doi: 10.1038/emboj.2008.276

Berthold, A. D., Babcock, G. T., and Yocum, C. (1981). A highly resolved, oxygen evolving photosystem II preparation from spinach thylakoid membranes. *FEBS Lett.* 134, 231–234. doi: 10.1016/0014-5793(81)80608-4

Boekema, E. J., Hankamer, B., Bald, D., Kruip, J., Nield, J., Boonstra, A. F., et al. (1995). Supramolecular structure of the photosystem II complex from green plants and cyanobacteria. *Proc. Natl. Acad. Sci. U.S.A.* 92, 175–179. doi: 10.1073/pnas.92.1.175

Boekema, E. J., Nield, J., Hankamer, B., and Barber, J. (1998). Localization of the 23-kDa subunit of the oxygen-evolving complex of photosystem II by electron microscopy. *Eur. J. Biochem.* 252, 268–276. doi: 10.1046/j.1432-1327.1998.2520268.x

Boekema, E. J., van Breemen, J. F. L., van Roon, H., and Dekker, J. P. (2000). Conformational changes in photosystem II supercomplexes upon removal of extrinsic subunits. *Biochemistry* 39, 12907–12915. doi: 10.1021/bi0009183

Boekema, E. J., Van Roon, H., Van Breemen, J. F., and Dekker, J. P. (1999). Supramolecular organization of photosystem II and its light-harvesting antenna in partially solubilized photosystem II membranes. *Eur. J. Biochem.* 266, 444–452. doi: 10.1046/j.1432-1327.1999.00876.x

Bricker, T. M., and Frankel, L. K. (2011). Auxiliary functions of the PsbO, PsbP and PsbQ proteins of higher plant Photosystem II: a critical analysis. *J. Photochem. Photobiol. B Biol.* 104, 165–178. doi: 10.1016/j.jphotobiol.2011.01.025

Caffarri, S., Croce, R., Cattivelli, L., and Bassi, R. (2004). A look within LHCII: differential analysis of the Lhcb1-3 complexes building the major trimeric antenna complex of higher-plant photosynthesis. *Biochemistry* 43, 9467–9476. doi: 10.1021/bi036265i

Caffarri, S., Kouril, R., Kereïche, S., Boekema, E. J., and Croce, R. (2009). Functional architecture of higher plant photosystem II supercomplexes. *EMBO J.* 28, 3052–3063. doi: 10.1038/emboj.2009.232

Cardona, T., Sedoud, A., Cox, N., and Rutherford, A. W. (2012). Charge separation in photosystem II: a comparative and evolutionary overview. *Biochim. Biophys. Acta* 1817, 26–43. doi: 10.1016/j.bbabio.2011.07.012

Che, Y., Fu, A., Hou, X., McDonald, K., Buchanan, B. B., Huang, W., et al. (2013). C-terminal processing of reaction center protein D1 is essential for the function and assembly of photosystem II in *Arabidopsis*. *Proc. Natl. Acad. Sci. U.S.A.* 110, 16247–16252. doi: 10.1073/pnas.1313894110

Danielsson, R., Suorsa, M., Paakkarinen, V., Albertsson, P., Styring, S., Aro, E. M., et al. (2006). Dimeric and monomeric organization of photosystem II. *J. Biol. Chem.* 281, 14241–14249. doi: 10.1074/jbc.M600634200

de Bianchi, S., Dall'Osto, L., Tognon, G., Morosinotto, T., and Bassi, R. (2008). Minor antenna proteins CP24 and CP26 affect the interactions between photosystem II subunits and the electron transport rate in grana membranes of *Arabidopsis*. *Plant Cell* 20, 1012–1028. doi: 10.1105/tpc.107.055749

Dekker, J. P., and Boekema, E. J. (2005). Supramolecular organization of thylakoid membrane proteins in green plants. *Biochim. Biophys. Acta* 1706, 12–39. doi: 10.1016/j.bbabio.2004.09.009

Edelman, M., and Mattoo, A. K. (2008). D1-protein dynamic in photosystem II: the lingering enigma. *Photosynth. Res.* 98, 609–620. doi: 10.1007/s11120-008-9342-x

Eshaghi, S., Andersson, B., and Barber, J. (1999). Isolation of a highly active PSII-LHCII supercomplex from thylakoid membranes by a direct method. *FEBS Lett.* 446, 23–26. doi: 10.1016/S0014-5793(99)00149-0

Farci, D., Bowler, M. W., Kirkpatrick, J., McSweeney, S., Tramontano, E., and Piano, D. (2014). New features of the cell wall of the radio-resistant bacterium *Deinococcus radiodurans*. *Biochim. Biophys. Acta* 1838, 1978–1984. doi: 10.1016/j.bbamem.2014.02.014

Ferreira, K. N., Inverson, T. M., Maghlaoui, K., Barber, J., and Iwata, S. (2004). Architecture of the photosynthetic oxygen-evolving center. *Science* 303, 1831–1838. doi: 10.1126/science.1093087

Fey, H., Piano, D., Horn, R., Fischer, D., Schröder, W. P., Bock, R., et al. (2008). Isolation of highly active photosystem II core complexes with a His-tagged Cyt b559 subunit from transplastomic tobacco plants. *Biochim. Biophys. Acta* 1777, 1501–1509. doi: 10.1016/j.bbabio.2008.09.012

Gerotto, C., Franchin, C., Arrigoni, G., and Morosinotto, T. (2015). In vivo identification of Photosystem II Light Harvesting Complexes interacting with Photosystem II Subunit S. *Plant Physiol.* 168, 1747–1761. doi: 10.1104/pp.15.00361

Ghanotakis, D. F., Demetriou, D. M., and Yocum, C. F. (1987). Isolation and characterization of an oxygen-evolving photosystem II reaction center core preparation and a 28 kDa Chl a binding protein. *Biochim. Biophys. Acta* 891, 15–21. doi: 10.1016/0005-2728(87)90078-8

Haniewicz, P., De Sanctis, D., Büchel, C., Schröder, W. P., Loi, M. C., Kieselbach, T., et al. (2013). Isolation of monomeric photosystem II that retains the subunit PsbS. *Photosynth. Res.* 118, 199–207. doi: 10.1007/s11120-013-9914-2

Kihara, S., Hartzler, D. A., and Savikhin, S. (2014). Oxygen concentration inside a functioning photosynthetic cell. *Biophys. J.* 106, 1882–1889. doi: 10.1016/j.bpj.2014.03.031

Liu, Z., Yan, H., Wang, K., Kuang, T., Zhang, J., Gui, L., et al. (2004). Crystal structure of spinach major light-harvesting complex at 2.72 Å resolution. *Nature* 428, 287–292. doi: 10.1038/nature02373

Loll, B., Kern, J., Saenger, W., Zouni, A., and Biesiadka, J. (2005). Towards complete cofactor arrangement in the 3.0 Å resolution structure of photosystem II. *Nature* 438, 1040–1044. doi: 10.1038/nature04224

Mummadisetti, M. P., Frankela, L. K., Bellamyb, H. D., Sallansc, L., Goettertb, J. S., Brylinskia, M., et al. (2014). Use of protein cross-linking and radiolytic footprinting to elucidate PsbP and PsbQ interactions within higher plant Photosystem II. *Proc. Natl. Acad. Sci. U.S.A.* 111, 16178–16183. doi: 10.1073/pnas.1415165111

Munekage, Y., Hojo, M., Meurer, J., Endo, T., Tasaka, M., and Shikanai, T. (2002). PGR5 is involved in cyclic electron flow around photosystem I and II essential for photoprotection in *Arabidopsis*. *Cell* 110, 361–371. doi: 10.1016/S0092-8674(02)00867-X

Nield, J., and Barber, J. (2006). Refinement of the structural model for the Photosystem II supercomplex of higher plants. *Biochim. Biophys. Acta* 1757, 353–361. doi: 10.1016/j.bbabio.2006.03.019

Nield, J., Orlova, E. V., Morris, E. P., Gowen, B., van Heel, M., and Barber, J. (2000). 3D map of the plant photosystem II supercomplex obtained by cryoelectron microscopy and single particle analysis. *Nat. Struct. Biol.* 7, 44–47. doi: 10.1038/71242

Nixon, P. J., Michoux, F., Yu, J., Boehm, M., and Komenda, J. (2010). Recent advances in understanding the assembly and repair of photosystem II. *Ann. Bot.* 106, 1–16. doi: 10.1093/aob/mcq059

Pagliano, C., Chimirri, F., Saracco, G., Marsano, F., and Barber, J. (2011). Onestep isolation and biochemical characterization of highly active plant PSII monomeric core. *Photosynth. Res.* 108, 33–46. doi: 10.1007/s11120-011-9650-4

Piano, D., El Alaoui, S., Korza, H. J., Filipek, R., Sabala, I., Haniewicz, P., et al. (2010). Crystallization of the photosystem II core complex and its chlorophyll binding subunit CP43 from transplastomic plants of *Nicotiana tabacum*. *Photosynth. Res.* 106, 221–226. doi: 10.1007/s11120-010-9597-x

Pietrzykowska, M., Suorsa, M., Semchonok, D. A., Tikkanen, M., Boekema, E. J., Aro, E. M., et al. (2014). The light-harvesting chlorophyll a/b binding proteins Lhcb1 and Lhcb2 play complementary roles during state transitions in *Arabidopsis*. *Plant Cell* 26, 3646–3660. doi: 10.1105/tpc.114.127373

Pokorska, B., Zienkiewicz, M., Powikrowska, M., Drozak, A., and Romanowska, E. (2009). Differential turnover of the photosystem II reaction centre D1 protein in mesophyll and bundle sheath chloroplast of maize. *Biochim. Biophys. Acta* 1787, 1161–1169. doi: 10.1016/j.bbabio.2009.05.002

Porra, R. J., Thompson, W. A., and Kriedmann, P. E. (1989). Determination of accurate extinction coefficients and simultaneous equations for assaying chlorophylls a and b with four different solvents: verifications of the concentration of chlorophyll standards by atomic absorption spectroscopy. *Biochim. Biophys. Acta* 975, 384–394. doi: 10.1016/S0005-2728(89)80347-0

Puthiyaveetil, S., Tsabari, O., Lowry, T., Lenhert, S., Lewis, R. R., Reich, Z., et al. (2014). Compartmentalization of the protein repair machinery in photosynthetic membranes. *Proc. Natl. Acad. Sci. U.S.A.* 111, 15839–15844. doi: 10.1073/pnas.1413739111

Schägger, H., and von Jagow, G. (1991). Blue native electrophoresis for isolation of membrane protein complexes in enzymatically active form. *Anal. Biochem.* 199, 223–231. doi: 10.1016/0003-2697(91)90094-A

Shi, L. X., Hall, M., Funk, C., and Schröder, W. P. (2012). Photosystem II, a growing complex: updates on newly discovered components and low molecular mass proteins. *Biochim. Biophys. Acta* 1817, 13–25. doi: 10.1016/j.bbabio.2011.08.008

Suorsa, M., Regel, R. E., Paakkarinen, V., Battchikova, N., Herrmann, R. G., and Aro, E. M. (2004). Protein assembly of photosystem II and accumulation of subcomplexes in the absence of low molecular mass subunits PsbL and PsbJ. *Eur. J. Biochem.* 271, 96–107. doi: 10.1046/j.1432-1033.2003.03906.x

Takahashi, T., Inoue-Kashino, N., Ozawa, S., Takahashi, Y., Kashino, Y., and Satoh, K. (2009). Photosystem II complex in vivo is a monomer. *J. Biol. Chem.* 284, 15598–15606. doi: 10.1074/jbc.M109.000372

Tomizioli, M., Lazar, C., Brugière, S., Burger, T., Salvi, D., Gatto, L., et al. (2014). Deciphering thylakoid sub-compartments using a mass spectrometry-based approach. *Mol. Cell. Proteomics* 13, 2147–2167. doi: 10.1074/mcp.M114.040923

Watanabe, M., Iwai, M., Narikawa, R., and Ikeuchi, M. (2009). Is the photosystem II complex a monomer or a dimer? *Plant Cell Physiol.* 50, 1674–1680. doi: 10.1093/pcp/pcp112

Conflict of Interest Statement: The authors declare that the research was conducted in the absence of any commercial or financial relationships that could be construed as a potential conflict of interest.

10

The Receptor for Activated C Kinase in Plant Signaling: Tale of a Promiscuous Little Molecule

Tania Islas-Flores[1], *Ahasanur Rahman*[2], *Hemayet Ullah*[2] *and Marco A. Villanueva*[1*]

[1] *Unidad Académica de Sistemas Arrecifales, Instituto de Ciencias del Mar y Limnología, Universidad Nacional Autónoma de México, Puerto Morelos, México,* [2] *Department of Biology, Howard University, Washington, DC, USA*

Two decades after the first report of the plant homolog of the Receptor for Activated C Kinase 1 (RACK1) in cultured tobacco BY2 cells, a significant advancement has been made in the elucidation of its cellular and molecular role. The protein is now implicated in many biological functions including protein translation, multiple hormonal responses, developmental processes, pathogen infection resistance, environmental stress responses, and miRNA production. Such multiple functional roles are consistent with the scaffolding nature of the plant RACK1 protein. A significant advance was achieved when the β-propeller structure of the *Arabidopsis* RACK1A isoform was elucidated, thus revealing that its conserved seven WD repeats also assembled into this typical topology. From its crystal structure, it became apparent that it shares the structural platform for the interaction with ligands identified in other systems such as mammals. Although RACK1 proteins maintain conserved Protein Kinase C binding sites, the lack of a *bona fide* PKC adds complexity and enigma to the nature of the ligand partners with which RACK1 interacts in plants. Nevertheless, ligands recently identified using the split-ubiquitin based and conventional yeast two-hybrid assays, have revealed that plant RACK1 is involved in several processes that include defense response, drought and salt stress, ribosomal function, cell wall biogenesis, and photosynthesis. The information acquired indicates that, in spite of the high degree of conservation of its structure, the functions of the plant RACK1 homolog appear to be distinct and diverse from those in yeast, mammals, insects, etc. In this review, we take a critical look at the novel information regarding the many functions in which plant RACK1 has been reported to participate, with a special emphasis on the information on its currently identified and missing ligand partners.

Edited by:
Christopher J. Staiger, Purdue University, USA

Reviewed by:
Viktor Zarsky, Charles University, Czech Republic
Rui Malhó, Universidade de Lisboa, Portugal

***Correspondence:**
Marco A. Villanueva
marco@cmarl.unam.mx

Keywords: hormonal regulation, RACK1, signal transduction, WD-repeat protein, WD40 domain

INTRODUCTION

Receptors for activated C kinase (RACKs) were initially described as ~33 kDa proteins from particulate fractions of rat heart, which fulfilled the criteria of intracellular receptors for the activated enzyme (Mochly-Rosen et al., 1991a,b). The first cloned sequence encoding one of such receptors was obtained from a rat brain cDNA expression library and termed RACK1 due to its ability to bind activated protein kinase C (PKC) (Ron et al., 1994). RACK1 is now known to be a highly conserved intracellular adaptor protein belonging to the WD-repeat family only present in eukaryotic organisms. Its constitutive expression and ubiquitous presence in both unicellular and multicellular organisms highlights its functional importance. The protein sequence contains seven WD-40 domains which assemble into a typical seven-bladed β-propeller structure. Since this assembly provides an interactive platform for the binding of potential ligand partners in proximity, the notion that RACK1 only functions as a receptor for active PKC isoforms has changed to that of a versatile protein that also provides a scaffold for direct or indirect interaction with many different ligands. This ability has positioned RACK1 as a central hub for integration of multiple pathways that impinge key cellular functions (Reviewed in Adams et al., 2011).

Much of our current knowledge about RACK1 function has arisen from mammalian cell studies; however, new emerging models such as yeast, slime molds, and worms have also provided clues to the variety of biochemical pathways in which RACK1 participates. For example, the *Saccharomyces cerevisiae* homolog Ascl has been reported to regulate protein translation critical for cell wall integrity near the yeast budding sites (Melamed et al., 2010). DdRACK1 from *Dyctiostelium discoideum*, has been reported to interact with heterotrimeric G proteins and impact growth and developmental processes (Omosigho et al., 2014). Finally, the *Caenorhabditis elegans*, RACK1 homolog interacts with the actin-binding protein UNC-115/abLIM for axon pathfinding and lamellipodia and filopodia formation, while in an independent and novel signaling pathway, it participates in gonadal tip cell migration (Demarco and Lundquist, 2010).

Besides the many model organisms in which RACK1 has been studied, plant RACK1 homologs have also been described. The first report was that of the auxin-regulated RACK1 homolog arcA (*ArcA* for auxin-regulated gene from cultured cells) from tobacco BY2 cells (Ishida et al., 1993). Since then, plant RACK1 homologs were reported from *Oryza sativa* (Iwasaki et al., 1995; Nakashima et al., 2008), *Brassica napus* (Kwak et al., 1997), *Medicago sativa* (McKhann et al., 1997), *Solanum lycopersicum* (Kiyosue and Ryan, 1999), *Glycine max* (Nielsen et al., 2001); *Arabidopsis thaliana* (van Nocker and Ludwig, 2003), *Phaseolus vulgaris* (Islas-Flores et al., 2009), and *Zea mays* (Wang et al., 2014), among others. In animal cells, a single copy of *RACK1* is encoded; however, this number is variable in plant genomes. For example, the genome of *A. thaliana* encodes three *RACK1* genes termed *AtRACK1A* (At1g18080), *AtRACK1B* (At1g48630), and *AtRACK1C* (At1g18130) (Chen et al., 2006), while two genes are present in *O. sativa* (Zhang et al., 2014). To date, the evidence acquired indicates that plant RACK1 does not have a unique function but rather, it is emerging as a multi-functional protein playing a central role in critical biological responses (**Figure 1**). Through cryo-electron microscopy (Cryo-EM) studies using ribosomes from the fungi *Thermomyces lanuginosus* and *Saccharomyces cereviseae*, it was revealed that RACK1 is a component of the ribosomal 40S subunit (Sengupta et al., 2004) that anchors PKC, a family of serine/threonine kinases, to phosphorylate the eukaryotic Initiation Factor6 (eIF6). In addition to binding with PKC, RACK1 plays crucial role in protein translation, tissue development, circadian clock, neural responses and tumorigenesis in mammals (McCahill et al., 2002; Robles et al., 2010). In plants, RACK1 regulates various signaling pathways ranging from developmental processes such as seed germination, flowering and leaf production, to immune and stress responses against pathogen and environmental stimuli (Chen et al., 2006; Ullah et al., 2006). Intriguing however, despite the recent findings of new plant RACK1 ligands, has been the search for signal-transduction and/or biochemical pathways in which plant RACK1 is involved. In this review, we cover and critically discuss past and present findings regarding the various signaling and biochemical pathways in which plant RACK1 has been reported to participate. In addition, we take a critical look at the various binding partners with which it has been reported to interact, and discuss the present trends and directions of the field.

WD-REPEAT PROTEIN RACK1: A SMALL INBUILT β-PROPELLER MOLECULE WITH LARGE RESPONSIBILITIES

WD-40 domains assemble into β sheets to form blade-like structures, and proteins containing seven WD-40 repeats such as the β subunit of the heterotrimeric G protein can assemble into a seven-bladed β-propeller-like structure (Reviewed in McCahill et al., 2002). The β-propeller is a molecular platform that facilitates interaction with several ligands from distinct signaling pathways at the same time. RACK1 is a 36 kDa cytosolic protein with high sequence identity with the β subunit of heterotrimeric G proteins, and highly conserved in eukaryotes where it shares 43–76% of sequence identity (Ullah et al., 2008). Consequently, the WD-40 repeats are highly conserved in all RACK1 homologs characterized so far, indicating that all of them share the same structure. This assumption was supported by the first crystallographic resolution of the AtRACK1A isoform structure at 2.4 Å (Ullah et al., 2008). Subsequent determinations of the crystallographic structure for RACK1 from different organisms, such as, *S. cerevisiae* (Yatime et al., 2011), *Tetrahymena thermophila* (Rabl et al., 2011), and *Homo sapiens* (Ruiz Carrillo et al., 2012) at 2.1, 2.4, 3.9, and 2.45 Å resolution, also confirmed the β-propeller structure platform. Thus, the available crystal structures have shown that its seven-bladed structural dynamics presents the features to scaffold a large number of proteins to facilitate numerous cellular signaling pathways. As of now, more than 130 different interacting partners of plant RACK1 homologs have been reported (**Table 1** and **Figure 1**), suggesting its involvement in various physiological functions in this kingdom (Ullah et al., 2006; Kundu et al., 2013).

HORMONAL SIGNALING: SPECIES- AND TISSUE-SPECIFIC RESPONSES

Since the discovery of *arcA*, the first plant RACK1 homolog sequence from a subtraction library of BY2 cultured cells subjected to auxin treatment (Ishida et al., 1993), it was reported that its expression was regulated (induced) by the hormone. Interestingly, this expression was exclusively stimulated by auxins but not by cytokinins, abscisic acid (ABA), ethylene or heat

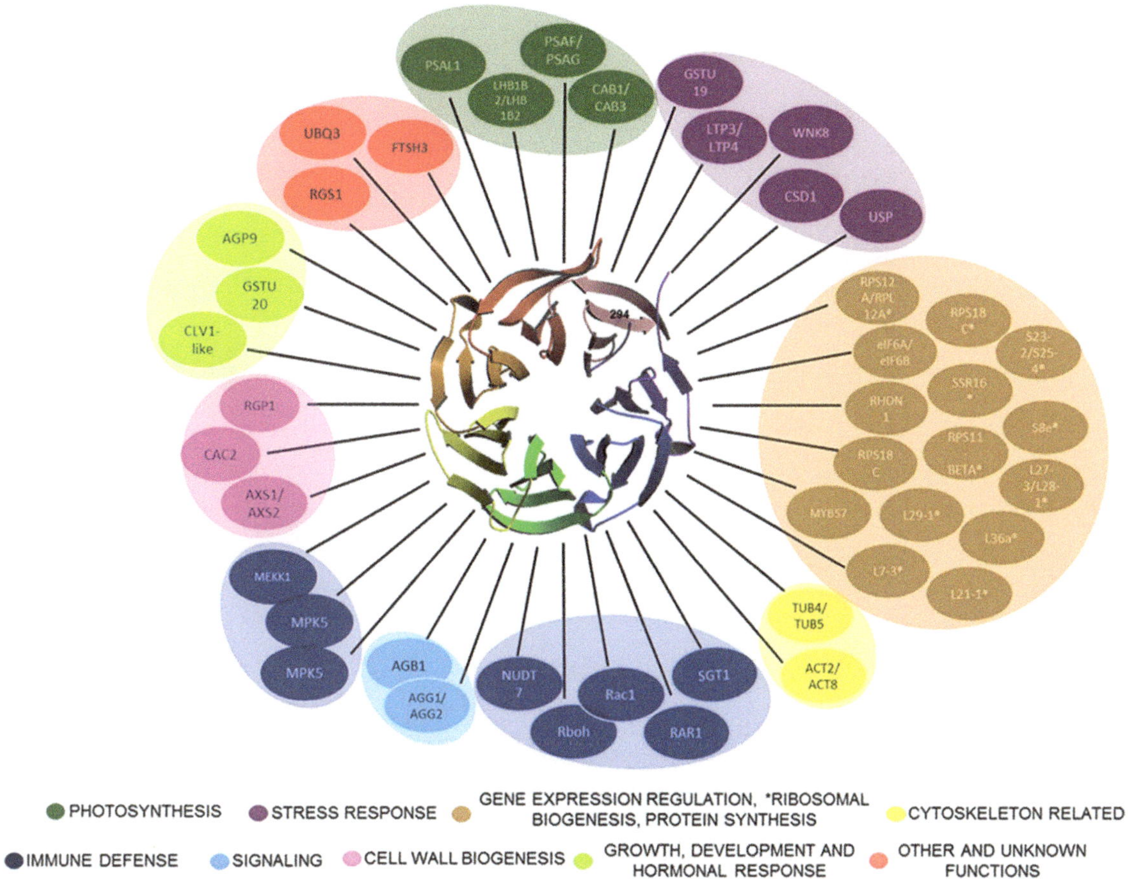

FIGURE 1 | Plant RACK1 is related with several cellular processes through its interacting ligands. Only some of the proteins demonstrated to interact directly with RACK1 are outlined. The β-propeller structure at the center corresponds to the top view of AtRACK1 (Ullah et al., 2008). The RACK1 interactors belong to different cellular processes ranging from Cell wall biogenesis, Photosynthesis, Cytoskeletal, Immune defense, Stress response, Signaling, Gene expression regulation, *Ribosomal biogenesis, Protein synthesis, Growth, Development and hormonal response, to Other and unknown.

(Ishida et al., 1996). It is interesting to note that RACK1 has been found in stoichiometric quantities on the small subunit of crystallized eukaryotic ribosomes (Gibson, 2012), and that it has also been implicated in protein translation and ribosome binding in plants (see below Chang et al., 2005; Ullah et al., 2008; Guo et al., 2011a,b). Thus, it is tempting to suggest that plant RACK1 may be able to regulate its own expression via hormonal regulation. After Ishida et al. (1993) reported auxin-regulated expression of *arcA*, several other reports implicated RACK1 in hormonal signaling pathways. For example, Nakashima et al. (2008) reported that auxin, ABA and methyl jasmonate, induced rice *RACK1A* expression. As in tobacco cultured cells, in rice cultured cells the auxin was also an inducer of Os*RACK1*; however, unlike the case of *ArcA*, ABA did stimulate *OsRACK1A* expression (Ishida et al., 1996; Nakashima et al., 2008). In a separate similar report on *Zea mays*, growing seedlings were subjected to ABA treatment while methyl jasmonate was sprayed to the expanding leaves of growing plants. In both cases, *ZmRACK1* expresssion was induced and only a slight difference between both hormonal treatments was observed (Wang et al., 2014). These results further suggested that RACK1 indeed participated in hormonal signaling, albeit, under different regulation from species to species.

Early reports included the implication of Msgbl, the RACK1 homolog from *M. sativa*, in hormone-mediated cell division since the transcript was preferentially located to dividing cells of nodule primordia and meristem (McKhann et al., 1997). However, contrary to *arcA*, which was exclusively induced by auxin (Ishida et al., 1996), Mgsbl was induced by cytokinin treatment of roots (McKhann et al., 1997). This work suggested that plant RACK1 was involved in several hormone-mediated pathways. Most recent work has been carried out on *A. thaliana RACK1* isogenes *AtRACK1A*, *B*, and *C*, although a functional redundancy derived from the high sequence conservation was expected and subsequently confirmed (Guo and Chen, 2008). The involvement of RACK1 in multiple plant hormonal pathways was confirmed by the study of hormonal treatments and developmental processes of loss-of-function *Arabidopsis rack1a* mutants (Chen et al., 2006; Fennell et al., 2012). The *rack1a* mutants displayed defects

TABLE 1 | Plant RACK1 ligand proteins.

	RACK1 isoform	Ligand ID	Name of the ligand	Process	Assay	Specie	References
CELL WALL BIOGENESIS							
1	AtRACK1C	AT1G08200	AXS2 (UDP-D-APIOSE/UDP-D-XYLOSE SYNTHASE 2); UDP-glucuronate decarboxylase	Cell wall organization	c	*A. thaliana*	Klopffleisch et al., 2011
2	AtRACK1A	AT1G20220	Alba DNA/RNA binding protein (NP 564108.1)	Cell wall related	e	*A. thaliana*	Kundu et al., 2013
3	AtRACK1A/ AtRACK1C	AT2G27860	AXS1 (UDP-D-apiose/UDP-D-xylose synthase 1); NAD or NADH Binding/UDP-glucuronate decarboxylase	Cell wall organization	c, g	*A. thaliana*	Klopffleisch et al., 2011
4	AtRACK1C	AT3G02230	RGP1, Reversibly glycosylated polypeptide 1; UDP-arabinose mutase 1	Cell wall biogenesis	c	*A. thaliana*	Klopffleisch et al., 2011
5	AtRACK1C	AT3G25150	NTF2, Nuclear transport factor 2 family; protein with RNA Binding (RRM-RBD-RNP motifs) domain	Cell wall organization	c	*A. thaliana*	Klopffleisch et al., 2011
6	AtRACK1C	AT5G14430	PMT9, S-adenosyl-L-methionine-dependent methyltransferases superfamily protein	Cell wall related	c	*A. thaliana*	Klopffleisch et al., 2011
7	AtRACK1A	AT5G35360	CAC2, Biotin carboxylase subunit	Cell wall organization/Fatty acid biosynthesis	e	*A. thaliana*	Kundu et al., 2013
PHOTOSYNTHESIS							
8	AtRACK1A	AT1G20340	Plastocyanin major isoform, DNA-damage resistance protein (DRT112)	Light response	e, f	*A. thaliana*	Kundu et al., 2013
9	AtRACK1A	AT1G29910	CAB3, Chlorophyll A/B binding protein 3	Photosynthesis	e	*A. thaliana*	Kundu et al., 2013
10	AtRACK1A	AT1G29930	CAB1, Chlorophyll A/B binding protein1	Photosynthesis	e, f	*A. thaliana*	Kundu et al., 2013
11	AtRACK1A	AT1G31330	PSAF, Photosystem I subunit F	Photosynthesis	e	*A. thaliana*	Kundu et al., 2013
12	AtRACK1A	AT1G55670	PSAG, Photosystem I subunit G	Electron transport in photosystem I	e	*A. thaliana*	Kundu et al., 2013
13	AtRACK1A	AT1G59840	CCB4, Cofactor assembly of complex C	Photosystem II assembly	e	*A. thaliana*	Kundu et al., 2013
14	AtRACK1A	AT1G60950	FED A, Ferredoxin-2	Light response	e	*A. thaliana*	Kundu et al., 2013
15	AtRACK1A	AT1G61520	LHCA3, Photosystem I Light Harvesting Complex Gene 3	Light harvesting	e, f	*A. thaliana*	Kundu et al., 2013
16	AtRACK1A	AT1G67090	RuBisCO1A, Rubisco small subunit 1A	Light response/Photosyntesis	e, f	*A. thaliana*	Kundu et al., 2013
17	AtRACK1A	AT2G34420	LHB1B2, Photosystem II light harvesting complex protein B1B2	Light harvesting	e	*A. thaliana*	Kundu et al., 2013
18	AtRACK1A	AT2G34430	LHB1B1, Light-harvesting chlorophyll protein complex II subunit B1	Light harvesting	e	*A. thaliana*	Kundu et al., 2013
19	AtRACK1A	AT2G43560	FKBP16-3, Peptidyl-prolyl cis-trans isomerase	Protein folding/Photosynthesis	e	*A. thaliana*	Kundu et al., 2013
20	AtRACK1A	AT3G56940	CRD1, Magnesium-protoporphyrin IX monomethyl ester [oxidative] cyclase	Chlorophyll biosynthesis	e	*A. thaliana*	Kundu et al., 2013
21	AtRACK1A	AT3G61470	LHCA2, Photosystem I Light Harvesting Complex Gene 2	Light harvesting	e, f	*A. thaliana*	Kundu et al., 2013
22	AtRACK1A	AT3G62410	CP12-2, Domain-containing protein 1	Light response	e	*A. thaliana*	Kundu et al., 2013
23	AtRACK1A	AT4G12800	PSAL, Photosystem I subunit L; Photosystem I reaction center subunit XI	Light reaction/Photosystem II assembly	e	*A. thaliana*	Kundu et al., 2013
24	AtRACK1A	AT5G38430	RuBisCO1B, Small chain 1B	Light response	e	*A. thaliana*	Kundu et al., 2013
25	AtRACK1A	AT5G64040	PSAN, Photosystem I reaction center subunit N	Electron transport in photosystem I	e	*A. thaliana*	Kundu et al., 2013

(Continued)

TABLE 1 | Continued

	RACK1 isoform	Ligand ID	Name of the ligand	Process	Assay	Specie	References
CYTOSKELETON RELATED							
26	AtRACK1A	AT1G20010	TUB5, Tubulin beta-5 chain	Cytoskeleton organization	e	*A. thaliana*	Kundu et al., 2013
27	AtRACK1A	AT1G49240	ACT8, Actin 8	Cytoskeleton organization	e	*A. thaliana*	Kundu et al., 2013
28	AtRACK1A	AT3G18780	ACT2, Actin 2	Cytoskeleton organization	e	*A. thaliana*	Kundu et al., 2013
29	AtRACK1A	AT4G20890	TUB9, Tubulin beta-9 chain	Cytoskeleton organization	e	*A. thaliana*	Kundu et al., 2013
30	AtRACK1A	AT5G23860	TUB8, Tubulin beta-8 chain	Cytoskeleton organization	e	*A. thaliana*	Kundu et al., 2013
31	AtRACK1A	AT5G44340	TUB4, Tubulin beta-4 chain	Cytoskeleton organization	e	*A. thaliana*	Kundu et al., 2013
IMMUNE DEFENSE							
32	OsRACK1A	AAF18438	SGT1, Suppressor of G2 allele of Skp1	Disease resistance	b, d, f	*O. sativa*	Nakashima et al., 2008
33	OsRACK1A/ OsRACK1B	AB029508	Rac1, Small GTP-binding protein 1	Disease resistance/cell death	a, b, d, f	*O. sativa*	Nakashima et al., 2008
34	OsRACK1A	AB029510	Rac3, Small GTP-binding protein 3; Rho family small GTPases	Disease resistance/cell death	f	*O. sativa*	Nakashima et al., 2008
35	OsRACK1A	AK058414	Rac7, Small GTP-binding protein 7; Rho family small GTPases	Disease resistance/cell death	f	*O. sativa*	Nakashima et al., 2008
36	OsRACK1A	AK067504	Rac5, Small GTP-binding protein 5; Rho family small GTPases	Disease resistance/cell death	f	*O. sativa*	Nakashima et al., 2008
37	OsRACK1A	AK100842	Rac6, Small GTP-binding protein 7; Rho family small GTPases	Disease resistance/cell death	f	*O. sativa*	Nakashima et al., 2008
38	OsRACK1A	AK111881	RAR1, Required for *Mla12* resistance; zinc-binding protein	Disease resistance	b, d, f	*O. sativa*	Nakashima et al., 2008
39	AtRACK1A	AT1G07790	H2B Histone	Defense response	e	*A. thaliana*	Kundu et al., 2013
40	AtRACK1A	AT2G16870	TIR-NBS-LRR class disease resistance protein	Defense response	e	*A. thaliana*	Kundu et al., 2013
41	AtRACK1A	AT2G43790	MPK6, MAP kinase 6	Defense response	b, g, h	*A. thaliana*	Cheng et al., 2015
42	AtRACK1A	AT3G01500	CA1, Carbonic anhydrase 1	Defense response	e, f	*A. thaliana*	Kundu et al., 2013
43	AtRACK1A	AT3G12780	PGK1, Phosphoglycerate kinase 1	Defense and stress response	e	*A. thaliana*	Kundu et al., 2013
44	AtRACK1A	AT3G21220	MKK5, Mitogen-activated protein kinase kinase 5	Defense response	b, g, h	*A. thaliana*	Cheng et al., 2015
45	AtRACK1A	AT3G45640	MPK3, Mitogen-activated protein kinase 3	Defense response	b, g, h	*A. thaliana*	Cheng et al., 2015
46	AtRACK1A	AT3G62030	ROC4, Peptidyl-prolyl cis-trans isomerase CYP20-3	Defense response	e, f	*A. thaliana*	Kundu et al., 2013
47	AtRACK1A	AT4G00870	Transcription factor bHLH14	Defense response	e	*A. thaliana*	Kundu et al., 2013
48	AtRACK1A/ AtRACK1b/ AtRACK1C	AT4G08500	MEKK1, Mitogen-activated protein kinase kinase kinase 1 in a inactive form (K361M)	Stress and defense response	b, g, h	*A. thaliana*	Cheng et al., 2015
49	AtRACK1A	AT4G12720	NUDT7, Nudix hydrolase 7	Defense and oxidative stress response	b, d, g	*A. thaliana*	Olejnik et al., 2011
50	AtRACK1A	AT5G04140	GLU1 Ferredoxin-dependent glutamate synthase 1 (Fd-GOGaT)	Defense and stress response	e	*A. thaliana*	Kundu et al., 2013
51	AtRACK1C	AT5G09650	PPa6, Soluble inorganic pyrophosphatase 1	Defense and stress response	c	*A. thaliana*	Klopffleisch et al., 2011
52	AtRACK1A	AT5G20630	GER3, Germin-like protein subfamily 3 member 3	Defense and stress response	e	*A. thaliana*	Kundu et al., 2013
53	AtRACK1A	AT5G24780	VSP1, Vegetative storage protein 1 acid phosphatase	Defense response	e	*A. thaliana*	Kundu et al., 2013
54	OsRACK1A	AY603975	Rboh, NADPH oxidase	Defense response	f	*O. sativa*	Nakashima et al., 2008

(Continued)

TABLE 1 | Continued

	RACK1 isoform	Ligand ID	Name of the ligand	Process	Assay	Specie	References
STRESS RESPONSE							
55	AtRACK1A	AT1G08830	CSD1, Superoxide dismutase [cu-Zn]	Response to oxidative stress	e, f	*A. thaliana*	Kundu et al., 2013
56	AtRACK1A	AT1G32640	MYC2 Transcription factor	Response to drought stress and ABA treatment	e	*A. thaliana*	Kundu et al., 2013
57	AtRACK1A	AT1G52420	UDP-Glycosyltransferase-like protein	Response to hypoxia	e	*A. thaliana*	Kundu et al., 2013
58	AtRACK1A/ AtRACK1C	AT1G78380	GSTU19, TAU 19 Glutathione S-transferase	Response to drought and oxidative stress	c, e	*A. thaliana*	Kundu et al., 2013; Klopffleisch et al., 2011
59	AtRACK1A/ AtRACK1C	AT2G01140	Fructose-bisphosphate aldolase 3	Response to oxidative and saline stress	c, e	*A. thaliana*	Kundu et al., 2013; Klopffleisch et al., 2011
60	AtRACK1A	AT3G52300	ATPQ, ATP synthase subunit D	Response to saline stress	e	*A. thaliana*	Kundu et al., 2013
61	AtRACK1A/ AtRACK1C	AT3G53990	USP, Universal stress protein family protein/Adenine nucleotide alpha hydrolases-like superfamily protein	Cold stress response	c, g	*A. thaliana*	Klopffleisch et al., 2011
62	AtRACK1A	AT5G20010	RAN-1 GTP-binding nuclear protein Ran-1; RAS-related nuclear protein-1	Response to saline stress/Protein import into the nucleus	e	*A. thaliana*	Kundu et al., 2013
63	AtRACK1C	AT5G20150	SPX1 domain-containing protein 1	Response to stress starvation	c, g	*A. thaliana*	Klopffleisch et al., 2011
64	AtRACK1A/ AtRACK1B/ AtRACK1C	AT5G41990	WNK8, WITH NO LYSINE (K) KINASE 8; Serine/threonine-protein kinase	Response to saline and osmotic stress/ Response to glucose	c, d, g	*A. thaliana*	Klopffleisch et al., 2011; Urano et al., 2015
65	AtRACK1A	AT5G59310	LTP4, Non-specific lipid-transfer protein 4	Response to drought stress	e	*A. thaliana*	Kundu et al., 2013
66	AtRACK1A	AT5G59320	LTP3, Non-specific lipid transfer protein 3	Response to drought stress	e	*A. thaliana*	Kundu et al., 2013
SIGNALING							
67	AtRACK1A	AT1G05000	Tyrosine phosphatase family; Atypical dual specificity phosphatase 1	Signaling	c, d	*A. thaliana*	Klopffleisch et al., 2011
68	AtRACK1B	AT3G18130	AtRACK1C, Receptor for Activated C Kinase 1C	Signaling	c, g	*A. thaliana*	Klopffleisch et al., 2011
69	AtRACK1A	AT3G22942	AGG2, G-protein gamma-subunit 2	Signaling	b, d, g	*A. thaliana*	Olejnik et al., 2011
70	AtRACK1A	AT3G63420	AGG1, G-protein gamma-subunit 1	Signaling	b, d, g	*A. thaliana*	Olejnik et al., 2011
71	AtRACK1A/ AtRACK1B/ AtRACK1C	AT4G34460	AGB1, G-protein beta-subunit 1	Signaling	b, g, h	*A. thaliana*	Cheng et al., 2015
72	AtRACK1A	AT4G34870	ROC5, Rotamase cyclophilin 5; Peptidyl-prolyl cis-trans isomerase CYP18-4	Signal transduction	e	*A. thaliana*	Kundu et al., 2013
GENE EXPRESSION REGULATION							
73	AtRACK1A	AT1G06190	RHON1, Rho termination factor, N-terminal domain; Ribonucleic acid-binding protein	Transcription termination	e	*A. thaliana*	Kundu et al., 2013
74	AtRACK1A	AT1G09250	bHLH149, Basic helix-loop-helix DNA-binding superfamily protein	Transcription	e	*A. thaliana*	Kundu et al., 2013
75	AtRACK1A	AT1G09590	L21-1, 60S Ribosomal protein	Translation	e	*A. thaliana*	Kundu et al., 2013
76	AtRACK1A	AT1G15930	RPS12A, 40S Ribosomal protein S12-1	Translation/Stress response	e	*A. thaliana*	Kundu et al., 2013
77	AtRACK1A	AT2G18090	PHD Finger, SWIB/MDM2 and GYF domain-containing protein	Transcription	e	*A. thaliana*	Kundu et al., 2013
78	AtRACK1A	AT2G19730	L28-1, 60S Ribosomal protein	Translation	e	*A. thaliana*	Kundu et al., 2013
79	AtRACK1A/ AtRACK1B/ AtRACK1C	AT2G27100	SE, Serrate RNA effector molecule; miRNA factor	Pri-miRNA processing	b, c, d, g	*A. thaliana*	Speth et al., 2013

(Continued)

TABLE 1 | Continued

	RACK1 isoform	Ligand ID	Name of the ligand	Process	Assay	Specie	References
80	AtRACK1A/ AtRACK1B/ AtRACK1C	AT2G39820	eIF6B, 60S Ribosomal protein	Translation/Ribosome biogenesis	d, g	*A. thaliana*	Guo et al., 2011b
81	AtRACK1A	AT2G44120	L7-3, 60S Ribosomal protein	Translation	e	*A. thaliana*	Kundu et al., 2013
82	AtRACK1A	AT3G01530	MYB57, Myb domain protein 57	Transcription	e	*A. thaliana*	Kundu et al., 2013
83	AtRACK1A	AT3G06700	L29-1, 60S Ribosomal protein	Translation/Ribosome biogenesis	e	*A. thaliana*	Kundu et al., 2013
84	AtRACK1A	AT3G23390	L36a, 60S Ribosomal protein	Translation	e	*A. thaliana*	Kundu et al., 2013
85	AtRACK1A	AT3G27830	RPL12-A, 50S Ribosomal protein L12-1	Translation	e	*A. thaliana*	Kundu et al., 2013
86	AtRACK1A/ AtRACK1B/ AtRACK1C	AT3G55620	eIF6A, 60S Ribosomal protein	Translation/Ribosome biogenesis	d, g	*A. thaliana*	Guo et al., 2011b
87	AtRACK1A	AT4G09800	RPS18C, 40S Ribosomal protein S18	Translation	e	*A. thaliana*	Kundu et al., 2013
88	AtRACK1C	AT4G13940	HOG1, Homology-dependent gene silencing 1; MEE58 adenosylhomocysteinase 1; SAHH1, EMB1395, SAH1	Post-transcriptional gene silencing	c	*A. thaliana*	Klopffleisch et al., 2011
89	AtRACK1A	AT4G15000	L27-3, 60S Ribosomal protein	Translation	e	*A. thaliana*	Kundu et al., 2013
90	AtRACK1A	AT4G21660	Splicing factor 3B subunit 4	mRNA Processing	e	*A. thaliana*	Kundu et al., 2013
91	AtRACK1A	AT4G34620	SSR16, Small subunit ribosomal protein 16	Translation/Ribosome biogenesis	e	*A. thaliana*	Kundu et al., 2013
92	AtRACK1A	AT4G39200	S25-4, 40S Ribosomal protein	Translation	e	*A. thaliana*	Kundu et al., 2013
93	AtRACK1A	AT5G02960	S23-2, 40S Ribosomal protein	Translation	e	*A. thaliana*	Kundu et al., 2013
94	AtRACK1A	AT5G06360	S8e, Ribosomal protein family protein	Translation/Ribosome biogenesis	e	*A. thaliana*	Kundu et al., 2013
95	AtRACK1A	AT5G23740	RPS11-BETA, 40S Ribosomal protein S11-3	Translation	e	*A. thaliana*	Kundu et al., 2013
96	AtRACK1B	AT5G43960	NTF2, Nuclear transport factor 2 family protein with RNA Binding (RRM-RBD-RNP motifs) domain	Nucleocytoplasmic transport	d	*A. thaliana*	Dreze et al., 2011
97	AtRACK1B	AT5G48650	NTF2, Nuclear transport factor 2 family protein with RNA Binding (RRM-RBD-RNP motifs) domain	Nucleocytoplasmic transport	d	*A. thaliana*	Dreze et al., 2011
98	AtRACK1C	AT5G51190	ERF105, Ethylene-responsive transcription factor	Transcription	c	*A. thaliana*	Klopffleisch et al., 2011
GROWTH, DEVELOPMENT, AND HORMONAL RESPONSE							
99	AtRACK1A	AT1G08590	CLV1-Like leucine rich repeat transmembrane receptor-like protein kinase	Vascular development	e	*A. thaliana*	Kundu et al., 2013
100	AtRACK1A	AT1G78370	GSTU20, TAU 20 Glutathione S-transferase	Regulation of growth	e	*A. thaliana*	Kundu et al., 2013
101	AtRACK1A	AT2G02850	ARPN Plantacyanin	Development	e, f	*A. thaliana*	Kundu et al., 2013
102	AtRACK1A	AT2G14890	AGP9, Arabinogalactan protein 9	Meristem growth	e	*A. thaliana*	Kundu et al., 2013
103	AtRACK1A	AT3G07900	O-fucosyltransferase-like protein	Root hair cell differentiation	e	*A. thaliana*	Kundu et al., 2013
104	AtRACK1C	AT3G20830	AGC, cAMP-dependent, cGMP-dependent and protein kinase C family protein	Response to brassinosteroid and auxin	c	*A. thaliana*	Klopffleisch et al., 2011
105	AtRACK1B/ AtRACK1C	AT4G35470	PIRL4, Plant intracellular ras group-related LRR 4	Gibberellin signaling	d	*A. thaliana*	Dreze et al., 2011
106	AtRACK1C	AT5G06110	DNAJ Domain; Myb-like DNA-binding domain	Cell division/Protein folding	c	*A. thaliana*	Klopffleisch et al., 2011
OTHER AND UNKNOWN FUNCTIONS							
107	AtRACK1A	AT1G18210	CML27, Putative calcium-binding protein	Unknown	e	*A. thaliana*	Kundu et al., 2013
108	AtRACK1C	AT1G22920	CSN5A, COP9 Signalosome complex subunit 5a, JAB1, AJH1	Photomorphogenesis	c	*A. thaliana*	Klopffleisch et al., 2011

(Continued)

TABLE 1 | Continued

	RACK1 isoform	Ligand ID	Name of the ligand	Process	Assay	Specie	References
109	AtRACK1A	AT1G23100	GroES-like protein	Protein folding	e	*A. thaliana*	Kundu et al., 2013
120	AtRACK1A	AT1G64230	UBC28, Ubiquitin-conjugating enzyme E2 28	Protein catabolic process	e	*A. thaliana*	Kundu et al., 2013
111	AtRACK1C	AT1G68410	PP2Cc, Protein phosphatase 2C 15 family protein	Unknown	c	*A. thaliana*	Klopffleisch et al., 2011
112	AtRACK1A	AT2G04900	Uncharacterized protein	Unknown	e	*A. thaliana*	Kundu et al., 2013
113	AtRACK1A	AT2G10940	Bifunctional inhibitor/lipid-transfer protein/seed storage 2S albumin superfamily protein	Lipid transport	e	*A. thaliana*	Kundu et al., 2013
114	AtRACK1A	AT2G18510	RRM1_SF3B4; RNA recognition motif 1 in splicing factor 3B subunit 4 (SF3B4)	Embryo dormancy	e	*A. thaliana*	Kundu et al., 2013
115	AtRACK1A	AT2G21045	Rhodanese-like domain-containing protein	Ion transport	e	*A. thaliana*	Kundu et al., 2013
116	AtRACK1C	AT2G22880	VQ Motif-containing protein	Response to UVB	c	*A. thaliana*	Klopffleisch et al., 2011
117	AtRACK1B/ AtRACK1C	AT2G29080	FTSH3 protease/AAA+ ATPase	Protein catabolic process	c	*A. thaliana*	Klopffleisch et al., 2011
118	AtRACK1A	AT2G30105	LRR-UBQ, Leucine-rich repeats and ubiquitin-like domain-containing protein	Unknown	e	*A. thaliana*	Kundu et al., 2013
119	AtRACK1C	AT2G44310	Calcium-binding EF hand containing protein	Unknown	c	*A. thaliana*	Klopffleisch et al., 2011
120	AtRACK1A	AT2G44500	O-fucosyltransferase family protein	Unknown	e	*A. thaliana*	Kundu et al., 2013
121	AtRACK1A	AT2G46000	Uncharacterized protein	Unknown	e	*A. thaliana*	Kundu et al., 2013
122	AtRACK1C	AT2G47090	Zinc ion binding/nucleic acid binding	Unknown	c	*A. thaliana*	Klopffleisch et al., 2011
123	AtRACK1A	AT2G47590	PHR2, Photolyase/blue-light receptor 2	DNA repair	e	*A. thaliana*	Kundu et al., 2013
124	AtRACK1A	AT3G07565	Uncharacterized protein	Unknown	e	*A. thaliana*	Kundu et al., 2013
125	AtRACK1A	AT3G08690	UBC11, Ubiquitin-conjugating enzyme E2 11	Protein catabolic process	e	*A. thaliana*	Kundu et al., 2013
126	AtRACK1A	AT3G13520	AGP12, Arabinogalactan protein 12	Modified amino acid biosynthesis	e	*A. thaliana*	Kundu et al., 2013
127	AtRACK1A/ AtRACK1B	AT3G26090	RGS1, Regulator of G-protein signaling 1	Sugar response	c, d, g	*A. thaliana*	Klopffleisch et al., 2011
128	AtRACK1C	AT3G56410	Uncharacterized protein	Unknown	c	*A. thaliana*	Klopffleisch et al., 2011
129	AtRACK1A	AT4G27960	UBC9, Ubiquitin conjugating enzyme	Protein catabolic process	e	*A. thaliana*	Kundu et al., 2013
130	AtRACK1A	AT4G28030	GCN5-Related N-acetyltransferase (GNAT) family protein	Metabolism	e	*A. thaliana*	Kundu et al., 2013
131	AtRACK1C	AT4G37540	LBD39, LOB domain-containing protein 39	Membrane fluidity	c	*A. thaliana*	Klopffleisch et al., 2011
132	AtRACK1B/ AtRACK1C	AT5G03240	UBQ3, Polyubiquitin 3	Protein catabolic process	a	*A. thaliana*	Kim et al., 2013
133	AtRACK1A	AT5G04750	Putative F1F0-ATPase inhibitor protein	Unknown	e	*A. thaliana*	Kundu et al., 2013
134	AtRACK1A	AT5G11500	Uncharacterized protein (DUF814)	Unknown	e	*A. thaliana*	Kundu et al., 2013
135	AtRACK1A	AT5G48180	NSP5, Nitrile specifier protein 5	Catabolic processes	e	*A. thaliana*	Kundu et al., 2013
136	AtRACK1A	AT5G52430	Hydroxyproline-rich glycoprotein family protein	Unknown	e	*A. thaliana*	Kundu et al., 2013
137	AtRACK1A	AT5G53300	UBC10, Ubiquitin-conjugating enzyme E2 10	Protein catabolic process	e	*A. thaliana*	Kundu et al., 2013
138	AtRACK1C	AT5G65780	ATBCAT-5, Branched-chain-amino-acid aminotransferase 5	Metabolism	c	*A. thaliana*	Klopffleisch et al., 2011

In order to obtain homogeneous data, the ligand names and related process were assigned based on the gene information from each ID at the NCBI database. a, Affinity Capture; b, Co-Immunoprecipitation; c, cDNA Library screen by YTH assay; d, One by one YTH assay; e, cDNA Library screen by suYTH system; f, One by one suYTH assay; g, in planta or protoplasts BiFC; and h, Split firefly luciferase complementation (SFLC).

in seed germination, flowering, and production of leaves. The mutants also displayed a number of altered responses to hormones such as: (a) reduced sensitivity to gibberellin and brassinosteroid during seed germination; (b) hypersensitivity to ABA during seed germination and early seedling development; and (c) hyposensitivity to auxin in adventitious and lateral root formation (Chen et al., 2006). The authors concluded that AtRACK1A participated in multiple signal transduction pathways during plant developmental processes. These data were further substantiated by the report that *Arabidopsis rack1a single, and rack1a/rack1b or rack1a/rack1c* double mutants, were hypersensitive to ABA in developmental processes that included seed germination, cotyledon greening and root growth (Guo et al., 2009a). In addition, the mutants lost water more slowly from the rosettes than the wild type, and were hypersensitive to high concentrations of NaCl during seed germination. On the other hand, plants overexpressing AtRACK1A displayed ABA insensitivity. The authors concluded that *Arabidopsis* RACK1 was a negative regulator of ABA responses and that the three *AtRACK1* genes act redundantly to regulate such ABA responses. Interestingly, the expression of all three *AtRACK1* genes was down-regulated by ABA, an effect contrary to the stimulation of expression by auxin (Ishida et al., 1993) or cytokinin (McKhann et al., 1997). In a later study, it was found that all three AtRACK1 proteins interacted physically with the two eIF6 protein isoforms in *Arabidopsis* (Guo et al., 2011a), and that ABA down-regulated the expression of both *AtRACK1* and *eIF6* transcripts. These data further suggested that plant RACK1 may be able to regulate its own expression in a hormonally-regulated feedback loop.

Interestingly, it was found in a different work that the AtRACK1A isoform seems to specifically mediate gibberellin signaling, at least in a D-allose mediated inhibition pathway (Fennell et al., 2012). When gibberellin was applied to *Arabidopsis* seeds, a significant up-regulation of AtRACK1A:GFP expression in the embryo root tip region of seedlings was observed after 72 h. The opposite effect was observed by treatment with the rare sugar D-allose, which is an inhibitor of seed germination (Fennell et al., 2012). Further analysis on *Arabidopsis rack1a* knockout mutants showed a significantly higher hypersensitivity to D-allose inhibition of germination compared to wild type seeds, and this inhibition was not counteracted by the addition of gibberellin. Finally, in seeds harboring a functional RACK1A in a *rack1b/rack1c* double mutant background, neither D-allose nor D-allose plus gibberellin significantly affected seed germination. These data suggested that D-allose negatively regulated seed germination and gibberellin-mediated early seedling development through the inhibition of RACK1A expression (Fennell et al., 2012). These results also indicated that specific plant RACK1 isoforms may be involved in hormonal responses regulating developmental processes in different fashion.

Hormonal relationships to RACK1 in symbiotic legumes have been also carried out (McKhann et al., 1997; Islas-Flores et al., 2009, 2011, 2012). In the smaller *P. vulgaris RACK1* (*PvRACK1*) gene family, treatment of bean seeds with the synthetic auxins 2,4-D (2, 4-Dichlorophenoxyacetic acid) and IAA (Indole-3-acetic acid) showed that auxin negatively controlled *PvRACK1* transcript accumulation during germination. It was found that the maximum transcript accumulation of *PvRACK1* was at 32 h of germination without treatment. However, this accumulation was delayed 8 h from the control with synthetic auxin treatment (Islas-Flores et al., 2009). These findings suggested a weak negative regulation of *PvRACK1* transcription mediated by auxin. This conclusion was supported by the fact that the auxin transport blocker Naphthylphthalamic Acid (NPA) did not have any effect on the level of *PvRACK1* transcript, which was consistent with a putative blocking of the auxin transport to target sites (Islas-Flores et al., 2009). In a subsequent work, it was demonstrated that the *PvRACK1* transcript accumulation was induced by ABA, cytokinin, and gibberellin during root development. However, contrary to the above effect seen on *P. vulgaris* germination, the transcript was also induced by auxin on developing roots. It was shown that the transcript induction was higher with auxin, gibberellin and ABA than with cytokinin (Islas-Flores et al., 2011, 2012). Cytokinin induction of *RACK1* expression in roots of other legume was also documented in the previous report by McKhann et al. (1997), in *M. sativa*. Thus, hormonal signaling mediated by RACK1 on developing legume roots suggested its involvement in symbiotic processes (see below). Some of the provided data on the hormonal signaling-RACK1 relationship has also been discussed in a previous review (Zhang et al., 2013). Taken together, the results reported so far provide compelling evidence that plant RACK1 participates in hormonal signaling through regulation of protein expression. Its own expression, in a likely array of feedback loops, would also be regulated by phytohomones in tightly regulated signaling networks critical for fundamental plant developmental processes.

CELL PROLIFERATION AND PLANT DEVELOPMENT

RACK1 has been related and even suggested as marker for mammalian cell proliferative processes since early reports of its overexpression in breast, lung and hepatocellular carcinoma (reviewed in Li and Xie, 2015). Following this association, Islas-Flores et al. (2011) inferred that plant RACK1 could be involved in cell proliferation processes in symbiotic root nodules. Using the model of *P. vulgaris* root nodules, and with previous evidence that suggested that its genome only harbored one *PvRACK1* gene (Islas-Flores et al., 2009), they used an RNAi approach to silence its expression and follow the root nodule phenotype during development. It was observed that, as expected, silencing of *PvRACK1* expression led to a reduced number and size of the nodules compared to controls (Islas-Flores et al., 2011). They also observed a reduced red-brown coloration of *PvRACK1*-knockdown nodules with respect to controls. A closer examination of the cell ultrastructure revealed that *PvRACK1* knockdown prevented a proper formation of the symbiosome and impaired cell expansion, which resulted in a reduced cell size of both infected and non-infected cells (Islas-Flores et al., 2011). These data were also consistent with earlier findings where the *RACK1* transcript was preferentially located to dividing cells of nodule primordia and meristem in *M. sativa* (McKhann et al., 1997). These results suggested that plant RACK1 is also involved in cell proliferation and expansion directly related to symbiotic processes.

RACK1-induced developmental defects were studied in *A. thaliana* with T-DNA insertion mutants. The *rack1a* loss-of-function mutation in *Arabidopsis* led to the impairment of multiple developmental processes that included seed germination, leaf production, and flowering (Chen et al., 2006). It was later reported (Guo and Chen, 2008), that loss-of-function mutations in either *rack1b* or *rack1c* did not confer the same developmental defects and phenotypes were apparently normal, indicating that *AtRACK1A* could functionally complement those mutations. Interestingly, either *rack1b* or *rack1c* mutation enhanced the developmental defects caused by the *rack1a* mutation in double mutant analyses. Furthermore, severe effects on development and lethality were observed in the *rack1a/rack1b/rack1c* triple mutants. The authors concluded that *RACK1* genes function in an unequally redundant manner, critically and tightly regulating plant development (Guo and Chen, 2008). More recently, Zhang et al. (2014) reported an interesting work linking OsRACK1 to rice seed germination. They first observed that in wild type rice seeds, the *OsRACK1A* gene was highly expressed while *OsRACK1B* was poorly expressed. They assessed its role in the control of seed germination and observed that OsRACK1A underexpressing seeds showed a significant delay in germination and a decrease in their germination rate compared to the wild type. These knockdown mutants also displayed an increased sensitivity to an ABA-induced germination delay compared to wild type or OsRACK1A overexpressing seeds (Zhang et al., 2014). These results indicated that OsRACK1A positively regulated seed germination in rice, similar to the first reports on the effect of RACK1 on *Arabidopsis* germination (Chen et al., 2006). Taken together, these reports clearly highlight a critical role of RACK1 in cell proliferation and developmental processes in plants. Zhang et al. (2014) proposed a mechanism in which OsRACK1A positively regulates seed germination through enhancing ABA catabolism and stimulation of H_2O_2 production, and that they both interact to regulate seed germination. More recently, a direct interaction between AtRACK1 and Gβ in *Arabidopsis* upon activation of a pathogen defense signaling pathway was reported (Cheng et al., 2015). Although, some direct RACK1 interactors have been found and these have revealed some clues regarding the impact they may have on developmental processes, many more are predicted yet to be discovered. These RACK1 interactors will be discussed in detail below.

INNATE IMMUNITY AND ROS PRODUCTION

RACK1 is involved in innate immunity responses in plants through the formation of immune complexes. This aspect of RACK1 function is discussed in detail in a parallel review within this topic and we will only mention the most relevant and recent findings. Innate immunity in rice is regulated by a complex of regulatory proteins located at the plasma membrane. This assembly is composed of OsRACK1, Rac1, RAR1, SGT1, Rboh and the two HSP proteins, HSP90 and HSP70 (Thao et al., 2007; Nakashima et al., 2008). It was found that RACK1 participates in this complex by binding to Rac1, RAR1, SGT1, and Rboh but not HSP90 (Nakashima et al., 2008). Rac1 transcriptionally and post-transcriptionally regulates RACK1 and vice versa in a feedback loop. A model has been proposed in which, when a rice plant is attacked by pathogens (i.e., rice blast fungus), Rac1 and/or RACK1 is/are activated and the immune complex is formed. This newly formed immune complex can then directly regulate the immune response through HSP70 and HSP90, or interact with the N-terminus of RbohB to trigger the production of ROS and fight the pathogen (Nakashima et al., 2008; Kawano et al., 2014). More recently, Cheng et al. (2015) reported that in *Arabidopsis*, all three RACK1 subunits also serve as scaffolds for the MAPK pathway through binding to the Gβ subunit upon activation of the MAPK cascade by a pathogen-secreted protease. This cascade resulted in the activation of the immune response although further downstream defense responses were not investigated. It is possible that in this instance, ROS are elicited to fight the invader, as it has been proposed to occur in rice (Nakashima et al., 2008). The authors concluded that a mechanistically distinct immune signaling must occur in plants compared to yeast or mammals. In the latter, Ste5 or β-arrestin scaffold MAPK's, after upstream GPCRs activation, whereas plants do not possess the corresponding orthologs (Cheng et al., 2015). Recently, RACK1 expression was linked to pathogen responses that led to ROS production in Maize (Wang et al., 2014). These authors observed an increased level of *ZmRACK1* expression upon ABA or methyl jasmonate treatment and hypothesized that in maize, a similar innate immunity response to that occurring in rice (Nakashima et al., 2008), could exist. In agreement, *ZmRACK1* overexpression resulted in a 2.5–3-fold increase in expression levels of the pathogenesis-related protein genes *PR-1* and *PR-5*, and the protein was shown to interact with RAC1, RAR1 and SGT1. These proteins were present in the Racl (a ROP/RAC small GTPase) immune complex as determined by the Yeast Two Hybrid (YTH) assay (Wang et al., 2014). It was also observed that ROS production was higher in seedlings and leaves from overexpressing lines than in the wild-type. These results were in direct analogy with the immune response in rice where an equivalent complex is formed and ROS production ensues (Kawano et al., 2014).

G-protein signaling has been linked to ROS generation in *Arabidopsis* guard cells as *gpa1* mutants were impaired in the production of ROS in response to ABA (Zhang et al., 2011). In addition, Gudesblat et al. (2007) reported that MPK3 kinase acts downstream of ROS production in guard cell ABA signaling. Thus, pathogen-activated defense responses in plants may occur through similar mechanisms to those utilized for hormone-mediated ROS production. This is consistent with the model proposed by Kawano et al. (2014), where G-protein functions upstream of OsRac1 in the early steps of rice defense signaling.

RACK1 ALSO APPEARS ON THE SCENE OF STRESS RESPONSES

RACK1 has been implicated in a variety of responses elicited by both biotic and abiotic stress in yeast (Núñez et al., 2009), worms (Ziegler et al., 2009), mammalian cells (Arimoto et al., 2008), and plants. Since ABA plays a major role in drought and saline stress, and Guo et al. (2009a) had shown that *rack1a* single, *rack1a*/*rack1b*, *rack1a*/*rack1c*, and *rack1b*/*rack1c* double loss-of-function mutants exhibited hypersensitivity to ABA, they tested seed germination on the same mutant lines subjected to salt stress. The results showed a direct relationship between ABA sensitivity and salt stress sensitivity in seed germination. In their assays, *rack1a* single and *rack1a*/*rack1b* or *rack1a*/*rack1c* double mutants, were severely affected during seed germination under salt stress. On the other hand, the *RACK1A* overexpressing plants showed a hyposensitivity to the salt treatment in the germination assay. These results directly linked AtRACK1 to stress responses in *Arabidopsis* (Guo et al., 2009a). Unfortunately, no determination of the effect of salt or other types of stress on the expression of the *AtRACK1* genes was carried out on germinating *Arabidopsis* wild type seeds to directly test the stress effects on *AtRACK1* regulation.

In a separate work, using *P. vulgaris* plant roots transformed with *A. rhizogenes* carrying a *PvRACK1* overexpression construct, severe damage and necrosis was observed when these were subjected to heat stress. Heat-shocked transformed seedlings showed systemic necrosis at 4–5 days post-inoculation, no callus formation at the inoculation zone, and interrupted progression to transgenic root formation (Islas-Flores, 2011; Islas-Flores et al., 2012). These observations suggested that the overexpression of the *PvRACK1* gene in *P. vulgaris*, caused a severe imbalance in the RACK1-mediated signaling cross-talk leading to an enhanced effect of heat stress, which resulted in lethality. The exact mechanisms and RACK1 interactors underpinning this enhanced stress effect remain to be studied.

Subsequently, Kundu et al. (2013) identified oxidative, drought and saline stress RACK1 ligands when AtRACK1A was used as a bait to screen an *A. thaliana* inflorescence cDNA library by the split-ubiquitin Yeast Two-Hybrid system (suYTH). Identified ligands corresponded to CSD1 (Superoxide dismutase [cu-Zn]), MYC2 (Transcription factor), GSTU19 (TAU 19 Glutathione S-transferase), Fructose-bisphosphate aldolase 3, ATPQ (ATP synthase subunit D), RAN-1 (GTP-binding nuclear protein Ran-1; RAS-related nuclear protein-1), WNK8 (WITH NO LYSINE (K) KINASE 8; Serine/threonine-protein kinase), and LTP3 and LTP4 (Non-specific lipid-transfer proteins 3 and 4; Kundu et al., 2013; **Table 1**). These results also linked RACK1 function to diverse types of stress including hypoxia, cold, and starvation.

More recently, Olejnik et al. (2011) found that AtRACK1A binds to AtNUDT7. AtNUDT7 is a pyrophosphatase that hydrolyzes NADH and ADP-ribose *in vitro*, and plays a role in the response to biotic and abiotic stress. The authors proposed that the AtRACK1A-AtNUDT7 interaction negatively regulates the cellular level of ROS and the cellular defense pathway. This work uncovered new players in a novel pathway in which RACK1 scaffolded interactors that are involved in stress responses in *Arabidopsis*. Similarly, RACK1 expression was linked to biotic stress imposed by pathogen responses in rice and maize (Nakashima et al., 2008; Wang et al., 2014). Rice and maize plants overexpressing OsRACK1A or ZmRACK1, respectively led to a reduction in leaf symptoms caused by the fungi *Magnaporthe grisea* and *Exserohilum turcicum* (Pass.), respectively. The extension of the leaf chlorosis and necrosis was statistically and significantly lower in overexpressing OsRACK1A or ZmRACK1 leaves than in the wild type. Unfortunately, no assessment of the stress effect imposed by the pathogens on *OsRACK1A* or *ZmRACK1* expression was carried out, although the *ZmRACK1* transcript was up-regulated in leaves pre-treated with ABA and methyl jasmonate (Wang et al., 2014). The above results implicate both, plant RACK1 in the activation of biotic and abiotic stress responses, and ABA in the regulation of these responses. The exact mechanisms of action and clarification of the opposing effects in different plant species await further study.

DOES RACK1 REGULATE THE microRNA PATHWAY?

MicroRNAs (miRNAs) play significant roles in living systems by modifying most protein coding transcripts at the post-transcriptional level. miRNAs are a class of 21 nucleotide non-coding small RNAs, simultaneously targeting many transcripts and fine-tuning the expression of genes (Carrington and Ambros, 2003; Bartel, 2004; Nunez et al., 2013). Although the process of miRNA-mediated regulation of gene expression shows similarities in plant and animal systems, there are clear distinctions in terms of the identity of the regulatory proteins and the cellular sites of miRNA biogenesis. In animals, these tiny (~21 nucleotide) miRNAs are involved in developmental and pathological processes. In plants, miRNAs participate in growth, flowering and development by regulating hormone signaling, nutrient sensing, stress responses and immunity against pathogen invasion (Ding et al., 2013; Jin et al., 2013). While plant and non-plant miRNAs are both transcribed by the RNA polymerase II, their biogenesis differs in terms of location. Metazoan miRNAs are processed at two locations, the cytoplasm and the nucleus. In the nucleus, pri-miRNAs are processed by DROSHA and DIGEORGE SYNDROME CRITICAL REGION 8 (DGCR8), and the processed pre-miRNAs are then further processed by DICER in the cytoplasm (Lee et al., 2003; Kim, 2005). Unlike their animal counterpart, plants do not have DROSHA and DGCR8 to process the pri-miRNA to the pre-miRNA. Instead, another RNase III-like protein known as DICER-LIKE 1 (DCL1), in conjunction with the SERRATE (SE) or HYPONASTIC LEAVES 1 (HYL1), process the pri-miRNAs into the mature miRNAs, which occur exclusively in the nucleus. The mature miRNAs are loaded onto the ARGONAUTE (AGO) effector complexes-miRNA-induced complex (miRISC), which regulates the target mRNAs for degradation or repression of translation (Kurihara and Watanabe, 2004).

To date, three different reports have implicated RACK1 in the miRNA pathway in *C. elegans*, humans, and *Arabidopsis*

(Jannot et al., 2011; Speth et al., 2013; Speth and Laubinger, 2014). In plants it was found, through a YTH assay, that RACK1 interacts with the SE protein to regulate the mature miRNA biogenesis in the nucleus (Speth et al., 2013; Speth and Laubinger, 2014). Subsequently, the interaction was confirmed *in vivo* by the Bimolecular Fluorescence Complementation (BiFC) assay. It was found that such interaction took place in distinct sub-nuclear foci D-bodies, where SE was previously reported to localize (Fang and Spector, 2007). Further support came with the report that miRNA accumulation in *rack1* loss-of-function mutants was globally decreased causing de-repression of the target mRNAs (Speth et al., 2013; Speth and Laubinger, 2014). It was also reported that some pri-miRNAs accumulated at higher levels in *rack1* mutants, suggesting that RACK1 affects the processing and transcription/stability of certain pri-miRNAs in plants (Speth et al., 2013; Speth and Laubinger, 2014). Moreover, aberrant non-canonical miRNAs were more abundant in *rack1* mutants, indicating that RACK1 is also involved in ensuring precise processing (Speth et al., 2013; Speth and Laubinger, 2014). Plant pri-miRNAs and pre-miRNAs are cleaved in the nucleus by the same RNase III enzyme DICER-LIKE 1 (DCL1). Therefore, *Arabidopsis* RACK1 may be required for steps upstream of, or with DCL1, to affect pri-miRNA processing or stability.

Although RACK1 interaction with SE indicates a role in miRNA biogenesis, it is not difficult to envision a RACK1 role independent of miRNA biogenesis. For example, Otsuka et al. (2011) reported, in hepatocellular carcinoma, that a few miRNAs displayed impaired silencing upon RACK1 depletion with no detectable changes in their overall levels (Otsuka et al., 2011). However, lower amounts of these miRNAs were detected in Ago2-containing complexes, leading the authors to propose that RACK1 functions after miRNA maturation and is required to load mature miRNAs into miRISCs. Similarly, Jannot et al. (2011) observed that in *C. elegans* and human cells depletion of RACK1 impaired miRNA regulation and reduced the amount of AGO associated with the polysome (Jannot et al., 2011). They proposed that RACK1, as an AGO-interacting protein, facilitates the recruitment of miRISC to the ribosome. Interestingly, Speth and Laubinger (2014) have shown that *Arabidopsis* AtRACK1 and AGO1 are part of a common complex outside the ribosome, and that they co-localize in the nucleus and the cytoplasm. As AtRACK1 has been shown to regulate pri-miRNA processing in the nucleus, it is likely that free, rather than ribosome-bound RACK1, is involved in the regulation of miRNA biogenesis (Speth et al., 2013; Speth and Laubinger, 2014). Differences in localization and binding partners may explain the distinct functions of RACK1 in miRNA biogenesis and miRISC function. Ribosome-bound RACK1 may help recruit miRISCs to the ribosome, while free RACK1, perhaps with associated AGO, may play roles in miRNA biogenesis in the nucleus and/or the cytoplasm (Chu et al., 2014).

Whether RACK1 binds miRNA directly is not resolved yet. Although RACK1 is predicted to bind the ribosome by primarily anchoring to the 18S ribosomal RNA, the mechanism of binding to the miRNA may be mediated by interaction with other anchoring protein(s) like SE or HYL1.

Given the evidence that RACK1 is a negative regulator of the stress hormone ABA signaling, it is not far-fetched to suggest that RACK1 regulates this pathway possibly through miRNA regulation. Indications that miRNAs participate in the ABA response were first provided by the isolation of ABA-hypersensitive mutants impaired in any of the several key genes of the miRNA biogenesis pathway, *HYL1*, *DCL1*, *HEN1*, *SE*, and *HASTY* (Ding et al., 2013). The *hyl1* mutant was shown to be hypersensitive to ABA during *Arabidopsis* germination (Lu and Fedoroff, 2000). Drought-induced miRNAs down-regulate their target mRNAs which may encode negative functional proteins involved in the drought response. Conversely, other miRNAs are down-regulated, leading to the accumulation of their target mRNAs that contribute positively to stress adaptation. The precise identification of the miRNAs and their targets in the *rack1* mutants would definitely allow to pinpoint the exact role of RACK1 in the miRNA-mediated drought responses.

ARE THE RIBOSOME-BOUND RACK1 LIGANDS LOST IN TRANSLATION?

Although RACK1 has been implicated in many stress signaling pathways, the *in silico* study of RACK1 inside the GENEVESTIGATOR database indicates that RACK1 predominantly regulates protein translation and ribosome biogenesis (Guo et al., 2011b). Through a careful analysis of the online data and co-expression studies, Guo et al. (2011b) found that the protein synthesis and ribosome biogenesis function of RACK1 is regulated by the stress hormone ABA. Loss in the relative abundance of 60S ribosome subunits and 80S ribosome in the *rack1a/rack1b* double mutants indicated a role of RACK1 in the ribosome biogenesis. Such predominant role of RACK1 in protein translation and ribosome biogenesis is, therefore, represented by a thicker arrow in **Figure 2**. In addition, by virtue of the RACK1 predominant localization on the ribosome in different species, a role for the protein in the regulation of global translation through ribosome biogenesis and interaction with regulatory proteins has been envisioned (Shor et al., 2003; Gerbasi et al., 2004; Nilsson et al., 2004; Sengupta et al., 2004; Chang et al., 2005; Giavalisco et al., 2005; Yu et al., 2005; Regmi et al., 2008; Coyle et al., 2009; Armache et al., 2010; Guo et al., 2011a,b). In a recent report, it was also revealed that *Arabidopsis* RACK1A, in conjunction with other ribosomal proteins, controls the upstream Open Reading Frame (uORF)-mediated protein translation of the key transcription factor SAC51 to regulate growth and development (Kakehi et al., 2015). Using Cryo-EM, Sengupta et al. (2004) found that fungus RACK1 was located on the head region of the 40S subunit, in the immediate vicinity of the mRNA exit channel. Later, Coyle et al. (2009) provided evidence that *S. cerevisiae* RACK1, Ascl functions on the ribosome, implying a physical link between the eukaryotic ribosome and cell signaling pathways *in vivo*. Chang et al. (2005), in a proteomic analysis of ribosomal proteins, first reported the identification of RACK1 protein from the *Arabidopsis* ribosome, where it was associated with both the 40S ribosome subunit and polysomes. It has been

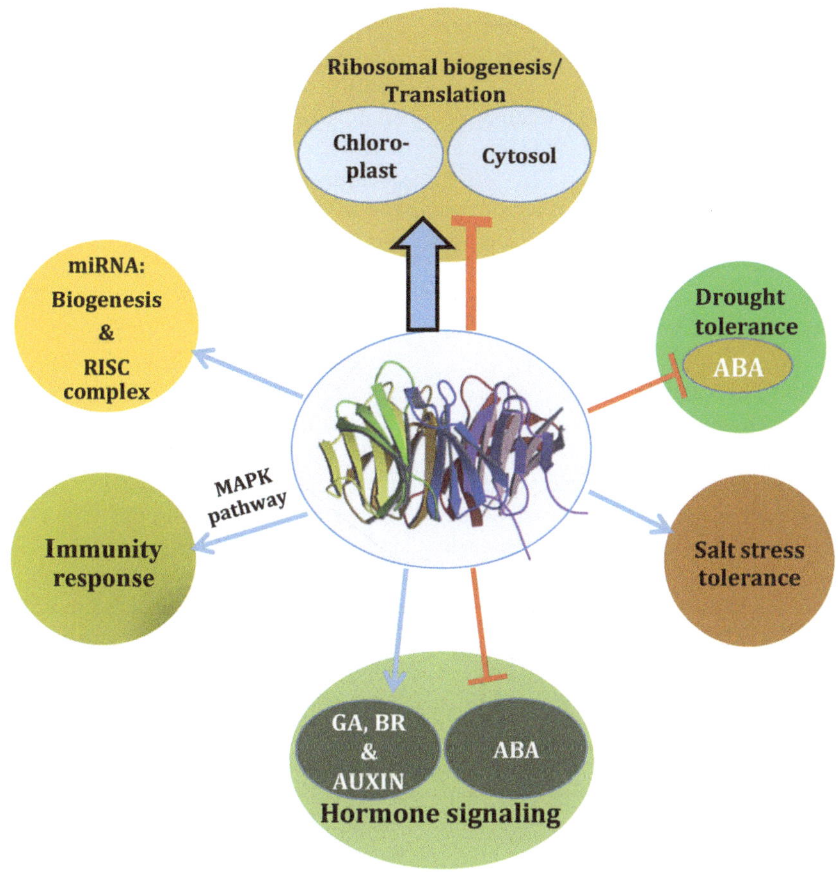

FIGURE 2 | As an integrator of signaling, RACK1 regulates diverse cellular signaling pathways. Based on the available literature, a comprehensive view is presented. Depending on the pathways, RACK1 acts either as a positive regulator (blue arrow) or as a negative regulator (red inhibitory arrow), most likely through mediating protein-protein interaction. The thicker arrow denotes a predominant role of RACK1 in protein translation and ribosome biogenesis. Note that the pathways may not always act in a unidirectional way as they often cross-talk. See text for further detail.

suggested that, although ribosome-bound RACK1 might directly regulate translation *per se*, its function as a ribosomal protein was likely linked to its capacity to recruit a particular cohort of RACK1-associated proteins such as activated PKCβII in animals (Adams et al., 2011). However, the lack of *bona fide* PKC isoforms from the plant kingdom indicate that other kinases may play a similar role in regulating RACK1-mediated ribosome translation processes. On the other hand, a suggested model for the role of RACK1 protein in ribosome assembly in non-plants entails phosphorylation of eIF6 by PKC. The latter as a result of interaction with eIF6 and PKC bound to a nearby 40S subunit via RACK1 (Sengupta et al., 2004). The report that *Arabidopsis* RACK1 directly interacts with eIF6 implicates plant RACK1 in mediating the ribosome assembly in a similar way as observed with their counterparts in mammalian cells (Guo et al., 2011a). The superposition of the deduced crystal structure of *Arabidopsis* RACK1A onto the yeast RACK1-40S ribosome model (PDB ID code 1ARI) supports the notion that the top rim of RACK1A is in contact with the ribosome allowing the bottom rim to interact with regulatory proteins (Ullah et al., 2008). The lack of any structural data on plant ribosome-bound RACK1 makes difficult to support the claim that both, the mammalian and plant RACK1 proteins function in the same mechanistic way during the regulation of protein translation. It appears even more difficult, to extrapolate similar pathways of RACK1-ribosomal complexes that are modulated under stress in mammalian cells (Arimoto et al., 2008). When these cells enter stress by hypoxia or heat shock (type 1 stress), the formation of stress granules is induced. Stress granules are molecular aggregates of stalled translation pre-initiation complexes (that prevent the accumulation of mis-folded proteins), which also sequester RACK1. This RACK1 immobilization, in turn, suppresses the activation of the MTK1-SAPK pathway leading to apoptosis, which would otherwise be induced under conditions of type 2 stress (X-ray or genotoxic drug exposure; Arimoto et al., 2008). Plants do not display, at least phenotypically, any similar response under the various stress conditions. This is most likely due to the plasticity and ancestry of plant genomes brought about by the evolution pressure that the incapability of movement has imposed. The mechanisms of regulation of translation in plants under stress are more likely to be under hormonal control, in which RACK1 also actively participates. For example, since RACK1 is known to be a negative regulator of the stress hormone ABA, it is possible that plants are also able to modulate global protein translation through RACK1-ribosomal associations under stress conditions. The putative critical kinases and phosphatases key for regulation through phosphorylation/dephosphorylation events, whether or not present in multimolecular complexes with RACK1, remain unidentified.

Finally, another feature that sets apart plant RACK1 proteins from their non-plant counterparts, is their putative role in the regulation of the chloroplastic ribosome-based translation process. Kundu et al. (2013) reported that RACK1A, the predominant isoform in *Arabidopsis*, physically interacts with dozens of cytosolic and chloroplastic ribosomal proteins. Important among them is the RPL12-A protein which contains a signal sequence that exclusively localizes the protein to the chloroplast. In addition, RACK1A is also found to interact with a member of Ribosomal protein L36a/L7-3 family protein that is also reported to be localized in the chloroplast (Kundu et al., 2013). Interestingly, it was reported that the spinach chloro-ribosome L7-3/L12-A proteins in combination with L1, form the base stalks of the 70S chloro-ribosome as observed by Cryo-EM. These base stalks were larger than those of the *Escherichia coli* 70S ribosome, one of various features that distinguished them from one another (Sharma et al., 2007). In view of the RACK1 associations with putative chloroplast-localized ribosomal proteins, it is tempting to speculate that RACK1 may regulate the translation process by either affecting the chloro-ribosome assembly and/or stability, or by directly regulating the translation process.

RACK1A is also reported to interact with the major components of photosynthetic proteins translated by the chloro-ribosomes which include photosystem I and II components and the large subunit of RuBisCo-associated proteins. Since chloro-ribosomes control chloroplastic translation in a light dependent manner, it will be interesting to investigate whether plant RACK1 is able to regulate the photosynthetic protein abundance of the chloroplast. This would have a direct effect on regulating the photosynthesis-based total growth and development of a plant.

RACK1-INTERACTING LIGANDS, SOME SOLID EVIDENCE, AND SOME ENIGMA

Structurally, RACK1 presents multiple conserved surface residues contained on the WD-40 repeats, where these binding sites allow multiple interactions and set the bases of its scaffolding characteristics. Although, it is capable to form homodimers, it is also found as a free monomer allowing to form hetero-complexes with an expanding number of partners. It is well known that RACK1 has a large number of direct ligands; for example, in metazoans over 90 ligands have been reported as RACK1 interactors (Adams et al., 2011). RACK1 binding partners have been identified in several intracellular fractions, where they are transported and placed in proximity to their modifying enzymes or substrates.

Plant RACK1 presents two conserved surface regions (region 1, blades 1–4; and region 2, blades 5 and 6) believed to mediate potential protein–protein interaction sites (Ullah et al., 2008). Probable interacting partners were proposed based on the comparison of residues contained in these regions, with known residues that mediate interaction with characterized ligands (Dell et al., 2002; Chen et al., 2004; Ullah et al., 2008). In addition, RACK1 has a structural and translational function as it has been documented that it forms part of the ribosome in metazoans and plants, and it is coexpressed with up to 80% of ribosome proteins (Guo et al., 2011b). RACK1 has been identified on the head of the 40S ribosomal subunit were its conserved region 1 directly interacts with the helices 39 and 40 of the 18S rRNA, and the ribosomal protein S2p, while the conserved region 2 is exposed for interaction and recruiting of binding partners such as PKC (Nilsson et al., 2004; Sengupta et al., 2004; Armache et al., 2010). Consequently, plant RACK1 should also be found in the ribosome for recruiting PKC to phosphorylate eIF6 as in animal cells (Ceci et al., 2003). However, it is well documented that plant genomes lack this type of kinase (Guo et al., 2011b) and so far no equivalent plant kinase to fulfill a parallel function has been found. Therefore, since direct comparison of RACK1 interacting ligands in animal cells cannot be extrapolated to plant cells, alternative strategies to unequivocally identify plant RACK1 interactors have been necessary. For example, approaches that demonstrate direct physical interaction such as the yeast two hybrid (YTH) and BiFC assays, confirmed interaction of AtRACK1A, B, and C with eIF6A and B (Guo et al., 2011b). These strategies confirmed, in part, what was observed in metazoans, as well as the functional involvement of plant RACK1 in protein translation and ribosome stability. It is important to highlight that, to date, still no plant kinase equivalent to animal PKC has been identified in those complexes so how the initiation of translation is regulated in plants remains an open question. The identification of *Oryza sativa* RACK1A (OsRACK1A) as part of the immune complex interacting with Rac1, Rboh, RAR1, and SGT1 (Nakashima et al., 2008) revealed the first plant RACK1 ligands, which had remained elusive for many years. This was achieved by means of a Rac1 affinity column in a search for Rac1 ligands in an assay using prior activation by a sphingolipid elicitor. OsRACK1A interaction was further confirmed by the YTH and suYTH assays. While OsRACK1B bound to a less extent to Rac1, the binding of the other Rac isoforms, Rac3, Rac5, Rac6, and Rac7 to OsRACK1A was also confirmed. Furthermore, the direct interaction of OsRACK1A with the other immune response related proteins RAR1 and SGT1, and with N-Rboh was also shown by YTH and suYTH. Finally, it was also demonstrated that the interaction of OsRACK1A with Rac1, RAR1, and SGT1 occurred on blades 1 and 2, indicating that the conserved region 1 was mediating this interaction. Subsequently, the same group also reported that the key transcription factor OsRap2.6 contributes to rice innate immunity through its interaction with RACK1A in compatible interactions (Wamaitha et al., 2012). These results established the role of plant RACK1 in innate immunity and resistance to rice blast fungus infection.

After the first rice report of RACK1-interacting ligands, Klopffleisch et al. (2011) reported an *Arabidopsis* signal-transduction interactome that included the heterotrimeric G protein subunits and the three *Arabidopsis* RACK1 isoforms.

This was achieved by a high-throughput YTH system screening prey cDNA libraries from nine *Arabidopsis* tissues. After ten screens, seven ligands for AtRACK1A, four for AtRACK1B, and twenty four for AtRACK1C were identified (**Table 1**). Eight of these were confirmed by BiFC *in planta* and interestingly, novel unexpected functions were found. The nature of the identified AtRACK1 ligands revealed an involvement in cell wall formation and stress (Klopffleisch et al., 2011). Earlier work had suggested a more important function for AtRACK1A than AtRACK1B and AtRACK1C because AtRACK1A loss-of-function resulted in multiple defects in plant development while the loss of function of AtRACK1B and AtRACK1C did not (Guo and Chen, 2008). Therefore, it was unexpected that the largest proportion of ligands was obtained for AtRACK1C. Interestingly, no direct interaction among RACK1 isoforms and G protein subunits was found in this interactome (Klopffleisch et al., 2011). More plant RACK1 ligands were identified when AtRACK1A was used as a bait to systematically screen a suYTH-based *Arabidopsis* inflorescence cDNA library (Kundu et al., 2013). Ninety potential ligands were reported as involved in several functional plant processes such as oxidative, drought and saline stress, photosynthesis, and protein biosynthetic pathways (Kundu et al., 2013; **Table 1**). These data also demonstrated that the suYTH assay was more efficient than the YTH since more ligands were identified. It is likely that in this assay, the activation of the reporters was carried out in the cytoplasm where the interactors are found, and the probability to get positive interactions was higher. Kundu et al. (2013) also showed that phosphorylation on tyrosine 248 (Y248) of AtRACK1A was critical for the interactions since when a Y248F-RACK1A mutant was used in the suYTH assay, no positive interactions were observed. Importantly, this Y248 residue is conserved among *A. thaliana*, *H. sapiens*, *D. melanogaster*, and *S. cerevisiae*, and is located on blade 6 of the conserved region 2, at the bottom of the propeller (Ullah et al., 2008). These results also confirmed that tyrosine phosphorylation occurs on plant RACK1, and that it is required for both, homodimerization of AtRACK1A, and binding of interactors. Within the identified ligands in this report, four kinases were found: CLV1-like leucine-rich repeat transmembrane receptor-like protein kinase, AGC (cAMP-dependent, cGMP-dependent and protein kinase C family protein), PGK1 (Phosphoglycerate kinase 1), and WNK8 (WITH NO LYSINE (K) KINASE 8) protein kinase (see below). From these kinases, only WNK8 has been characterized as an atypical serine/threonine kinase that directly interacts with RACK1B and RACK1C, but not with RACK1A (Klopffleisch et al., 2011). Furthermore, YTH assays revealed that all three AtRACK1 proteins physically interacted with WNK8, and confirmation of this interaction was carried out on tobacco leaf epidermal cells by BiFC. Indeed, the three AtRACK1 isoforms were phosphorylated by WNK8 at threonine 162 and serine 122 (T162 and S122) without substrate specificity. The RACK1 residue S122 is also conserved in several species (*A. thaliana*, *O. sativa*, *H. sapiens*, *D. melanogaster*, and *S. cerevisiae*) and localizes on blade 3, while T162 residue is on blade 4 and is plant-specific. Blades 3 and 4 are in the conserved region 1, on the top rim of the propeller (Ullah et al., 2008). AtRACK1 was confirmed as a substrate of WNK8 and this phosphorylation negatively affected its stability. The phosphomimetic mutations on S122 and T162 abolished its expression at the protein level but the accumulation of the *AtRACK1* transcript was not affected. This suggested that AtRACK1 is controlled by phosphorylation and subsequent protein degradation (Urano et al., 2015). From these data, more information has been obtained regarding the binding properties of RACK1, new interacting ligands, and how phosphorylation affects this binding. However, the responsible tyrosine kinase critical for binding, and whether its association in a complex with AtRACK1 occurs, remain unknown and new strategies will have to be implemented to solve this enigma.

Growing evidences show that RACK1 is also involved in MAPK pathways. For example, in mammalian systems it has been reported that RACK1 induces cell proliferation via the MAPK cascade in association with the ERK, JNK and p38 families (López-Bergami et al., 2005; Vomastek et al., 2007). While significant knowledge of the RACK1-MAPK cascade relationship in mammalian and yeast systems over the past two decades has emerged, much less is known on how these pathways function in plants. This is further hampered by the facts that plant cells do not have *bona fide* G protein-coupled receptors (GPCRs), canonical PKC, or any of the RACK1, G protein and MAPK-linking chaperone orthologs of yeast *ste5* or mammalian beta arrestin (Witzel et al., 2012; Urano et al., 2013). Although, it was completely unknown how plant RACK1 contributed to the MAPK pathways, recently, Cheng et al. (2015) provided the first intriguing evidence that *Arabidopsis* RACK1 scaffolds the MAPK pathway through binding to the Gβ subunit. This finding links RACK1 to upstream G-protein signaling and downstream activation of a MAPK cascade in a pathogen-secreted, protease-mediated immune signaling pathway. They hypothesized that the 25% amino-acid sequence identity and similar seven-bladed β-propeller structure between RACK1 and Gβ might play a role in linking the G protein complex to downstream MAPK pathways. Their study showed that *Arabidopsis* RACK1 forms a complex with Gβ, MEKK1 (a MAPKKK), MKK4/MKK5 (two redundant MAPKKs), and MPK3/MPK6 in the presence of a pathogen-secreted protease treatment. Furthermore, they found that all three RACK1 protein isoforms bind the Gβ subunit. MPK3/MPK6 induced immune responses against pathogen infection through phosphorylation activation of their substrates. In a separate study in which RACK1 association was not investigated, Lieberherr et al. (2005) showed that attenuation of *OsRac1* by RNAi-mediated knockdown or loss of function of Gα (termed as *dwarf1*), drastically decreased OsMAPK6 levels. They found that OsMAPK6 and active but not inactive OsRac1 formed a protein complex in rice immunity. It is worth noting that Ras-MAPK or G protein-MAPK cascades occur in response to various stimuli, such as hormones or environmental stresses (Kawano et al., 2010). Since OsMAPK6 and OsRac1 proteins are in the same protein complex in the immune signaling pathway, it can be envisaged that OsRACK1 might have a functional link in this complex. However, unlike *Arabidopsis*, a direct evidence showing that OsRACK1 scaffolds the rice immunity signaling pathway with any member of the MAPK superfamily has not been reported so far and remains to be elucidated.

The heterotrimeric G protein is composed of three different canonical subunits (α, β, and γ), which in *Arabidopsis* are named GPA1, AGB1, and AGG1/AGG2 (there are two genes encoding the Gγ subunit in *Arabidopsis*). Before 2011, all the screens identifying one of the heterotrimeric G protein subunits interacting with RACK1 had been in animals. However, the essential residues for this interaction were already identified on blades 1–4 of plant RACK1 and thus, the conserved region 1 could mediate binding to the Gβγ dimer and the Gαβγ trimer (Dell et al., 2002; Ullah et al., 2008). Earlier reports documented that AtRACK1A, AtRACK1B, and AtRACK1C did not interact directly with AGB1 as tested by YTH, suYTH, and Co-IP assays (Guo et al., 2009b). However, a recent study demonstrated that all three AtRACK1 isoforms interacted with AGB1 by three methods (Cheng et al., 2015). First, BiFC in *Nicotiana benthamiana* leaves demonstrated that RACK1A, RACK1B, and RACK1C interacted with AGB1, MEKK1(K361M), MKK4, MKK5, MPK3, and MPK6, but not GPA1 or MPK4. Second, the interactions were further confirmed only for AtRACK1A by means of BiFC and split firefly luciferase complementation (SFLC) assays in *Arabidopsis* protoplasts. Third, binding between all three *Arabidopsis* RACK1 proteins and AGB1 was observed in co-immunoprecipitation experiments with *Arabidopsis* mesophyll protoplasts using Flag-tagged RACK1 proteins as the bait and HA-tagged AGB1 as the prey. In contrast, no binding was observed when HA-tagged GPA1 was used as prey (Cheng et al., 2015). This further supported the central role of RACK1 in innate immunity, as previously shown for *O. sativa* (Nakashima et al., 2008). In addition, Olejnik et al. (2011) later reported that AtRACK1A bound to AGG1, AGG2, and AtNUDT7. The latter is a pyrophosphatase capable of hydrolyzing NADH and ADP-ribose *in vitro*, and plays a role in the response to biotic and abiotic stresses. These data suggested a correlation between heterotrimeric G protein function and stress responses.

The data indicates that GPA1does not directly interact with RACK1 in any of the implemented assays (YTH, suYTH, Co-IP, *in planta* BiFC, SFLC, etc.). Moreover, although the interaction between RACK1 and the AGG1/AGG2 subunits was not analyzed in these reports, it was previously shown that AtRACK1A consistently interacted with both AGG1 and AGG2 using *in vitro* (pull-down), and *in vivo* (YTH and BiFC in protoplasts) assays (Olejnik et al., 2011). Therefore, the accumulating evidence leads to the notion that plant RACK1 can form complexes with the Gβ and Gγ subunits, but not with Gα. Thus, it is likely to participate in the regulation of some heterotrimeric G protein functions such as pathogen-activated immune response and oxidative stress, which are processes where other plant RACK1 interactors have been identified. Nevertheless, plant RACK1 participation in G protein-mediated developmental processes as found in other models (Omosigho et al., 2014), should not be ruled out.

CONCLUDING REMARKS AND FUTURE DIRECTIONS

It is clear that plant RACK1, in analogous manner to its animal homolog, has earned itself a name in the WD-repeat protein family as a promiscuous but multi-interactive protein key for many critical plant processes and responses.

Plant RACK1 ligands remained elusive for almost two decades after the report of the first homolog in tobacco; however due the growing interest in RACK1 function, and the development of new analytical techniques, a total of 138 ligands have been identified so far (**Table 1**). This identification allowed to assign RACK1 function to a wide range of biological processes in which RACK1 is the central scaffolding molecule.

The particular feature of RACK1 to scaffold multiprotein complexes, also render it with the capability to participate in a wide variety of cell processes. Its properties result in the spatio-temporal regulation of critical signal-transduction events in plants including hormonal control, stress responses, development, immune defense, protein translation regulation, miRNA production, photosynthesis, and cell wall biogenesis. A central participation in all these processes require its constitutive and ubiquitous expression in order to carry out its varied and exquisite functions.

Although, significant progress has been made, the mechanisms that control its expression and functions are not yet well understood. It is now known that its expression is, in part, affected in a species- and tissue-specific dependent fashion by hormones (auxins, gibberellins, cytokinin, etylene, ABA, brassinosteroids, and methyl jasmonate), phosphorylation (Tyr248, S122, and T162), stress, and some ligands. An example of the species-specific differential regulation of RACK1 in plants is the distinct marked effect produced by ABA, which functions as a negative regulator of *RACK1* expression in *Arabidopsis*, but does not have any effect on the expression of *arcA* in *N. tabacum*. On the other hand, it induces the expression of *RACK1* in *O. sativa*, *M. truncatula*, *P. vulgaris,* and *Z. mays*. It is important to point out that the variation of the plant RACK1 expression patterns in different species could be the result of tissue-specific regulation. For example, these assays were carried out in different tissues and organs such as seedlings (*A. thaliana*), roots (*M. truncatula* and *P. vulgaris*), leaves (*Z. mays*), and cultured cells (*N. tabacum* and *O. sativa*). Furthermore, a differential regulation of RACK1 expression induced by auxin was observed between germination (slight down-regulation) and root development (up-regulation) in the same plant, *P. vulgaris*. In addition, it was shown recently that RACK1 becomes unstable and undergoes protein degradation upon phosphorylation by WNK8 on serine 122 and threonine 162 residues. Furthermore, RACK1 expression is also controlled by some ligands such as Rac1, which associates in an immune complex, thereby regulating its expression at transcriptional and post-transcriptional levels. RACK1, in turn, regulates Rac1 transcription in a feedback loop. In summary, RACK1 regulation in plants is carried out at several levels which involve responses to hormones, post-translational modifications such as phophorylation, and direct interactions with ligands.

Another salient feature of its functional versatility is that RACK1 forms a structural part of the 40S ribosome and thus, it is involved in regulation of protein expression. This is substantiated by the facts that: (a) several ribosomal proteins were identified as ligands (**Figure 1**; **Table 1**); (b) it binds to the translation initiation factor eIF6 A and B to maintain the 60S ribosome biogenesis and whose phosphorylation leads to 80S monosome assembly; and (c) it participates in the pre-miRNA processing via interaction with the SE protein, increasing the processivity and transcription/stability of certain pri-miRNAs.

Finally, defense against pathogens has emerged as one of the main processes in which RACK1 is involved in plants (**Table 1**). As shown for *O. sativa* and *Z. mays*, it is involved in the formation of the immune complex that participates in the resistance against the rice blast fungus, where it directly interacts with Racl, RAR1, SGT1, and Rboh. RACK1 also interacts directly with the pyrophosphatase NUDT7, which is induced by bacterial pathogens and abiotic stressors. In addition, the relationship between RACK1 and processes of stress, photosynthesis and cell wall biogenesis became apparent since a dozen, eighteen and seven interactors, respectively, were indentified for each process (**Table 1**). These are novel processes in which RACK1 had not been implicated until recently. These recent findings highlight the importance to study and identify new interactors and their mechanisms that regulate and dictate when and where RACK1 will bind to channel its participation in a particular process.

Even though *A. thaliana* has been used traditionally as a plant model and copious research on RACK1 has been carried out on this species, future studies will likely move into the direction of searching for RACK1 functions in plant systems with smaller RACK1 gene families and/or in which one isogene predominantly expresses such as *P. vulgaris*, *O. sativa*, or *Z. mays*.

In conclusion, significant light has been shed regarding the role of plant RACK1 function in critical plant physiological pathways so far (**Figure 2**). With the ever flowing advent of new technologies and the availability of simpler plant models, the fine dissection of the interaction mechanistics and the unequivocal identification of critical RACK1 interactors should be achieved in order to unveil the whole of plant RACK1 function.

AUTHOR CONTRIBUTIONS

TI wrote part of hormonal signaling, interacting ligands, part of concluding remarks, assembled table and drew **Figure 1**, provided critical comments and reviewed all the manuscript. AR wrote part of miRNA, drew **Figure 2**, HU wrote part of miRNA, protein translation, drew **Figure 2**, provided images, provided critical comments and reviewed all the manuscript. MV wrote introduction, part of hormonal signaling, immunity, stress, part of concluding remarks, reviewed, and coordinated writing of the manuscript.

REFERENCES

Adams, D. R., Ron, D., and Kiely, P. A. (2011). RACK1, A multifaceted scaffolding protein: structure and function. *Cell Comm. Signal.* 9:22. doi: 10.1186/1478-811X-9-22

Arimoto, K., Fukuda, H., Imajoh-Ohmi, S., Saito, H., and Takekawa, M. (2008). Formation of stress granules inhibits apoptosis by suppressing stress-responsive MAPK pathways. *Nat. Cell Biol.* 10, 1324–1332. doi: 10.1038/ncb1791

Armache, J. P., Jarasch, A., Anger, A. M., Villa, E., Becker, T., Bhushan, S., et al. (2010). Localization of eukaryote specific ribosomal proteins in a 5.5-Å cryo-EM map of the 80S eukaryotic ribosome. *Proc. Natl. Acad. Sci. U.S.A.* 107, 19754–19759. doi: 10.1073/pnas.1010005107

Bartel, D. P. (2004). MicroRNAs: genomics, biogenesis, mechanism, and function. *Cell* 116, 281–297. doi: 10.1016/S0092-8674(04)00045-5

Carrington, J. C., and Ambros, V. (2003). Role of microRNAs in plant and animal development. *Science* 301, 336–338. doi: 10.1126/science.1085242

Ceci, M., Gaviraghi, C., Gorrini, C., Sala, L. A., Offenhäuser, N., Marchisio, P. C., et al. (2003). Release of eIF6 ($p27^{BBP}$) from the 60S subunit allows 80S ribosome assembly. *Nature* 426, 579–584. doi: 10.1038/nature02160

Chang, I.-F., Szick-Miranda, K., Pan, S., and Bailey-Serres, J. (2005). Proteomic characterization of evolutionarily conserved and variable proteins of *Arabidopsis* cytosolic ribosomes. *Plant Physiol.* 137, 848–862. doi: 10.1104/pp.104.053637

Chen, J.-G., Ullah, H., Temple, B., Liang, J., Guo, J., Alonso, J., et al. (2006). RACK1 mediates multiple hormone responsiveness and developmental processes in *Arabidopsis*. *J. Exp. Bot.* 57, 2697–2708. doi: 10.1093/jxb/erl035

Chen, S., Spiegelberg, B. D., Lin, F., Dell, E. J., and Hamm, H. E. (2004). Interaction of Gβγ with RACK1 and other WD40 repeat proteins. *J. Mol. Cell. Cardiol.* 37, 399–406. doi: 10.1016/j.yjmcc.2004.04.019

Cheng, Z., Li, J.-F., Niu, Y., Zhang, X.-C., Woody, O. Z., Xiong, Y., et al. (2015). Pathogen-secreted proteases activate a novel plant immune pathway. *Nature* 521, 213–216. doi: 10.1038/nature14243

Chu, Y.-D., Wang, W.-C., Chen, S.-A. A., Hsu, Y.-T., Yeh, M.-W., Slack, F. J., et al. (2014). RACK-1 regulates let-7 microRNA expression and terminal cell differentiation in *Caenorhabditis elegans*. *Cell Cycle* 13, 1995–2009. doi: 10.4161/cc.29017

Coyle, S. M., Gilbert, W. V., and Doudna, J. A. (2009). Direct link between RACK1 function and localization at the ribosome *in vivo*. *Mol. Cell Biol.* 29, 1626–1634. doi: 10.1128/MCB.01718-08

Dell, E. J., Connor, J., Chen, S., Stebbins, E. G., Skiba, N. P., Mochly-Rosen, D., et al. (2002). The βγ subunit of heterotrimeric G proteins interacts with RACK1 and two other WD repeat proteins. *J. Biol. Chem.* 277, 49888–49895. doi: 10.1074/jbc.M202755200

Demarco, R. S., and Lundquist, E. A. (2010). RACK-1 acts with Rac GTPase signaling and UNC-115/abLIM in *Caenorhabditis elegans* axon pathfinding and cell migration. *PLoS Genet.* 6:e1001215. doi: 10.1371/journal.pgen.1001215

Ding, Y., Tao, Y., and Zhu, C. (2013). Emerging roles of microRNAs in the mediation of drought stress response in plants. *J. Exp. Bot.* 64, 3077–3086. doi: 10.1093/jxb/ert164

Dreze, M., Carvunis, A. R., Charloteaux, B., Galli, M., Pevzner, S. J., Tasan, M., et al. (2011). Evidence for network evolution in an *Arabidopsis* interactome map. *Science* 333, 601–607. doi: 10.1126/science.1203877

Fang, Y., and Spector, D. L. (2007). Identification of nuclear dicing bodies containing proteins for microRNA biogenesis in living *Arabidopsis* plants. *Curr Biol.* 17, 818–823. doi: 10.1016/j.cub.2007.04.005

Fennell, H., Olawin, A., Mizanur, R. M., Izumori, K., Chen, J. G., and Ullah, H. (2012). *Arabidopsis* scaffold protein RACK1A modulates rare sugar D-allose regulated gibberellin signaling. *Plant Signal. Behav.* 7, 1407–1410. doi: 10.4161/psb.21995

Gerbasi, V. R., Weaver, C. M., Hill, S., Friedman, D. B., and Link, A. J. (2004). Yeast Asc1p and mammalian RACK1 are functionally orthologous core 40S ribosomal proteins that repress gene expression. *Mol. Cell. Biol.* 24, 8276–8287. doi: 10.1128/MCB.24.18.8276-8287.2004

Giavalisco, P., Wilson, D., Kreitler, T., Lehrach, H., Klose, J., Gobom, J., et al. (2005). High heterogeneity within the ribosomal proteins of the *Arabidopsis thaliana* 80S ribosome. *Plant Mol. Biol.* 57, 577–591. doi: 10.1007/s11103-005-0699-3

Gibson, T. J. (2012). RACK1 research-ships passing in the night? *FEBS Lett.* 586, 2787–2789. doi: 10.1016/j.febslet.2012.04.048

Gudesblat, G. E., Iusem, N. D., and Morris, P. C. (2007). Guard cell-specific inhibition of *Arabidopsis* MPK3 expression causes abnormal stomatal responses to abscisic acid and hydrogen peroxide. *New Phytol.* 173, 713–721. doi: 10.1111/j.1469-8137.2006.01953.x

Guo, J., and Chen, J.-G. (2008). RACK1 genes regulate plant development with unequal genetic redundancy in *Arabidopsis. BMC Plant Biol.* 8:108. doi: 10.1186/1471-2229-8-108

Guo, J., Jin, Z., Yang, X., Li, J.-F., and Chen, J.-G. (2011a). Eukaryotic Initiation Factor 6, an evolutionarily conserved regulator of ribosome biogenesis and protein translation. *Plant Signal. Behav.* 6, 766–771. doi: 10.4161/psb.6.5.15438

Guo, J., Wang, J., Xi, L., Huang, W.-D., Liang, J., and Chen, J.-G. (2009a). RACK1 is a negative regulator of ABA responses in *Arabidopsis J. Exp. Bot.* 60, 3819–3833. doi: 10.1093/jxb/erp221

Guo, J., Wang, S., Valerius, O., Hall, H., Zeng, Q., Li, J.-F., et al. (2011b). Involvement of *Arabidopsis* RACK1 in protein translation and its regulation by abscisic acid. *Plant Physiol.* 155, 370–383. doi: 10.1104/pp.110.160663

Guo, J., Wang, S., Wang, J., Huang, W.-D., Liang, J., and Chen, J.-G. (2009b). Dissection of the relationship between RACK1 and heterotrimeric G-proteins in *Arabidopsis. Plant Cell Physiol.* 50, 1681–1694. doi: 10.1093/pcp/pcp113

Ishida, S., Takahashi, Y., and Nagata, T. (1993). Isolation of a cDNA of an auxin-regulated gene encoding a G protein β-subunit-like protein from tobacco BY-2 cells. *Proc. Natl. Acad. Sci. U.S.A.* 90, 11152–11156. doi: 10.1073/pnas.90.23.11152

Ishida, S., Takahashi, Y., and Nagata, T. (1996). The mode of expression and promoter analysis of the arcA gene, an auxin-regulated gene in tobacco BY-2 cells. *Plant Cell Physiol.* 37, 439–448. doi: 10.1093/oxfordjournals.pcp.a028965

Islas-Flores, T. (2011). *Functional Analysis of the Role of PvRACK1 During Nodulation in Phaseolus Vulgaris (Análisis Funcional del Papel que Desempeña PvRACK1 Durante la Nodulación de Phaseolus Vulgaris).* PhD thesis. National Autonomous University of Mexico (UNAM). Available online at: http://132.248.9.195/ptb2011/junio/0670514/Index.html

Islas-Flores, T., Guillén, G., Alvarado-Affantranger, X., Lara-Flores, M., Sánchez, F., and Villanueva, M. A. (2011). PvRACK1 loss-of-function impairs cell expansion and morphogenesis in *Phaseolus vulgaris* L. root nodules. *Mol. Plant Microbe Interact.* 24, 819–826. doi: 10.1094/MPMI-11-10-0261

Islas-Flores, T., Guillén, G., Islas-Flores, I., San Román-Roque, C., Sánchez, F., Loza-Tavera, H., et al. (2009). Germination behavior, biochemical features and sequence analysis of the RACK1/arcA homolog from *Phaseolus vulgaris. Physiol. Plant.* 137, 264–280. doi: 10.1111/j.1399-3054.2009.01280.x

Islas-Flores, T., Guillén, G., Sanchez, F., and Villanueva, M. A. (2012). Changes in RACK1 expression induce defects in nodulation and development in *Phaseolus vulgaris. Plant Signal. Behav.* 7, 132–134. doi: 10.4161/psb.7.1.18485

Iwasaki, Y., Komano, M., Ishikawa, A., Sasaki, T., and Asahi, T. (1995). Molecular cloning and characterization of cDNA for a rice protein that contains seven repetitive segments of the Trp-Asp forty-amino-acid repeat (WD-40 repeat). *Plant Cell Physiol.* 3, 505–551.

Jannot, G., Bajan, S., Giguère, N. J., Bouasker, S., Banville, I. H., Piquet, S., et al. (2011). The ribosomal protein RACK1 is required for microRNA function in both *C. elegans* and humans. *EMBO Rep.* 12, 581–586. doi: 10.1038/embor.2011.66

Jin, D., Wang, Y., Zhao, Y., and Chen, M. (2013). MicroRNAs and their cross-talks in plant development. *J. Genet. Genom.* 40, 161–170. doi: 10.1016/j.jgg.2013.02.003

Kakehi, J.-I., Kawano, E., Yoshimoto, K., Cai, Q., Imai, A., and Takahashi, T. (2015). Mutations in ribosomal proteins, RPL4 and RACK1, suppress the phenotype of a thermospermine-deficient mutant of *Arabidopsis thaliana. PLoS ONE* 10:e0117309. doi: 10.1371/journal.pone.0117309

Kawano, Y., Chen, L., and Shimamoto, K. (2010). The function of Rac small GTPase and associated proteins in rice innate immunity. *Rice* 3, 112–121. doi: 10.1007/s12284-010-9049-4

Kawano, Y., Kaneko-Kawano, T., and Shimamoto, K. (2014). Rho family GTPase-dependent immunity in plants and animals. *Front. Plant Sci.* 5:522. doi: 10.3389/fpls.2014.00522

Kim, D.-Y., Scalf, M., Smith, L. M., and Vierstra, R. D. (2013). Advanced proteomic analyses yield a deep catalog of ubiquitylation targets in *Arabidopsis. Plant Cell* 25, 1523–1540. doi: 10.1105/tpc.112.108613

Kim, V. N. (2005). MicroRNA biogenesis: coordinated cropping and dicing. *Nature Rev. Mol. Cell Biol.* 6, 376–385. doi: 10.1038/nrm1644

Kiyosue, T., and Ryan, C. (1999). The electronic plant gene register. *Plant Physiol.* 119, 1567–1568. doi: 10.1104/pp.119.4.1567

Klopffleisch, K., Phan, N., Augustin, K., Bayne, R. S., Booker, K. S., Botella, J. R., et al. (2011). *Arabidopsis* G-protein interactome reveals connections to cell wall carbohydrates and morphogenesis. *Mol. Syst. Biol.* 7, 532. doi: 10.1038/msb.2011.66

Kundu, N., Dozier, U., Deslandes, L., Somssich, I. E., and Ullah, H. (2013). *Arabidopsis* scaffold protein RACK1A interacts with diverse environmental stress and photosynthesis related proteins. *Plant Signal. Behav.* 8:e24012. doi: 10.4161/psb.24012

Kurihara, Y., and Watanabe, Y. (2004). *Arabidopsis* micro-RNA biogenesis through Dicer-like 1 protein functions. *Proc. Natl. Acad. Sci. U.S.A.* 101, 12753–12758. doi: 10.1073/pnas.0403115101

Kwak, J., Kim, S., Lee, S., Oh, S.-A., Byoun, C.-H., Han, J.-K., et al. (1997). Insulin-induced maturation of *Xenopus* oocytes is inhibited by microinjection of a *Brassica napus* cDNA clone with high similarity to a mammalian receptor for activated protein kinase C. *Planta* 201, 245–251. doi: 10.1007/s004250050063

Lee, Y., Ahn, C., Han, J., Choi, H., Kim, J., Yim, J., et al. (2003). The nuclear RNase III Drosha initiates microRNA processing. *Nature* 425, 415–419. doi: 10.1038/nature01957

Li, J.-J., and Xie, D. (2015). RACK1, a versatile hub in cancer. *Oncogene* 34, 1890–1898. doi: 10.1038/onc.2014.127

Lieberherr, D., Thao, N. P., Nakashima, A., Umemura, K., Kawasaki, T., and Shimamoto, K. (2005). A sphingolipid elicitor-inducible mitogen-activated protein kinase is regulated by the small GTPase OsRac1 and heterotrimeric G-protein in rice. *Plant Physiol.* 138, 1644–1652. doi: 10.1104/pp.104.057414

López-Bergami, P., Habelhah, H., Bhoumik, A., Zhang, W., Wang, L. H., and Ronai, Z. E. (2005). Receptor for RACK1 mediates activation of JNK by protein kinase C. *Mol. Cell* 19, 309–320. doi: 10.1016/j.molcel.2005.06.025

Lu, C., and Fedoroff, N. (2000). A mutation in the *Arabidopsis HYL1* gene encoding a dsRNA binding protein affects responses to abscisic acid, auxin, and cytokinin. *Plant Cell* 12, 2351–2366. doi: 10.1105/tpc.12.12.2351

McCahill, A., Warwicker, J., Bolger, G. B., Houslay, M. D., and Yarwood, S. J. (2002). The RACK1 scaffold protein: a dynamic cog in cell response mechanisms. *Mol. Pharmacol.* 62, 1261–1273. doi: 10.1124/mol.62.6.1261

McKhann, H. I., Frugier, F., Petrovics, G., de la Peña, T. C., Jurkevitch, E., Brown, S., et al. (1997). Cloning of a WD-repeat-containing gene from alfalfa (*Medicago sativa*): a role in hormone-mediated cell division? *Plant Mol. Biol.* 34, 771–780.

Melamed, D., Bar-Ziv, L., Truzman, Y., and Arava, Y. (2010). Asc1 supports cell-wall integrity near bud sites by a Pkc1 independent mechanism. *PLoS ONE* 5:e11389. doi: 10.1371/journal.pone.0011389

Mochly-Rosen, D., Khaner, H., and Lopez, J. (1991a). Identification of intracellular receptor proteins for activated protein kinase C. *Proc. Natl. Acad. Sci. U.S.A.* 88, 3997–4000. doi: 10.1073/pnas.88.9.3997

Mochly-Rosen, D., Khaner, H., Lopez, J., and Smith, B. L. (1991b). Intracellular receptors for activated protein kinase C. Identification of a binding site for the enzyme. *J. Biol. Chem.* 266, 14866–14868.

Nakashima, A., Chen, L., Thao, N. P., Fujiwara, M., Wong, H. L., Kuwano, M., et al. (2008). RACK1 functions in rice innate immunity by interacting with the Rac1 immune complex. *Plant Cell* 20, 2265–2279. doi: 10.1105/tpc.107.054395

Nielsen, N. C., Beilinson, V., Bassüner, R., and Reverdatto, S. (2001). A G_β-like protein from soybean. *Physiol. Plant.* 111, 75–82. doi: 10.1034/j.1399-3054.2001.1110110.x

Nilsson, J., Sengupta, J., Frank, J., and Nissen, P. (2004). Regulation of eukaryotic translation by the RACK1 protein: a platform for signalling molecules on the ribosome. *EMBO Rep.* 5, 1137–1141. doi: 10.1038/sj.embor.7400291

Núñez, A., Franco, A., Madrid, M., Soto, T., Vicente, J., Gacto, M., et al. (2009). Role for RACK1 orthologue Cpc2 in the modulation of stress response in fission yeast. *Mol. Biol. Cell.* 20, 3996–4009. doi: 10.1091/mbc.E09-05-0388

Nunez, Y. O., Truitt, J. M., Gorini, G., Ponomareva, O. N., Blednov, Y. A., Harris, R. A., et al. (2013). Positively correlated miRNA-mRNA regulatory networks in mouse frontal cortex during early stages of alcohol dependence. *BMC Genomics.* 14:725. doi: 10.1186/1471-2164-14-725

Olejnik, K., Bucholc, M., Anielska-Mazur, A., Lipko, A., Kujawa, M., Modzelan, M., et al. (2011). *Arabidopsis thaliana* Nudix hydrolase AtNUDT7 forms complexes with the regulatory RACK1A protein and Ggamma subunits of the signal transducing heterotrimeric G protein. *Acta Biochim. Pol.* 58, 609–616. Available online at: http://www.actabp.pl/pdf/4_2011/609.pdf

Omosigho, N. N., Swaminathan, K., Plomann, M., Müller-Taubenberger, A., Noegel, A. A., and Riyahi, T. Y. (2014). The *Dictyostelium discoideum* RACK1 orthologue has roles in growth and development. *Cell Commun. Signal.* 12, 37. doi: 10.1186/1478-811X-12-37

Otsuka, M., Takata, A., Yoshikawa, T., Kojima, K., Kishikawa, T., Shibata, C., et al. (2011). Receptor for activated protein kinase C: requirement for efficient microRNA function and reduced expression in hepatocellular carcinoma. *PLoS ONE* 6:e24359. doi: 10.1371/journal.pone.0024359

Rabl, J., Leibundgut, M., Ataide, S. F., Haag, A., and Ban, N. (2011). Crystal structure of the eukaryotic 40S ribosomal subunit in complex with initiation factor 1. *Science* 331, 730–736. doi: 10.1126/science.1198308

Regmi, S., Rothberg, K. G., Hubbard, J. G., and Ruben, L. (2008). The RACK1 signal anchor protein from *Trypanosoma brucei* associates with eukaryotic elongation factor 1A: a role for translational control in cytokinesis. *Mol. Microbiol.* 70, 724–745. doi: 10.1111/j.1365-2958.2008.06443.x

Robles, M. S., Boyault, C., Knutti, D., Padmanabhan, K., and Weitz, C. J. (2010). Identification of RACK1 and protein kinase Cα as integral components of the mammalian circadian clock. *Science* 327, 463–466. doi: 10.1126/science.1180067

Ron, D., Chen, C. H., Caldwell, J., Jamieson, L., Orr, E., and Mochly-Rosen, D. (1994). Cloning of an intracellular receptor for protein kinase C: a homolog of the beta subunit of G proteins. *Proc. Natl. Acad. Sci. U.S.A.* 91, 839–843. doi: 10.1073/pnas.91.3.839

Ruiz Carrillo, D., Chandrasekaran, R., Nilsson, M., Cornvik, T., Liew, C. W., Tan, S. M., et al. (2012). Structure of human Rack1 protein at a resolution of 2.45Å. *Acta Crystallogr. Sect. F Struct. Biol. Cryst. Commun.* 68, 867–872. doi: 10.1107/S1744309112027480

Sengupta, J., Nilsson, J., Gursky, R., Spahn, C. M., Nissen, P., and Frank, J. (2004). Identification of the versatile scaffold protein RACK1 on the eukaryotic ribosome by cryo-EM. *Nat. Struct. Mol. Biol.* 11, 957–962. doi: 10.1038/nsmb822

Sharma, M. R., Wilson, D. N., Datta, P. P., Barat, C., Schluenzen, F., Fucini, P., et al. (2007). Cryo-EM study of the spinach chloroplast ribosome reveals the structural and functional roles of plastid-specific ribosomal proteins. *Proc. Natl. Acad. Sci. U.S.A.* 104, 19315–19320. doi: 10.1073/pnas.0709856104

Shor, B., Calaycay, J., Rushbrook, J., and McLeod, M. (2003). Cpc2/RACK1 is a ribosome-associated protein that promotes efficient translation in *Schizosaccharomyces pombe*. *J. Biol. Chem.* 278, 49119–49128. doi: 10.1074/jbc.M303968200

Speth, C., and Laubinger, S. (2014). RACK1 and the microRNA pathway. Is it déjà-vu all over again? *Plant Signal. Behav.* 9:e27909. doi: 10.4161/psb.27909

Speth, C., Willing, E. M., Rausch, S., Schneeberger, K., and Laubinger, S. (2013). RACK1 scaffold proteins influence miRNA abundance in *Arabidopsis*. *Plant J.* 76, 433–445. doi: 10.1111/tpj.12308

Thao, N. P., Chen, L., Nakashima, A., Hara, S., Umemura, K., Takahashi, A., et al. (2007). RAR1 and HSP90 form a complex with Rac/Rop GTPase and function in innate-immune responses in rice. *Plant Cell* 19, 4035–4045. doi: 10.1105/tpc.107.055517

Ullah, H., Scappini, E. L., Moon, A. F., Williams, L. V., Armstrong, D. L., and Pedersen, L. C. (2008). Structure of a signal transduction regulator, RACK1, from *Arabidopsis thaliana*. *Protein Sci.* 17, 1771–1780. doi: 10.1110/ps.035121.108

Ullah, H., Temple, B., Alonso, J. M., Ecker, J. R., and Jones, A. M. (2006). "RACK1 mediates multiple hormone responsiveness and developmental processes in *Arabidopsis*," in *Proceedings of the Second International Symposium on Plant Neurobiology* (Beijing).

Urano, D., Chen, J.-G., Botella, J. R., and Jones, A. M. (2013). Heterotrimeric G protein signalling in the plant kingdom. *Open Biol.* 3:120186. doi: 10.1098/rsob.120186

Urano, D., Czarnecki, O., Wang, X., Jones, A. M., and Chen, J.-G. (2015). *Arabidopsis* RACK1 Phosphorylation by WNK8 Kinase. *Plant Physiol.* 167, 507–516. doi: 10.1104/pp.114.247460

van Nocker, S., and Ludwig, P. (2003). The WD-repeat protein superfamily in *Arabidopsis*: conservation and divergence in structure and function. *BMC Genomics* 4:50. doi: 10.1186/1471-2164-4-50

Vomastek, T., Iwanicki, M. P., Schaeffer, H. J., Tarcsafalvi, A., Parsons, J. T., and Weber, M. J. (2007). RACK1 targets the extracellular signal-regulated kinase/mitogen-activated protein kinase pathway to link integrin engagement with focal adhesion disassembly and cell motility. *Mol. Cell. Biol.* 27, 8296–8305. doi: 10.1128/MCB.00598-07

Wamaitha, M. J., Yamamoto, R., Wong, H. L., Kawasaki, T., Kawano, Y., and Shimamoto, K. (2012). OsRap2.6 transcription factor contributes to rice innate immunity through its interaction with Receptor for Activated Kinase-C 1 (RACK1). *Rice* 5:35. doi: 10.1186/1939-8433-5-35

Wang, B., Yu, J., Zhu, D., Chang, Y., and Zhao, Q. (2014). Maize ZmRACK1 Is Involved in the plant response to fungal phytopathogens. *Int. J. Mol. Sci.* 15, 9343–9359. doi: 10.3390/ijms15069343

Witzel, F., Maddison, L., and Blüthgen, N. (2012). How scaffolds shape MAPK signaling: what we know and opportunities for systems approaches. *Front. Physiol.* 3:475. doi: 10.3389/fphys.2012.00475

Yatime, L., Hein, K. L., Nilsson, J., and Nissen, P. (2011). Structure of the RACK1 dimer from *Saccharomyces cerevisiae*. *J. Mol. Biol.* 411, 486–498. doi: 10.1016/j.jmb.2011.06.017

Yu, Y., Ji, H., Doudna, J. A., and Leary, J. A. (2005). Mass spectrometric analysis of the human 40S ribosomal subunit: native and HCV IRES-bound complexes. *Protein Sci.* 14, 1438–1446. doi: 10.1110/ps.041293005

Zhang, D., Chen, L., Li, D., Lv, B., Chen, Y., Chen, J., et al. (2014). *OsRACK1* Is involved in abscisic acid and H_2O_2-mediated signaling to regulate seed germination in rice (*Oryza sativa*, L.). *PLoS ONE* 9:e97120. doi: 10.1371/journal.pone.0097120

Zhang, D., Chen, L., Lv, B., and Liang, J. (2013). The scaffolding protein RACK1: a platform for diverse functions in the plant kingdom. *J. Plant Biol. Soil Health* 1, 7. doi: 10.13188/2331-8996.1000003

Zhang, W., Jeon, B. W., and Assmann, S. A. (2011). Heterotrimeric G-protein regulation of ROS signalling and calcium currents in *Arabidopsis* guard cells. *J. Exp. Bot.* 62, 2371–2379. doi: 10.1093/jxb/erq424

Ziegler, K., Kurz, C. L., Cypowyj, S., Couillault, C., Pophillat, M., Pujol, N., et al. (2009). Antifungal innate immunity in *C. elegans*: PKCδ links G protein signaling and a conserved p38 MAPK cascade. *Cell Host Microbe* 5, 341–352. doi: 10.1016/j.chom.2009.03.006

Conflict of Interest Statement: The authors declare that the research was conducted in the absence of any commercial or financial relationships that could be construed as a potential conflict of interest.

11

Dynamic epigenetic states of maize centromeres

*Yalin Liu[1,2], Handong Su[1,2], Jing Zhang[1], Yang Liu[1,2], Fangpu Han[1] and James A. Birchler[3]**

[1] State Key Laboratory of Plant Cell and Chromosome Engineering, Institute of Genetics and Developmental Biology, Chinese Academy of Sciences, Beijing, China, [2] University of Chinese Academy of Sciences, Beijing, China, [3] Division of Biological Sciences, University of Missouri at Columbia, Columbia, MO, USA

The centromere is a specialized chromosomal region identified as the major constriction, upon which the kinetochore complex is formed, ensuring accurate chromosome orientation and segregation during cell division. The rapid evolution of centromere DNA sequence and the conserved centromere function are two contradictory aspects of centromere biology. Indeed, the sole presence of genetic sequence is not sufficient for centromere formation. Various dicentric chromosomes with one inactive centromere have been recognized. It has also been found that *de novo* centromere formation is common on fragments in which centromeric DNA sequences are lost. Epigenetic factors play important roles in centromeric chromatin assembly and maintenance. Non-disjunction of the supernumerary B chromosome centromere is independent of centromere function, but centromere pairing during early prophase of meiosis I requires an active centromere. This review discusses recent studies in maize about genetic and epigenetic elements regulating formation and maintenance of centromere chromatin, as well as centromere behavior in meiosis.

Keywords: centromere inactivation, centromere pairing, non-disjunction, *de novo* centromere, neocentromere, epigenetics, maize

Edited by:
Ingo Schubert,
IPK Gatersleben – Leibniz Institute of Plant Genetics and Crop Plant Research, Germany

Reviewed by:
Jiri Macas,
Biology Centre CAS, Czech Republic
Michael Sandmann,
IPK Gatersleben – Leibniz Institute of Plant Genetics and Crop Plant Research, Germany

Correspondence:
James A. Birchler
birchlerj@missouri.edu

INTRODUCTION

The centromeres are the control centers of chromosomes and are essential for correct orientation and segregation in cell division. The kinetochore complex formation requires a functional centromere, so that the spindle can attach to the centromere regions for accurate orientation and segregation of chromosomes (Cleveland et al., 2003; Allshire and Karpen, 2008).

Centromeres can be divided into three types according to their structural organization: the point centromere with a 125 bp single nucleosome in *Saccharomyces cerevisiae*, the regional centromeres with several kilobases to megabases of repeat sequences in most organisms, and the holocentromeres that are spread throughout the chromosome as in *Caenorhabditis elegans* (Allshire and Karpen, 2008) and the genus Luzula (Heckmann et al., 2013). In plants, most centromeres are regional centromeres; the sizes are larger and the repeat sequences are more complicated than other species (Zhang and Dawe, 2012; Feng et al., 2015).

In maize (*Zea mays*), there are mainly two kinds of centromeric repeat sequences, namely a 156 bp tandem repeat CentC and centromeric retrotransposon of maize (CRM; Ananiev et al., 1998; Zhong et al., 2002). During evolution, maize centromere sizes and the arrangement of centromere sequences have experienced insertion, deletion, duplication, and other changes (Wang and Bennetzen, 2012). The B chromosome centromere in maize has an additional specific centromere repeat sequence called the B-repeat (Alfenito and Birchler, 1993) that surrounds and is interspersed with the CentC and CRM arrays (Jin et al., 2005; Lamb et al., 2005). The pericentromere sequence, Cent4, on chromosome 4 in maize has some sequence similarity to the B-repeat sequence (Page et al., 2001). In the second pollen mitosis, the functional B centromere undergoes non-disjunction, leading to both sister-chromatids segregating to the same pole (Carlson, 2007). Several elements encoded by various regions on the B chromosome act on the B centromere to perform non-disjunction (Lin, 1978). Centromere misdivision of the B centromere of the TB-9Sb translocation line in maize, which is a reciprocal translocation between the B chromosome and the short arm of chromosome 9, can produce derivatives with changed centromere sizes and DNA sequences (Kaszás and Birchler, 1998).

The functional centromere associates with a specific histone H3 variant, which is a heritable epigenetic marker of centromere identity. It was first identified in human and called centromere protein (CENP)-A in 1985 (Earnshaw and Rothfield, 1985). In plants, the centromeric H3 variants (CENH3) have also been identified in maize and other plants including *Arabidopsis* (Talbert et al., 2002; Zhong et al., 2002). There are two copies of CENH3 in barley, wheat, *Pisum*, and *Lathyrus*, which may have functions in polyploidy formation and chromosome evolution (Ishii et al., 2015; Neumann et al., 2015; Yuan et al., 2015). Another histone modification marker for functional centromeres in maize is phosphorylation at Thr133 in histone H2A (Dong and Han, 2012). The centromeric sequences CentC and CRM in maize interact with CENH3-nucleosomes (Zhong et al., 2002).

Centromere function is not determined by DNA sequence. As described below, several examples of inactive centromeres have been documented in maize as well as several examples of *de novo* centromere formation over unique sequences. Taken together, this evidence suggests an epigenetic basis of centromere specification in maize. In this review, we will discuss current studies and potential mechanisms in centromere formation, centromere activity maintenance and special centromere behavior in maize.

CENTROMERE INACTIVATION

Dicentric chromosomes can be produced through chromosome translocation. Chromosomes with two active centromeres are usually unstable during the cell cycle. Therefore stable dicentric chromosomes have only one functional centromere with the other one inactive.

In maize, multiple dicentric chromosomes have been produced by chromosome translocation involving A and B centromeres, B and B centromeres as well as A and A centromeres (**Figure 1**). The B-A translocation chromosome B9-Dp9 was produced by the short arm of chromosome 9 (9S) being translocated to the B centromere, on to which the inverted duplication of 9S was recombined (Zheng et al., 1999; Kato et al., 2005). Recombination within B9-Dp9 can generate dicentric chromosomes with two identical B centromeres, and then this dicentric chromosome can undergo a Breakage-Fusion-Bridge

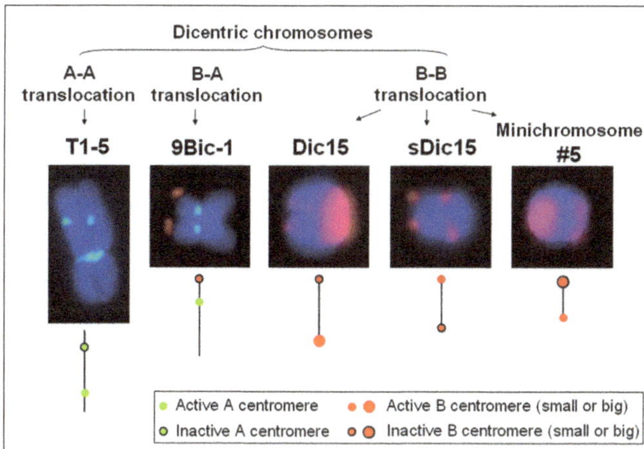

FIGURE 1 | Different kinds of dicentric chromosomes in maize. Dicentric chromosome T1-5 is derived from an A-A translocation with an active centromere 1. 9Bic-1 is derived from a B-A translocation with active centromere 9. Dic15, sDic15, and minichromosome #5 are derived from combining B centromeres using a reverse duplication of 9S on the TB-9Sb translocation as described in the text. CRM is in green; B-repeat is red.

(BFB) cycle. After several rounds of breakage and fusion, different kinds of newly formed dicentric mini-chromosomes are produced (Han et al., 2006; **Figure 2**). The configurations are distinct in these dicentric chromosomes: mini-chromosomes 2, 3, and 13 have two changed B centromere regions with similar sizes with one active; mini-chromosome 10 has multiple regions with centromeric DNA sequences while the functional centromere does not occupy all regions; and in mini-chromosome 5 the smaller centromere is functional rather than the larger one (**Figure 1**; Han et al., 2006). These mini chromosomes show patterns of centromere cohesion and disjunction during meiosis different from the normal ten pairs of A centromeres (Han et al., 2007a). The sister chromatids separate early at meiosis I in plants containing one copy of the mini chromosome, and the cohesion-mediating histone phosphorylation of H3S10 as well as the cohesion protector Shugoshin protein are still located on the separated mini chromosome in anaphase I (Han et al., 2007a). Among the dicentric derivatives of B9-Dp9, the B-A translocation chromosome 9Bic-1 was detected, with the B centromere transferred to the short arm of chromosome 9. The centromere 9 is active while the B centromere is inactive; this B-A translocation is an unexpected product from B-B dicentric formation that initiates a chromosome type BFB cycle (Han et al., 2006). In 9Bic-1, the process of non-disjunction of the B centromere in the presence of a whole B chromosome will cause chromosome breakage because the chromosome nine centromeres separate but the terminal B centromere sisters remain adhered to each other. This releases a fragment with the inactive B centromere. One such broken B centromere containing fragment attached to the short arm of chromosome 7 to generate a new dicentric chromosome 7Bic-1 (Han et al., 2007b; **Figure 2**). Apart from these dicentric chromosomes derived from the B centromere, an A-A translocation between chromosomes 1 and 5 in maize produced a structurally dicentric chromosome

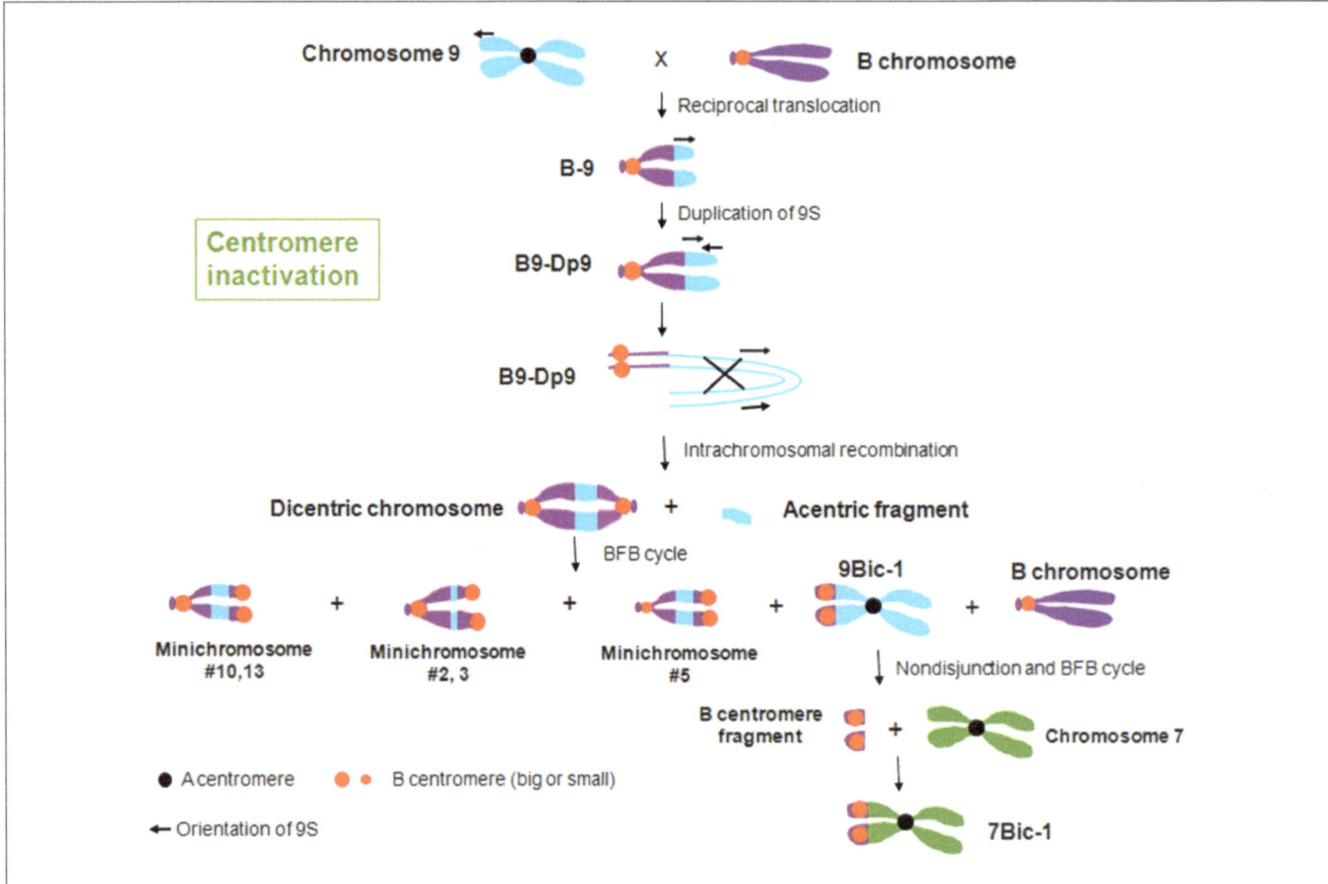

FIGURE 2 | Centromere inactivation through the process of BFB cycle in maize. Reciprocal translocation between chromosome 9 and the B chromosome produced chromosome B-9, with an active B centromere and 9S. Duplication of 9S on B-9 generated chromosome B9-Dp9. The arrows indicate the orientation of the duplicated regions. Intrachromosomal recombination of B9-Dp9 produced dicentric chromosomes with two B centromere regions and a chromosomal fragment without a centromere region. This dicentric chromosome undergoes a BFB cycle, and produced different kinds of new dicentric chromosomes. Most of these dicentric chromosomes are with two B centromere regions, while in 9Bic-1, centromere 9 is active and the B centromere is inactive. The inactive B centromere region on 9Bic-1 remains adhered at the second pollen mitosis, which causes the chromosome to break. The broken piece formed a new A-B translocated chromosome 7Bic-1 with an inactive B centromere.

T1-5 (8041), which possesses an inactive centromere (Gao et al., 2011; **Figure 1**). In wheat, hybridization between wheat and rye generated a translocation line with two centromere regions (Fu et al., 2012). The same holds true for wheat-barley translocations, where one of the two centromeric regions on the translocated chromosomes became inactive (Nasuda et al., 2005).

The inactive centromeres have no function during cell division, and there is no CENH3 loading on the inactive centromeres (Han et al., 2006; Gao et al., 2011). As the original centromere sequences are unchanged in inactive centromeres, there are apparently epigenetic factors regulating CENH3-nucleosome loading only on active centromere regions (Birchler and Han, 2009). Epigenetic factors can also operate when centromere function is regained on a chromosome formerly with only inactive centromere regions. For example, when plants with the translocation chromosome B9-Dp9 (with a big B centromere region) were crossed with another carrying centromere misdivision derivative T3-5(+) (with a small B centromere region), dicentric chromosome Dic15 was produced and transmitted to the next generation with one big active centromere and one inactive small centromere (**Figure 3**). The joining of the two centromeres resulted in inactivation of the smaller one. Then intrachromosomal recombination of Dic15 produced new dicentric chromosomes with two small inactive centromeres or two large active centromeres. On the chromosome with two small centromeres, one of the originally inactive small centromeres appeared to be reactivated (Han et al., 2009; **Figure 3**). Among these new chromosomes with two sites of small centromere regions, sDic15 was found to have *de novo* centromere sequences as determined by CENH3 ChIP-seq. It is therefore possible that the regaining of centromere function on this chromosome is from a *de novo* centromere rather than reactivation of the B centromere sequence. There is no B centromere reference sequence to test whether the formerly inactive centromere sequences were involved in a reactivation process. The analysis showed that a 723 kb genome DNA sequence from the short arm of chromosome 9 was involved in the *de novo* centromere formation (Zhang et al., 2013a; **Figure 3**).

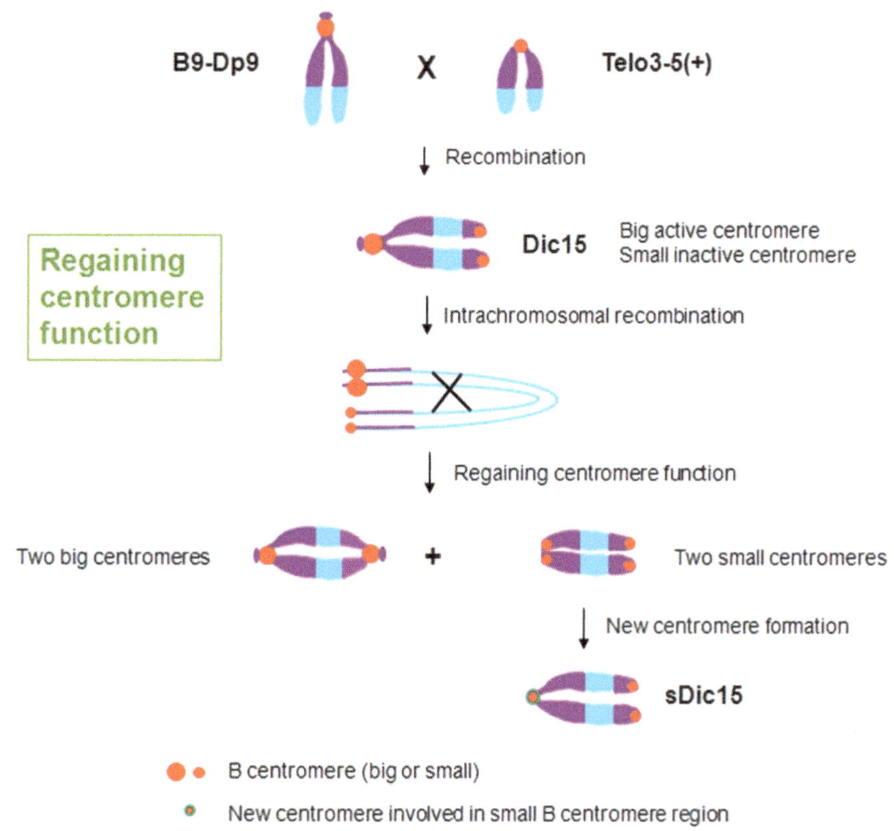

FIGURE 3 | Regaining centromere activity in maize. The hybridization between plants with translocation chromosome B9-Dp9 and those with B centromere misdivision derivative T3-5(+) produced dicentric chromosome Dic15 with one big centromere and one small centromere with the small one inactive. Intrachromosomal recombination of Dic15 can generate new dicentric chromosomes with two big centromeres or two small centromeres. sDic15 is a dicentric chromosome with small centromeres that regained centromere activity via a 723 kb genomic DNA sequence *de novo* centromere.

From study of dicentric chromosomes with one big centromere and one small centromere as determined by FISH for centromeric DNA amounts, it can be realized that centromere size does not determine centromere activity. In Dic15, the small centromere is inactive, while in mini-chromosome 5 derived from intrachromosomal recombination and BFB cycle of B9-Dp9, the small centromere is active (Han et al., 2006, 2009; **Figure 1**). For these two dicentric chromosomes, all centromeres have B-repeat containing regions. Minimal centromere size for function has been studied in maize using the system of centromere misdivison of the B centromere on chromosome B-9 (the chromosome containing the B centromere in TB-9Sb). It was revealed that chromosomes with estimated small centromeres have low transmission rate (Kaszás and Birchler, 1996, 1998). However, it is important to realize that the size of a centromere as determined by DNA amount might not necessarily reflect the centromere chromatin domain size. At present, it is not possible to make generalization about centromere inactivation, which potentially could be stochastic.

In an active centromere, the chromatin maintains an open state for CENH3-nucleosome deposition dynamically in the cell cycle, but in the inactive centromere, the chromatin state is closed. The arrangement order and higher structure of centromeric DNA sequence may provide the basic structure for centromere establishment, while epigenetic elements including histone modifications and chromatin assembly factors are the determinants for CENH3-nucleosome loading (Ekwall, 2007; Allshire and Karpen, 2008; Black and Cleveland, 2011). These epigenetic elements may create on or off states of centromeric chromatin, so that CENH3-nucleosomes can load or fail to load onto the centromere region (Svensson and Ekwall, 2014). In yeast and human, several factors involved in centromere assembly have been reported, including histone deacetylation (Sato et al., 2012), centromere histone H2B monoubiquitination (Sadeghi et al., 2014), CENP-A Ser68 phosphorylation (Yu et al., 2015), and K124 ubiquitylation (Niikura et al., 2015). However, in maize, such factors have not been identified.

CENTROMERE NON-DISJUNCTION

The centromere of the maize B chromosome can undergo non-disjunction at the second pollen mitosis, and the sperm with B chromosomes will preferentially fertilize the egg in the process of double fertilization (Roman, 1947, 1948; Carlson, 2007). There are factors on the B chromosome that are required for non-disjunction of the B centromere, apparently including the B centromere adjacent heterochromatin, a site in proximal euchromatin and another at the very distal tip of the long arm (Roman, 1947; Lin, 1978). In the B-A translocation chromosomes 9Bic-1 and 7Bic-1, the inactive B centromere is transferred to the short arm of chromosome 9 or chromosome 7, respectively. These inactive B centromeres can still perform non-disjunction in the presence of a whole B chromosome; thus non-disjunction does not rely on centromere function. Furthermore, the knob heterochromatin region near the B centromere is deleted in 9Bic-1 and 7Bic-1, so it is likely that the B specific centromere sequence is responsible for non-disjunction given that it is the only major unique repeat unit remaining on these chromosomes (Han et al., 2007b). The inactive B centromere state does not affect non-disjunction, indicating the sequence plays a role in non-disjunction. However in rye, the non-disjunction of the B chromosome relies on B centromere function; the pericentromere cohesion is related to B centromere non-disjunction (Banaei-Moghaddam et al., 2012).

CENTROMERE PAIRING

The function of the centromere in chromosome orientation and segregation during the cell cycle has been studied in detail. In maize, homologous chromosome pairing is initiated from centromeres, and centromere pairing occurs prior to telomere bouquet formation (Zhang et al., 2013b). It was revealed that function rather than DNA sequence is responsible of centromere pairing in maize, suggesting that dynamic chromosome binding factors at active centromeres may take part in the homologous pairing process. In 7Bic-1, the inactive B centromere regions are not paired, when other active centromeres are paired completely at the leptotene stage (**Figure 4**). Compared to the chromosome arm regions and pericentromere sites, functional centromere regions are the earliest recognition points for homologous chromosomes. The synaptonemal complex (SC) and sister-chromatid cohesion are required for centromere pairing, suggesting that they are involved in homologous centromere recognition (Zhang et al., 2013b; **Figure 5**).

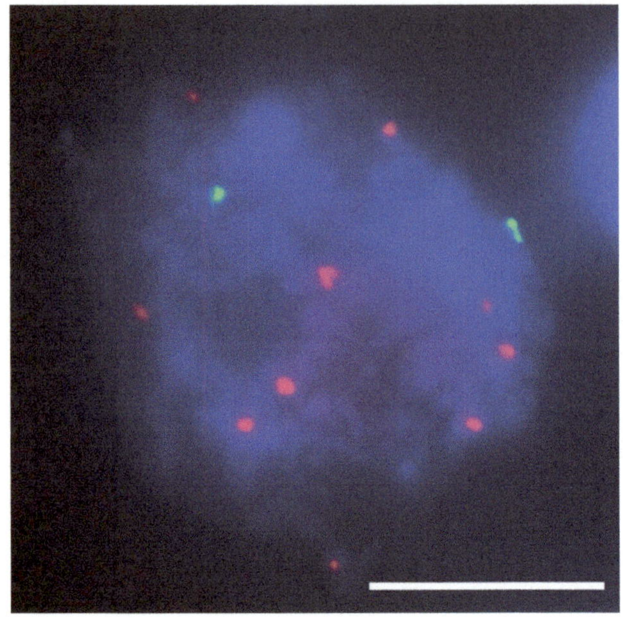

FIGURE 4 | Inactive centromeres (7Bic-1 in green) can not pair at the leptotene stage, when the ten pairs of functional centromeres pair completely. Immuno-FISH using an antibody against centromeric histone CENH3 in red and B repeat probe in green; chromosomes are counterstained with DAPI in blue. The inactive B centromeres are separated but the ten active A centromeres are all paired. Bar = 10 μm.

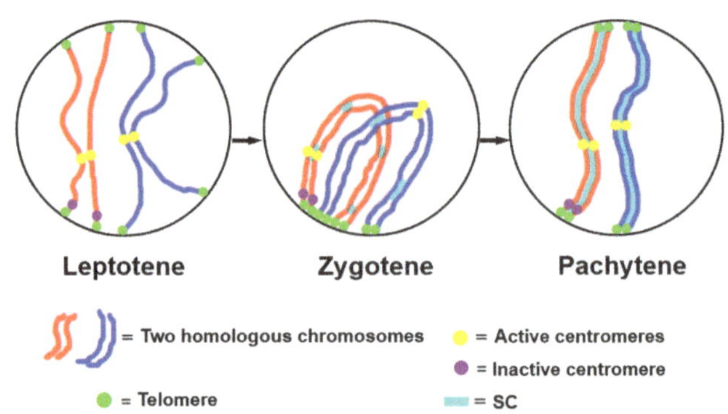

FIGURE 5 | A model for centromere pairing in meiotic early prophase I in maize. We suggest a model of homologous chromosome initiation in maize in which homologous chromosomes find each other at the leptotene stage and are recognized at centromeres via a centromere function dependent mechanism. The inactive centromeres do not participate in this pairing. Then chromosome pairing is further facilitated by telomere bouquet formation and synaptonemal complex (SC)-dependent pathways.

DE NOVO CENTROMERE FORMATION

Inactive centromeres show that DNA sequence alone can not determine centromere establishment. Furthermore, a specific DNA sequence is not always required for centromere assembly. Additional evidence for epigenetic effects on centromere specification comes from the recognition that *de novo* centromeres can be formed on acentric chromosomal fragments without canonical centromeric repeat sequences produced by chromosome breakage. It has been proposed that a tug of war between typically larger endogenous centromeres and smaller *de novo* ones inactivate the latter (Liu et al., 2015), clearing it from the arm in analogous fashion to the generation of Dic15 described above when a small centromere was placed on a dicentric with a normal sized endogenous centromere and became inactive (Han et al., 2009).

De novo centromeres have been found in several plant species. In a wheat background, a stable 7HS telosome with no detectable barley or wheat centromeric sequences was generated (Nasuda et al., 2005). In maize, several *de novo* centromeres have been found. On the chromosomal fragment Duplication 3a (Dp3a), a *de novo* centromere appeared with a 350 kb CENH3 binding region from the long arm of chromosome 3 (Fu et al., 2013;

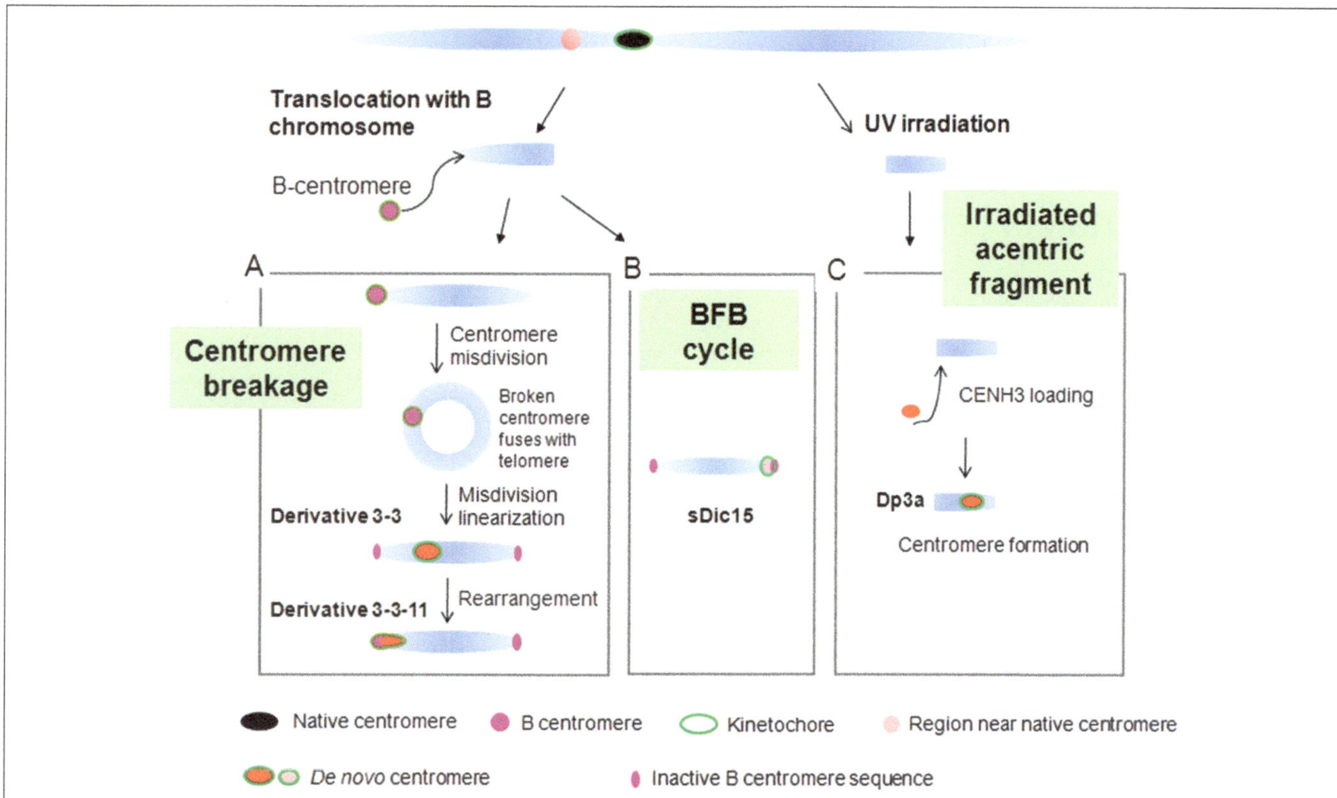

FIGURE 6 | *De novo* centromere formation in maize. (A) *De novo* centromere generated by centromere misdivision. B centromere misdivision on the B-9 chromosome causes a broken centromere that fuses with the telomere region of the same chromosome to produce a ring chromosome. The ring chromosome broke and became linear and gained a 288 kb *de novo* centromere in derivative 3-3. In the derivative of 3-3, this *de novo* centromere was inactivated and another 200 kb *de novo* centromere arose. (B) sDic15 has a *de novo* centromere derived from a 723 kb sequence. (C) Irradiated acentric chromosomal fragments can acquire a *de novo* centromere by CENH3 seeding at that position. Chromosome fragment Dp3a has a 350 kb *de novo* centromere from a sequence on the long arm of chromosome 3.

Figure 6). The dicentric chromosome sDic15, mentioned above, contains 723 kb of genomic DNA sequence from the short arm of chromosome 9 in the active centromere, with a similar DNA methylation level and DNA composition to the native centromere regions (Zhang et al., 2013a; **Figure 6**). By the process of the B centromere misdivision in maize translocation line TB-9Sb (Kaszás and Birchler, 1998), chromosome derivative 3-3 was produced with a 288 kb *de novo* centromere derived from the distal region of the short arm of chromosome 9. In subsequent derivatives of 3-3, the *de novo* centromere of 3-3 was inactivated with another *de novo* centromere formed in 3-3-11 with 200 kb sequence on the short arm of chromosome 9. These *de novo* centromeres are derived from different regions and occupy different sequences (Liu et al., 2015; **Figure 6**). Two isolates of maize chromosome 3 contained neocentromeres near the original centromere 3 when normal maize chromosomes were transfered to an oat background (Wang et al., 2014).

The above-mentioned cases indicate that *de novo* centromere formation is common in maize. The large native centromere appears capable of suppressing potential *de novo* centromeres forming on chromosome arms. Potential *de novo* centromeres be established only on fragments without native centromeres. Any process that produces chromosomal fragments can potentially promote *de novo* centromere formation on the otherwise acentric chromosome fragments.

The sites for neocentromere formation are not related and the mechanism of centromere formation at ectopic regions is still a mystery (Scott and Sullivan, 2014). In maize, *de novo* centromeres derived from the sequences near active centromere regions or on the distal regions of chromosomal arms have been found (Fu et al., 2013; Zhang et al., 2013a). The sizes of *de novo* centromeres vary as noted above, but the smallest, which is only 200 kb, can transmit stably in meiosis (Liu et al., 2015). Maize centromeres sizes were dramatically expanded and adopt a uniform size in the genetic background of oat, which have larger centromere chromatin domains than maize (Wang et al., 2014).

For many *de novo* centromeres in various species, no common DNA features or DNA motifs for centromere deposition have been found (Birchler et al., 2011). In sDic15, there are two DNA motifs enriched in the *de novo* centromere that are also in native centromeres (Zhang et al., 2013a), but in other *de novo* centromeres, such motifs are not recognized. After the native centromere is deleted on a chromosomal fragment, which kinds of factors are involved in choosing a specific region for CENH3 seeding and centromere establishment are still not known. Most acentric fragments are lost during cell division because there is no CENH3 loading.

DNA METHYLATION IN CENTROMERIC REGION

DNA methylation level is important for centromere identity and function through regulating the centromeric chromatin state (Choo, 2001; Schueler and Sullivan, 2006). The DNA methylation level has been determined in *de novo*, active and inactive centromeres in maize. The CENH3-nucleosome binding DNA sequences in native centromeres of maize is hypomethylated compared to the DNA sequences associated with the flanking pericentromere, as is the case in *Arabidopsis thaliana* (Zhang et al., 2008). For the B chromosome centromere, hypomethylated DNA also exists in the active B centromere, and the inactive B centromere has hypermethylated DNA (Koo et al., 2011). Core centromere chromatin is not heterochromatin; to some degree it has a loose structure for CENH3-nucleosome deposition. The hypomethylated DNA in centromere regions may provide a relaxed chromatin environment to allow centromeric transcription and also serve as a marker recognized by other factors for centromere assembly.

As centromere formation and maintenance of centromeric chromatin are regulated by epigenetic factors, the methylation modification on centromere DNA may take part in centromere chromatin assembly (Gopalakrishnan et al., 2009). The DNA methylation in centromere regions may influence the transcription state, and transcription in a centromere may play a role in CENH3-nucleosome assembly (Allshire and Karpen, 2008).

TRANSCRIPTION IN CENTROMERE REGIONS

Centromere transcripts from the repeat sequences have been found in many species. These transcripts are essential for regulating CENH3 nucleosomes loading in the centromere region (Chan and Wong, 2012) and promoting kinetochore complex assembly (Scott, 2013). In maize, transcripts from centromere repeat sequences CentC and CRM2 have been detected (Topp et al., 2004; Du et al., 2010) and the CentC RNA can interact with CENP-C (Du et al., 2010). In the *de novo* centromere sDic15, several genes inside the 723 kb centromere region are transcribed (Zhang et al., 2013a). In a neocentromere of human, the transcript of retrotransposon LINE 1 was required for stable transmission of the neocentromere during the cell cycle (Chueh et al., 2009). As *de novo* centromeres have no traditional centromere repeat sequences, the transcripts in *de novo* centromeres should be different from the native centromeres.

The function of centromere transcription in CENH3-nucleosome assembly has been studied. In *Drosophila melanogaster*, non-coding RNA of satellite III in the centromere region is required for deposition of CENP-A nucleosomes and CENP-C (Rosic et al., 2014). The work using an ectopic centromere system in *Drosophila* showed that centromere transcription is required for CENP-A deposition, and the production of these transcripts is dependent on CENP-A chaperone recruiting chromosome assembly factor and RNA polymerase II to the centromere region (Chen et al., 2015). The transcripts of centromere satellite in human interacts with CENP-A and its chaperone Holliday junction recognition protein (HJURP; (Quenet and Dalal, 2014).

The function of centromere transcripts in CENH3-nucleosome loading can be divided into two parts. Firstly, the chromosome assembly factor as well as histone chaperone and RNA polymerase II are recruited to the centromere region to produce an open chromatin state for CENH3-nucleosome loading. Then the transcribed RNA can interact with CENH3 and the chaperone as well as kinetochore proteins to promote CENH3 assembly and maintain the centromeric chromatin. In dicentric chromosomes, the transcripts from an active centromere cannot work in the inactive centromere region, and may result from the relatively closed chromatin state in the inactive centromere.

SUMMARY

DNA sequence and histone modification are involved in formation and maintenance of centromere activity. Centromere DNA is the carrier for centromere chromatin. *De novo* centromere formation is common in maize. Centromere inactivation shows that apart from histone variants and histone modifications, other elements including histone chaperones and chromatin assembly factors as well as transcription factors are working in regulation of centromere chromatin. The transcription products may be combined in the centromere region controlling CENH3-nucleosome loading.

Centromere DNA can play roles independent of centromere function in non-disjunction of the B centromere in maize, in that specific products emanating from a whole B chromosome can act in trans on the inactive B centromere to induce non-disjunction (Han et al., 2007b). When the higher structure of centromeric chromatin is studied in more depth, the nature of non-repeat DNA sequence in *de novo* centromeres and the highly repetitive DNA sequences in native centromeres as well as how the centromere DNA sequence is involved in centromere formation and centromere non-disjunction will be understood better.

An active centromere is required for homologous chromosome pairing in maize (Zhang et al., 2013b). The key factors controlling homologous chromosome pairing may locate in centromere regions in early meiosis. Thus, the function of the centromere is reflected not only in chromosome orientation and segregation, but also in chromosome recognition and genome stability.

KEY CONCEPTS OF CENTROMERE FUNCTION IN MAIZE

Concept One: Centromere

The centromere is the major constriction on a metaphase chromosome. In most species, it is composed of repeat sequences. Accurate assembly of the kinetochore complex is dependent on functional centromeric chromatin, and the correct spindle attachment on the centromere region ensures accurate chromosome orientation and segregation.

Concept Two: Dicentric Chromosomes

The chromosomes with two sites of centromere sequences are called dicentric chromosomes. In order to remain stable, one centromere must be active while the other centromere must become inactive. Otherwise chromosomes with two functional centromeres would be unstable during cell division if the two centromeres separate in opposite directions.

Concept Three: Centromere Inactivation

The process of centromere state change from active to inactive is called centromere inactivation. The inactive centromere loses CENH3, while the centromeric DNA sequence is still present. There is no spindle contact to the inactive centromere during cell division, and the dicentric chromosome can transmit stably.

Concept Four: Centromere Non-disjunction

At the second pollen mitosis, the maize B sister centromeres are held together so that they are transferred to the same pole at anaphase. Such non-disjoined chromatids preferentially transmit to the egg cell. In the presence of the whole B chromosome, an inactive B centromere region can perform non-disjunction.

Concept Five: Centromere Pairing

In maize, during early meiotic prophase, centromere associations occur between homologous chromosomes before telomere bouquet formation and chromosome arm pairing. Homologous chromosomes recognize each other through centromere regions. An active centromere is necessary to participate in this process.

Concept Six: *De Novo* Centromere

De novo centromeres are newly formed centromeres in otherwise non-centromeric regions on chromosomes without traditional centromeric repeat sequences. *De novo* centromeres form on chromosomal fragments generated through different processes, such as chromosome rearrangements from centromere misdivision, BFB cycle and so on, in which the native centromere has been deleted or inactivated.

ACKNOWLEDGMENTS

Research in the Birchler lab was funded by NSF grant IOS1444514 and NSFC grant (0.31320103912) for FH and JB.

REFERENCES

Alfenito, M. R., and Birchler, J. A. (1993). Molecular characterization of a maize B chromosome centric sequence. *Genetics* 135, 589–597.

Allshire, R. C., and Karpen, G. H. (2008). Epigenetic regulation of centromeric chromatin: old dogs, new tricks? *Nat. Rev. Genet.* 9, 923–937. doi: 10.1038/nrg2466

Ananiev, E. V., Phillips, R. L., and Rines, H. W. (1998). Chromosome-specific molecular organization of maize (*Zea mays* L.) centromeric regions. *Proc. Natl. Acad. Sci. U.S.A.* 95, 13073–13078. doi: 10.1073/pnas.95.22.13073

Banaei-Moghaddam, A. M., Schubert, V., Kumke, K., Weibeta, O., Klemme, S., Nagaki, K., et al. (2012). Nondisjunction in favor of a chromosome: the mechanism of rye B chromosome drive during pollen mitosis. *Plant Cell* 24, 4124–4134. doi: 10.1105/tpc.112.105270

Birchler, J. A., Gao, Z., Sharma, A., Presting, G. G., and Han, F. (2011). Epigenetic aspects of centromere function in plants. *Curr. Opin. Plant Biol.* 14, 217–222. doi: 10.1016/j.pbi.2011.02.004

Birchler, J. A., and Han, F. (2009). Maize centromeres: structure, function, epigenetics. *Annu. Rev. Genet.* 43, 287–303. doi: 10.1146/annurev-genet-102108-134834

Black, B. E., and Cleveland, D. W. (2011). Epigenetic centromere propagation and the nature of CENP-a nucleosomes. *Cell* 144, 471–479. doi: 10.1016/j.cell.2011.02.002

Carlson, W. R. (2007). Locating a site on the maize B chromosome that controls preferential fertilization. *Genome* 50, 578–587. doi: 10.1139/g07-035

Chan, F. L., and Wong, L. H. (2012). Transcription in the maintenance of centromere chromatin identity. *Nucleic Acids Res.* 40, 11178–11188. doi: 10.1093/nar/gks921

Chen, C.-C., Bowers, S., Lipinszki, Z., Palladino, J., Trusiak, S., Bettini, E., et al. (2015). Establishment of centromeric chromatin by the CENP-A assembly factor CAL1 requires FACT-mediated transcription. *Dev. Cell* 34, 73–84. doi: 10.1016/j.devcel.2015.05.012

Choo, K. H. A. (2001). Domain organization at the centromere and neocentromere. *Dev. Cell* 1, 165–177. doi: 10.1016/S1534-5807(01)00028-4

Chueh, A. C., Northrop, E. L., Brettingham-Moore, K. H., Choo, K. H., and Wong, L. H. (2009). LINE retrotransposon RNA is an essential structural and functional epigenetic component of a core neocentromeric chromatin. *PLoS Genet.* 5:e1000354. doi: 10.1371/journal.pgen.1000354

Cleveland, D. W., Mao, Y., and Sullivan, K. F. (2003). Centromeres and kinetochores: from epigenetics to mitotic checkpoint signaling. *Cell* 112, 407–421. doi: 10.1016/S0092-8674(03)00115-6

Dong, Q., and Han, F. (2012). Phosphorylation of histone H2A is associated with centromere function and maintenance in meiosis. *Plant J.* 71, 800–809. doi: 10.1111/j.1365-313X.2012.05029.x

Du, Y., Topp, C. N., and Dawe, R. K. (2010). DNA binding of centromere protein C (CENPC) is stabilized by single-stranded RNA. *PLoS Genet.* 6:e1000835. doi: 10.1371/journal.pgen.1000835

Earnshaw, W. C., and Rothfield, N. (1985). Identification of a family of human centromere proteins using autoimmune sera from patients with scleroderma. *Chromosoma* 91, 313–321. doi: 10.1007/bf00328227

Ekwall, K. (2007). Epigenetic control of centromere behavior. *Annu. Rev. Genet.* 41, 63–81. doi: 10.1146/annurev.genet.41.110306.130127

Feng, C., Liu, Y., Su, H., Wang, H., Birchler, J., and Han, F. (2015). Recent advances in plant centromere biology. *Sci. China Life Sci.* 58, 240–245. doi: 10.1007/s11427-015-4818-3

Fu, S., Gao, Z., Birchler, J., and Han, F. (2012). Dicentric chromosome formation and epigenetics of centromere formation in plants. *J Genet. Genomics* 39, 125–130. doi: 10.1016/j.jgg.2012.01.006

Fu, S., Lv, Z., Gao, Z., Wu, H., Pang, J., Zhang, B., et al. (2013). De novo centromere formation on a chromosome fragment in maize. *Proc. Natl. Acad. Sci. U.S.A.* 110, 6033–6036. doi: 10.1073/pnas.1303944110

Gao, Z., Fu, S., Dong, Q., Han, F., and Birchler, J. A. (2011). Inactivation of a centromere during the formation of a translocation in maize. *Chromosome Res.* 19, 755–761. doi: 10.1007/s10577-011-9240-5

Gopalakrishnan, S., Sullivan, B. A., Trazzi, S., Della Valle, G., and Robertson, K. D. (2009). DNMT3B interacts with constitutive centromere protein CENP-C to modulate DNA methylation and the histone code at centromeric regions. *Hum. Mol. Genet.* 18, 3178–3193. doi: 10.1093/hmg/ddp256

Han, F., Gao, Z., and Birchler, J. A. (2009). Reactivation of an inactive centromere reveals epigenetic and structural components for centromere specification in maize. *Plant Cell* 21, 1929–1939. doi: 10.1105/tpc.109.066662

Han, F., Gao, Z., Yu, W., and Birchler, J. A. (2007a). Minichromosome analysis of chromosome pairing, disjunction, and sister chromatid cohesion in maize. *Plant Cell* 19, 3853–3863. doi: 10.1105/tpc.107.055905

Han, F., Lamb, J. C., Yu, W., Gao, Z., and Birchler, J. A. (2007b). Centromere function and nondisjunction are independent components of the maize B chromosome accumulation mechanism. *Plant Cell* 19, 524–533. doi: 10.1105/tpc.106.049577

Han, F., Lamb, J. C., and Birchler, J. A. (2006). High frequency of centromere inactivation resulting in stable dicentric chromosomes of maize. *Proc. Natl. Acad. Sci. U.S.A.* 103, 3238–3243. doi: 10.1073/pnas.0509650103

Heckmann, S., Macas, J., Kumke, K., Fuchs, J., Schubert, V., Ma, L., et al. (2013). The holocentric species Luzula elegans shows interplay between centromere and large-scale genome organization. *Plant J.* 73, 555–565. doi: 10.1111/tpj.12054

Ishii, T., Karimi-Ashtiyani, R., Banaei-Moghaddam, A. M., Schubert, V., Fuchs, J., and Houben, A. (2015). The differential loading of two barley CENH3 variants into distinct centromeric substructures is cell type- and development-specific. *Chromosome Res.* 23, 277–284. doi: 10.1007/s10577-015-9466-8

Jin, W., Lamb, J. C., Vega, J. M., Dawe, R. K., Birchler, J. A., and Jiang, J. (2005). Molecular and functional dissection of the maize B chromosome centromere. *Plant Cell* 17, 1412–1423. doi: 10.1105/tpc.104.030643

Kaszás, E., and Birchler, J. A. (1996). Misdivision analysis of centromere structure in maize. *EMBO J.* 15, 5246–5255.

Kaszás, E., and Birchler, J. A. (1998). Meiotic transmission rates correlate with physical features of rearranged centromeres in maize. *Genetics* 150, 1683–1692.

Kato, A., Zheng, Y. Z., Auger, D. L., Phelps-Durr, T., Bauer, M. J., Lamb, J. C., et al. (2005). Minichromosomes derived from the B chromosome of maize. *Cytogenet. Genome Res.* 109, 156–165. doi: 10.1159/000082395

Koo, D. H., Han, F., Birchler, J. A., and Jiang, J. (2011). Distinct DNA methylation patterns associated with active and inactive centromeres of the maize B chromosome. *Genome Res.* 21, 908–914. doi: 10.1101/gr.116202.110

Lamb, J. C., Kato, A., and Birchler, J. A. (2005). Sequences associated with A chromosome centromeres are present throughout the maize B chromosome. *Chromosoma* 113, 337–349. doi: 10.1007/s00412-004-0319-z

Lin, B. (1978). Regional control of nondisjunction of B-chromosome in maize. *Genetics* 90, 613–627.

Liu, Y., Su, H., Pang, J., Gao, Z., Wang, X. J., Birchler, J. A., et al. (2015). Sequential de novo centromere formation and inactivation on a chromosomal fragment in maize. *Proc. Natl. Acad. Sci. U.S.A.* 112, E1263–E1271. doi: 10.1073/pnas.1418248112

Nasuda, S., Hudakova, S., Schubert, I., Houben, A., and Endo, T. R. (2005). Stable barley chromosomes without centromeric repeats. *Proc. Natl. Acad. Sci. U.S.A.* 102, 9842–9847. doi: 10.1073/pnas.0504235102

Neumann, P., Pavlikova, Z., Koblizkova, A., Fukova, I., Jedlickova, V., Novak, P., et al. (2015). Centromeres off the hook: massive changes in centromere size and structure following duplication of CenH3 gene in fabeae species. *Mol. Biol. Evol.* 32, 1862–1879. doi: 10.1093/molbev/msv070

Niikura, Y., Kitagawa, R., Ogi, H., Abdulle, R., Pagala, V., and Kitagawa, K. (2015). CENP-A K124 ubiquitylation is required for CENP-A deposition at the centromere. *Dev. Cell* 32, 589–603. doi: 10.1016/j.devcel.2015.01.024

Page, B. T., Wanous, M. K., and Birchler, J. A. (2001). Characterization of a maize chromosome 4 centromeric sequence: evidence for an evolutionary relationship with the B chromosome centromere. *Genetics* 159, 291–302.

Quenet, D., and Dalal, Y. (2014). A long non-coding RNA is required for targeting centromeric protein A to the human centromere. *Elife* 3:e03254. doi: 10.7554/eLife.03254

Roman, H. (1947). Mitotic nondisjunction in the case of interchanges involving the B-type chromosomes in maize. *Genetics* 32, 391–409.

Roman, H. (1948). Directed fertilization in maize. *Proc. Natl. Acad. Sci. U.S.A.* 34, 36–42. doi: 10.1073/pnas.34.2.36

Rosic, S., Kohler, F., and Erhardt, S. (2014). Repetitive centromeric satellite RNA is essential for kinetochore formation and cell division. *J. Cell Biol.* 207, 335–349. doi: 10.1083/jcb.201404097

Sadeghi, L., Siggens, L., Svensson, J. P., and Ekwall, K. (2014). Centromeric histone H2B monoubiquitination promotes noncoding transcription and chromatin integrity. *Nat. Struct. Mol. Biol.* 21, 236–243. doi: 10.1038/nsmb.2776

Sato, H., Masuda, F., Takayama, Y., Takahashi, K., and Saitoh, S. (2012). Epigenetic inactivation and subsequent heterochromatinization of a centromere stabilize dicentric chromosomes. *Curr. Biol.* 22, 658–667. doi: 10.1016/j.cub.2012.02.062

Schueler, M. G., and Sullivan, B. A. (2006). Structural and functional dynamics of human centromeric chromatin. *Annu. Rev. Genomics Hum. Genet.* 7, 301–313. doi: 10.1146/annurev.genom.7.080505.115613

Scott, K. C. (2013). Transcription and ncRNAs: at the cent(rome)re of kinetochore assembly and maintenance. *Chromosome Res.* 21, 643–651. doi: 10.1007/s10577-013-9387-3

Scott, K. C., and Sullivan, B. A. (2014). Neocentromeres: a place for everything and everything in its place. *Trends Genet.* 30, 66–74. doi: 10.1016/j.tig.2013.11.003

Svensson, J. P., and Ekwall, K. (2014). Epigenetic activation and inactivation of centromeres. *Epigenomics* 6, 451–454. doi: 10.2217/epi.14.33

Talbert, P. B., Masuelli, R., Tyagi, A. P., Comai, L., and Henikoff, S. (2002). Centromeric localization and adaptive evolution of an *Arabidopsis* histone H3 variant. *Plant Cell* 14, 1053–1066. doi: 10.1105/tpc.010425

Topp, C. N., Zhong, C. X., and Dawe, R. K. (2004). Centromere-encoded RNAs are integral components of the maize kinetochore. *Proc. Natl. Acad. Sci. U.S.A.* 101, 15986–15991. doi: 10.1073/pnas.0407154101

Wang, H., and Bennetzen, J. L. (2012). Centromere retention and loss during the descent of maize from a tetraploid ancestor. *Proc. Natl. Acad. Sci. U.S.A.* 109, 21004–21009. doi: 10.1073/pnas.1218668109

Wang, K., Wu, Y., Zhang, W., Dawe, R. K., and Jiang, J. (2014). Maize centromeres expand and adopt a uniform size in the genetic background of oat. *Genome Res.* 24, 107–116. doi: 10.1101/gr.160887.113

Yu, Z., Zhou, X., Wang, W., Deng, W., Fang, J., Hu, H., et al. (2015). Dynamic phosphorylation of CENP-A at Ser68 orchestrates its cell-cycle-dependent deposition at centromeres. *Dev. Cell* 32, 68–81. doi: 10.1016/j.devcel.2014.11.030

Yuan, J., Guo, X., Hu, J., Lv, Z., and Han, F. (2015). Characterization of two CENH3 genes and their roles in wheat evolution. *New Phytol.* 206, 839–851. doi: 10.1111/nph.13235

Zhang, B., Lv, Z., Pang, J., Liu, Y., Guo, X., Fu, S., et al. (2013a). Formation of a functional maize centromere after loss of centromeric sequences and gain of ectopic sequences. *Plant Cell* 25, 1979–1989. doi: 10.1105/tpc.113.110015

Zhang, J., Pawlowski, W. P., and Han, F. (2013b). Centromere pairing in early meiotic prophase requires active centromeres and precedes installation of the synaptonemal complex in maize. *Plant Cell* 25, 3900–3909. doi: 10.1105/tpc.113.117846

Zhang, H., and Dawe, R. K. (2012). Total centromere size and genome size are strongly correlated in ten grass species. *Chromosome Res.* 20, 403–412. doi: 10.1007/s10577-012-9284-1

Zhang, W., Lee, H. R., Koo, D. H., and Jiang, J. (2008). Epigenetic modification of centromeric chromatin: hypomethylation of DNA sequences in the CENH3-associated chromatin in *Arabidopsis thaliana* and maize. *Plant Cell* 20, 25–34. doi: 10.1105/tpc.107.057083

Zheng, Y. Z., Roseman, R. R., and Carlson, W. R. (1999). Time course study of the chromosome-type breakage-fusion-bridge cycle in maize. *Genetics* 153, 1435–1444.

Zhong, C. X., Marshall, J. B., Topp, C., Mroczek, R., Kato, A., Nagaki, K., et al. (2002). Centromeric retroelements and satellites interact with maize kinetochore protein CENH3. *Plant Cell* 14, 2825–2836. doi: 10.1105/tpc.006106

Conflict of Interest Statement: The authors declare that the research was conducted in the absence of any commercial or financial relationships that could be construed as a potential conflict of interest.

12

Photosystem II Assembly from Scratch

Thilo Rühle[1] *and Dario Leister*[1,2*]

[1] Plant Molecular Biology, Department of Biology, Ludwig-Maximilians-University Munich,, Munich, Germany, [2] Department of Plant and Environmental Sciences, Copenhagen Plant Science Centre, University of Copenhagen, Copenhagen, Denmark

Keywords: PSII assembly, PSII, PSII complex, assembly factor, synthetic bacterium, forward genetic screen, reverse genetic, chloroplast

Construction of a functional Photosystem II (PSII) in cyanobacteria and chloroplasts depends on the action of auxiliary factors, which transiently interact with PSII intermediates during assembly. In addition to a common PSII structure and a conserved set of PSII assembly factors, cyanobacteria, and higher plants have evolved additional, clade-specific assembly factors. Most such factors in cyanobacteria and chloroplasts have been identified by "top-down" approaches (forward and reverse genetics), which involved genetic disruption of individual components in the assembly process and subsequent characterization of the ensuing phenotypic effects on the respective mutant lines/strains. In contrast, a "bottom-up" strategy, based on the engineering of a synthetic bacterium with a plant-type PSII, has the potential to identify all assembly factors sufficient to make a functional plant PSII.

Edited by:
Julian Eaton-Rye,
University of Otago, New Zealand

Reviewed by:
Eva-Mari Aro,
University of Turku, Finland
Johnna Roose,
Louisiana State University, USA

****Correspondence:***
Dario Leister
leister@lmu.de

Photosystem II (PSII) is a water-plastoquinone photo-oxidoreductase, which is found in cyanobacteria and their endosymbiotic descendants, the chloroplasts. Light-driven water splitting and subsequent electron transfer steps are carried out with the assistance of non-proteinaceous cofactors. Thus, the PSII monomer harbors a Mn_4CaO_5 cluster, chloride, bicarbonate, 1-2 hemes, 1 nonheme iron, 35 chlorophyll *a* molecules, 2 pheophytins, 11 β-carotenes, and 2 plastoquinones (Umena et al., 2011), all of which are embedded in a shell made up of at least 20 proteins (Shen, 2015) that determine their correct positioning and relative orientation. Several PSII-associated lipids have been identified in crystal structures and might also be important for functionality (Mizusawa and Wada, 2012; Kansy et al., 2014). The structural core of PSII is conserved between chloroplasts and cyanobacteria (Allen et al., 2011). However, the oxygen-evolving complex in cyanobacteria contains subunits U and V, which are replaced by Q, R, P, and Tn in higher plants (Bricker et al., 2012). Furthermore, in contrast to the soluble, peripherally attached phycobilisomes found in cyanobacteria, green photosynthetic eukaryotes have evolved integrated light-harvesting complexes and lack phycobilisomes (Hohmann-Marriott and Blankenship, 2011; **Figure 1A**).

In accordance with its structural complexity, the assembly of PSII is an elaborate and highly coordinated process, which depends on the action of a network of assembly factors and the fabrication of distinct metastable modules during the course of assembly (see for reviews on this topic: Nixon et al., 2010; Nickelsen and Rengstl, 2013; and papers in this special issue of Frontiers in Plant Science). Several modules common to cyanobacterial and chloroplast PSII assembly processes have been described and are characterized by transient binding of specific assembly factors. An additional level of complexity arises from the fact that PSII is susceptible to photodamage, with subunit D1 being the primary target (Kato and Sakamoto, 2009), as replacement of damaged D1 entails partial disassembly of PSII and reassembly of a functional complex. This repair mechanism features some distinct intermediates, but otherwise involves steps and assembly factors that are shared with the *de novo* assembly pathway (Järvi et al., 2015).

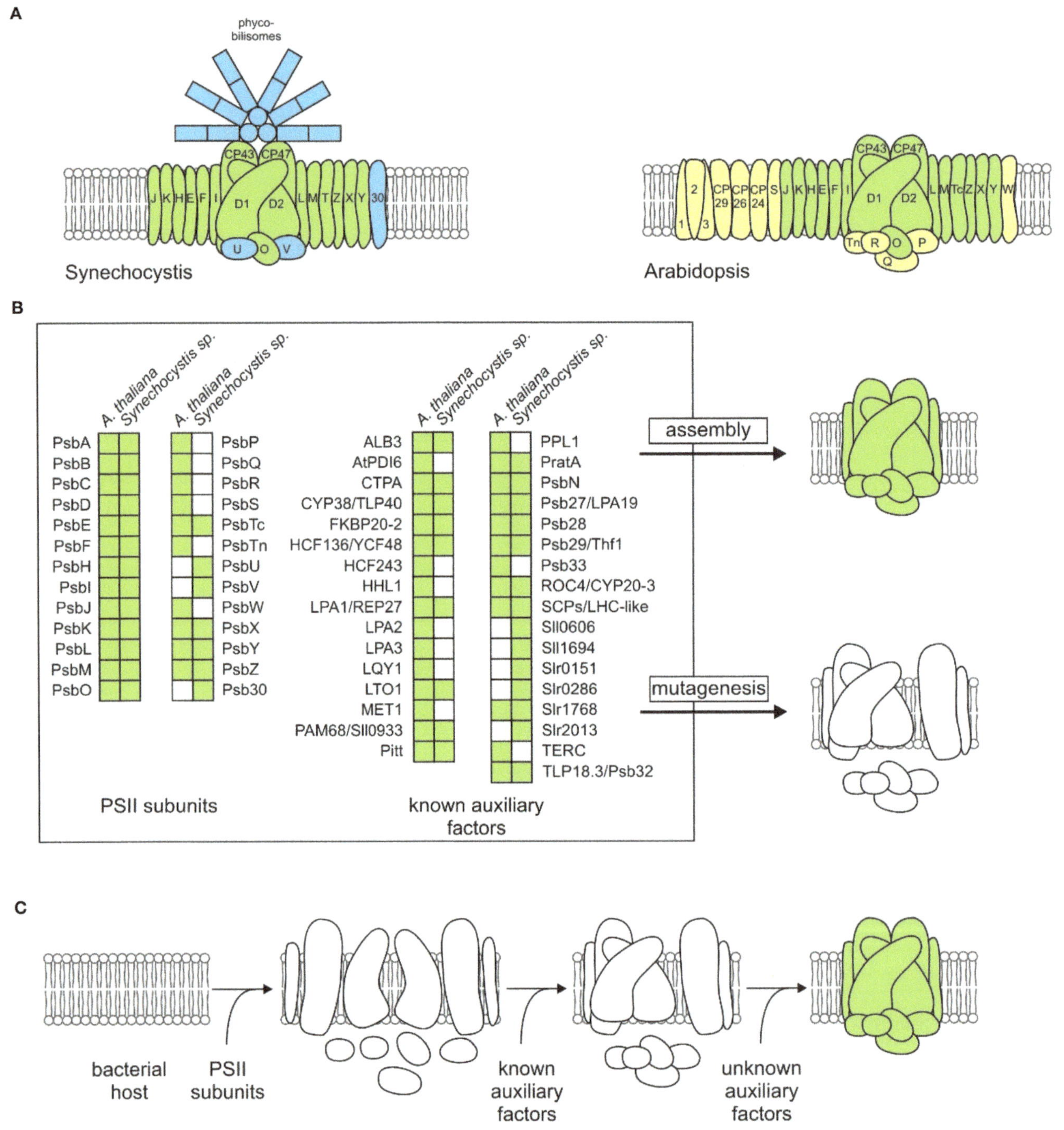

FIGURE 1 | Strategies for identifying auxiliary factors in PSII assembly. (A) Subunit compositions of cyanobacterial (*Synechocystis*) and plant thylakoid (*Arabidopsis*) PSII complexes. Components shown in blue or yellow are only found in *Synechocystis* or *Arabidopsis*, respectively. Light-harvesting proteins Lhcb1, 2 and 3 are labeled with numbers (1, 2, and 3). **(B)** PSII subunits and known PSII auxiliary factors. The list of auxiliary factors has been adapted from Nickelsen and Rengstl (2013) and Järvi et al. (2015), and the recently identified auxiliary factors MET1 (Bhuiyan et al., 2015) and Slr0151 (Yang et al., 2014) have been added. Note that for simplicity, no distinction has been made between factors necessary for assembly, repair or supercomplex formation, and Deg/FtsH proteases are not listed. The classical strategy for study of the assembly process is based on mutagenesis of single components in an intact system. Subsequent analyses focus on the identification of a specific loss of function in a photosynthetic organism. Green and white squares indicate presence and absence of a component listed for each organism, respectively. **(C)** Generation of a synthetic, bacterial host strain carrying a functional plant-type PSII. The design of the synthetic approach largely depends on the choice of the host organism. In the case of a cyanobacterial host, endogenous PSII subunits or auxiliary factors must be deleted or replaced by the plant-type subunits/factors, whereas the presence of essential cofactors (see text), in addition to subunits and auxiliary factors, must be assured for the assembly of a functional, plant-type PSII in a non-photosynthetic bacterium. White and green PSII cartoons in **(B,C)** indicate un- or partially assembled and fully assembled PSII complexes, respectively.

It has become increasingly clear in recent years that the set of assembly factors is largely conserved between cyanobacteria and chloroplasts (Nickelsen and Rengstl, 2013). However, higher plants have extended their inventory during evolution, giving rise to new, plant-specific factors, such as the D1 stabilization factor HCF243 (Zhang et al., 2011) or the repair factors PPL1 and LQY1 (Ishihara et al., 2007; Lu et al., 2011; **Figure 1B**). Furthermore, the consequences of disruption of conserved auxiliary factors sometimes differ between the two systems. For instance, *Arabidopsis* PAM68 and its cyanobacterial counterpart (Armbruster et al., 2010), as well as HCF136 (Meurer et al., 1998) and the cyanobacterial homolog YCF48 (Komenda et al., 2008), participate in the early steps of PSII assembly, but their absence has more severe effects on photosynthesis in plants. Thus, to understand the evolutionary diversification of assembly factors, it is crucial to study chloroplast and cyanobacterial PSII assembly in parallel.

CLASSICAL OR "TOP-DOWN" METHODS OF IDENTIFYING PSII ASSEMBLY FACTORS

Extensive efforts have been made over the past several decades to identify auxiliary factors involved in PSII assembly, and a large number have been found by screening *Synechocystis* sp. PCC 6803 (hereafter *Synechocystis*), *Chlamydomonas reinhardtii*, or *Arabidopsis thaliana* (hereafter *Arabidopsis*) mutant collections for PSII-defective mutants. This type of genetic approach has emerged as a powerful strategy, since single components of an initially intact system can be functionally deleted without directly altering other elements in the assembly process. Subsequent studies then allow the in-depth characterisation of the effects of the loss of a specific function in the assembly process.

Forward Genetic Screens

A classical way to screen for mutants defective in PSII is based on the observation of a high-chlorophyll fluorescence (HCF) phenotype in mutagenized *Arabidopsis* plants, which can be recognized when leaves are exposed to UV light in the dark (Meurer et al., 1996). Several assembly factors, including HCF136 and HCF243 (stabilization of D1 and assembly of the reaction center), LPA1 (D1 integration), LPA2 and LPA3 (integration of CP43) were identified in this manner (Meurer et al., 1998; Peng et al., 2006; Ma et al., 2007; Cai et al., 2010; Zhang et al., 2011). The sensitivity of this screening method for PSII mutants can be further increased by measuring photosynthetic parameters of large mutant collections under varying growth conditions using automatic pulse-amplitude modulation (PAM) of chlorophyll fluorescence (Varotto et al., 2000) or imaging PAM technologies (Ajjawi et al., 2010). This type of approach has led to the identification of several novel factors such as LPA19, PAM68, LQY1, HHL1, and PSB33 (Armbruster et al., 2010; Wei et al., 2010; Lu et al., 2011; Jin et al., 2014; Fristedt et al., 2015). However, forward screens are both labor-intensive and time-consuming, and more than 90% of nuclear genes in *Arabidopsis* have at least one homolog (Armisén et al., 2008). In many cases, duplicated genes code for proteins with redundant functions, and for that reason are inaccessible to classical forward screening techniques.

Reverse Genetic Screens

The increasing accumulation of information on entire genomes, transcriptomes, and proteomes in public databases, together with the availability of indexed libraries of *Arabidopsis* mutant lines with mutations in almost every gene, and the establishment of efficient methods for gene silencing, and genetic engineering by means of homologous recombination in cyanobacteria, have made it possible to apply reverse genetic strategies to both cyanobacterial and eukaryotic photosynthetic model organisms and selectively knock out whole gene families. In one such case, the *Synechocystis* genome database was searched for TPR (tetratricopeptide repeat) genes (Klinkert et al., 2004) and the ORFs were systematically mutagenized. Of the 22 putative TPR proteins identified, PratA and Pitt were shown to be important for Mn^{2+} transport to D1 (Stengel et al., 2012) and for early steps in photosynthetic pigment-protein complex formation (Schottkowski et al., 2009), respectively. Interestingly, the green lineage-specific TPR-domain-containing protein MET1 was recently shown to assist in PSII supercomplex formation and in PSII repair (Bhuiyan et al., 2015). A further example of the power of reverse genetics is the screen of the lumenal immunophilin family in *Arabidopsis*, which comprises at least 16 FKBP (FK-506 binding proteins) and cyclophilins that are known to function as protein chaperones or foldases. Two of these, CYP38/TLP40 and FKBP20-2, were found to be involved in early PSII biogenesis and PSII-LHCII assembly, respectively (Lima et al., 2006; Fu et al., 2007; Sirpiö et al., 2008).

Potential and Limitations of Top-down Approaches

It is obvious that additional PSII assembly auxiliary factors will be identified in forward as well as reverse genetic screens. But the latter are becoming more important owing to major advances in rates of data generation and data mining brought by "omics" technologies. Additionally, methods for large-scale quantification of transient protein interactions of PSII subunits or known PSII auxiliary factors with unknown proteins (Braun et al., 2013) will provide a more thorough understanding of the components and their function in the assembly process.

However, an inherent limitation of top-down approaches is that they allow only those factors that are *required* for a given process to be identified. To define the suite of assembly factors and additional auxiliary factors *sufficient* for PSII assembly, an entirely different strategy is needed, in which all structural PSII proteins and auxiliary factors are brought together and the final outcome of their interplay (PSII assembly) can be monitored. This "bottom-up" concept cannot be optimally implemented in commonly used model organisms like *Arabidopsis*, maize, barley or tobacco, as "brute force" genetic approaches (like shot-gun complementation) are not feasible in these organisms and non-photosynthetic propagation of mutants is difficult to achieve, if at all. Consequently, new tactics have to be adopted to speed up and complement research on PSII assembly. Such efforts must be directed at the discovery of the complete set of auxiliary factors sufficient to make a functional plant-type photosystem and are outlined below.

A NOVEL "BOTTOM-UP" APPROACH: GENERATION OF A SYNTHETIC BACTERIUM WITH A FUNCTIONAL PLANT-TYPE PSII

In this context, microbial cell systems possess unsurpassed advantages over plants in several respects. They are fast growing (with doubling times as short as 20 min), they have small genomes that are easy to manipulate, and do not show genetic compartmentalization. The ideal host system for study of the assembly of a plant-type PSII via a "bottom-up" approach should fulfill the following requirements: (i) be able to synthesize such essential non-proteinaceous cofactors as Chl *a*, pheophytin, heme and β-carotene, (ii) possess an appropriate electron acceptor for PSII, and (iii) a lipid composition similar to that found in thylakoids, (iv) be capable of heterotrophic growth, and (v) intermediates should accumulate in mutants that are dysfunctional in the assembly process. Notably several cyanobacteria meet virtually all these criteria, which is a reflection of their evolutionary relationship to chloroplasts. However, in the case of a cyanobacterial host like *Synechocystis* endogenous PSII subunits and known, conserved auxiliary factors must be deleted, because they could interfere with the plant-type PSII subunits or auxiliary factors to be tested. This would entail the use of either a sequential, marker-less deletion strategy (Viola et al., 2014 and references within) or a time-saving, multiplex editing strategy recently described for bacterial strains, but not yet established in polyploid cyanobacteria (Ramey et al., 2015). Alternatively, cyanobacterial PSII genes can be directly replaced by their (appropriately optimized) plant counterparts in order to keep untranslated regulatory regions and operon structures intact (Viola et al., 2014). A plant-type PSII can then be constructed stepwise by mimicking the chloroplast assembly process. Known and normally short-lived assembled modules, for instance the D2-Cyt b_{559} complex, the reaction-center complex lacking both CP47 and CP43 or the reaction-center complex lacking CP43 alone (Müller and Eichacker, 1999; Komenda et al., 2004; Dobáková et al., 2007; Nixon et al., 2010), can serve as checkpoints after introducing known plant-type assembly factors into a cyanobacterial strain expressing the respective plant-type subunits found in each module. As the ultimate goal, unknown factors required for the building of each PSII module can be identified by complementation assays with cDNA libraries derived from higher plants (**Figure 1C**). An initial benchmark for this kind of approach would be the successful assembly of the minimal PSII core of the five subunits D1, D2, α, and β subunits of Cyt b_{559} and PsbI, which is able to perform charge separation (Nanba and Satoh, 1987).

Critical Aspects

The bottom-up strategy is inherently appealing, but there are significant uncertainties. While top-down strategies rely on the disruption of a single component of a complex system followed by the characterisation of the effects attributable to loss of that component, the success of the bottom-up strategy critically depends on the ability to monitor the completion of each step in the assembly process. This is the vital prerequisite for the identification of novel assembly factors introduced by shot-gun complementation approaches by perhaps only incremental advances in the assembly process. Nevertheless, in light of the tremendously fast progress in sequencing, gene-synthesis and gene-editing technologies, such synthetic biology-related approaches will sooner or later enrich the field of thylakoid assembly research.

FUNDING

This work was funded by a grant from the Ludwig-Maximilians-University (LMUexcellent) and by the German Science Foundation (DFG, grant LE 1265/20-1).

ACKNOWLEDGMENTS

We thank Paul Hardy for critical comments on the manuscript.

REFERENCES

Ajjawi, I., Lu, Y., Savage, L. J., Bell, S. M., and Last, R. L. (2010). Large-scale reverse genetics in *Arabidopsis*: case studies from the Chloroplast 2010 Project. *Plant Physiol.* 152, 529–540. doi: 10.1104/pp.109.148494

Allen, J. F., de Paula, W. B. M., Puthiyaveetil, S., and Nield, J. (2011). A structural phylogenetic map for chloroplast photosynthesis. *Trends Plant Sci.* 16, 645–655. doi: 10.1016/j.tplants.2011.10.004

Armbruster, U., Zühlke, J., Rengstl, B., Kreller, R., Makarenko, E., Rühle, T., et al. (2010). The *Arabidopsis* thylakoid protein PAM68 is required for efficient D1 biogenesis and Photosystem II assembly. *Plant Cell* 22, 3439–3460. doi: 10.1105/tpc.110.077453

Armisén, D., Lecharny, A., and Aubourg, S. (2008). Unique genes in plants: specificities and conserved features throughout evolution. *BMC Evol. Biol.* 8:280. doi: 10.1186/1471-2148-8-280

Bhuiyan, N. H., Friso, G., Poliakov, A., Ponnala, L., and Van Wijk, K. J. (2015). MET1 is a thylakoid-associated TPR protein involved in Photosystem II supercomplex formation and repair in *Arabidopsis*. *Plant Cell* 27, tpc.114.132787. doi: 10.1105/tpc.114.132787

Braun, P., Aubourg, S., Van Leene, J., De Jaeger, G., and Lurin, C. (2013). Plant protein interactomes. *Annu. Rev. Plant Biol.* 64, 161–187. doi: 10.1146/annurev-arplant-050312-120140

Bricker, T. M., Roose, J. L., Fagerlund, R. D., Frankel, L. K., and Eaton-Rye, J. J. (2012). The extrinsic proteins of Photosystem II. *Biochim. Biophys. Acta* 1817, 121–142. doi: 10.1016/j.bbabio.2011.7.006

Cai, W., Ma, J., Chi, W., Zou, M., Guo, J., Lu, C., et al. (2010). Cooperation of LPA3 and LPA2 is essential for Photosystem II assembly in *Arabidopsis*. *Plant Physiol.* 154, 109–120. doi: 10.1104/pp.110.159558

Dobáková, M., Tichy, M., and Komenda, J. (2007). Role of the PsbI protein in Photosystem II assembly and repair in the cyanobacterium *Synechocystis* sp. PCC 6803. *Plant Physiol.* 145, 1681–1691. doi: 10.1104/pp.107.107805

Fristedt, R., Herdean, A., Blaby-Haas, C. E., Mamedov, F., Merchant, S. S., Last, R. L., et al. (2015). PHOTOSYSTEM II PROTEIN33, a protein conserved in the plastid lineage, is associated with the chloroplast thylakoid membrane and provides stability to Photosystem II supercomplexes in *Arabidopsis*. *Plant Physiol.* 167, 481–492. doi: 10.1104/pp.114.253336

Fu, A., He, Z., Cho, H. S., Lima, A., Buchanan, B. B., and Luan, S. (2007). A chloroplast cyclophilin functions in the assembly and maintenance of Photosystem II in *Arabidopsis thaliana*. *Proc. Natl. Acad. Sci. U.S.A.* 104, 15947–15952. doi: 10.1073/pnas.0707851104

Hohmann-Marriott, M. F., and Blankenship, R. E. (2011). Evolution of photosynthesis. *Annu. Rev. Plant Biol.* 62, 515–548. doi: 10.1146/annurev-arplant-042110-103811

Ishihara, S., Takabayashi, A., Ido, K., Endo, T., Ifuku, K., and Sato, F. (2007). Distinct functions for the two PsbP-like proteins PPL1 and PPL2 in the chloroplast thylakoid lumen of *Arabidopsis*. *Plant Physiol.* 145, 668–679. doi: 10.1104/pp.107.105866

Järvi, S., Suorsa, M., and Aro, E.-M. (2015). Photosystem II repair in plant chloroplasts—Regulation, assisting proteins and shared components with Photosystem II biogenesis. *Biochim. Biophys. Acta* 1847, 900–909. doi: 10.1016/j.bbabio.2015.01.006

Jin, H., Liu, B., Luo, L., Feng, D., Wang, P., Liu, J., et al. (2014). HYPERSENSITIVE TO HIGH LIGHT1 Interacts with LOW QUANTUM YIELD OF PHOTOSYSTEM II1 and Functions in Protection of Photosystem II from Photodamage in *Arabidopsis*. *Plant Cell* 26, 1213–1229. doi: 10.1105/tpc.113.122424

Kansy, M., Wilhelm, C., and Goss, R. (2014). Influence of thylakoid membrane lipids on the structure and function of the plant Photosystem II core complex. *Planta* 240, 781–796. doi: 10.1007/s00425-014-2130-2

Kato, Y., and Sakamoto, W. (2009). Protein quality control in chloroplasts: a current model of D1 protein degradation in the Photosystem II repair cycle. *J. Biochem.* 146, 463–469. doi: 10.1093/jb/mvp073

Klinkert, B., Ossenbühl, F., Sikorski, M., Berry, S., Eichacker, L., and Nickelsen, J. (2004). PratA, a periplasmic tetratricopeptide repeat protein involved in biogenesis of Photosystem II in *Synechocystis* sp. PCC 6803. *J. Biol. Chem.* 279, 44639–44644. doi: 10.1074/jbc.M405393200

Komenda, J., Nickelsen, J., Tichi, M., Prášil, O., Eichacker, L. A., and Nixon, P. J. (2008). The cyanobacterial homologue of HCF136/YCF48 is a component of an early Photosystem II assembly complex and is important for both the efficient assembly and repair of Photosystem II in *Synechocystis* sp. PCC 6803. *J. Biol. Chem.* 283, 22390–22399. doi: 10.1074/jbc.M801917200

Komenda, J., Reisinger, V., Müller, B. C., Dobáková, M., Granvogl, B., and Eichacker, L. A. (2004). Accumulation of the D2 Protein is a key regulatory step for assembly of the Photosystem II reaction center complex in *Synechocystis* PCC 6803. *J. Biol. Chem.* 279, 48620–48629. doi: 10.1074/jbc.M405725200

Lima, A., Lima, S., Wong, J. H., Phillips, R. S., Buchanan, B. B., and Luan, S. (2006). A redox-active FKBP-type immunophilin functions in accumulation of the Photosystem II supercomplex in *Arabidopsis thaliana*. *Proc. Natl. Acad. Sci. U.S.A.* 103, 12631–12636. doi: 10.1073/pnas.0605452103

Lu, Y., Hall, D. A., and Last, R. L. (2011). A small zinc finger thylakoid protein plays a role in maintenance of Photosystem II in *Arabidopsis thaliana*. *Plant Cell* 23, 1861–1875. doi: 10.1105/tpc.111.085456

Ma, J., Peng, L., Guo, J., Lu, Q., Lu, C., and Zhang, L. (2007). LPA2 is required for efficient assembly of Photosystem II in *Arabidopsis thaliana*. *Plant Cell* 19, 1980–1993. doi: 10.1105/tpc.107.050526

Meurer, J., Meierhoff, K., and Westhoff, P. (1996). Isolation of high-chlorophyll-fluorescence mutants of *Arabidopsis thaliana* and their characterisation by spectroscopy, immunoblotting and northern hybridisation. *Planta* 198, 385–396. doi: 10.1007/BF00620055

Meurer, J., Plücken, H., Kowallik, K. V., and Westhoff, P. (1998). A nuclear-encoded protein of prokaryotic origin is essential for the stability of Photosystem II in *Arabidopsis thaliana*. *EMBO J.* 17, 5286–5297. doi: 10.1093/emboj/17.18.5286

Mizusawa, N., and Wada, H. (2012). The role of lipids in Photosystem II. *Biochim. Biophys. Acta* 1817, 194–208. doi: 10.1016/j.bbabio.2011.04.008

Müller, B., and Eichacker, L. A. (1999). Assembly of the D1 precursor in monomeric Photosystem II reaction center precomplexes precedes chlorophyll *a*-triggered accumulation of reaction center II in barley etioplasts. *Plant Cell* 11, 2365–2377. doi: 10.1105/tpc.11.12.2365

Nanba, O., and Satoh, K. (1987). Isolation of a Photosystem II reaction center consisting of D-1 and D-2 polypeptides and cytochrome b-559. *Proc. Natl. Acad. Sci. U.S.A.* 84, 109–112. doi: 10.1073/pnas.84.1.109

Nickelsen, J., and Rengstl, B. (2013). Photosystem II assembly: from cyanobacteria to plants. *Annu. Rev. Plant Biol.* 64, 609–635. doi: 10.1146/annurev-arplant-050312-120124

Nixon, P. J., Michoux, F., Yu, J., Boehm, M., and Komenda, J. (2010). Recent advances in understanding the assembly and repair of Photosystem II. *Ann. Bot.* 106, 1–16. doi: 10.1093/aob/mcq059

Peng, L., Ma, J., Chi, W., Guo, J., Zhu, S., Lu, Q., et al. (2006). LOW PSII ACCUMULATION1 is involved in efficient assembly of Photosystem II in *Arabidopsis thaliana*. *Plant Cell* 18, 955–969. doi: 10.1105/tpc.105.037689

Ramey, C. J., Barón-Sola, Á., Aucoin, H. R., and Boyle, N. R. (2015). Genome engineering in cyanobacteria: where we are and where we need to go. *ACS Synth. Biol.* 4, 1186–1196. doi: 10.1021/acssynbio.5b00043

Schottkowski, M., Ratke, J., Oster, U., Nowaczyk, M., and Nickelsen, J. (2009). Pitt, a novel tetratricopeptide repeat protein involved in light-dependent chlorophyll biosynthesis and thylakoid membrane biogenesis in *Synechocystis* sp. PCC 6803. *Mol. Plant* 2, 1289–1297. doi: 10.1093/mp/ssp075

Shen, J.-R. (2015). The structure of Photosystem II and the mechanism of water oxidation in photosynthesis. *Annu. Rev. Plant Biol.* 66, 23–48. doi: 10.1146/annurev-arplant-050312-120129

Sirpiö, S., Khrouchtchova, A., Allahverdiyeva, Y., Hansson, M., Fristedt, R., Vener, A. V., et al. (2008). AtCYP38 ensures early biogenesis, correct assembly and sustenance of Photosystem II. *Plant J.* 55, 639–651. doi: 10.1111/j.1365-313X.2008.03532.x

Stengel, A., Gügel, I. L., Hilger, D., Rengstl, B., Jung, H., and Nickelsen, J. (2012). Initial steps of Photosystem II de novo assembly and preloading with manganese take place in biogenesis centers in *Synechocystis*. *Plant Cell* 24, 660–675. doi: 10.1105/tpc.111.093914

Umena, Y., Kawakami, K., Shen, J.-R., and Kamiya, N. (2011). Crystal structure of oxygen-evolving Photosystem II at a resolution of 1.9Å. *Nature* 473, 55–60. doi: 10.1038/nature09913

Varotto, C., Pesaresi, P., Maiwald, D., Kurth, J., Salamini, F., and Leister, D. (2000). Identification of photosynthetic mutants of *Arabidopsis* by automatic screening for altered effective quantum yield of Photosystem 2. *Photosynthetica* 38, 497–504. doi: 10.1023/A:1012445020761

Viola, S., Rühle, T., and Leister, D. (2014). A single vector-based strategy for marker-less gene replacement in *Synechocystis* sp. PCC 6803. *Microb. Cell Fact.* 13:4. doi: 10.1186/1475-2859-13-4

Wei, L., Guo, J., Ouyang, M., Sun, X., Ma, J., Chi, W., et al. (2010). LPA19, a Psb27 homolog in *Arabidopsis thaliana*, facilitates D1 protein precursor processing during PSII biogenesis. *J. Biol. Chem.* 285, 21391–21398. doi: 10.1074/jbc.M110.105064

Yang, H., Liao, L., Bo, T., Zhao, L., Sun, X., Lu, X., et al. (2014). Slr0151 in *Synechocystis* sp. PCC 6803 is required for efficient repair of Photosystem II under high-light condition. *J. Integr. Plant Biol.* 56, 1136–1150. doi: 10.1111/jipb.12275

Zhang, D., Zhou, G., Liu, B., Kong, Y., Chen, N., Qiu, Q., et al. (2011). HCF243 encodes a chloroplast-localized protein involved in the D1 protein stability of the *Arabidopsis* Photosystem II complex. *Plant Physiol.* 157, 608–619. doi: 10.1104/pp.111.183301

Conflict of Interest Statement: The authors declare that the research was conducted in the absence of any commercial or financial relationships that could be construed as a potential conflict of interest.

13

WD40-Repeat Proteins in Plant Cell Wall Formation: Current Evidence and Research Prospects

*Gea Guerriero[1], Jean-Francois Hausman[1] and Inés Ezcurra[2]**

[1] Environmental Research and Innovation, Luxembourg Institute of Science and Technology, Esch-sur-Alzette, Luxembourg,
[2] School of Biotechnology, Division of Industrial Biotechnology, KTH Royal Institute of Technology, AlbaNova University Center, Stockholm, Sweden

The metabolic complexity of living organisms relies on supramolecular protein structures which ensure vital processes, such as signal transduction, transcription, translation and cell wall synthesis. In eukaryotes WD40-repeat (WDR) proteins often function as molecular "hubs" mediating supramolecular interactions. WDR proteins may display a variety of interacting partners and participate in the assembly of complexes involved in distinct cellular functions. In plants, the formation of lignocellulosic biomass involves extensive synthesis of cell wall polysaccharides, a process that requires the assembly of large transmembrane enzyme complexes, intensive vesicle trafficking, interactions with the cytoskeleton, and coordinated gene expression. Because of their function as supramolecular hubs, WDR proteins could participate in each or any of these steps, although to date only few WDR proteins have been linked to the cell wall by experimental evidence. Nevertheless, several potential cell wall-related WDR proteins were recently identified using *in silico* approaches, such as analyses of co-expression, interactome and conserved gene neighborhood. Notably, some WDR genes are frequently genomic neighbors of genes coding for GT2-family polysaccharide synthases in eukaryotes, and this WDR-GT2 collinear microsynteny is detected in diverse taxa. In angiosperms, two WDR genes are collinear to cellulose synthase genes, *CesAs*, whereas in ascomycetous fungi several WDR genes are adjacent to chitin synthase genes, *chs*. In this Perspective we summarize and discuss experimental and *in silico* studies on the possible involvement of WDR proteins in plant cell wall formation. The prospects of biotechnological engineering for enhanced biomass production are discussed.

Edited by:
Hemayet Ullah,
Howard University, USA

Reviewed by:
Ian S. Wallace,
The University of Nevada, Reno, USA
Biswapriya Biswavas Misra,
University of Florida, USA

***Correspondence:**
Inés Ezcurra
ines@biotech.kth.se

Keywords: WDR proteins, protein–protein interaction, plant cell wall, genomic collinearity, lignocellulose

PLANT CELL WALL BIOSYNTHESIS: A MULTIPLAYER MATTER

Plant lignocellulosic biomass is one of the most abundant renewable resources on Earth, an important feedstock for the supply of polymers and chemicals and an alternative to the use of petroleum (Guerriero et al., 2014a, 2015a). It is mainly composed of secondary cell walls, heterogeneous composites of polysaccharides (cellulose, hemicelluloses) and proteins impregnated with the aromatic polymer lignin, which confers mechanical support and recalcitrance to degradation (Guerriero et al., 2015b). Cell walls are dynamic structures which are continuously synthesized and remodeled during plant development to accommodate the needs of the growing plant cells and the broad variety of cell shapes (Guerriero et al., 2014b). The biosynthesis of plant cell walls is a complex and highly regulated process and several aspects have been thoroughly investigated. This has been possible thanks to the large amount of data generated via transcriptomics, proteomics, metabolomics on both woody and herbaceous species (Lin et al., 2013; Printz et al., 2015) and to functional studies in model plants, i.e., *Arabidopsis thaliana* (Anderson et al., 2015). The increasing number of sequenced plant genomes and genome-wide analyses of cellulose and lignin-related genes (Peng et al., 2013; Myburg et al., 2014) has further contributed to our understanding of plant cell wall biosynthesis. Moreover, the availability of plant cell culture systems differentiating tracheary elements (Oda et al., 2005) has provided a valuable tool to study the sequential steps of secondary cell wall biosynthesis using high-resolution imaging techniques (Lacayo et al., 2010).

Plant cell wall biosynthesis is regulated during different stages of gene expression, namely at the transcriptional and post-translational level (Hijazi et al., 2014; Zhong and Ye, 2014). A transcriptional wiring composed of master and downstream regulators determines the regulation of cell wall structural genes, typically encoding cellulose synthases and other carbohydrate-active enzymes, or enzymes in lignin biosynthesis. This hierarchical organization is conserved across different species, from woody to herbaceous, from monocots to dicots (Winzell et al., 2010; Zhao and Bartley, 2014).

At the post-translational level, plant wall biosynthesis requires the assembly of large protein complexes at the membranes, a process which relies upon intensive intracellular trafficking (Wightman and Turner, 2010) and cytoskeleton interactions (Gutierrez et al., 2009). A key example entails the cellulose synthase complex (CSC, a.k.a the "rosette terminal complex") which is pre-assembled in the Golgi and then delivered in vesicles to the plasma membrane (Wightman and Turner, 2010). The complex was recently shown to be composed of six particles, each containing a trimer of cellulose synthase catalytic subunits (CESA4, CESA7, CESA8) in equimolar stoichiometry (1:1:1), in *A. thaliana* secondary cell walls (Hill et al., 2014). CSC establishes interactions with other proteins during its vesicle trafficking to and from the membrane (Gutierrez et al., 2009) and at the plasma membrane (Vain et al., 2014). Protein–protein interactions are likewise crucial for the biosynthesis of non-cellulosic polysaccharides in the Golgi (reviewed by Oikawa et al., 2013) and for the organization of multienzyme complexes, or "metabolons" (Jørgensen et al., 2005), which maximize shunting of metabolites into branches of the plant secondary metabolism that are relevant to the cell wall, such as the phenylpropanoid pathway (Chen et al., 2014).

It is therefore clear that the formation of supramolecular protein complexes is crucial for cell wall biosynthesis in plants, and in this perspective we aim at understanding more about the scaffolding components mediating protein–protein interaction, to inspire additional biotechnological strategies to tailor plant cell wall biosynthesis. We focus on a widely distributed family of scaffolding proteins, the WD40-repeat proteins (WDRs), and discuss their role in cell wall biosynthesis by presenting evidence linking these proteins to cell wall-related processes. We also present potential biotechnological uses of plant WDRs to modulate lignocellulose synthesis.

PLANT WDR PROTEINS HAVE PLEIOTROPIC FUNCTIONS

Plants are among the most complicated organisms to study from a systems biology perspective. Their metabolic redundancy, a consequence of their sessile lifestyle (Mishra et al., 2012), determines intricate interactome maps (Morsy et al., 2008). Protein–protein interactions coordinate the formation of supramolecular complexes which ensure the correct execution of sequential steps in a specific metabolic pathway. The association of proteins, either temporal or stable, requires usually the presence of ancillary scaffolding proteins. In eukaryotes a family of scaffolding proteins, the WDR proteins, participates in assembling protein complexes (Stirnimann et al., 2010). The structure of these proteins is a β-propeller, where several repeating units, composed of ca 40–60 amino acids (among which conserved GH and WD residues), fold into four-stranded anti-parallel β-sheets (Stirnimann et al., 2010; Mishra et al., 2012). The β-sheets arrange circularly around a central axis, thereby creating a multi-bladed toroidal structure which can establish molecular interactions via the top, bottom, and circumference (**Figures 1A,B**). WDR proteins are rigid platforms that can interact with many partners, a phenomenon known as "moonlighting." This promiscuity explains the wide range of cellular processes in which they are implicated (Stirnimann et al., 2010; Mishra et al., 2012), which span from (a)biotic stress response to intracellular trafficking and transcriptional regulation, just to name a few.

PLANT WDR PROTEINS WITH ROLES IN CELL WALL-RELATED PROCESSES

Given their functional versatility and scaffolding properties, WDR proteins are plausible candidates for roles in cell wall biosynthesis, although to date only three WDR proteins have been associated to the cell wall through experimental evidence, namely FRAGILE FIBER3 (FRA3), LEUNIG Homolog (LUH) and TWD40-2. In *Arabidopsis,* FRA3 is a type II inositol polyphosphate 5-phosphatases containing an N-terminal WDR domain (Zhong et al., 2004). Mutations in the *FRA3* gene cause phenotypes related to a reduced secondary cell wall thickness in fibers and xylem vessels, with consequent decrease in stem strength (Zhong et al., 2004). Analysis of cell wall composition showed that the mutants have a 37% reduction in glucose, thereby suggesting that cellulose is mainly affected by the mutations (Zhong et al., 2004). Moreover, the actin organization in fiber cells is compromised with consequences on trafficking of vesicles containing CESA complexes and/or cell wall polysaccharides. The second *Arabidopsis* WDR protein, LUH, is a transcriptional co-repressor belonging to the Gro/TLE

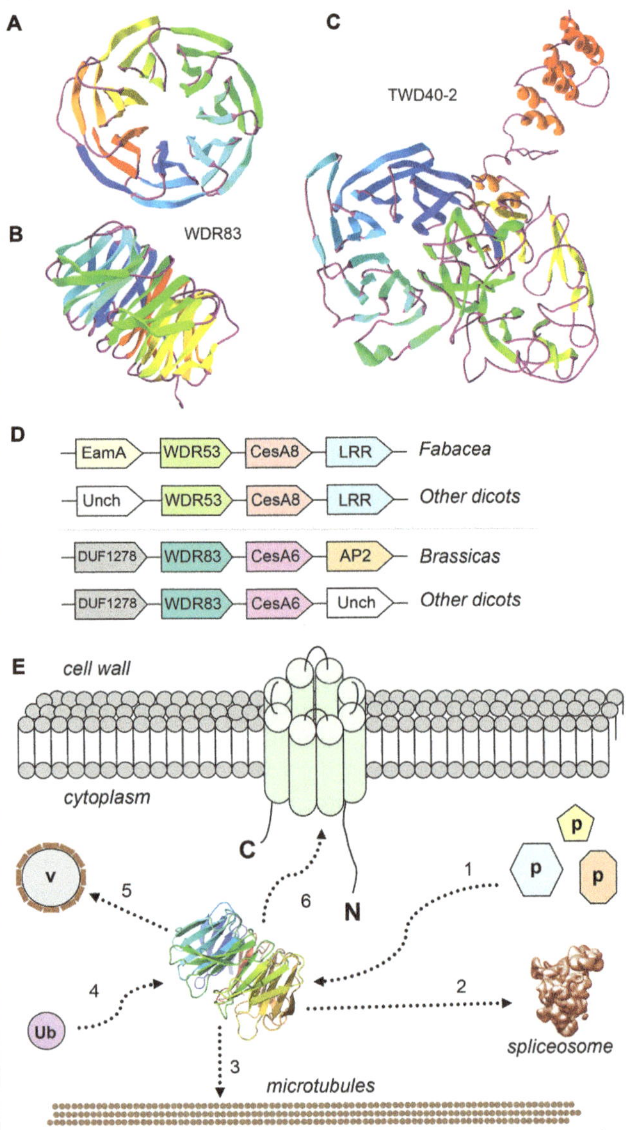

FIGURE 1 | Cell wall-related WDRs: predicted 3D structure, conserved genomic neighbors of *CesAs*, and model of their possible roles.
(A) Predicted 3D structure of AtWDR83 (AT5G64730), view from the top of the propeller, showing the seven blades; **(B)** view from the side, showing the toroidal structure with top, bottom and circumferential sites of interactions; **(C)** View of predicted 3D structure of TWD40-2 (AT5G24710), showing the two seven-bladed propellers; The models were obtained with Phyre2 (Kelley et al., 2015; available at http://www.sbg.bio.ic.ac.uk/phyre2/html/page.cgi?id=index) and visualized with Swiss-PdbViewer (Guex and Peitsch, 1997; available at http://www.expasy.org/spdbv/). Templates for the AtWDR83 model were multiple WDR proteins, including Gbeta5/RGS9, ribosomal Rsa4, and pre-mRNA-splicing factor prp5, whereas the TWD40-2 template was the coatomer subunit alpha, alpha-COP; **(D)** Conserved microcolinearity of *WDR* and *CesA* genes in angiosperms. Unch, Uncharacterized; EamA, EamA-like transporter; LRR, Leucine-rich repeat; **(E)** Hypothetical model showing the possible functions of WDR53/WDR83-like proteins in cellulose biosynthesis. 1: WDR53/WDR83 represent a "platform" for the binding of unknown cell wall-related proteins; 2: WDR53/WDR83 interact with the spliceosome and splices *CesA* transcripts; 3: WDR53/WDR83 interact with the cytoskeleton; 4: WDR53/WDR83 bind ubiquitin; 5: WDR53/WDR83 mediate trafficking of Golgi vesicles (v); 6: WDR53/WDR83 interact with CESA complexes at the plasma membrane.

family (Bui et al., 2011). *Arabidopsis luh-1* mutants have defects associated with mucilage extrusion, which are reminiscent of the *mum2* mutants (*mucilage modified 2*, a β-galactosidase involved in mucilage maturation). LUH and (possibly) the closely related WDR protein LEUNIG (LUG) regulate the expression of cell wall-related genes associated with mucilage maturation (Bui et al., 2011). The penetrance of the *luh-1* phenotype indicates that this WDR protein is a key regulator of cell wall modifications associated with a critical phase of seed coat maturation (Bui et al., 2011). The third *Arabidopsis* WDR protein, TWD40-2, was recently linked to clathrin-mediated endocytosis of cellulose synthases during cellulose biosynthesis (Bashline et al., 2015). Mutation of the *TWD40-2* gene caused impaired endocytosis, unregulated overaccumulation of CESAs at the plasma membrane, decreased cellulose content and stunted plant growth. This highlights the importance of controlled CESA endocytosis in regulating cellulose biosynthesis, and the TWD40-2 WDR protein is key in this process. TWD40-2 is unique to plants and some protozoan species, and is similar to the alpha and beta coatomer WDR subunits, alpha- and beta-COP (Zhang et al., 2015). Although TWD40-2 is structurally related to alpha- and beta-COP, with its two N-terminal propellers and a C-terminal alpha-helical region (**Figure 1C**), it is unclear whether it coats endocytic vesicles as alpha- and beta-COP do (Dodonova et al., 2015).

WDR PROTEINS LINKED TO THE CELL WALL BY *IN SILICO* PREDICTIONS

Although only three WDR proteins have been linked to the cell wall by biochemical and/or genetic evidence, several potential cell wall-related WDR proteins were recently identified using *in silico* approaches, such as analyses of co-expression, interactome and conserved gene neighborhood. A WDR gene (At5g53500) was found to be highly co-regulated with the primary cell wall *CesA1*, *CesA3*, and *CesA6* by regression analysis of public microarray data sets (Persson et al., 2005). More recently, a computational framework was developed for the prediction of proteins related to cell wall synthesis, based on known cell wall synthesis-related proteins and their predicted protein-protein interactions in the *Arabidopsis* interactome, which is inferred from multiple indirect lines of evidence, including coexpression, colocalization, coevolution, annotation similarity, domain interaction, and homologous interactions in other species (Geisler-Lee et al., 2007; Zhou et al., 2010). The study identified three WDR proteins, AT1G04510, AT2G43770, and AT4G34460, among the 42 most likely cell wall-related candidate proteins. Other studies show that AT1G04510 codes for MAC3A, a WDR protein with roles in plant immunity and activation of the spliceosome (Monaghan et al., 2009). The computational analysis predicts that MAC3A/AT1G04510 interacts with a rhamnose synthase, RHM1 (AT1G78570), involved in accumulation of the pectic polysaccharide rhamnogalacturonan in the cell wall. Both functions of MAC3A, spliceosome activation and plant defense, are consistent with a possible role in the cell wall, as both processes may entail the cell wall (Huang et al., 2013; Bellincampi et al., 2014). Further, cell wall pectin is an important factor in plant defense (Bellincampi et al., 2014), and thus the predicted interaction of MAC3A/AT1G04510 with RHM1 is in agreement with its function. Then, MAC3A/AT1G04510 could potentially regulate plant defense through splicing of the RHM1/AT1G78570 transcript, in analogy to regulation by splicing of Callose Synthase5 (Huang et al., 2013), although this remains to be shown experimentally. The second predicted cell wall-related WDR protein, AT2G43770, is the plant homolog of SNRNP40, also a spliceosome component, and is predicted to interact with a UDP-D-xylose synthase, UXS6 (AT2G28760). Finally, the third predicted cell wall WDR protein, AT4G34460, is AGB1, a G-Protein β subunit, and the study predicts its interaction with a GDP-D-mannose 3′,5′-epimerase (AT5G28840). The role of AGB1 in the cell wall was experimentally confirmed in later studies showing its involvement in cell wall defense (Delgado-Cerezo et al., 2012), suggesting a fair level of accurate predictions of cell wall-involved WDRs by the computational framework.

CONSERVED COLLINEAR MICROSYNTENY BETWEEN WDR AND GLYCAN SYNTHASE GENES IN DIVERSE EUKARYOTE TAXA

The clustering of functionally related genes is prevalent in prokaryotes but was conventionally assumed to be less common in eukaryotes. However, this notion is shifting as clusters of functionally related genes are increasingly detected in eukaryotes (Hurst et al., 2004; Yi et al., 2007) including, recently, plants (Nützmann and Osbourn, 2014). Although gene clusters in plant and fungi concern secondary metabolism, a recent bioinformatic study identified a cell wall gene cluster in fungi (Pacheco-Arjona and Ramirez-Prado, 2014), suggesting that conserved genomic neighborhood may provide a means to identifying genes involved in processes other than secondary metabolism. The function of physically clustering functionally related genes in eukaryotes is unclear, but could involve coordinated gene expression or keeping genes together to avoid toxic effects of their individual loss.

We previously identified two cases of conserved genomic collinearity between a WDR gene and a *CesA* in the genomes of several dicots (Guerriero et al., 2012). Specifically, a plant *WDR53* gene (represented by At5g45760 in *Arabidopsis*) is collinear with the secondary cell wall *CesA8*, while a *WDR83* (At5g64730) is adjacent to the primary cell wall *CesA6* (schematized in **Figure 1D**). We showed that *WDR53* and *CesA8* are partially co-expressed in apple and *Populus* (Guerriero et al., 2012). Further, we identified a bicistronic *WDR53-CesA8* transcript in phytoplasma-infected apple tissues, where the *CesA8* cistron is alternatively spliced and lacks a portion corresponding the RING-type zinc finger domain. We speculate on whether this alternatively spliced *CesA8* cistron is translated in order to alter cellulose biosynthesis in response to phytoplasma infection.

We have also showed that the genomic association between a WDR gene and a gene encoding a glycosyltransferase from family 2 (GT2) is a conserved feature in eukaryotes, since, besides dicots, it is reported in metazoans (hyaluronan synthase-WDR microsynteny) and in fungi (chitin synthase-WDR microsynteny; Guerriero et al., 2012). In fungi, the genomic collinearity between a chitin synthase (*chs*) and a WDR gene is conserved among several species (Guerriero et al., in revision). This WDR gene (AN10216 in *Aspergillus nidulans*) occurs mainly in fungi and plants, and is therefore named Fungal-Plant WDR (*FPWD*). *FPWD* is localized within a recently identified cell wall metabolism gene cluster involving among others, besides *chsD*, a Chs activator, a myosin V, a GH17 cell wall glucanase, scw11 (Pacheco-Arjona and Ramirez-Prado, 2014). Deletion of *FPWD* and a tightly linked neighboring beta-flanking gene causes cell wall and chitin-related phenotypes in *A. nidulans*, confirming *FPWD*'s role in the cell wall (Guerriero et al., in revision). These results suggest that identifying conserved gene neighbors may be a useful bioinformatic approach to mining for gene targets, in addition to more traditional approaches based on gene expression, proteomics or interactome inference.

In light of the conserved genomic microsynteny *WDR53-CesA8* and *WDR83-CesA6*, it is then plausible to assume a cellulose biosynthesis-related role for these WD40 proteins. Given the moonlighting properties of many plant WDR proteins, they could be involved in one or several processes involving the delivery of the CSC to membrane microdomains, e.g., via regulating transcript splicing, vesicle trafficking, the formation of platforms mediating protein-protein interactions and/or protein turnover via ubiquitination (schematized in **Figure 1E**).

TABLE 1 | Phylogenetic cluster location of cell wall-related WDR proteins (upper section), listed together with representative members of each cluster (lower section).

Gene ID	Annotation	Phylo-genetic cluster	Function
AT1G04510	MAC3A	I	Splicing, plant defense
AT2G43770	–	I	Unknown
AT4G34460	AGB1	I	Cell wall defense
AT5G64730	WDR83	I	Unknown
AT1G78070	FPWD-1	II	Unknown
AT1G36070	FPWD-2	II	Unknown
AT1G55680	FPWD-3	II	Unknown
AT3G13340	FPWD-4	II	Unknown
AT5G56190	FPWD-5	II	Unknown
AT5G24710	TWD40-2	II	Cell wall
AT1G65580	FRA3	III	Cell wall
AT3G50590	TWD40-1	III	Cell wall
AT5G45760	WDR53	III	Unknown
AT3G49660	WDR5	I	Histone methylation
AT5G27030	TPR3	II	Chromatin regulation
AT3G21060	RBL	II	Histone methylation
AT3G49180	RID3	III	Pre-rRNA processing
AT1G03060	SPIRRIG, SPI	III	Membrane; RNA processing

In flowering plants, the *FPWD* gene occurs as a small gene family of around five genes of unknown function, represented in *Arabidopsis* by At1g78070, At1g36070, At1g55680, At3g13340, and At5g56190. We noticed that the plant *FPWDs* are conserved genomic neighbors of cell wall or carbohydrate-related genes, coding for trehalose-6-phosphate phosphatase, GH3 and actin ACT7 (in higher plants), as well as nucleotide sugar transporter and GT57 glucosyltransferase (in green algae). Examination of their expression using publicly available microarray databases in *Arabidopsis* and *Populus* suggests that they may be either largely xylem-specific, or pollen-specific (data not shown), consistent with a possible role in the cell wall. Clearly, the precise functions of these potentially wall-involved WDRs, and of their gene neighbors, remain to be further elucidated.

PHYLOGENETIC RELATIONSHIPS OF CELL WALL CANDIDATE WDR PROTEINS

We here identify a number of plant WDR proteins with confirmed or potential roles in the lignocellulosic cell wall. A recent phylogenetic analysis of all *Arabidopsis* (comprising 237 gene ids) and rice (comprising 200 gene ids) WDRs established that the plant WDR phylogenetic tree consists of five major branches, numbered I–V (Ouyang et al., 2012). To better understand the function and evolution of the cell wall candidate WDRs, we analyzed their cluster location within the phylogenetic tree (**Table 1**). We observe that cell wall-related WDR proteins belong to distinct phylogenetic clusters: WDR83, MAC3A, AT2G43770 and AGB1 belong to cluster I, together with WDR5, involved in histone methylation (Jiang et al., 2011). WDR53 is in cluster III with FRA3, involved in secondary cell wall formation, ROOT INITIATION DEFECTIVE 3, RID3, with roles in pre-rRNA processing (Shinohara et al., 2014), and SPIRRIG, a BEACH domain WDR which regulates mRNA stability (Steffens et al., 2015). Further, the FPWDs branch in cluster II, together with the TOPLESS-related, TPR, protein cluster, which are involved in chromatin-mediated repression by histone deacetylation (Pi et al., 2015), and RBBP5 LIKE, RBL, involved in histone methylation (Jiang et al., 2011). Finally, TWD40-2, involved in clathrin mediated endocytosis of CESA complexes, also branches in cluster II. In summary, the cell wall candidate WDR proteins do not cluster together but are rather widely distributed across the plant WDR superfamily.

BIOTECHNOLOGICAL PROSPECTS: WDR PROTEINS TO BOOST PLANT BIOMASS PRODUCTION

The depletion of fossil fuel pushes toward the identification and use of renewable energy sources to promote the development of a circular economy. In this perspective plant biomass is very important, since it constitutes one of the most abundant sustainable resources for humanity and it can provide biopolymers, fibers, biofuel, thereby promoting economic growth, while minimizing environmental impacts. Besides woody species like *Populus* and *Eucalyptus*, fast growing herbaceous species, like the fiber-crops flax and hemp, have also considerable importance for the bio-economy, because of their shorter growth cycles, their high biomass yield (e.g., 24 tons/hectare for hemp; Kolaříková et al., 2013) and the presence of cellulosic fibers in their stems (the so-called bast fibers; Guerriero et al., 2013). Because of these advantageous characteristics and the properties of their fibers (length, strength, renewability), these crops are widely used in industry to provide fibers for the reinforcement of composites. For the bio-economy high biomass-production is a sought-after character and biotechnological strategies aimed at boosting biomass formation are attracting research efforts (Nieminen et al., 2012; Dubouzet et al., 2013). Here we identify a list of WDR proteins with confirmed or potential roles in the lignocellulosic cell wall, which are interesting targets for engineering plants with enhanced biomass production. Such effort would involve the use of promoters to drive altered expression in the cambium or xylem (Ji et al., 2010; Ratke et al., 2015), to influence xylem proliferation and/or lignocellulose biosynthesis. As one possible example, seedlings of *twd40-2* mutants display reduced endocytosis, cellulose content, and plant growth. The *TWD40-2* gene is expressed in most tissues, but its expression is highest in stem or xylem in *Arabidopsis* and *Populus* according to the BAR eFT gene expression browser (http://bar.utoronto.ca/; data not shown). Then, enhancing *TWD40-2* expression in the xylem could increase cellulose content. A further approach to enhance the function of target WDR proteins could potentially involve mutation of selected surface residues involved in protein interaction, a.k.a. "interaction hotspots," in order to decrease the target's eventual pleiotropy. This would require identification of interaction hotspots for example using tools for *in silico* prediction, such as modeling their 3D structure, combined with experimental confirmation. The ultimate goal of such mutagenesis, which could be generated through genome editing, would be to reduce pleiotropy or moonlighting of the target WDR, and obtain more specialized enzymes, to avoid undesired side effects.

AUTHOR CONTRIBUTIONS

GG and IE conceived the idea of writing the Perspective paper. GG and IE obtained the bioinformatic data. GG, J-FH, and IE interpreted the data and wrote the manuscript.

ACKNOWLEDGMENTS

GG gratefully acknowledges the Austrian Science Fund (FWF; http://www.fwf.ac.at/en/) Grant No. M1315. GG and JFH wish to thank the support by the Fonds National de la Recherche, Luxembourg, (Project CANCAN C13/SR/5774202). IE gratefully acknowledges the Swedish Governmental Agency for Innovation Systems, VINNOVA and the Swedish Energy Agency, Energimyndigheten for financial support.

REFERENCES

Anderson, N., Bonawitz, N. D., Nyffeler, K. E., and Chapple, C. (2015). Loss of ferulate 5-hydroxylase leads to Mediator-dependent inhibition of soluble phenylpropanoid biosynthesis in *Arabidopsis*. *Plant Physiol.* 169, 1557–1567. doi: 10.1104/pp.15.00294

Bashline, L., Li, S., Zhu, X., and Gu, Y. (2015). The TWD40-2 protein and the AP2 complex cooperate in the clathrin-mediated endocytosis of cellulose synthase to regulate cellulose biosynthesis. *Proc. Natl. Acad. Sci. U.S.A.* 112, 12870–12875. doi: 10.1073/pnas.1509292112

Bellincampi, D., Cervone, F., and Lionetti, V. (2014). Plant cell wall dynamics and wall-related susceptibility in plant–pathogen interactions. *Front. Plant Sci.* 5:228. doi: 10.3389/fpls.2014.00228

Bui, M., Lim, N., Sijacic, P., and Liu, Z. (2011). LEUNIG_HOMOLOG and LEUNIG regulate seed mucilage extrusion in *Arabidopsis*. *J. Integr. Plant Biol.* 53, 399–408. doi: 10.1111/j.1744-7909.2011.01036.x

Chen, H. C., Song, J., Williams, C. M., Shuford, C. M., Liu, J., Wang, J. P., et al. (2014). Systems biology of lignin biosynthesis in *Populus trichocarpa*: heteromeric 4-coumaric acid:coenzyme A ligase protein complex formation, regulation, and numerical modeling. *Plant Cell* 26, 876–893. doi: 10.1105/tpc.113.119685

Delgado-Cerezo, M., Sánchez-Rodríguez, C., Escudero, V., Miedes, E., Fernández, P. V., Jordá, L., et al. (2012). *Arabidopsis* heterotrimeric G-protein regulates cell wall defense and resistance to necrotrophic fungi. *Mol. Plant* 5, 98–114. doi: 10.1093/mp/ssr082

Dodonova, S. O., Diestelkoetter-Bachert, P., von Appen, A., Hagen, W. J., Beck, R., Beck, M., et al. (2015). VESICULAR TRANSPORT. A structure of the COPI coat and the role of coat proteins in membrane vesicle assembly. *Science* 349, 195–198. doi: 10.1126/science.aab1121

Dubouzet, J. G., Strabala, T. J., and Wagner, A. (2013). Potential transgenic routes to increase tree biomass. *Plant Sci.* 212, 72–101. doi: 10.1016/j.plantsci.2013.08.006

Geisler-Lee, J., O'Toole, N., Ammar, R., Provart, N. J., Millar, A. H., and Geisler, M. (2007). A predicted interactome for *Arabidopsis*. *Plant Physiol.* 145, 317–329. doi: 10.1104/pp.107.103465

Guerriero, G., Hausman, J. F., Strauss, J., Ertan, H., and Siddiqui, K. S. (2015a). Lignocellulosic biomass: biosynthesis, degradation and industrial utilization. *Eng. Life Sci.* doi: 10.1002/elsc.201400196

Guerriero, G., Hausman, J. F., Strauss, J., Ertan, H., and Siddiqui, K. S. (2015b). Destructuring plant biomass: focus on fungal and extremophilic cell wall hydrolases. *Plant Sci.* 234, 180–193. doi: 10.1016/j.plantsci.2015.02.010

Guerriero, G., Sergeant, K., and Hausman, J. F. (2013). Integrated -omics: a powerful approach to understanding the heterogeneous lignification of fibre crops. *Int. J. Mol. Sci.* 14, 10958–10978. doi: 10.3390/ijms140610958

Guerriero, G., Sergeant, K., and Hausman, J. F. (2014a). Wood biosynthesis and typologies: a molecular rhapsody. *Tree Physiol.* 34, 839–855. doi: 10.1093/treephys/tpu031

Guerriero, G., Hausman, J. F., and Cai, G. (2014b). No stress! Relax! Mechanisms governing growth and shape in plant cells. *Int. J. Mol. Sci.* 15, 5094–5114. doi: 10.3390/ijms15035094

Guerriero, G., Spadiut, O., Kerschbamer, C., Giorno, F., Baric, S., and Ezcurra, I. (2012). Analysis of cellulose synthase genes from domesticated apple identifies collinear genes WDR53 and CesA8A: partial co-expression, bicistronic mRNA, and alternative splicing of CESA8A. *J. Exp. Bot.* 63, 6045–6056. doi: 10.1093/jxb/ers255

Guex, N., and Peitsch, M. C. (1997). SWISS-MODEL and the Swiss-PdbViewer: an environment for comparative protein modeling. *Electrophoresis* 18, 2714–2723. doi: 10.1002/elps.1150181505

Gutierrez, R., Lindeboom, J. J., Paredez, A. R., Emons, A. M., and Ehrhardt, D. W. (2009). *Arabidopsis* cortical microtubules position cellulose synthase delivery to the plasma membrane and interact with cellulose synthase trafficking compartments. *Nat. Cell Biol.* 11, 797–806. doi: 10.1038/ncb1886

Hijazi, M., Velasquez, S. M., Jamet, E., Estevez, J. M., and Albenne, C. (2014). An update on post-translational modifications of hydroxyproline-rich glycoproteins: toward a model highlighting their contribution to plant cell wall architecture. *Front. Plant Sci.* 5:395. doi: 10.3389/fpls.2014.00395

Hill, J. L. Jr., Hammudi, M. B., and Tien, M. (2014). The *Arabidopsis* cellulose synthase complex: a proposed hexamer of CESA trimers in an equimolar stoichiometry. *Plant Cell* 26, 4834–4842. doi: 10.1105/tpc.114.131193

Huang, X. Y., Niu, J., Sun, M. X., Zhu, J., Gao, J. F., Yang, J., et al. (2013). CYCLIN-DEPENDENT KINASE G1 is associated with the spliceosome to regulate CALLOSE SYNTHASE5 splicing and pollen wall formation in *Arabidopsis*. *Plant Cell* 25, 637–648. doi: 10.1105/tpc.112.107896

Hurst, L. D., Pál, C., and Lercher, M. J. (2004). The evolutionary dynamics of eukaryotic gene order. *Nat. Rev. Genet.* 5, 5299–5310. doi: 10.1038/nrg1319

Ji, J., Strable, J., Shimizu, R., Koenig, D., Sinha, N., and Scanlon, M. J. (2010). WOX4 promotes procambial development. *Plant Physiol.* 152, 1346–1356. doi: 10.1104/pp.109.149641

Jiang, D., Kong, N. C., Gu, X., Li, Z., and He, Y. (2011). *Arabidopsis* COMPASS-like complexes mediate histone H3 lysine-4 trimethylation to control floral transition and plant development. *PLoS Genet.* 7:e1001330. doi: 10.1371/journal.pgen.1001330

Jørgensen, K., Rasmussen, A. V., Morant, M., Nielsen, A. H., Bjarnholt, N., Zagrobelny, M., et al. (2005). Metabolon formation and metabolic channeling in the biosynthesis of plant natural products. *Curr. Opin. Plant Biol.* 8, 280–291. doi: 10.1016/j.pbi.2005.03.014

Kelley, L. A., Mezulis, S., Yates, C. M., Wass, M. N., and Sternberg, M. J. (2015). The Phyre2 web portal for protein modeling, prediction and analysis. *Nat. Protoc.* 10, 845–858. doi: 10.1038/nprot.2015.053

Kolaříková, M., Ivanova, T., and Havrland, B. (2013). "Energy balance of briquettes made of hemp (*Cannabis sativa* L.) cultivars (Ferimon, Bialobrzeskie) from autumn harvest to produce heat for household use," in *Proceedings of the Engineering for Rural Development* (Jelgava: Latvia University of Agriculture), 504–508.

Lacayo, C. I., Malkin, A. J., Holman, H. Y., Chen, L., Ding, S. Y., Hwang, M. S., et al. (2010). Imaging cell wall architecture in single *Zinnia elegans* tracheary elements. *Plant Physiol.* 154, 121–133. doi: 10.1104/pp.110.155242

Lin, Y. C., Li, W., Sun, Y. H., Kumari, S., Wei, H., Li, Q., et al. (2013). SND1 transcription factor-directed quantitative functional hierarchical genetic regulatory network in wood formation in *Populus trichocarpa*. *Plant Cell* 25, 4324–4341. doi: 10.1105/tpc.113.117697

Mishra, A. K., Puranik, S., and Prasad, M. (2012). Structure and regulatory networks of WD40 protein in plants. *J. Plant Biochem. Biotechnol.* 21, S32–S39. doi: 10.1111/j.1744-7909.2012.01159.x

Monaghan, J., Xu, F., Gao, M., Zhao, Q., Palma, K., Long, C., et al. (2009). Two Prp19-like U-box proteins in the MOS4-associated complex play redundant roles in plant innate immunity. *PLoS Pathog.* 5:e1000526. doi: 10.1371/journal.ppat.1000526

Morsy, M., Gouthu, S., Orchard, S., Thorneycroft, D., Harper, J. F., Mittler, R., et al. (2008). Charting plant interactomes: possibilities and challenges. *Trends Plant Sci.* 13, 183–191. doi: 10.1016/j.tplants.2008.01.006

Myburg, A. A., Grattapaglia, D., Tuskan, G. A., Hellsten, U., Hayes, R. D., Grimwood, J., et al. (2014). The genome of *Eucalyptus grandis*. *Nature* 510, 356–362. doi: 10.1038/nature13308

Nieminen, K., Robischon, M., Immanen, J., and Helariutta, Y. (2012). Towards optimizing wood development in bioenergy trees. *New Phytol.* 194, 46–53. doi: 10.1111/j.1469-8137.2011.04011.x

Nützmann, H. W., and Osbourn, A. (2014). Gene clustering in plant specialized metabolism. *Curr. Opin. Biotechnol.* 26, 91–99. doi: 10.1016/j.copbio.2013.10.009

Oda, Y., Mimura, T., and Hasezawa, S. (2005). Regulation of secondary cell wall development by cortical microtubules during tracheary element differentiation in *Arabidopsis* cell suspensions. *Plant Physiol.* 137, 1027–1036. doi: 10.1104/pp.104.052613

Oikawa, A., Lund, C. H., Sakuragi, Y., and Scheller, H. V. (2013). Golgi-localized enzyme complexes for plant cell wall biosynthesis. *Trends Plant Sci.* 18, 49–58. doi: 10.1016/j.tplants.2012.07.002

Ouyang, Y., Huang, X., Lu, Z., and Yao, J. (2012). Genomic survey, expression profile and co-expression network analysis of OsWD40 family in rice. *BMC Genomics* 20:100. doi: 10.1186/1471-2164-13-100

Pacheco-Arjona, J. R., and Ramirez-Prado, J. H. (2014). Large-scale phylogenetic classification of fungal chitin synthases and identification of a putative cell-wall metabolism gene cluster in *Aspergillus* genomes. *PLoS ONE* 9:e104920. doi: 10.1371/journal.pone.0104920

Peng, Z., Lu, Y., Li, L., Zhao, Q., Feng, Q., Gao, Z., et al. (2013). The draft genome of the fast-growing non-timber forest species moso bamboo (*Phyllostachys heterocycla*). *Nat. Genet.* 45, 456–461. doi: 10.1038/ng.2569

Persson, S., Wei, H., Milne, J., Page, G. P., and Somerville, C. R. (2005). Identification of genes required for cellulose synthesis by regression analysis of public microarray data sets. *Proc. Natl. Acad. Sci. U.S.A.* 102, 8633–8638. doi: 10.1073/pnas.0503392102

Pi, L., Aichinger, E., van der Graaff, E., Llavata-Peris, C. I., Weijers, D., Hennig, L., et al. (2015). Organizer-derived WOX5 signal maintains root columella stem cells through chromatin-mediated repression of CDF4 expression. *Dev. Cell.* 33, 576–588. doi: 10.1016/j.devcel.2015.04.024

Printz, B., Guerriero, G., Sergeant, K., Renaut, J., Lutts, S., and Hausman, J. F. (2015). Ups and downs in alfalfa: proteomic and metabolic changes occurring in the growing stem. *Plant Sci.* 238, 13–25. doi: 10.1016/j.plantsci.2015.05.014

Ratke, C., Pawar, P. M., Balasubramanian, V. K., Naumann, M., Duncranz, M. L., Derba-Maceluch, M., et al. (2015). *Populus* GT43 family members group into distinct sets required for primary and secondary wall xylan biosynthesis and include useful promoters for wood modification. *Plant Biotechnol. J.* 13, 26–37. doi: 10.1111/pbi.12232

Shinohara, N., Ohbayashi, I., and Sugiyama, M. (2014). Involvement of rRNA biosynthesis in the regulation of CUC1 gene expression and pre-meristematic cell mound formation during shoot regeneration. *Front. Plant Sci.* 5:159. doi: 10.3389/fpls.2014.00159

Steffens, A., Bräutigam, A., Jakoby, M., and Hülskamp, M. (2015). The BEACH domain protein SPIRRIG is essential for *Arabidopsis* salt stress tolerance and functions as a regulator of transcript stabilization and localization. *PLoS Biol.* 13:e1002188. doi: 10.1371/journal.pbio.1002188

Stirnimann, C. U., Petsalaki, E., Russell, R. B., and Müller, C. W. (2010). WD40 proteins propel cellular networks. *Trends Biochem. Sci.* 35, 565–574. doi: 10.1016/j.tibs.2010.04.003

Vain, T., Crowell, E. F., Timpano, H., Biot, E., Desprez, T., Mansoori, N., et al. (2014). The cellulase KORRIGAN is part of the cellulose synthase complex. *Plant Physiol.* 165, 1521–1532. doi: 10.1104/pp.114.241216

Wightman, R., and Turner, S. (2010). Trafficking of the plant cellulose synthase complex. *Plant Physiol.* 153, 427–432. doi: 10.1093/aob/mcu040

Winzell, A., Aspeborg, H., Wang, Y., and Ezcurra, I. (2010). Conserved CA-rich motifs in gene promoters of Pt x tMYB021-responsive secondary cell wall carbohydrate-active enzymes in *Populus. Biochem. Biophys. Res. Commun.* 394, 848–853. doi: 10.1016/j.bbrc.2010.03.10

Yi, G., Sze, S. H., and Thon, M. R. (2007). Identifying clusters of functionally related genes in genomes. *Bioinformatics* 23, 1053–1060. doi: 10.1093/bioinformatics/btl673

Zhang, Y., Persson, S., Hirst, J., Robinson, M. S., van Damme, D., and Sánchez-Rodríguez, C. (2015). Change your TPLATE, change your fate: plant CME and beyond. *Trends Plant Sci.* 20, 41–48. doi: 10.1016/j.tplants.2014.09.002

Zhao, K., and Bartley, L. E. (2014). Comparative genomic analysis of the R2R3 MYB secondary cell wall regulators of *Arabidopsis*, poplar, rice, maize, and switchgrass. *BMC Plant Biol.* 14:135. doi: 10.1186/1471-2229-14-135

Zhong, R., Burk, D. H., Morrison, W. H. III, and Ye, Z. H. (2004). FRAGILE FIBER3, an *Arabidopsis* gene encoding a type II inositol polyphosphate 5-phosphatase, is required for secondary wall synthesis and actin organization in fiber cells. *Plant Cell* 16, 3242–3259. doi: 10.1105/tpc.104.02 7466

Zhong, R., and Ye, Z. H. (2014). Complexity of the transcriptional network controlling secondary wall biosynthesis. *Plant Sci.* 229, 193–207. doi: 10.1016/j.plantsci.2014.09.009

Zhou, C., Yin, Y., Dam, P., and Xu, Y. (2010). Identification of novel proteins involved in plant cell-wall synthesis based on protein-protein interaction data. *J. Proteome Res.* 9, 5025–5037. doi: 10.1021/pr100249c

Conflict of Interest Statement: The authors declare that the research was conducted in the absence of any commercial or financial relationships that could be construed as a potential conflict of interest.

14

Ternary WD40 Repeat-Containing Protein Complexes: Evolution, Composition and Roles in Plant Immunity

Jimi C. Miller[1†], William R. Chezem[2†] and Nicole K. Clay[2]*

[1] Department of Molecular Biophysics and Biochemistry, Yale University, New Haven, CT, USA, [2] Department of Molecular, Cellular and Developmental Biology, Yale University, New Haven, CT, USA

Plants, like mammals, rely on their innate immune system to perceive and discriminate among the majority of their microbial pathogens. Unlike mammals, plants respond to this molecular dialog by unleashing a complex chemical arsenal of defense metabolites to resist or evade pathogen infection. In basal or non-host resistance, plants utilize signal transduction pathways to detect "non-self," "damaged-self," and "altered-self"-associated molecular patterns and translate these "danger" signals into largely inducible chemical defenses. The WD40 repeat (WDR)-containing proteins Gβ and TTG1 are constituents of two independent ternary protein complexes functioning at opposite ends of a plant immune signaling pathway. They are also encoded by single-copy genes that are ubiquitous in higher plants, implying the limited diversity and functional conservation of their respective complexes. In this review, we summarize what is currently known about the evolutionary history of these WDR-containing ternary complexes, their repertoire and combinatorial interactions, and their downstream effectors and pathways in plant defense.

Keywords: heterotrimeric G-protein, MYB-bHLH-WDR complex, TTG1, Gβ, plant innate immunity, RACK1, flavonoid metabolism

Edited by:
Marco A. Villanueva,
Instituto de Ciencias del Mar y Limnología, UNAM, Mexico

Reviewed by:
Viktor Zarsky,
Charles University, Czech Republic
Tania Islas-Flores,
Instituto de Ciencias del Mar y Limnología, UNAM, Mexico

****Correspondence:***
Nicole K. Clay
nicole.clay@yale.edu

† These authors have contributed equally to this work.

INTRODUCTION

WD40 repeat (WDR)-containing proteins are prevalent in eukaryotes, but rarely present in prokaryotes (Janda et al., 1996; Stirnimann et al., 2010). Plant genomes typically encode more than 200 putative WDR-containing proteins (van Nocker and Ludwig, 2003; Ouyang et al., 2012), which is slightly less than the human genome (349; Letunic et al., 2014). The basic function of WDR-containing proteins is to serve as rigid scaffolds for protein–protein and protein-DNA interactions. WDR-containing proteins are involved in fundamental mechanisms such as signal transduction, chromatin modification and transcriptional regulation. They are also involved in a wide variety of plant processes, including cell division, meristem organization, light signaling, floral development, secondary metabolism, and innate immunity (Smith et al., 1999; van Nocker and Ludwig, 2003; Perfus-Barbeoch et al., 2004).

Plants, unlike mammals, lack mobile defender cells and an adaptive immune system. Instead, they rely on the innate immunity of each cell, systemic peptide and chemical signals emanating from infection sites, and preformed and inducible chemical defenses at infection sites to ward off invading pathogens (Dixon, 2001; Jones and Dangl, 2006; Zipfel, 2014). Plants, like mammals, have a multi-tiered pathogen-detection system. The first layer is evolutionarily more ancient and involves the cell-surface perception of conserved microbial or "non self" molecular signatures known as microbe-/pathogen-associated molecular patterns (or MAMPs/PAMPs) and pathogen-generated "damaged/altered-self" molecular signatures known as damage-associated molecular patterns (or DAMPs). These "danger" signals are recognized by pattern recognition receptors (or PRRs), which in plants are plasma membrane-localized receptor-like proteins (RLPs) or receptor-like kinases (RLKs). MAMPs, *inter alia*, are also thought to be the molecular determinants of induced systemic resistance (ISR) that is activated by beneficial plant-microbe interactions in the roots (Van Wees et al., 1997; Meziane et al., 2005; Bakker et al., 2007). The second layer of immunity involves the cytosolic perception of pathogen-specific effector proteins by intracellular nucleotide binding leucine-rich repeat (or NB-LRR)-containing resistance proteins to trigger programmed cell death at infection sites and, in many cases, systemic acquired resistance in the host plant (Jones and Dangl, 2006).

Plant immunity, in particular, boasts two distinct but structurally similar classes of WDR-containing proteins: Gβ, and TRANSPARENT TESTA GLABRA1 (TTG1). The Gβ and TTG1 proteins are constituents of ternary regulatory complexes (**Figures 1A, 2A**). While the Gβ is widely conserved across a diversity of eukaryotes (Adams et al., 2011; Bradford et al., 2013), TTG1 is only present in higher plants (**Figure 1B**). The Gβ and TTG1 proteins are coupled to type-I membrane receptors and transcription factors, respectively, in a plant innate immune signaling pathway that convert extracellular signals into a subset of intracellular chemical defense responses (see below).

Structural Conservation of WDR-Containing Proteins Gβ and TTG1

The common and defining structural feature of WDR-containing proteins is the seven-tandem WDR motif sequence, which adopts a seven-bladed β-propeller-like structure with three potential surfaces for molecular interactions—the top, bottom and circumference (**Figure 3A**; Lambright et al., 1996; Smith et al., 1999; Ullah et al., 2008; Adams et al., 2011; Ruiz Carrillo et al., 2012). Each blade of the propeller-like structure consists of four antiparallel β strands; the first three strands of one blade and the fourth strand of the next are formed by a single WDR motif; the overlap between two adjacent propeller blades provides an interlocking architecture that holds the propeller-like structure in a closed, rigid ring structure (Smith et al., 1999; Adams et al., 2011). The seven-bladed β-propeller structure is best demonstrated for the WDR-containing receptor for activated C kinase 1 (RACK1) protein, for which there are four known structures from a diversity of eukaryotes (**Figure 3A**; Ullah et al., 2008; Coyle et al., 2009; Rabl et al., 2011; Ruiz Carrillo et al., 2012).

Unlike TTG1, the Gβ protein additionally contains an *N*-terminal α-helix (**Figure 3A**) that forms a coiled-coil structure with the Gγ protein, as indicated by the crystal structure of the human Gβ HsGNB1 partially encircled by the Gα HsGNAT1 (Sondek et al., 1996). However, HsGNB1 remains the sole Gβ with a solved crystal structure, which serves as the foundation (along with a handful of solved RACK1 structures) for the predicted Gβ structures generated by structural bioinformatics. Within these confines, there is some evidence that the Gβ-specific structure mediating the Gβγ interaction may not be conserved across eukaryotes. For example, Gβ proteins from more primitive eukaryotes (e.g., alveolate *Tetrahymena thermophila* Gβ and the green alga *Chara braunii* Gβ) are predicted to lack the *N*-terminal helix (**Figure 3A**) but still retain the Gβγ interaction (Hackenberg et al., 2013), presumably through a novel Gβγ interaction domain(s) within the β-propeller structure. Additional crystal structures of non-metazoan Gβ sequences are needed to provide structural details on the Gβγ interaction across eukaryotes.

HETEROTRIMERIC G-PROTEIN COMPLEX

The most extensively studied WDR-containing protein to date is the Gβ subunit of the heterotrimeric G-protein complex, which is one of the most conserved and elaborate receptor-effector signaling mechanisms in eukaryotes. The Gβ reversibly interacts with the GDP-bound Gα subunit and forms an obligate heterodimer (Gβγ) with the Gγ subunit. While the interaction between the Gα and the Gβγ dimer serves as a molecular switch, the Gβ serves as a scaffold for effector proteins (**Figure 1A**). In animals and fungi, ligand perception by the heptahelical membrane receptors, G-protein-coupled receptors (GPCRs), leads to replacement of GDP with GTP in the Gα subunit and activation of the heterotrimer (Li et al., 2007; Oldham and Hamm, 2008). Upon activation, the GTP-bound Gα and Gβγ dimer dissociate from each other and from the receptor complex, releasing their bound effectors to activate various signaling cascades. Signaling terminates when the intrinsic GTPase activity of the Gα hydrolyzes GTP to GDP and the inactive heterotrimer reforms at the receptor.

Elusive Receptor-Effector Signaling Mechanism

Although signal transduction through a heterotrimeric G-protein complex is common to animals and plants, there are some mechanistic differences between the evolutionary branches. For example, in plants and basal eukaryotes, the canonical Gα subunit isoform is self-activating, and thus does not require GPCR-like proteins for its activation (Jones et al., 2011a,b; Bradford et al., 2013). Plants also contain non-canonical Gα subunit isoforms, which have a slower rate of GTP hydrolysis (Heo et al., 2012), but it is not yet known whether they are also self-activating. In addition, canonical GPCR-like sequences are absent or rare in plants (Urano et al., 2013; Taddese et al., 2014). Instead, plants have several families of non-canonical GPCR-like sequences, three of which (GCR1, GTG1, and GTG2) have been shown to interact *in planta* with the *Arabidopsis* canonical Gα GPA1 and modulate an ABA-mediated drought response (Pandey and Assmann, 2004; Pandey et al., 2009). It remains controversial whether the GPCR-like proteins are *bona fide* GPCRs, although a recent structural bioinformatics study has found GCR1 to be a strong GPCR candidate based on its predicted heptahelical scaffold and GPCR fold (Taddese et al., 2014). Plants also contain hundreds of membrane RLP and RLK sequences (Shiu and Bleecker, 2001, 2003; Fritz-Laylin et al., 2005), two of which (the

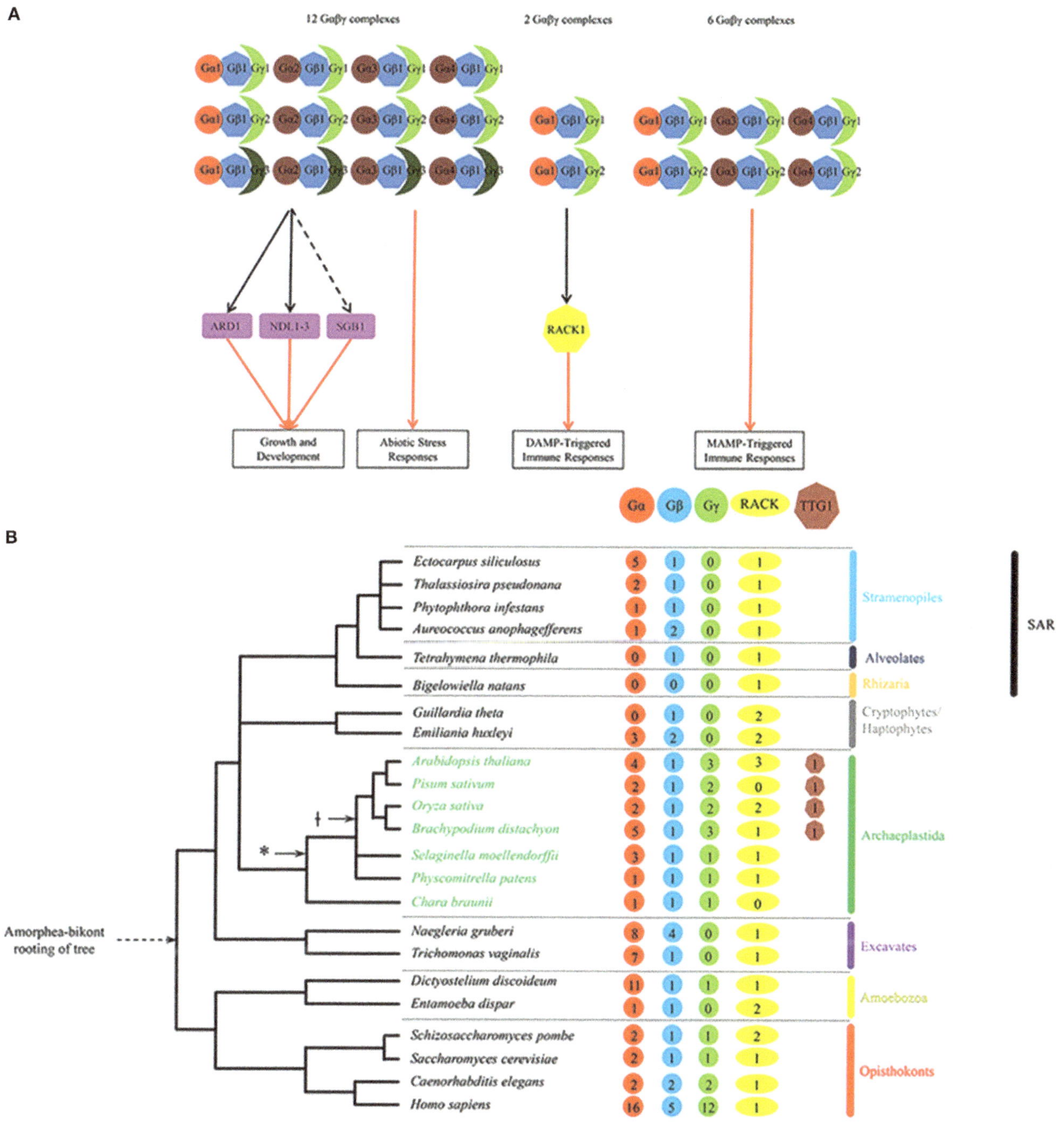

FIGURE 1 | (A) Regulatory network of known Gβ-dependent pathways in *Arabidopsis* illustrating the interactions between G-protein subunits and between Gβ1 and its effectors for various regulated plant processes. Growth and development processes include stomatal density and opening, seed germination, hypocotyl elongation, and organ (i.e., leaf, silique, seed) morphology. Abiotic stress responses include salt stress, chemical-induced endoplasmic reticulum stress, and sugar stress. DAMP-triggered immune responses include MAP kinase activation and ROS generation. MAMP-triggered immune responses include aforementioned immune responses as well as modulation of the flavonoid anthocyanin pathway. Unbroken and broken black lines indicate indirect and direct interaction, respectively; red arrow indicates positive regulation. Shapes: heptagons, WDR-containing proteins Gβ and AtRACK1; circles, Gα proteins; moons, Gγ proteins; rectangles, downstream Gβγ effectors. For a given shape, different colors denote different classes of G-protein subunit isoforms or WDR-containing proteins. *Arabidopsis* proteins: Gβ1, AGB1; Gα1, GPA1; Gα2, XLG1; Gα3, XLG2; Gα4, XLG3; Gγ1, AGG1; Gγ2, AGG2; and Gγ3, AGG3. Note that the expanded diversity of the non-WDR-containing subunits in the complex likely provides functional specificity within plant innate immune signaling. Also note the paucity of identified effectors downstream of the G-protein complexes for all known regulated processes. **(B)** Tree depicting the number of Gβ, RACK1 and TTG1 sequences identified to date for representative species of the five eukaryote supergroups as well as two major plant evolutionary milestones: plant colonization of land (*) and the first appearance of the bHLH interaction motif present on constituent MYBs in the MYB-bHLH-TTG1 complex (ǂ). Note that no Gβ or TTG1 sequences were identified in the Rhizaria group.

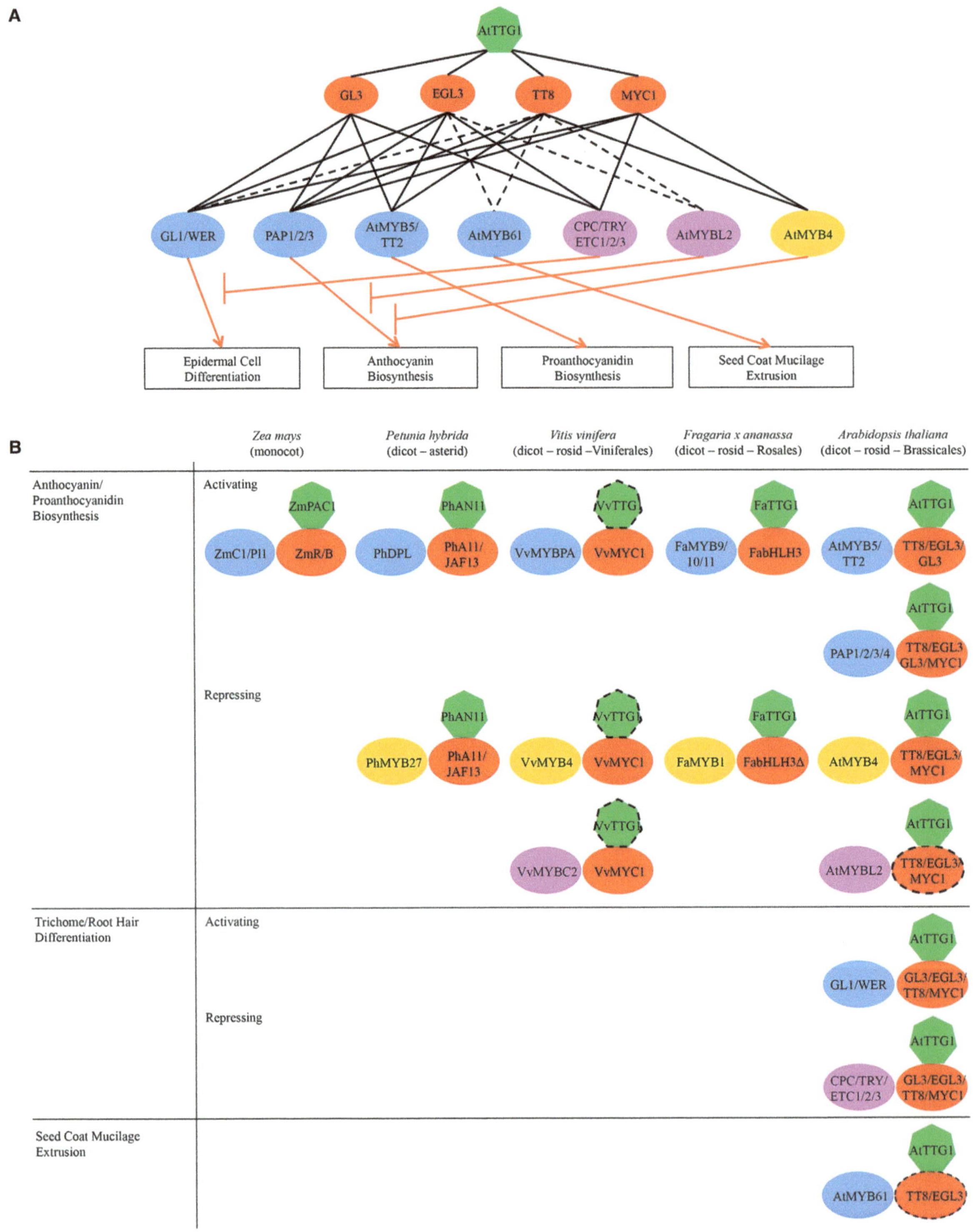

FIGURE 2 | (A) Regulatory network of AtTTG1-dependent pathways in *Arabidopsis* illustrating the potential/indirect (broken black lines) and demonstrated (unbroken black lines) interactions among proteins/genes. Boxed are regulated processes. Key: green heptagons, WDR-containing protein AtTTG1; red ovals, bHLH transcription factors from subgroup IIIf; blue ovals, activator-type R2R3-type MYB subfamilies; yellow ovals, repressor-type R2R3-type MYBs; purple ovals, repressor-type R3-type MYB subfamilies. Note the narrowing of regulatory specificity across the tiers from top to bottom. **(B)** Schematic representation of MYB-bHLH-TTG1 (MBW) complexes regulating flavonoid defense metabolism and other processes in representative plant species of major clades within flowering plants (monocot vs. dicot, asterid vs. rosid, different rosid orders). Dotted lines indicate potential genetic/physical interactions. Note the functional conservation of MBW complexes in the regulation of flavonoid biosynthesis across different plant species.

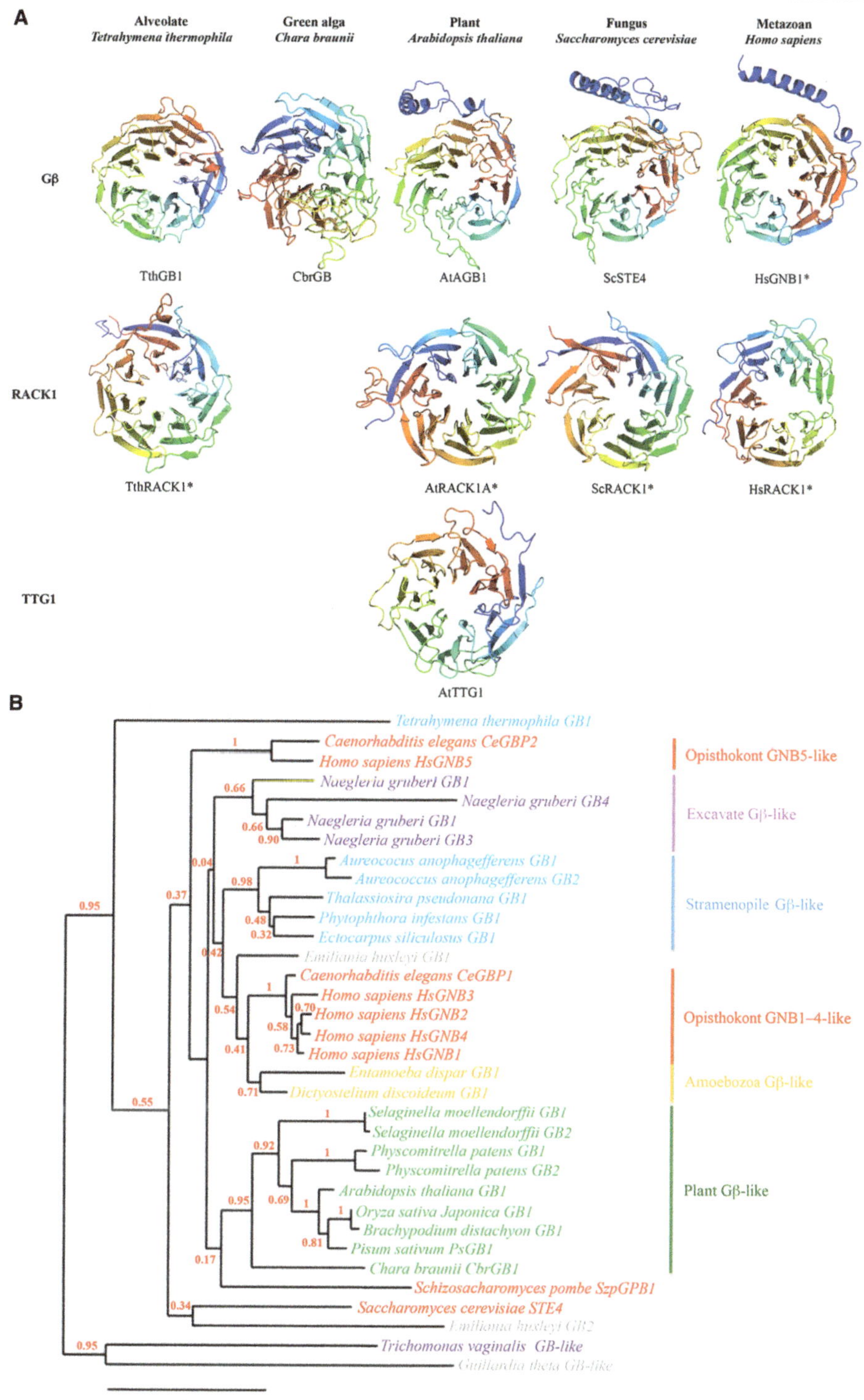

FIGURE 3 | (A) Seven-bladed propeller-like structures of WDR-containing proteins Gβ (top row), RACK1 (middle row) and TTG1 (bottom row) proteins from an alveolate, green algal, plant, fungal, and metazoan species. Homology models were based on known structures of TthRACK1, AtRACK1A, ScRACK1, HsGNB1, and HsRACK1 (marked by asterisks) as well as predicted structures from multiple sequence templates using the PHYRE2 protein fold recognition server (www.sbg.bio.ic.ac.uk/phyre2/; Kelley and Sternberg, 2009). Acronyms: Tth, *Tetrahymena thermophila*; Cbr, *Chara braunii*; At, *Arabidopsis thaliana*; Sc, *Saccharomyces cerevisiae*; Hs, *Homo sapiens*. Note the presence of an *N*-terminal alpha helix on the plant, fungal and metazoan Gβ proteins but not on the alveolate and green algal Gβ proteins. **(B)** Maximum likelihood phylogenetic tree of Gβ sequences from representative species in the five eukaryotic supergroups. Tree was generated using MUSCLE multiple sequence alignment, PhyML phylogeny, and TreeDyn tree viewer programs (http://phylogeny.lirmm.fr; Dereeper et al., 2008). Bootstrap value ($n = 100$ replicate trees) is shown in red at the nodes. Note that the plant Gβ sequences cluster as a well-supported monophyletic group.

maize RLP FEA2 and the *Arabidopsis* RLK RPK2) have been shown to interact *in planta* with the maize canonical Gα CT2 and *Arabidopsis* canonical Gβ AGB1, respectively, to regulate stem cell proliferation (meristem organization; Bommert et al., 2013; Ishida et al., 2014). Finally, while all three constituents of the mammalian G-protein complex interact with the GPCR (Taylor et al., 1994, 1996), only the Gα and Gγ subunits of the *Arabidopsis* heterotrimer have been shown to interact with the receptor complex (Aranda-Sicilia et al., 2015). In the absence of receptor interaction, it remains unclear how the Gβ subunit participates in the receptor signaling complex.

Downstream, the plant Gβγ dimer has been shown to regulate the MAPK cascade by interacting directly with a MAPK protein (Bhardwaj et al., 2011; Xu et al., 2015) or by recruiting RACK1 proteins as MAPK cascade scaffolds (Cheng et al., 2015). By contrast, mammalian and fungal Gβγ dimers instead recruit the MAPK scaffolding proteins β-arrestin2 and Ste5, respectively (Witzel et al., 2012), while mammalian RACK1 proteins serve as Jun N-terminal kinase (JNK) MAPK cascade scaffolds for the protein kinase C (PKC) signaling pathway (Ron et al., 1994; López-Bergami et al., 2005). Although RACK1 is highly conserved between plants and animals (**Figure 1B**), β-arrestin2, Ste5 and second-messenger-regulated PKC proteins are absent in plants (Stone and Walker, 1995; Witzel et al., 2012). Despite the diversity of MAPK cascade scaffolds between plants and animals, the use of scaffolding proteins in signal transduction pathways appears universal.

Evolutionary History of the Plant Gαβγ Trimer

Gβ sequences (and those of the other two G-protein subunits) are present in the genomes of all five eukaryotic supergroups Archaeplastida, Excavata, Opisthokonta, Amoebozoa, and Stramenopila/Alveolata/Rhizaria (or SAR), and are absent only in the Rhizaria subgroup of SAR (**Figure 1B**). Although each supergroup consists of a diversity of eukaryotes, most of which are microbial (e.g., protists and algae; Keeling et al., 2005; Burki, 2014), the best-characterized Gβ sequences are from animals/metazoans and fungi in the Opisthokonta supergroup. The oldest extant Gβ sequence in the Archaeplastida supergroup (e.g., land plants and green/red algae) is a single-copy gene found in the green alga *Chara braunii* (Hackenberg et al., 2013; **Figure 1B**). This green algal Gβ sequence is not distinct from the Gβ sequences present in the genomes of basal plant lineages (e.g., bryophytes and lycophytes) and the diploid genomes of higher plant lineages (**Figure 3B**; Urano et al., 2013), indicating that they descended from a single ancestral plant Gβ sequence. In contrast, phylogenetic analysis of metazoan Gβ sequences identified three distinct Gβ classes (GNB1-4-like, GNB5-like, and Gbe-like); the first two are found in humans, and the third is specific for arthropods (de Mendoza et al., 2014; Krishnan et al., 2015). GNB1-4-like and GNB5-like sequences are likely present in the last common metazoan ancestor and are confined within metazoans (de Mendoza et al., 2014; Krishnan et al., 2015; **Figure 3B**).

Previous phylogenetic analysis for ancestral plant Gβ sequences suggested that plant Gβ sequences are more closely related to Gβ sequences from the SAR (e.g., diatom) and Amoebozoa (e.g., entamoeba) supergroups than those of Excavata (Friedman et al., 2009). Although it is still not clear how the eukaryotic supergroups relate to one another, the most popular hypothesis (Amorphea-bikont rooting) places the root of the eukaryotic tree between the last common ancestor of the amoebozoans and opisthokonts and the remaining eukaryotes (Keeling et al., 2005; Burki, 2014). The Amorphea-bikont rooting positions the Gβ sequences in the Excavata supergroup between the plant Gβ sequences and those of the amoebozoans and opisthokonts (**Figure 1B**). Phylogenetic analysis of a representative sampling of Gβ sequences from all five supergroups supports this hypothesis by sandwiching the Gβ sequences in the Excavata supergroup between the plant Gβ sequences and the animal GNB1-4-like sequences (**Figure 3B**).

Combinatorial Diversity of Plant G-Proteins

Although the heterotrimeric G-protein complex consists of three subunits, subunit isoforms can give rise to many heterotrimeric combinations, limited in part by amino acid sequence differences in the contact regions that lead to selective interactions. Given the large number of known G-protein-mediated signaling pathways, a diversity of G-protein isoforms is needed for signaling specificity (Wettschureck and Offermanns, 2005). For example, the human genome encodes 16 Gα, 5 Gβ, and 12 Gγ subunit isoforms, allowing for approximately 700 potential Gαβγ combinations (Hillenbrand et al., 2015; **Figure 1B**). By contrast, *Arabidopsis thaliana*, like most diploid plants, contains 4 Gα (GPA1, XLG1–3), one Gβ (AGB1), and 3 Gγ (AGG1–3) subunit isoforms, allowing for 12 potential Gαβγ combinations (**Figure 1**; Ma et al., 1990; Weiss et al., 1994; Mason and Botella, 2000, 2001; Zhu et al., 2009; Thung et al., 2012; Chakravorty et al., 2015; Maruta et al., 2015). This number falls short of the specificity needed for the large number of known G-protein-mediated signaling pathways regulating fundamental processes in plants, and remains a bottleneck issue in plant G-protein signaling (Urano et al., 2013).

The sole Gβ subunit isoform is a limiting factor for plant G-protein combinatorial diversity. There are different complex models on how one Gβ subunit isoform is able to transduce so many diverse signals (Urano and Jones, 2014). In addition, the ubiquitous presence of Gβ-like sequences across plant genomes has led to a hypothesis on the existence of additional non-canonical classes of plant Gβ subunits that have yet to be discovered, a situation analogous to the recent discoveries of new classes of plant Gα sequences (XLG1–3-like) and plant Gγ subunits (AGG3-like; Lee and Assmann, 1999; Thung et al., 2012). The XLG1–3-like Gα subunit differs from the canonical Gα subunit in its possession of a long N-terminal extension of unknown function and its nuclear- and plasma membrane-localization (Ding et al., 2008; Chakravorty et al., 2015; Maruta et al., 2015). Similarly, the AGG3-like Gγ subunit differs from the canonical Gγ subunit in its possession of a C-terminal extension that is cysteine-rich and of unknown function (Chakravorty et al., 2011; Trusov et al., 2012).

Aside from genetic interaction data, there is physical interaction evidence from yeast three-hybrid studies supporting interaction specificity within the heterotrimer and its putative coupled receptor/adaptor. For example, the *Arabidopsis* Gα subunit isoforms, XLG1 and XLG2 have been shown to strongly interact with the Gβγ heterodimers AGB1-AGG1/2, while the Gα subunit isoform GPA1 strongly interacts with the Gβγ heterodimer AGB1-AGG3 and XLG3 strongly interacts with all three Gβγ heterodimers AGB1-AGG1/2/3 (Chakravorty et al., 2015; Maruta et al., 2015), suggesting that all three Gγ isoforms are somewhat selective of their interaction partners, each preferring two of the four Gα isoforms. In addition, yeast split-ubiquitin and Bimolecular Fluorescence Complementation (BIFC) studies indicate that the other two Gγ isoforms AGG1/2 mediate the interaction between the plant heterotrimer and the co-receptor proteins BAK1 and CERK1 (Aranda-Sicilia et al., 2015). These reports are consistent with similar reports of animal Gγ isoforms conferring specificity to the G-protein complex-GPCR interaction (Im et al., 1988; Kisselev and Gautam, 1993).

G-Protein Complexes in Defense

One of the best-characterized functions of the *Arabidopsis* heterotrimeric G-protein complex is in plant innate immunity, where it participates in multiple immune signaling pathways and defense responses (e.g., reactive oxygen species (ROS) production, mitogen-activated protein kinase (MAPK) activation, defense gene activation, callose deposition, and programmed cell death) against a variety of fungal (Llorente et al., 2005; Trusov et al., 2006, 2007, 2009; Delgado-Cerezo et al., 2012; Torres et al., 2013) and bacterial pathogens (Zhang et al., 2008; Ishikawa, 2009; Zeng and He, 2010; Lee et al., 2013; Liu et al., 2013; Torres et al., 2013). Evidence of a physical interaction between a plant heterotrimer and a ligand-binding innate immune receptor (e.g., FLS2, EFR, LYK4/5) is still elusive, although a recent report showed a direct interaction between the canonical Gα GPA1 and the Gγ isoforms AGG1/2 (but not the Gβ AGB1) with the co-receptor proteins BAK1 and CERK1 by yeast split-ubiquitin assay and BiFC studies (Aranda-Sicilia et al., 2015). If validated, this is the first report of a novel plant-specific interaction between a heterotrimer and a receptor complex via co-receptor adaptors. If the plant heterotrimer is coupled directly to the receptor complex, then further research is needed to understand how the plant heterotrimer converts MAMP and/or DAMP signals from the receptors into intracellular defense responses, especially if the heterotrimer is self-activating.

Nearly all of the *Arabidopsis* G-protein subunit isoforms (save two—XLG1 and AGG3) participate in plant defense (**Figure 1A**; Maruta et al., 2015) and an even smaller subset of G-protein subunit isoforms in a bacterial DAMP-triggered immune pathway involving RACK1 proteins as MAPK cascade scaffolds (**Figure 1A**; Cheng et al., 2015). The sole Gβ AGB1 participates in all G-protein-mediated processes, and positively contribute to all tested immune responses, including ROS production, callose deposition, MAPK activation, defense gene activation and programmed cell death (Llorente et al., 2005; Maeda et al., 2009; Liu et al., 2013; Torres et al., 2013). Among the Gα subunit isoforms, XLG2 is the major contributor to resistance against the hemibiotrophic bacterium *Pseudomonas syringae*, necrotrophic fungi *Alternaria brassicicola* and *Plectosphaerella cucumerina*, and the hemibiotrophic fungus *Fusarium oxysporum*. The loss-of-function *xlg2* mutant most closely recapitulates the phenotypes of the loss-of-function *agb1* mutant in its pathogen susceptibility (Llorente et al., 2005; Trusov et al., 2006; Zhu et al., 2009; Torres et al., 2013; Maruta et al., 2015), and the XLG2 protein was shown to interact with the AGB1 protein *in planta* by co-immunoprecipitation of overexpressed proteins (Zhu et al., 2009). In addition, the Gα isoform GPA1 contributes to bacterial resistance by mediating stomatal closure, a MAMP-triggered immune response that retards pathogen entry through the stomata (natural openings in the plant surface; Zhang et al., 2008), while the Gα isoform XLG3 contributes partly to resistance against *Fusarium oxysporum* (Maruta et al., 2015) through an unknown mechanism. Among the Gγ subunit isoforms, AGG1 and AGG2 are mostly redundant in their contribution to plant immunity (Trusov et al., 2007; Thung et al., 2013). The loss-of-function *aggl/agg2* double mutant recapitulates the phenotypes of the *agb1* mutant in pathogen susceptibility (Trusov et al., 2007; Liu et al., 2013; Torres et al., 2013; Maruta et al., 2015). These results suggest that multiple G-protein combinations are required for a successful plant defense response against a majority of pathogens, with the Gβ playing a central but undefined role in immune signaling.

Search for G-Protein Complexes in Pathogenesis

Although compatibility between plants and their pathogens leads to disease and symptom development, it is rarely found in nature due to the effectiveness of the plant innate immune system. One exceptional case is the small family of GPCR-like mildew resistance locus O (MLO) receptor proteins, which are found throughout flowering plants. A subset of MLO proteins has been shown to be a conserved requirement for the compatibility of monocots and dicots with their adapted powdery mildew pathogens (Devoto et al., 2003; Consonni et al., 2006; Humphry et al., 2006). While it remains controversial whether GPCR-like proteins are *bona fide* GPCRs, some *Arabidopsis* MLO proteins have predicted heptahelical scaffolds, GPCR folds, and G-protein coupling (Taddese et al., 2014). However, the three MLO proteins involved in fungal pathogenesis (AtMLO2/6/12) do not contain these GPCR hallmarks and thus are unlikely to function as canonical GPCRs (Taddese et al., 2014). In addition, attempts to couple the two plant heterotrimers (GPA1-AGB1-AGG1/2) to the MLO receptor were not successful (Lorek et al., 2013). Still, further research is needed to discern whether other heterotrimeric combinations are involved in regulating the compatibility between plants and their pathogens.

TERNARY MYB-bHLH-TTG1 COMPLEXES

Flavonoids in Defense

Plant secondary (or defense) metabolites are essential for plant survival in complex environments, and are the primary means of defense against microbial pathogens as well as insect herbivores and competitor plants (Dixon, 2001; Kliebenstein, 2013). The phenylpropanoids represent one of the largest and

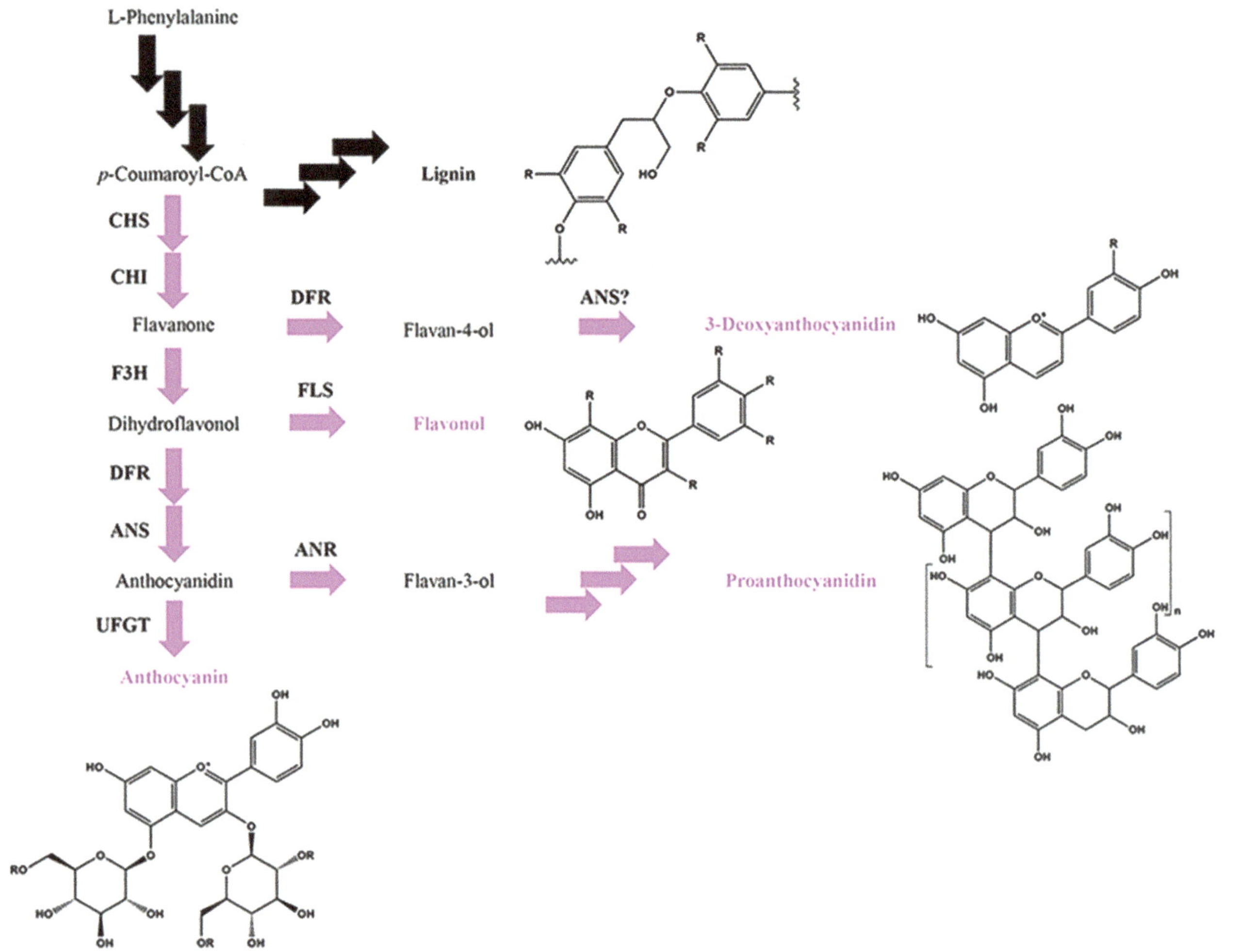

FIGURE 4 | Simplified phenylpropanoid pathway depicting the flavonoid (in purple) and lignin branch pathways, major flavonoid classes and basic flavonoid structures. Multiple arrows indicate multiple enzymatic steps. Key flavonoid steps: CHS, chalcone synthase; CHI, chalcone isomerase; F3H, flavanone 3-hydroxylase; FLS, flavonol synthase, DFR, dihydroflavonol 4-reductase; ANS, anthocyanidin synthase; ANR, anthocyanidin reductase; UFGT, UDP-glucose flavonoid 3-*O*-glucosyltransferase.

most ancient families of defense metabolites in land plants. The two largest subfamilies within the phenylpropanoids are the flavonoids and the lignins, whose partially overlapping, competing branch pathways share the first three steps in the phenylpropanoid biosynthetic pathway (**Figure 4**). Among the flavonoid classes, the anthocyanins, proanthocyanidins and 3-deoxyanthocyanidins have been shown to be critical for disease resistance in a number of plant species. For example, the 3-deoxyanthocyanidin phytoalexins in *Sorghum* contribute to localized resistance against the *Sorghum* anthracnose disease fungus *Colletotrichum sublineolum*, the southern leaf blight fungus *Cochliobolus heterostrophus* and head smut fungus *Sporisorium reilianum* (Nicholson et al., 1987; Snyder and Nicholson, 1990; Lo et al., 1999; Zuther et al., 2012). In grapevine, proanthocyanidins (also known as condensed tannins) contribute to resistance against the gray mold fungus *Botrytis cinerea* by competitively inhibiting fungal laccases involved in plant cell wall degradation and phytoalexin detoxification (Pezet et al., 1992; van Baarlen et al., 2007). In *Arabidopsis*, the colorful anthocyanins may be the molecular basis of the ISR that is induced by the biocontrol fungus *Trichoderma hamatum* against *Botrytis cinerea* (Mathys et al., 2012; Kottb et al., 2015). Additionally, the *Trichoderma*-synthesized volatile 6-pentyl-α-pyrone may be the MAMP signal responsible for triggering ISR (Kottb et al., 2015). Although the identities of the flavonoid metabolite(s) involved in ISR remain unknown, mutant analysis confirmed that flavonoid biosynthesis is required for ISR (Mathys et al., 2012).

Anthocyanins, proanthocyanidins and 3-deoxyanthocyanidins represent partially overlapping, competing pathways within the flavonoid biosynthetic pathway (**Figure 4**; Hipskind et al., 1996; Winkel-Shirley, 2001). In addition, the flavonoid pathway itself competes with the lignin pathway for shared resources (**Figure 4**). The immediate biochemical and physiological needs for the defense response are best illustrated during the *de novo* synthesis of defense metabolites critical to disease resistance, where the synthesis of more common/less essential phenylpropanoids are repressed in order to maintain metabolic balance between the competing pathways (Kombrink and Hahlbrock, 1990). For example, in *Arabidopsis*, the bacterial MAMP flg22 transcriptionally upregulates the synthesis of lignins, while repressing the light-dependent production of anthocyanins (Saijo et al., 2009; Adams-Phillips et al., 2010; Schenke et al., 2011). Similarly, in *Sorghum*, unknown fungal-derived signals transcriptionally upregulate the synthesis of the 3-deoxyanthocyanidins, while repressing light-dependent production of anthocyanins (Weiergang et al., 1996; Lo and Nicholson, 1998; Wharton and Nicholson, 2000; Shih et al., 2006). Defense metabolites can also be synthesized from two or more overlapping pathways, requiring more complex pathway regulation. For example, most monocots, including maize, rice, and wheat, incorporate the flavonoid tricin into the lignin polymer (Lan et al., 2015), although it is not yet known whether the flavonoid content contributes to the defense function of lignins.

Elusive Transcription Factor-Regulon Mechanism

In plants, the majority of the non-homologous gene constituents of well-characterized multi-gene pathways in defense metabolism and cell differentiation are not genetically clustered as they are in fungi and bacteria, but are distributed throughout the plant genome (Lappin et al., 2006; Kliebenstein and Osbourn, 2012). Instead, to facilitate their common regulation, co-regulated genes along a pathway or sections within typically share common or overlapping sets of promoter elements that are bound by specific transcription factor, and thus are transcriptionally clustered into an operon- like gene module known as a regulon. This functional gene organization allows for a limited set of transcriptional regulators to combinatorially control a regulon for pathway function. The best-studied transcriptional regulator of pathways in defense metabolism in higher plants is the MYB-bHLH-WDR (MBW) complex, wherein the sole constituent WDR-containing protein is known as TTG1 and the MYB and bHLH constituents are paralogous transcription factors from small subfamilies (analogous to classes of G-protein subunit isoforms; **Figure 2A**). MYB constituents of the MBW complexes are either R2R3-type or R3-type MYBs, where R stands for the number of adjacent repeats of the MYB DNA-binding, helix-turn-helix domain; plant MYBs lack the first repeat (Kranz et al., 1998; Stracke et al., 2001; Dubos et al., 2010).

The pathway specificity of the MBW complex is determined by the diversity of the MYB constituent, limited in part by the overlapping functions of paralogous members within a MYB subfamily (Stracke et al., 2001). However, *in vitro* promoter binding studies to date indicate that R2R3-type MYBs from different subfamilies appear to bind to the same or very similar *cis*-regulatory elements, which are described generally as "AC-like" sequences (e.g., TACC(T/A)A(C/A)C (MBSIIG1–4 motif), CACC(T/A)A(C/A)C (MBSIIG5–8 motif), ACCTACC (AC-I or SMRE-8 motif), ACCAACC (AC-II or SMRE-4 motif), ACCTAAC (AC-III or SMRE-7 motif), ACC(T/A)ACC (AC-I/SMRE-8 or AC-II/SMRE-4 motif), and ACCCGCC) (Grotewold et al., 1994; Williams and Grotewold, 1997; Romero et al., 1998; Zhao et al., 2007; Zhou et al., 2009; Fornalé et al., 2010; Prouse and Campbell, 2013). However, *in vitro*-defined consensus motifs may not be sufficient and/or present in the majority of their *in vivo* DNA target sequences, as was demonstrated for several animal transcription factors (Carr and Biggin, 1999; Yang et al., 2006; Rabinovich et al., 2008) and the maize R2R3-type MYB ZmMYBP1 from subgroup 7, which was found to preferentially bind to sequences containing 6 to 8 repeats of CxxC (where X corresponds to any nucleotide) with no preference for A or T between the C base pairs (Morohashi et al., 2012). Currently, it remains unclear how the different MYB subfamilies confer pathway specificity to the MBW complex.

Because AC-like sequences are also typically present in the promoters of activator- and repressor-type R2R3-type MYB genes, constituent MYBs not only participate in MBW complexes but also can transcriptionally regulate other components in the complexes and/or competing MYBs via autoregulatory, feedback or feedforward loops (Baudry et al., 2006; Zhao et al., 2007). In addition, the expression of repressor-type MYBs is typically development-specific or stress-responsive and thus can serve to fine-tune pathway regulation (Jin et al., 2000; Preston et al., 2004; Fornalé et al., 2014). For example, chromatin immunoprecipitation experiments in *Arabidopsis* have shown that the WDR-containing TTG1 and the bHLH constituent GL3 are associated with the promoters of pathway genes in anthocyanin and proanthocyanidin metabolism as well as the promoters of repressor- and R3-type MYBs CPC, TRY, and ETC1 (Morohashi et al., 2007; Zhao et al., 2008).

Role of TTG1 in Defense

The WDR-containing TTG1 has both direct and indirect roles in plant defense. For example, heterologous expression of the *Sorghum* R2R3-type MYB SbY1 in maize induced 3-deoxyanthocyanidin synthesis and enhanced resistance against leaf blights (Ibraheem et al., 2010, 2015). Although maize contains the SbY1 ortholog ZmP1 (Grotewold et al., 1994) and the necessary flavonoid biosynthetic genes, ZmP1 does not participate in MBW complexes nor induce 3-deoxyanthocyanidin synthesis in response to fungal attack (Ibraheem et al., 2015). These observations suggest that SbY1 requires additional regulatory factors to induce 3-deoxyanthocyanidin synthesis in *Sorghum* (and in maize). One of these factors may be the *Sorghum* TTG1 ortholog Tan1 (Mizuno et al., 2014). QTL analysis identified *Tan1* gene as a major determinant of the relative proportions of 3-deoxyanthocyanidins responsible for the pathogen-induced color variation in *Sorghum* (Mizuno et al., 2014). TTG1 has also been found to play a direct role in plant defense in dicots. For example, in tobacco, TTG1 physically interacts with the oomycete-specific effector ParA1 to activate plant immune responses, such as ROS generation and programmed cell death (Wang et al., 2009).

Combinatorial MYB-bHLH Interactions

TTG1 serves as a scaffold for the combinatorial interactions between the MYB and bHLH constituent isoforms, which are essential for regulating the binding specificity and transcriptional activity of the MBW complex. Both the MYB and bHLH constituents can be sequestered by other bHLH and MYB proteins, respectively, to form non-functional MYB-bHLH complexes and thus disrupt the MBW complex (Burr et al., 1996; Sawa, 2002; Esch et al., 2003; Kirik et al., 2004; Simon et al., 2007; Tominaga et al., 2007; Dubos et al., 2008; Matsui et al., 2008). In addition, the bHLH constituent can form functional homo- or heterodimers with other bHLH proteins through their hydrophobic helix-loop-helix regions and/or ACT domains (Feller et al., 2006; Kong et al., 2012), whereas the MYB constituent is functionally dependent on their its bHLH partners (Lee and Schiefelbein, 1999; Zhang et al., 2003; Hernandez et al., 2007).

The MYB constituent determines the transcriptional activity of the complex (i.e., activating or repressing). There are two repressor-type MYB subfamilies: R2R3-type MYBs from subgroup 4 (e.g., *Arabidopsis* AtMYB4, petunia PhMYB27, grapevine VvMYBC2-L1–3) and a single group of R3-type MYBs (e.g., *Arabidopsis* CPC, TRY, ETC1, AtMYBL2). Members of both subfamilies contain bHLH interaction motifs as well as one or more repressor motifs, including the ethylene response factor (ERF)-associated amphiphilic repressor (EAR) motif LNLxL (Kranz et al., 1998; Jin et al., 2000; Cavallini et al., 2015). Genetic and *in vitro* physical interaction studies in *Arabidopsis*, petunia and grapevine suggest that MBW complexes may be able to switch between activating and repressive modes through a modular exchange of MYB components between activator- and repressor-type MYB subfamilies (Jin et al., 2000; Zimmermann et al., 2004; Albert et al., 2014; Cavallini et al., 2015). It remains to be seen whether repressor-type MBW complexes exist *in planta*.

Evolutionary History of the MYB-bHLH-TTG1 Complex

Not all MYB and bHLH subfamilies participate in MBW complexes. To date, only bHLHs from subgroup IIIf and MYBs containing the bHLH interaction motif [DE]Lx_2[RK]x_3Lx_6Lx_3R (Zimmermann et al., 2004) have been shown to directly interact with TTG1 and participate in MBW complexes (Payne et al., 2000; Stracke et al., 2001; Heim et al., 2003; Zhang et al., 2003; Baudry et al., 2004; Zimmermann et al., 2004). bHLHs from subgroup IIIf are the best-studied bHLH proteins in plants and function only in complex with R2R3-type MYBs from subgroups 4, 5, 6, and 15 as well as R3-type MYBs. Although the bHLH interaction motif arose early in land plant evolution and is conserved between angiosperms and gymnosperms (**Figure 1B**; Bedon et al., 2007), the bHLHs from subgroup IIIf themselves are evolutionarily older, arising before the origin of the mosses approximately 400 million years ago (Carretero-Paulet et al., 2010; Pires and Dolan, 2010). In contrast, TTG1 sequences are present in angiosperms, but not in gymnosperms or more ancient plant lineages (**Figure 1B**; Humphries et al., 2005; Brueggemann et al., 2010; Ben-Simhon et al., 2011). This suggests that the bHLHs from subgroup IIIf functioned as dimers in transcriptional regulation long before the MBW complex, and that ancestral MBW complexes may have originated as MYB-bHLH dimers that then recruited the WDR-containing protein TTG1 to aid in complex stability and/or subnuclear localization (Zhao et al., 2008).

The role of MBW complexes in mediating the synthesis of anthocyanins and proanthocyanidins appears to be conserved in diverse species of flowering plants, including *Arabidopsis* and related species, *Medicago*, strawberry, petunia and maize (**Figure 2B**; Grotewold et al., 2000; Carey et al., 2004; Dressel and Hemleben, 2009; Pang et al., 2009; Zhang et al., 2009; Schaart et al., 2013; Albert et al., 2014; Chopra et al., 2014). For example, maize uses the MBW complexes ZmC1/PL1-ZmR/B-ZmPAC1 to regulate the anthocyanin pathway (Goff et al., 1990; Tuerck and Fromm, 1994; Selinger and Chandler, 1999; Walker et al., 1999; Carey et al., 2004). In addition, *Arabidopsis* and other dicots have evolved two different classes of MBW complexes to separately regulate the anthocyanin and proanthocyanidin pathways (**Figure 2A**). *Arabidopsis* MBW complexes PAP1/2/3-EGL3/GL3/TT8-AtTTG1 activate anthocyanin biosynthesis genes (e.g., *DFR*; **Figures 2A, 4**; Walker et al., 1999; Borevitz et al., 2000; Nesi et al., 2000, 2001; Zhang et al., 2003; Zimmermann et al., 2004; Teng et al., 2005; Gonzalez et al., 2008); whereas, MBW complexes AtMYB5/TT2-GL3/EGL3/TT8-AtTTG1 activate genes specific for proanthocyanidin biosynthesis (e.g., *BAN/ANR*; **Figures 2A, 4**). Both classes of MBW complexes share the same bHLH and WDR constituents, leaving their MYB constituents to determine the promoter target activation and pathway specificity of the MBW complex. In addition, the MYB constituents AtMYB5/TT2 from subgroup 5 and PAP1-3 from subgroup 6 are closely related. They are in fact so close that the sequence variation at the R/G^{39} motif in their R2 motifs and at the A/SNDV or $DNEI^{90-93}$ motif in their R3 motifs (also known as the A2 box; Cavallini et al., 2015) has been shown to be sufficient to confer pathway specificity (Heppel et al., 2013). The same A2 box also confers pathway specificity for the repressor-type MYBs (Cavallini et al., 2015).

CONCLUSION

Plants are the basis for human nutrition and a renewable source for fuel and chemical feedstocks. Diminishing food security from plant disease/pests, climate instability and population growth, concomitant with rising energy costs and dwindling petrochemical-based fossil fuel supplies, have placed high demands on the productivity of food crops and other crops of economic importance (Krattiger, 1997; Lobell et al., 2011; Lobell and Gourdji, 2012; UN-DESA, 2013). Because WDR-containing trimeric complexes are at the heart of immune signaling and transcriptional regulation of chemical defenses, continued basic and translational research on these complexes in plant immunity will certainly improve agriculture and food security as well as our understanding of fundamental processes of signal transduction and gene regulation.

Plant Gβ sequences are ubiquitous across all five eukaryotic supergroups, with only a handful of species having more than two Gβ sequences (**Figure 1B**). Its signaling mechanism has evolved very slowly and yet pervasively so that it cannot be easily extricated from multiple immune signaling pathways (**Figure 1A**). In contrast, TTG1 is plant-specific and arose after the functional dimerization of its MYB-bHLH partners (**Figure 2A**). Evolution has added several layers of specialization to its regulatory mechanism, including multiple partially redundant bHLH and MYB partners, regulatory loops, and repressor-type MYBs to fine-tune the activation of its target pathways (**Figure 2B**). While its regulatory mechanism has evolved more quickly, TTG1 and its complex are more dispensable and limited in reach, affecting only a handful of pathways in the most evolved plants (**Figure 2B**).

Despite differences in evolutionary history, mechanism and target pathways, the plant Gβ and TTG1 proteins represent the apex of the hierarchy of network interactions in their respective pathways (**Figures 1A, 2A**), and are the sole constituents of their respective complexes to preside over all signaling and regulatory pathways in plant immunity that are mediated by WDR-containing ternary complexes. While there are still many open questions concerning the dynamics of these complexes and the specificity of their interactions with other protein partners and their downstream effectors or target DNA sequences, the large and still-growing body of research on these proteins and their complexes underscores the importance of these signaling and regulatory complexes.

AUTHOR CONTRIBUTIONS

JM wrote GB section. WC wrote TTG1 section. NC wrote introduction and edited manuscript. All three wrote conclusion.

ACKNOWLEDGMENTS

This work was supported by the National Institutes of Health under award numbers T32 GM007223 to JM and T32 GM007499 to WC.

REFERENCES

Adams, D. R., Ron, D., and Kiely, P. A. (2011). RACK1, A multifaceted scaffolding protein: structure and function. *Cell Commun. Signal.* 9, 22. doi: 10.1186/1478-811X-9-22

Adams-Phillips, L., Briggs, A. G., and Bent, A. F. (2010). Disruption of poly(ADP-ribosyl)ation mechanisms alters responses of *Arabidopsis* to biotic stress. *Plant Physiol.* 152, 267–280. doi: 10.1104/pp.109.148049

Albert, N. W., Davies, K. M., Lewis, D. H., Zhang, H., Montefiori, M., Brendolise, C., et al. (2014). A conserved network of transcriptional activators and repressors regulates anthocyanin pigmentation in eudicots. *Plant Cell* 26, 962–980. doi: 10.1105/tpc.113.122069

Aranda-Sicilia, M. N., Trusov, Y., Maruta, N., Chakravorty, D., Zhang, Y., and Botella, J. R. (2015). Heterotrimeric G proteins interact with defense-related receptor-like kinases in *Arabidopsis*. *J. Plant Physiol.* 188, 44–48. doi: 10.1016/j.jplph.2015.09.005

Bakker, P. A., Pieterse, C. M., and van Loon, L. C. (2007). Induced systemic resistance by fluorescent *Pseudomonas* spp. *Phytopathology* 97, 239–243. doi: 10.1094/PHYTO-97-2-0239

Baudry, A., Caboche, M., and Lepiniec, L. (2006). TT8 controls its own expression in a feedback regulation involving TTG1 and homologous MYB and bHLH factors, allowing a strong and cell-specific accumulation of flavonoids in *Arabidopsis thaliana*. *Plant J.* 46, 768–779. doi: 10.1111/j.1365-313X.2006.02733.x

Baudry, A., Heim, M. A., Dubreucq, B., Caboche, M., Weisshaar, B., and Lepiniec, L. (2004). TT2, TT8, and TTG1 synergistically specify the expression of BANYULS and proanthocyanidin biosynthesis in *Arabidopsis thaliana*. *Plant J.* 39, 366–380. doi: 10.1111/j.1365-313X.2004.02138.x

Bedon, F., Grima-Pettenati, J., and Mackay, J. (2007). Conifer R2R3-MYB transcription factors: sequence analyses and gene expression in wood-forming tissues of white spruce (*Picea glauca*). *BMC Plant Biol.* 7:17. doi: 10.1186/1471-2229-7-17

Ben-Simhon, Z., Judeinstein, S., Nadler-Hassar, T., Trainin, T., Bar-Ya'akov, I., Borochov-Neori, H., et al. (2011). A pomegranate (*Punica granatum* L.) WD40-repeat gene is a functional homologue of *Arabidopsis* TTG1 and is involved in the regulation of anthocyanin biosynthesis during pomegranate fruit development. *Planta* 234, 865–881. doi: 10.1007/s00425-011-1438-4

Bhardwaj, D., Sheikh, A. H., Sinha, A. K., and Tuteja, N. (2011). Stress induced β subunit of heterotrimeric G-proteins from *Pisum sativum* interacts with mitogen activated protein kinase. *Plant Signal. Behav.* 6, 287–292. doi: 10.4161/psb.6.2.14971

Bommert, P., Je, B. II, Goldshmidt, A., and Jackson, D. (2013). The maize Gα gene COMPACT PLANT2 functions in CLAVATA signalling to control shoot meristem size. *Nature* 502, 555–558. doi: 10.1038/nature12583

Borevitz, J. O., Xia, Y., Blount, J., Dixon, R. A., and Lamb, C. (2000). Activation tagging identifies a conserved MYB regulator of phenylpropanoid biosynthesis. *Plant Cell* 12, 2383–2394. doi: 10.1105/tpc.12.12.2383

Bradford, W., Buckholz, A., Morton, J., Price, C., Jones, A. M., and Urano, D. (2013). Eukaryotic G protein signaling evolved to require G protein-coupled receptors for activation. *Sci. Signal.* 6, ra37. doi: 10.1126/scisignal.2003768

Brueggemann, J., Weisshaar, B., and Sagasser, M. (2010). A WD40-repeat gene from Malus × domestica is a functional homologue of *Arabidopsis thaliana* TRANSPARENT TESTA GLABRA1. *Plant Cell Rep.* 29, 285–294. doi: 10.1007/s00299-010-0821-0

Burki, F. (2014). The eukaryotic tree of life from a global phylogenomic perspective. *Cold Spring Harb. Perspect. Biol.* 6, a016147. doi: 10.1101/cshperspect.a016147

Burr, F. A., Burr, B., Scheffer, B. E., Blewitt, M., Wienand, U., and Matz, E. C. (1996). The maize repressor-like gene intensifier1 shares homology with the r1/b1 multigene family of transcription factors and exhibits missplicing. *Plant Cell* 8, 1249–1259. doi: 10.1105/tpc.8.8.1249

Carey, C. C., Strahle, J. T., Selinger, D. A., and Chandler, V. L. (2004). Mutations in the pale aleurone color1 regulatory gene of the *Zea mays* anthocyanin pathway have distinct phenotypes relative to the functionally similar TRANSPARENT TESTA GLABRA1 gene in *Arabidopsis thaliana*. *Plant Cell* 16, 450–464. doi: 10.1105/tpc.018796

Carr, A., and Biggin, M. D. (1999). A comparison of *in vivo* and *in vitro* DNA-binding specificities suggests a new model for homeoprotein DNA binding in *Drosophila* embryos. *EMBO J.* 18, 1598–1608. doi: 10.1093/emboj/18.6.1598

Carretero-Paulet, L., Galstyan, A., Roig-Villanova, I., Martinez-Garcia, J. F., Bilbao-Castro, J. R., and Robertson, D. L. (2010). Genome-wide classification and evolutionary analysis of the bHLH family of transcription factors in *Arabidopsis*, poplar, rice, moss, and algae. *Plant Physiol.* 153, 1398–1412. doi: 10.1104/pp.110.153593

Cavallini, E., Matus, J. T., Finezzo, L., Zenoni, S., Loyola, R., Guzzo, F., et al. (2015). The phenylpropanoid pathway is controlled at different branches by a set of R2R3-MYB C2 repressors in grapevine. *Plant Physiol.* 167, 1448–1470. doi: 10.1104/pp.114.256172

Chakravorty, D., Gookin, T. E., Milner, M. J., Yu, Y., and Assmann, S. (2015). Extra-large G proteins (XLGs) expand repertoire of subunits in *Arabidopsis* heterotrimeric G protein signaling. *Plant Physiol.* 169, 512–529. doi: 10.1104/pp.15.00251

Chakravorty, D., Trusov, Y., Zhang, W., Acharya, B. R., Sheahan, M. B., McCurdy, D. W., et al. (2011). An atypical heterotrimeric G-protein γ-subunit is involved in guard cell K^+-channel regulation and morphological development in *Arabidopsis thaliana*. *Plant J.* 67, 840–851. doi: 10.1111/j.1365-313X.2011.04638.x

Cheng, Z., Li, J.-F., Niu, Y., Zhang, X.-C., Woody, O. Z., Xiong, Y., et al. (2015). Pathogen-secreted proteases activate a novel immune pathway. *Nature* 521, 213–216. doi: 10.1038/nature14243

Chopra, D., Wolff, H., Span, J., Schellmann, S., Coupland, G., Albani, M. C., Schrader, A., et al. (2014). Analysis of TTG1 function in *Arabis alpina*. *BMC Plant Biol.* 14:16. doi: 10.1186/1471-2229-14-16

Consonni, C., Humphry, M. E., Hartmann, H. A., Livaja, M., Durner, J., and Westphal, L. (2006). Conserved requirement for a plant host cell protein in powdery mildew pathogenesis. *Nat. Genet.* 38, 716–720. doi: 10.1038/ng1806

Coyle, S. M., Gilbert, W. V., and Doudna, J. A. (2009). Direct link between RACK1 function and localization at the ribosome *in vivo*. *Mol. Cell Biol.* 29, 1626–1634. doi: 10.1128/MCB.01718-08

de Mendoza, A., Sebé-Pedrós, A., and Ruiz-Trillo, I. (2014). The evolution of the GPCR signaling system in eukaryotes: modularity, conservation, and the transition to metazoan multicellularity. *Genome Biol. Evol.* 6, 606–619. doi: 10.1093/gbe/evu038

Delgado-Cerezo, M., Sánchez-Rodríguez, C., Escudero, V., Miedes, E., Fernández, P. V., Jordá, L., et al. (2012). *Arabidopsis* heterotrimeric G-protein regulates cell wall defense and resistance to necrotrophic fungi. *Mol. Plant* 5, 98–114. doi: 10.1093/mp/ssr082

Dereeper, A., Guignon, V., Blanc, G., Audic, S., Buffet, S., Chevenet, F., et al. (2008). Phylogeny.fr: robust phylogenetic analysis for the non-specialist. *Nucleic Acids Res.* 1, W465–W469. doi: 10.1093/nar/gkn180

Devoto, A., Hartmann, H. A., Piffanelli, P., Elliott, C., Simmons, C., Taramino, G., et al. (2003). Molecular phylogeny and evolution of the plant-specific seven-transmembrane MLO family. *J. Mol. Evol.* 56, 77–88. doi: 10.1007/s00239-002-2382-5

Ding, L., Pandey, S., and Assmann, S. M. (2008). *Arabidopsis* etra-large G proteins (XLGs) regulate root morphogenesis. *Plant J.* 53, 248–263. doi: 10.1111/j.1365-313X.2007.03335.x

Dixon, R. A. (2001). Natural products and plant disease resistance. *Nature* 411, 843–847. doi: 10.1038/35081178

Dressel, A., and Hemleben, V. (2009). Transparent Testa Glabra 1 (TTG1) and TTG1-like genes in *Matthiola incana* R. Br. and related Brassicaceae and mutation in the WD-40 motif. *Plant Biol. (Stuttg)* 11, 204–212. doi: 10.1111/j.1438-8677.2008.00099.x

Dubos, C., Le Gourrierec, J., Baudry, A., Huep, G., Lanet, E., Debeaujon, I., et al. (2008). MYBL2 is a new regulator of flavonoid biosynthesis in *Arabidopsis thaliana*. *Plant J.* 55, 940–953. doi: 10.1111/j.1365-313X.2008.03564.x

Dubos, C., Stracke, R., Grotewold, E., Weisshaar, B., Martin, C., and Lepiniec, L. (2010). MYB transcription factors in *Arabidopsis*. *Trends Plant Sci.* 15, 573–581. doi: 10.1016/j.tplants.2010.06.005

Esch, J. J., Chen, M., Sanders, M., Hillestad, M., Ndkium, S., Idelkope, B., et al. (2003). A contradictory GLABRA3 allele helps define gene interactions controlling trichome development in *Arabidopsis*. *Development* 130, 5885–5894. doi: 10.1242/dev.00812

Feller, A., Hernandez, J. M., and Grotewold, E. (2006). An ACT-like domain participates in the dimerization of several plant basic-helix-loop-helix transcription factors. *J. Biol. Chem.* 39, 28964–28974. doi: 10.1074/jbc.M603262200

Fornalé, S., Shi, X., Chai, C., Encina, A., Irar, S., Capellades, M., et al. (2010). ZmMYB31 directly represses maize lignin genes and redirects the phenylpropanoid metabolic flux. *Plant J.* 64, 633–644. doi: 10.1111/j.1365-313X.2010.04363.x

Fornalé, S., Lopez, E., Salazar-Henao, J. E., Fernandez-Nohales, P., Rigau, J., and Caparros-Ruiz, D. (2014). AtMYB7, a new player in the regulation of UV-sunscreens in *Arabidopsis thaliana*. *Plant Cell Physiol.* 55, 507–516. doi: 10.1093/pcp/pct187

Friedman, E. J., Temple, B. R. S., Hicks, S. N., Sondek, J., Jones, C. D., and Jones, A. M. (2009). Prediction of protein–protein interfaces on G-protein β subunits reveals a novel phospholipase C β2 binding domain. *J. Mol. Biol.* 392, 1044–1054. doi: 10.1016/j.jmb.2009.07.076

Fritz-Laylin, L. K., Krishnamurthy, N., Tör, M., Sjölander, K. V., and Jones, J. D. G. (2005). Phylogenomic analysis of the receptor-like proteins of rice and *Arabidopsis*. *Plant Physiol.* 138, 611–623. doi: 10.1104/pp.104.054452

Goff, S. A., Klein, T. M., Roth, B. A., Fromm, M. E., Cone, K. C., Radicella, J. P., et al. (1990). Transactivation of anthocyanin biosynthetic genes following transfer of B regulatory genes into maize tissues. *EMBO J.* 9, 2517–2522.

Gonzalez, A., Zhao, M., Leavitt, J. M., and Lloyd, A. M. (2008). Regulation of the anthocyanin biosynthetic pathway by the TTG1/bHLH/Myb transcriptional complex in *Arabidopsis* seedlings. *Plant J.* 53, 814–827. doi: 10.1111/j.1365-313X.2007.03373.x

Grotewold, E., Drummond, B. J., Bowen, B., and Peterson, T. (1994). The myb-homologous P gene controls phlobaphene pigmentation in maize floral organs by directly activating a flavonoid biosynthetic gene subset. *Cell* 76, 543–553. doi: 10.1016/0092-8674(94)90117-1

Grotewold, E., Sainz, M. B., Tagliani, L., Hernandez, J. M., Bowen, B., and Chandler, V. L. (2000). Identification of the residues in the *Myb domain* of maize C1 that specify the interaction with the bHLH cofactor R. *Proc. Natl. Acad. Sci. U.S.A.* 97, 13579–13584. doi: 10.1073/pnas.250379897

Hackenberg, D., Sakayama, H., Nishiyama, T., and Pandey, S. (2013). Characterization of the heterotrimeric G-protein complex and its regulator from the green alga *Chara braunii* expands the evolutionary breadth of plant G-protein signaling. *Plant Physiol.* 163, 1510–1517. doi: 10.1104/pp.113.230425

Heim, M. A., Jakoby, M., Werber, M., Martin, C., Weisshaar, B., and Bailey, P. C. (2003). The basic helix-loop-helix transcription factor family in plants: a genome-wide study of protein structure and functional diversity. *Mol. Biol. Evol.* 20, 735–747. doi: 10.1093/molbev/msg088

Heo, J. B., Sung, S., and Assmann, S. M. (2012). Ca^{2+}-dependent GTPase, Extra-large G protein 2 (XLG2), promotes activation of DNA-binding protein Related to Vernalization 1 (RTV1), leading to activation of floral integrator genes and early flowering in *Arabidopsis*. *J. Biol. Chem.* 287, 8242–8253. doi: 10.1074/jbc.M111.317412

Heppel, S. C., Jaffe, F. W., Takos, A. M., Schellmann, S., Rausch, T., Walker, A. R., et al. (2013). Identification of key amino acids for the evolution of promoter target specificity of anthocyanin and proanthocyanidin regulating MYB factors. *Plant Mol. Biol.* 82, 457–471. doi: 10.1007/s11103-013-0074-8

Hernandez, J. M., Feller, A., Morohashi, K., Frame, K., and Grotewold, E. (2007). The basic helix loop helix domain of maize R links transcriptional regulation and histone modifications by recruitment of an EMSY-related factor. *Proc. Natl. Acad. Sci. U.S.A.* 104, 17222–17227. doi: 10.1073/pnas.0705629104

Hillenbrand, M., Schori, C., Schöppe, J., and Plückthun, A. (2015). Comprehensive analysis of heterotrimeric G-protein complex diversity and their interactions with GPCRs in solution. *Proc. Natl. Acad. Sci. U.S.A.* 112, E1181–E1190. doi: 10.1073/pnas.1417573112

Hipskind, J. D., Goldsbrough, P. B., Urmeev, F., and Nicholson, R. L. (1996). Synthesis of 3-deoxyanthocyanidin phytoalexins in *Sorghum* does not occur via the same pathway as 3-hydroxylated anthocyanidins and phlobaphenes. *Maydica* 41, 155–166.

Humphries, J. A., Walker, A. R., Timmis, J. N., and Orford, S. J. (2005). Two WD-repeat genes from cotton are functional homologues of the *Arabidopsis thaliana* TRANSPARENT TESTA GLABRA1 (TTG1) gene. *Plant Mol. Biol.* 57, 67–81. doi: 10.1007/s11103-004-6768-1

Humphry, M., Consonni, C., and Panstruga, R. (2006). mlo-based powdery mildew immunity: silver bullet or simply non-host resistance? *Mol. Plant Pathol.* 7, 605–610. doi: 10.1111/j.1364-3703.2006.00362.x

Ibraheem, F., Gaffoor, I., and Chopra, S. (2010). Flavonoid phytoalexin-dependent resistance to anthracnose leaf blight requires a functional yellow seed1 in *Sorghum bicolor*. *Genetics* 184, 915–926. doi: 10.1534/genetics.109.111831

Ibraheem, F., Gaffoor, I., Tan, Q., Shyu, C. R., and Chopra, S. (2015). A *Sorghum* MYB transcription factor induces 3-deoxyanthocyanidins and enhances resistance against leaf blights in maize. *Molecules* 20, 2388–2404. doi: 10.3390/molecules20022388

Im, M. J., Holzhöfer, A., Böttinger, H., Pfeuffer, T., and Helmreich, E. J. (1988). Interactions of pure β_{γ}-subunits of G-proteins with purified β_1-adrenoceptor. *FEBS Lett.* 227, 225–229. doi: 10.1016/0014-5793(88)80903-7

Ishida, T., Tabata, R., Yamada, M., Aida, M., Mitsumasu, K., Fujiwara, M., et al. (2014). Heterotrimeric G proteins control stem cell proliferation through CLAVATA signaling in *Arabidopsis*. *EMBO Rep.* 15, 1202–1209. doi: 10.15252/embr.201438660

Ishikawa, A. (2009). The *Arabidopsis* G-protein β-subunit is required for defense response against *Agrobacterium tumefaciens*. *Biosci. Biotechnol. Biochem.* 73, 47–52. doi: 10.1271/bbb.80449

Janda, L., Tichy, P., Spízek, J., and Petrícek, M. (1996). A deduced *Thermomonospora curvata* protein containing serine/threonine protein kinase and WD-repeat domains. *J. Bacteriol.* 178, 1487–1489.

Jin, H., Cominelli, E., Bailey, P., Parr, A., Mehrtens, J. J., Tonelli, C., Weisshaar, B., et al. (2000). Transcriptional repression by AtMYB4 controls production of UV-protecting sunscreens in *Arabidopsis*. *EMBO J.* 19, 6150–6161. doi: 10.1093/emboj/19.22.6150

Jones, J. C., Duffy, J. W., Machius, M., Temples, B. R. S., Dohlman, H. G., and Jones, A. M. (2011a). The crystal structure of a self-activating G protein α subunit reveals its distinct mechanism of signal initiation. *Sci. Signal.* 4, ra8. doi: 10.1126/scisignal.2001446

Jones, J. C., Temple, B. R., Jones, A. M., and Dohlman, H. G. (2011b). Functional reconstitution of an atypical G protein heterotrimer and regulator of G protein signaling protein (RGS1) from *Arabidopsis thaliana*. *J. Biol. Chem.* 286, 13143–13150. doi: 10.1074/jbc.M110.190355

Jones, J. D. G., and Dangl, J. L. (2006). The plant immune system. *Nature* 444, 323–329. doi: 10.1038/nature05286

Keeling, P. J., Burger, G., Durnford, D. G., Lang, B. F., Lee, R. W., Pearlman, R. E., et al. (2005). The tree of eukaryotes. *Trends Ecol. Evol.* 20, 670–676. doi: 10.1016/j.tree.2005.09.005

Kelley, L. A., and Sternberg, M. J. E. (2009). Protein structure prediction on the Web: a case study using the Phyre server. *Nat. Protoc.* 4, 363–371. doi: 10.1038/nprot.2009.2

Kirik, V., Simon, M., Wester, K., Schiefelbein, J., and Hülskamp, M. (2004). ENHANCER of TRY and CPC 2 (ETC2) reveals redundancy in the region-specific control of trichome development of *Arabidopsis*. *Plant Mol. Biol.* 55, 389–398. doi: 10.1007/s11103-004-0893-8

Kisselev, O., and Gautam, N. (1993). Specific interaction with rhodopsin is dependent on the γ subunit type in a G protein. *J. Biol. Chem.* 268, 24519–22422.

Kliebenstein, D. J. (2013). Making new molecules—evolution of structures for novel metabolites in plants. *Curr. Opin. Plant Biol.* 16, 112–117. doi: 10.1016/j.pbi.2012.12.004

Kliebenstein, D. J., and Osbourn, A. (2012). Making new molecules—evolution of pathways for novel metabolites in plants. *Curr. Opin. Plant Biol.* 15, 415–423. doi: 10.1016/j.pbi.2012.05.005

Kombrink, E., and Hahlbrock, K. (1990). Rapid, systemic repression of the synthesis of ribulose 1,5-bisphosphate carboxylase small-subunit mRNA in fungus-infected or elicitor-treated potato leaves. *Planta* 181, 216–219. doi: 10.1007/BF02411541

Kong, Q., Pattanaik, S., Feller, A., Werkman, J. R., Chai, C., Wang, Y., et al. (2012). Regulatory switch enforced by basic helix-loop-helix and ACT-domain mediated dimerizations of the maize transcription factor R. *Proc. Natl. Acad. Sci.* 109, E2091–E2097. doi: 10.1073/pnas.1205513109

Kottb, M., Gigolashvili, T., Großkinsky, D. K., and Piechulla, B. (2015). Trichoderma volatiles effecting *Arabidopsis*: from inhibition to protection against phytopathogenic fungi. *Front. Microbiol.* 6:995. doi: 10.3389/fmicb.2015.00995

Kranz, H. D., Denekamp, M., Greco, R., Jin, H., Leyva, A., Meissner, R. C., et al. (1998). Towards functional characterisation of the members of the R2R3-MYB gene family from *Arabidopsis thaliana*. *Plant J.* 16, 263–276. doi: 10.1046/j.1365-313x.1998.00278.x

Krattiger, A. F. (1997). *Insect Resistance in Crops: A Case Study of Bacillus thuringiensis (Bt) and its Transfer to Developing Countries.* Ithaca, NY: ISAAA Briefs.

Krishnan, A., Mustafa, A., Almén, M. S., Fredriksson, R., Williams, M. J., and Schiöth, H. B. (2015). Evolutionary hierarchy of vertebrate-like heterotrimeric G protein families. *Mol. Phylogenet. Evol.* 91, 27–40. doi: 10.1016/j.ympev.2015.05.009

Lambright, D. G., Sondek, J., Bohm, A., Skiba, N. P., Hamm, H. E., and Sigler, P. B. (1996). The 2.0 A crystal structure of a heterotrimeric G protein. *Nature* 379, 311–319. doi: 10.1038/379311a0

Lan, W., Lu, F., Regner, M., Zhu, Y., Rencoret, J., Ralph, S. A., and et al. (2015). Tricin, a flavonoid monomer in monocot lignification. *Plant Physiol.* 167, 1284–1295. doi: 10.1104/pp.114.253757

Lappin, T. R., Grier, D. G., Thompson, A., and Halliday, H. L. (2006). HOX genes: seductive science, mysterious mechanisms. *Ulster Med. J.* 75, 23–31.

Lee, M. M., and Schiefelbein, J. (1999). WEREWOLF, a MYB-related protein in *Arabidopsis*, is a position-dependent regulator of epidermal cell patterning. *Cell* 99, 473–483. doi: 10.1016/S0092-8674(00)81536-6

Lee, S., Rojas, C. M., Ishiga, Y., Pandey, S., and Mysore, K. S. (2013). *Arabidopsis* heterotrimeric G-proteins play a critical role in host and nonhost resistance against *Pseudomonas syringae* pathogens. *PLoS ONE* 8:e82445. doi: 10.1371/journal.pone.0082445

Lee, Y. R., and Assmann, S. M. (1999). *Arabidopsis thaliana* 'extra-large GTP-binding protein' (AtXLG1): a new class of G-protein. *Plant Mol. Biol.* 40, 55–64. doi: 10.1023/A:1026483823176

Letunic, I., Doerks, T., and Bork, P. (2014). SMART: recent updates, new developments and status in 2015. *Nucleic Acids Res.* 43, D257–D260. doi: 10.1093/nar/gku949

Li, L., Wright, S. J., Krystofova, S., Park, G., and Borkovich, K. A. (2007). Heterotrimeric G protein signaling in filamentous fungi. *Annu. Rev. Microbiol.* 61, 423–452. doi: 10.1146/annurev.micro.61.080706.093432

Liu, J., Ding, P., Sun, T., Nitta, Y., Dong, O., Huang, X., et al. (2013). Heterotrimeric G proteins serve as a converging point in plant defense signaling activated by multiple receptor-like kinases. *Plant Physiol.* 161, 2146–2158. doi: 10.1104/pp.112.212431

Llorente, F., Alonso-Blanco, C., Sánchez-Rodriguez, C., Jorda, L., and Molina, A. (2005). ERECTA receptor-like kinase and heterotrimeric G protein from *Arabidopsis* are required for resistance to the necrotrophic fungus *Plectosphaerella cucumerina*. *Plant J.* 43, 165–180. doi: 10.1111/j.1365-313X.2005.02440.x

Lo, S. C., De Verdier, K., and Nicholson, R. L. (1999). Accumulation of 3-deoxyanthocyanidin phytoalexins and resistance to *Colletotrichum sublineolum* in *Sorghum bicolor*. *Physiol. Mol. Plant Path.* 61, 179–188. doi: 10.1006/pmpp.2002.0428

Lo, S. C., and Nicholson, R. L. (1998). Reduction of light-induced anthocyanin accumulation in inoculated *Sorghum* mesocotyls. *Plant Physiol.* 116, 979–989. doi: 10.1104/pp.116.3.979

Lobell, D. B., and Gourdji, S. M. (2012). The influence of climate change on global crop productivity. *Plant Physiol.* 160, 1686–1697. doi: 10.1104/pp.112.208298

Lobell, D. B., Schlenker, W., and Costa-Roberts, J. (2011). Climate trends and global crop production since 1980. *Science* 333, 616–620. doi: 10.1126/science.1204531

López-Bergami, P., Habelhah, H., Bhoumik, A., Zhang, W., Wang, L. H., and Ronai, Z. (2005). Receptor for RACK1 mediates activation of JNK by protein kinase C. *Mol. Cell* 19, 309–320. doi: 10.1016/j.molcel.2005.06.025

Lorek, J., Griebel, T., Jones, A. M., Kuhn, H., and Panstruga, R. (2013). The role of *Arabidopsis* heterotrimeric G-protein subunits in MLO2 function and MAMP-triggered immunity. *Mol. Plant Microbe Interact.* 26, 991–1003. doi: 10.1094/MPMI-03-13-0077-R

Ma, H., Yanofsky, M. F., and Meyerowitz, E. M. (1990). Molecular cloning and characterization of GPA1, a G protein α subunit gene from *Arabidopsis thaliana*. *Proc. Natl. Acad. Sci. U.S.A.* 87, 3821–3825. doi: 10.1073/pnas.87.10.3821

Maeda, K., Houjyou, Y., Komatsu, T., Hori, H., Kodaira, T., and Ishikawa, A. (2009). AGB1 and PMR5 contribute to PEN2-mediated preinvasion resistance to *Magnaporthe oryzae* in *Arabidopsis thaliana*. *Mol. Plant Microbe Interact.* 22, 1331–1340. doi: 10.1094/MPMI-22-11-1331

Maruta, N., Trusov, Y., Brenya, E., Parekh, U., and Botella, J. R. (2015). Membrane-localized extra-large G proteins and Gβγ of the heterotrimeric G proteins form functional complexes engaged in plant immunity in *Arabidopsis*. *Plant Physiol.* 167, 1004–1016. doi: 10.1104/pp.114.255703

Mason, M. G., and Botella, J. R. (2000). Completing the heterotrimer: isolation and characterization of an *Arabidopsis thaliana* G protein γ-subunit cDNA. *Proc. Natl. Acad. Sci. U.S.A.* 97, 1478–1488. doi: 10.1073/pnas.97.26.14784

Mason, M. G., and Botella, J. R. (2001). Isolation of a novel G-protein γ-subunit from *Arabidopsis thaliana* and its interaction with Gβ. *Biophys. Acta* 1520, 147–153. doi: 10.1016/s0167-4781(01)00262-7

Mathys, J., De Cremer, K., Timmermans, P., Van Kerckhove, S., Lievens, B., Vanhaecke, M., et al. (2012). Genome-Wide Characterization of ISR induced in *Arabidopsis thaliana* by *Trichoderma hamatum* T382 against *Botrytis cinerea* infection. *Front. Plant Sci.* 3:108. doi: 10.3389/fpls.2012.00108

Matsui, K., Umemura, Y., and Ohme-Takagi, M. (2008). AtMYBL2, a protein with a single MYB domain, acts as a negative regulator of anthocyanin biosynthesis in *Arabidopsis*. *Plant J.* 55, 954–967. doi: 10.1111/j.1365-313X.2008.03565.x

Meziane, H., Van der Sluis, I., van Loon, L. C., Hofte, M., and Bakker, P. A. (2005). Determinants of *Pseudomonas putida* WCS358 involved in inducing systemic resistance in plants. *Mol. Plant Pathol.* 6, 177–185. doi: 10.1111/j.1364-3703.2005.00276.x

Mizuno, H., Yazawa, T., Kasuga, S., Sawada, Y., Ogata, J., Ando, T., et al. (2014). Expression level of a flavonoid 3′-hydroxylase gene determines pathogen-induced color variation in *Sorghum*. *BMC Res. Notes* 7:761. doi: 10.1186/1756-0500-7-761

Morohashi, K., Casas, M. I., Ferreyra, M. L. F., Mejía-Guerra, M. K., Pourcel, L., Yilmaz, A., et al. (2012). A genome-wide regulatory framework identifies maize pericarp color1 controlled genes. *Plant Cell* 24, 2745–2764. doi: 10.1105/tpc.112.098004

Morohashi, K., Zhao, M., Yang, M., Read, B., Lloyd, A., Lamb, R., et al. (2007). Participation of the *Arabidopsis* bHLH factor GL3 in trichome initiation regulatory events. *Plant Physiol.* 145, 736–746. doi: 10.1104/pp.107.104521

Nicholson, R. L., Kollipara, S. S., Vincent, J. R., Lyons, P. C., and Cadena-Gomez, G. (1987). Phytoalexin synthesis by the *Sorghum* mesocotyl in response to infection by pathogenic and nonpathogenic fungi. *Proc. Natl. Acad. Sci. U.S.A.* 84, 5520–5524. doi: 10.1073/pnas.84.16.5520

Nesi, N., Debeaujon, I., Jond, C., Pelletier, G., Caboche, M., and Lepiniec, L. (2000). The TT8 gene encodes a basic helix-loop-helix domain protein required for expression of DFR and BAN genes in *Arabidopsis* siliques. *Plant Cell* 12, 1863–1870. doi: 10.1105/tpc.12.10.1863

Nesi, N., Jond, C., Debeaujon, I., Caboche, M., and Lepiniec, L. (2001). The *Arabidopsis* TT2 gene encodes an R2R3 MYB domain protein that acts as a key determinant for proanthocyanidin accumulation in developing seed. *Plant Cell* 13, 2099–2114. doi: 10.1105/tpc.13.9.2099

Oldham, W. M., and Hamm, H. E. (2008). Heterotrimeric G protein activation by G-protein-coupled receptors. *Nat. Rev. Mol. Cell Biol.* 9, 60–71. doi: 10.1038/nrm2299

Ouyang, Y., Huang, X., Lu, Z., and Yao, J. (2012). Genomic survey, expression profile and co-expression network analysis of OsWD40 family in rice. *BMC Genomics* 13:100. doi: 10.1186/1471-2164-13-100

Pandey, S., and Assmann, S. M. (2004). The *Arabidopsis* putative G protein-coupled receptor GCR1 interacts with the G protein α subunit GPA1 and regulates abscisic acid signalling. *Plant Cell* 16, 1616–1632. doi: 10.1105/tpc.020321

Pandey, S., Nelson, D. C., and Assmann, S. M. (2009). Two novel GPCR-type G proteins are abscisic acid receptors in *Arabidopsis*. *Cell* 136, 136–148. doi: 10.1016/j.cell.2008.12.026

Pang, Y., Wenger, J. P., Saathoff, K., Peel, G. J., Wen, J., Huhman, D., et al. (2009). A WD40 repeat protein from *Medicago truncatula* is necessary for tissue-specific anthocyanin and proanthocyanidin biosynthesis but not for trichome development. *Plant Physiol.* 151, 1114–1129. doi: 10.1104/pp.109.144022

Payne, C. T., Zhang, F., and Lloyd, A. M. (2000). GL3 encodes a bHLH protein that regulates trichome development in *Arabidopsis* through interaction with GL1 and TTG1. *Genetics* 156, 1349–1362.

Perfus-Barbeoch, L., Jones, A. M., and Assmann, S. M. (2004). Plant heterotrimeric G protein function: insights from *Arabidopsis* and rice mutants. *Curr. Opin. Plant Biol.* 7, 719–731. doi: 10.1016/j.pbi.2004.09.013

Pezet, R., Pont, V., and Hoang-Van, K. (1992). "Enzymatic detoxification of stilbenes by *Botrytis cinerea* and inhibition by grape berries proanthocyanidins," in *Recent Advances in Botrytis Research*, eds K. Verhoeff, N. E. Malathrakis, and B. Williamson (Wageningen: Pudoc Scientific Publishers), 87–92.

Pires, N., and Dolan, L. (2010). Origin and diversification of basic-helix-loop-helix proteins in plants. *Mol. Biol. Evol.* 27, 862–874. doi: 10.1093/molbev/msp288

Preston, J., Wheeler, J., Heazlewood, J., Li, S. F., and Parish, R. W. (2004). AtMYB32 is required for normal pollen development in *Arabidopsis thaliana*. *Plant J.* 40, 979–995. doi: 10.1111/j.1365-313X.2004.02280.x

Prouse, M. B., and Campbell, M. M. (2013). Interactions between the R2R3-MYB transcription factor, AtMYB61, and target DNA binding sites. *PLoS ONE* 8:e65132. doi: 10.1371/journal.pone.0065132

Rabinovich, A., Jin, V. X., Rabinovich, R., Xu, X., and Farnham, P. J. (2008). E2F *in vivo* binding specificity: comparison of consensus versus non-consensus binding sites. *Genome Res.* 18, 527–544. doi: 10.1101/gr.080622.108

Rabl, J., Leibundgut, M., Ataide, S. F., Haag, A., and Ban, N. (2011). Crystal structure of the eukaryotic 40S ribosomal subunit in complex with initiation factor 1. *Science* 331, 730–736. doi: 10.1126/science.1198308

Romero, I., Fuertes, A., Benito, M., Malpica, J. M., Leyva, A., and Paz-Ares, J. (1998). More than 80R2R3-MYB regulatory genes in the genome of *Arabidopsis thaliana*. *Plant J.* 14, 273–284. doi: 10.1046/j.1365-313X.1998.00113.x

Ron, D., Chen, C. H., Caldwell, J., Jamieson, L., and Orr, E. Mochly-Rosen, D. (1994). Cloning of an intracellular receptor for protein kinase C: a homolog of the β subunit of G proteins. *Proc. Natl. Acad. Sci. U.S.A.* 91, 839–843. doi: 10.1073/pnas.91.3.839

Ruiz Carrillo, D., Chandrasekaran, R., Nilsson, M., Cornvik, T., Liew, C. W., Tan, S. M., et al. (2012). Structure of human Rack1 protein at a resolution of 2.45Å. *Acta Crystallogr. Sect. F: Struct. Biol. Cryst. Commun.* 68, 867–872. doi: 10.1107/S1744309112027480

Saijo, Y., Tinto, N., Lu, X., Rauf, P., Pajerowska-Mukhtar, K., Häweker, H., et al. (2009). Receptor quality control in the endoplasmic reticulum for plant innate immunity. *EMBO J.* 28, 3439–3449. doi: 10.1038/emboj.2009.263

Sawa, S. (2002). Overexpression of the AtmybL2 gene represses trichome development in *Arabidopsis*. *DNA Res.* 9, 31–34. doi: 10.1093/dnares/9.2.31

Schaart, J. G., Dubos, C., Romero De La Fuente, I., van Houwelingen, A. M., de Vos, R. C., et al. (2013). Identification and characterization of MYB-bHLH-WD40 regulatory complexes controlling proanthocyanidin biosynthesis in strawberry (Fragaria × ananassa) fruits. *New Phytol.* 197, 454–467. doi: 10.1111/nph.12017

Schenke, D., Boettcher, C., and Scheel, D. (2011). Crosstalk between abiotic ultraviolet-B stress and biotic (flg22) stress signalling in *Arabidopsis* prevents flavonol accumulation in favor of pathogen defence compound production. *Plant. Cell Environ.* 34, 1849–1864. doi: 10.1111/j.1365-3040.2011.02381.x

Selinger, D. A., and Chandler, V. L. (1999). A mutation in the pale aleurone color1 gene identifies a novel regulator of the maize anthocyanin pathway. *Plant Cell* 11, 5–14. doi: 10.1105/tpc.11.1.5

Shih, C. H., Chu, I. K., Yip, W. K., and Lo, C. (2006). Differential expression of two flavonoid 3′-hydroxylase cDNAs involved in biosynthesis of anthocyanin pigments and 3-deoxyanthocyanidin phytoalexins in *Sorghum*. *Plant Cell Physiol.* 47, 1412–1419. doi: 10.1093/pcp/pcl003

Shiu, S. H., and Bleecker, A. B. (2001). Plant receptor-like kinase gene family: diversity, function, and signaling. *Sci. STKE* 2001, 22. doi: 10.1126/stke.2001.113.re22

Shiu, S. H., and Bleecker, A. B. (2003). Expansion of the receptor-like kinase/Pelle gene family and receptor-like proteins in *Arabidopsis*. *Plant Physiol.* 132, 530–543. doi: 10.1104/pp.103.021964

Simon, M., Lee, M. M., Lin, Y., Gish, L., and Schiefelbein, J. (2007). Distinct and overlapping roles of single-repeat MYB genes in root epidermal patterning. *Dev. Biol.* 311, 566–578. doi: 10.1016/j.ydbio.2007.09.001

Smith, T. F., Gaitatzes, C., Saxena, K., and Neer, E. J. (1999). The WD repeat: a common architecture for diverse functions. *Trends Biochem. Sci.* 24, 181–185. doi: 10.1016/S0968-0004(99)01384-5

Snyder, B. A., and Nicholson, R. L. (1990). Synthesis of phytoalexins in *Sorghum* as a site-specific response to fungal ingress. *Science* 248, 1637–1639. doi: 10.1126/science.248.4963.1637

Sondek, J., Bohm, A., Lambright, D. G., Hamm, H. E., and Sigler, P. B. (1996). Crystal structure of a G-protein βγdimer at 2.1 Å resolution. *Nature* 379, 369–374. doi: 10.1038/379369a0

Stirnimann, C. U., Petsalaki, E., Russell, R. B., and Muller, C. W. (2010). WD40 proteins propel cellular networks. *Trends Biochem. Sci.* 35, 565–574. doi: 10.1016/j.tibs.2010.04.003

Stone, J. M., and Walker, J. C. (1995). Plant protein kinase families and signal transduction. *Plant Physiol.* 108, 451–457. doi: 10.1104/pp.108.2.451

Stracke, R., Werber, M., and Weisshaar, B. (2001). The R2R3-MYB gene family in *Arabidopsis thaliana*. *Curr. Opin. Plant Biol.* 4, 447–456. doi: 10.1016/S1369-5266(00)00199-0

Taddese, B., Upton, G. J. G., Bailey, G. R., Jordan, S. R. D., Abdulla, N. Y., Reeves, P. J., et al. (2014). Do plants contain G protein-coupled receptors? *Plant Phyisol.* 164, 287–307. doi: 10.1104/pp.113.228874

Taylor, J. M., Jacob-Mosier, G. G., Lawton, R. G., Remmers, A. E., and Neubig, R. R. (1994). Binding of an alpha 2 adrenergic receptor third intracellular loop peptide to G β and the amino acid terminus of G alpha. *J. Biol. Chem.* 269, 27618–27624.

Taylor, J. M., Jacob-Mosier, G. G., Lawton, R. G., Van Dort, M., and Neubig, R. R. (1996). Receptor and membrane interaction sites on Gβ. A receptor- derived peptide binds to the carboxyl terminus. *J. Biol. Chem.* 271, 3336–3339.

Teng, S., Keurentjes, J., Bentsink, L., Koornneef, M., and Smeekens, S. (2005). Sucrose-specific induction of anthocyanin biosynthesis in *Arabidopsis* requires the MYB75/PAP1 gene. *Plant Physiol.* 139, 1840–1852. doi: 10.1104/pp.105.066688

Thung, L., Chakravorty, D., Trusov, Y., Jones, A. M., and Botella, J. R. (2013). Signaling specificity provided by the *Arabidopsis thaliana* heterotrimeric G-protein g subunits AGG1 and AGG2 is partially but not exclusively provided through transcriptional regulation. *PLoS ONE* 8:e58503. doi: 10.1371/journal.pone.0058503

Thung, L., Trusov, Y., Chakravorty, D., and Botella, J. R. (2012). Gγ1+Gγ2+Gγ3 = Gβ: the search for heterotrimeric G-protein γ subunits in *Arabidopsis* is over. *J. Plant Physiol.* 169, 542–545. doi: 10.1016/j.jplph.2011.11.010

Tominaga, R., Iwata, M., Okada, K., and Wada, T. (2007). Functional analysis of the epidermal-specific MYB genes CAPRICE and WEREWOLF in *Arabidopsis. Plant Cell* 19, 2264–2277. doi: 10.1105/tpc.106.045732

Torres, M. A., Morales, J., Sánchez-Rodríguez, C., Molina, A., and Dangl, J. L. (2013). Functional interplay between *Arabidopsis* NADPH oxidases and heterotrimeric G protein. *Mol. Plant Microbe Interact.* 26, 686–694. doi: 10.1094/MPMI-10-12-0236-R

Trusov, Y., Chakravorty, D., and Botella, J. R. (2012). Diversity of heterotrimeric G-protein γ subunits in plants. *BMC Res. Notes* 5:608. doi: 10.1186/1756-0500-5-608

Trusov, Y., Rookes, J. E., Chakravorty, D., Armour, D., Schenk, P. M., and Botella, J. R. (2006). Heterotrimeric G proteins facilitate *Arabidopsis* resistance to necrotrophic pathogens and are involved in jasmonate signaling. *Plant Physiol.* 140, 210–220. doi: 10.1104/pp.105.069625

Trusov, Y., Rookes, J. E., Tilbrook, K., Chakravorty, D., Mason, M. G., Anderson, D., et al. (2007). Heterotrimeric G protein γ subunits provide functional selectivity in Gβγ dimer signaling in *Arabidopsis. Plant Cell* 19, 1235–1250. doi: 10.1105/tpc.107.050096

Trusov, Y., Sewelam, N., Rookes, J. E., Kunkel, M., Nowak, E., Schenk, P. M., et al. (2009). Heterotrimeric G proteins-mediated resistance to necrotrophic pathogens includes mechanisms independent of salicylic acid-, jasmonic acid/ethylene- and abscisic acid-mediated defense signaling. *Plant J.* 58, 69–81. doi: 10.1111/j.1365-313X.2008.03755.x

Tuerck, J. A., and Fromm, M. E. (1994). Elements of the maize A1 promoter required for transactivation by the anthocyanin B/C1 or phlobaphene P regulatory genes. *Plant Cell* 6, 1655–1663. doi: 10.1105/tpc.6.11.1655

Ullah, H., Scappini, E. L., Moon, A. F., Williams, L. V., Armstrong, D. L., and Pedersen, L. C. (2008). Structure of a signal transduction regulator, RACK1, from *Arabidopsis thaliana. Protein Sci.* 17, 1771–1780. doi: 10.1110/ps.035121.108

UN-DESA. (2013). *World Population Prospects: The 2012 Revision, Press Release.* New York: United Nations, Department of Economic and Social Affairs, Population Division (accessed June 13, 2013).

Urano, D., Chen, J. G., Botella, J. R., and Jones, A. M. (2013). Heterotrimeric G protein signalling in the plant kingdom. *Open Biol.* 3, 120186. doi: 10.1098/rsob.120186

Urano, D., and Jones, A. M. (2014). Heterotrimeric G protein-coupled signaling in plants. *Annu. Rev. Plant Biol.* 65, 365–384. doi: 10.1146/annurev-arplant-050213-040133

van Baarlen, P., Legendre, L., and van Kan, J. A. L. (2007). "Plant defence compounds against botrytis infection," in *Botrytis: Biology, Pathology and Control*, eds Y. Elad, B. Williamson, P. Tudzynski, and N. Delen (Dordrecht: Kluwer Academic Publishers), 143–155. doi: 10.1007/978-1-4020-2626-3_9

van Nocker, S., and Ludwig, P. (2003). The WD-repeat protein super family in *Arabidopsis*: conservation and divergence in structure and function. *BMC Genom.* 4:50. doi: 10.1186/1471-2164-4-50

Van Wees, S. C., Pieterse, C. M., Trijssenaar, A., Van 't Westende, Y. A., Hartog, F., and van Loon, L. C. (1997). Differential induction of systemic resistance in *Arabidopsis* by biocontrol bacteria. *Mol. Plant-Microbe Interact.* 10, 716–724. doi: 10.1094/MPMI.1997.10.6.716

Walker, A. R., Davison, P., Bolognesi-Winfield, A., James, C. N., Srinivasan, N., Blundell, T. L., et al. (1999). The TRANSPARENT TESTA GLABRA1 locus, which regulates trichome differentiation and anthocyanin biosynthesis in *Arabidopsis*, encodes a WD40 repeat protein. *Plant Cell* 11, 1337–1350. doi: 10.1105/tpc.11.7.1337

Wang, Y., Liu, R., Chen, L., Wang, Y., Liang, Y., Wu, X., et al. (2009). *Nicotiana tabacum* TTG1 contributes to ParA1-induced signalling and cell death in leaf trichomes. *J. Cell Sci.* 122, 2673–2685. doi: 10.1242/jcs.049023

Weiergang, I., Hipskind, J. D., and Nicholson, R. L. (1996). Synthesis of 3-deoxyanthocyanidin phytoalexins in *Sorghum* occurs independent of light. *Physiol. Mol. Plant Path.* 49, 377–388. doi: 10.1006/pmpp.1996.0060

Weiss, C. A., Gamaat, C. W., Mukai, K., Hu, Y., and Ma, H. (1994). Isolation of cDNAs encoding guanine nucleotide-binding protein β-subunit homologues from maize (ZGB1) and *Arabidopsis* (AGB1). *Proc. Natl. Acad. Sci. U.S.A.* 91, 9554–9558. doi: 10.1073/pnas.91.20.9554

Wettschureck, N., and Offermanns, S. (2005). Mammalian G proteins and their cell type specific functions. *Physiol. Rev.* 85, 1159–1204. doi: 10.1152/physrev.00003.2005

Wharton, P. S., and Nicholson, R. L. (2000). Temporal synthesis and radiolabelling of the *Sorghum* 3-deoxyanthocyanidin phytoalexins and the anthocyanin, cyanidin 3-dimalonyl glucoside. *New Phytol.* 145, 457–469. doi: 10.1046/j.1469-8137.2000.00600.x

Williams, C. E., and Grotewold, E. (1997). Differences between plant and animal Myb domains are fundamental for DNA binding activity, and chimeric Myb domains have novel DNA binding specificities. *J. Biol. Chem.* 272, 563571.

Winkel-Shirley, B. (2001). Flavonoid biosynthesis. A colorful model for genetics, biochemistry, cell biology, and biotechnology. *Plant Physiol.* 126, 485–493. doi: 10.1104/pp.126.2.485

Witzel, F., Maddison, L., and Blüthgen, N. (2012). How scaffolds shape MAPK signaling: what we know and opportunities for systems approaches. *Front. Physiol.* 3:475. doi: 10.3389/fphys.2012.00475

Xu, D., Chen, M., Ma, Y., Xu, Z., Li, L., Chen, Y., et al. (2015). A G-protein β subunit, AGB1, negatively regulates the ABA response and drought tolerance by down-regulating AtMPK6-related pathway in *Arabidopsis. PLoS ONE* 10:e0116385. doi: 10.1371/journal.pone.0116385

Yang, A., Zhu, Z., Kapranov, P., McKeon, F., Church, G. M., Gingeras, T. R., et al. (2006). Relationships between p63 binding, DNA sequence, transcription activity, and biological function in human cells. *Mol. Cell* 24, 593–602. doi: 10.1016/j.molcel.2006.10.018

Zeng, W., and He, S. Y. (2010). A prominent role of the flagellin receptor FLAGELLIN-SENSING2 in mediating stomatal response to *Pseudomonas syringae* pv tomato DC3000 in *Arabidopsis. Plant Physiol.* 153, 1188–1198. doi: 10.1104/pp.110.157016

Zhang, F., Gonzalez, A., Zhao, M., Payne, C. T., and Lloyd, A. (2003). A network of redundant bHLH proteins functions in all TTG1-dependent pathways of *Arabidopsis. Development* 130, 4859–4869. doi: 10.1242/dev.00681

Zhang, J., Lu, Y., Yuan, Y., Zhang, X., Geng, J., Chen, Y., et al. (2009). Map-based cloning and characterization of a gene controlling hairiness and seed coat color traits in *Brassica rapa. Plant Mol. Biol.* 69, 553–563. doi: 10.1007/s11103-008-9437-y

Zhang, W., He, S. Y., and Assmann, S. M. (2008). The plant innate immunity response in stomatal guard cells invokes G-protein-dependent ion channel regulation. *Plant J.* 56, 984–996. doi: 10.1111/j.1365-313X.2008.03657.x

Zhao, J., Zhang, W., Zhao, Y., Gong, X., Guo, L., Zhu, G., et al. (2007). SAD2, an importin -like protein, is required for UV-B response in *Arabidopsis* by mediating MYB4 nuclear trafficking. *Plant Cell* 19, 3805–3818. doi: 10.1105/tpc.106.048900

Zhao, M., Morohashi, K., Hatlestad, G., Grotewold, E., and Lloyd, A. (2008). The TTG1-bHLH-MYB complex controls trichome cell fate and patterning through direct targeting of regulatory loci. *Development* 135, 1991–1999. doi: 10.1242/dev.016873

Zhou, J., Lee, C., Zhong, R., and Ye, Z. H. (2009). MYB58 and MYB63 are transcriptional activators of lignin biosynthetic pathway during secondary cell wall formation in *Arabidopsis. Plant Cell* 21, 248–266. doi: 10.1105/tpc.108.063321

Zhu, H., Li, G. J., Ding, L., Cui, X., Berg, H., Assmann, S. M., et al. (2009). *Arabidopsis* extra large G-protein 2 (XLG2) interacts with the Gβ subunit of heterotrimeric G protein and functions in disease resistance. *Mol. Plant* 2, 513–525. doi: 10.1093/mp/ssp001

Zimmermann, I. M., Heim, M. A., Weisshaar, B., and Uhrig, J. F. (2004). Comprehensive identification of *Arabidopsis thaliana* MYB transcription factors interacting with R/B-like BHLH proteins. *Plant J.* 40, 22–34. doi: 10.1111/j.1365-313X.2004.02183.x

Zipfel, C. (2014). Plant pattern-recognition receptors. *Trends Immunol.* 35, 345–351. doi: 10.1016/j.it.2014.05.004

Zuther, K., Kahnt, J., Utemark, J., Imkampe, J., Uhse, S., and Schirawski, J. (2012). Host specificity of *Sporisorium reilianum* is tightly linked to generation of the phytoalexin luteolinidin by *Sorghum bicolor*. *Mol. Plant Microbe Interact.* 25, 1230–1237. doi: 10.1094/MPMI-12-11-0314

Conflict of Interest Statement: The authors declare that the research was conducted in the absence of any commercial or financial relationships that could be construed as a potential conflict of interest.

15

Immunoprofiling of Rice Root Cortex Reveals Two Cortical Subdomains

*Sophia Henry, Fanchon Divol, Mathilde Bettembourg, Charlotte Bureau, Emmanuel Guiderdoni, Christophe Périn * and Anne Diévart **

CIRAD, UMR AGAP, Montpellier, France

The formation and differentiation of aerenchyma, i.e., air-containing cavities that are critical for flooding tolerance, take place exclusively in the cortex. The understanding of development and differentiation of the cortex is thus an important issue; however, studies on this tissue are limited, partly because of the lack of available molecular tools. We screened a commercially available library of cell wall antibodies to identify markers of cortical tissue in rice roots. Out of the 174 antibodies screened, eight were cortex-specific. Our analysis revealed that two types of cortical tissues are present in rice root seedlings. We named these cell layers "inner" and "outer" based on their location relative to the stele. We then used the antibodies to clarify cell identity in lateral roots. Without these markers, previous studies could not distinguish between the cortex and sclerenchyma in small lateral roots. By immunostaining lateral root sections, we showed that the internal ground tissue in small lateral roots has outer cortical identity.

Keywords: rice root, cortex, markers, antibodies, lateral roots, tissue identity, confocal microscopy, confocal imaging

Edited by:
Elison B. Blancaflor, The Samuel Roberts Noble Foundation, USA

Reviewed by:
David Domozych, Skidmore College, USA
Laura Elizabeth Bartley, University of Oklahoma, USA

****Correspondence:***
Christophe Périn
perin@cirad.fr;
Anne Diévart
anne.dievart@cirad.fr

INTRODUCTION

Rice has a complex root architecture with a mix of embryonic and post-embryonic roots. The radicle emerges first during germination, followed soon thereafter by embryonic coronary roots (Rebouillat et al., 2009; Coudert et al., 2010). The first adventitious post-embryonic roots resulting from shoot apical meristem activity appear a few days after seedling germination. Adventitious roots are also called coronary or crown roots and are produced throughout the life of a plant.

The internal rice root anatomy is well-described, particularly at the seedling stages (Rebouillat et al., 2009). Several concentric tissues (from the periphery to the stele) are located between the epidermis and the vascular tissues (**Figure 1**). These tissues, collectively referred to as "ground tissue," are sometimes also referred to as the "cortex" (Lux et al., 2004). Here, we retain the generic term, i.e., ground tissue, to describe the exodermal, sclerenchyma, cortical, and endodermal tissues. We refer to the "cortex" as the tissue located between the endodermis and the sclerenchyma.

Ground tissue is composed of several layers of tissues that have the same parenchymatous origin but different specific characteristics that play various and important structural and functional roles in roots. For example, the exodermis and endodermis are apoplastic barriers that control the radial transport of water and nutrients from the soil to the photosynthetic organs, a feature mostly provided by suberin lamellae deposition and secondary cell wall differentiation (Gregory, 2006). Sclerenchyma are lignified cells that represent a critical barrier to some metals and possibly play a mechanical support role (Huang et al., 2012). In contrast to these three monolayer

Abbreviations: ae, aerenchyma; c, cortex; ed, endodermis; ep, epidermis; ex, exodermis; gt, ground tissue; ic, inner cortex; L-LLR, Large-type Large Lateral Root; oc, outer cortex; sc, sclerenchyma; SLR, Small Lateral Root; st, stele; T-LLR, Thin-type Large Lateral Root; x, xylem.

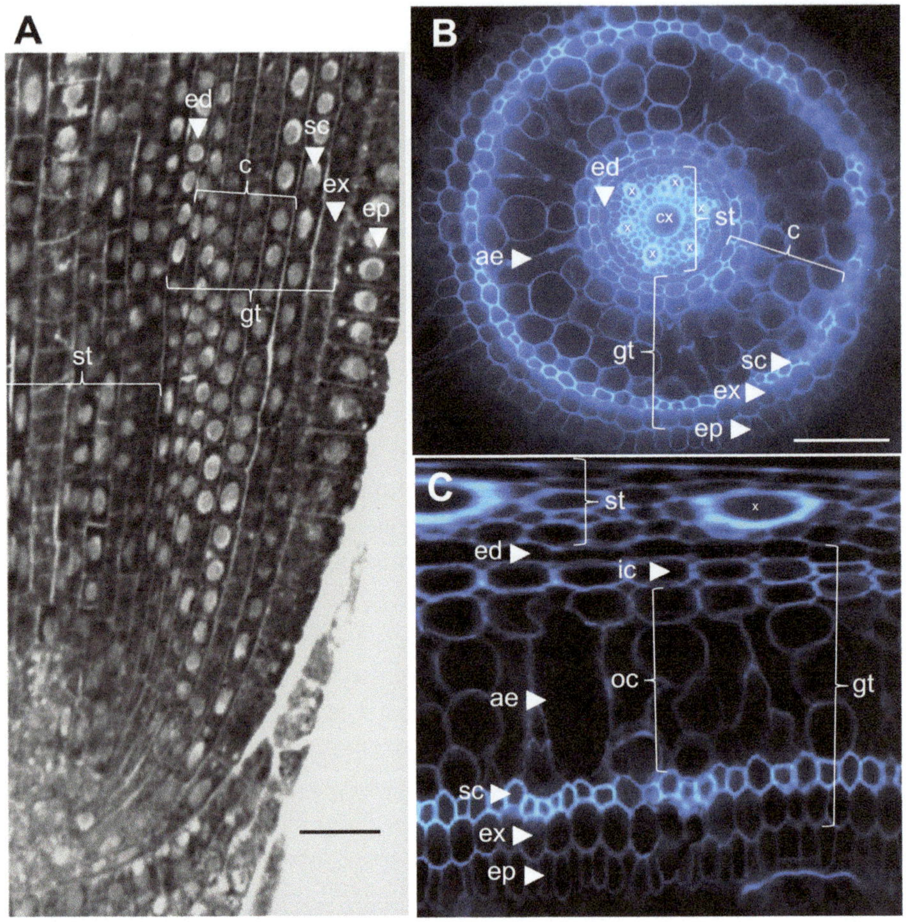

FIGURE 1 | Anatomy of a 6-day-old Nipponbare radicle. (A) Longitudinal section of a radicle fixed and paraplast embedded. Bar = 100 μm. **(B)** Transverse section of a fresh radicle observed under UV light. Bar = 50 μm. **(C)** Polar transformed view of B. Layers of tissues are labeled as follows: stele (st), ground tissue (gt) composed of several cell layers: one layer of endodermis (ed), several layers of cortex (c), one layer of sclerenchyma (sc), and one layer of exodermis (ex). The endodermal (ed) and exodermal (ex) layers are easily identified by the absence/reduction of fluorescence in the cell wall centers of these cells. The radicle is protected by one layer of epidermis (ep). Two types of cortical layers can be identified: inner cortex (ic), which is next to the endodermis (ed), and outer cortex (oc), which will eventually form aerenchyma (ae). In the stele, note the presence of a large central metaxylem (cx) surrounded by 6 metaxylem (x) vessels.

tissues, the cortical tissue in the rice radicle is composed of four or five layers. This tissue, also named the mid-cortex or mesodermis, is characterized by the schizogenous formation of spaces, leading to the formation of aerenchyma in rice (Justin and Armstrong, 1991; Kawai et al., 1998). Aerenchyma plays a major role in tolerance to flooding conditions and ensures continuous oxygen flow from the shoot to the root, allowing root growth under anoxic conditions. The cortex has an important but variable function in root physiology of numerous species and is involved in storage, flooding tolerance, or symbiotic interactions (Lux et al., 2004).

The radicle and crown roots produce three types of lateral roots: small lateral roots (SLRs) and two types of large lateral roots [Thin (T-LLRs) and Large (L-LLRs)] (Kawata and Shibayama, 1965; Kono et al., 1972; Rebouillat et al., 2009). This division is based on (i) root diameter: SLRs are the smallest (50–60 μm) and L-LLRs are the largest (120–150 μm), (ii) ramifications, which are only observed on LLRs, (iii) the presence of a central metaxylem vessel only in LLRs, and (iv) the presence of a sclerenchyma layer only in L-LLRs (Kono et al., 1972; Kawata et al., 1977; Rebouillat et al., 2009). Despite this classification, the identity of internal root tissues in lateral roots, particularly SLRs, remains uncertain. Indeed, Rebouillat et al. (2009) and Kawata et al. (1977) both claimed that SLR ground tissue was comprised of the endodermis, sclerenchyma and exodermis (from the inside toward the outside), while in Kono et al. (1972), the inner-cortex sclerenchymatous cell layer was described as absent in SLRs and T-LLRs.

To solve this dilemma, and to study the development and function of particular tissues in general, the use of identity markers is required. Most of the tissue markers used until now are morphological markers or histochemical stains, which have provided a useful way to characterize cell identity. Berberine reveals the presence of suberin in cell walls, which is found almost exclusively in the Casparian strips that surround each individual endodermis cell and is thus a good indirect marker of endodermal identity (Brundett et al., 1988). In *A. thaliana,* tissue-specific markers were also developed using tissue-specific promoters fused to reporter genes (GUS and/or GFP) in genetically modified plants. For instance, the *SCARECROW* promoter is used as an endodermal identity marker (Sugimoto et al., 2010). QC25 and QC46 enhancer trap GUS lines (Sabatini et al., 2003) and QHB in rice (Kamiya et al., 2003) were also identified as quiescent center (QC) markers based on their QC-specific expression. Despite interest in these markers, only a few are available and almost exclusively in *A. thaliana.* They also have drawbacks. For instance, the suitability of using the *SCARECROW* (*SCR*) promoter as an "endodermis" marker is debatable when specific questions on SCR and SHORT ROOT (SHR) roles in ground tissue differentiation are addressed. Indeed, the *SCR* and *SHR* genes are involved *per se* in endodermis/cortex formation (Wu and Gallagher, 2014). In *A. thaliana,* the only available root cortex markers are the *Co2* and *Co3* promoters (Heidstra et al., 2004; Ten Hove et al., 2010). In rice, these cortex markers are not yet used because there are no clear orthologs for the *Co2* and *Co3* genes. Only indirect morphological markers have been used to date in rice, such as aerenchyma formation for cortical identity (Rebouillat et al., 2009).

Another class of markers is commercially available antibodies directed against plant cell walls (http://www.plantprobes.net/index.php). These markers have been used scarcely in the past, but their efficiency has been demonstrated. For instance, the CCRC-M2 and JIM13 cell wall antibodies were used in *A. thaliana* to show that the single layer of internal tissue present in the *shr* mutant had cortical identity while the ground tissue layer of the *scr* mutant exhibited multiple identities (Di Laurenzio et al., 1996; Helariutta et al., 2000). These markers have many advantages: their simplicity and the possibility of combining secondary antibodies coupled to different fluorochromes to limit overlapping with auto-fluorescence, they do not require genetic transformation and they can be used to complement other classes of markers.

In this paper, we first describe a simple medium-throughput protocol for immunolabeling fresh vibratome tissue sections. Using this protocol, we screened radial sections of rice radicles with a large library of 174 cell wall antibodies. Among these, we identified eight cortex-specific markers and demonstrated that rice possesses two types of cortical tissues with distinct identity. We called these layers the inner and outer cortex based on their respective anatomical position in the root sections. To demonstrate the usefulness of these cortex-specific markers, we clarified the ground tissue identities of lateral roots using three of these antibodies. In SLRs in particular, we show that the internal tissue layer has an outer cortical identity. Our protocol is generic enough to be used for the development of tissue markers in any species without the need to generate transgenic plants. This protocol can be easily adapted with fresh shoot, leaf, root, or any tissues in other plants and represents a simple and easy way to identify tissue-specific markers. Moreover, these markers can be used in conjunction with other molecular markers.

RESULTS

Immunoprofiling of Cell Wall Antibodies in Rice Root Radial Sections Reveals Specific Antibodies for Cortex Layers

With the aim of identifying antibody-based markers for cortical cells, we first established a protocol for medium throughput immunolabeling of radial sections of fresh rice roots (see Materials and Methods and Supplementary Figure 1 for details). Then, we performed a screen with monoclonal antibodies raised against cell wall components (from the Complex Carbohydrate Research Center (CarboSource Services, Athens Georgia, USA) and Paul Knox Cell Wall lab (university of Leeds, UK), see Materials and Methods for details)) to identify cortex markers. From the 174 antibodies tested, only 12 were retained as putatively cortex specific. We tested their repeatability by performing two additional experiments and concluded that eight antibodies displayed similar and robust profiles among the three replicates (**Figure 2**). Five antibodies (JIM7, M14, M38, M130, and M131) provided the same immunolabeling pattern in the cortex. Only non-adjacent cell walls were labeled, resulting in a singular "diamond shape" labeling of all cortical cells. For the three other antibodies (LM5, M133, and M107), the labeling was detected in the cortex and sometimes the sclerenchyma cells, particularly with the LM5 and M107 antibodies. The signal was homogeneous on the cortex cell wall, but was much weaker in the inner cortex layers (the layer of cortex adjacent to the endodermal layer) than in the outer cortex layers. Thus, all of these antibodies with profiles that are complementary to cortical tissues are specific markers that can be used to analyze cortical cell identity. Moreover, the two different cortical labels, defining two subcortical areas, suggest that the outer and inner layers have a distinct molecular identity that was revealed by these antibodies.

Some specific features of these two subcortical layers were highlighted when the "polar transformer" plugin of the ImageJ software (http://imagej.nih.gov/ij/) was used to view the images of the root sections (**Figures 1B,C**; Schneider et al., 2012; Lartaud et al., 2014). The inner cells were characterized by a thick cell wall compared with the outer cells, which exhibited various air-containing cavities resulting from cell wall fusion. The inner cortical cells also exhibited a flattened shape resembling endodermal cells, in contrast to the outer cells, which were rounder. In this transformed picture, the radial cell wall that is wall running from the inner side of the cell to the outer side, autofluorescence of the endodermal and exodermal cells are not visible, presumably due to the presence of Casparian strips. Similarly, but to a lesser extent, small holes exempt of autofluorescence were also visible on the radial cell walls of inner cortical cells.

All Lateral Root Types Possess Only Outer Cortical Tissue

To demonstrate the usefulness of the cortex markers and to characterize cell identities, we labeled the three rice lateral root types (L-LLR, T-LLR and SLR) with three of these antibodies

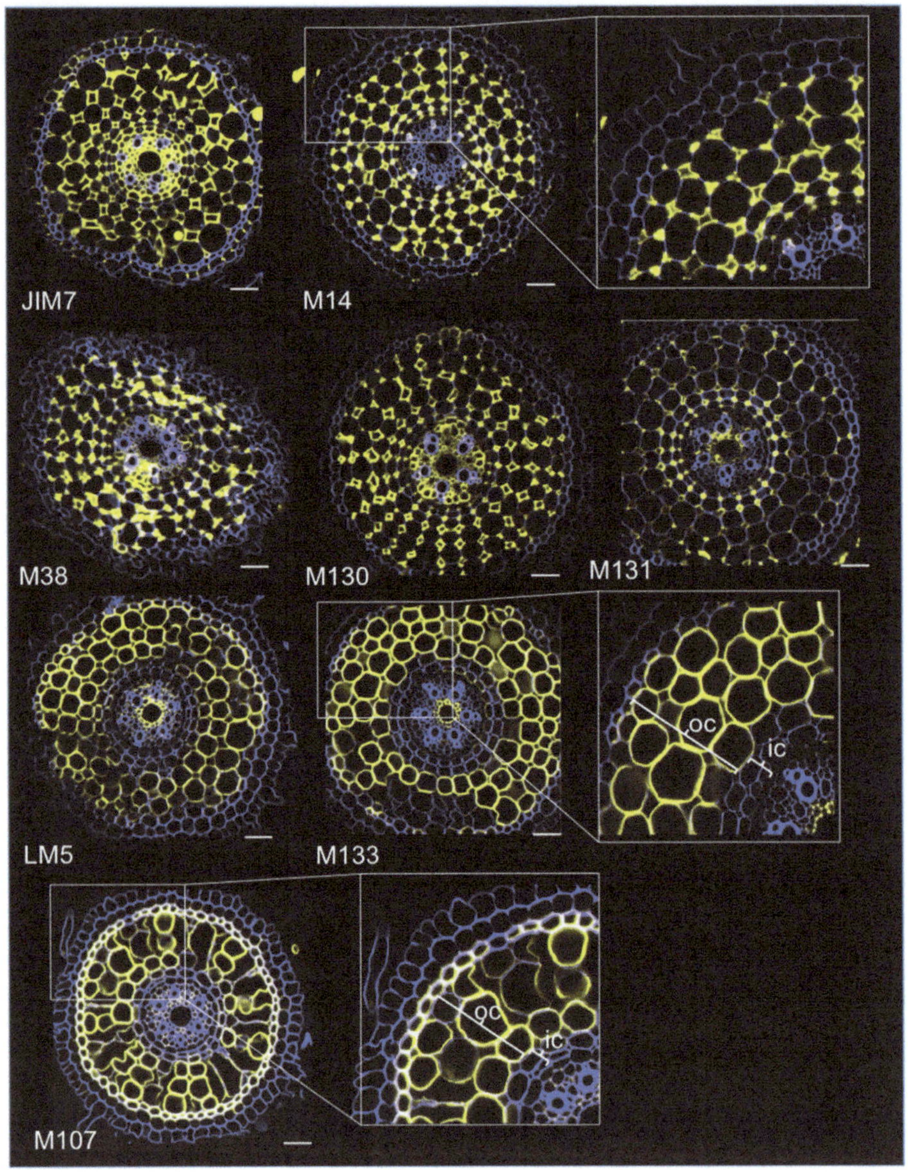

FIGURE 2 | Identification of eight specific cortex antibodies (JIM7, M14, M38, M130, M131, LM5, M133, and M107) among the 174 screened from the monoclonal cell wall antibody library. For each antibody, merged images of transverse sections were obtained under a confocal microscope after immunohistological labeling: in yellow, antibody pattern at 561 nm, and in blue, cell wall autofluorescence under UV light. Five antibodies (JIM7, M14, M38, M130, and M131) show a similar profile within all cortical cell layers stained (diamond shapes). Three antibodies (LM5, M133, and M107) homogeneously labeled cortical cell walls. Note that for these antibodies, inner cortical (ic) layers are not stained compared with outer cortical (oc) layers. Insets are close-ups of the corresponding images. Bars = 20 μm.

(M107, M133, and M14). For this experiment, because L-LLRs do not often develop on Petri dishes, lateral roots were collected from rice seedlings grown under hydroponic conditions for 2 weeks. We first analyzed the ground tissue organization of each type of lateral root by observing the autofluorescence of the cell walls in the transverse and polar views of the root sections (**Figure 3**). This first analysis revealed slightly distinct internal radial anatomy for the three types of lateral roots. Similar to the radicle sections viewed under UV light, the endodermal, and exodermal tissues can be identified by the fluorescence extinction in the center of the radial cell walls. These two tissues were present in all lateral roots. Between the endodermis and exodermis, several tissue layers were present in the L-LLRs and T-LLRs while the SLRs possessed only a single (unidentified) cell layer. The anatomy of the L-LLRs looked like seminal roots and external to the endodermis, several layers of cortical tissue were easily identified when aerenchyma tissue was formed. In addition, a sclerenchyma layer formed of closely packed cells

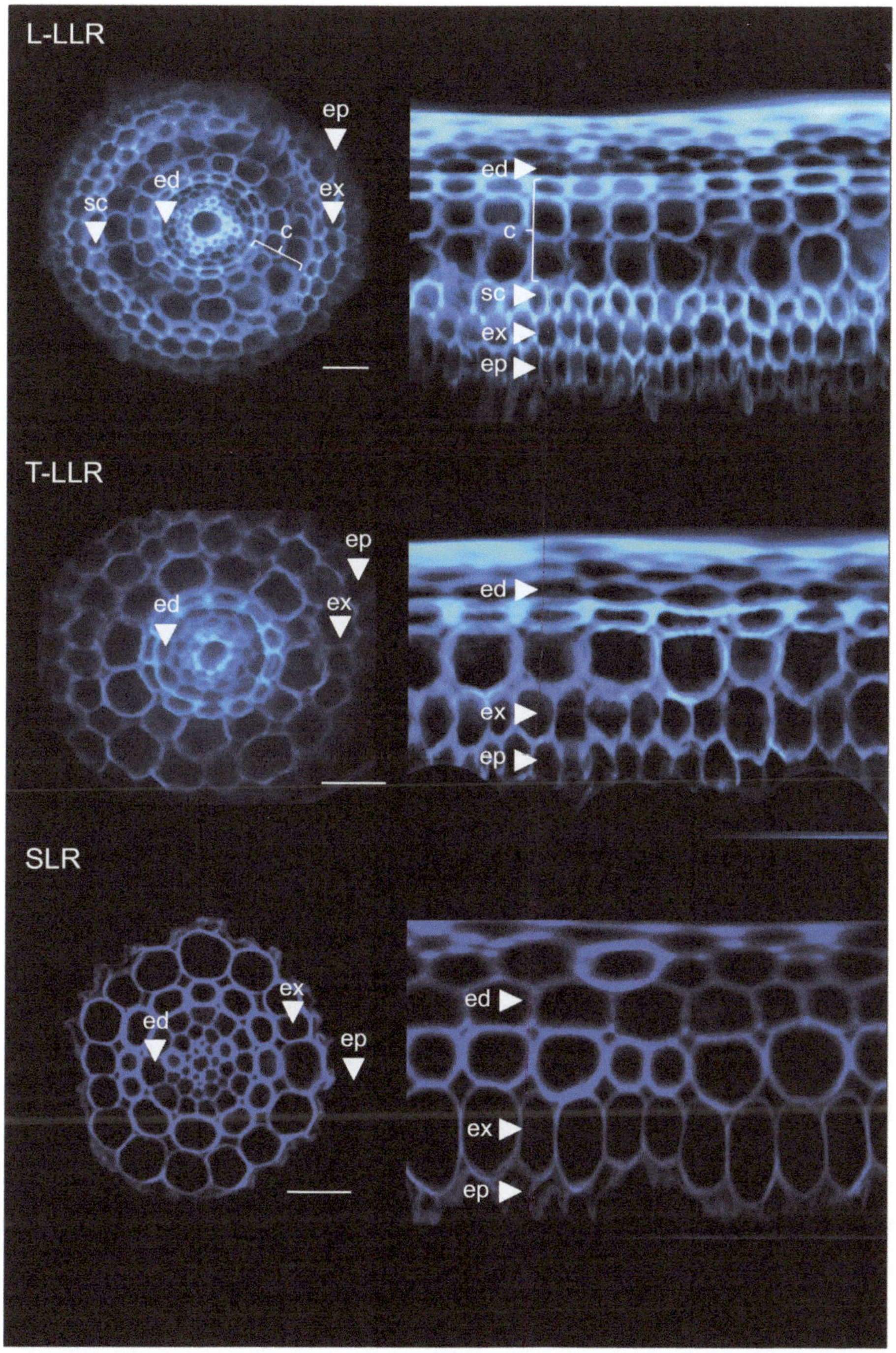

FIGURE 3 | Tissue composition of the Large Lateral Root types (Large-LLR and Thin-LLR) and Small Lateral Root (SLR). Seedlings were grown under hydroponic conditions for 2 weeks. Cell wall autofluorescence of transverse sections and polar transformed view under UV light. Tissues are labeled as follows: epidermis (ep), exodermis (ex), cortex (c), endodermis (ed). Note that the epidermis (ep) was often fragmented during the course of the experiment. Bars = 20 μm.

with wide cell walls was also observed. Depending on the experiment, this cell layer was not always completely developed. In the T-LLR sections, the external layer resembled cortical tissue and the internal layer resembled the single (unidentified) SLR layer.

To identify the cell layers present in the SLR and T-LLR ground tissues and to confirm the cortical identity in the L-LLRs, we immunolabeled lateral root sections with three cortex-specific antibodies (M107, M133 and M14; **Figure 4**). The M107 and M133 antibodies labeled all cell

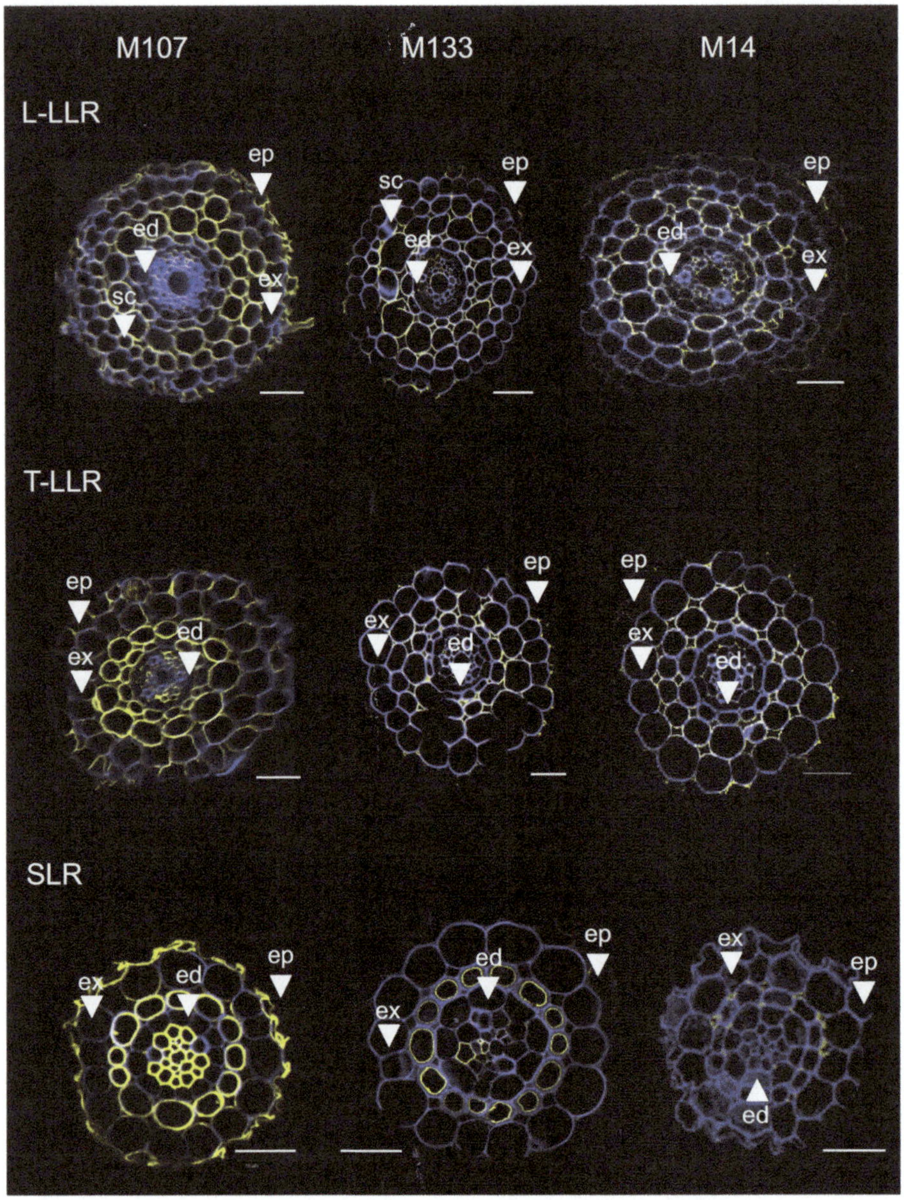

FIGURE 4 | Merged images of the Large Lateral Root types (Large-LLR and Thin-LLR) and Small Lateral Root (SLR) sections immunolabeled with M107, M133, and M14. Seedlings were grown under hydroponic conditions for 2 weeks. In yellow, antibody labeling at 561 nm, and in blue, autofluorescence under UV light. Tissues are labeled as follows: epidermis (ep), exodermis (ex), cortex (c), endodermis (ed). Note that the epidermis (ep) was often fragmented during the course of the experiment. Bars = 20 μm.

layers located between the endodermis and exodermis in all lateral roots. Based on radicle staining for these antibodies, this observation indicated that these cell layers could have sclerenchyma or cortical identity. However, the fact that M14 also immunolabeled these layers, particularly in the SLR sections, clearly demonstrates that these layers had outer cortical identity.

Variations in the number of outer cortical cell layers (one in the SLRs, two in the T-LLRs, three in the L-LLRs and four to five in the radicle) play a major role in rice root diameters (**Figure 5**).

CONCLUSIONS

Different terms have already been used to qualify subcortical areas in plants. For example, the presence of an "inner" lignified cortex layer, comparable to the one studied here, has already been described in wild rice (Yang et al., 2014). In maize, the two cortical domains are defined differently (Baluška et al., 1993). The "inner" domain forms aerenchyma, corresponding to the outer cortex here, and the "middle-outer" cortex is a lignified support tissue, which could be sclerenchyma. In *Arabidopsis*, a "middle cortex" cell layer, located between the cortical monolayer

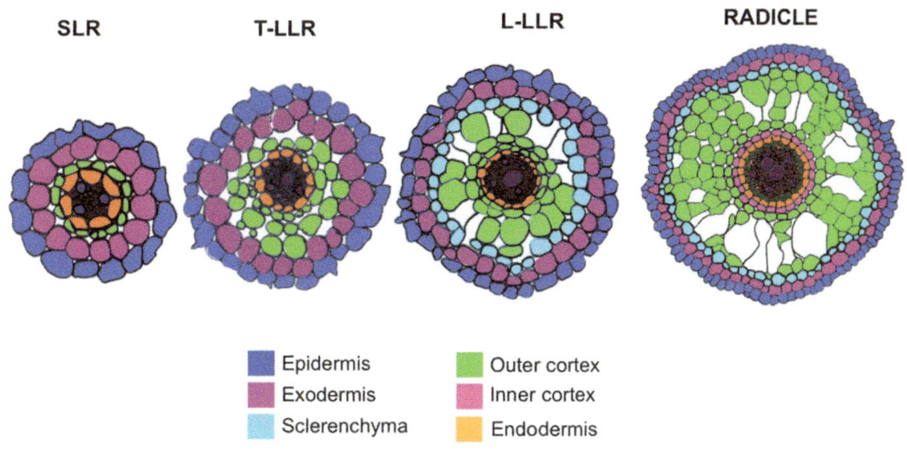

FIGURE 5 | Schematic representation of the lateral root and radicle tissues.

and the endodermis, has also been described (Baum et al., 2002; Paquette and Benfey, 2005). Recent results suggest that this cortex layer, whose formation is regulated by GA, oxidative stress, and SHR and SCR among others, plays a role in abiotic tolerance (Paquette and Benfey, 2005; Cui and Benfey, 2009; Pauluzzi et al., 2012; Wu and Gallagher, 2014). Interestingly, this layer is also described as a helix-like layer with a typical shape (Baum et al., 2002). Could this *Arabidopsis* "middle cortex" layer be functionally and structurally similar to the inner cortex layer of the rice radicle? In longitudinal and transverse sections of rice, the inner cortical layer also usually forms a helix, suggesting these two tissue layers could be comparable. In *Arabidopsis*, the "middle cortex" layer is derived from an endodermis periclinal division, which appears away from the QC between 7 and 14 days after germination (Paquette and Benfey, 2005). New tools will be necessary to verify these important parameters in rice. For example, no mutant affecting the timing of formation of this layer has been described up to now. Moreover, technical advances in imaging the rice QC are also required. Nevertheless, our cortex-specific markers offer a new tool to compare the function and molecular network involved in inner cortex (rice) vs. "middle cortex" (*Arabidopsis*) formation. Work is ongoing in our laboratory to identify the molecular determinants of inner and outer cortex formation, with SCR and SHR rice orthologs as first candidates for the regulation of cortex formation in rice roots (Pauluzzi et al., 2012).

MATERIALS AND METHODS

Plant Material

Six-Day-Old Seedlings

Seeds of the *Oryza sativa* L. ssp. *japonica* Nipponbare cultivar were grown vertically in sterile Petri dishes (Corning, 431301; 20 × 20 cm) under controlled conditions (day/night rhythm: 12/12 h, 28/25°C, light intensity: 500 μE m-2 s-1). First, the seeds were surface-sterilized by rinsing in 70% ethanol for ~1 min. Then, ethanol was replaced by a solution composed of 40% bleach in distilled water containing three drops of tween 80 (Sigma-Aldrich P4780-500 mL). The seeds were soaked in this solution for 30 min with gentle agitation, and then rinsed at least four times with sterile distilled water. Hot (~50°C) autoclave sterilized half strength Murashige and Skoog (MS/2) medium (250 mL) is poured in Petri dishes, and let solidify for ~ 45 min. The MS/2 solid medium is composed of 2.15 g.L-1 of MS medium basal salt mixture (Duchefa Biochemie, M0221), 75 mg.L-1 of MS vitamin mixture (Duchefa Biochemie, M0409), and 8 g.L-1 of agarose type II (Sigma-Aldrich, A6877). Sterile seeds are then pushed into the solidified MS/2 medium with the radicle oriented downwards. Roots were harvested after 6 days of growth.

Four-Week-Old Plants

After 3 days of germination in water, the seedlings were transferred to a hydroponic system under controlled conditions (day/night rhythm: 12/12 h, 28/25°C, light intensity: 500 μE m-2 s-1, relative humidity: 55%). Hydroponic system consists of a 50 L plastic box containing 30 L of hydroponic solution on which a 1 cm thick foam mattress is laid. The foam mattress is pierced with gaps to maintain seedling stem bases. The hydroponic medium is composed of $(NH_4)_2SO_4$ (0.5 mM), $MgSO_4.7H_2O$ (1.6 mM), $Ca(NO_3)_2.4H_2O$ (1.2 mM), KNO_3 (0.7 mM), $FeSO_4$ (0.1 mM), Na_2EDTA (0.1 mM), $MnSO_4.H_2O$ (1.7 μM), $(NH_4)_6Mo_7O_{24}.4H_2O$ (0.2 μM), $ZnSO_4.7H_2O$ (0.2 μM), $CuSO_4.5H_2O$ (0.2 μM), H_3BO_3 (1.4 μM), and KH_2PO_4 (0.6 mM). The solution is aerated by a pump placed at the bottom of the plastic box and changed every 10 days. The pH was adjusted and maintained at 5.4 ± 0.2.

Fresh Root Sectioning

The tips of growing radicles or lateral roots (2 cm) were cut using a sharp blade and placed parallel to each other to align the root tips (Supplementary Figures 1A,B). They were embedded in one drop of 3% melted agarose (50°C; Supplementary Figure 1C). Patches containing root tips were inserted into a 3 × 1 × 1-cm well filled with 3% melted agarose (Supplementary Figure 1D).

After solidification, blocks were resized and glued on a vibratome plate to be sliced. The vibratome (Hm650v (Thermo Scientific Microm)) parameters were speed 30, frequency 70, amplitude 0.8, and thickness 60 μm (Supplementary Figure 1E). Sections were transferred either onto chamber slides (Lab-teak 177402) for immunostaining (Supplementary Figure 1F), or onto slides humidified with 1X alkaline phosphate buffer (PBS, Sigma-Aldrich P3813) for observation.

Immunolocalization

The sections placed on the chamber slides (three per chamber) were first rinsed in 0.1 M glycine supplemented with 1X PBS and then twice in 1X PBS, each for 10 min. The tissues were then immersed in a PBS solution containing 5% bovine fetal serum (blocking solution, Thermo Fisher 37520) at 4°C overnight under agitation. Primary antibodies, diluted 1/10 in this blocking solution, were applied overnight at 4°C under agitation. The sections were then rinsed 3 times in PBS 1X for 10 min. The root sections were incubated for 2 h with the secondary antibody diluted 500-fold in blocking solution under agitation. These antibodies were coupled to a fluorophore, Alexa 546 anti-rat antibody (Invitrogen A11081) or Alexa 546 anti-mouse antibody (Invitrogen A11060). Then, the sections were rinsed again three times in 1X PBS under agitation, for 10 min each. The chambers were removed and few drops of mowiol mounting media (Sigma-Aldrich 81381) were added. A coverslip was placed on the root sections, which were allowed to dry for 36 h at 4°C in the dark.

Paraplast Fixation

Radicles grown on Petri dishes for 6 days were fixed in 4% paraformaldehyde (in PBS 1X) overnight at 4°C and rinsed twice using PBS 1X (Jackson, 1991). The fixed tissues were dehydrated in ethanol, cleared in Histochoice Clearing Agent (HistoClear, Sigma Aldrich), and embedded in Paraplast (Fisher). The tissues were sectioned (6 μm thick) on a Leica RM2255 microtome and mounted on SuperfrostPlus slides (Fisher).

Microscopy

Bright field and autofluorescence observations were performed using a Leica DM4500 microscope. For autofluorescence, images were taken with the "A" filter cube (excitation range: UV; excitation filter: BP 340–380; suppression filter: LP 425). Immunostained sections were observed with confocal microscopes: Zeiss LSM 510 or Leica SP8. The cell walls were first visualized using autofluorescence [720-nm (chameleon laser) or 405-nm respectively]. The secondary antibody was visualized using a Helium/Neon laser at 543 or 561 nm respectively. Pictures were taken with a color Retiga 2000R camera (QIMAGING, Canada) running Volocity image acquisition software (Improvision, UK).

AUTHOR CONTRIBUTIONS

SH, FD, MB, CB: acquisition of data; SH, CP, AD analysis of data; SH, EG, CP, AD: drafting of the manuscript.

ACKNOWLEDGMENTS

We thank MRI (Montpellier RIO Imaging plateform), PHIV and La Gaillarde Plateforms for their technical support for immunolocalization and/or confocal and multiphoton imaging. SH is supported by a CIRAD PhD fellowship.

REFERENCES

Baluška, F., Brailsford, R. W., Hauskrecht, M., Jackson, M. B., and Barlow, P. W. (1993). Cellular dimorphism in the maize root cortex: involvement of microtubules, ethylene and gibberellin in the differentiation of cellular behaviour in postmitotic growth zones. *Bot. Acta* 106, 394–403. doi: 10.1111/j.1438-8677.1993.tb00766.x

Baum, S. F., Dubrovsky, J. G., and Rost, T. L. (2002). Apical organization and maturation of the cortex and vascular cylinder in *Arabidopsis thaliana* (Brassicaceae) roots. *Am. J. Bot.* 89, 908–920. doi: 10.3732/ajb.89.6.908

Brundett, M. C., Enstone, D. E., and Peterson, C. A. (1988). A Berberine-aniline blue fluorescent staining procedure for suberin, lignin, and callose in plant tissue. *Protoplasma* 146, 133–142. doi: 10.1007/BF01405922

Coudert, Y., Périn, C., Courtois, B., Khong, N. G., and Gantet, P. (2010). Genetic control of root development in rice, the model cereal. *Trends Plant Sci.* 15, 219–226. doi: 10.1016/j.tplants.2010.01.008

Cui, H., and Benfey, P. N. (2009). Interplay between SCARECROW, GA and LIKE HETEROCHROMATIN PROTEIN 1 in ground tissue patterning in the Arabidopsis root. *Plant J.* 58, 1016–1027. doi: 10.1111/j.1365-313X.2009.03839.x

Di Laurenzio, L., Wysocka-Diller, J., Malamy, J. E., Pysh, L., Helariutta, Y., Freshour, G., et al. (1996). The SCARECROW gene regulates an asymmetric cell division that is essential for generating the radial organization of the Arabidopsis root. *Cell* 86, 423–433. doi: 10.1016/S0092-8674(00)80115-4

Gregory, P. J. (2006). *Plant Roots: Growth, Activity, and Interaction with Soils.* Oxford; Ames, IA: Blackwell Pub.

Heidstra, R., Welch, D., and Scheres, B. (2004). Mosaic analyses using marked activation and deletion clones dissect Arabidopsis SCARECROW action in asymmetric cell division. *Genes Dev.* 18, 1964–1969. doi: 10.1101/gad.305504

Helariutta, Y., Fukaki, H., Wysocka-Diller, J., Nakajima, K., Jung, J., Sena, G., et al. (2000). The SHORT-ROOT gene controls radial patterning of the Arabidopsis root through radial signaling. *Cell* 101, 555–567. doi: 10.1016/S0092-8674(00)80865-X

Huang, C. F., Yamaji, N., Ono, K., and Ma, J. F. (2012). A leucine-rich repeat receptor-like kinase gene is involved in the specification of outer cell layers in rice roots. *Plant J.* 69, 565–576. doi: 10.1111/j.1365-313X.2011.04824.x

Jackson, D. (1991). "*In-situ* hybridization in plants" in *Molecular Plant Pathology: A Practical Approach*, eds D. J. Bowles, M. McPherson, and S. J. Gurr (Totowa, NJ: Humana Press), 163–174.

Justin, S. H. F. W., and Armstrong, W. (1991). Evidence for the involvement of ethylene in aerenchyma formation in adventitious roots of rice (*Oryza sativa* L.). *New Phytol.* 118, 49–62. doi: 10.1111/j.1469-8137.1991.tb00564.x

Kamiya, N., Nagasaki, H., Morikami, A., Sato, Y., and Matsuoka, M. (2003). Isolation and characterization of a rice WUSCHEL-type homeobox gene that is specifically expressed in the central cells of a quiescent center in the root apical meristem. *Plant J.* 35, 429–441. doi: 10.1046/j.1365-313X.2003.01816.x

Kawai, M., Samarajeewa, P. K., Barrero, R. A., Nishiguchi, M., and Uchimiya, H. (1998). Cellular dissection of the degradation pattern of cortical cell death during aerenchyma formation of rice roots. *Planta* 204, 277–287. doi: 10.1007/s004250050257

Kawata, S., Sasaki, O., and Yamazaki, K. (1977). On the structure of the crown root and the lateral root, and the vessel connection between them, in rice plants. *Jpn. J. Crop. Sci.* 46, 569–579. doi: 10.1626/jcs.46.569

Kawata, S., and Shibayama, B. (1965). On the lateral root primordia formation in the crown roots of rice plant. *Proc. Crop Sci. Soc. Jpn.* 33, 423–431. doi: 10.1626/jcs.33.423

Kono, Y., Igeta, M., and Yamada, N. (1972). Studies on the developmental physiology of the lateral roots in rice seminal roots. *Jpn. J. Crop. Sci.* 41, 192–204. doi: 10.1626/jcs.41.192

Lartaud, M., Perin, C., Courtois, B., Thomas, E., Henry, S., Bettembourg, M., et al. (2014). PHIV-RootCell: a supervised image analysis tool for rice root anatomical parameter quantification. *Front. Plant Sci.* 5:790. doi: 10.3389/fpls.2014.00790

Lux, A., Luxova, M., Abe, J., and Morita, S. (2004). Root cortex: structural and functional variability and responses to environmental stress. *Root Res.* 13, 117–131. doi: 10.3117/rootres.13.117

Paquette, A. J., and Benfey, P. N. (2005). Maturation of the ground tissue of the root is regulated by gibberellin and SCARECROW and requires SHORT-ROOT. *Plant Physiol.* 138, 636–640. doi: 10.1104/pp.104.058362

Pauluzzi, G., Divol, F., Puig, J., Guiderdoni, E., Dievart, A., and Périn, C. (2012). Surfing along the root ground tissue gene network. *Dev. Biol.* 365, 14–22. doi: 10.1016/j.ydbio.2012.02.007

Rebouillat, J., Dievart, A., Verdeil, J., Escoute, J., Giese, G., Breitler, J., et al. (2009). Molecular genetics of rice root development. *Rice* 2, 15–34. doi: 10.1007/s12284-008-9016-5

Sabatini, S., Heidstra, R., Wildwater, M., and Scheres, B. (2003). SCARECROW is involved in positioning the stem cell niche in the Arabidopsis root meristem. *Genes Dev.* 17, 354–358. doi: 10.1101/gad.252503

Schneider, C. A., Rasband, W. S., and Eliceiri, K. W. (2012). NIH Image to ImageJ: 25 years of image analysis. *Nat. Methods* 9, 671–675. doi: 10.1038/nmeth.2089

Sugimoto, K., Jiao, Y., and Meyerowitz, E. M. (2010). Arabidopsis regeneration from multiple tissues occurs via a root development pathway. *Dev. Cell* 18, 463–471. doi: 10.1016/j.devcel.2010.02.004

Ten Hove, C. A., Willemsen, V., de Vries, W. J., Van Dijken, A., Scheres, B., and Heidstra, R. (2010). SCHIZORIZA encodes a nuclear factor regulating asymmetry of stem cell divisions in the Arabidopsis root. *Curr. Biol.* 20, 452–457. doi: 10.1016/j.cub.2010.01.018

Wu, S., and Gallagher, K. L. (2014). The movement of the non-cell-autonomous transcription factor, SHORT-ROOT relies on the endomembrane system. *Plant J.* 80, 396–409. doi: 10.1111/tpj.12640

Yang, C., Zhang, X., Li, J., Bao, M., Ni, D., and Seago, J. L. (2014). Anatomy and histochemistry of roots and shoots in wild rice (*Zizania latifolia* Griseb.). *J. Bot.* 2014, 9. doi: 10.1155/2014/181727

Conflict of Interest Statement: The authors declare that the research was conducted in the absence of any commercial or financial relationships that could be construed as a potential conflict of interest.

16

Mitochondrial pleomorphy in plant cells is driven by contiguous ER dynamics

Erica-Ashley Jaipargas, Kiah A. Barton, Neeta Mathur and Jaideep Mathur *

Laboratory of Plant Development and Interactions, Department of Molecular and Cellular Biology, University of Guelph, ON, Canada

Mitochondria are pleomorphic, double membrane-bound organelles involved in cellular energetics in all eukaryotes. Mitochondria in animal and yeast cells are typically tubular-reticulate structures and several micro-meters long but in green plants they are predominantly observed as 0.2–1.5 μm punctae. While fission and fusion, through the coordinated activity of several conserved proteins, shapes mitochondria, the endoplasmic reticulum (ER) has recently been identified as an additional player in this process in yeast and mammalian cells. The mitochondria-ER relationship in plant cells remains largely uncharacterized. Here, through live-imaging of the entire range of mitochondria pleomorphy we uncover the underlying basis for the predominantly punctate mitochondrial form in plants. We demonstrate that mitochondrial morphology changes in response to light and cytosolic sugar levels in an ER mediated manner. Whereas, large ER polygons and low dynamics under dark conditions favor mitochondrial fusion and elongation, small ER polygons result in increased fission and predominantly small mitochondria. Hypoxia also reduces ER dynamics and increases mitochondrial fusion to produce giant mitochondria. By observing elongated mitochondria in normal plants and fission-impaired Arabidopsis *nmt1-2* and *drp3a* mutants we also establish that thin extensions called matrixules and a beads-on-a-string mitochondrial phenotype are direct consequences of mitochondria-ER interactions.

Keywords: mitochondria, pleomorphy, sugar, light, hypoxia, fission, endoplasmic reticulum

Edited by:
Elison B. Blancaflor, The Samuel Roberts Noble Foundation, USA

Reviewed by:
Diane C. Bassham, Iowa State University, USA
Frantisek Baluska, University of Bonn, Germany

****Correspondence:***
Jaideep Mathur, Laboratory of Plant Development and Interactions, Department of Molecular and Cellular Biology, University of Guelph, 50 Stone Road, Guelph, ON N1G2W1, Canada
jmathur@uoguelph.ca

Introduction

The term mitochondrion (Greek-mitos = thread; chondrion = grain) introduced by Benda (1899) recognized the peculiar thread-grain nature of these organelles and their tendency to morph quickly from one form to another. Lewis and Lewis (1915) provided meticulous descriptions of the dynamic behavior of mitochondria involving continuous contortions and undulatory movements that lead to changes in their morphology and position within the living cell. Today mitochondria are recognized as double membrane-bound organelles of an endo-symbiogenic origin that constitute major sites of oxygen consumption and ATP production in all eukaryotes. It has been unequivocally established that mitochondria undergo fission and fusion in order to maintain the optimal gradients required for driving energy production (Cavers, 1914; Bereiter-Hahn and Vöth, 1994; Nunnari et al., 1997; Bleazard et al., 1999; Welchen et al., 2014). Alterations in mitochondrial form thereby reflect the continuous flux in cellular energy that occurs due to fluctuations in temperature, oxygen, and carbon availability, and various metabolic processes (Hackenbrock, 1968; Hackenbrock et al., 1971; Novikoff and Holtzman, 1976; Bereiter-Hahn and Vöth, 1994).

The inner mitochondrial membrane (IMM) is the seat of the electron transport chain (ETC) and oxidative phosphorylation reactions and responds to fluctuating ATP/ADP levels. The relative arrangements of mitochondrial cristae and matrix change under different conditions as well. There are two general states that these arrangements fall under: orthodox or condensed. The orthodox state is characteristic of ATP-enriched, motile mitochondria and is indicated by an expanded matrix. By contrast the condensed state is exhibited by relatively immobile ADP-enriched mitochondria with electron dense matrix and dilated inter-cristal regions (Hackenbrock et al., 1971; Bereiter-Hahn and Vöth, 1994; Logan and Leaver, 2000; Logan, 2006a). Whereas, different eukaryotic cell types and even similar cells within an organism exhibit mitochondria of different shapes and sizes (Youle and van der Bliek, 2012) the predominant mitochondrial form observed in a majority of cells in an organism is usually presented as being typical for that particular organism. For example, the typical mitochondrion is described as long and filamentous in animal fibroblasts, whereas hepatocytes have a predominance of spherical or ovoid mitochondria (Youle and van der Bliek, 2012). While in yeast cells a tubular-reticulate form comprising of 1–10 mitochondria is considered typical (Hoffman and Avers, 1973; Bereiter-Hahn and Vöth, 1994; Nunnari et al., 1997; Sesaki and Jensen, 1999), in green plants they predominantly appear as discrete, spherical to ovoid, punctate organelles with diameters ranging from 0.2 to 1.5 μm (Matzke and Matzke, 1986; Köhler et al., 1997; Logan and Leaver, 2000; Arimura et al., 2004; Logan, 2006a). Interestingly, the first depiction of mitochondria in plant cells had shown elongated forms (Meves, 1904) which was reinforced by several subsequent light microscopy based studies (Gunning and Steer, 1986; Lichtscheidl and Url, 1990). Other investigations have revealed that mitochondria become abnormally enlarged and vermiform under certain conditions (Stickens and Verbelen, 1996; Logan and Leaver, 2000; Van Gestel and Verbelen, 2002; Logan, 2006b; Seguí-Simarro et al., 2008). These observations clearly indicate that as in other organisms, mitochondria in plants are also capable of becoming elongated and vermiform and some specific but as yet undefined conditions must cause and maintain their fission into small units. The first objective for our investigations was therefore to identify a common condition that might explain the punctate morphology of mitochondria in plant cells.

In animal cells, normally elongated mitochondria become small or fragmented under hyperglycemic conditions (Yu et al., 2008; Jhun et al., 2013). This led to the hypothesis that a similar sugar-induced phenomenon occurs in plant cells. In addition, a contextual requirement was to understand the cell-biological mechanism through which an elongated mitochondrion in a plant cell may become small. The widely accepted present-day view on the mechanism underlying mitochondria shape and motility involves several highly conserved genes (Logan et al., 2004; Scott et al., 2006; Arimura et al., 2008). Whereas, homologs for all the proteins implicated in mitochondrial fission in yeast and animal cells have not been identified so far in plants, at least two major factors have been well-characterized. These are Fission1, a tail-anchored membrane protein (hFis1 in mammals: Yoon et al., 2003; Stojanovski et al., 2004; Fis1p in yeast: Mozdy et al., 2000; Tieu and Nunnari, 2000; Fis1/BIGYIN in plants: Scott et al., 2006) and a mechano-chemical GTPase Dnm1 (Dnm1 in yeast: Bleazard et al., 1999; Dlp1 in mammals: Pitts et al., 1999; DRP3/ADL2 in plants: Arimura et al., 2004; Logan et al., 2004). The possibility of NETWORK MITOCHONDRIA/ELONGATED MITOCHONDRIA1 (NMT1/ELM1) being a functional plant homolog of the yeast Mdv1/Caf4 proteins has also been suggested (Logan et al., 2003; Arimura et al., 2008; Logan, 2010). Our present mechanistic understanding of mitochondria dynamics relies heavily on the relative frequency with which mitochondrial fission vs. fusion takes place in a cell. Frequent fission results in small mitochondria whereas a reduced fission frequency or an increased tendency to fuse leads to large mitochondria. Demonstrations of speculated stoichiometric relationships between the different mitochondria-associated proteins discovered so far greatly support this viewpoint but do not immediately suggest how two or more portions of an elongated mitochondrion might actually be forced apart.

The recent demonstration of tightly coupled dynamics and extensive membrane contacts between mitochondria and tubules of the endoplasmic reticulum (ER) in yeast and mammalian cells has been a major breakthrough in this context (Friedman et al., 2011; Friedman and Nunnari, 2014). Close associations between mitochondria and the ER have been known in plant cells for a long time (Gunning and Steer, 1986; Lichtscheidl and Url, 1990; Stickens and Verbelen, 1996) and have been used to support their biochemical relationship. However, the idea that the ER might be directly involved in the fission of elongated mitochondria into the predominant, punctate form in plant cells has not been explored to date. A major objective for our investigations was to assess a possible role of contiguous ER tubules and cisternae in mediating mitochondrial fission in plant cells.

The work on mitochondria-ER interactivity has focused largely on the phenomenon of mitochondrial fission and on identifying the protein complexes that aid inter-organelle interactions (Friedman et al., 2011). However, in a diversity of organisms including plants abnormally large mitochondria, called mega mitochondria and giant mitochondria have been observed under suboptimal conditions of growth and development including hypoxia, metabolic disorders, and chemical-induction (Tandler and Hoppel, 1986; Ramonell et al., 2001; Wakabayashi, 2002; Van Gestel and Verbelen, 2002). In addition, both animal (Grainger and James, 1969; Bowes and Gupta, 2008) and plant (Logan et al., 2004) cells exhibit sporadic thin extensions from and between mitochondria that have been called matrixules (Logan et al., 2004; Logan, 2006b; Scott et al., 2007). Elongated mitochondria also display a rather perplexing phenotype described as beads-on-a-string. It is unknown whether the ER plays a role in shaping all these different mitochondrial forms too. Whereas, earlier investigations have described the effect that specific conditions (Ramonell et al., 2001; Van Gestel and Verbelen, 2002) have on mitochondrial morphology, our study considers the entire range of mitochondrial pleomorphy in plant cells in relation to the organization and behavior of contiguous ER tubules and cisternae.

Our investigations involve time-lapse imaging of living cells expressing fluorescent proteins targeted to mitochondria (mito-GFP: Logan and Leaver, 2000; mito-mEosFP: Mathur et al., 2010) and the ER [ER-lumen retained RFP (RER); Sinclair et al., 2009]. Physiological manipulations on transgenic Arabidopsis plants have revealed that shifting plants between light and dark conditions of growth, changing cytosolic sugar levels through exogenous feeding of sucrose and maintaining seedlings transiently under hypoxia produce significant alterations in mitochondrial morphology. Our investigations involving a wide range of transient shapes including small spherical mitochondria, rings, and coiled, branched, and beaded tubules displayed by elongated mitochondria are reinforced through observations on abnormally long, fission-impaired mitochondria in the Arabidopsis *drp3a/adl2a* (Arimura et al., 2004; Logan et al., 2004) and *nmt1-2/elm1* (Logan et al., 2003; Feng et al., 2004; Arimura et al., 2008; Logan, 2010) mutants. In addition the *pah1pah2* double mutant in the phosphatidic acid phosphohydrolase1 and 2 that contains over-expanded ER cisternal membranes (Eastmond et al., 2010) was used to test possible association between mitochondrial form and ER. Our observations suggest a strong correlation between mitochondrial pleomorphy and the ER.

Results

Earlier reports on transgenic tobacco and Arabidopsis plants, protoplasts, single cells, and detached organs (Logan and Leaver, 2000; Van Gestel and Verbelen, 2002; Logan, 2006a) that between them describe the complete range of mitochondrial pleomorphy in plants formed the basis for our optimization of procedures for 7–9 days old seedlings of stable transgenic *Arabidopsis thaliana* (Columbia ecotype) expressing mito-GFP. Subsequent visualization of punctate, elongated, and giant mitochondria and the ER simultaneously in double transgenic plants co-expressing mito-GFP and RFP-HDEL (RER; Sinclair et al., 2009) used our optimized methods.

Cytosolic Sugar Levels and Exposure to Light Affect Mitochondrial Size in Plant Cells

Photosynthesis in green plant cells causes cytosolic sugar levels to fluctuate considerably between cells and during the day and night cycle (Azcón-Bieto and Osmond, 1983; Azcón-Bieto et al., 1983). This was the rationale behind our using a more controlled, *in vitro*, exogenous sugar-feeding approach. Arabidopsis mito-GFP seeds were sown on MS medium with and without sucrose and kept in the dark to first assess if exogenous sugar in the MS medium actually altered cytosolic sugar levels in seedlings. Cytosolic sugar estimations (**Figure 1**; Supplementary Information) for 7 day old plants using a phenol-sulfuric acid colorimetric method (Buysse and Merckx, 1993) established that plants grown on MS medium with 3% sucrose had higher soluble sugar levels [10.14 $\pm$ 1.10 μg/mg of fresh weight (f.w.)] as compared to plants grown without sucrose in the medium (1.92 $\pm$ 0.77 μg/mg f.w.).

Confocal laser scanning microscopy of hypoctyl cells of mito-GFP revealed that although all cells had a mixture of punctate ($\leq$0.85 μm long) and elongated (>0.85 μm long) mitochondria, the relative frequency between these two broad categories varied. Eighty-six percent of mitochondria were elongated when plants were grown in the dark without sucrose (**Figures 1A1,B**), which decreased to 74% for plants grown in

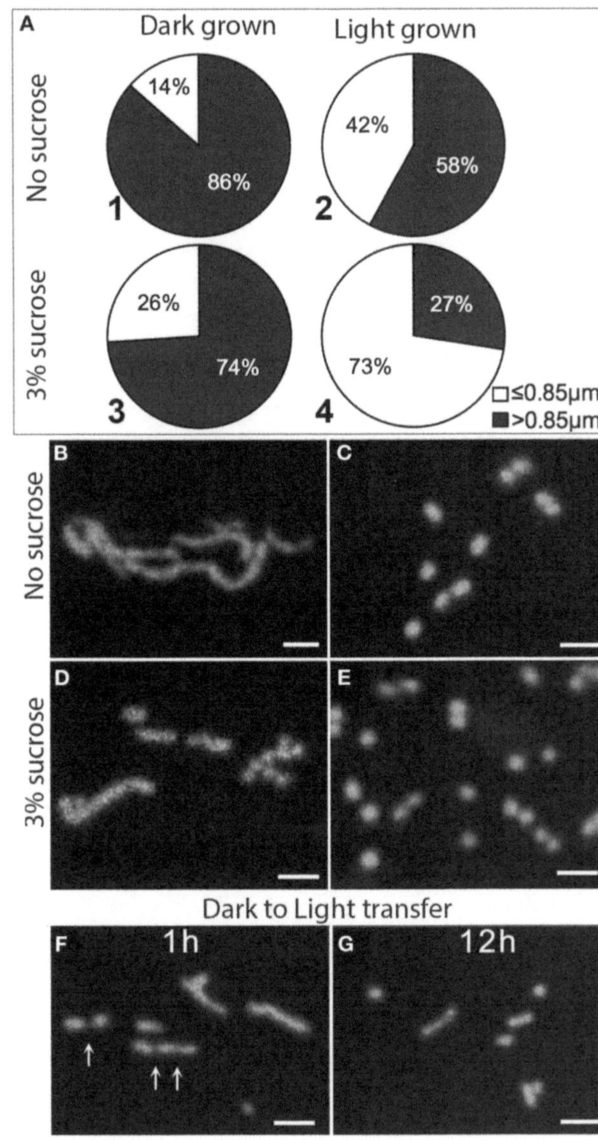

FIGURE 1 | Effects of light, dark and sugar on mitochondria length. **(A–E)** Pie diagram showing the relative percentage of small (less than 0.85 μm) vs. elongated (longer than 0.85 μm) mitochondria in a cellular population for under different conditions. Data based on Arabidopsis mito-GFP transgenics grown for 7 days in light (70 μmol m^{-2} s^{-1}) and dark on MS medium with no sucrose **(A1,A2,B,C)** and with 3% sucrose **(A3,A4,D,E)**. Mitochondria in seedlings grown in the dark were predominantly elongated **(B,D)** as compared to the punctate mitochondria in plants grown in light **(C,E)**. **(F,G)** Dark grown plants grown without sucrose were transferred to light (70 μmol $m^{-2}s^{-1}$) for 1 h **(F)** and 12 h **(G)** without sucrose. Putative fission sites on tubular mitochondria are indicated (arrows).

the dark with 3% sucrose (**Figures 1A3,D**). The observations on dark grown plants had to be taken quickly since rapid fission of elongated mitochondria occurred within minutes and changed the relative number of small vs. elongated mitochondria in a cell. Whereas, 58% of mitochondria were elongated in plants grown in light without sucrose (**Figures 1A2,C**), only 27% were elongated in plants grown in the light with sucrose. Furthermore, the light and sucrose grown plants conversely had the highest proportion of small, round mitochondria (**Figures 1A4,E**).

The effect of light on mitochondrial fission was confirmed in a separate experiment by taking plants grown in the dark without sugar with clearly elongated mitochondria and transferring them to 70 μmol $m^{-2}s^{-1}$ light for 1 and 12 h (**Figures 1F,G**). In both cases the predominant mitochondrial size approached ≤0.85 μm.

We concluded that darkness and low sugar levels favor an elongated mitochondrial morphology whereas high light with and without high cytosolic sugar drives mitochondrial fission to create the punctate mitochondria considered typical of plant cells. Having observed that elongated mitochondria can form in normal plant cells under sugar-depleted conditions, we turned to other ways of obtaining a change in mitochondrial form.

Oxygen Limited Conditions Induce Mitochondrial Fusion and Alter Mitochondrial Form

In agreement with earlier reports (Van Gestel and Verbelen, 2002; Logan, 2006a) seedlings with approximately 1.5–2 μm long mitochondria, when kept between a glass slide and coverslip for more than 30 min showed an increase in mitochondrial lengths of up to 5–8 μm (**Figures 2A,B**). Mitochondrial elongation took place irrespective of whether the seedlings were grown in the dark or light, with or without sucrose. The considerably elongated mitochondria displayed normal motility rates of 1.5 ± 0.4 μms^{-1} and morphed continuously, coiling, and forming branched and reticulate shapes (**Figure 2B**). Time-lapse imaging of tubular mitochondria showed sporadic dilations and constrictions forming along their length resulting in a beads-on-a-string appearance (**Figure 2C**). Exposing the plants to 800 ± 50 μmol $m^{-2}s^{-1}$ light for approximately

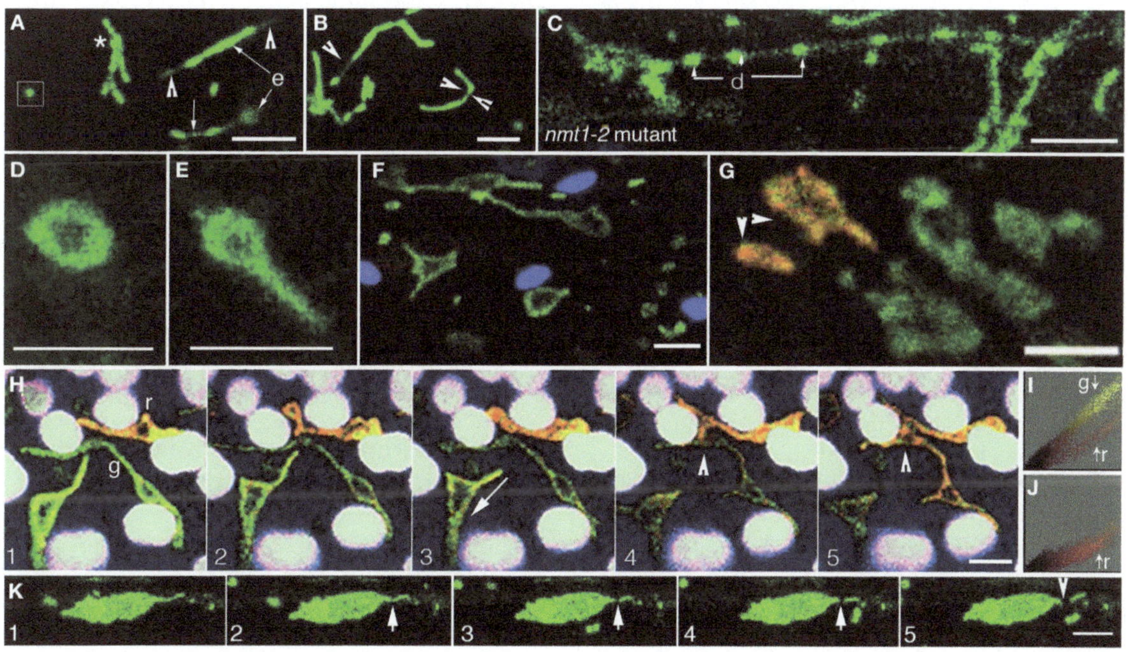

FIGURE 2 | The range of mitochondrial forms obtained in Arabidopsis plants under different conditions. (A) Mitochondria in dark grown hypocotyl cells could appear punctate (box), clustered (*), or elongated (e). An elongated mitochondrion could exhibit terminal (arrowheads) or medial matrixules (small arrow). The medial matrixule was often observed just before the fission of an elongated mitochondrion. **(B)** Elongated mitochondria displayed shape contortions that eventually led to mitochondrial fission. Thin, stretched out regions, as shown in an elongated mitochondrion (single arrowhead) and a possible break-point shown in a sickle-shaped mitochondrion (facing arrowheads) were commonly observed. **(C)** Elongated mitochondria from dark-grown plants often exhibited a beads-on-a-string form consisting of dilated (d) and narrow intervening regions in the nmt1-2 mutant. Whereas, in elongated mitochondria in the wild type this form often preceded fission, it persisted in the fission-impaired *adl2a* and *nmt1-2* mutants. **(D,E)** Single, expanded torus-like mitochondrial forms were observed under oxygen-limited conditions. As suggested by the extension in **(E)** this shape is not fixed and changes over time. **(F)** Under prolonged hypoxia enlarged mitochondria expanded into flattened discs and sheet-like forms. **(G)** Hypoxia in a plant expressing mitochondrial targeted mEosFP resulted in expanded mitochondria, some of which (arrowheads) were photo-converted from green to orange-red. **(H)** A time-lapse image series showed the interaction (frames 1–4) and eventual fusion (frame 5-arrowhead) of a green (g) and a photo-converted, red (r) mitochondrion to produce an even larger giant mitochondrion (frame 5) with an intermediate orange fluorescence. **(I,J)** A color distribution map of frame1 compared to frame 5 in **(H)** provided an estimate of redistribution of mitochondrial matrix through the prominent color mixing that occurred upon fusion of the green non-photo-converted with the red, photo-converted mitochondrion. **(K)** A time-lapse image sequence obtained after irradiating a giant mitochondrion formed under low oxygen conditions **(1)** with bright light for 6 min initiated a break point (arrowhead in **2**), progressive separation of a small fragment **(3,4)** and its eventual moving away from the large mitochondrion. Size bars = 5 μm.

30 s using the bright-field illumination on our upright epi-fluorescence microscope led to the rapid fission of such mitochondria into smaller units. In nearly all cases the beads-on-a-string form preceded a fission event with the actual break occurring at the narrow neck region (**Figures 2A,B**). Occasionally, single mitochondria displayed thin transient extensions that have been called matrixules (Scott et al., 2007; **Figure 2A**).

Under the conditions described above it was difficult to observe elongated mitochondria in mito-GFP for more than a few minutes without their either undergoing fission or progressing into relatively inactive enlarged mitochondria. In order to reliably observe the numerous shapes that elongated mitochondria adopt transiently, we used the fission impaired *nmt1-2/elm1-1* mutant (Logan et al., 2003; Arimura et al., 2008; Logan, 2010). Eight to ten days old mutant seedlings grown in a 16 h light/8 h dark cycle were used and displayed 8 ± 3 μm long mitochondria in cotyledon cells. Interestingly nearly 70% of the mitochondria in hypocotyl cells initially appeared normal. However, within 2 min mitochondria commenced fusing and elongating rapidly. Similarly in cotyledon cells of the *nmt1-2* mutant, there was an increase in mitochondrial length up to 20–25 μm and the formation of tubular, branched networks.

It was concluded that while light promoted fission mitochondrial fusion was favored under a low oxygen environment. This led to mitochondrial elongation in both wild type and *nmt1-2* and *drp3a* mutants. Contrary to the light-triggered fission of elongated mitochondria in wild type plants, the greatly elongated mitochondria in *nmt1-2/elm1-1* and *drp3a* mutants continued morphing for longer periods. Mitochondria in the mutants also displayed a high frequency of matrixule formation and the beads-on-a-string phenotype (**Figure 2C**; *nmt1-2* mutant).

However, if mito-GFP and *nmt1-2/elm1-1* seedlings were left immersed in water for more than 45 min (maximum time assessed was 3 h) mitochondrial motility decreased from 1.5 ± 0.4 $\mu m s^{-1}$ to approximately 0.5 ± 0.2 $\mu m s^{-1}$ or displayed an oscillatory motion. All mitochondria exhibited isotropic expansion around this time (**Figures 2D–F**). Average numbers and sizes could not be obtained due to the wide variability displayed even within a single cell. In both wild type and mutant plants, some mitochondria remained as relatively immobile expanded discs (**Figures 2D,E**) while others enlarged into flat, expanded, irregular shapes to form giant mitochondria ranging in length from 10 to 16 μm (**Figure 2F**).

Earlier studies have suggested that giant mitochondria form due to hypoxia (Ramonell et al., 2001; Van Gestel and Verbelen, 2002; Logan, 2006a). This observation was confirmed in wild type seedlings by overlaying seedlings on the slide with mineral oil, a standard microbiological procedure for creating anoxic conditions (Edwards et al., 1947; Jacobson et al., 2004). Since the polygonal shapes adopted by giant mitochondria (**Figure 2F**) appeared quite different from the annulus-like, iso-tropically expanded mitochondria (**Figure 2D**) we investigated whether giant mitochondria formation involves fusion.

Formation of Giant Mitochondria Involves Fusion of Several Expanded Mitochondria

A stable transgenic Arabidopsis line expressing a mitochondrial-targeted green to red photo-convertible monomeric Eos fluorescent protein (mito-mEosFP; Mathur et al., 2010) was used for these experiments (**Figure 2G**). Seedlings were immersed in water on a slide and single mitochondria were observed progressing through the elongated phase into the large, expanded forms. Single flattened mitochondria within different clusters were photo-converted (**Figure 2G**; arrowheads). Close interactions between photo-converted (red) and non-photo-converted mitochondria (green) over several minutes resulted in their fusion and the formation of a giant mitochondrion (**Figure 2H**; Supplementary Movie 1) with an intermediate, yellow-orange color (**Figures 2H5,I,J**). Whereas, observations of more than 20 such events clearly established that the formation of giant mitochondria involves the fusion of already expanded mitochondria, the initial distance between the fusion pair and the time taken for fusion were very variable and made it difficult to predict the degree of interactivity required for the event. We wondered whether the giant mitochondria represented a terminal phase in mitochondrial morphology or could be made to undergo fission again. This was investigated next.

Whereas, tubular mitochondria undergo rapid fission upon exposure to high light (800 ± 50 $\mu mol\ m^{-2}s^{-1}$) for less than a minute achieving a similar degree of fission in giant mitochondria required a longer exposure time of more than 10 min. However, as shown (**Figure 2K**) giant mitochondria do undergo fission and produce small mitochondria with irregular forms.

Having observed the range of mitochondrial pleomorphy and optimized conditions for obtaining mitochondrial sizes ranging from punctate (≤0.85 μm long) to giant (>8 μm long) mitochondria in 7 day old Arabidopsis seedlings we undertook the simultaneous visualization of mitochondria and the ER to assess their possible relationship.

Mitochondrial Fission Correlates with the Reorganization of ER Tubules

Simultaneous imaging of mitochondria and the ER in Arabidopsis transgenic seedlings co-expressing mito-GFP and RER revealed that mitochondria were nestled within polygons created by ER tubules. Time-lapse images showed further that the seemingly erratic movement of mitochondria correlated with the motility and organization patterns of ER tubules (Supplementary Movie 2). Immersion of seedlings in water under a coverslip for 30–45 min allowed mitochondria to elongate without a noticeable change in ER motility. Subsequent exposure to bright light for 30 s triggered mitochondrial fission. Through highly reproducible time-lapse series taken in over 80 different cells from 10 plants and from iso-surface volume rendering of confocal image stacks it became apparent that single elongated mitochondria were encircled by multiple ER tubules at several positions (**Figures 3A–C**; Supplementary Movie 2). The position of these ER bands around mitochondria often changed during the reorganization of the dynamic ER mesh. Changes in the organization of ER tubules and the consequent narrowing or broadening of ER polygons resulted in the stretching and

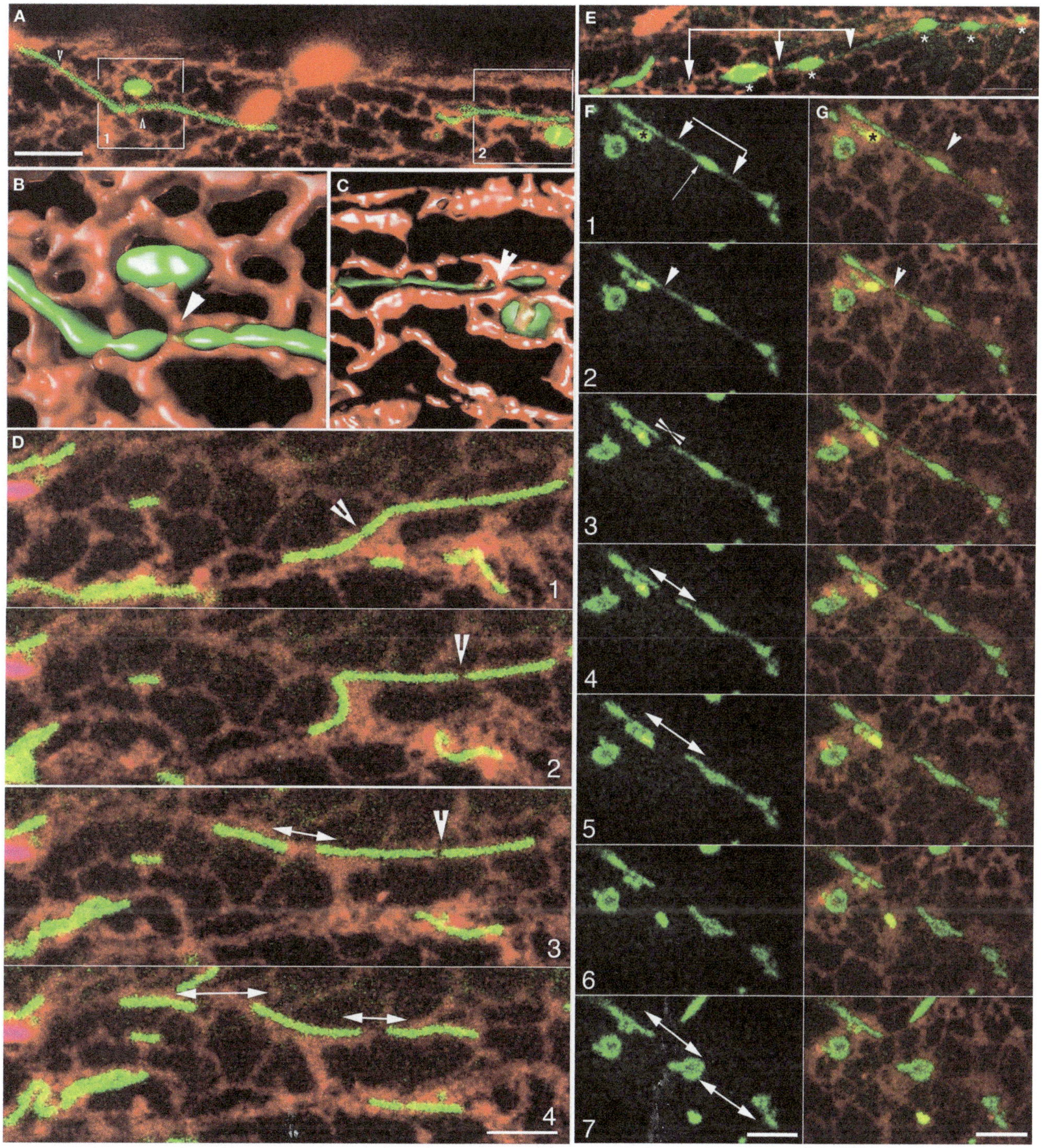

FIGURE 3 | Contortions and fission of elongated mitochondria occur through physical interactions with the ER mesh. (A) A collapsed confocal image stack shows two elongated and two swollen, oval mitochondria embedded in the ER. Arrowheads indicate ER bands that encircle the tubular mitochondria. **(B)** Iso-surface volume rendering of boxed-in region 1 in **(A)** depicts the 3-dimensionality of the mitochondrion-ER spatial relation. An oval mitochondrion is located within an ER-cup while a tubular mitochondria appears threaded between ER polygons. Note the constrictions in the elongated mitochondrion due to encircling ER bands (e.g., arrowhead). **(C)** 3-D volume rendering of boxed-in region 2 in **(A)** shows a tubular mitochondrion enmeshed in the channel created by contiguous ER polygons of variable diameter. Increasing the transparency of the green channel has allowed an appreciation of the ER tubules lying behind the mitochondria. **(D)** Four snapshots selected from a time-lapse series (Supplementary Movie 2) show an elongated mitochondrion (panel 1) with a putative fission site (arrowhead) contorting (panel 2) before breaking (panel 3). The distance between the longer portion of the mitochondria and the broken off segment widens in panels 3 and 4 while another break appears (arrowhead in panel 3) and complete fission of one elongated mitochondria into three is achieved in panel 4. Note that the contiguous ER also stretches and rearranges during this process. **(E)** An image suggesting that different portions of mitochondria can become enmeshed within different ER polygons and thus get

(Continued)

FIGURE 3 | Continued
stretched during the dynamic reorganization of the ER reorganizes. The stretching creates dilations (*) and narrow regions (arrows) in the elongated mitochondrion suggesting the beads-on-a-string form. The terminal, thin regions may also be interpreted as matrixules. **(F,G)** Snapshots acquired from a time-lapse image sequence show the stretching of an elongated mitochondrion and the formation of thin regions (arrowheads in **F**1) suggesting matrixules. Note that the dilated region (**G**1; arrowhead) locates to enlarged ER cistern. Panels **2–5** show fission and expanding ER cisternae in G that increase the gap between the different mitochondrial segments. Panels **5–7** suggest a snap-back phenomenon where the mitochondrial segments curl up on themselves as the neighboring ER completes their separation. A single yellow fluorescent peroxisome (*in panel 1) was also observed in this image series. Size bars = 5 μm.

twisting of enmeshed mitochondria. Whereas, an average time for a fission event could not be estimated due to the rather asynchronous interactions of mitochondria and the ER, a medial narrowing or constriction of the mitochondrion preceded every such event (**Figures 3D,E**; Supplementary Movie 3). This narrowing usually formed the point of fission (**Figures 3F,G**). A certain amount of force to separate the mitochondrial fragments was indicated in the way that the two portions of the mitochondrion and the ER pulled away into opposite directions after a fission event (**Figures 3F,G**). The pulling force often created a thin region between two fragments, very much like a medial matrixule, prior to their actual separation (**Figures 3F,G**).

Since the formation of matrixules (Logan et al., 2004; Logan, 2006b; Scott et al., 2007) and the beads-on-a-string form are interesting transient phenomenon the role of the ER in such occurrences was investigated next.

Transient Matrixules and the "Beads-on-a-string" form Are Created When Elongated Mitochondria Navigate through the ER Mesh

Our assessment of the pleomorphy of elongated mitochondria in mito-GFP plants as well as the *nmt1-2/elm1-1* mutant revealed a wide range of transient shapes including polygons, rings, coiling, branched, and beaded tubules (**Figure 2**). As mentioned earlier, in wild type plants the mitochondrial contortions invariably led to stretching and breaking at weak points. However, elongated mitochondria such as those found in the mitochondrial fission-impaired *nmt1-2/elm1-1* plants co-expressing mito-GFP/Mt-GFP and RER appeared to be threaded through the different-sized polygons making up the ER mesh. During their trajectory through the mesh the elongated mitochondria sporadically displayed thin regions and dilated areas that conveyed an impression of beads-on-a-string (Supplementary Movie 4). We obtained a similar impression upon observing mitochondria stained with MitoTracker in the *adl2/drp3a* mutant (Arimura et al., 2004; Logan et al., 2004), which is also impaired in fission.

The observation that mitochondrial fission is aided by the activity of neighboring ER tubules suggested that the variation in the size of mitochondria might be correlated with the size of ER polygons in a cell. This was investigated next.

Mitochondrial Length Correlates with the Size of Surrounding ER Polygons

Our earlier observations had established that while mitochondria in plants grown under light are small, there is a significant increase in the sub-population of elongated mitochondria in the dark (**Figure 1**). A possible correlation with the ER was sought by comparing mito-GFP-RER seedlings grown under dark and light conditions. In comparison to the meshwork of small ER polygons under light-growth conditions (**Figure 4A**), the polygons were significantly larger in dark grown seedlings (**Figures 4B,C**). In general, the increased ER-polygon size correlated well with increased mitochondrial length in the dark (**Figure 4C**). Therefore, it was perplexing to find some small mitochondria too in dark grown plants. The reason for this apparent discrepancy was traced to the presence of numerous small ER polygons that are formed between large ER-polygons (**Figure 4D**). Based on our observation of numerous small mitochondria trapped in these small-ER-polygon enriched regions it appeared that elongated mitochondria enmeshed within such pockets, named "corrals," broke up to form small mitochondria. Time-lapse imaging of dark grown plants following their exposure to light revealed that the formation of ER corrals increased over time so that after a few hours in light, the ER network comprised predominantly of small polygons. Notably there was a concomitant increase in the population of small mitochondria within the cell. Thus, our observations clearly indicated that the length of a mitochondrion in wild type plants depends upon the size of contiguous ER polygons and small polygons correlate with increased mitochondrial fission.

We concluded that all transient forms exhibited by elongated mitochondria were in response to their physical interactions with neighboring ER and frequently led to fission. However, these contorted forms differed considerably from the expanded, sheet-like mitochondria observed under hypoxia. We asked whether the mitochondria-ER relationship continues under oxygen limited conditions and investigated this next.

Hypoxia-induced Mitochondrial Forms Correlate with Expanded ER Cisternae and Reduced Polygon Formation by ER Tubules

As optimized earlier, the immersion of seedlings in water for about 45 min to an hour resulted in a general expansion of mitochondria. For light grown seedlings the simultaneous visualization of the ER and mitochondria at this stage showed a gradual reduction in the motility of both organelles with concomitant expansion of ER cisternae and single mitochondria (**Figure 5A** vs. **Figure 5B**; arrowheads in **Figure 5B**). For dark grown plants with small mitochondria clustered in ER corrals (**Figure 5C**-box) the expanded ER cisternae did not become as apparent as in light-grown plants. However, ER motility and the rearrangement of ER polygons did slow down and enlarged, flattened mitochondria became evident within the same duration as light grown plants (**Figure 5D**). Time-lapse observations suggested that reduced ER dynamics increased interaction

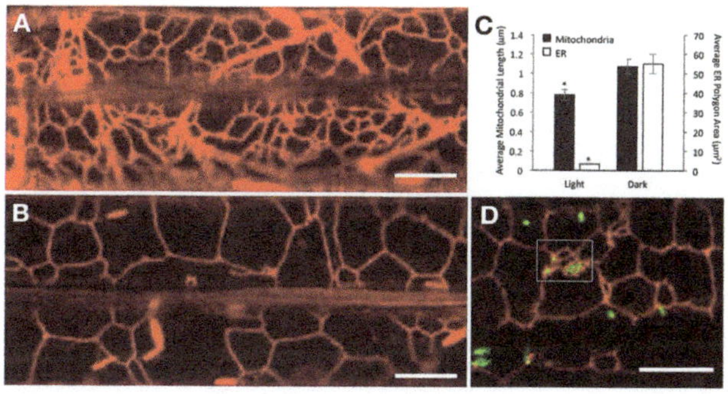

FIGURE 4 | Average size of ER polygons and mitochondria correlates under light and dark growth conditions. (A,B) Representative images from 8 day old plants of RER Arabidopsis transgenics grown in the light (164 µmol $m^{-2}s^{-1}$) **(A)** and complete darkness **(B)** show the visible difference in the size of ER polygons. **(C)** A comparison of mitochondria and ER size from plants grown in the light and dark. Mitochondria from mito-GFP plants ($n = 200$ mitochondria per treatment) were significantly longer when plants were grown in the dark than in the light (1.07 ± 0.34 and 0.78 ± 0.15 µm, dark and light, respectively; $p < 0.01$). The average ER polygon area was also significantly larger in dark grown plants than those grown in the dark (55.07 ± 39.44 and 3.27 ± 2.05 µm, dark and light respectively; $n = 120$ ER polygons per treatment; $p < 0.01$). Standard error bars are shown. **(D)** "Corrals," regions with small ER polygons formed between large ER polygons (boxed in area) were observed in dark grown plants. Mitochondria enmeshed in these regions were small while mitochondria elsewhere in the cell appeared elongated. The number of small ER polygons increased upon exposure to light and coincided with the increase in population of small mitochondria. Scale bars in **(A,B,D)** = 10 µm.

time between mitochondria and promoted their fusion. These observations were tested in the *nmt1-2/elm1-1* mutant transformed with RER. While the mutant has characteristically elongated mitochondria we were able to observe punctate mitochondria, by growing plants on 3% sugar containing MS medium and exposing them to light for 6 h before subjecting them to hypoxia. Punctate mitochondria clustered rapidly, fused and converted into long tubules within 15 min (Supplementary Movie 5). While the fusion in *nmt1-2/elm1-1* mutant took place quickly and the ER appeared slightly diffuse and gave the impression of having expanded cisternae we used another mutant to reinforce this point. An Arabidopsis double mutant *pah1pah2* in the phosphatidic acid phosphohydrolase1 and 2 contains over-expanded ER cisternal membranes (Eastmond et al., 2010) and was used to address this question. RER and mito-GFP dual probe expression in the *pah1pah2* double mutant background showed expanded mitochondria with a diffuse outline in regions with large cisternae (**Figure 5E**).

In order to complete our investigation on ER involvement in mitochondrial pleomorphy we considered that reinvigorating the ER into producing and rearranging tubules should break up the giant mitochondria. Indeed exposing seedlings with a rather moribund ER and giant mitochondria to light for up to 10 min resulted in resumption of ER tubule extension and polygon formation. In over 25 observations this reactivation coincided with changes in the shape of giant mitochondria and their fission into smaller mitochondrial units. Similar patterns of dilations and thinning and stretching as described earlier were observed (**Figure 5F**; Supplementary Movie 6).

Discussion

A High Energy Status Creates and Maintains the Predominantly Small Mitochondria in Green Plant Cells

Mitochondria are routinely described as dynamic, pleomorphic organelles (Cavers, 1914; Lewis and Lewis, 1915; Bereiter-Hahn and Vöth, 1994; Nunnari et al., 1997; Logan and Leaver, 2000; Youle and van der Bliek, 2012; Friedman and Nunnari, 2014). Whereas, elongated mitochondria have been described in green algae (McFadden and Wetherbee, 1982 and references therein), characean internodal cells (Foissner, 2004), in leaves of Ficus (Duckett and Toth, 1977) and Arabidopsis (Ramonell et al., 2001), and in tobacco cells (Stickens and Verbelen, 1996; Van Gestel and Verbelen, 2002) several years of live-imaging using vital dyes and mitochondria-targeted fluorescent proteins have led to the general view that their predominant form is small and punctate in most green plants (Matzke and Matzke, 1986; Köhler et al., 1997; Logan and Leaver, 2000). However, the basis for the formation and maintenance of the discrete, punctate mitochondrial form in plants is unclear and therefore formed the focus of our investigations.

Our observations reveal that significant changes in mitochondrial size occur in response to alterations in the cytosolic sugar levels in plant cells. We demonstrate that whereas cells in plants kept in the dark and starved of sugar exhibit elongated mitochondria, sugar-replete cells predominantly exhibit small mitochondria. These observations agree with an energy dependent internal arrangement of mitochondrial cristae that is believed to underlie mitochondrial dynamics (Bereiter-Hahn and Vöth, 1994; Van der Klei et al., 1994). In green, photosynthesizing plants there are considerable fluctuations in the sugar status of a plant cell between the day and night periods (Azcón-Bieto and Osmond, 1983; Azcón-Bieto et al., 1983; Taiz et al., 2015). In addition we found that exposure to light also results in small mitochondria. While light is the major driver of photosynthesis it is also responsible for changes in the redox status of a plant cell (Douce, 1985; Siedow and Umbach, 1995; Noctor et al., 2007). In plants chloroplasts are major contributors to subcellular reactive oxygen species (ROS). At day break the photo-conversion of proto-chlorophyllide, enriched in chloroplasts during the dark period, into chlorophyll generates different ROS (Meskauskiene et al., 2001) and might trigger rapid mitochondrial fission. Here we have not investigated ROS involvement in mitochondrial fission in depth but subcellular ROS production might also be augmented through reactions in the mitochondrial electron transport chain (ETC) (Logan (2006a). Observations on hyperglycemic animal cells with small, punctate mitochondria similar in appearance to those found in the cells of green plants strongly suggest that a similar sugar-ROS link might operate in green autotrophic plants. Clearly the high-energy state of plant cells during the day favors the formation and maintenance of a population of small-sized mitochondria while a relatively lower energy status at night makes mitochondria more elongated.

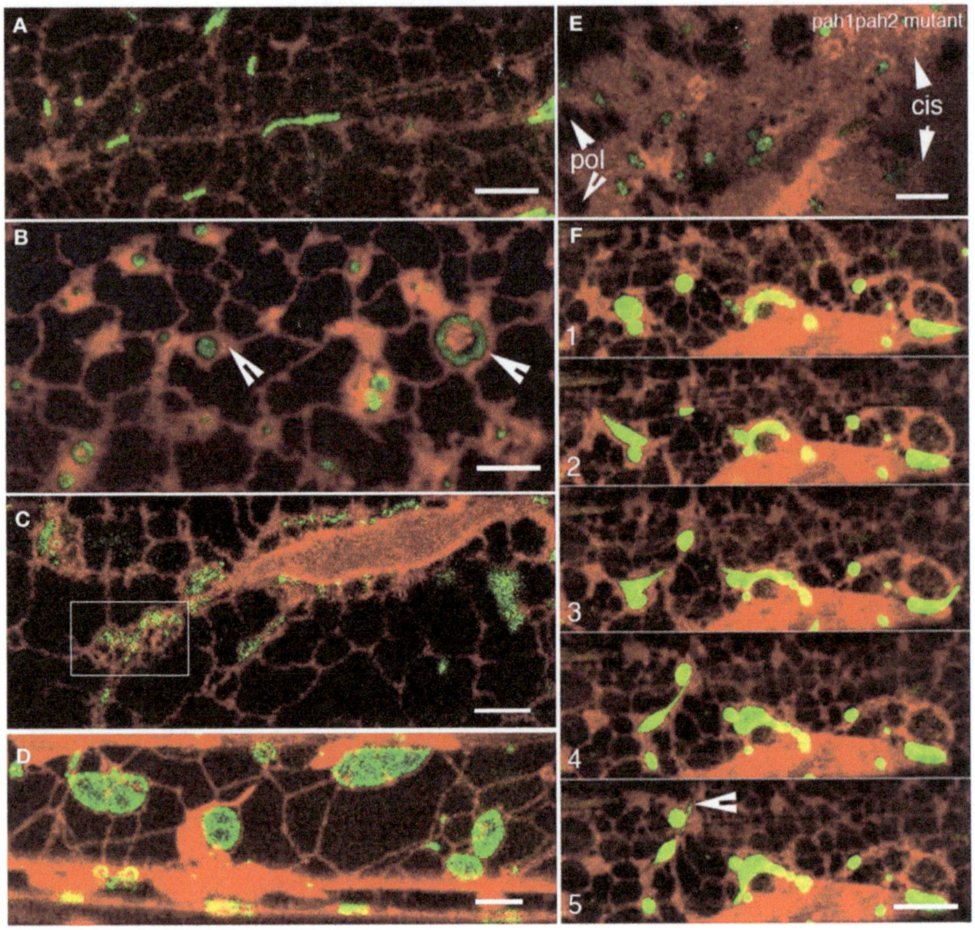

FIGURE 5 | Reduced ER dynamics promote mitochondrial fusion to form giant mitochondria. (A) Representative view of the ER organization and assorted mitochondrial size in normal, light-grown hypocotyl cell. **(B)** Representative view of expanded ER cisternae surrounding almost isotropically expanded mitochondria after 1 h under oxygen-limited conditions. **(C)** Mitochondrial clusters in ER corrals (e.g., boxed in region) surrounded by relatively large ER polygons were observed in dark-grown plants. **(D)** A representative image from a dark plant similar to **(C)** after an hour of oxygen limited conditions shows flat, giant mitochondria that appear to have formed through the fusion of mitochondria caught up in ER corrals. Note that there are no unduly expanded cisternae. **(E)** Representative image showing rather diffuse, enlarged mitochondria embedded in greatly expanded ER cisternae (cis) and not very well defined ER polygons (pol) in the *pah1pah2* double mutant. **(F)** Snapshots from a time-lapse series of a cell with expanded mitochondria and ER cisternae exposed to light showed the resumption of dynamic behavior for both organelles. Mitochondrial blobs started extending and stretching and eventually breaking up as contiguous ER polygons got reorganized. A matrixule-like projection was extended from an expanded mitochondrion resuming dynamic behavior (arrowhead in panel F5). Size bars **(A–E)** = 5 μm; **(F)** =10 μm.

The ER Has a Role in Making Elongated Mitochondria Small

Following our understanding of cellular energy status as the major factor affecting mitochondrial size we investigated the mechanism through which elongated mitochondria broke into smaller units. Active fission of elongated mitochondria was indicated. Proteins such as Fis1p/hFis1/FIS1 and Dnm1/Drp1/DRP3 are firmly implicated in maintaining the fission-fusion cycle in different organisms including plants (Hoppins et al., 2007; Logan, 2010; Rafelski, 2013). In addition ER tubules have been identified as key players that exercise pulling force to promote mitochondrial splitting (Friedman et al., 2011; Lackner et al., 2013; Friedman and Nunnari, 2014). While ER-mitochondrion co-visualization has been achieved for dividing tobacco protoplasts and expanding and regenerating cells (Stickens and Verbelen, 1996) to the best of our knowledge, no investigations have approached the possible involvement of the ER in mediating mitochondrial fission and pleomorphy in plants. Our observations show that dark grown plants have large ER polygons compared to plants grown in light. This observation correlates well with the presence of elongated mitochondria in the dark and small mitochondria in light. Further, using time-lapse imaging we found that in response to light the tubules making up the ER mesh rearrange into smaller sized-polygons and their remodeling increases the frequency of mitochondrial fission. Our observations concur with findings on ER-mitochondrial co-operation in animal and yeast cells (Friedman et al., 2011; Kornmann, 2014). However, our work has gone one step further by investigating the role of the ER in mitochondrial fusion.

The ER Has a Role in Mitochondrial Fusion

The mechanism for mitochondrial fusion in plants remains unknown as many of the proteins involved in this process in animal and yeast cells appear to have no clear homologs in plants (Zhang and Chan, 2007). A study by Sheahan et al. (2005) on mitochondrial fusion in plants utilized protoplasts and several inhibitors to conclude that mitochondrial elongation results from massive mitochondrial fusion and requires IMM electric potential, cytoplasmic protein synthesis, microtubules, and functional kinesin but not ATP or actin filaments. Interestingly, inhibitors used to disrupt actin filaments stopped all mitochondrial movement except Brownian oscillations but did not hinder fusion or fission whereas the myosin inhibitors 2,3-butanedione monoxime (BDM) and N-ethylmaleimide (NEM) affected both mitochondrial motility and mitochondrial fusion negatively (Sheahan et al., 2005). These findings on myosin inhibitors reinforce our own observations since these chemicals directly affect ER dynamics (Liebe and Menzel, 1995; Ostap, 2002). Our findings that mitochondria elongate in the dark and on low sugar levels suggested that fusion rather than fission was being favored under these conditions. A correlation was found in concomitant decrease in ER motility and enlarged polygon size. Further support for increased fusion was obtained under increasing hypoxia when both mitochondrial motility and ER polygon rearrangement slowed down considerably. In addition, the ratio between tubular ER and flattened cisternae changed and most single mitochondria started exhibiting a characteristic isotropic swelling. Mitochondria often became clustered between expanded ER cisternae and over time fused to form giant mitochondria. While we have concluded that a moribund state of the ER might be responsible for creating giant mitochondria, it is equally possible that the low energy status of mitochondria under hypoxia causes the surrounding ER to flatten out first and then subsequently affects the mitochondria as well. Our observations match the commonly observed formation of giant or mega mitochondria in response to oxygen deprivation and inhibition of respiration (Tandler and Hoppel, 1986; Bereiter-Hahn and Vöth, 1994 and references therein; Karbowski et al., 1999; Teranishi et al., 1999; Vartapetian et al., 2003). In aerobic organisms, the efficient generation of ATP during oxidative phosphorylation requires oxygen (O_2). Limiting the supply of O_2 thereby interferes with the efficiency of mitochondrial ATP generation, ultimately limiting the cellular energy status. Similar observations that plant cells maintained under a cover-slip in water start exhibiting elongated/expanded mitochondria after about 30 min have been made by Van Gestel and Verbelen (2002) and Logan (2006a). Therefore, an important consideration to be kept in mind while viewing mitochondria is that their shapes may change during and due to the process of imaging itself.

Contorted Mitochondrial Forms Result from Close Alignment with the ER

Another manifestation of the ER-mitochondria cooperation observed by us was the frequent morphing of elongated mitochondria to form apparent networks, sinous forms, loops, circles, a beads-on-a-string phenotype, and the formation of terminal and median matrixules that are characteristic of mitochondria in living plant cells. Conventionally these shapes have been attributed to rearrangements of internal mitochondrial membranes (Bereiter-Hahn and Vöth, 1994). However, Logan (2006a) has questioned this interpretation based on intrinsic factors only and pointed out that the activities of molecular motors and the cytoskeleton should also be considered. While this work has not investigated the involvement of motor proteins and the cytoskeleton *per se* our observations strongly suggest that all mitochondrial contortions result from their alignment with contiguous ER tubules. However, the dynamic behavior of the ER and mitochondria is intimately associated with both the actin cytoskeleton and myosin motors (Van Gestel et al., 2002; Avisar et al., 2008; Ueda et al., 2010; Peremyslov et al., 2012; Joensuu et al., 2014). Thus, there is a possibility that actin-myosin dependent mitochondrial dynamics also influence the behavior of contiguous ER. A general organelle interactivity model where each subcellular compartment influences the other is being actively investigated by us.

ER-mitochondrial Association Explains Matrixules and the Beads-on-a-string-phenotype

The use of *nmt1-2/elm1-1* and *adl2a/drp3a* mutants allowed us to explore the behavior of elongated mitochondria in more detail and revealed a key role for the ER in creating two interesting morphologies described in mitochondrial literature on plants. Several elongated mitochondria adopt a transient beads-on-a-string shape comprising of narrow and swollen areas while others display matrixules (Logan et al., 2004; Logan, 2006b; Scott et al., 2007). The matrixule has been described as a thin, several micro-meters long protuberance that extends from individual mitochondrion (Logan et al., 2004). According to Logan et al. (2004) matrixules are rarely observed in wild type plants but observed frequently in the *adl2a/drp3A* mutants and might form as a mitochondrion is being pushed through a constrictive collar such as a mitochondrial division ring comprising of DRP3/ADL2 and other proteins (Logan, 2006b). Matrixules may also have a terminal or a medial location (Logan: Supplementary Movie 4: http://www.plantmitochondria.net/Plant_Mitochondria/Movies.html). Whereas, observations of matrixules were made primarily in the *adl2a/drp3A* mutants (Logan et al., 2004), the *nmt1-2/elm1-1* mutant also provided succinct examples of the beads-on-a-string morphology (Arimura et al., 2008). Our observations suggest that both matrixule formation and the beads-on-a-string phenotype are not limited to the two mutants but can be observed in all elongated mitochondria. Further we demonstrate that both transient forms result as mitochondria move through the continuously rearranging ER mesh. We agree with Logan's (2006b) view that matrixules form as a mitochondria passes through a constrictive collar and based on our time-lapse images and 3D volume renditions (**Figure 3**) submit that the collar also includes ER tubules and not just proteins implicated in mitochondrial fission. The resultant view also suggests that as a flexible, elongated mitochondrion encounters the different-sized openings in the ER mesh during its motor-driven, ER-aided motility, the tubule becomes squeezed in some places and dilated in others to produce the beads-on-a-string form. Our view does not reduce the importance of the DRP3A and NMT1/ELM1 protein localizations (Arimura et al., 2004, 2008; Logan et al., 2004) but points to the involvement of the ER-membrane scaffolding for the mitochondrial fission complex.

Questions that remain unanswered in our study relate to why and how a certain degree of alignment or attachment between mitochondria and the ER is created. The presence of membrane contact sites (MCS) between the two organelles and protein complexes localized to these MCS have been described in other organisms (Friedman et al., 2011; Prinz, 2014) and can readily explain such coordinated behavior. While proteins with similar activity and localization patterns have yet to be identified in plants our observations certainly lay down the basis for such investigations. In a more general eukaryotic cell scenario our live-imaging based view points to the ER mesh as a physical barrier with which mitochondria and possibly other organelles interact as they move around the cell. As demonstrated by us these physical interactions mold organelle morphology. The involvement of cytoskeletal elements and motor proteins during the interactive processes poses interesting questions that require further work.

Materials and Methods

Transgenic Lines and Plant Growth Conditions

Arabidopsis thaliana transgenic plants expressing a mitochondrial targeting signal sequence of the beta-ATPase subunit fused to Green Fluorescent Protein (mito-GFP; Logan and Leaver, 2000) or to photo-convertible mEosFP (Mathur et al., 2010) were used for observing mitochondrial morphology and dynamics. The *Agrobacterium* floral dip method (Clough and Bent, 1998) was used transform both mito-GFP and *nmt1-2/elm1-1* (At5g22350; Arimura et al., 2008; Logan, 2010) to create double transgenics lines expressing both RFP-HDEL (RER; Sinclair et al., 2009) and mito-GFP. The *adl2a/drp3a/apm1-1* mutant used has been described earlier (Arimura et al., 2004; Logan et al., 2004; Mano et al., 2004). The *pah1pah2* mutant (Eastmond et al., 2010) was transformed separately with mito-GFP and RER constructs and resultant transgenic plants were crossed to obtain the double transgenics in the double mutant background. Seeds were sterilized using commercial bleach, and unless stated otherwise, plated on Murashige and Skoog (1962) medium (MS) with and without 3% sucrose, stratified at 4°C for 2 days and then grown at 21°C for 7 days at 70 μmol $m^{-2}s^{-1}$ under a 16/8 h light dark cycle.

To determine the effects of sugar on mitochondrial morphology and dynamics, seedlings were grown on MS medium with 0 and 3% sucrose. For light-dark comparisons, seedlings on MS medium with and without sucrose were grown in light (70 μmol m^{-2} s^{-1}, unless stated otherwise) and in complete darkness, while maintaining all other growth conditions as described earlier. Plants grown in the dark without sucrose were transferred to light (70 μmol m^{-2} s^{-1}) for 1 and 12 h. Unless stated otherwise, all observations of mitochondrial and ER morphology were made in cells within the mid region of the hypocotyl. Cells around the vasculature were not considered as they usually exhibit longer mitochondria than epidermal and cortical cells.

Sugar Quantification

The phenol-sulfuric acid colorimetric method for quantifying the total soluble sugar of plant tissue described by Buysse and Merckx (1993) was implemented to determine the relative levels of sugar of plants grown in the light and dark. Plants were grown in the light (155 μmol $m^{-2}s^{-1}$) or dark for 7 days on MS medium with or without 3% sucrose. Fifty plants were harvested from each treatment either in light 1.5 h into the light cycle or in a room in near darkness. The plants were placed into pre-weighed 2 mL microfuge tubes and immediately frozen in liquid nitrogen and the tubes weighed again. Tissue fresh-weight was calculated by subtracting the post-harvest weight from the pre-weighed weight of the tube.

Sample Preparation

The tissue was ground using magnetic beads and a bead beater (Retsch MM301) for 2 min at a frequency of 30.01/s and placed on ice immediately. After removing the magnetic beads, 500 μL of 80% ethanol was added to the tissue and vortexed to homogenize the tissue in solution. The tubes were centrifuged at 14,000 rpm for 5 min at 4°C. The supernatant was transferred to a new tube on ice. This elution in ethanol was repeated two more times. After the final elution, tubes were centrifuged one more time to remove residual tissue that could hamper quantification.

Quantification

A stock of 100 μg/mL of glucose in 80% ethanol was diluted to 10–90 μg/mL. These were used to create a standard line. Five hundred microliters of the standard or sugar elution was added to a cuvette. To each cuvette, 500 μL of 28% phenol (w/w) was added. 2.5 mL of concentrated sulfuric acid was added directly to the liquid surface in a steady stream. After 15 min, the absorbance was read at 490 nm. The blank was prepared in the same manner, except that 80% ethanol was used as the sample.

Creating a Low Oxygen Environment during Microscopic Observations

Whole seedlings were left submerged in water between a slide and coverslip to create an oxygen-limited environment for 45 min and up to 3 h. Observations were taken periodically during this time. Caution was also taken to make sure that no air pockets or air bubbles were trapped around the seedlings, as these hindered the hypoxic effects by prolonging O_2 availability.

Fluorescence Microscopy

A Leica TCS SP5 confocal microscope was used for visual analysis. Argon (488 nm) and Helium-Neon (543 nm) excitation lasers were used to simultaneously visualize GFP labeled mitochondria (emission band 503–524 nm) and RFP labeled ER (emission band 566–643) while chlorophyll auto-fluorescence was collected between 660 and 750 nm in green tissues as an internal control for cell health. Mitotracker® Orange CMTMRos (Life technologies; cat. No. M-7510) used to stain mitochondria in the *adl2a/drp3a/apm1-1* mutant was excited using 543 nm laser and visualized at 555–622 nm. A 50 W halogen bright field lamp on the epi-fluorescence microscope was used for high light (800 ± 50 μmol $m^{-2}s^{-1}$) treatments to induce the fission of elongated and giant mitochondria.

Photo-conversion of mito-mEosFP was carried out as reported earlier (Mathur et al., 2010) using epi-fluorescent lighting through a D filter cube (Leica UV/violet; Ex, BP 355–425; dichroic, 455; Em, LP 470 nm) and a 40 X water-immersion lens (N.A. 0.8). Photo-converting only a subset of mitochondria from green to red permitted assessment of subsequent fusion, which was indicated by the mixing of green and red matrix to form an intermediate orange color.

All confocal images were acquired at a box-size of 1024 × 512 pixels. Three-D (X,Y,Z) stacks maintained a 0.99 μm distance between images whereas the difference between successive images (x,y,t) was maintained at 3.79 s.

Image Analysis and Statistics

For mitochondrial length analysis, 10 mitochondria per cell were measured randomly. All experiments consisted of at least four biological and two technical replications. When there was an unequal sample size per treatment, a two-tailed *t*-test was used to determine the significance of results. For equal sample sizes, a Tukey's test of significance was used (Tukey, 1949). Significance was predetermined as having a $p < 0.05$ (95% confidence interval).

The image analysis software, ImageJ (http://imagej.nih.gov/ij/) was used to measure the length of mitochondria and create color profiles using the 3D inspector plugin. All images were cropped and processed for brightness/contrast as complete montages using Adobe Photoshop CS6 (http://www.adobe.com). The layer function in Photoshop was used to introduce text, regions of interest, and color overlays. The 3-dimensional (3D) iso-surface volume rendering of images of mito-GFP RER double transgenics was carried out using IMARIS (v. 6.4.0; Bitplane AG).

Author Contributions

EA and JM co-designed and carried out experiments and wrote the manuscript while KB and NM provided strong support in generating images and plant materials.

Author Note

The author responsible for distribution of materials integral to the findings presented in this article in accordance with the policy described in the Instructions for Authors (www.plantcell.org) is: JM (jmathur@uoguelph.ca).

Acknowledgments

We thank David Logan for the gift of mito:GFP seeds, Shin-ichi Arimura for seeds of the *nmt1-2/elm1-1*, Shoji Mano and Mikio Nishimura for the *apm1-1* mutant seeds and Peter Eastmond for *pah1pah2* double mutant seeds. JM and KB gratefully acknowledge funding by Natural Sciences and Engineering Research Council of Canada (NSERC).

References

Arimura, S., Aida, G. P., Fujimoto, M., Nakazono, M., and Tsutsumi, N. (2004). Arabidopsis dynamin-like protein 2a (ADL2a), like ADL2b, is involved in plant mitochondrial division. *Plant Cell Physiol.* 45, 236–242. doi: 10.1093/pcp/pch024

Arimura, S., Fujimoto, M., Doniwa, Y., Kadoya, N., Nakazono, M., Sakamoto, W., et al. (2008). Arabidopsis ELONGATED MITOCHONDRIA1 is required for localization of DYNAMIN-RELATED PROTEIN3A to mitochondrial fission sites. *Plant Cell* 20, 1555–1566. doi: 10.1105/tpc.108.058578

Avisar, D., Prokhnevsky, A. I., Makarova, K. S., Koonin, E. V., and Dolja, V. V. (2008). Myosin XI-K Is required for rapid trafficking of Golgi stacks, peroxisomes, and mitochondria in leaf cells of *Nicotiana benthamiana*. *Plant Physiol.* 146, 1098–1108. doi: 10.1104/pp.107.113647

Azcón-Bieto, J., Lambers, H., and Day, D. A. (1983). Effects of photosynthesis and carbohydrate status on respiratory rates and the involvement of the alternative pathway in leaf respiration. *Plant Physiol.* 71, 598–603. doi: 10.1104/pp.72.3.598

Azcón-Bieto, J., and Osmond, C. B. (1983). Relationship between photosynthesis and respiration. The effects of carbohydrate status on the rate of CO_2 production by respiration in darkened and illuminated wheat leaves. *Plant Physiol.* 71, 574–581. doi: 10.1104/pp.71.3.574

Benda, D. (1899). Weitere Beobachlungen uber die Mitochondria und ihr Verhaltnis zu Secretguanulationen, nebst kritischen Bemerkungen. *Verh. d. Phys. Ges. 1903 Die Mitochondria. Ergeb. d.Anat. und Entw.* 12.

Bereiter-Hahn, J., and Vöth, M. (1994). Dynamics of mitochondria in living cells: shape changes, dislocations, fusion, and fission of mitochondria. *Microscop. Res. Tech.* 27, 198–219. doi: 10.1002/jemt.1070270303

Bleazard, W., McCaffery, J. M., King, E. J., Bale, S., Mozdy, A., Tieu, Q., et al. (1999). The dynamin-related GTPase Dnm1 regulates mitochondrial fission in yeast. *Nat. Cell Biol.* 1, 298–304. doi: 10.1038/13014

Bowes, T., and Gupta, R. S. (2008). Novel mitochondrial extensions provide evidence for a link between microtubule-directed movement and mitochondrial fission. *Biochem. Biophys. Res. Commun.* 376, 40–45. doi: 10.1016/j.bbrc.2008.08.120

Buysse, J., and Merckx, R. (1993). An improved colorimetric method to quantify sugar content of plant tissue. *J. Exp. Bot.* 44, 1627–1629. doi: 10.1093/jxb/44.10.1627

Cavers, F. (1914). Chondriosomes (mitochondria) and their significance. *New Phytol.* 13, 96–106. doi: 10.1111/j.1469-8137.1914.tb05742.x

Clough, S. J., and Bent, A. F. (1998). Floral dip: a simplified method for Agrobacterium-mediated transformation of *Arabidopsis thaliana*. *Plant J.* 16, 735–743 doi: 10.1046/j.1365-313x.1998.00343.x

Douce, R. (1985). *Mitochondria in Higher Plants: Structure, Function and Biogenesis*. Orlando, FL: Academic Press.

Duckett, J. G., and Toth, R. (1977). Giant mitochondria in a periclinal chimera, Ficus elastica Roxb. forma variegata. *Ann. Bot.* 41, 903–912.

Eastmond, P. J., Quettier, A. L., Kroon, J. T., Craddock, C., Adams, N., and Slabas, A. R. (2010). Phosphatidic acid phosphohydrolase 1 and 2 regulate phospholipid synthesis at the endoplasmic reticulum in Arabidopsis. *Plant Cell.* 22, 2796–2811. doi: 10.1105/tpc.109.071423

Edwards, G. A., Buell, C. B., and Weston, W. H. (1947). The influence of mineral oil upon the oxygen consumption of *Sordaria fimicola*. *Am. J. Bot.* 34, 551–555. doi: 10.2307/2437336

Feng, X. G., Arimura, S., Hirano, H. Y., Sakamoto, W., and Tsutsumi, N. (2004). Isolation of mutants with aberrant mitochondrial morphology from *Arabidopsis thaliana*. *Genes Genet. Syst.* 79, 301–305. doi: 10.1266/ggs.79.301

Foissner, I. (2004). Microfilaments and microtubules control the shape, motility, and subcellular distribution of cortical mitochondria in characean internodal cells. *Protoplasma* 224, 145–157. doi: 10.1007/s00709-004-0075-1

Friedman, J. R., Lackner, L. L., West, M., DiBenedetto, J. R., Nunnari, J., and Voeltz, G. K. (2011). ER tubules mark sites of mitochondrial division. *Science* 334, 358–362. doi: 10.1126/science.1207385

Friedman, J. R., and Nunnari, J. (2014). Mitochondrial form and function. *Nature* 505, 335–343. doi: 10.1038/nature12985

Grainger, F., and James, D. W. (1969). Mitochondrial extensions associated with microtubules in outgrowing processes from chick spinal cord *in vitro*. *J. Cell Sci.* 4, 729–737.

Gunning, B. E. S., and Steer, M. W. (1986). *Plant Cell Biology: An Ultrastructural Approach*. Dublin, OH: Steer, M.W. University College.

Hackenbrock, C. R. (1968). Chemical and physical fixation of isolated mitochondria in low-energy and high-energy states. *Proc. Natl. Acad. Sci. U.S.A.* 61, 598–605. doi: 10.1073/pnas.61.2.598

Hackenbrock, C. R., Rehn, T. G., Weinbach, E. C., and Lemasters, J. J. (1971). Oxidative phosphorylation and ultrastructural transformation in mitochondria in the intact ascites tumor cell. *J. Cell Biol.* 51, 123–137. doi: 10.1083/jcb.51.1.123

Hoffman, H. P., and Avers, C. J. (1973). Mitochondrion of yeast: ultrastructural evidence for one giant, branched organelle per cell. *Science* 181, 749–751. doi: 10.1126/science.181.4101.749

Hoppins, S., Lackner, L., and Nunnari, J. (2007). The Machines that Divide and Fuse Mitochondria. *Annu. Rev. Biochem.* 76, 751–780. doi: 10.1146/annurev.biochem.76.071905.090048

Jacobson, A. P., Thunberg, R. L., Johnson, M. L., Hammack, T. S., and Andrews, W. H. (2004). Alternative anaerobic enrichments to the bacteriological analytical manual culture method for isolation of *Shigella sonnei* from selected types of fresh produce. *J. AOAC Int.* 87, 1115–1122.

Jhun, B. S., Lee, H., Jin, Z. G., and Yoon, Y. (2013). Glucose stimulation induces dynamic change of mitochondrial morphology to promote insulin secretion in the insulinoma cell line INS-1E. *PLoS ONE* 8:e60810. doi: 10.1371/journal.pone.0060810

Joensuu, M., Belevich, I., Rämö, O., Nevzorov, I., Vihinen, H., Puhka, M., et al. (2014). ER sheet persistence is coupled to myosin1c-regulated dynamic actin filament arrays. *Mol Biol Cell.* 25, 1111–1126. doi: 10.1091/mbc.E13-12-0712

Karbowski, M., Kurono, C., Wozniak, M., Ostrowski, M., Teranishi, M., Nishizawa, Y., et al. (1999). Free radical-induced megamitochondria formation and apoptosis. *Free Radic. Biol. Med.* 26, 396–409. doi: 10.1016/S0891-5849(98)00209-3

Köhler, R. H., Zipfel, W. R., Webb, W. W., and Hanson, M. R. (1997). The green fluorescent protein as a marker to visualize plant mitochondria *in vivo*. *Plant J.* 11, 613–621. doi: 10.1046/j.1365-313X.1997.11030613.x

Kornmann, B. (2014). Quality control in mitochondria: use it, break it, fix it, trash it. *F1000Prime Rep.* 6:15. doi: 10.12703/P6-15

Lackner, L. L., Ping, H., Graef, M., Murley, A., and Nunnari, J. (2013). Endoplasmic reticulum-associated mitochondria-cortex tether functions in the distribution and inheritance of mitochondria. *Proc. Natl. Acad. Sci. U.S.A.* 110, E458–E467. doi: 10.1073/pnas.1215232110

Lewis, M. R., and Lewis, W. H. (1915). Mitochondria (and other cytoplasmic structures) in tissue cultures. *Amer. J. Anatomy* 17, 339–401.

Lichtscheidl, I. K., and Url, W. G. (1990). Organization and dynamics of cortical endoplasmic reticulum in inner epidermal cells of onion bulb scales. *Protoplasma* 157, 203–215. doi: 10.1007/BF01322653

Liebe, S., and Menzel, D. (1995). Actomyosin-based motility of endoplasmic reticulum and chloroplasts in Vallisneria mesophyll cells. *Biol. Cell* 85, 207–222. doi: 10.1016/0248-4900(96)85282-8

Logan, D. C. (2006a). The mitochondrial compartment. *J. Expt. Bot.* 57, 1225–1243. doi: 10.1093/jxb/erj151

Logan, D. C. (2006b). Plant mitochondrial dynamics. *Biochim. Biophys. Acta* 1763, 430–441. doi: 10.1016/j.bbamcr.2006.01.003

Logan, D. C. (2010). Mitochondrial fusion, division and positioning in plants. *Biochem. Soc. Trans.* 38, 789–795. doi: 10.1042/BST0380789

Logan, D. C., and Leaver, C. J. (2000). Mitochondria-targeted GFP highlights the heterogeneity of mitochondrial shape, size and movement within living plant cells. *J. Exp. Bot.* 51, 865–871. doi: 10.1093/jexbot/51.346.865

Logan, D. C., Scott, I., and Tobin, A. K. (2003). The genetic control of plant mitochondrial morphology and dynamics. *Plant J.* 36, 500–509. doi: 10.1046/j.1365-313X.2003.01894.x

Logan, D. C., Scott, I., and Tobin, A. K. (2004). ADL2a, like ADL2b, is involved in the control of higher plant mitochondrial morphology. *J. Exp. Bot.* 55, 783–785. doi: 10.1093/jxb/erh073

Mano, S., Nakamori, C., Kondo, M., Hayashi, M., and Nishimura, M. (2004). An Arabidopsis dynamin-related protein, DRP3A, controls both peroxisomal and mitochondrial division. *Plant J.* 38, 487–498. doi: 10.1111/j.1365-313X.2004.02063.x

Mathur, J., Radhamony, R., Sinclair, A. M., Donoso, A., Dunn, N., Roach, E., et al. (2010). mEosFP-based green-to-red photoconvertible subcellular probes for plants. *Plant Physiol.* 154, 1573–1587. doi: 10.1104/pp.110.165431

Matzke, M. A., and Matzke, A. J. M. (1986). Visualization of mitochondria and nuclei in living plant cells by the use of a potential-sensitive fluorescent dye. *Plant Cell Environ.* 9, 73–77.

McFadden, G. I., and Wetherbee, R. (1982). Serial reconstruction of the mitochondrial reticulum in the Antarctic flagellate, *Pyramimonas gelidicola* (*Parsinophyceae, Chlorophyta*) *Protoplasma* 111, 79–82. doi: 10.1007/BF01287648

Meskauskiene, R., Nater, M., Goslings, D., Kessler, F., op den Camp, R., and Apel, K. (2001). FLU: a negative regulator of chlorophyll biosynthesis in *Arabidopsis thaliana*. *Proc. Natl. Acad. Sci. U.S.A.* 98, 12826–12831. doi: 10.1073/pnas.221252798

Meves, F. (1904). Über das vorkommen von mitochondrion bezw. Chrondromiten der Pflanzen zellen. *Ber. Deutsch Bot. Ges.* 22, 264–286.

Mozdy, A. D., McCaffery, J. M., and Shaw, J. M. (2000). Dnm1p GTPase-mediated mitochondrial fission is a multi-step process requiring the novel integral membrane component Fis1p. *J. Cell Biol.* 151, 367–380. doi: 10.1083/jcb.151.2.367

Murashige, T., and Skoog, F. (1962). A revised medium for rapid growth and bioassays with tobacco tissue cultures. *Physiol. Plant.* 15, 473–797. doi: 10.1111/j.1399-3054.1962.tb08052.x

Noctor, G., de Paepe, R., and Foyer, C. H. (2007). Mitochondrial redox biology and homeostasis in plants. *Trends Plant Sci.* 12, 125–134. doi: 10.1016/j.tplants.2007.01.005

Novikoff, A. B., and Holtzman, E. (1976). *Cells and organelles, 2nd Edn.* Holt: Rinehart and Winston.

Nunnari, J., Marshall, W. F., Straight, A., Murray, A., Sedat, J. W., and Walter, P. (1997). Mitochondrial transmission during mating in Saccharomyces cerevisiae is determined by mitochondrial fusion and fission and the intra-mitochondrial segregation of mitochondrial DNA. *Mol. Biol. Cell* 8, 1233–1242. doi: 10.1091/mbc.8.7.1233

Ostap, E. M. (2002). 2,3-Butanedione monoxime (BDM) as a myosin inhibitor. *J. Muscle Res. Cell Motil.* 23, 305–308. doi: 10.1023/A:1022047102064

Peremyslov, V. V., Klocko, A. L., Fowler, J. E., and Dolja, V. V. (2012). Arabidopsis myosin XI-K localizes to the motile endomembrane vesicles associated with F-actin. *Front. Plant Sci.* 3:184. doi: 10.3389/fpls.2012.00184

Pitts, K. R., Yoon, Y., Krueger, E. W., and McNiven, M. A. (1999). The dynamin-like protein DLP1 is essential for normal distribution and morphology of the endoplasmic reticulum and mitochondria in mammalian cells. *Mol. Biol. Cell* 10, 4403–4417. doi: 10.1091/mbc.10.12.4403

Prinz, W. (2014). Bridging the gap: membrane contact sites in signaling, metabolism, and organelle dynamics. *J. Cell Biol.* 205, 759–769. doi: 10.1083/jcb.201401126

Rafelski, S. M. (2013). Mitochondrial network morphology: building an integrative, geometrical view. *BMC Biology* 11:71. doi: 10.1186/1741-7007-11-71

Ramonell, K. M., Kuang, A., Porterfield, D. M., Crispi, M. L., Xiao, Y., McClure, G., et al. (2001). Influence of atmospheric oxygen on leaf structure and starch deposition in *Arabidopsis thaliana*. *Plant Cell Environ.* 24, 419–428. doi: 10.1046/j.1365-3040.2001.00691.x

Scott, I., Sparkes, I. A., and Logan, D. C. (2007). The missing link: inter-organellar connections in mitochondria and peroxisomes? *Trends Plant Sci.* 12, 380–381. doi: 10.1016/j.tplants.2007.08.010

Scott, I., Tobin, A. K., and Logan, D. C. (2006). BIGYIN, an orthologue of human and yeast FIS1 genes functions in the control of mitochondrial size and number in *Arabidopsis thaliana*. *J. Exp. Bot.* 57, 1275–1280. doi: 10.1093/jxb/erj096

Seguí-Simarro, J. M., Coronado, M. J., and Staehelin, L. A. (2008). The mitochondrial cycle of Arabidopsis shoot apical meristem and leaf primordium meristematic cells is defined by a perinuclear tentaculate/cage-like mitochondrion. *Plant Physiol.* 148, 1380–1393. doi: 10.1104/pp.108.126953

Sesaki, H., and Jensen, R. E. (1999). Division versus fusion: Dnm1p and Fzo1p antagonistically regulate mitochondrial shape. *J. Cell Biol.* 147, 699–706. doi: 10.1083/jcb.147.4.699

Sheahan, M. B., McCurdy, D. W., and Rose, R. J. (2005). Mitochondria as a connected population: ensuring continuity of the mitochondrial genome during plant cell dedifferentiation through massive mitochondrial fusion. *Plant J.* 44, 744–755. doi: 10.1111/j.1365-313X.2005.02561.x

Siedow, J. N., and Umbach, A. L. (1995). Plant mitochondrial electron transfer and molecular biology. *Plant Cell* 7, 821–831. doi: 10.1105/tpc.7.7.821

Sinclair, A. M., Trobacher, C. P., Mathur, N., Greenwood, J. S., and Mathur, J. (2009). Peroxule extension over ER-defined paths constitutes a rapid subcellular response to hydroxyl stress. *Plant J.* 59, 231–242. doi: 10.1111/j.1365-313X.2009.03863.x

Stickens, D., and Verbelen J. P. (1996). Spatial structure of mitochondria and ER denotes changes in cell physiology of cultured tobacco protoplasts. *Plant J.* 9, 85–92. doi: 10.1046/j.1365-313X.1996.09010085.x

Stojanovski, D., Koutsopoulos, O. S., Okamoto, K., and Ryan, M. T. (2004). Levels of human Fis1 at the mitochondrial outer membrane regulate mitochondrial morphology. *J. Cell Sci.* 117, 1201–1210. doi: 10.1242/jcs.01058

Taiz, L., Zeiger, E., Møller, I. M., and Murphy, A. (2015). *Plant Physiology and Development, 6th Edn.* Sunderland, MA: Sinauer Associates.

Tandler, B., and Hoppel, C. L. (1986). Studies on giant mitochondria. *Ann. N.Y. Acad. Sci.* 488, 65–81. doi: 10.1111/j.1749-6632.1986.tb46548.x

Teranishi, M., Karbowski, M., Kurono, C., Soji, T., and Wakabayashi, T. (1999). Two types of the enlargement of mitochondria related to apoptosis: simple swelling and the formation of megamitochondria. *J. Electron Microsc.* 48, 637–651. doi: 10.1093/oxfordjournals.jmicro.a023730

Tieu, Q., and Nunnari, J. (2000). Mdv1p is a WD repeat protein that interacts with the dynamin-related GTPase, Dnm1p, to trigger mitochondrial division. *J. Cell Biol.* 151, 353–365. doi: 10.1083/jcb.151.2.353

Tukey, J. W. (1949). Comparing individual means in the analysis of variance. *Biometrics* 5, 99–114. doi: 10.2307/3001913

Ueda, H., Yokota, E., Kutsuna, N., Shimada, T., Tamura, K., Shimmen, T., et al. (2010). Myosin-dependent endoplasmic reticulum motility and F-actin organization in plant cells. *Proc. Natl. Acad. Sci. U.S.A.* 107, 6894–6899. doi: 10.1073/pnas.0911482107

Van der Klei, I. J., Veenhuis, M., and Neupert, W. (1994). A morphological view on mitochondrial protein targeting. *Microsc. Res. Techn.* 27, 284–293. doi: 10.1002/jemt.1070270404

Van Gestel, K., Köhler, R. H., and Verbelen, J. P. (2002). Plant mitochondria move on F-actin, but their positioning in the cortical cytoplasm depends on both F-actin and microtubules. *J. Exp. Bot.* 53, 659–667. doi: 10.1093/jexbot/53.369.659

Van Gestel, K., and Verbelen, J. P. (2002). Giant mitochondria are a response to low oxygen pressure in cells of tobacco (*Nicotiana tabacum* L.). *J. Exp. Bot.* 53, 1215–1218. doi: 10.1093/jexbot/53.371.1215

Vartapetian, B. B., Andreeva, I. N., Generozova, I. P., Polyakova, L. I., Maslova, I. P., Dolgikh, Y. I., et al. (2003). Functional electron microscopy in studies of plant response and adaptation to anaerobic stress. *Ann. Bot.* 91, 155–172. doi: 10.1093/aob/mcf244

Wakabayashi, T. (2002). Megamitochondria formation—physiology and pathology. *J. Cell Mol. Med.* 6, 497–538. doi: 10.1111/j.1582-4934.2002.tb00452.x

Welchen, E., Garciìa, L., Mansilla, N., and Gonzalez, D. H. (2014). Coordination of plant mitochondrial biogenesis: keeping pace with cellular requirements. *Front. Plant Sci.* 4:551. doi: 10.3389/fpls.2013.00551

Yoon, Y., Krueger, E. W., Oswald, B. J., and McNiven, M. A. (2003). The mitochondrial protein hFis1 regulates mitochondrial fission in mammalian cells through an interaction with the dynamin-like protein DLP1. *Mol. Cell Biol.* 23, 5409–5420. doi: 10.1128/MCB.23.15.5409-5420.2003

Youle, R. J., and van der Bliek, A. M. (2012). Mitochondrial fission, fusion, and stress. *Science* 337, 1062–1065. doi: 10.1126/science.1219855

Yu, T., Sheu, S. S., Robotham, J. L., and Yoon, Y. (2008). Mitochondrial fission mediates high glucose-induced cell death through elevated production of reactive oxygen species. *Cardiovasc. Res.* 79, 341–351. doi: 10.1093/cvr/cvn104

Zhang, Y., and Chan, D. C. (2007). New insights into mitochondrial fusion. *FEBS Lett.* 581, 2168–2173. doi: 10.1016/j.febslet.2007.01.095

Conflict of Interest Statement: The authors declare that the research was conducted in the absence of any commercial or financial relationships that could be construed as a potential conflict of interest.

17

Cooperative control between AtRGS1 and AtHXK1 in a WD40-repeat protein pathway in *Arabidopsis thaliana*

Jian-Ping Huang[1,2], Meral Tunc-Ozdemir[2], Ying Chang[1]* and Alan M. Jones[2,3]*

[1] College of Life Science, Northeast Agricultural University, Harbin, China, [2] Department of Biology, University of North Carolina, Chapel Hill, NC, USA, [3] Department of Pharmacology, University of North Carolina, Chapel Hill, NC, USA

HEXOKINASE 1 (AtHXK1) and Regulator of G-protein Signaling 1 (AtRGS1) pathways, mediate D-glucose signaling in *Arabidopsis*. However, it is not known the degree, if any, that these pathways overlap and how. We show modest signaling crosstalk between these pathways, albeit complex with both epistatic interactions and additive effects that may be indirect. The action of HXK1 on AtRGS1 signaling lies downstream of the primary step in G protein-mediated sugar signaling in which the WD-repeat protein, AGB1, is the propelling signaling element. RHIP1, a previously unknown protein predicted here to have a 3-stranded helical structure, interacts with both AtRGS1 and AtHXK1 *in planta* and is required for some glucose-regulated gene expression, providing a physical connection between these two proteins in sugar signaling. The *rhip1* null mutant displays similar seedling growth phenotypes as *rgs1-2* in response to glucose, further suggesting a role for RHIP1 in glucose signaling. In conclusion, glucose signaling is a complex hierarchical relationship which is specific to the target gene and sugar phenotype and suggests that there are two glycolysis-independent glucose signaling sensors: AtRGS1 and AtHXK1 that weakly communicate with each other via feed-back and feed-forward loops to fine tune the response to glucose.

Edited by:
Christopher J. Staiger,
Purdue University, USA

Reviewed by:
Ben Matthew Abell,
Sheffield Hallam University, UK
Ling Zhu,
The University of Texas at Austin, USA

***Correspondence:**
Ying Chang
changying@neau.edu.cn;
Alan M. Jones
alan_jones@unc.edu

Keywords: plant heterotrimeric G protein, sugar signaling, *Arabidopsis* regulator of G-protein signaling 1 protein (AtRGS1), hexokinase 1 (HXK1)

INTRODUCTION

Glucose affects plant growth and development by acting as both a metabolite and a signaling molecule (Sheen, 2014). In *Arabidopsis thaliana*, glucose is sensed by (1) a plasma membrane G-protein-coupled pathway involving *REGULATOR OF G-PROTEIN SIGNALING 1* (AtRGS1), (2) a *HEXOKINASE 1* (AtHXK1) pathway, and (3) a glycolysis-dependent *SNF1-RELATED KINASE 1/TARGET OF RAPAMYCIN* (SnRK1/TOR) pathway. On the *Arabidopsis* plasma membrane, the G protein complex contains the core G-protein subunits Gα (AtGPA1), Gβ (AGB1) and Gγ (AGG1), three WNK kinases (AtWNK1, AtWNK8, and AtWNK10) and the 7-transmembrane AtRGS1 protein. AtRGS1 keeps the core G protein complex in its inactive (GDP-bound) state and physical decoupling of AtRGS1 from AtGPA1 allows this G protein to self-activate by spontaneously binding GTP. Sustained activation of glucose signaling depends, in part, on removal of the inhibiting AtRGS1 protein by endocytosis of this plasma membrane-anchored glucose sensor (Urano et al., 2012; Fu et al., 2014). The WD40 repeat subunit of the G protein AGB1 recruits the WNK kinases which phosphorylate AtRGS1 consequently triggering the endocytosis event.

AtRGS1-dependent G-protein-coupled signaling regulates many cellular biological processes, such as cell growth and proliferation (Chen et al., 2006a; Booker et al., 2010; Urano et al., 2012), abiotic stress tolerance (Chen et al., 2006b; Colaneri et al., 2014), stomatal density (Zhang et al., 2008), chloroplast development (Zhang et al., 2009), seed germination (Chen et al., 2006c), morphogenesis and development (Chen and Jones, 2004; Chen et al., 2006b). Specifically, loss of *AtRGS1* confers increased growth and high glucose tolerance in *Arabidopsis*. Loss-of-function mutations in the WD40-repeat subunit AGB1 have the most severe phenotypes.

AtHXK1 regulates glucose-responsive gene transcription directly by forming a nuclear co-repressor complex with the vacuolar H+-ATPase B1 (VHA-B1) and the 19S regulatory particle of the proteasome subunit (RPT5B) (Cho et al., 2006). The *hxk1* (aka *gin2*-1), *vha-b1* and *rpt5b* null mutants display similar seedling and adult plant phenotypes (e.g., reduced growth in roots, leaves and inflorescences), and glucose response defects (namely, insensitivity to high glucose-mediated developmental arrest) (Moore et al., 2003; Cho et al., 2006).

The glycolysis-dependent SnRK1/TOR pathway contains two protein kinases SnRK1 and TOR that may function as cellular energy sensors instead of sugar sensors (Lastdrager et al., 2014; Tome et al., 2014). TOR is activated and promotes growth in response to favorable nutritional and energy conditions, while SnRK1 is stimulated upon nutrient and energy starvation conditions (Lastdrager et al., 2014; Tome et al., 2014).

Until now, the mechanisms within each of these glucose signal transduction pathways were studied independently of each other, yet, analogous to the integrative HXK/G protein mechanism for glucose sensing in yeast (Rolland et al., 2000; Kim et al., 2013), it is likely that an integration of these pathways is essential for plants to make a fast and coordinated response to extracellular and intracellular glucose conditions. In this study, we focused on functional crosstalk between AtRGS1- and AtHXK1-dependent sugar signaling (Moore et al., 2003; Chen and Jones, 2004; Urano et al., 2012). By comparing the expression of a selected set of glucose-regulated genes as well as a series of physiological sugar-related phenotypes, we determined that the two glucose sensors AtRGS1 and AtHXK1 communicate via cooperative loops in determining the final response to glucose. This occurs downstream of AtRGS1 endocytosis. RGS1-HXK1 INTERACTING PROTEIN 1 (RHIP1) serves as the physical scaffold for these two sensors.

MATERIALS AND METHODS

Plant Materials

The T-DNA insertion mutant *rgs1*-2 was described by Chen et al. (2003), and the ethylmethane sulfonate (EMS)-mutagenized mutant *gin2*-1 was described by Moore et al. (2003). The *35S::RGS1-YFP* over-expression line was previously described by Urano et al. (2012). *hxk1*-1 (*SALK_034233*), *hxk1*-2 (*CS864200*), *hxk1*-3 (*CS861759*), *rhip1*-1 (*SALK_091518*), and *rhip1*-2 (*SALK_061002*) mutants were obtained from the *Arabidopsis* Biological Resource Center (ABRC) stock. Plants homozygous for *hxk1*-1, *hxk1*-2, *hxk1*-3, *rhip1*-1, and *rhip1*-2 were isolated by PCR using genomic DNA with T-DNA insertion verification primers and the insertion was confirmed by sequencing. Reverse transcriptase polymerase chain reaction (PCR) was used to analyze the *AtHXK1* or *RHIP1* transcripts in these five mutants. The primers used are shown in Supplementary Table S1. Our analyses disproved an earlier claim by Reda (2013) that *hxk1*-1 (*SALK_034233C*) lacked expression of *HXK1* (Supplementary Figure S1); *hxk1*-1 is not a null mutant as previously reported and the reported phenotypes of the *hxk1*-1 mutant should be re-interpreted. The open reading frame of *AtRGS1* in pEarleyGate101 vector (C-terminal *YFP-HA*) was previously described by Urano et al. (2012). The *35S::AtRGS1-YFP* over-expression lines in the *hxk1*-3, *rhip1*-1 or *rhip1*-2 mutant background were generated by *Agrobacterium*-mediated transformation (Bechtold et al., 1993), and T2 generation seed was used. The *pCDKA;1::GUS* reporter was crossed to *rgs1*-2 and *hxk1*-3 single mutant plants, and F3 and F4 generation plants were used for phenotype analysis. All mutants used are in the ecotype Columbia (Col-0) except *gin2*-1 which is in the Landsberg (Ler) background.

Quantitative Real-time PCR (qRT-PCR)

Quantitative real-time PCR was performed exactly as described by Grigston et al. (2008) and Urano et al. (2012). Briefly, ~150 seeds were cultured in 75 ml 1/2 × MS liquid media with 1% sucrose at 23°C shaking (~140 rpm) under constant low light (70 μmol $s^{-1}m^{-2}$) conditions. After 7 days, the seedlings were washed in sterilized water three times, and then starved in 1/2 × MS liquid media lacking sugar for 2 days in the dark. The seedlings were treated with fresh 1/2 × MS liquid media containing 0% (control), or 3% (w/v) D-glucose for 3 h in dark conditions, then harvested by flash freezing in liquid N_2. The mRNA and cDNA were prepared with RNAeasy^{TM} (Qiagen) and Superscript III (Invitrogen), respectively, according to the manufacturer's instructions. Expression of *TBL26*, *HXK1*, *RGS1*, *CA2*, *CAB2*, *DIN1*, and *TUB4* was analyzed by qRT-PCR with SYBRgreen (Invitrogen). The comparative CT (Threshold Cycle defined as the cycle number at which the fluorescence generated within a reaction crosses the threshold line) method described in detail in Livak and Schmittgen (2001), was performed for qRT-PCR. First, validation experiments demonstrated that the targets (*TBL26*, *HXK1*, *RGS1*, *CA2*, *CAB2*, *DIN1*) and endogenous control (*TUB4*) have single amplification products determined by gel electrophoresis and relatively equivalent PCR efficiencies (relative standard curve). Second, samples from each biological repeat were run in triplicate on a single plate. The CT mean and standard deviation values of the three replicate sample results were calculated. The ΔCT value was calculated by:

$$\Delta CT = CT_{\text{target gene}} - CT_{\text{endogenous control gene}}$$

The standard deviation of the ΔCT was calculated by:

$$S = (S^2_{\text{target gene}} + S^2_{\text{endogenous control gene}})^{1/2};$$

(S = standard deviation)

The $\Delta\Delta CT$ was calculated by:

$$\Delta\Delta CT = \Delta CT_{\text{test sample}} - \Delta CT_{\text{control sample}}$$

The standard deviation of the $\Delta\Delta CT$ value is the same as the standard deviation of the ΔCT value. Fold-differences between test sample and control sample was calculated by $2^{-\Delta\Delta Ct}$ with $\Delta\Delta CT + S$ and $\Delta\Delta CT - S$, where S is the standard deviation of the $\Delta\Delta CT$ value. Fold-differences calculated using $\Delta\Delta CT$ method are shown as a range from $2^{-\Delta\Delta Ct + s}$ to $2^{-\Delta\Delta Ct - s}$.

Plant Growth Assays

Seeds were sown onto soil and stratified at 4°C for 3 days, and then transferred to a 23°C growth chamber in 16/8 h (L/D). Two or 3-week-old seedlings were photographed. Alternatively, seeds were surface-sterilized with 70% ethanol and 95% ethanol for 10 min each, then sown onto 1/4 × MS with 0.5% phytoagar (pH adjusted to 5.75 with 5 N KOH). Seeds were stratified at 4°C in the dark for 3 days, germinated and grown horizontally at 23°C with a 16/8 h (L/D) cycle (70 μmol $s^{-1}m^{-2}$). Three days later, the seedlings were transferred to 1/4 × MS with 0.5% phytoagar (pH adjusted to 5.75 with 5 N KOH) containing 0, 0.25, 0.5, 0.75, 1, 2, or 3% (w/v) D-glucose, and grown vertically for root growth analysis. Every 24 h, the root tips were marked. The root lengths were calculated using ImageJ software. The "green seedling" assay was performed as described by Moore and coworkers (Moore et al., 2003). Briefly, surface-sterilized seeds of Col-0, *rgs1*-2, *hxk1*-3, *hxk1*-3/*rgs1*-2, and *rhip1*-2 mutant were sown onto 1/4 × MS with 0.5% phytoagar containing 6% (w/v) D-glucose or D-mannitol, stratified at 4°C in the dark for 3 days, germinated and grown horizontally at 23°C with continuous light (70 μmol $s^{-1}m^{-2}$) for 10 days. The average percentage of seedlings showing green cotyledons was determined. The "green seedling" assay in response to saline stress was performed as described by Colaneri et al. (2014). Briefly, surface-sterilized seeds were germinated and seedlings grown on 1/4 MS agar media supplemented with 125 mM NaCl and 0.5% D-glucose (w/v). Green seedlings per plate (36 seeds per plate) were counted 10 days after germination.

β-Glucuronidase (GUS) Staining Assay

Arabidopsis seeds were surface-sterilized and stratified as described above. Seeds germinated in 24-well plates containing 1 ml 1/2 MS liquid medium (pH adjusted to 5.75 with 5 N KOH) on a shaker (~30 rpm) at 23°C under continuous low light (70 μmol $s^{-1}m^{-2}$). 3-day-old seedlings were then transferred to 1 ml fresh 1/2 MS liquid medium supplemented with 3% (w/v) D-glucose or D-mannitol (pH adjusted to 5.75 with 5 N KOH) for 3 more days in culture.

GUS staining was performed following the method described by Malamy and Benfey (1997). Seedlings were infiltrated in 100 mM Tris-HCl (pH 7.5) buffer containing 2.9 mg/ml NaCl, 20% (v/v) methanol, 0.001% (v/v) Triton X-100 and 0.5 mg/ml X-gluc (5-Bromo-4-chloro-3-indoxyl-beta-D-glucuronide cyclohexyl ammonium salt; Gold Biotechnology, Inc.). After incubation for 12–18 h at 37°C in dark, seedlings were cleared in 70% ethanol and photographed on a Nikon inverted microscope DIAPHOT-TMD. ImageJ software was used for analysis.

Informatic Methods

The protein sequence of RHIP1 (At4g26410) was obtained by searching against the *Arabidopsis thaliana* protein database supported by The *Arabidopsis* Information Resource (TAIR). The RHIP1 protein sequence was submitted to I-TASSER (Iterative Threading ASSEmbly Refinement) online[1], to generate the predicted RHIP1 protein structure and the GO biological process of RHIP1 protein based on global and local protein similarity. PyMOL(TM) Molecular Graphics System (Version 1.7.0.0.) was used to generate the final high-quality protein structure model image based on the data of model 1 from I-TASSER. GO annotations of proteins directly interacting with RHIP1 were retrieved from TAIR GO Annotation Search[2], and functional categorization by their TAIR loci for GO molecular function or GO biological process were expressed as percent of total.

AtRGS1-YFP Internalization Analysis

Fluorescence quantification for AtRGS1-YFP internalization was performed as described by Urano et al. (2012) and Fu et al. (2014). Sterilized, stratified seeds were germinated in 6-well plates containing 2 ml 1/2 × MS liquid medium (pH adjusted to 5.75 with 5 N KOH) at 23°C under darkness. Seedlings (7-day-old) were treated with 0 or 3% D-glucose (w/v) for 30 min. Hypocotyl epidermal cells located 2–4 mm below the cotyledon were imaged (Z stacks obtained) using a Zeiss LSM710 confocal laser scanning microscope equipped with a 20× Plan-NeoFluor (N.A. = 0.5) objective and a 40× C-Apochromat (N.A. = 1.20) water immersion objective. YFP fluorescence was excited by a 514 nm argon laser and detected at 526–569 nm by a photomultiplier detector. At least 10 sets of images from 5 seedlings were obtained for internalization quantification analysis by ImageJ software.

Bimolecular Fluorescence Complementation (BiFC)

Bimolecular fluorescence complementation was performed as described in Klopffleisch et al. (2011). The entire open reading frames of *RHIP1* or *AtHXK1* were amplified by PCR from a cDNA library made from seedlings in qRT-PCR analysis. Open reading frames were subcloned into the p-ENTR/D-TOPO vector (Invitrogen, Carlsbad, CA, USA) and recombined into the BiFC vectors pBatTL-sYFP-N and pBatTL-sYFP-C (C-terminal split-nYFP and cYFP tag, respectively), and pCL112 and pCL113 (N-terminal split-nYFP and cYFP tag, respectively) by LR recombination reaction. The open reading frame of *AtRGS1* in BiFC vectors was previously described by Grigston et al. (2008). Split nYFP- and cYFP-tagged protein pairs, p19 (gene silencing suppressor), and Mt-rk (mitochondrial RFP marker, an internal transformation control) were co-expressed in 4- to 5-week-old *Nicotiana benthamiana* leaves by *Agrobacterium tumifaciens*-mediated infection. Tobacco leaf epidermal cells

[1] http://zhanglab.ccmb.med.umich.edu/I-TASSER/

[2] http://www.arabidopsis.org/tools/bulk/go/index.jsp

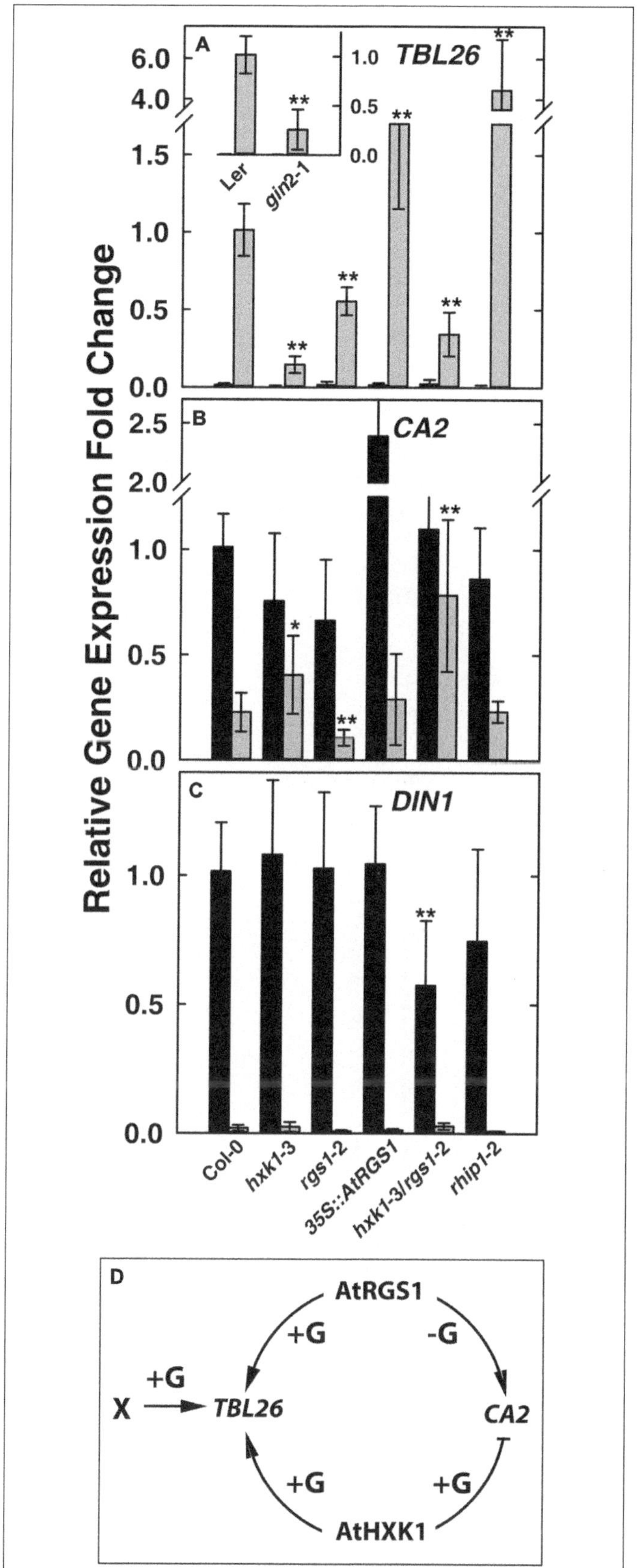

FIGURE 1 | Continued

FIGURE 1 | Continued

Quantitative Real-time PCR (qPCR) analysis of glucose-reporter genes *TBL26*, *CA2* and *DIN1* in *Arabidopsis* seedlings and the corresponding mechanism models. Seven-day-old seedlings with the indicated genotypes indicated at the bottom of panels **(A–C)** were starved for 2 days then treated with 1/2 × MS medium (0% D-glucose, shown by black bars) or 3%D-glucose (gray bars) for 3 h as described in Section "Materials and Methods." Transcripts of *TBL26* **(A)**, *CA2* **(B)**, and *DIN1* **(C)** were quantitated using qRT-PCR. Values are the means of the fold changes ± SD of 3–5 independent biological replicates. Each biological replication had at least three technical replications. ANOVA single factor analysis ($\alpha = 0.05$) was conducted to compare the relative fold-change in gene expression of different plant lines with wild type control plants. $^*P < 0.05$; $^{**}P < 0.01$. **(D)** Simplest model accounting for the gene expression data based on the mutant behaviors. +G, addition of D-glucose; –G, no glucose.

were imaged using a Zeiss LSM710 confocal laser scanning microscope equipped with an Apochromat 40 × water-emersion objective (N.A. = 1.2). YFP fluorescence was excited by a 514 nm argon laser and detected at 526–569 nm by a photomultiplier detector, and RFP fluorescence was excited by a diode laser and detected at 565–621 nm.

RESULTS AND DISCUSSION

AtRGS1 and AtHXK1 Co-regulate the Expression of a Group of Glucose-regulated Genes

Glucose induces in an AtRGS1-dependent manner the expression of *TRICHOMELESS 26* (*TBL26*, At4g01080), *JACALIN-RELATED LECTIN 7* (*JAL7*, At1g52040), *JACALIN-RELATED LECTIN23* (*JAL23*, At2g39330), At1g54020 (*JAL* gene) among ~30 genes (Grigston et al., 2008; Urano et al., 2012). The time course and dose responsiveness for regulation of *TBL26* is the best documented among the small set of genes regulated by AtRGS1. Even fewer genes to date are documented to be regulated by HXK1 (Sheen et al., 1999). The best characterized is the involvement of AtHXK1 in glucose repression of the *CARBONIC ANHYDRASE 2* (*CA2*, At5g14740) and the photosynthetic gene *CHLOROPHYL A/B-BINDING 2* (*CAB2*, At1g29920) (Moore et al., 2003; Cho et al., 2006). As shown in **Figure 1A** inset, the D-glucose induced expression of *TBL26* was severely depressed in the *hxk1* mutant (*gin2*-1 in the Ler background) with ~80% difference between *TBL26* expression in *Ler* and in *gin2*-1, suggesting an active role of AtHXK1 in glucose-induced *TBL26* expression. All other studies on *TBL26* were conducted using the Col ecotype. In order to exclude the influence of different ecotypes, it was necessary to generate a transcript-null allele (*hxk1*-3) in the *Columbia* ecotype (Supplementary Figure S1). Glucose-induced expression of *TBL26* was impaired in seedlings lacking either *AtRGS1* or *AtHXK1* (**Figure 1A**), and this impairment was not the result of lowered expression of *AtRGS1* in the *hxk1*-3 mutant, or of *AtHXK1* in the *rgs1*-2 mutant (Supplementary Figure S2), suggesting an overlapping or additive function between AtRGS1 and AtHXK1 in glucose-regulated gene expression.

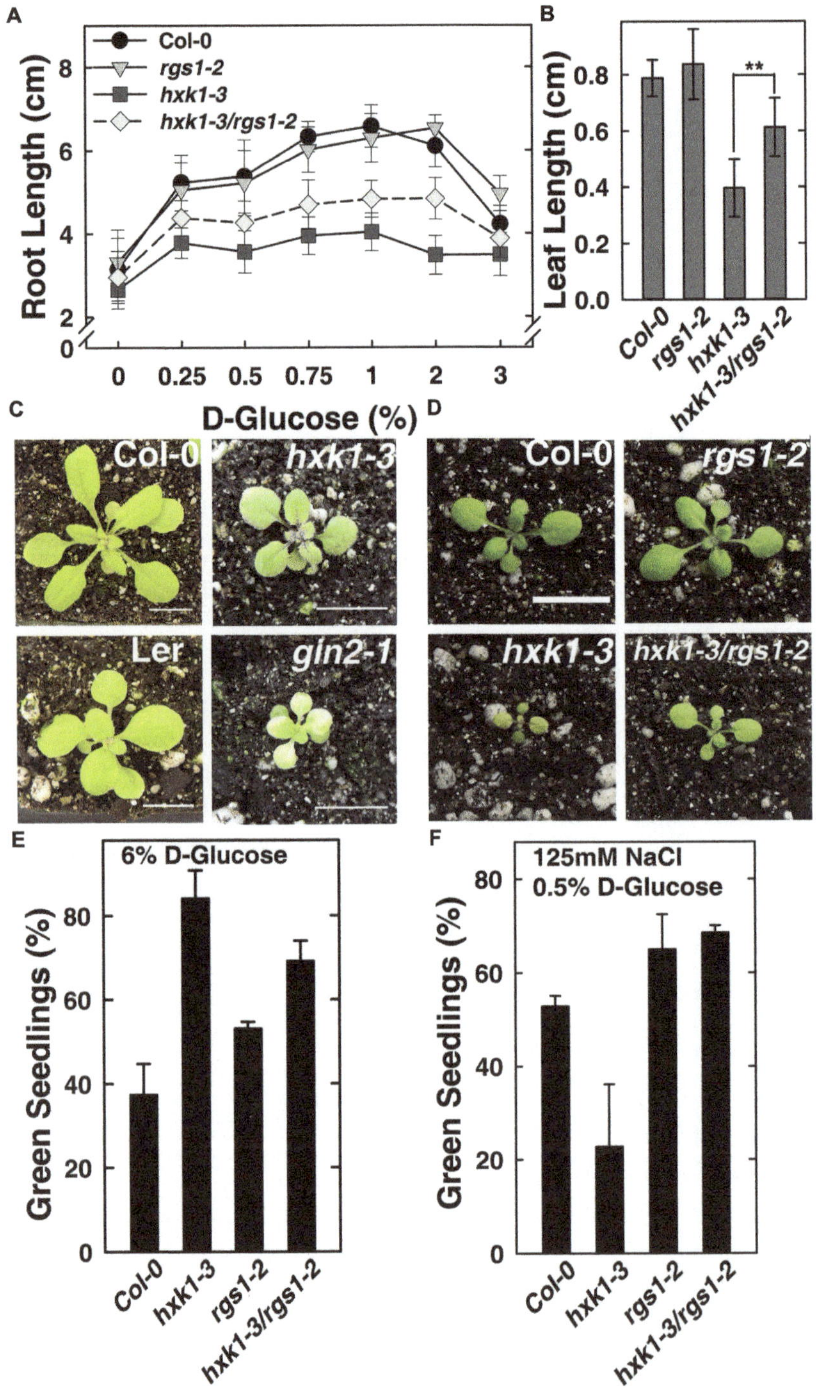

FIGURE 2 | *hxk1-3/rgs1-2* double mutants show intermediate growth phenotype between *hxk1-3* and *rgs1-2* sngle mutants. (A) Comparison of root elongation of Col-0, *rgs1-2*, *hxk1-3*, and *hxk1-3/rgs1-2* mutant seedlings transferred to1/4 MS medium supplemented with 0, 0.25, 0.5, 0.75, 1, 2 or 3% D-glucose (w/v) under low intensity light (70 μmol s^{-1} m^{-2}) in long-day chamber in 16/8 h (L/D). Values are means ± SD (n = 10–18). **(B)** Comparison of leaf size (the third leaf) of 2-week-old Col-0, *rgs1-2*, *hxk1-3*, and *hxk1-3/rgs1-2* mutant seedlings grown under light (160 μmol s^{-1} m^{-2}) in 16/8 h (L/D). Values are means ± SD (n = 10–12). ANOVA single factor analysis ($\alpha = 0.05$) was conducted to compare the leaf size in *hxk1-3*and *hxk1-3/rgs1-2*. $^{**}P < 0.01$. **(C)** 3-week-old Col-0, *hxk1-3*, Ler, and *gin2-1* plants grown under high intensity light (160 μmol s^{-1} m^{-2}) in 16/8 h (L/D). Scale bar = 1 cm. **(D)** 2-week-old Col-0, *hxk1-3*, *rgs1-2*, and *hxk1-3/rgs1-2* mutant plants grown under high intensity light (160 μmol s^{-1} m^{-2}) in 16/8 h (L/D). Scale bar = 1 cm. **(E)** Green seedling assay of the Col-0, *rgs1-2*, *hxk1-3*, and *hxk1-3/rgs1-2* mutant. The assay was performed as described in Section "Materials and Methods." The average percentage of seedlings showing green cotyledons was determined and presented with means ± SD from one representative experiment of 4 biological replications. **(F)** Col, *rgs1-2*, *hxk1-3,* and *rgs1-2/hxk1-3* have altered responses to saline stress. The assay was performed as described in Section "Material and Methods." Values are means ± SD from quintuplicate.

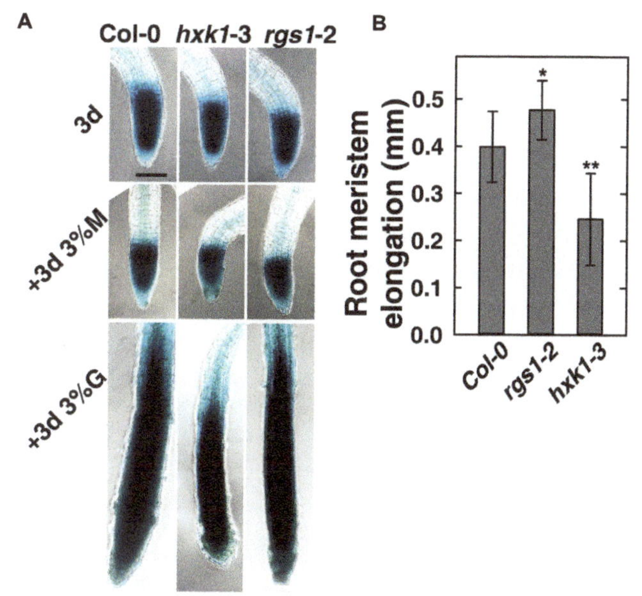

FIGURE 3 | Cell cycle progress in *rgs1-2* or *hxk1-3* mutant showed changed response to D-glucose. (A) Meristematic activity in the primary root tip of Col-0, *hxk1-3* and *rgs1-2* plants as visualized by GUS-staining under the indicated sugar conditions as described in Section "Materials and Methods." Seeds were germinated and cultured in 1/2 MS liquid medium for 3 days (3d); and then transferred to 1/2 MS liquid medium supplemented with 3% (w/v) D-glucose(+3d 3%G) or 3% (w/v) D-mannitol (+3d 3% M) for 3 more days. Scale bar = 0.1 mm. **(B)** The root meristem elongation as determined by the length-change of GUS-staining region in the primary root tip of Col-0, *hxk1-3* and *rgs1-2* seedlings. Values are means ± SD ($n = 8$–12) from one representative experiment of 3 biological replications. Pairwise Student's *t*-test was used to compare values to the Col-0. *$P < 0.05$; **$P < 0.01$.

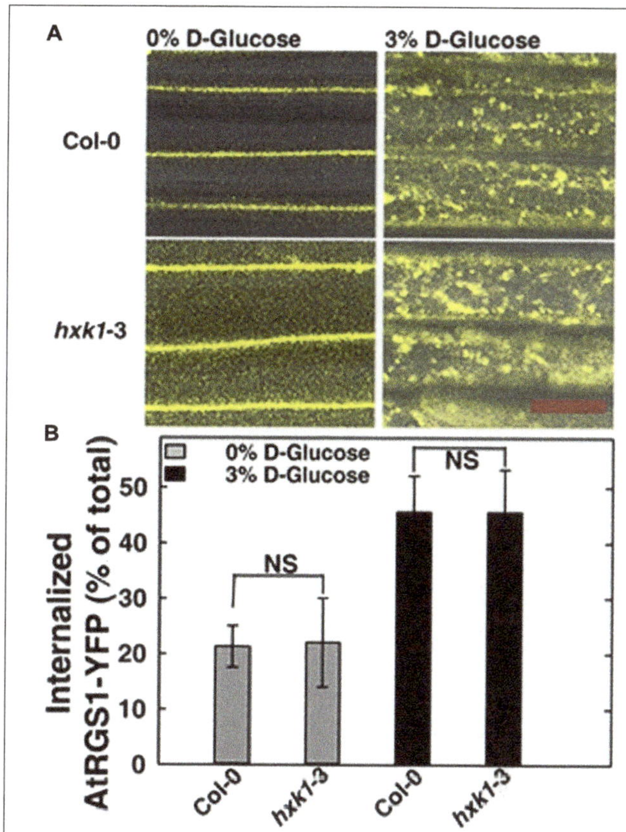

FIGURE 4 | AtRGS1 internalizes in response to D-glucose even in the absence of AtHXK1. (A) Hypocotyls cells of 7-day-old Col-0 and *hxk1-3* *Arabidopsis* seedlings stably expressing AtRGS1-YFP imaged after treatment with 3% D-glucose (w/v) for 30 min. Scale bar = 20 μm. **(B)** Quantization of the percentage of AtRGS1-YFP internalized in epidermal cells of Col-0 and *hxk1-3* mutant seedlings hypocotyls before and after glucose stimulation. Values are percentage means ± SD, from a representative experiment ($n = 10$) of 3 biological replications. ANOVA single factor analysis ($\alpha = 0.05$) was conducted to compare the difference between plant lines with the same glucose treatment.

The level of *TBL26* expression in the *hxk1*-3 mutant was lower than in the *rgs1*-2 mutant ($P < 0.01$) but combining these alleles conferred statistically similar level as the *rgs1*-2 mutant indicating that the *rgs1*-2 allele is epistatic to the *hxk1*-3 allele. Note, however, that loss of both AtRGS1 and AtHXK1 did not completely eliminate glucose induction of *TBL26* expression indicating that a third (glucose) input must exist. AtRGS1 slightly inhibits *TBL26* expression but also positively reinforced AtHXK1 although with a different weight than inhibition.

Glucose represses some gene expression in an AtHXK1-dependent manner (Sheen, 1990; Koch, 1996). We next examined whether AtRGS1 influences AtHXK1-mediated gene repression of *CA2* gene expression. Glucose-repression of *CA2* was slightly attenuated in the *hxk1* mutants (both Col and Ler alleles), consistent with previous studies (Moore et al., 2003; Cho et al., 2006). The *rgs1*-2 mutant displayed slightly lower *CA2* gene expression with ($P = 0.003236$) or without ($P = 0.00223$) glucose treatment (**Figure 1B**), suggesting that AtRGS1 promotes *CA2* basal gene expression. To test this hypothesis, we examined the effect of over-expression of AtRGS1 on the basal expression of *CA2*. The basal *CA2* expression without glucose treatment was strongly increased to nearly 2.5-fold in the *35S::AtRGS1* plants (P = 1.25E-05), while the glucose-repressed level of expression was similar to wild type. The *hxk1-3/rgs1-2* double mutant had statistically the same basal level of *CA2* expression ($P > 0.05$) but the glucose repression was attenuated although statistically to no more or less degree than the in the *hxk1*-3 mutant ($P < 0.025$) (**Figure 1B**). The epistasis analysis did not provide strong evidence for a heirarchial role of AtRGS1 in AtHXK1 regulation of *CA2* expression, however, the large effect of over-expression of AtRGS1 is suggestive of cooperation between these two pathways although with the accompanying caveats associated with interpreting gain-of-function alleles. We also examined another AtHXK1-mediated gene repression, namely *CAB2*, and obtained a similar result as for *CA2* (Supplementary Figure S3). AtHXK1 mediates the main pathway toward *CA2* repression. AtRGS1, based on the gain-of-function data, induces *CA2* gene expression in the absence of glucose. An inhibitory genetic interaction by AtHXK1 on AtRGS1 fits the observed result. There is no need to implicate an independent glucose pathway since loss of both AtRGS1

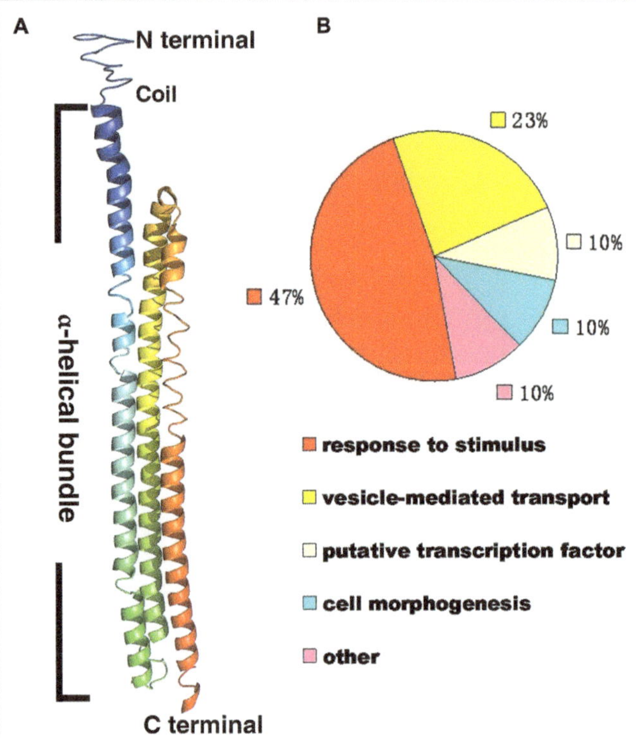

FIGURE 5 | Bioinformational analysis of RHIP1 (At4g26410). **(A)** A modeled structure of *Arabidopsis thaliana* RHIP1. The structure was predicted by the I-TASSER program based on the RHIP1 protein sequence, and processed by PyMOL software. The three α-helical bundleis shown with rainbow colors, and the top is the N-terminal head composed of coils. **(B)** Functional analysis of 21 RHIP1 interacting proteins are annotated as "response to stimulus" (47%). These 21 proteins are AT2G32670, AT5G05760, AT1G35720, At2g26400, At2g42400, AT1G28520, AT1G20100, AT2G26300, AT4G17730, AT1G05500, AT3G26090, AT3G49290, At5g42030, AT3G60600, AT5G14240, AT5G43850, AT5G03540, AT3G59220, AT4g13640, At5g67380, and AT4G29130. The function cluster analysis was performed based on TAIR GO annotations.

and AtHXK1 statistically eliminated glucose repression of *CA2* expression.

As discussed above, there are at least three sugar-sensing pathways in *Arabidopsis*, the two examined above and the third being the SnRK1/TOR pathway tested next. Glucose-induced repression of *DARK INDUCIBLE* 1 (*DIN1*, At4g35770) is regulated by glycolysis-dependent SnRK1(Baena-Gonzalez et al., 2007; Baena-Gonzalez and Sheen, 2008). As shown in **Figure 1C**, there was no obvious difference in glucose repression of *DIN1* among Col-0, *hxk1*-3, *rgs1*-2, and *35S::AtRGS1* over-expression lines, suggesting no role for either HXK1 or AtRGS1 in glucose-regulated *DIN1* expression. However, the *hxk1-3/rgs1*-2 double mutants had a lower basal level of *DIN1* expression ($P = 0.007405$), implying redundancy or, more likely, an indirect effect on the glucose economy.

The behavior of the *hxk1*, *rgs1*, and *hxk1/rgs1* mutants in glucose-regulated gene expression paints a complex relationship between two glucose sensing pathways in *Arabidopsis* and clearly indicate that these two pathways do not operate independently without at least modest influence on each other through direct or indirect mechanisms. The epistasis analyses suggest that AtHXK1 and AtRGS1 operate cooperatively for *TBL26* and *CA2* regulation (**Figure 1D**) and redundantly or not at all in *DIN1* regulation. This unusual genetic relationship is born out in the developmental phenotypes of the mutants described below.

AtRGS1 Genetically Interacts with AtHXK1 in Modulating Early Seedling Development

The feed-back regulation of AtHXK1 and AtRGS1 on each other in glucose-regulated gene expression predicts that loss of either would confer opposite physiological phenotypes while loss of both would confer a phenotype shared by neither (e.g., an intermediate phenotype). Previous reports indicate that genetic ablation of *AtHXK1* reduces growth of roots, leaves and inflorescences in the Ler ecotype (Moore et al., 2003; Cho et al., 2006), whereas, *rgs1*-2 null mutants have longer roots, hypocotyls, and greater rosette diameter than wild type (Chen et al., 2003; Urano et al., 2013). In order to study the different roles between AtHXK1 and AtRGS1 in glucose-regulated primary root growth, 3-day-old *Arabidopsis* seedlings germinated in 1/4 MS solid medium containing 0% glucose were transferred to different glucose conditions, and the lengths of primary roots on the 10th day was measured. As shown in **Figure 2A**, compared to wild type plants, *hxk1*-3 mutants had shorter roots under all glucose conditions; *rgs1*-2 mutants had similar root lengths to Col-0; *hxk1*-3/*rgs1*-2 double mutants had intermediate root elongation between the *rgs1*-2 and *hxk1*-3 mutants. The *rgs1*-2 allele rescued the *hxk1*-3 allele in leaf and rosette size (**Figures 2B–D**). Given that root elongation and leaf extension are all based on cell division in the meristem and subsequent cell expansion, the intermediate phenotype of *hxk1-3/rgs1*-2 double mutants imply that the two glucose sensors—AtRGS1 and AtHXK1 have opposite roles in regulation of cell proliferation. Glucose and salt stress responsiveness were also measured (**Figures 2E,F**). Both *hxk1*-3 and *rgs1*-2 were hyposensitive to a high dose of glucose but this hyposensitivity was not additive. The *hxk1*-3 allele conferred hypersensitivity to NaCl while the *rgs1*-2 allele and the double mutant were less sensitive. The *rgs1*-2 allele is epistatic to the *hxk1*-3 allele indicating that AtRGS1 and AtHXK1 operate in the same genetic pathway.

To verify the role of AtRGS1and AtHXK1 in meristematic cell activity, and to extend a previous study on the behavior of glucose sensors in sugar promoting meristem activation (Xiong et al., 2013), we compared the cell cycle activity in the primary roots of Col-0, *hxk1*-3 and *rgs1*-2 single mutant lines (**Figure 3**). The steady-state level of A-type CDK—CDKA;1, which is constitutively high in actively dividing cells was investigated using existing GUS reporter genes which were stably integrated into the wild-type (West et al., 2004; Skylar et al., 2011) or mutant backgrounds (this study). In order to minimize the influence of endogenous sugar from the endosperm or photosynthesis, seeds were germinated in sugar-free liquid medium, then 3-day-old seedlings were transferred to fresh liquid medium supplemented with 3% D-glucose or D-mannitol for

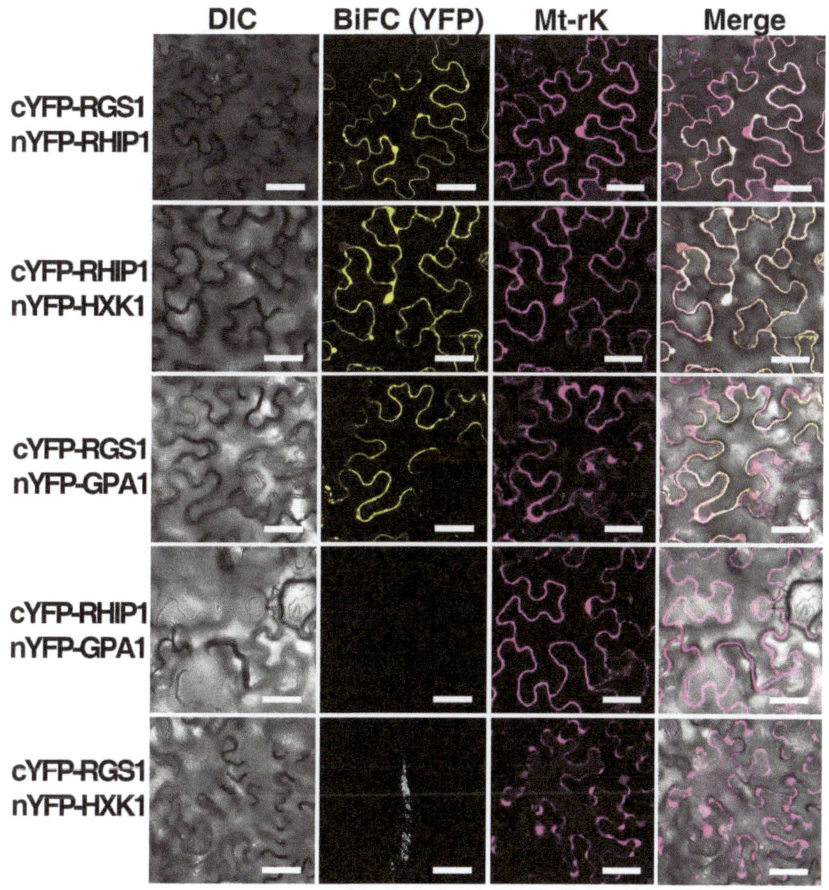

FIGURE 6 | *In vivo* interaction between AtRGS1 and RHIP1, and AtHXK1 and RHIP1 determined by physical complementation of YFP using Bimolecular Fluorescence complementation (BiFC). The test pairs are indicated on the left. **(Column 1)** Differential interference contrast (DIC) images of transformed cells. **(Column 2)** BIFC (YFP). cYFP-tagged proteins were co-transformed with nYFP-tagged proteins into tobacco leaves as described in Section "Materials and Methods." **(Column 3)** Mt-rk is an RFP mitochondria marker used for the transformation control. Fluorescence complementation of split YFP and expression of Mt-rk were observed by confocal fluorescence microscopy. **(Column 4)** DIC, BiFC, and Mt-rk are merged. Scale bars = 50 μm.

another 3 days treatment as previously described (Zhang et al., 2010; Xiong et al., 2013). As shown in **Figure 3A**, D-glucose, but not D-mannitol, activated the *CDKA;1* promoter in all three genotypes. Quantitative analysis showed that the *rgs1*-2 mutant had increased ($P < 0.05$) root meristem size in response to glucose as determined by GUS-staining, whereas the *hxk1*-3 mutant had decreased ($P < 0.01$) size (**Figure 3B**), consistent with the phenotypes we observed in glucose-promoted root growth (**Figure 2A**). Moreover, it indicated an overlap in function between AtRGS1 and AtHXK1 in root meristem modulation in response to glucose.

AtHXK1 Operates Downstream of Rapid Glucose-induced AtRGS1 Endocytosis

In *Saccharomyces cerevisiae*, glucose activation of cAMP synthesis (cAMP/PKA pathway) requires both the presence of a G-protein-coupled receptor system for extracellular glucose sensing, and phosphorylation of the sugar by hexokinases (HXKs, Hxk2/Snf1/Mig1 pathway) for cytoplasmic glucose sensing (Rolland et al., 2000). While the exact mechanism in this process is unknown, the data indicate a functional connection between G-proteins and hexokinases. To test if an analogous relationship between extracellular and cytoplasmic glucose perception exists in *Arabidopsis*, we quantitated glucose-induced activation of G signaling. The most rapid *in vivo* reporter for G protein activation currently available is glucose-induced endocytosis of AtRGS1 which occurs in minutes (Urano et al., 2012; Fu et al., 2014). Specifically, AtGPA1 (Gα) is self-activating through spontaneous binding of GTP but is held in the resting state by the inhibitory action of its physical partner, AtRGS1. Glucose relieves this inhibition by causing physical de-coupling of AtRGS1 and AtGPA1 through WNK kinase-dependent phosphorylation at the C terminus of AtRGS1 (Urano et al., 2012). Phosphorylation of AtRGS1 drives its endocytosis thus physically uncoupling AtRGS1 from AtGPA1 and allowing activation of G protein signaling (Urano et al., 2012; Fu et al., 2014). As shown in **Figure 4**, loss of *AtHXK1* did not affect the level of glucose activated AtRGS1 re-localization, confirming that the regulatory role of AtHXK1 in AtRGS1-dependent *TBL26* induction shown in **Figure 1A**

FIGURE 7 | *rhip1* mutants share growth phenotypes with the *rgs1*-2 null mutant. (A) 2-week-old Col-0, *hxk1*-3, *rgs1*-2, *rhip1*-1, and *rhip1*-2 plants grown in 1/4 × MS medium supplemented with 1% D-glucose (w/v) and grown at 200 μmol s^{-1} m^{-2} in 8/16 h (L/D). **(B)** 5-week-old Col-0, *hxk1*-3, *rgs1*-2, *rhip1*-1, and *rhip1*-2 plants grown at 160 μmol s^{-1} m^{-2} in 16/8 h (L/D). **(C)** Comparison of root elongation of Col-0, *hxk1*-3, *rgs1*-2, *rhip1*-1, and *rhip1*-2 seedlings transferred to 1/4 × MS medium supplemented with 0.25% (black) or 3% (gray) w/v D-glucose and grown at 70 μmol s^{-1} m^{-2} in 16/8 h (L/D). Values indicate means ± SD (n = 10–18) from a representative experiment. For the 3% D-glucose treatment, ANOVA single factor analysis ($\alpha = 0.05$) was conducted to compare values to the Col-0. $^{**}P < 0.01$. **(D)** 2-week-old Col-0 and *35S::YFP-RHIP1* plants grown in 1/4 × MS medium supplemented with 3% D-glucose (w/v). **(E)** Green seedling assay of the *rhip1*-2 mutant. The assay was performed as described in Section "Material and Methods." The average percentage of seedlings showing green cotyledons was determined and presented with means ± SD from one representative experiment of 4 biological replications. ANOVA single factor analysis ($\alpha = 0.05$) was conducted to compare values to the Col-0. $^{*}P < 0.05$.

lies downstream of the apical event in G protein coupled signaling.

RHIP1 is a Nexus in AtRGS1- and HXK1-dependent Sugar Sensing

A deep screen for *Arabidopsis* G-protein interacting proteins (Klopffleisch et al., 2011) yielded an uncharacterized conserved protein encoded by *At4g26410* that interacts with both AtRGS1 and AtHXK1 (Supplementary Figure S4). We investigated the possibility that this protein provides a functional connection between AtHXK1 and AtRGS1 glucose signaling. We designated this protein ***R***GS1 and ***H***XK1 ***I***nteracting ***P***rotein 1 (RHIP1). RHIP1 contains an *U*ncharacterized *C*onserved "*P*rotein" domain (UCP022280) with only one other protein (At2g45060) containing this domain in *Arabidopsis*. To our knowledge, these two proteins have not previously been identified in any study. The predicted protein structure of RHIP1 is a three-stranded helix with an N-terminal head composed of coils (**Figure 5A**). The predicted GO biological process of RHIP1 protein based on global and local protein similarity includes response to fungicide, receptor internalization, and regulation of receptor recycling (I-TASSER server) (Zhang, 2008; Roy et al., 2010, 2012). The top 1 enzyme homolog of RHIP1 protein in the Protein Database is the *Escherichia coli* histidine kinase sensor TorS sensor domain involved in signal transduction (Zhang, 2008; Moore and Hendrickson, 2009; Roy et al., 2010, 2012). RHIP1 interacts with 21 other proteins in the G protein interactome (Klopffleisch et al., 2011). Based on TAIR gene ontology annotations, 47% of these proteins are predicted to be involved in response to stimulus (**Figure 5B**)[3]. This finding prompted us to hypothesize that RHIP1 may have a role in sugar signaling through interactions with HXK1 and AtRGS1. To confirm interactions of RHIP1 with both AtRGS1 and AtHXK1 in

[3]http://www.arabidopsis.org/tools/bulk/go/index.jsp

plant cells, we used BiFC performed as described by Klopffleisch et al. (2011). As shown in **Figure 6**, AtRGS1 and AtHXK1 interacted with RHIP1 *in vivo*. Two negative and one positive control were included. The positive control was AtRGS1 and AtGPA1. The two negative controls were RHIP1-AtGPA1 and AtRGS1-AtHXK1 pairs. While *in vivo* interaction is confirmed, we do not reach a conclusion on the subcellular location of this interaction. Tobacco pavement cells, due to the large vacuole, are not suitable for subcellular location analysis. The dynamics of the confirmed interactions is not clear. While the data show that RHIP1 interacts with both AtRGS1 and AtHXK1, they do not show that this interaction is simultaneous; neither do the data preclude a scaffold role of RHIP1 to bring together AtRGS1 and AtHXK1. This is because that while AtRGS1 is located on the plasma membrane and the endomembrane system, AtHXK1 subcellular location is broad, including the cytoplasm, the mitochondrion and the nucleus and therefore it is plausible that RHIP1 is a protein complex scaffold.

RHIP1 Participates in Glucose Signaling in *Arabidopsis*

In order to elucidate the physiological role of RHIP1 in glucose signaling, two *RHIP1* T-DNA insertion mutants, *rhip1*-1 and *rhip1*-2 were generated (Supplementary Figure S5). Plants homozygous for *rhip1*-1 and *rhip1*-2 were confirmed to be transcript null. Both *rhip1*-1 and *rhip1*-2 mutants had longer roots in young seedlings and qualitatively larger inflorescence in adult plants (**Figures 7A–C**); these are phenotypes shared by the *rgs1*-2 mutant. Over-expression of *RHIP1* in Col-0 resulted in defects in root elongation (**Figure 7D**). Acute doses of glucose (6%, w/v) cause *Arabidopsis* seedling development to arrest and this phenotype is attenuated in both *HXK1* and *RGS1* loss-of-function mutants (Moore et al., 2003; Chen and Jones, 2004). We used this standardized green seedling assay to explore the role of RHIP1. As shown in **Figure 7E**, *rhip1*-2 mutants showed less sensitivity to 6% glucose than Col-0 seedlings. These *rhip1* phenotypes were not the consequences of altered transcript level for *RGS1* or *HXK1* (Supplementary Figure S2). As shown in **Figures 1A–C**, genetic ablation of *RHIP1* increased glucose-induced *TBL26* expression (around 4.5 fold), while there was no effect on the expression of *CA2* or *DIN1*.

Since it is predicted that AtHXK1 operates on AtRGS1 signaling downstream of the rapid glucose-induced AtRGS1 endocytosis (**Figures 1A** and **4**), we explored the role of their common interactor —AtRHIP1 in this process. As expected, loss of *AtRHIP1* did not affect the level of glucose activated AtRGS1 re-localization (Supplementary Figure S6), consistent with the regulatory role of AtHXK1 localized downstream of the apical event in G protein coupled signaling.

In summary, a partial functional conversation occurring between AtRGS1 and AtHXK1 was shown by glucose-regulated gene expression, epistasis analyses, and protein–protein interaction: (1) AtRGS1 and AtHXK1 showed hierarchical relationships in regulation of *TBL26*, *CA2*, and *CAB2* expression, suggesting collaboration between AtRGS1 and AtHXK1 in regulating gene expression. (2) *hxk1* mutant exhibited the opposite root elongation and leaf expansion phenotype to the *rgs1* mutants, while the double mutant showed an intermediate phenotype, implying a feed-back relationship between AtRGS1 and AtHXK1 in controlling cell activity. This was confirmed by the opposite behavior in glucose promotion of root meristem activation. The intermediate response in the double mutant was not additive since the phenotype of the single mutants were opposite. (3) Loss of *AtHXK1* abrogated AtRGS1-dependent signaling as measured by *TBL26* expression without changing the level of glucose activated AtRGS1 endocytosis, indicating that AtHXK1 acts downstream of AtRGS1 in sugar-activated AtRGS1 signaling. (4) AtRGS1 and AtHXK1 have a common protein interactor, RHIP1, a predicted signaling component, providing a direct nexus in AtRGS1- and AtHXK1-dependent sugar sensing. (5) Gene expression and seedling growth phenotype analyses of the *rhip1* mutant confirmed a potential role in AtRGS1 and AtHXK1 signaling.

Accession Numbers

AtRGS1, At3g26090; TBL26, At4g01080; AtHXK1, At4g29130; AtCA2, At5g14740; AtTUB4, At5g44340; AtDIN1, At4g35770; RHIP1, At4g26410; CAB2, At1g29920. Mutant alleles for the genes were generated from the indicated T-DNA insertion lines: *hxk1-1*, SALK_034233; *hxk1-2*, CS864200; *hxk1-3*, CS861759; *rhip1-1*, SALK_091518C; *rhip1-2*, SALK_061002.

FUNDING

This work was supported by the National Institute of General Medical Sciences (R01GM065989), and the National Science Foundation (MCB- 0718202) to AJ. The Division of Chemical Sciences, Geosciences, and Biosciences, Office of Basic Energy Sciences of the US Department of Energy through the grant DE-FG02-05er15671 to AJ funded technical support in this study.

ACKNOWLEDGMENTS

J-PH thanks The University of North Carolina, where these experiments were conducted in entirety for hosting her from 2012 to 2014 as an international scholar supported by an exchange scholarship from the Chinese Scholarship Council. We thank Jing Yang for technical assistance, Alejandro Colaneri for discussions, and Dirk Inzé for providing the *pCDKA;1::GUS* reporter line in the Col background.

REFERENCES

Baena-Gonzalez, E., Rolland, F., Thevelein, J. M., and Sheen, J. (2007). A central integrator of transcription networks in plant stress and energy signalling. *Nature* 448, 938–942. doi: 10.1038/nature06069

Baena-Gonzalez, E., and Sheen, J. (2008). Convergent energy and stress signaling. *Trends Plant Sci.* 13, 474–482. doi: 10.1016/j.tplants.2008.06.006

Bechtold, N., Ellis, J., and Pelletier, G. (1993). In planta *Agrobacterium* mediated gene transfer by infiltration of adult *Arabidopsis thaliana* plants. *C. R. Acad. Sci. Paris Life Sci.* 316, 1194–1199.

Booker, K. S., Schwarz, J., Garrett, M. B., and Jones, A. M. (2010). Glucose attenuation of auxin-mediated bimodality in lateral root formation is partly coupled by the heterotrimeric G protein complex. *PLoS ONE* 5:e12833. doi: 10.1371/journal.pone.0012833

Chen, J. G., Gao, Y., and Jones, A. M. (2006a). Differential roles of *Arabidopsis* heterotrimeric G-protein subunits in modulating cell division in roots. *Plant Physiol.* 141, 887–897. doi: 10.1104/pp.106.079202

Chen, Y., Ji, F., Xie, H., and Liang, J. (2006b). Overexpression of the regulator of G-protein signalling protein enhances ABA-mediated inhibition of root elongation and drought tolerance in *Arabidopsis. J. Exp. Bot.* 57, 2101–2110. doi: 10.1093/jxb/erj167

Chen, Y., Ji, F., Xie, H., Liang, J., and Zhang, J. (2006c). The regulator of G-protein signaling proteins involved in sugar and abscisic acid signaling in *Arabidopsis* seed germination. *Plant Physiol.* 140, 302–310. doi: 10.1104/pp.105.069872

Chen, J. G., and Jones, A. M. (2004). AtRGS1 function in *Arabidopsis thaliana. Methods Enzymol.* 389, 338–350. doi: 10.1016/S0076-6879(04)89020-7

Chen, J. G., Willard, F. S., Huang, J., Liang, J., Chasse, S. A., Jones, A. M., et al. (2003). A seven-transmembrane RGS protein that modulates plant cell proliferation. *Science* 301, 1728–1731. doi: 10.1126/science.1087790

Cho, Y. H., Yoo, S. D., and Sheen, J. (2006). Regulatory functions of nuclear hexokinase1 complex in glucose signaling. *Cell* 127, 579–589. doi: 10.1016/j.cell.2006.09.028

Colaneri, A. C., Tunc-Ozdemir, M., Huang, J. P., and Jones, A. M. (2014). Growth attenuation under saline stress is mediated by the heterotrimeric G protein complex. *BMC Plant Biol.* 14:129. doi: 10.1186/1471-2229-14-129

Fu, Y., Lim, S., Urano, D., Tunc-Ozdemir, M., Phan, N. G., Elston, T. C., et al. (2014). Reciprocal encoding of signal intensity and duration in a glucose-sensing circuit. *Cell* 156, 1084–1095. doi: 10.1016/j.cell.2014.01.013

Grigston, J. C., Osuna, D., Scheible, W. R., Liu, C., Stitt, M., and Jones, A. M. (2008). D-Glucose sensing by a plasma membrane regulator of G signaling protein, AtRGS1. *FEBS Lett.* 582, 3577–3584. doi: 10.1016/j.febslet.2008.08.038

Kim, J. H., Roy, A., Jouandot, D., and Cho, K. H. (2013). The glucose signaling network in yeast. *Biochim. Biophys. Acta* 1830, 5204–5210. doi: 10.1016/j.bbagen.2013.07.025

Klopffleisch, K., Phan, N., Augustin, K., Bayne, R. S., Booker, K. S., Botella, J. R., et al. (2011). *Arabidopsis* G-protein interactome reveals connections to cell wall carbohydrates and morphogenesis. *Mol. Syst. Biol.* 7, 532. doi: 10.1038/msb.2011.66

Koch, K. E. (1996). Carbohydrate-modulated gene expression in plants. *Annu. Rev. Plant Physiol. Plant Mol. Biol.* 47, 509–540. doi: 10.1146/annurev.arplant.47.1.509

Lastdrager, J., Hanson, J., and Smeekens, S. (2014). Sugar signals and the control of plant growth and development. *J. Exp. Bot.* 65, 799–807. doi: 10.1093/jxb/ert474

Livak, K. J., and Schmittgen, T. D. (2001). Analysis of relative gene expression data using real-time quantitative PCR and the 2(-Delta Delta C(T)) Method. *Methods* 25, 402–408. doi: 10.1006/meth.2001.1262

Malamy, J. E., and Benfey, P. N. (1997). Organization and cell differentiation in lateral roots of *Arabidopsis thaliana. Development* 124, 33–44.

Moore, B., Zhou, L., Rolland, F., Hall, Q., Cheng, W. H., Liu, Y. X., et al. (2003). Role of the *Arabidopsis* glucose sensor HXK1 in nutrient, light, and hormonal signaling. *Science* 300, 332–336. doi: 10.1126/science.1080585

Moore, J. O., and Hendrickson, W. A. (2009). Structural analysis of sensor domains from the TMAO-responsive histidine kinase receptor TorS. *Structure* 17, 1195–1204. doi: 10.1016/j.str.2009.07.015

Reda, M. (2013). Regulation of nitrate reduction in *Arabidopsis* WT and hxk1 mutant under C and N metabolites. *Physiol. Plant.* 149, 260–272. doi: 10.1111/ppl.12045

Rolland, F., De Winde, J. H., Lemaire, K., Boles, E., Thevelein, J. M., and Winderickx, J. (2000). Glucose-induced cAMP signalling in yeast requires both a G-protein coupled receptor system for extracellular glucose detection and a separable hexose kinase-dependent sensing process. *Mol. Microbiol.* 38, 348–358. doi: 10.1046/j.1365-2958.2000.02125.x

Roy, A., Kucukural, A., and Zhang, Y. (2010). I-TASSER: a unified platform for automated protein structure and function prediction. *Nat. Protoc.* 5, 725–738. doi: 10.1038/nprot.2010.5

Roy, A., Yang, J., and Zhang, Y. (2012). COFACTOR: an accurate comparative algorithm for structure-based protein function annotation. *Nucleic Acids Res.* 40, W471–W477. doi: 10.1093/nar/gks372

Sheen, J. (1990). Metabolic repression of transcription in higher plants. *Plant Cell* 2, 1027–1038. doi: 10.2307/3869242

Sheen, J. (2014). Master regulators in plant glucose signaling networks. *J. Plant Biol.* 57, 67–79. doi: 10.1007/s12374-014-0902-7

Sheen, J., Zhou, L., and Jang, J.-C. (1999). Sugars as signaling molecules. *Curr. Opin. Plant Biol.* 2, 410–418. doi: 10.1016/S1369-5266(99)00014-X

Skylar, A., Sung, F., Hong, F., Chory, J., and Wu, X. (2011). Metabolic sugar signal promotes *Arabidopsis* meristematic proliferation via G2. *Dev. Biol.* 351, 82–89. doi: 10.1016/j.ydbio.2010.12.019

Tome, F., Nagele, T., Adamo, M., Garg, A., Marco-Llorca, C., Nukarinen, E., et al. (2014). The low energy signaling network. *Front. Plant Sci.* 5:353. doi: 10.3389/fpls.2014.00353

Urano, D., Chen, J. G., Botella, J. R., and Jones, A. M. (2013). Heterotrimeric G protein signalling in the plant kingdom. *Open Biol.* 3, 120186. doi: 10.1098/rsob.120186

Urano, D., Phan, N., Jones, J. C., Yang, J., Huang, J., Grigston, J., et al. (2012). Endocytosis of the seven-transmembrane RGS1 protein activates G-protein-coupled signalling in *Arabidopsis. Nat. Cell Biol.* 14, 1079–1088. doi: 10.1038/ncb2568

West, G., Inze, D., and Beemster, G. T. (2004). Cell cycle modulation in the response of the primary root of *Arabidopsis* to salt stress. *Plant Physiol.* 135, 1050–1058. doi: 10.1104/pp.104.040022

Xiong, Y., Mccormack, M., Li, L., Hall, Q., Xiang, C., and Sheen, J. (2013). Glucose-TOR signalling reprograms the transcriptome and activates meristems. *Nature* 496, 181–186. doi: 10.1038/nature12030

Zhang, L., Hu, G., Cheng, Y., and Huang, J. (2008). Heterotrimeric G protein alpha and beta subunits antagonistically modulate stomatal density in *Arabidopsis thaliana. Dev. Biol.* 324, 68–75. doi: 10.1016/j.ydbio.2008.09.008

Zhang, L., Wei, Q., Wu, W., Cheng, Y., Hu, G., Hu, F., et al. (2009). Activation of the heterotrimeric G protein alpha-subunit GPA1 suppresses the ftsh-mediated inhibition of chloroplast development in *Arabidopsis. Plant J.* 58, 1041–1053. doi: 10.1111/j.1365-313X.2009.03843.x

Zhang, Y. (2008). I-TASSER server for protein 3D structure prediction. *BMC Bioinformatics* 9:40. doi: 10.1186/1471-2105-9-40

Zhang, Z. W., Yuan, S., Xu, F., Yang, H., Zhang, N. H., Cheng, J., et al. (2010). The plastid hexokinase pHXK: a node of convergence for sugar and plastid signals in *Arabidopsis. FEBS Lett.* 584, 3573–3579. doi: 10.1016/j.febslet.2010.07.024

Conflict of Interest Statement: The authors declare that the research was conducted in the absence of any commercial or financial relationships that could be construed as a potential conflict of interest.

18

A phylogenetic approach to study the origin and evolution of the CRINKLY4 family

Natalia Nikonorova[1,2], *Lam D. Vu*[1,2,3,4], *Nathan Czyzewicz*[5], *Kris Gevaert*[3,4] and *Ive De Smet*[1,2,5,6*]

[1] Department of Plant Systems Biology, Flanders Institute for Biotechnology (VIB), Ghent, Belgium, [2] Department of Plant Biotechnology and Bioinformatics, Ghent University, Ghent, Belgium, [3] Department of Medical Protein Research, Flanders Institute for Biotechnology (VIB), Ghent University, Ghent, Belgium, [4] Department of Biochemistry, Ghent University, Ghent, Belgium, [5] Division of Plant and Crop Sciences, School of Biosciences, University of Nottingham, Loughborough, UK, [6] Center for Plant Integrative Biology, University of Nottingham, Loughborough, UK

Edited by:
Ikram Blilou,
Wageningen University, Netherlands

Reviewed by:
Biswapriya Biswavas Misra,
University of Florida, USA
Luca Santuari,
Wageningen University, Netherlands

***Correspondence:**
Ive De Smet
ive.desmet@psb.vib-ugent.be

Cell–cell communication plays a crucial role in plant growth and development and relies to a large extent on peptide ligand–receptor kinase signaling mechanisms. The CRINKLY4 (CR4) family of receptor-like kinases is involved in a wide range of developmental processes in plants, including mediating columella stem cell identity and differentiation in the *Arabidopsis thaliana* root tip. Members of the CR4 family contain a signal peptide, an extracellular part, a single-pass transmembrane helix and an intracellular cytoplasmic protein kinase domain. The main distinguishing features of the family are the presence of seven "crinkly" repeats and a TUMOR NECROSIS FACTOR RECEPTOR (TNFR)-like domain in the extracellular part. Here, we investigated the evolutionary origin of the CR4 family and explored to what extent members of this family are conserved throughout the green lineage. We identified members of the CR4 family in various dicots and monocots, and also in the lycophyte *Selaginella moellendorffii* and the bryophyte *Physcomitrella patens*. In addition, we attempted to gain insight in the evolutionary origin of different CR4-specific domains, and we could detect "crinkly" repeat containing proteins already in single celled algae. Finally, we related the presence of likely functional CR4 orthologs to its best described signaling module comprising CLAVATA3/EMBRYO SURROUNDING REGION-RELATED 40 (CLE40), WUSCHEL RELATED HOMEOBOX 5 (WOX5), CLAVATA 1 (CLV1), and ARABIDOPSIS CR4 (ACR4), and established that this module likely is already present in bryophytes and lycophytes.

Keywords: CR4, WOX5, CLV1, CLE40, stem cells, evolution, phylogenetics, crinkly repeat

INTRODUCTION

Stem cells are fundamental for post-embryonic plant growth, and the stem cell niche is subject to a tight regulation (ten Hove et al., 2015). In plants, stem cells are located in the central zone of the shoot apical meristem and around the quiescent center in the root apical meristem. Initiated during early embryogenesis they constantly produce cells during the entire plant life (Scheres, 2007). Maintaining the balance between cell proliferation, specification, and differentiation requires the coordinated activity of, for example, hormones (Pacifici et al., 2015), peptides (Murphy et al., 2012), (mobile) transcription factors (Han et al., 2014), and receptor kinases

(De Smet et al., 2009). Despite the obvious involvement of receptor kinases in animals, our understanding of their role in plant growth and development remains very incomplete.

Over 600 receptor-like kinase genes have been identified in *Arabidopsis thaliana* and similar or even higher numbers were found in other plant species, such as *Physcomitrella patens* (329), soybean (605), tomato (647), rice (977–1132), and poplar (1192) (Shiu and Bleecker, 2001b; Shiu et al., 2004; Lehti-Shiu et al., 2009; Liu et al., 2009). Mutations in receptor kinases often lead to obvious and crucial developmental defects as they are, for example, involved in root and shoot apical meristem maintenance (De Smet et al., 2009; Stahl and Simon, 2012; Wierzba and Tax, 2013).

One of these receptor kinases is ARABIDOPSIS CRINKLY4 (ACR4), which belongs to the CRINKLY4 (CR4) family of receptor-like kinases. CR4 was first identified in maize (*Zea mays*), where the *cr4* mutation affects leaf epidermis differentiation (Becraft et al., 1996). Epidermal cells of maize *cr4* mutants have irregularities in shape, cell wall thickness and structure, cuticle formation, and vesicle trafficking, and often display tumor-like sections with large undifferentiated and disorganized cells (Becraft et al., 1996, 2001; Jin et al., 2000). The *cr4* mutant plants exhibit very short stature and also display defects in floral organs, including the glumes, anthers, and silks, and CR4 functions in aleurone formation in seeds and may be involved in sex-determination (Becraft et al., 1996; Jin et al., 2000; Kang et al., 2002; Tian et al., 2007). Expression studies indicated that the maize *cr4* transcript is present in vegetative and floral organs, but absent in the root (Jin et al., 2000; Becraft et al., 2001; Kang et al., 2002). Following the initial characterization in maize, orthologs of *CR4* were identified and characterized in rice (*Oryza sativa*) and *A. thaliana* (Shiu and Bleecker, 2001b; Cao et al., 2005). Rice *cr4* mutants also show tumor-like cell growth in the outer epidermis, wart-like cell masses in the inner epidermis of palea and lemma, and abnormal cells with discontinuous cuticles and uneven cell walls (Pu et al., 2012). In *A. thaliana*, *ACR4* is strongly expressed in the protodermal cells of the embryo, in the L1 layer of the shoot apical meristem, in the epidermis of leaf primordia, in the small daughter cells after the first asymmetric pericycle cell division and in the root stem cell niche (Tanaka et al., 2002; Gifford, 2003; Watanabe et al., 2004; De Smet et al., 2008). Subcellular localization of transiently expressed *ACR4* in *Nicotiana benthamiana* leaf epidermal cells shows that ACR4 preferentially associates with plasmodesmata (Stahl et al., 2013). The *ACR4* expression pattern corresponds with *A. thaliana acr4* mutant phenotypes, such as seed development, embryo morphogenesis, various abnormalities related to epidermal differentiation—which include disorganized cell layers in the ovule integument—, loss of coordination of pericycle cell divisions during lateral root initiation, and irregular columella stem cell division and differentiation in the root apex (Gifford, 2003; Watanabe et al., 2004; De Smet et al., 2008).

In the *A. thaliana* genome four CRINKLY4-RELATED (CCR) proteins were found (Cao et al., 2005). Members of the CR4 family contain a signal peptide, an extracellular part, a single-pass transmembrane helix, and a cytoplasmic serine/threonine protein kinase domain (Jin et al., 2000). Kinase assays demonstrated that ACR4 is an active serine/threonine kinase, while CCR1 and CCR2 are nearly inactive in autophosphorylation assays (Gifford, 2003; Cao et al., 2005; Meyer et al., 2011, 2015). The distinguishing feature of the family is the presence of seven "crinkly" repeats in the extracellular part which is required both for signaling and for normal protein internalization (Gifford et al., 2005). A core C(X~10)CWG sequence motif is highly conserved amongst the repeats, and the regularly spaced Cys residues in the extracellular "crinkly" repeat domain are likely to form stabilizing disulfide bridges. Another feature of the CR4 family extracellular part is the presence of a domain with homology to the three Cys-rich repeats of the TUMOR NECROSIS FACTOR RECEPTOR (TNFR) extracellular domain (Becraft et al., 1996; Gifford et al., 2005). Finally, ACR4 and CCRs can interact through the conserved "KD/ESAF" motif in the juxtamembrane domain (Meyer et al., 2013, 2015).

The initial idea that the CR4 ligand may be a peptide (Becraft et al., 1996) was supported in *A. thaliana,* since the *acr4* mutant is largely insensitive to CLAVATA3/EMBRYO SURROUNDING REGION-RELATED 40 (CLE40) peptide application with respect to columella stem cell phenotypes (Stahl et al., 2009). In addition, ACR4 and CLAVATA 1 (CLV1) form homo- and heterodimers that regulate root meristem maintenance in response to the CLE40 signaling peptide (Stahl et al., 2013). Thus, ACR4, CLV1, CLE40 peptide, and the downstream WUSCHEL RELATED HOMEOBOX 5 (WOX5) transcription factor were proposed as a signaling module controlling the stem cell niche in root meristems (Stahl et al., 2009). The involvement of the latter is further supported by the fact that the intracellular domain of ACR4 could interact with and phosphorylate WOX5 (Meyer et al., 2015).

Here, we investigated the phylogenetic relationships between CR4 family members in the green plant lineage and explored when members of this family (or key motifs) appeared. In addition, we searched for putative orthologs of other components of the CLE40-ACR4-CLV1-WOX5 signaling module in the green plant lineage.

MATERIALS AND METHODS

Data Sources and Candidate Retrieval

In our analysis we included genomes of 23 species of single-celled (*Ostreococcus lucimarinus*, *Ostreococcus tauri*, *Micromonas* sp. RCC299, *Micromonas pusilla* CCMP1545, *Chlorella vulgaris*, *Chlorella variabilis*, and *Chlamydomonas reinhardtii*), colony-forming (*Volvox carteri*), and multicellular green algae (*Spirogyra pratensis* and *Coleochaete orbicularis*), the bryophyte *Physcomitrella patens*, the lycophyte *Selaginella moellendorffii* and various higher plants, including the monocots *Sorghum bicolor*, *Zea mays*, *Hordeum vulgare*, *Triticum aestivum*, and *Orysa sativa* and the dicots *Populus trichocarpa*, *Cucumis sativus*, *Medicago truncatula*, *Glycine max*, *Vitis vinifera*, and *Solanum tuberosum*. The *A. thaliana* genome encodes ACR4 and four CRINKLY4-RELATED (CCR) proteins (Cao et al., 2005). We used the respective *A. thaliana* full-length protein sequences as well as their extracellular domain sequences (Supplementary data set 1) for BLASTP analyses on www.uniprot.org, http://blast.st-va.ncbi.nlm.nih.gov/Blast.cgi, and www.phytozome.net with an *E*-value = 1 as filter, and retrieved a large number of putative CR4 family members (data not shown). Subsequently, reciprocal BLASTP analyses (where we BLASTed the highest-scoring candidate protein sequence against an organism database with well-annotated and described proteins) were performed for putative CR4 family members against the *A. thaliana* protein database TAIR10 Proteins (www.arabidopsis.org/Blast/index.jsp). Putative CR4 orthologs were considered when the protein of interest was retrieved in the top five for the reciprocal BLASTP analyses in *A. thaliana*. A similar approach was used for CLE40, WOX5, and CLV1 (Supplementary data set 1). The complete workflow is summarized in Supplementary Figure 1.

Alignment and Phylogenetic Analyses

Retrieved sequences were aligned using a progressive alignment algorithm (Feng and Doolittle, 1987) in order to create multiple sequence alignments with CLC DNA Workbench 7 using the following settings: Gap open cost (10), Extension cost (1), End gap cost (as any other), Alignment (very accurate). To visualize conserved motifs and domains we also used WebLogo (weblogo.berkeley.edu; Crooks, 2004). The phylogenetic trees for the *A. thaliana* CR4 family and for Viridiplantae were built using the Neighbor Joining and UPGMA tree construction method, respectively. For protein distance measurement a Jukes–Cantor method was used and a bootstrap analysis with 1000 replicates within the CLC DNA Workbench 6 (CLC Bio-Qiagen, Aarhus, Denmark) was performed. Putative CR4 orthologs were considered further when they grouped in the clade of the protein of interest and matched to the results obtained after the reciprocal BLASTP analysis. The taxonomy tree was generated with the phyloT online tool (phylot.biobyte.de).

Domain Prediction

For the prediction of signal peptides, the PSORT prediction tool for plants (http://psort.hgc.jp/form.html) together with the SignalP 4.1 server (www.cbs.dtu.dk/services/SignalP) were used with the standard settings for PSORT and the following parameters for SignalP: organism group (Eukaryotes), D-cutoff values (Default), method (input sequences may include TM regions). For the transmembrane and kinase domain prediction, TMHMM 2.0 (www.cbs.dtu.dk/services/TMHMM) and the Pfam database (pfam.xfam.org/) were used, respectively.

3D Structure Modeling

Structure prediction for the putative extracellular domain of ACR4 from *A. thaliana* (NP_191501.1), *Zea mays* (NP_001105395.1), *Selaginella moellendorffii* (XP_002960988.1), and *Physcomitrella patens* (XP_001754560.1) was carried out in the SWISS-MODEL workspace (http://swissmodel.expasy.org/workspace/; Arnold et al., 2006; Guex et al., 2009; Kiefer et al., 2009). For the same purpose, full length sequences of ACR4 homologs from *Micromonas* sp. RCC299 (XP_002508396.1) and *Chlorella variabilis* (XP_005845729.1) were used. Templates for the modeling studies were identified in automated mode against the SWISS-MODEL template library (Supplementary Table 1). Structure representations were generated using the PyMOL Molecular Graphics System, Version 1.7.4, Schrödinger, LLC (www.pymol.org).

Data Mining

Analyses of available transcriptome data was done through the Genevestigator platform (genevestigator.com/gv/index.jsp; Zimmermann et al., 2004, 2005; Hruz et al., 2008) on 27/07/2015.

RESULTS AND DISCUSSION

Features of the CR4 Family in *A. thaliana*

In *A. thaliana*, the CR4 family consists of five members distributed over two clades; one that includes ARABIDOPSIS CRINKLY 4 (ACR4)/CRINKLY 4 (CR4) (AT3G59420–Q9LX29), CRINKLY 4 RELATED 1 (CCR1 or CRR1) (AT3G09780–Q9S7D9), and CRINKLY 4 RELATED 2 (CCR2 or CRR2) (AT2G39180–O80963) and one with CRINKLY 4 RELATED 3 (CCR3 or CRR3) (AT3G55950–Q9LY50) and CRINKLY 4 RELATED 4 (CCR4)/CRINKLY 4-RELATED KINASE 1 or CYTOKININ-REGULATED KINASE 1 (CRK1) (AT5G47850–Q9FIJ6) (**Figure 1A**). All the family members share common features, such as a signal peptide, an extracellular part (with characteristic features such as seven "crinkly" repeats and a TNFR-Cys domain), a transmembrane domain and an intracellular part (including a juxtamembrane and protein kinase domain; **Figure 1B**). The ACR4 extracellular part is required for protein function, and the "crinkly" repeats include several cysteine residues in C(X∼10)CWG motifs that are likely important for stabilizing disulfide bridges and generating the β-propeller structure of the extracellular part (Gifford et al., 2005) (**Figure 2A**). The ACR4 extracellular part also contains a domain with homology to the three Cys-rich repeats of the TNFR extracellular domain (Becraft et al., 1996; Gifford et al., 2005) (**Figure 2B**). With respect to the kinase domain, it appears that this is not essential for ACR4 (Gifford et al., 2005) and is likely inactive for CCR1 and CCR2 (Gifford, 2003; Cao et al., 2005). Finally, ACR4 and CCRs can interact through the conserved "KD/ESAF" motif in the juxtamembrane domain (Meyer et al., 2013, 2015).

Identification of Putative CR4 Family Members in the Green Lineage

While there is some information on CR4 family members in a few plant species, such as *A. thaliana*, maize, and rice (Becraft et al., 1996; Gifford, 2003; Watanabe et al., 2004; Cao et al., 2005; Pu et al., 2012), a comprehensive overview is lacking. In addition, available web-resources such as Plaza (Proost et al., 2009), Phytozome (Goodstein et al., 2012), and Ensembl Plants (Kersey et al., 2014) appeared to be incomplete and do not always apply sufficient criteria to identify true orhologs. To gain insight into the size and evolutionary origin of the CR4 family, we analyzed available genomes from the green plant lineage. In our analysis we included genomes of single-celled, colony-forming

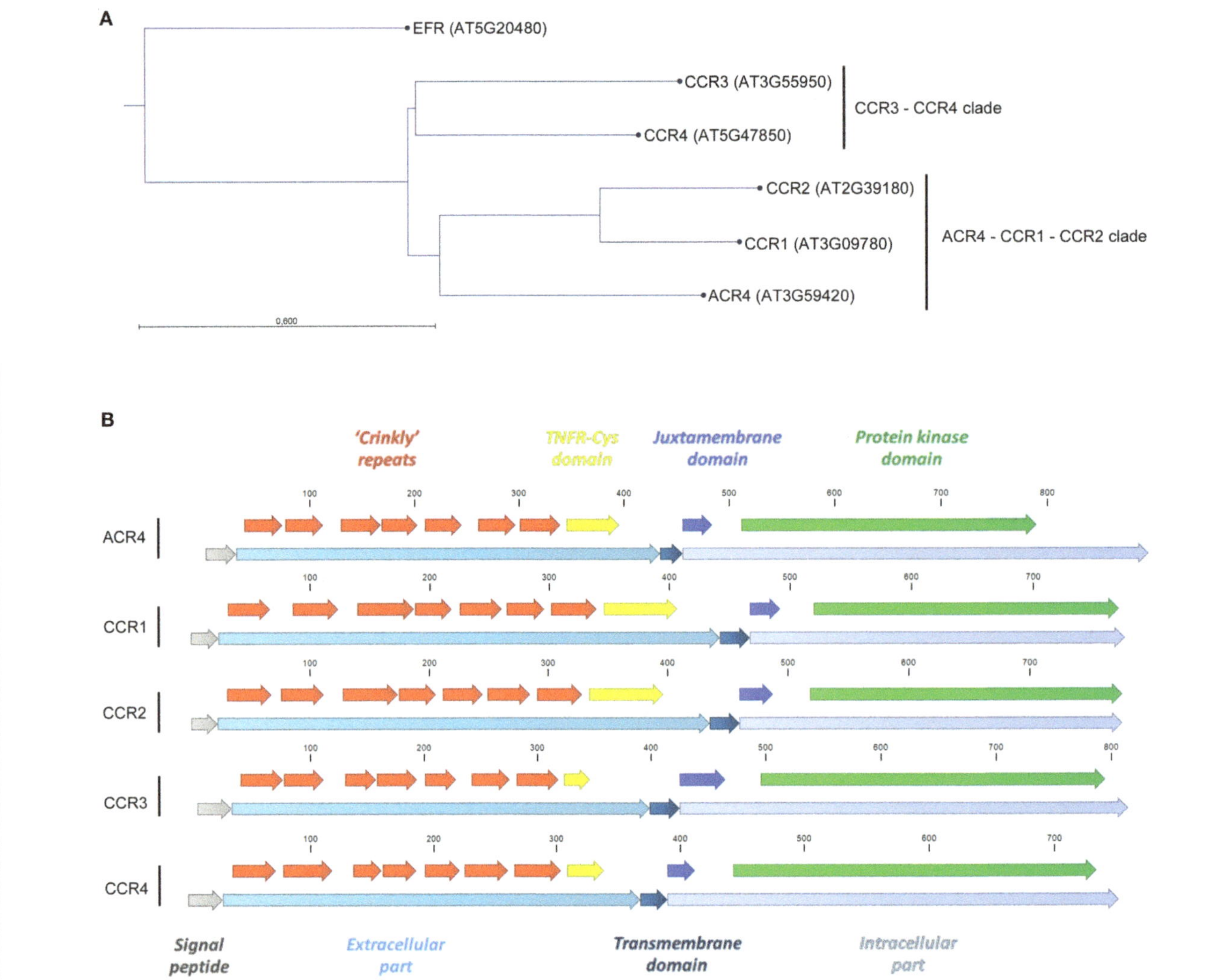

FIGURE 1 | *A. thaliana* CR4 family members. (A) Neighbor joining tree construction method with Jukes–Cantor protein distance measure and bootstrap analysis with 1000 replicates. **(B)** Common features for *A. thaliana* CR4 family members.

and multicellular green algae, the bryophyte *Physcomitrella patens*, the lycophyte *Selaginella moellendorffii*, and higher plants, totaling 23 species. Through extensive BLASTP and reciprocal BLASTP analyses, we generated a set of possible CR4 family members (Supplementary Table 2).

Our analyses showed that putative members of the CR4 family are present in algae, bryophytes, lycophytes, and in all higher plant species selected for our analysis (Supplementary Table 2). Among the 10 algae species included in our search, BLASTP analyses revealed only three putative orthologs of CR4 family proteins encoded in the genomes of *Chlorella variabilis*, *Micromonas pusilla*, and *Micromonas* sp. RCC299. However, due to the lack of complete genome information for the more multicellular algae, such as *S. pratensis* and *C. orbicularis*, we cannot rule out that CR4 family members are present in those species. In addition, there appeared to be no putative orthologs for CCR1 and CCR2 in monocots, which is consistent with a previous study (Cao et al., 2005). Furthermore, we could not identify putative CCR1 and CCR2 orthologs in *Medicago truncatula*.

As proteins are generally composed of one or more functional regions or domains that can provide insight into their function and evolutionary relationship (Lee et al., 2007), we explored to what extent the key domains and motifs are conserved in the putative CR4 family members (see below). Taking these data into account will allow us to generate a conclusive list of likely functional CR4 family members.

"Crinkly" Repeats and TNFR Domain in Putative CR4 Family Members

Since the extracellular part is required for protein function and normal protein internalization (Gifford et al., 2005), the set of putative CR4 family members was first investigated in terms of presence and conservation level of key extracellular domains,

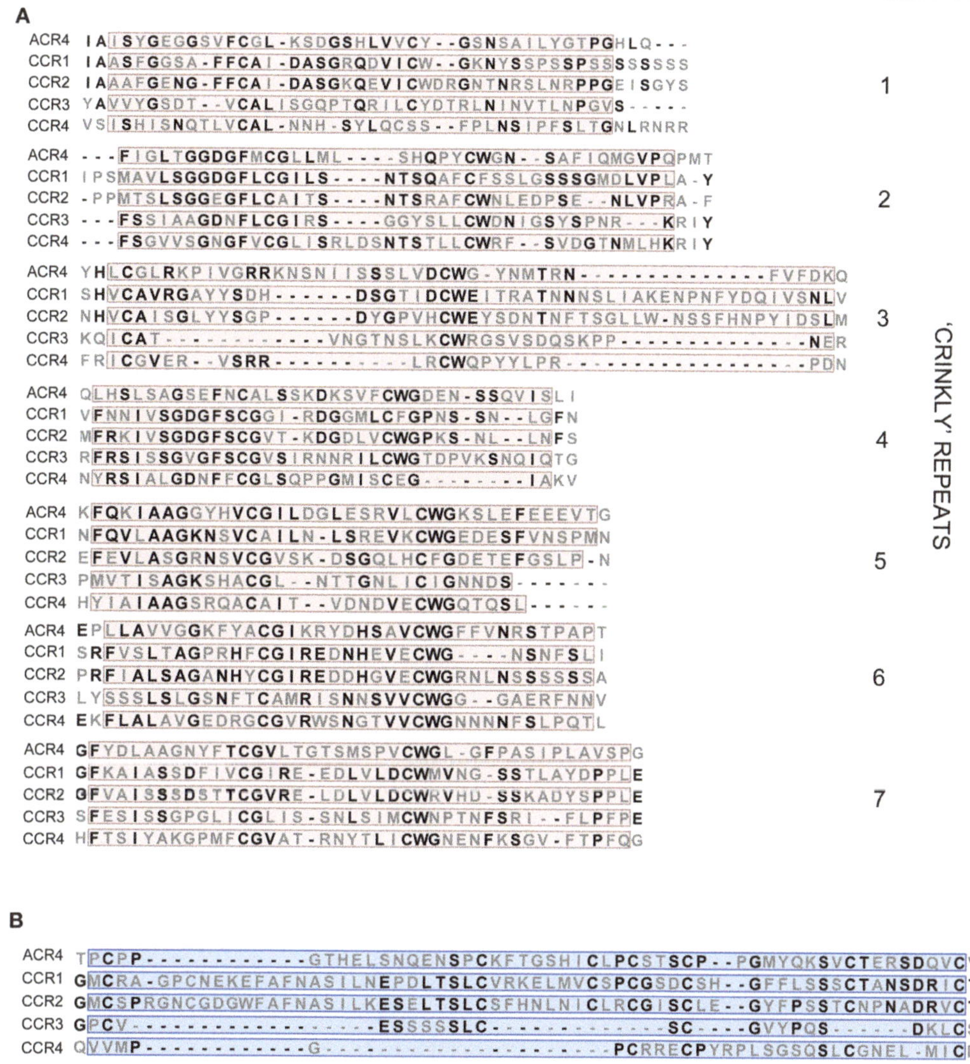

FIGURE 2 | Key domains in extracellular part of *A. thaliana* CR4 family members. (A,B) Alignment of amino acid sequences of seven "crinkly" repeats **(A)** and TNFR-Cys domain **(B)** of *A. thaliana* CR4 family members. Conserved residues are in bold.

including "crinkly" repeats and the TNFR domain. Sequence alignment indicated that the extracellular part is highly conserved within monocots and dicots (**Figure 3** and Supplementary Table 3). Although putative ACR4 orthologs in algae, *P. patens*, and *S. moellendorffii* have a much lower conservation level, the key residues remained highly conserved within the "crinkly" repeats and TNFR domain. With respect to putative ACR4 orthologs, the first "crinkly" repeat is the most variable, since only two Cys, two Gly, and one Pro residue were conserved and for algae sequences only Pro matched to *A. thaliana* sequence. The other six "crinkly" repeats were relatively more conserved and almost in all algae, *P. patens*, and *S. moellendorffii* sequences the C(X∼10)CWG motif was detected in each repeat (**Figure 3** and Supplementary Table 3). However, only Cys, Gly, and Trp residues of this motif were found in all of repeats. In ACR4 and its putative orthologs in *P. patens* and *S. moellendorffii*, the TNFR domain included seven Cys residues; however, algae species contained none (*Micromonas* sp. RCC299) or two Cys residues (*Chlorella variabilis*; **Figure 4** and Supplementary Table 3). Although the TNFR domain of dicots and monocots is conserved, it has some characteristics that distinguish the taxonomic groups. For example, for putative ACR4 orthologs, the Pro at position 346 in ACR4 is highly conserved in all dicot species, but is replaced by Ala or Ser in monocots. Similarly, the Ser at position 360 in ACR4 is highly variable in dicots and is replaced by Lys in monocots. Interestingly, the Gln at position 383 in ACR4, which is conserved in dicots, is substituted for Glu in monocots and also in *P. patens* and *S. moellendorffii* sequences. At the moment, it is not clear if this has structural implications and would affect functionality.

3D Structure Modeling of Selected Putative CR4 Family Members

For ACR4, the extracellular part is predicted to fold into a β-propeller type structure (McCarty and Chory, 2000; Cao et al., 2005; Gifford et al., 2005). In the case of ACR4—given

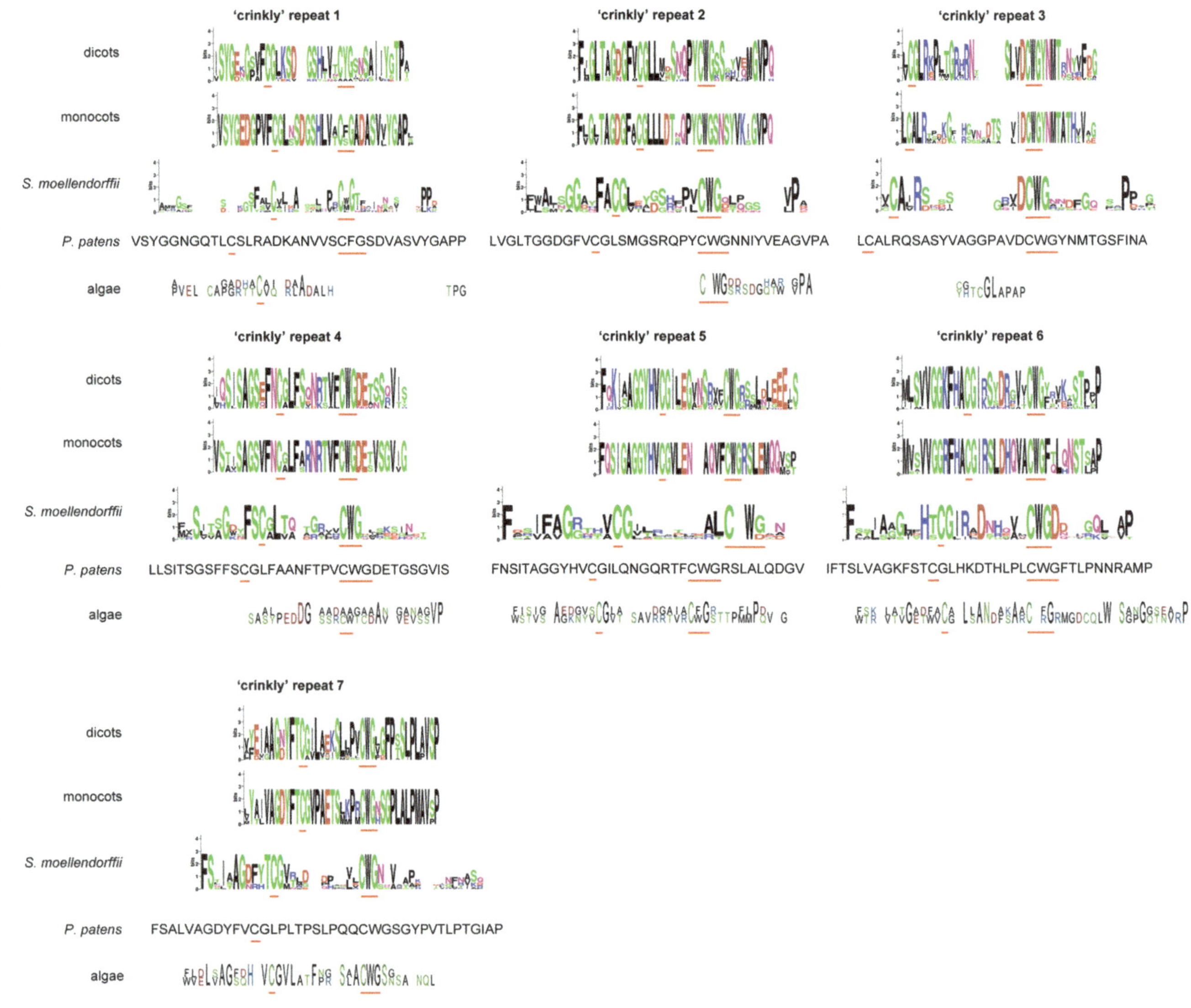

FIGURE 3 | "Crinkly" repeats sequences of (putative) CR4 family orthologs. WebLogo or sequence of "crinkly" repeats in dicots (*A. thaliana*, *Populus trichocarpa*, *Cucumis sativus*, *Medicago truncatula*, *Glycine max*, *Vitis vinifera*, *Solanum tuberosum*), monocots (*Sorghum bicolor*, *Zea mays*, *Hordeum vulgare*, *Triticum aestivum*, *Orysa sativa*), the bryophyte *Physcomitrella patens*, the lycophyte *Selaginella moellendorffii*, and algae (*Chlorella variabilis* and *Micromonas* sp. RCC299). Red lines indicate conserved C(X~10)CWG motif.

the protein–protein interactions formed by similarly structured proteins—one would expect that the β-propeller formed by the extracellular part would facilitate binding of a ligand, which is presumably of peptide/protein origin (McCarty and Chory, 2000; Gifford et al., 2005) and which would result in the activation of the intracellular kinase domain.

Beta-propellers are tertiary structures formed from a number of tandem repeat domains, forming a tilted disc-like structure similar to that of propeller blades (Chaudhuri et al., 2008). These blade domains are mostly thought to originate from single-bladed ancestor proteins in a number of both pro- and eukaryotic organisms (Kopec and Lupas, 2013). These ancestor proteins were subsequently duplicated multiple times, and further tandem repeats of the blade domain were incorporated, resulting in a functionally diverse range of proteins containing β-propeller domains, with varying numbers of blades (Fülöp and Jones, 1999). Usually, each blade consists of four β-sheets which are arranged antiparallel to each other in a "β-meander," forming a triangular structure (Baker et al., 1997; Chaudhuri et al., 2008). The blades are themselves arranged in a circular structure, and are slightly tilted, forming propeller-like structures containing between 4 and 10 blades (Chen et al., 2011). In order to form the final propeller structure, β-sheets at N- and C-termini interact with each other, to form a closed circle (Baker et al., 1997; Chaudhuri et al., 2008). A common configuration of closing is the 3+1 structure where one β-sheet is present at the C-terminus, and forms the innermost part of the final blade, and three sheets are contributed by the N-terminus, forming a

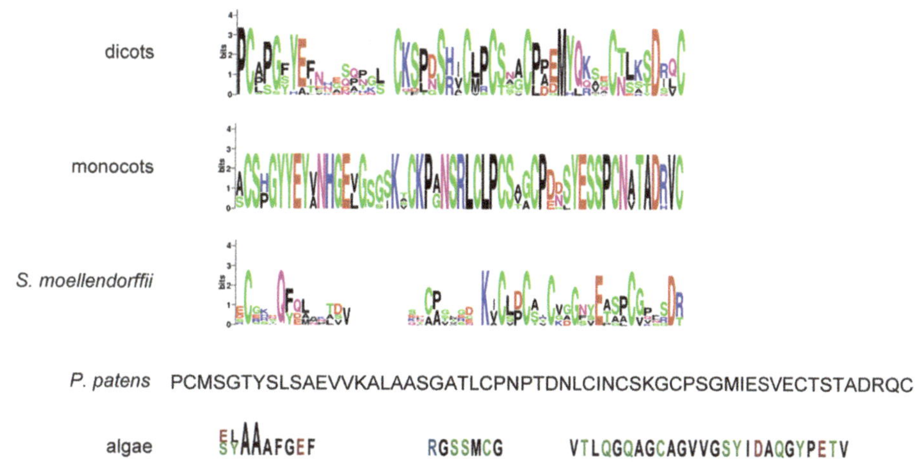

FIGURE 4 | TNFR domain sequences of (putative) CR4 family orthologs. WebLogo or sequence of TNFR domain in dicots (*A. thaliana*, *Populus trichocarpa*, *Cucumis sativus*, *Medicago truncatula*, *Glycine max*, *Vitis vinifera*, *Solanum tuberosum*), monocots (*Sorghum bicolor*, *Zea mays*, *Hordeum vulgare*, *Triticum aestivum*, *Orysa sativa*), the bryophyte *Physcomitrella patens*, the lycophyte *Selaginella moellendorffii*, and algae (*Chlorella variabilis* and *Micromonas* sp. RCC299).

"velcro closing" mechanism by the formation of hydrogen and/or disulfide bonds along the length of two β-sheets (Neer and Smith, 1996; Chaudhuri et al., 2008). There are variations on this theme amongst proteins containing β-propeller domains, namely velcro closing domains can also be formed from 1 + 3 and 2 + 2 β-sheets from N- and C-termini, respectively, and in the case of four-bladed β-propellers, disulfide bridges commonly form between the first and last blade domains, replacing the velcro closing mechanism altogether (Fülöp and Jones, 1999). The resulting propeller structure serves to stabilize the protein, and an array of hydrogen donors and acceptors arranged on the edges of the β-strands simultaneously allows for a diverse range of functions, from protein-protein interactions to enzymatic activity (Neer and Smith, 1996; Baker et al., 1997; Fülöp and Jones, 1999). The active site for the imparted function of β-propeller forming proteins is commonly found in the central "tunnel" region, and substrates are determined by size exclusion and interactions with exposed amino acid side-chains (Neer and Smith, 1996; Baker et al., 1997; Fülöp and Jones, 1999).

CR4-like receptors contain seven copies of a ~39 amino acid repeat that is distantly related to REGULATOR OF CHROMOSOME CONDENSATION 1 (RCC1; Renault et al., 1998). Modeling of the ACR4 extracellular "crinkly" repeats was previously undertaken based on the x-ray structures of two seven-bladed β-propeller folds of the β-LACTAMASE INHIBITORY PROTEIN II) (BLIP-II) from *Streptomyces exfoliates* (Lim et al., 2001) and RCC1 from human (Renault et al., 1998). Molecular modeling studies support the structural homology of the CRINKLY4 domain to the RCC1 family of seven-bladed propeller proteins from three, rather than four, antiparallel β-sheets per blade (McCarty and Chory, 2000; Cao et al., 2005; Gifford et al., 2005). According to the earlier analyses, the regularly spaced Cys residues in the extracellular "crinkly" repeat domain, which are absent in template structures, are held close in space on neighboring antiparallel β-strands within each repeat, and thus, are likely to form stabilizing disulfide bridges which act to stabilize the β-propeller structure in the oxidizing extracellular environment (Gifford et al., 2005).

To gain insight in the 3D structure of the extracellular part of potential CR4 family members, we modeled this part from ACR4 from selected species, such as *A. thaliana, Zea mays, Selaginella moellendorffii, Physcomitrella patens, Chlorella variabilis,* and *Micromonas* sp. RCC299 (Supplementary data set 2). While various methods to predict protein structure exist, including the *ab initio* approach (which predicts the protein structure from the physicochemical parameters based on the protein sequence), or the protein threading approach (which uses a specific energy function to determine and evaluate the fitting between the target sequence and the protein folds in the library; Kelley, 2009; Lee et al., 2009), we employed the structure modeling webserver SWISS-MODEL which uses a comparative homology-based modeling algorithm (Arnold et al., 2006) and does not require a large computational resource and is less time-consuming. Despite the low sequence identity and similarity (Supplementary Table 1), largely due to the unspecific sequence of the loops between the repeats, this method sufficiently provided a first insight in the overall structure of the crinkly repeat domain of CR4 family proteins.

The model structure of the ACR4 extracellular domain was based on the newly available 1.7 Å crystal structure of the plant photoreceptor UV RESISTANCE LOCUS 8 (UVR8) (Christie et al., 2012; also see Materials and Methods). Consistent with the sequence of the protein, the generated ACR4 structure model exhibits an "open" β-propeller fold, similarly to UVR8, with seven structural repeats, each composed of a blade-shaped β-sheet contiguous in sequence (**Figure 5**). The motifs CGX (X = a hydrophobic amino acid, X = L in the first four repeats, X = I for the next two repeats, and V in the last one, the glycine is exchanged to an alanine the fourth repeat) and CWG (with exception in the first repeat where W is exchange by Y, an amino acid with similar properties) are present in each repeat

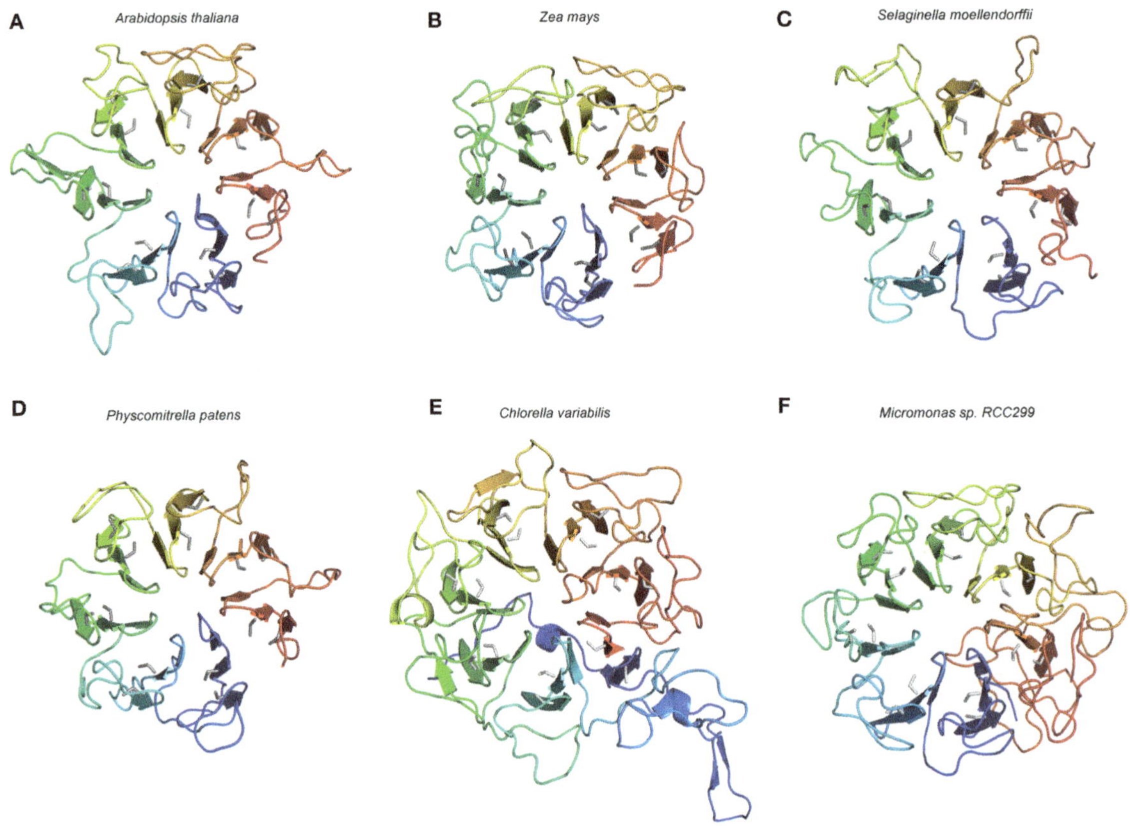

FIGURE 5 | Structural model of the extracellular part of selected putative CR4 family orthologs. Models are shown for the dicot *A. thaliana* **(A)**, the monocot *Zea mays* **(B)**, the lycophyte *Selaginella moellendorffii* **(C)**, the bryophyte *Physcomitrella patens* **(D)**, the single-celled algae *Chlorella variabilis* **(E)**, and *Micromonas* sp. RCC299 **(F)**. The color gradient from blue to red represents the (selected) sequence from N-terminus to C-terminus. The conserved cysteine is colored in gray and the seven repeats of blade-shaped β-sheets are visible.

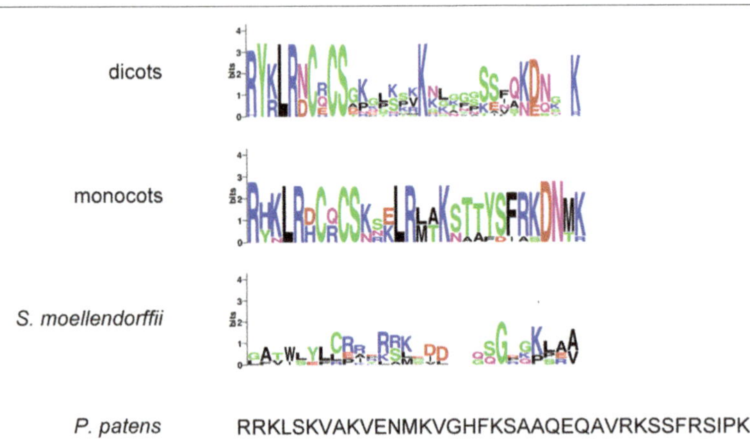

FIGURE 6 | Juxtamembrane domain sequences of (putative) CR4 family orthologs. WebLogo or sequence of TNFR domain in dicots (*A. thaliana*, *Populus trichocarpa*, *Cucumis sativus*, *Medicago truncatula*, *Glycine max*, *Vitis vinifera*, *Solanum tuberosum*), monocots (*Sorghum bicolor*, *Zea mays*, *Hordeum vulgare*, *Triticum aestivum*, *Orysa sativa*), the bryophyte *Physcomitrella patens*, the lycophyte *Selaginella moellendorffii*, and algae (*Chlorella variabilis* and *Micromonas* sp. RCC299).

(**Figure 3**). These motifs are also found in all the β-bladed repeats of all putative CR4 family orthologs, while not being consistently found in protein structure templates used for the modeling (**Figure 3**). The predicted disulfide bridges are restrained in the modeled structure. However, the approximate location of the conserved cysteine residues indicates the formation of a disulfide bond, which can be crucial for the folding of the protein and implies the existence of these "crinkly" repeats in putative CR4 family orthologs. An extended sequence in the third repeat of ACR4 is not covered by any similarities with the sequence template used for the modeling and may form an additional loop. Considering their exposed position on the protein surface, the variation in length and composition of these loops across the species may reflect the diversity of the input signaling molecules and may have an impact on the function of the protein. Overall, the structural core elements of *A. thaliana, Z. mays, S. moellendorffii,* and *P. patens* were very similar and suggested a functional extracellular domain (**Figures 5A–D**).

In the potential CR4 family ortholog from *Chlorella variabilis,* six crinkly repeats with conserved Cys residues are found in the structural model of the retrieved full length protein. Interestingly, a seventh motif with CGV and CWG sub-motifs is found at both ends of the protein sequence, at the N- and C-terminus, respectively. Similar to its template, in this case the human RCC1, the structural model of *C. variabilis* CR4 displays a distinct "velcro closing," in contrast to other potential CR4 family orthologs that are lacking this feature and represent "open" propeller structures. The structure model also displays longer linking loops between repeats with the greatest length of >30 amino acids (in four loops), compared to 28 (in one loop) in the ACR4 model (**Figure 5E**). This may indicate distinct structural dynamics and functions of this protein, and since this protein lacks a kinase domain it was not retained as a member of the CR4 family. The potential CR4 family ortholog from *Micromonas* sp. RCC299, of which the model is based on the W285A variant of UVR8, displayed the consistent cysteine-rich motifs and can be considered as the intermediate case between higher plant CR4 and *S. moellendorffii* CR4 (**Figure 5F**). While representing the seven "crinkly" repeats element and the open propeller structure similarly to the orthologs in higher plant species, the protein is predicted to lack a transmembrane and a kinase domain, and was not retained as a member of the CR4 family.

Signal Peptide, and Transmembrane, Juxtamembrane, and Kinase Domains in Putative CR4 Family Members

In addition to the extracellular domain, CR4 family members hold other features that can be used to identify them. We used SignalP 4.1 (Petersen et al., 2011), PSORT (Nakai and Horton, 1999), TMHMM 2.0 (Sonnhammer et al., 1998; Krogh et al., 2001), and the Pfam database (Finn et al., 2014), in addition to sequence alignment, to determine the presence of a signal peptide and transmembrane, juxtamembrane, and kinase domains in putative CR4 family members (Supplementary Table 3).

First, all *A. thaliana* CR4 family members have an N-terminal signal peptide. Commonly signal peptides are 15–30 amino acids long and cleaved off during translocation of a protein across the membrane. The presence of a signal peptide might thus suggest that a protein is targeted to the endoplasmic reticulum and eventually destined to be either secreted, retained in the lumen of the endoplasmic reticulum, lysosome, or any other organelle along the secretory pathway, or to be a single-pass membrane protein (Izard and Kendall, 1994; Mackenzie, 2005). The SignalP 4.1 analysis gave ambiguous results, as for many (potential) CR4 family protein sequences, including *A. thaliana* CCR4, the signal peptide was not found (Supplementary Table 3). Therefore, we also analyzed sequences based on the alignment with annotated *A. thaliana* sequences. This approach allowed us to identify a signal peptide for almost all putative CR4 family members (Supplementary Table 3).

Second, the highly conserved "KD/ESAF" motif from the juxtamembrane domain within the CR4 family intracellular part is important, for example for protein–protein interaction between family members (Meyer et al., 2015). Our data showed that this motif was not very conserved within species (**Figure 6**). The total length of this motif in *A. thaliana* is five amino acids, in other dicots nine amino acids and in all monocots seven amino acids. Specifically, only the Lys is highly conserved among both dicots and monocots, the Ala in the ACR4 motif is substituted by Ser in other species, and the Phe is highly conserved only in monocots. In algae, *P. patens* and *S. moellendorffii* a "KD/ESAF" motif could not be detected. The juxtamembrane domain itself is highly conserved in monocots (30 aa), has some variation in dicots (28–32 aa), and is not very conserved in *P. patens* and *S. moellendorffii* (**Figure 6** and Supplementary Table 3).

Third, another important feature of CR4 family members is the presence of a 21–26 amino acids long (in *A. thaliana*) transmembrane domain. Our theoretical data based on the TMHMM Server v. 2.0 prediction tool corresponded to the experimental data retrieved from sequence alignments in terms of positioning and length of the domain. In general, transmembrane domains of (potential) CR4 family members consist of 18–30 amino acids. The amino acid composition of transmembrane domains within the CR4 family is different. For example, the transmembrane domain of CCR1, CCR2, and CCR3 subgroups contained 1–2 Cys residue, while no Cys residues were found in CR4 and CCR4. Transmembrane domains are relatively conserved in monocots, dicots, *P. patens* and *S. moellendorffii* but absent in protein sequences of all algae species (*Micromonas sp., Micromonas pusilla,* and *Chlorella variabilis*; Supplementary Table 3).

Finally, kinase activity of some members of the CR4 family was demonstrated *in vitro*, and while one report showed that the kinase domain is not essential for ACR4 function, there are other reports suggesting otherwise (Gifford, 2003; Watanabe et al., 2004; Cao et al., 2005; Gifford et al., 2005; Meyer et al., 2011, 2015; Pu and Sun, 2012; Pu et al., 2012). Based on prediction tools and alignment analyses, it could be shown that in most cases a kinase domain was present in the putative CR4 family members. However, almost all candidates from *S. moellendorffii* had no predicted kinase domain except for two protein sequences belonging to the CCR1 clade. Strikingly, a kinase domain was also absent in protein sequences of algae species (*Micromonas sp.* RCC299, *Micromonas pusilla,* and *Chlorella variabilis).*

CR4 Family Members

Taken together, the above-described analyses allowed us to further refine the CR4 family with likely functional orthologs. Overall, land plants contain at least one member of this family

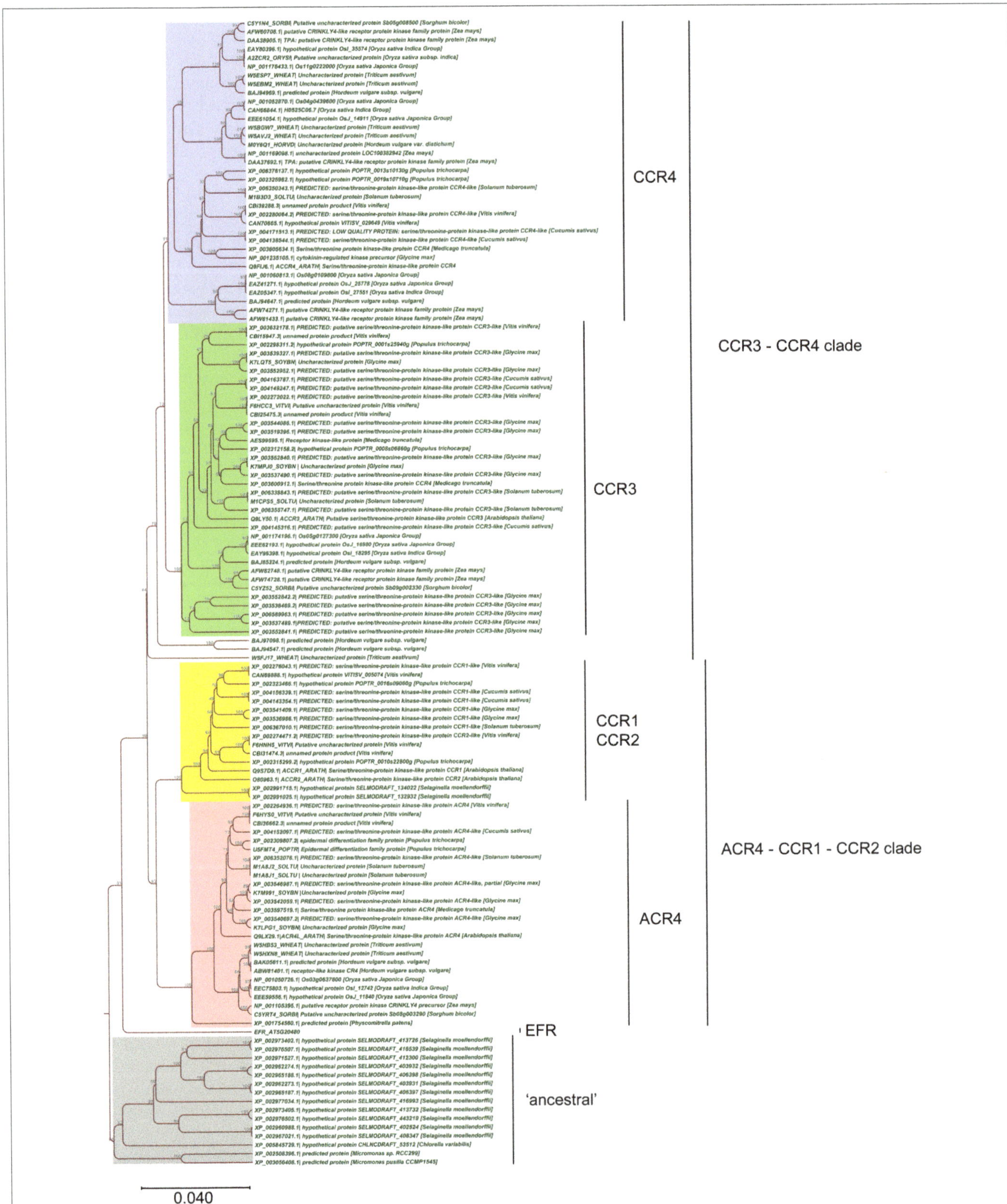

FIGURE 7 | Phylogenetic tree of CR4 family. The CR4 family clades are indicated, in addition to the EFR outgroup and possibly proteins "ancestral" to the CR4 family.

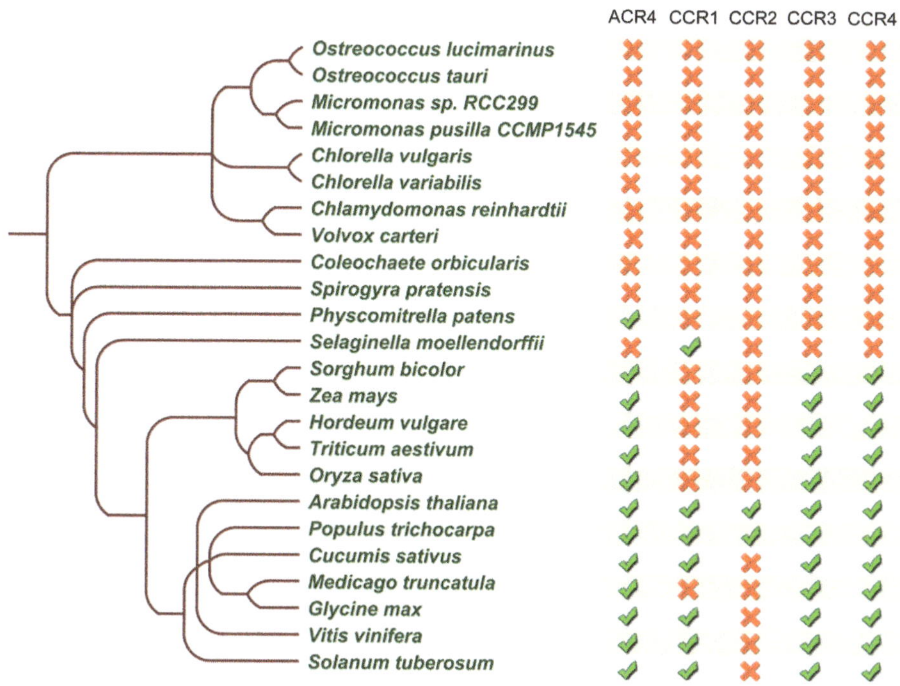

FIGURE 8 | Summary of CR4 family occurrence in the green lineage. Table built based on the results after reciprocal BLASTP searches and domain analysis of CR4 family members within selected species. The presence (✔) or absence (✖) of a likely functional ortholog is indicated.

(**Figures** 7, **8**), supporting their important role in plant growth and development. Combining the data resulted in a CR4 family tree that consists of two main clades similar to the *A. thaliana* tree. The first clade includes CR4, CCR1, and CCR2 orthologs and the second one includes CCR3 and CCR4 family members. A third, likely evolutionary more ancient one, contains algae and *S. moellendorffii* sequences, mainly potential orthologs of ACR4 and ACCR1 that, however, lack transmembrane and/or kinase domains (**Figure** 7). For the moss *P. patens* only one protein was found as a true CR4 family member containing all of characteristic domains. At the same time the lycophyte *S. moellendorffii* is represented by two candidates in the ACR4-CCR1-CCR2 clade. Finally, no sequences from algal species were retained based on the necessary presence of all CR4 family characteristics.

Based on the obtained data we concluded that although domains resembling parts of CR4 family members, such as the extracellular part, are present in the investigated non-land plants, these proteins did not appear to act as a receptor kinase on their own. However, due to the lack of complete genome information for multicellular algae, such as *S. pratensis* and *C. orbicularis*, we cannot rule out that CR4 family members are present in those species.

Evolutionary Insight into the CLE40-ACR4-CLV1-WOX5 Module

While ACR4 has been shown to play important roles in several processes (Becraft et al., 1996; Tanaka et al., 2002; Gifford, 2003; Watanabe et al., 2004; De Smet et al., 2008), the functional characterization of other members of the CR4 family in *A. thaliana* has been hardly addressed, and their characterization is largely limited to a redundant role in lateral root development (De Smet et al., 2008). Given that several receptor kinases—including ACR4—are involved in a multitude of physiological and developmental processes (Shiu and Bleecker, 2001a; De Smet et al., 2009; Stahl and Simon, 2012), it is likely that CR4 family members also play a broader role. To pinpoint new roles for *Arabidopsis* CR4 family members, we probed available micro-array data using Genevestigator (Zimmermann et al., 2004). Based on these results, only *ACR4* expression was found as differentially regulated during the first stages of lateral root initiation (**Figure 9A**; De Smet et al., 2008). This analysis further revealed that *ACR4* (AT3G59420) and *CCR1* (AT3G09780) are significantly up-regulated during the early stages of seed germination (**Figure 9A**). In addition, this meta-analysis also revealed a strong regulation of *CR4* family members, specifically *CCR3* and *CCR4*, upon biotic stress and elicitor treatment (**Figure 9B**). A connection for CR4 family members with biotic stress is further supported by an analysis of the response of *acr4* to *Botrytis cinerea*, which showed that *acr4* has reduced susceptibility to this necrotrophic pathogen (E-Zereen and Ingram, 2012). Furthermore, other environmental and endogenous triggers appeared to regulate expression levels of *CR4* family members in *A. thaliana* (Supplementary Table 4). This information provides interesting leads for future detailed functional characterization.

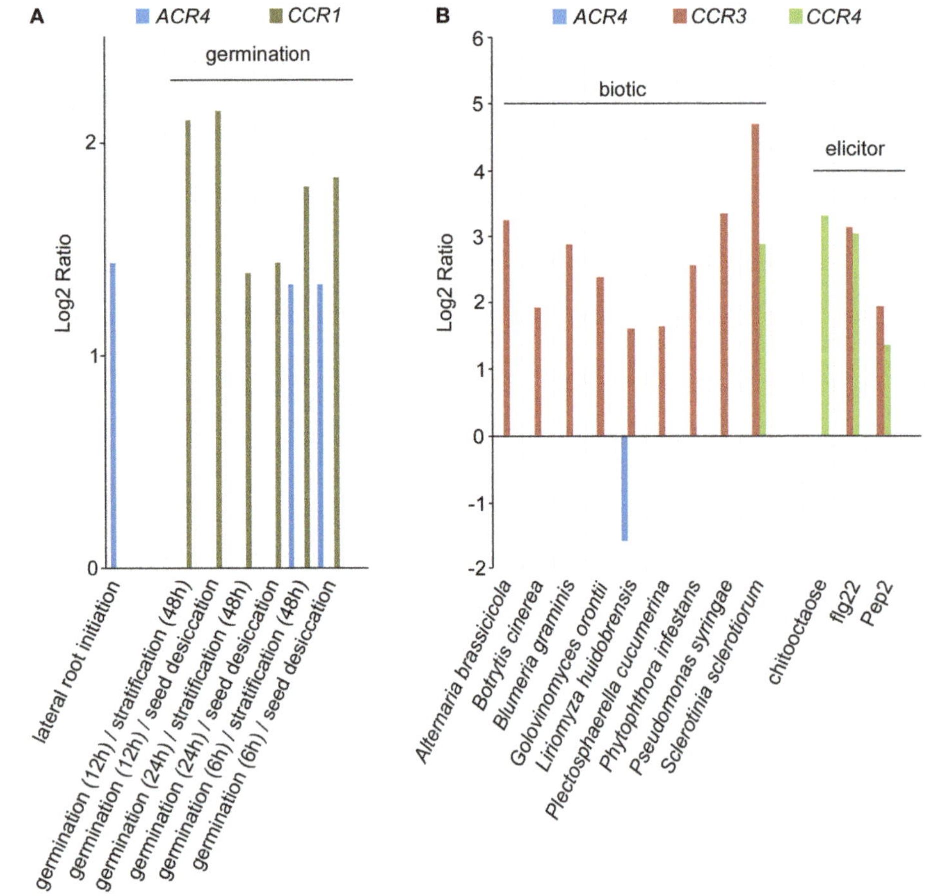

FIGURE 9 | Expression analysis of *A. thaliana CR4* family. (A,B) Summary of selected Genevestigator data (Log2 ratio) with respect to lateral root initiation and germination **(A)** and biotic stress and elicitor treatment **(B)** of *CR4* family members in *A. thaliana*. $P < 0.01$. Details on experiments in Supplementary Table 4. Pep2, Elicitor peptide 2; flg22, 22-amino acid sequence of the conserved N-terminal part of flagellin.

In addition to these observations, our mechanistic understanding of ACR4 action is mainly increasing through the analysis of the CLE40-ACR4-CLV1-WOX5 module in the *A. thaliana* root tip (Stahl et al., 2009, 2013; Meyer et al., 2015; Richards et al., 2015). Therefore, we probed if the other components of this module are also conserved. We performed extensive BLASTP searches and reciprocal BLASTP analyses on the same 23 species included in the CR4 family analysis using *A. thaliana* WOX5, CLV1, and CLE40 sequences as input sequences. This revealed that all components of this signaling module (or closely related proteins) could be identified in *P. patens*, *S. moellendorffii* and almost all higher plant species except for *Triticum aestivum* and *Solanum tuberosum*. With respect to the latter two, this might be due to incomplete genome sequence information and/or the function of missing components might be replaced by another protein or peptide (**Figure 10** and Supplementary Table 5). It is noteworthy that some of the components of this module are present in algae species, but no CLE40 orthologs were identified in these algae species. Taken together, a complete CLE40-ACR4-CLV1-WOX5 module could already be present in *P. patens*, and this can be explored in future.

CONCLUSION

Recent studies showed that in monocot and dicot plants, such as *Arabidopsis*, rice, and maize, CR4 receptor kinases play important roles (Becraft et al., 1996; Jin et al., 2000; Tanaka et al., 2002; Gifford, 2003; Watanabe et al., 2004; De Smet et al., 2008; Pu et al., 2012). Given how important CR4 receptor kinases are for root growth, floral organ formation, and seed development, our detailed analysis of this family in a wider range of (crop) plants, could impact on crop improvement. In addition, our results suggest that the origin of the CR4 family coincides with the colonization of terrestrial habitats, as the first CR4 family members with all key domains appear in *P. patens* and *S. moellendorffii*. In future, the function in these species will have to be investigated in detail to reveal if CR4 family members already play a key role in regulating e.g., formative cell divisions. On the other hand, in algae we found ancestral forms of CR4 family members, namely without kinase domain, which could indicate a different role of these proteins.

It is important to note that incomplete genome annotation in some species might explain the absence of some key players. We expect that continuous improvement in genome sequencing

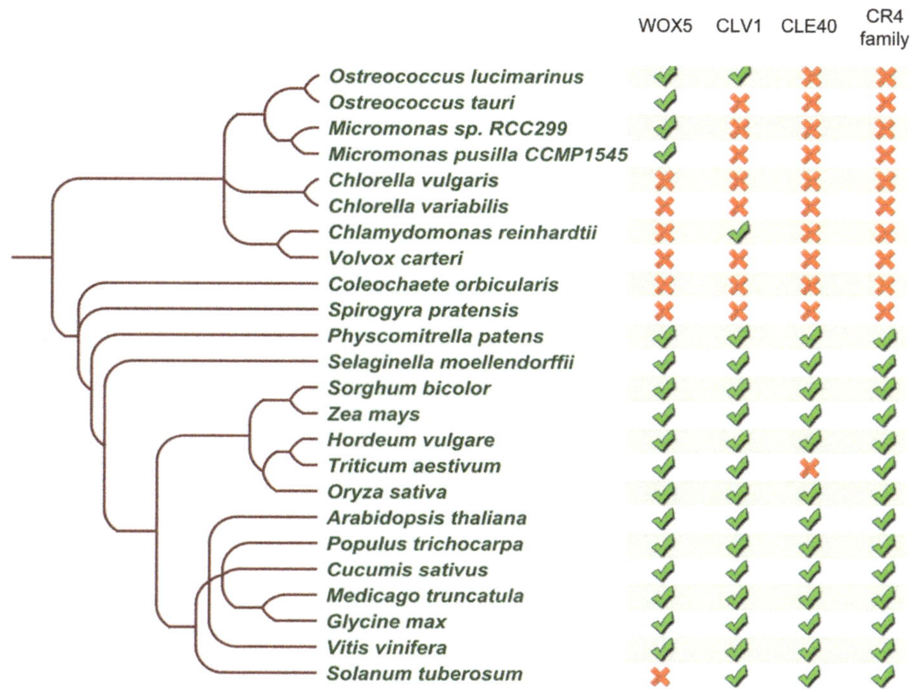

FIGURE 10 | Summary of CLE40-ACR4-CLV1-WOX5 module occurrence in the green lineage. Table built based on the results after reciprocal BLASTP searches for WOX5, CLE40, and CLV1 (no further analyses to identify true orthologs), and on a collapse of the CR4 family members. The presence (✔) or absence (✖) of a likely functional ortholog is indicated. Details can be found in Supplementary Tables 2, 5.

and gene/protein annotation will, in future, further increase our insight in the CR4 family and the associated signaling module. In this context, as this is a bio-informatics study, it remains essential to perform functional analyses of the orthologs proposed in this study in order to correctly annotate them and to avoid mis-annotations or over-confident annotations.

ACKNOWLEDGMENTS

We thank Remco Sprangers for useful discussions. This work was supported by a BBSRC David Phillips Fellowship (BB/H022457/1) and a Marie Curie European Reintegration Grant (PERG06-GA-2009-256354) to ID, and a BBSRC CASE Studentship co-funded by Bayer CropScience to NC (BB/J012726/1). LV received a VIB International Ph.D. Scholarship in life sciences.

REFERENCES

Arnold, K., Bordoli, L., Kopp, J., and Schwede, T. (2006). The SWISS-MODEL workspace: a web-based environment for protein structure homology modelling. *Bioinformatics* 22, 195–201. doi: 10.1093/bioinformatics/bti770

Baker, S. C., Saunders, N. F. W., Willis, A. C., Ferguson, S. J., Hajdu, J., and Fülöp, V. (1997). Cytochrome cd_1 Structure: unusual haem environments in a nitrite reductase and analysis of factors contributing to β-propeller folds. *J. Mol. Biol.* 269, 440–455. doi: 10.1006/jmbi.1997.1070

Becraft, P. W., Kang, S. H., and Suh, S. G. (2001). The maize CRINKLY4 receptor kinase controls a cell-autonomous differentiation response. *Plant Physiol.* 127, 486–496. doi: 10.1104/pp.010299

Becraft, P. W., Stinard, P. S., and McCarty, D. R. (1996). CRINKLY4: a TNFR-like receptor kinase involved in maize epidermal differentiation. *Science* 273, 1406–1409. doi: 10.1126/science.273.5280.1406

Cao, X., Li, K., Suh, S.-G., Guo, T., and Becraft, P. W. (2005). Molecular analysis of the CRINKLY4 gene family in *Arabidopsis thaliana*. *Planta* 220, 645–657. doi: 10.1007/s00425-004-1378-3

Chaudhuri, I., Söding, J., and Lupas, A. N. (2008). Evolution of the β-propeller fold. *Proteins* 71, 795–803. doi: 10.1002/prot.21764

Chen, C. K.-M., Chan, N.-L., and Wang, A. H.-J. (2011). The many blades of the β-propeller proteins: conserved but versatile. *Trends Biochem. Sci.* 36, 553–561. doi: 10.1016/j.tibs.2011.07.004

Christie, J. M., Arvai, A. S., Baxter, K. J., Heilmann, M., Pratt, A. J., O'Hara, A., et al. (2012). Plant UVR8 photoreceptor senses UV-B by tryptophan-mediated disruption of cross-dimer salt bridges. *Science* 335, 1492–1496. doi: 10.1126/science.1218091

Crooks, G. E. (2004). WebLogo: a sequence logo generator. *Genome Res.* 14, 1188–1190. doi: 10.1101/gr,.849004

De Smet, I., Vassileva, V., De Rybel, B., Levesque, M. P., Grunewald, W., van Damme, D., et al. (2008). Receptor-like kinase ACR4 restricts formative cell divisions in the *Arabidopsis* root. *Science* 322, 594–597. doi: 10.1126/science.1160158

De Smet, I., Voss, U., Jürgens, G., and Beeckman, T. (2009). Receptor-like kinases shape the plant. *Nat. Cell Biol.* 11, 1166–1173. doi: 10.1038/ncb1009-1166

E-Zereen, J., and Ingram, G. (2012). A possible involvement of ACR4, a receptor like kinase, in plant defence mechanism. *Bangladesh Pharm. J.* 15, 127–130. doi: 10.3329/bpj.v15i2.12576

Feng, D. F., and Doolittle, R. F. (1987). Progressive sequence alignment as a prerequisite to correct phylogenetic trees. *J. Mol. Evol.* 25, 351–360. doi: 10.1007/BF02603120

Finn, R. D., Bateman, A., Clements, J., Coggill, P., Eberhardt, R. Y., Eddy, S. R., et al. (2014). Pfam: the protein families database. *Nucleic Acids Res.* 42, 222–230. doi: 10.1093/nar/gkt1223

Fülöp, V., and Jones, D. T. (1999). β propellers: structural rigidity and functional diversity. *Curr. Opin. Struct. Biol.* 9, 715–721. doi: 10.1016/S0959-440X(99)00035-4

Gifford, M. L. (2003). The *Arabidopsis* ACR4 gene plays a role in cell layer organisation during ovule integument and sepal margin development. *Development* 130, 4249–4258. doi: 10.1242/dev.00634

Gifford, M. L., Robertson, F. C., Soares, D. C., and Ingram, G. C. (2005). *ARABIDOPSIS* CRINKLY4 function, internalization, and turnover are dependent on the extracellular crinkly repeat domain. *Plant Cell* 17, 1154–1166. doi: 10.1105/tpc.104.029975

Goodstein, D. M., Shu, S., Howson, R., Neupane, R., Hayes, R. D., Fazo, J., et al. (2012). Phytozome: a comparative platform for green plant genomics. *Nucleic Acids Res.* 40, 1178–1186. doi: 10.1093/nar/gkr944

Guex, N., Peitsch, M. C., and Schwede, T. (2009). Automated comparative protein structure modeling with SWISS-MODEL and Swiss-PdbViewer: a historical perspective. *Electrophoresis* 30, S162–S173. doi: 10.1002/elps.200900140

Han, X., Kumar, D., Chen, H., Wu, S., and Kim, J.-Y. (2014). Transcription factor-mediated cell-to-cell signalling in plants. *J. Exp. Bot.* 65, 1737–1749. doi: 10.1093/jxb/ert422

Hruz, T., Laule, O., Szabo, G., Wessendorp, F., Bleuler, S., Oertle, L., et al. (2008). Genevestigator V3: a reference expression database for the meta-analysis of transcriptomes. *Adv. Bioinformatics* 2008:420747. doi: 10.1155/2008/420747

Izard, J. W., and Kendall, D. A. (1994). Signal peptides: exquisitely designed transport promoters. *Mol. Microbiol.* 13, 765–773. doi: 10.1111/j.1365-2958.1994.tb00469.x

Jin, P., Guo, T., and Becraft, P. W. (2000). The maize CR4 receptor-like kinase mediates a growth factor-like differentiation response. *Genesis* 27, 104–116. doi: 10.1002/1526-968X(200007)27:3<104::AID-GENE30>3.0.CO;2-I

Kang, S.-G., Lee, H. J., and Suh, S.-G. (2002). The maize Crinkly4 gene is expressed spatially in vegetative and floral organs. *J. Plant Biol.* 45, 219–224. doi: 10.1007/BF03030363

Kelley, L. A. (2009). "Fold recognition," in *From Protein Structure to Function with Bioinformatics* (Dordrecht: Springer), 27–55.

Kersey, P. J., Allen, J. E., Christensen, M., Davis, P., Falin, L. J., Grabmueller, C., et al. (2014). Ensembl Genomes 2013: scaling up access to genome-wide data. *Nucleic Acids Res.* 42, 546–552. doi: 10.1093/nar/gkt979

Kiefer, F., Arnold, K., Künzli, M., Bordoli, L., and Schwede, T. (2009). The SWISS-MODEL repository and associated resources. *Nucleic Acids Res.* 37, D387–D392. doi: 10.1093/nar/gkn750

Kopec, K. O., and Lupas, A. N. (2013). β-Propeller blades as ancestral peptides in protein evolution. *PLoS ONE* 8:e77074. doi: 10.1371/journal.pone.0077074

Krogh, A., Larsson, B., von Heijne, G., and Sonnhammer, E. L. (2001). Predicting transmembrane protein topology with a hidden markov model: application to complete genomes. *J. Mol. Biol.* 305, 567–580. doi: 10.1006/jmbi.2000.4315

Lee, D., Redfern, O., and Orengo, C. (2007). Predicting protein function from sequence and structure. *Nat. Rev. Mol. Cell Biol.* 8, 995–1005. doi: 10.1038/nrm2281

Lee, J., Wu, S., and Zhang, Y. (2009). "*Ab initio* protein structure prediction," in *From Protein Structure to Function with Bioinformatics*, ed D. J. Rigden (Dordrecht: Springer Netherlands), 3–25.

Lehti-Shiu, M. D., Zou, C., Hanada, K., and Shiu, S.-H. (2009). Evolutionary history and stress regulation of plant receptor-like kinase/pelle genes. *Plant Physiol.* 150, 12–26. doi: 10.1104/pp.108.134353

Lim, D., Park, H. U., De Castro, L., Kang, S. G., Lee, H. S., Jensen, S., et al. (2001). Crystal structure and kinetic analysis of beta-lactamase inhibitor protein-II in complex with TEM-1 beta-lactamase. *Nat. Struct. Biol.* 8, 848–852. doi: 10.1038/nsb1001-848

Liu, P., Wei, W., Ouyang, S., Zhang, J.-S., Chen, S.-Y., and Zhang, W.-K. (2009). Analysis of expressed receptor-like kinases (RLKs) in soybean. *J. Genet. Genomics* 36, 611–619. doi: 10.1016/S1673-8527(08)60153-8

Mackenzie, S. A. (2005). Plant organellar protein targeting: a traffic plan still under construction. *Trends Cell Biol.* 15, 548–554. doi: 10.1016/j.tcb.2005.08.007

McCarty, D. R., and Chory, J. (2000). Conservation and innovation in plant signaling pathways. *Cell* 103, 201–209. doi: 10.1016/S0092-8674(00)00113-6

Meyer, M. R., Lichti, C. F., Townsend, R. R., and Rao, A. G. (2011). Identification of *in vitro* autophosphorylation sites and effects of phosphorylation on the Arabidopsis CRINKLY4 (ACR4) receptor-like kinase intracellular domain: insights into conformation, oligomerization, and activity. *Biochemistry* 50, 2170–2186. doi: 10.1021/bi101935x

Meyer, M. R., Shah, S., and Rao, A. G. (2013). Insights into molecular interactions between the juxtamembrane and kinase subdomains of the *Arabidopsis* Crinkly-4 receptor-like kinase. *Arch. Biochem. Biophys.* 535, 101–110. doi: 10.1016/j.abb.2013.03.014

Meyer, M. R., Shah, S., Zhang, J., Rohrs, H., and Rao, A. G. (2015). Evidence for intermolecular interactions between the intracellular domains of the *Arabidopsis* receptor-like kinase ACR4, its homologs and the WOX5 transcription factor. *PLoS ONE* 10:e0118861. doi: 10.1371/journal.pone.0118861

Murphy, E., Smith, S., and De Smet, I. (2012). Small signaling peptides in *Arabidopsis* development: how cells communicate over a short distance. *Plant Cell* 24, 3198–3217. doi: 10.1105/tpc.112.099010

Nakai, K., and Horton, P. (1999). PSORT: a program for detecting sorting signals in proteins and predicting their subcellular localization. *Trends Biochem. Sci.* 24, 34–35. doi: 10.1016/S0968-0004(98)01336-X

Neer, E. J., and Smith, T. F. (1996). G protein heterodimers: new structures propel new questions. *Cell* 84, 175–178. doi: 10.1016/S0092-8674(00)80969-1

Pacifici, E., Polverari, L., and Sabatini, S. (2015). Plant hormone cross-talk: the pivot of root growth. *J. Exp. Bot.* 66, 1113–1121. doi: 10.1093/jxb/eru534

Petersen, T. N., Brunak, S., von Heijne, G., and Nielsen, H. (2011). SignalP 4.0: discriminating signal peptides from transmembrane regions. *Nat. Methods* 8, 785–786. doi: 10.1038/nmeth.1701

Proost, S., Van Bel, M., Sterck, L., Billiau, K., Van Parys, T., Van de Peer, Y., et al. (2009). PLAZA: a comparative genomics resource to study gene and genome evolution in plants. *Plant Cell* 21, 3718–3731. doi: 10.1105/tpc.109.071506

Pu, C.-X., Ma, Y., Wang, J., Zhang, Y.-C., Jiao, X.-W., Hu, Y.-H., et al. (2012). Crinkly4 receptor-like kinase is required to maintain the interlocking of the palea and lemma, and fertility in rice, by promoting epidermal cell differentiation. *Plant J.* 70, 940–953. doi: 10.1111/j.1365-313X.2012.04925.x

Pu, C.-X., and Sun, Y. (2012). Rice Crinkly4 receptor-like kinase positively regulates culm elongation and amino acid K532 is not essential for its kinase activity. *Plant Signal. Behav.* 7, 1062–1064. doi: 10.4161/psb.21106

Renault, L., Nassar, N., Vetter, I., Becker, J., Klebe, C., Roth, M., et al. (1998). The 1.7 A crystal structure of the regulator of chromosome condensation (RCC1) reveals a seven-bladed propeller. *Nature* 392, 97–101. doi: 10.1038/32204

Richards, S., Wink, R. H., and Simon, R. (2015). Mathematical modelling of WOX5- and CLE40-mediated columella stem cell homeostasis in *Arabidopsis*. *J. Exp. Bot.* 66, 5375–5384. doi: 10.1093/jxb/erv257

Scheres, B. (2007). Stem-cell niches: nursery rhymes across kingdoms. *Nat. Rev. Mol. Cell Biol.* 8, 345–354. doi: 10.1038/nrm2164

Shiu, S. H., and Bleecker, A. B. (2001a). Plant receptor-like kinase gene family: diversity, function, and signaling. *Sci. STKE* 2001:re22. doi: 10.1126/stke.2001.113.re22

Shiu, S. H., and Bleecker, A. B. (2001b). Receptor-like kinases from *Arabidopsis* form a monophyletic gene family related to animal receptor kinases. *Proc. Natl. Acad. Sci. U.S.A.* 98, 10763–10768. doi: 10.1073/pnas.181141598

Shiu, S. H., Karlowski, W. M., Pan, R., Tzeng, Y. H., Mayer, K. F. X., and Li, W. (2004). Comparative analysis of the receptor-like kinase family in *Arabidopsis* and rice. *Plant Cell* 16, 1220–1234. doi: 10.1105/tpc.020834

Sonnhammer, E. L. L., von Heijne, G., and Krogh, A. (1998). "A hidden Markov model for predicting transmembrane helices in protein sequences," in *Proceedings of the Sixth International Conference on Intelligent Systems for Molecular Biology*, eds J. Glasgow, T. Littlejohn, F. Major, R. Lathrop, D. Sankoff, and C. Sensen (Menlo Park, CA: AAAI Press), 175–182.

Stahl, Y., Grabowski, S., Bleckmann, A., Kühnemuth, R., Weidtkamp-peters, S., Pinto, K. G., et al. (2013). Moderation of *Arabidopsis* root stemness by CLAVATA1 and *ARABIDOPSIS* CRINKLY4 receptor kinase complexes. *Curr. Biol.* 23, 362–371. doi: 10.1016/j.cub.2013.01.045

Stahl, Y., and Simon, R. (2012). "Receptor-like kinases in plants," in *Receptor-like Kinases in Plants, From Development to Defense*, eds F. Tax and B. Kemmerling (Berlin; Heidelberg: Springer Berlin Heidelberg), 23–39.

Stahl, Y., Wink, R. H., Ingram, G. C., and Simon, R. (2009). A signaling module controlling the stem cell niche in *Arabidopsis* root meristems. *Curr. Biol.* 19, 909–914. doi: 10.1016/j.cub.2009.03.060

Tanaka, H., Watanabe, M., Watanabe, D., Tanaka, T., Machida, C., and Machida, Y. (2002). ACR4, a putative receptor kinase gene of *Arabidopsis thaliana*, that is expressed in the outer cell layers of embryos and plants, is involved in proper embryogenesis. *Plant Cell Physiol.* 43, 419–428. doi: 10.1093/pcp/pcf052

ten Hove, C. A., Lu, K.-J., and Weijers, D. (2015). Building a plant: cell fate specification in the early *Arabidopsis* embryo. *Development* 142, 420–430. doi: 10.1242/dev.111500

Tian, Q., Olsen, L., Sun, B., Lid, S. E., Brown, R. C., Lemmon, B. E., et al. (2007). Subcellular localization and functional domain studies of DEFECTIVE KERNEL1 in maize and *Arabidopsis* suggest a model for aleurone cell fate specification involving CRINKLY4 and SUPERNUMERARY ALEURONE LAYER1. *Plant Cell* 19, 3127–3145. doi: 10.1105/tpc.106.048868

Watanabe, M., Tanaka, H., Watanabe, D., Machida, C., Machida, Y., and Agency, T. (2004). The ACR4 receptor-like kinase is required for surface formation of epidermis-related tissues in *Arabidopsis thaliana*. *Plant J.* 39, 298–308. doi: 10.1111/j.1365-313X.2004.02132.x

Wierzba, M. P., and Tax, F. E. (2013). Notes from the underground: receptor-like kinases in *Arabidopsis* root development. *J. Integr. Plant Biol.* 55, 1224–1237. doi: 10.1111/jipb.12088

Zimmermann, P., Hennig, L., and Gruissem, W. (2005). Gene-expression analysis and network discovery using Genevestigator. *Trends Plant Sci.* 10, 407–409. doi: 10.1016/j.tplants.2005.07.003

Zimmermann, P., Hirsch-Hoffmann, M., Hennig, L., and Gruissem, W. (2004). GENEVESTIGATOR. *Arabidopsis* microarray database and analysis toolbox. *Plant Physiol.* 136, 2621–2632. doi: 10.1104/pp.104.046367

Conflict of Interest Statement: The authors declare that the research was conducted in the absence of any commercial or financial relationships that could be construed as a potential conflict of interest.

19

cDNA Library Screening Identifies Protein Interactors Potentially Involved in Non-Telomeric Roles of *Arabidopsis* Telomerase

Ladislav Dokládal[1,2], David Honys[3], Rajiv Rana[3], Lan-Ying Lee[4], Stanton B. Gelvin[4] and Eva Sýkorová[1,2]*

[1] Mendel Centre for Plant Genomics and Proteomics, Central European Institute of Technology and Faculty of Science, Masaryk University, Brno, Czech Republic, [2] Institute of Biophysics – Academy of Sciences of the Czech Republic v.v.i., Brno, Czech Republic, [3] Institute of Experimental Botany – Academy of Sciences of the Czech Republic v.v.i., Prague, Czech Republic, [4] Department of Biological Sciences, Purdue University, West Lafayette, IN, USA

Edited by:
Inna Lermontova,
Leibniz Institute of Plant Genetics and Crop Plant Research, Germany

Reviewed by:
Biswapriya Biswavas Misra,
University of Florida, USA
Chee How Teo,
Taylor's University, Malaysia

****Correspondence:***
Eva Sýkorová
evin@ibp.cz

Telomerase-reverse transcriptase (TERT) plays an essential catalytic role in maintaining telomeres. However, in animal systems telomerase plays additional non-telomeric functional roles. We previously screened an *Arabidopsis* cDNA library for proteins that interact with the C-terminal extension (CTE) TERT domain and identified a nuclear-localized protein that contains an RNA recognition motif (RRM). This RRM-protein forms homodimers in both plants and yeast. Mutation of the gene encoding the RRM-protein had no detectable effect on plant growth and development, nor did it affect telomerase activity or telomere length *in vivo*, suggesting a non-telomeric role for TERT/RRM-protein complexes. The gene encoding the RRM-protein is highly expressed in leaf and reproductive tissues. We further screened an *Arabidopsis* cDNA library for proteins that interact with the RRM-protein and identified five interactors. These proteins are involved in numerous non-telomere-associated cellular activities. In plants, the RRM-protein, both alone and in a complex with its interactors, localizes to nuclear speckles. Transcriptional analyses in wild-type and *rrm* mutant plants, as well as transcriptional co-analyses, suggest that TERT, the RRM-protein, and the RRM-protein interactors may play important roles in non-telomeric cellular functions.

Keywords: telomerase, nuclear poly(A)-binding protein, telobox, metallothionein 2A, MODIFIER OF snc1, putative nuclear DNA-binding protein G2p, oxidation-related zinc finger 2 protein, BiFC

INTRODUCTION

Telomeres are nucleoprotein structures at the ends of eukaryotic chromosomes, distinguishing these ends from double strand DNA breaks (DDBs) and protecting them from the DNA damage repair (DDR) machinery. Due to the "end replication problem" (Olovnikov, 1971; Watson, 1972; Olovnikov, 1973) telomeres are shortened in each round of replication until they are too short to function, leading to cell senescence (Levy et al., 1992) and apoptosis (Harley et al., 1990; Counter et al., 1992). Thus, telomeres limit cellular proliferative capacity and act as a biological "clock." On the other hand, in cells with high proliferative need such as animal embryonic, stem, and cancer cells (reviewed in Blasco, 2005), or plant meristemic cells (Fitzgerald et al., 1996), telomere shortening is compensated by the action of telomerase, a conserved ribonucleoprotein complex with a reverse transcriptase subunit (Greider and Blackburn, 1985, 1987). Telomerase consists of two core subunits, telomerase RNA (TR) and telomerase reverse transcriptase (TERT), that are associated with several additional proteins not crucial for enzymatic activity. The TERT subunit has an evolutionary conserved primary structure which in most organisms can be further divided into N-terminal domains TEN (telomerase essential N-terminal) and TRBD (TR binding domain), a central reverse transcriptase (RT) domain, and a C-terminal extension (CTE; reviewed in Sykorova and Fajkus, 2009).

The fact that telomerase influences cellular life span and plays a role in various types of cancer intensified research in this field. Surprisingly, telomerase in mammalian cells influences tumourigenesis by additional mechanisms independent of telomere synthesis (reviewed in Majerska et al., 2011). These so-called non-telomeric functions of telomerase regulate processes such as apoptosis, cellular proliferation, and cell cycle regulation, generally by altering gene expression, or DDR by *de novo* telomere addition to the sites of DDB.

For the above reasons it is of great interest to study mechanisms and interactions through which telomerase is regulated, and by which telomerase regulates cellular functions other than telomere synthesis.

Telomerase from *Arabidopsis thaliana* represents a suitable model, especially because of the availability of viable T-DNA insertion mutants that are typically exploited in these types of studies. Classically, changes in telomere length and telomerase activity are measured in a particular mutant, which may lead to direct identification of important telomerase regulators. However, this approach may not detect interactors crucial for mediating non-telomeric activities of telomerase. For this purpose, methods such as tandem affinity purification or cDNA library screening may be more suitable.

The N- and C-terminal portions of TERT represent potential interacting targets for telomerase regulatory proteins. The CTE is highly conserved among vertebrates and plants and contains regions important for intracellular trafficking of human TERT, including a nuclear export signal, 14-3-3, and CRM1 binding sites (Seimiya et al., 2000). In our previous work we screened for *Arabidopsis* CTE protein–protein interactions against a cYFP-tagged *Arabidopsis* cDNA library in tobacco BY-2 protoplasts and identified two interacting partners, an armadillo/β-catenin-like repeat containing protein (encoded by At4g33945) interacting with CTE in the cytoplasm, and an RRM-containing protein (encoded by At5g10350; RRM) that interacts with the CTE in nuclei (Lee et al., 2012).

How telomerase executes its non-canonical activities and on which levels it regulates expression of its target genes are poorly understood. One possibility is regulation on the level of mRNA. The RRM protein belongs to a subfamily of *Arabidopsis* nuclear poly(A) binding proteins; that are characterized by a single RRM domain close to the C-terminus (reviewed in Eliseeva et al., 2013). The human nuclear poly(A) binding protein PABPN1 is implicated in a variety of mRNA stabilization and degradation processes, such as stimulation of poly(A) synthesis by poly(A) polymerase, protection of growing poly(A) chains from degradation, defining the length of growing poly(A) chains, and mRNA export (Wahle and Ruegsegger, 1999; Keller et al., 2000; Kuhn et al., 2009). In addition to RNA binding, the RRM domain may be responsible for interactions with other proteins or DNA (reviewed in Krietsch et al., 2013). These observations support the hypothesis that the interaction between TERT and RRM might be a mechanism by which telomerase could affect many cellular processes.

Here, we present further characterization of the RRM protein and discuss its potential physiological role in telomerase involvement in non-telomeric activities. We describe the interaction profile of the RRM protein and analyze telomere length, telomerase activity, and changes in gene expression in T-DNA insertion mutants that disrupt the *RRM* gene.

MATERIALS AND METHODS

Plant Material

Arabidopsis T-DNA insertion lines SALK_096285 (*rrm-1*) and SALK_116646C (*rrm-2*) were obtained from the Nottingham *Arabidopsis* Stock Centre. Both mutant and wild-type (Col-0) *A. thaliana* seeds were surface sterilized and germinated on 0.8% (w/v) agar plates supplemented with 1/2 Murashige and Skoog medium (MS; cat. n. M0255.0050; Duchefa[1]) and 1% (w/v) sucrose. Seedlings were potted after 7 days and further grown in the conditions of 16 h light, 21°C and 8 h dark, 19°C, illumination 150 μmol m^{-2} s^{-1}. Individual plants from each T-DNA insertion line were genotyped (see Supplementary Table S1 for primer sequences) and after selection of homozygous mutant plants, three subsequent generations were grown.

Telomere Length and Telomerase Activity Analyses

The terminal restriction fragment (TRF) analysis using Southern blot hybridization, the conventional TRAP (telomere-repeat-amplification-protocol) and the quantitative TRAP assays were performed as described (Fojtova et al., 2011). Mean telomere length values were calculated using TeloTool software (Gohring et al., 2014).

Entry Clone Generation

Sequences encoding full-length RRM (At5g10350) and G2p (At3g51800) proteins were amplified from 7-days-old seedling cDNA by Phusion HF DNA polymerase (Finnzymes[2]) according to the manufacturer's instructions. Sequences encoding RRM fragments [RRM-1(1–81); RRM-2(1–169); RRM-3(170–217); RRM-4(82–217); RRM-5(82–169)] were sub-cloned using KAPA Taq DNA polymerase (Kapabiosystems[3]) and a pGADT7-DEST::RRM construct as a template. Primers used for cloning are listed in Supplementary Table S1. PCR products were precipitated using PEG and cloned into pDONR/Zeo (Invitrogen[4]). The MT2A (At3g09390) coding sequence was sub-cloned into

[1] https://www.duchefa-biochemie.com/
[2] http://www.thermoscientificbio.com/finnzymes
[3] www.kapabiosystems.com
[4] http://www.lifetechnologies.com

pDONR/zeo from the cYFP cDNA library clone 212M1 (Lee et al., 2012). Entry clones encoding HSP70-1 (stock no. GC104920, At5g02500) and OZF2 (stock no. G10332, At4g29190; Yamada et al., 2003) were obtained from the ABRC[5]. Entry clones encoding AtTERT (At5g16850) fragments TEN(1–233), RID1(1–271), Fw3N-NLS(229–582), RT(597–987), and CTE2(958–1123) were prepared previously (Zachova et al., 2013).

Yeast Two Hybrid Analysis

Yeast two-hybrid experiments were performed using the Matchmaker^{TM} GAL4-based two-hybrid system (Clontech[6]). cDNA sequences encoding RRM protein (full-length and fragments), TERT fragments, G2p, MT2A, HSP70-1, and OZF2 were subcloned from their entry clones into the destination vectors pGADT7-DEST and pGBKT7-DEST. Each bait/prey combination was co-transformed into *Saccharomyces cerevisiae* PJ69-4a and yeast two hybrid analysis was performed as described in Schrumpfová et al. (2014). Protein expression was verified by immunoblotting using mouse anti-HA (kindly provided by Dr. Vojtěšek) or mouse anti-myc primary antibodies and HRP-conjugated anti-mouse secondary antibody (both Sigma-Aldrich[7]).

Bimolecular Fluorescence Complementation and Screening of cYFP cDNA Library

The constructs nYFP-TERT(CTE2), n/cYFP-TERT(RID1), and cYFP-RRM were created previously (Lee et al., 2012; Schrumpfová et al., 2014). The RRM, G2p, HSP70-1, and OZF2 coding sequences were subcloned from their entry clones into the destination vector pSAT4-DEST-nEYFP-C1 (Gelvin laboratory stock number pE3136). To visualize RRM subcellular localization, the RRM coding sequence was subcloned into the destination vector p2YGW7, generating a YFP tag at the N-terminus of the protein. The nYFP-RRM construct was screened against a cYFP cDNA library for protein–protein interactions in tobacco BY-2 protoplasts as described (Lee et al., 2012).

Tobacco BY-2 protoplasts were isolated and transfected as previously described (Tenea et al., 2009; Lee et al., 2012). *Arabidopsis* leaf protoplasts were isolated and transfected as described by Wu et al. (2009). To label cell nuclei, we co-transfected a plasmid expressing mRFP fused to the nuclear localization signal of the VirD2 protein from *Agrobacterium tumefaciens* (mRFP-VirD2NLS; Citovsky et al., 2006). To label nuclear speckles, a pSRp30-RFP nuclear speckles marker (Lorkovic et al., 2004) was co-transfected. Transfected protoplasts were incubated in the dark at room temperature overnight, and observed for fluorescence using a Zeiss AxioImager Z1 epifluorescence microscope (Tobacco BY-2) or a Leica SPE confocal scanning light microscope (*Arabidopsis*). As a negative control, we used the constructs nYFP- and cYFP-GAUT10 (At2g20810). Protein expression was tested by immunoblotting using mouse anti-GFP primary antibody (Roche[8]) and HRP-conjugated anti-mouse secondary antibody (Sigma-Aldrich). Proteins were extracted from protoplasts into an extraction buffer (50 mM Na_2HPO_4, 10 mM EDTA, 0.1% Triton X-100, 10 mM 2-Mercaptoethanol, 1x Proteinase inhibitors cocktail, 1 mM PMSF) by vortexing.

RNA Isolation and RT-qPCR Analysis

RNA from various *Arabidopsis* pollen developmental stages (Honys and Twell, 2004) was isolated using a Plant RNeasy Kit (Qiagen[9]) according to the manufacturer's instructions, and further purified by DNaseI treatment (TURBO DNA-free kit, Thermo Fisher Scientific[10]). RNA isolation from other tissues of mutant or wild-type plants and reverse transcription were performed as described (Fojtova et al., 2011; Ogrocka et al., 2012). Calli were derived from 7-days-old seedlings, propagated on cultivation medium with 1 $\mu g\ ml^{-1}$ 1-naphthaleneacetic acid and 1 $\mu g\ ml^{-1}$ 2,4-dichlorophenoxyacetic acid, and subcultured monthly onto fresh medium. Transcript levels relative to a ubiquitin reference gene were analyzed using FastStart SYBR Green Master (Roche) and a 7300 Real-Time PCR System (Applied Biosystems[11]). A 1 μl aliquot of cDNA was added to the 20 μl reaction mix; the final concentration of each forward and reverse primer (Supplementary Table S1) was 0.5 μM. Reactions were performed in triplicate; PCR cycle conditions consisted of 10 min of initial denaturation followed by 40 cycles of 20 s at 95°C, 30 s at 55°C, and 1 min at 72°C. SYBR Green I fluorescence was monitored after each extension step. The amount of the respective transcript was determined for at least two biological replicates using the $\Delta\Delta Ct$ method (Pfaffl, 2004).

Identification of Genes Co-regulated with *RRM*

GENEVESTIGATOR (Nebion AG[12]) application (Hruz et al., 2008) was used to identify *in silico* genes co-regulated with *RRM* and genes encoding its interacting partners TERT, G2P, MOS1, OZF2, HSP70-1, and MT2A. Using this tool, we first defined conditions under which any of these genes shows an at least twofold change in transcript levels. We then searched for genes responding either in a similar (score 0 to 1) or opposite (score −1 to 0) manner on the same subset of defined conditions. Genes with a co-regulation level score either higher than 0.5 or lower than −0.5 were considered for further analyses. Telobox motifs in candidate genes were identified in the literature or by manually searching for the motifs AAACCCT, AACCCTA and their corresponding reverse complements, in the genomic region 1000 bp upstream of the translation start (ATG) site using

[5]http://www.arabidopsis.org/

[6]www.clontech.com

[7]www.sigmaaldrich.com

[8]www.roche.com

[9]www.qiagen.com

[10]www.thermofisher.com

[11]www.appliedbiosystems.com

[12]https://genevestigator.com/gv/

publically available data at NCBI[13] and/or the gene datasets from Wang et al. (2011). Putative protein–protein-interaction networks were visualized using STRINGv10[14] (Szklarczyk et al., 2015).

RESULTS

Verification of *In vivo* Interaction between the RRM Protein and the CTE Domain of AtTERT

The RRM protein was identified as a putative interaction partner of AtTERT by screening a cYFP-tagged cDNA library using BiFC in tobacco BY-2 protoplasts (Lee et al., 2012). To test if this interaction is independent of the plant system used, we expressed both tagged partners in *Arabidopsis* leaf protoplasts and observed a positive BiFC signal in the nucleus (**Figure 1A**). We further employed the yeast-two-hybrid (Y2H) system, but no interaction was observed in yeast despite the expression of both proteins (Supplementary Figure S1). These results suggested that the *in vivo* interaction between the RRM and the AtTERT CTE domain was mediated by an additional plant protein or protein modification absent in yeast cells but present in telomerase-positive (BY-2) and telomerase-negative (*Arabidopsis* leaf) cells.

The C-terminus of the RRM Protein is Responsible for RRM Dimerization

We tested the ability of the At5g10350 protein containing a single RRM domain to form homodimers using Y2H and BiFC analyses (**Figures 1B,C**). RRM dimerization was observed using BiFC in tobacco BY-2 protoplasts, where the interaction provided a pattern similar to that of full length RRM-YFP fusion protein that co-localized with the pSRp30-RFP nuclear speckles marker (**Figure 1B**). We further tested this interaction using Y2H where the RRM protein showed strong self-interaction using both histidine and stringent adenine growth selection (**Figure 1C**). To determine which part of the RRM molecule is responsible for dimerization, we prepared five constructs corresponding to various structural domains of the RRM protein. We tested them for interaction with full length RRM and with each other (**Figure 1C**) by Y2H analysis. The RRM-4(82–217) fragment comprising the RRM domain and the C-terminus, and the RRM-3(170–217) fragment with the C-terminus only, interacted with the full length RRM protein, each other, and themselves using both histidine and stringent adenine growth selection, suggesting that the C-terminus is responsible for protein homodimerization. The fragments RRM-1(1–81) – N-terminus, RRM-2(1–169) – N-terminus and the RRM domain, and RRM-5(82–169) – the RRM domain did not show positive Y2H signals, although their successful expression was confirmed by immunoblotting (Supplementary Figure S1).

BiFC Screening of an *Arabidopsis* cDNA Library Identified Proteins that Interact with the RRM Protein

To obtain better insights into possible RRM cellular functions and its involvement in specific cellular processes, we screened in BY-2 protoplasts a nYFP-RRM fusion protein against a cYFP cDNA library (Lee et al., 2012). We identified one cDNA clone encoding the full length Metallothionein 2A protein (At3g09390; MT2A) and four additional cDNA clones encoding protein fragments that were in-frame with the YFP tag (**Figure 2**): (i) Modifier Of Snc1 (MOS1; At4g24680; fragment 1040–1427 aa); (ii) the putative nuclear DNA-binding protein G2p (At3g51800; 347–401 aa); (iii) Oxidation Related Zinc Finger 2 (OZF2; At4g29190; 1–68 aa); (iv) Heat Shock Cognate protein 70-1 (HSP70-1; At5g02500; 1–211 aa). In all cases, the interaction signal resembled nuclear speckles. We generated Y2H and BiFC constructs of G2p, OZF2, and HSP70-1 bearing the respective full length coding sequences to confirm interaction with the RRM protein and to test interaction with AtTERT fragments (**Figure 2B**). We were unable to obtain a full length MOS1 (1–1427 aa) construct, either by RT-PCR in our laboratory or from stock centers. Using the Y2H system we found strong interaction between OZF2 and RRM proteins, whereas MT2A, G2p, and HSP70-1 did not interact with RRM. None of the proteins interacted with any AtTERT fragments (not shown). Interestingly, using BiFC in tobacco BY-2 protoplasts we found interaction of the G2p and MT2A proteins with RRM and also with the N-terminal domain fragment RID1(1–271) of AtTERT (**Figure 2**, Supplementary Figure S2), with strong nucleolar and weak nucleoplasmic localization.

RRM is Highly Expressed in Leaves and Reproductive Tissues

To characterize RRM expression during plant development, we investigated the level of RRM transcripts (**Figure 3A**), including in telomerase-positive tissues. The transcripts were quantified in flower buds, calli, leaves, and 7-days-old seedlings of wild-type plants with a particular interest in detailed seedling analysis comprising whole seedlings, shoots, roots, and root tips. To quantify transcript levels in reproductive tissues, we included five pollen developmental stages (uninucleate microspores, early bicellular pollen, late bicellular pollen, immature tricellular pollen, and mature pollen). We observed RRM transcripts in all tissues tested. However, the greatest RRM transcript abundance was seen in proliferating tissues – young leaves and reproductive tissues. During pollen development, RRM transcript abundance peaked at the time of pollen mitosis I. Our RT-qPCR data confirmed previously published microarray data[15].

Telomere Length and Telomerase Activity in Homozygous *rrm* T-DNA Insertion Lines

To examine the role of RRM *in planta*, we analyzed *Arabidopsis* lines SALK_096285 (*rrm-1*) and SALK_116646C (*rrm-2*)

[13]http://www.ncbi.nlm.nih.gov/

[14]http://www.string-db.org/

[15]http://bbc.botany.utoronto.ca/

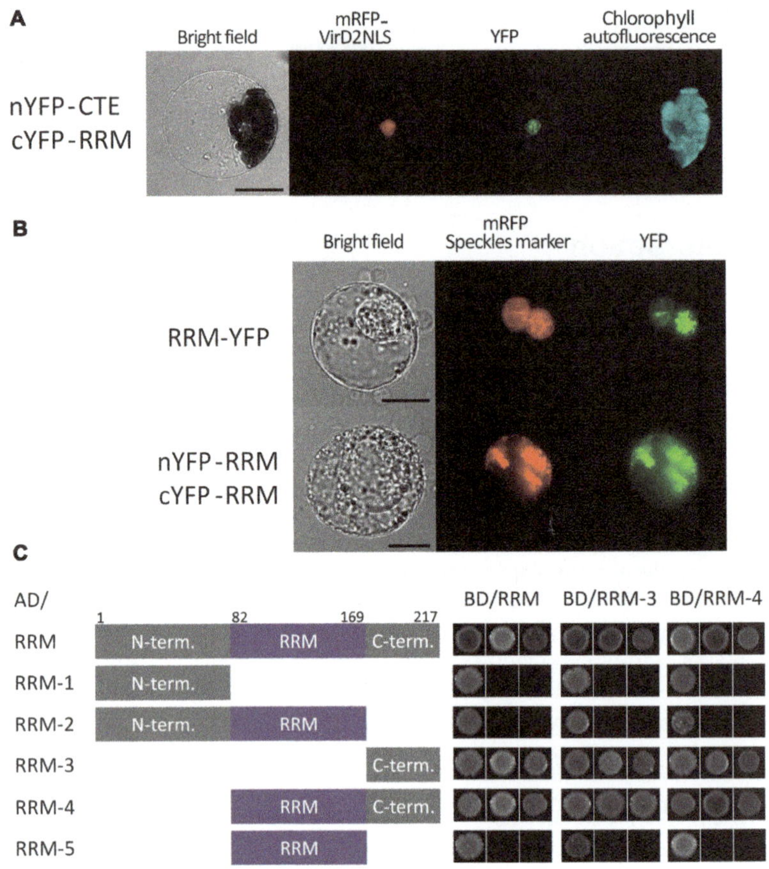

FIGURE 1 | Subcellular localization and dimerization of the RRM protein and verification of its interaction with the CTE domain of AtTERT. (A) BiFC in *Arabidopsis* leaf protoplasts confirmed a nuclear interaction of the RRM protein with AtTERT CTE domain. Protoplasts were co-transfected with plasmids encoding nYFP-tagged TERT(CTE2), cYFP-tagged RRM, and mRFP-VirD2(NLS) to label cell nuclei; nYFP- and cYFP-GAUT10 constructs served as negative control (not shown). YFP fluorescence is shown in green, mRFP fluorescence in red, and chlorophyll autofluorescence in blue pseudocolor. Scale bar indicates 10 μm. **(B)** RRM-YFP co-localizes with a pSRp30-RFP nuclear speckle marker in tobacco BY-2 protoplasts. The same localization pattern was observed for interaction of nYFP-RRM with cYFP-RRM. YFP fluorescence is shown in green, and mRFP fluorescence of the VirD2(NLS) marker is shown in red. Scale bars indicate 20 μm. **(C)** A Y2H system was used to assess RRM dimerization. Two sets of plasmids carrying full-length RRM or indicated RRM segments fused to either the GAL4 DNA-binding domain (BD) or the GAL4 activation domain (AD) were constructed and introduced into *Saccharomyces cerevisiae* PJ69-4a carrying *His3* and *Ade2* reporter genes. Co-transformation with an empty vector served as a negative control (not shown). Full-length RRM protein self-interacted on both histidine and stringent adenine selection plates. The same result was observed for interactions of the RRM-4(82–217) and the RRM-3(170–217) fragments with the full-length RRM protein, each other, and themselves, suggesting that the RRM-protein C-terminus is responsible for protein homodimerization. None of the other fragments [RRM-1(1–81), RRM-2(1–169), and RRM-5(82–169)] showed interaction, although their successful expression was confirmed by immunoblotting (Supplementary Figure S1).

harboring different T-DNA insertions in the *RRM* gene (**Figure 3B**). RT-qPCR results confirmed only *rrm-1* as a *null* allele, whereas the *rrm-2* allele caused only a partial knock-down of the *RRM* transcript (**Figure 3C**). No detectable morphological differences were observed in root length, rosette diameter, leaf number, flowering time, or silique number comparing soil-grown wild-type (Col-0) and three subsequent generations of homozygous *rrm/-* plants (not shown). Thus, RRM function does not appear to be essential for plant growth and development under these experimental conditions.

Telomere length was determined in three independent homozygous G3 mutant plants using the TRF analysis. Although telomeres in both *rrm-1/rrm-1* and *rrm-2/rrm-2* G3 generation plants were slightly longer when compared to wild-type plants (Supplementary Figure S3), a paired Student *t*-test indicated that these changes were not significant (the two-tailed *p*-values equal 0.0575 and 0.0656 for *rrm-1* and *rrm-2*, respectively). Telomerase activity in G3 generation homozygous *rrm-1* and *rrm-2* lines was tested by TRAP (telomere repeat amplification protocol) in 7-days-old seedlings. No changes in telomerase activity or processivity were observed using conventional TRAP analysis or quantitative TRAP analysis (not shown).

Changes in Transcript Levels in Homozygous *rrm* Lines

We analyzed the transcription profiles of genes identified by our cDNA library screen in homozygous *rrm-1* and *rrm-2 Arabidopsis* mutant lines (**Figure 3C**). *G2p*, *MOS1*, *OZF2*,

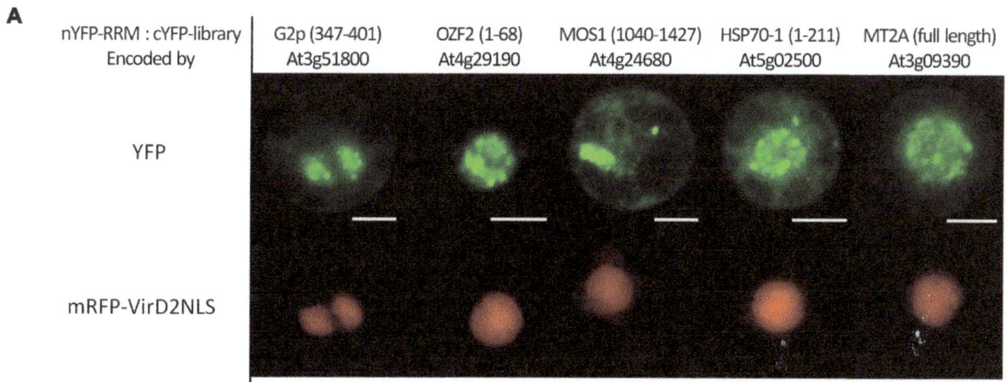

B

Full length protein	G2p	OZF2	MOS1	HSP70-1	MT2A
Interaction with RRM in Y2H	−	+	n.a.	−	−
Interaction with RRM in BiFC	+	n.e.	n.a.	n.e.	+
Interaction with TERT fragments in Y2H	−	−	n.a.	−	−
Interaction with TERT(RID1) in BiFC	+	n.e.	n.a.	n.e.	+
Co-purification with TERT fragments in TAP experiments*	+	−	+	c.b.d.	−

FIGURE 2 | BiFC screening of an *Arabidopsis* cDNA library identified interaction partners of the RRM protein. (A) Interactions of nYFP-RRM with cYFP-tagged protein fragments of G2p(347–401), OZF2(1–68), MOS1(1040–1427), HSP70-1(1–211) and full-length MT2A protein identified by screening a cYFP-tagged cDNA library. Tobacco BY-2 protoplasts were co-transfected with plasmids encoding a mRFP-VirD2NLS nuclear marker, nYFP-RRM, and one of the five interacting cYFP-tagged proteins. YFP fluorescence is shown in green, and mRFP fluorescence is shown in red. Scale bars indicate 20 μm. **(B)** Summary of investigated protein–protein interactions of full-length G2p, OZF2, HSP70-1, and MT2A proteins with full-length RRM protein and TERT fragments. One of the fragments (RID1) was used in BiFC; all other TERT fragments were investigated using a Y2H system. Using the GAL4-based Y2H system in *S. cerevisiae* PJ69-4a carrying *His3* and *Ade2* reporter genes, we confirmed interaction only between OZF2-AD and RRM-BD on both histidine and stringent adenine selection plates. Other investigated combinations were negative, excluding the OZF2-BD construct that showed false positive interactions, and the HSP70-1-AD construct that was not expressed. Protein expression was checked by immunoblotting (Supplementary Figure S1). In addition to interaction of MT2A with full-length RRM protein shown on **(A)**, BiFC analysis in tobacco BY-2 protoplasts revealed positive interactions of MT2A with the TERT(RID1) fragment and also of full-length G2p protein with both full-length RRM protein, and the TERT(RID1) fragment (Supplementary Figure S2). n.a., not analyzed, n.e., not expressed, c.b.d., cannot be determined. *G2p and MOS1 co-purified with TERT fragments in other work of our group (Majerska et al., manuscript in preparation).

HSP70-1, and *MT2A* transcripts were quantified in 21-days-old leaves, a tissue with high *RRM* expression (**Figure 3C**, right panel). *AtTERT* transcripts were quantified in 7-days-old seedlings (**Figure 3C**, left panel), as there is a very low *AtTERT* transcription in *Arabidopsis* leaves (Ogrocka et al., 2012). *G2p* and *TERT* transcript levels were significantly higher in both *rrm* T-DNA insertion lines, suggesting a possible role of RRM in the regulation of these genes and/or the stability of the mRNAs encoded by these genes. *MOS1*, *OZF2*, *HSP70-1*, and *MT2A* transcripts levels were similar in mutant and wild-type plants.

Using GENEVESTIGATOR software, we identified 2102 genes putatively transcriptionally co-regulated with *RRM* and/or with at least one of its interacting partners TERT, G2P, MOS1, OZF2, HSP70-1, and MT2A, using the same conditions subset in a similar or opposite manner. A narrow subset of 137 genes showed overlapping co-regulation with at least two of these genes. We observed that *RRM* and genes encoding its presumed interactors were co-regulated with numerous ribosomal protein genes. Interestingly, most ribosomal protein genes possess a telobox, a short regulatory motif over-represented in 5′ regions of *Arabidopsis* genes with sequences identical to the repeat (AAACCCT)n of plant telomeres (Regad et al., 1994). Telobox motifs are also found in promoters of genes involved in DNA replication (Tremousaygue et al., 1999; Wang et al., 2011). Because of a possible link between the cell cycle-dependent regulation of the expression of genes encoding ribosomal proteins and telomerase (Gaspin et al., 2010), we selected for transcription analysis a subset of identified co-regulated genes in addition to genes involved in DNA replication and translation-related genes with known telobox motifs (**Table 1**). Five genes showed a 2- to 4-fold increase of transcript levels in both mutant lines, and four genes displayed an increase in the homozygous *rrm-1* line only. Interestingly, none of the genes analyzed showed

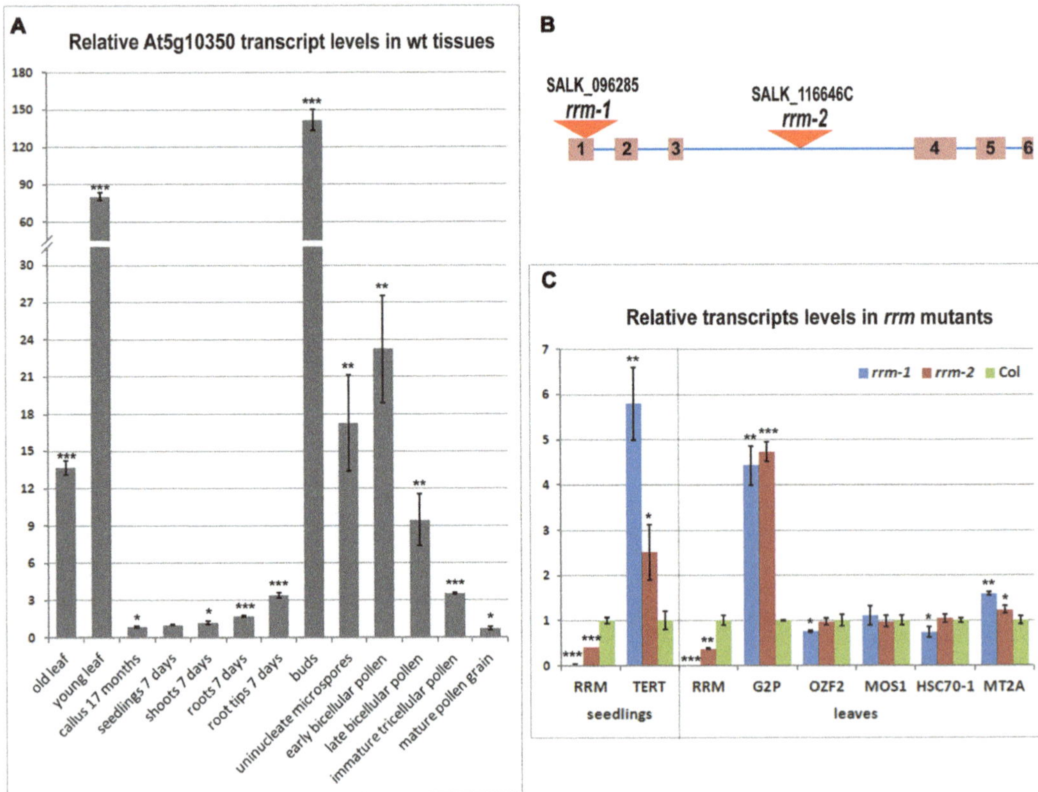

FIGURE 3 | Transcription profile of the RRM gene in tissues of wild-type plants and relative transcription of the RRM interactors in *rrm* mutants. **(A)** Level of *RRM* transcripts in various wild-type (wt) plant tissues and developmental stages calculated relative to 7-days-old seedlings (Col-0) using the $\Delta\Delta$Ct method and ubiquitin (*ubi10*) as a reference gene. **(B)** Schematic depiction of the *RRM* gene and position of the T-DNA insertion in homozygous mutant *Arabidopsis* lines *rrm-1* and *rrm-2*. **(C)** Level of transcripts of indicated genes in 7-days-old seedlings (left panel) and 21-days-old leaves (right panel) of *rrm-1* and *rrm-2* T-DNA insertion lines calculated relative to wild-type (Col-0) using the $\Delta\Delta$Ct method and *ubi10* as a reference gene. Two-tailed *p*-values were calculated using the unpaired *t*-test (*$p < 0.05$; **$p < 0.001$; ***$p = 0.0001$, see Supplementary Table S2 for details).

significantly decreased transcript levels in the mutant lines. Transcript levels of DNA replication-related genes were not altered in either mutant line, suggesting that the telobox in the 5′ region of these genes is not a critical determinant for RRM action.

DISCUSSION

The identification of an RRM protein as a nuclear interactor with the CTE domain of AtTERT in tobacco BY2 protoplasts (Lee et al., 2012) was somehow surprising. However, yeast genome-wide screens (Askree et al., 2004; Gatbonton et al., 2006; Ungar et al., 2009) revealed a number of proteins that influenced telomere length and which were involved in numerous cellular processes without a known link to telomere maintenance. Among these were human proteins involved in RNA metabolism and transcription pathways connected with non-telomeric functions of telomerase (see Majerska et al., 2011 for review). A conserved protein structure comprising the coiled-coil N-terminus, a single internal RRM domain, and the C-terminal region with a nuclear localization signal classified the RRM protein as a nuclear poly(A) binding protein (PABPN, Eliseeva et al., 2013). Recently, a RRM protein was identified as an interactor with AtCSP3 (COLD SHOCK DOMAIN PROTEIN 3; Kim et al., 2013), and hnRNP-like proteins (Arabidopsis Interactome Mapping Consortium, et al., 2011). Here, we verified RRM interaction with the CTE domain of AtTERT using BiFC in *Arabidopsis* protoplasts, but their direct interaction was not observed in yeast. Known technical differences of both screening systems suggest that the *in vivo* interaction is mediated by an additional protein absent in the yeast cell, or is facilitated by post-translational modification missing in yeast. Analysis of T-DNA insertion mutant lines showed no obvious changes in telomere lengths and telomerase activity, suggesting that the RRM protein is not essential for telomere maintenance. The observed interaction with telomerase may reflect possible non-telomeric functions.

The localization of the RRM-YFP signal in nuclear speckles and the observation that all described BiFC RRM interactions (and interaction with CSP3 protein, Kim et al., 2013) are nuclear-localized suggest that RRM is in fact a PABPN. Human PABPN1 localizes to nuclear speckles as a consequence of

TABLE 1 | Relative transcription levels of genes with identified telobox sequences and/or co-regulated with RRM interactors in homozygous *rrm* mutants.

AGI number	Gene name	*rrm-1**		*rrm-2**		Telobox	Reference	Co-regulation with RRM and its interactors (GENEVESTIGATOR score)[a]
		2^{-ddCt}	*SD*	2^{-ddCt}	*SD*			
Genes encoding cytoplasmic ribosomal proteins								
At1g23290	RPL27a	**2.08**	0.37	**2.56**	1.23	Yes	In this work	RRM (0,73), G2P (0,89), HSP70-1 (0,78)
At1g72370	RP40	**4.72**	0.13	**3.42**	0.22	Yes	Tremousaygue et al., 1999; Manevski et al., 2000	RRM (0,75), G2P (0,92), HSP70-1 (0,77)
At3g04840	RPS3Ae family	1.78	0.10	1.34	0.11	Yes	In this work	RRM (0,78), G2P (0,89), HSP70-1 (0,76)
At3g25520	RPL5	1.66	0.07	1.32	0.11	Yes	In this work	RRM (0,84), G2P (0,90), HSP70-1 (0,84)
At5g39740	RPL5b	1.41	0.01	1.12	0.13	Yes	In this work	RRM (0,87), G2P (0,90), HSP70-1 (0,75)
At3g47370	RPS10p/S20e family	**2.30**	0.01	1.45	0.10	Yes	In this work	RRM (0,73), MOS1 (−0,81), G2P (0,90), HSP70-1 (0,73)
At3g49010	BBC1	**3.70**	0.55	1.86	0.24	Yes	In this work	RRM (0,83), G2P (0,92), HSP70-1 (0,77)
At3g51190	RPL2 family	**2.41**	0.2	1.89	0.11	Yes	In this work	n.a.
At3g56340	RPS26e family	1.38	0.03	1.17	0.08	Yes	In this work	RRM (0,75), HSP70-1 (0,68)
At3g60770	RPS13/S15 family	1.95	0.12	1.44	0.07	Yes	In this work	RRM (0,78), G2P (0,90), HSP70-1 (0,80)
At4g00810	RPS60 family	**2.39**	0.09	**2.25**	0.10	Yes	Tremousayguo et al., 1999	MOS1 (−0,88)
At4g09800	RPS18C	**2.69**	0.23	**2.35**	0.05	Yes	Tremousaygue et al., 1999	No co-regulation
Genes encoding plastid ribosomal proteins								
At1g79850	PRPS17	0.68	0.02	1.68	0.28	Yes	Tremousaygue et al., 1999	No co-regulation
At2g33450	PRPL28	**2.62**	0.55	1.33	0.15	Yes	Tremousaygue et al., 1999	No co-regulation
Genes encoding translation factors								
At1g07940	EF1A family	1.88	0.01	0.87	0.01	Yes	Tremousaygue et al., 1999	No co-regulation
At1g54290	TIF SUI1 family	**2.77**	0.23	**2.43**	0.34	Yes	Tremousaygue et al., 1999	No co-regulation
DNA replication-related genes								
At1g07270	CDC6b	0.98	0.11	1.03	0.08	No	In this work	No co-regulation
At1g07370	PCNA1	1.13	0.07	1.13	0.11	Yes	Manevski et al., 2000	No co-regulation
At1g44900	MCM2	0.97	0.13	0.99	0.12	Yes	In this work	No co-regulation
At5g46280	MCM3	0.68	0.02	0.95	0.08	Yes	In this work	No co-regulation
At2g16440	MCM4	0.94	0.01	0.94	0.06	No	In this work	No co-regulation
At2g07690	MCM5	0.95	0.07	0.99	0.08	No	In this work	No co-regulation
At5g44635	MCM6	0.72	0.03	0.90	0.12	Yes	In this work	n.a.
At4g02060	MCM7	1.03	0.07	0.97	0.13	Yes	In this work	No co-regulation
Other co-regulated genes								
At2g19480	NAP1;2	1.85	0.15	1.42	0.15	Yes	In this work	RRM (0,79), HSP70-1 (0,75)
At3g54230	SUA	1.54	0.22	1.25	0.09	Yes	In this work	TERT (0,88), MOS1 (0,95)
At4g17520	Hyaluronan family	0.60	0.21	1.55	0.13	Yes	In this work	RRM (0,77), G2P (0,93)
At5g14790	ARM superfamily	0.76	0.05	0.87	0.08	No	In this work	TERT (−0,84)

*[a]n.a., data not available; *more than two-fold change in transcript level (2^{-ddCt}) is highlighted; SD, standard deviation.*

RNA poly(A) binding (Calado and Carmo-Fonseca, 2000). Our observations indicate a conserved structure-localization relationship of PABPNs across eukaryotic species. Nucleic acid binding of RRM-containing proteins is often mediated by a pair of RRM domains (Deo et al., 1999; Kwon and Chung, 2004). On the other hand, *Xenopus laevis* XlePABP2 and *Citrus sinensis* CsPABPN1, PABPNs that share similar structure with RRM protein, both bind poly(A) as monomers and undergo a dimer-monomer transition upon poly(A) binding (Song et al., 2008; Domingues et al., 2015). We visualized RRM protein dimerization in tobacco BY-2 protoplasts and observed the same nuclear speckle pattern as with RRM-YFP localization. Moreover, we demonstrated that RRM protein dimerizes through its C-terminal region. This last observation contradicts the published dimerization model of hPABPN1 (Ge et al., 2008), which identified the amino acid residues responsible for self-interaction within the RRM domain.

We revealed a possible connection between RRM and non-canonical telomerase functions by identifying interaction partners of RRM. Screening for nYFP-RRM protein–protein interactions against a cYFP-tagged cDNA library identified five putative RRM interactors with various annotated functions such as transcription regulation (OZF2), epigenetic regulation (MOS1), mRNA catabolism (MOS1), RNA methylation (G2p), protein nuclear import (G2p), protein folding (HSP70-1), proteolysis (G2p), cellular copper ion homeostasis (MT2A), or metabolism (MOS1, HSP70-1). In all five cases, the interaction localization pattern resembles nuclear speckles, as observed for RRM-YFP subcellular localization. A G2p-GFP fusion protein was previously localized in the nucleus (Zhang et al., 2005). However, the subcellular localization of other RRM interactors has not previously been described. Interestingly, other data from our group showed that G2p and MOS1 co-purify with TERT (**Figure 2**, Majerska et al., manuscript in preparation) and G2p and MT2A interact with TERT(RID1) using BiFC in tobacco BY-2 protoplasts (**Figure 2B**, Supplementary Figure S2), suggesting co-existence of TERT, RRM, G2p, MOS1, and MT2A in a multiprotein complex.

Analysis of telomere length and telomerase activity in homozygous *rrm-1* and *rrm-2* T-DNA insertion mutants indicated that the RRM protein was not important for the canonical telomeric functions of telomerase. On the other hand, *TERT* transcripts were elevated in homozygous *rrm* mutants, and TERT and RRM may share binding partners such as G2p, MOS1, and MT2A. These observations suggest that RRM plays a role in non-telomeric activities of telomerase. Interestingly, the G7 generation of a homozygous *tert* T-DNA insertion line showed increased *OZF2* and *MT2A* transcript levels (Amiard et al., 2014). PABPNs are implicated in processes that might be crucial for post-transcriptional regulation of gene expression. Our qPCR analyses indicated that RRM might generally function as a negative regulator of gene expression, because none of the 34 genes analyzed here showed a significant decrease in transcript levels in homozygous *rrm* mutants. Increased levels of *TERT* and *G2p* transcripts in homozygous *rrm* mutant lines indicated a possible feedback mechanism in RRM-TERT and RRM-G2p interactions. Moreover, nine ribosomal and translation-related genes also showed significantly increased transcript levels in a *rrm* mutant background. We have further analyzed these nine genes for transcript level perturbations across different conditions using GENEVESTIGATOR. *RPL2* transcripts were stable across various conditions. Interestingly, the transcript levels of the other eight genes changed more than two-fold in response to salt stress in the *myb44* T-DNA insertion line (Jung et al., 2008). RP40, RPL27A, RPS10p/S20e, BBC1, and RPL2 form a protein interaction network (STRING database[16], Szklarczyk et al., 2015) and are mutually co-regulated. The RRM interactome, subcellular localization, and co-regulation profile showing that the expression of the majority of its co-regulated genes contain telobox motifs in their promoters, further support the hypothesis that RRM may function in mediating non-telomeric (non-canonical) functions of telomerase. DNA replication-related genes were not co-regulated with genes encoding RRM or its interactors, and they also did not show changes in transcript abundance in a homozygous *rrm* background. These results suggest that the telobox in promoters of these genes are not a critical determinant of RRM action. Regulation of translation-related genes is generally important for the regulation of protein synthesis and consequently for cell growth. These genes regulate tumor onset and progression (reviewed in Loreni et al., 2014), further indicating a possible link between RRM and its interactors to TERT non-telomeric functions. Our results support a functional connection between RRM and its interaction partners in plant regulatory protein complex(es).

CONCLUSION

The RRM protein was previously identified as an interaction partner of AtTERT. However, telomere length shortening in RRM protein knockout mutant plants was not significant. By screening a cDNA library using cYFP-RRM as a bait, we identified five interaction partners; two of them interacted also with TERT fragments. Investigation of the subcellular localization and protein structure suggested that RRM-protein may function as a nuclear poly(A)-binding protein. Transcriptional profiling revealed a possible involvement of RRM-protein in the regulation of a subset of ribosomal and translation-related genes. Most of these genes contain a telobox motif in their promoters. *G2p* and *TERT* transcript levels were significantly higher in *rrm*/- knockout mutants, suggesting a possible role for RRM in the regulation of these genes and/or the stability of the mRNAs encoded by these genes. Overlaps of the RRM and TERT interactome, subcellular localization of protein–protein interactions, and co-regulation profiles support the hypothesis that RRM may be involved in mediating non-canonical telomerase functions.

AUTHOR CONTRIBUTION

LD performed most of experiments except pollen RT-qPCR performed by DH and RR. L-YL was involved in cDNA screening, ES was involved in cloning. ES and SBG designed the study.

---

[16]http://www.string-db.org/

FUNDING

The work was supported by the Czech Ministry of Education, Youth and Sports (project Interaktom, LH10352), the Grant Agency of the Czech Republic (13-06943S) and by the project "CEITEC - Central European Institute of Technology" (CZ.1.05/1.1.00/02.0068) from the European Regional Development Fund. Work in the Gelvin laboratory is supported by grants from the US National Science Foundation.

REFERENCES

Amiard, S., Da Ines, O., Gallego, M. E., and White, C. I. (2014). Responses to telomere erosion in plants. *PLoS ONE* 9:e86220. doi: 10.1371/journal.pone.0086220

Arabidopsis Interactome Mapping Consortium, Dreze, M., Carvunis, A., Charloteaux, B., Galli, M., Pevzner, S., et al. (2011). Evidence for network evolution in an *Arabidopsis* interactome map. *Science* 333, 601–607. doi: 10.1126/science.1203877

Askree, S. H., Yehuda, T., Smolikov, S., Gurevich, R., Hawk, J., Coker, C., et al. (2004). A genome-wide screen for *Saccharomyces cerevisiae* deletion mutants that affect telomere length. *Proc. Natl. Acad. Sci. U.S.A.* 101, 8658–8663. doi: 10.1073/pnas.0401263101

Blasco, M. A. (2005). Mice with bad ends: mouse models for the study of telomeres and telomerase in cancer and aging. *EMBO J.* 24, 1095–1103. doi: 10.1038/sj.emboj.7600598

Calado, A., and Carmo-Fonseca, M. (2000). Localization of poly(A)-binding protein 2 (PABP2) in nuclear speckles is independent of import into the nucleus and requires binding to poly(A) RNA. *J. Cell Sci.* 113(Pt 12), 2309–2318.

Citovsky, V., Lee, L.-Y., Vyas, S., Glick, E., Chen, M. H., Vainstein, A., et al. (2006). Subcellular localization of interacting proteins by bimolecular fluorescence complementation in planta. *J. Mol. Biol.* 362, 1120–1131. doi: 10.1016/j.jmb.2006.08.017

Counter, C. M., Avilion, A. A., Lefeuvre, C. E., Stewart, N. G., Greider, C. W., Harley, C. B., et al. (1992). Telomere shortening associated with chromosome instability is arrested in immortal cells which express telomerase activity. *EMBO J.* 11, 1921–1929.

Deo, R. C., Bonanno, J. B., Sonenberg, N., and Burley, S. K. (1999). Recognition of polyadenylate RNA by the poly(A)-binding protein. *Cell* 98, 835–845. doi: 10.1016/S0092-8674(00)81517-2

Domingues, M. N., Sforca, M. L., Soprano, A. S., Lee, J., De Souza Tde, A., Cassago, A., et al. (2015). Structure and mechanism of dimer-monomer transition of a plant poly(A)-binding protein upon RNA interaction: insights into its poly(A) tail assembly. *J. Mol. Biol.* 427, 2491–2506. doi: 10.1016/j.jmb.2015.05.017

Eliseeva, I. A., Lyabin, D. N., and Ovchinnikov, L. P. (2013). Poly(A)-binding proteins: structure, domain organization, and activity regulation. *Biochemistry Mosc.* 78, 1377–1391. doi: 10.1134/S00062979131 30014

Fitzgerald, M. S., Mcknight, T. D., and Shippen, D. E. (1996). Characterization and developmental patterns of telomerase expression in plants. *Proc. Natl. Acad. Sci. U.S.A.* 93, 14422–14427. doi: 10.1073/pnas.93.25.14422

Fojtova, M., Peska, V., Dobsakova, Z., Mozgova, I., Fajkus, J., and Sykorova, E. (2011). Molecular analysis of T-DNA insertion mutants identified putative regulatory elements in the AtTERT gene. *J. Exp. Bot.* 62, 5531–5545. doi: 10.1093/jxb/err235

Gaspin, C., Rami, J. F., and Lescure, B. (2010). Distribution of short interstitial telomere motifs in two plant genomes: putative origin and function. *BMC Plant Biol.* 10:283. doi: 10.1186/1471-2229-10-283

Gatbonton, T., Imbesi, M., Nelson, M., Akey, J. M., Ruderfer, D. M., Kruglyak, L., et al. (2006). Telomere length as a quantitative trait: genome-wide survey and genetic mapping of telomere length-control genes in yeast. *PLoS Genet.* 2:e35. doi: 10.1371/journal.pgen.0020035

Ge, H., Zhou, D., Tong, S., Gao, Y., Teng, M., and Niu, L. (2008). Crystal structure and possible dimerization of the single RRM of human PABPN1. *Proteins* 71, 1539–1545. doi: 10.1002/prot.21973

Gohring, J., Fulcher, N., Jacak, J., and Riha, K. (2014). TeloTool: a new tool for telomere length measurement from terminal restriction fragment analysis with improved probe intensity correction. *Nucleic Acids Res.* 42:e21. doi: 10.1093/nar/gkt1315

Greider, C. W., and Blackburn, E. H. (1985). Identification of a specific telomere terminal transferase activity in *Tetrahymena* extracts. *Cell* 43, 405–413. doi: 10.1016/0092-8674(85)90170-9

Greider, C. W., and Blackburn, E. H. (1987). The telomere terminal transferase of *Tetrahymena* is a ribonucleoprotein enzyme with two kinds of primer specificity. *Cell* 51, 887–898. doi: 10.1016/0092-8674(87)90576-9

Harley, C. B., Futcher, A. B., and Greider, C. W. (1990). Telomeres shorten during ageing of human fibroblasts. *Nature* 345, 458–460. doi: 10.1038/345 458a0

Honys, D., and Twell, D. (2004). Transcriptome analysis of haploid male gametophyte development in *Arabidopsis*. *Genome Biol.* 5:R85. doi: 10.1186/gb-2004-5-11-r85

Hruz, T., Laule, O., Sza bo, G., Wessendorp, F., Bleuler, S., Oertle, L., et al. (2008). Genevestigator v3: a reference expression database for the meta-analysis of transcriptomes. *Adv. Bioinformatics* 2008:420747. doi: 10.1155/2008/420747

Jung, C., Seo, J. S., Han, S. W., Koo, Y. J., Kim, C. H., Song, S. I., et al. (2008). Overexpression of AtMYB44 enhances stomatal closure to confer abiotic stress tolerance in transgenic *Arabidopsis*. *Plant Physiol.* 146, 623–635. doi: 10.1104/pp.107.110981

Keller, R. W., Kuhn, U., Aragon, M., Bornikova, L., Wahle, E., and Bear, D. G. (2000). The nuclear poly(A) binding protein, PABP2, forms an oligomeric particle covering the length of the poly(A) tail. *J. Mol. Biol.* 297, 569–583. doi: 10.1006/jmbi.2000.3572

Kim, M. H., Sonoda, Y., Sasaki, K., Kaminaka, H., and Imai, R. (2013). Interactome analysis reveals versatile functions of *Arabidopsis* COLD SHOCK DOMAIN PROTEIN 3 in RNA processing within the nucleus and cytoplasm. *Cell Stress Chaperones* 18, 517–525. doi: 10.1007/s12192-012-0398-3

Krietsch, J., Rouleau, M., Pic, E., Ethier, C., Dawson, T. M., Dawson, V. L., et al. (2013). Reprogramming cellular events by poly(ADP-ribose)-binding proteins. *Mol. Aspects Med.* 34, 1066–1087. doi: 10.1016/j.mam.2012.12.005

Kuhn, U., Gundel, M., Knoth, A., Kerwitz, Y., Rudel, S., and Wahle, E. (2009). Poly(A) tail length is controlled by the nuclear poly(A)-binding protein regulating the interaction between poly(A) polymerase and the cleavage and polyadenylation specificity factor. *J. Biol. Chem.* 284, 22803–22814. doi: 10.1074/jbc.M109.018226

Kwon, C., and Chung, I. K. (2004). Interaction of an *Arabidopsis* RNA-binding protein with plant single-stranded telomeric DNA modulates telomerase activity. *J. Biol. Chem.* 279, 12812–12818. doi: 10.1074/jbc.M3120 11200

Lee, L.-Y., Wu, F. H., Hsu, C. T., Shen, S. C., Yeh, H. Y., Liao, D. C., et al. (2012). Screening a cDNA library for protein-protein interactions directly in planta. *Plant Cell* 24, 1746–1759. doi: 10.1105/tpc.112.097998

Levy, M. Z., Allsopp, R. C., Futcher, A. B., Greider, C. W., and Harley, C. B. (1992). Telomere end-replication problem and cell aging. *J. Mol. Biol.* 225, 951–960. doi: 10.1016/0022-2836(92)90096-3

Loreni, F., Mancino, M., and Biffo, S. (2014). Translation factors and ribosomal proteins control tumor onset and progression: how? *Oncogene* 33, 2145–2156. doi: 10.1038/onc.2013.153

Lorkovic, Z. J., Hilscher, J., and Barta, A. (2004). Use of fluorescent protein tags to study nuclear organization of the spliceosomal machinery in transiently transformed living plant cells. *Mol. Biol. Cell* 15, 3233–3243. doi: 10.1091/mbc.E04-01-0055

Majerska, J., Sykorova, E., and Fajkus, J. (2011). Non-telomeric activities of telomerase. *Mol. Biosyst.* 7, 1013–1023. doi: 10.1039/c0mb00268b

Manevski, A., Bertoni, G., Bardet, C., Tremousaygue, D., and Lescure, B. (2000). In synergy with various cis-acting elements, plant insterstitial telomere motifs regulate gene expression in *Arabidopsis* root meristems. *FEBS Lett.* 483, 43–46. doi: 10.1016/S0014-5793(00)02056-1

Ogrocka, A., Sykorova, E., Fajkus, J., and Fojtova, M. (2012). Developmental silencing of the AtTERT gene is associated with increased H3K27me3 loading and maintenance of its euchromatic environment. *J. Exp. Bot.* 63, 4233–4241. doi: 10.1093/jxb/ers107

Olovnikov, A. M. (1971). [Principle of marginotomy in template synthesis of polynucleotides]. *Dokl. Akad. Nauk. SSSR* 201, 1496–1499.

Olovnikov, A. M. (1973). A theory of marginotomy. The incomplete copying of template margin in enzymic synthesis of polynucleotides and biological significance of the phenomenon. *J. Theor. Biol.* 41, 181–190. doi: 10.1016/0022-5193(73)90198-7

Pfaffl, M. W. (2004). "Quantification strategies in real-time PCR," in *A-Z of Quantitative PCR*, ed. S. A. Bustin (La Jolla, CA: International University Line), 87–112.

Regad, F., Lebas, M., and Lescure, B. (1994). Interstitial telomeric repeats within the *Arabidopsis thaliana* genome. *J. Mol. Biol.* 239, 163–169. doi: 10.1006/jmbi.1994.1360

Schrumpfová, P. P., Vychodilova, I., Dvorackova, M., Majerska, J., Dokladal, L., Schorova, S., et al. (2014). Telomere repeat binding proteins are functional components of *Arabidopsis* telomeres and interact with telomerase. *Plant J.* 77, 770–781. doi: 10.1111/tpj.12428

Seimiya, H., Sawada, H., Muramatsu, Y., Shimizu, M., Ohko, K., Yamane, K., et al. (2000). Involvement of 14-3-3 proteins in nuclear localization of telomerase. *EMBO J.* 19, 2652–2661. doi: 10.1093/emboj/19.11.2652

Song, J., Mcgivern, J. V., Nichols, K. W., Markley, J. L., and Sheets, M. D. (2008). Structural basis for RNA recognition by a type II poly(A)-binding protein. *Proc. Natl. Acad. Sci. U.S.A.* 105, 15317–15322. doi: 10.1073/pnas.0801274105

Sykorova, E., and Fajkus, J. (2009). Structure-function relationships in telomerase genes. *Biol. Cell* 101, 375–392. doi: 10.1042/BC20080205

Szklarczyk, D., Franceschini, A., Wyder, S., Forslund, K., Heller, D., Huerta-Cepas, J., et al. (2015). STRING v10: protein-protein interaction networks, integrated over the tree of life. *Nucleic Acids Res.* 43, D447–D452. doi: 10.1093/nar/gku1003

Tenea, G. N., Spantzel, J., Lee, L.-Y., Zhu, Y., Lin, K., Johnson, S. J., et al. (2009). Overexpression of several *Arabidopsis* histone genes increases *Agrobacterium*-mediated transformation and transgene expression in plants. *Plant Cell* 21, 3350–3367. doi: 10.1105/tpc.109.070607

Tremousaygue, D., Manevski, A., Bardet, C., Lescure, N., and Lescure, B. (1999). Plant interstitial telomere motifs participate in the control of gene expression in root meristems. *Plant J.* 20, 553–561. doi: 10.1046/j.1365-313X.1999.00627.x

Ungar, L., Yosef, N., Sela, Y., Sharan, R., Ruppin, E., and Kupiec, M. (2009). A genome-wide screen for essential yeast genes that affect telomere length maintenance. *Nucleic Acids Res.* 37, 3840–3849. doi: 10.1093/nar/gkp259

Wahle, E., and Ruegsegger, U. (1999). 3'-End processing of pre-mRNA in eukaryotes. *FEMS Microbiol. Rev.* 23, 277–295. doi: 10.1016/S0168-6445(99)00008-X

Wang, J., Wang, Y., Wang, Z., Liu, L., Zhu, X. G., and Ma, X. (2011). Synchronization of cytoplasmic and transferred mitochondrial ribosomal protein gene expression in land plants is linked to Telo-box motif enrichment. *BMC Evol. Biol.* 11:161. doi: 10.1186/1471-2148-11-161

Watson, J. D. (1972). Origin of concatemeric T7 DNA. *Nat. New Biol.* 239, 197–201. doi: 10.1038/newbio239197a0

Wu, F. H., Shen, S. C., Lee, L.-Y., Lee, S. H., Chan, M. T., and Lin, C. S. (2009). Tape-*Arabidopsis* sandwich - a simpler *Arabidopsis* protoplast isolation method. *Plant Methods* 5:16. doi: 10.1186/1746-4811-5-16

Yamada, K., Lim, J., Dale, J. M., Chen, H., Shinn, P., Palm, C. J., et al. (2003). Empirical analysis of transcriptional activity in the *Arabidopsis* genome. *Science* 302, 842–846. doi: 10.1126/science.1088305

Zachova, D., Fojtova, M., Dvorackova, M., Mozgova, I., Lermontova, I., Peska, V., et al. (2013). Structure-function relationships during transgenic telomerase expression in *Arabidopsis*. *Physiol. Plant* 149, 114–126. doi: 10.1111/ppl.12021

Zhang, W. K., Shen, Y. G., He, X. J., Du, B. X., Xie, Z. M., Luo, G. Z., et al. (2005). Characterization of a novel cell cycle-related gene from *Arabidopsis*. *J. Exp. Bot.* 56, 807–816. doi: 10.1093/jxb/eri075

Conflict of Interest Statement: The authors declare that the research was conducted in the absence of any commercial or financial relationships that could be construed as a potential conflict of interest.

20

The Roles of Cytochrome b_{559} in Assembly and Photoprotection of Photosystem II Revealed by Site-Directed Mutagenesis Studies

Hsiu-An Chu and Yi-Fang Chiu*

Institute of Plant and Microbial Biology – Academia Sinica, Taipei, Taiwan

Cytochrome b_{559} (Cyt b_{559}) is one of the essential components of the Photosystem II reaction center (PSII). Despite recent accomplishments in understanding the structure and function of PSII, the exact physiological function of Cyt b_{559} remains unclear. Cyt b_{559} is not involved in the primary electron transfer pathway in PSII but may participate in secondary electron transfer pathways that protect PSII against photoinhibition. Site-directed mutagenesis studies combined with spectroscopic and functional analysis have been used to characterize Cyt b_{559} mutant strains and their mutant PSII complex in higher plants, green algae, and cyanobacteria. These integrated studies have provided important *in vivo* evidence for possible physiological roles of Cyt b_{559} in the assembly and stability of PSII, protecting PSII against photoinhibition, and modulating photosynthetic light harvesting. This mini-review presents an overview of recent important progress in site-directed mutagenesis studies of Cyt b_{559} and implications for revealing the physiological functions of Cyt b_{559} in PSII.

Keywords: photosynthesis, photosystem II, cytochrome b_{559}, site-directed mutagenesis, photoprotection, photoinhibition

Edited by:
Julian Eaton-Rye,
University of Otago, New Zealand

Reviewed by:
Pavel Pospíšil,
Palacký University, Czech Republic
Gary Brudvig,
Yale University, USA

****Correspondence:***
Hsiu-An Chu
chuha@gate.sinica.edu.tw

INTRODUCTION

Cytochrome b_{559} is one of the essential components of Photosystem II in all oxygenic photosynthetic organisms (Whitmarsh and Pakrasi, 1996; Stewart and Brudvig, 1998; Guskov et al., 2009; Umena et al., 2011). Cyt b_{559} is a heme-bridged heterodimer protein comprising one α- and one β- subunit (encoded by the *psbE* and *psbF* genes) of 9 and 4 kDa, respectively (see **Figure 1**). Each subunit provides a His ligand (His-22 residue of the α- or β-subunit of Cyt b_{559} in *Synechocystis* sp. PCC 6803, corresponding to His-23 residue of α- or His-24 residue of the β-subunit of Cyt b_{559} in *Thermosynechococcus elongatus*) for the non-covalently bound heme, which is located near the stromal side of PSII. In addition, Cyt b_{559} has different redox potential forms depending on the type of PSII preparations and treatments: a HP form with a midpoint redox potential of about +400 mV, an IP form of about +200 mV, and a LP form with a midpoint redox potential of about 0–80 mV (Stewart and Brudvig, 1998; Roncel et al., 2001 and references therein). In intact PSII preparations, Cyt b_{559} is mostly in the reduced HP form under

Abbreviations: APC, allophycocyanin; Car, β-carotene; Cyt b_{559}, cytochrome b_{559}; HP, high potential; IP, intermediate potential; LP, low potential; NPQ, non-photochemical fluorescences quenching; PQ, plastoquinone; PQH_2, plastoquinol; PSII, Photosystem II; Q_A, primary quinone electron acceptor in PSII; Q_B, the secondary quinone electron acceptor in PSII; Q_C, the third plastoquinone-binding site in PSII.

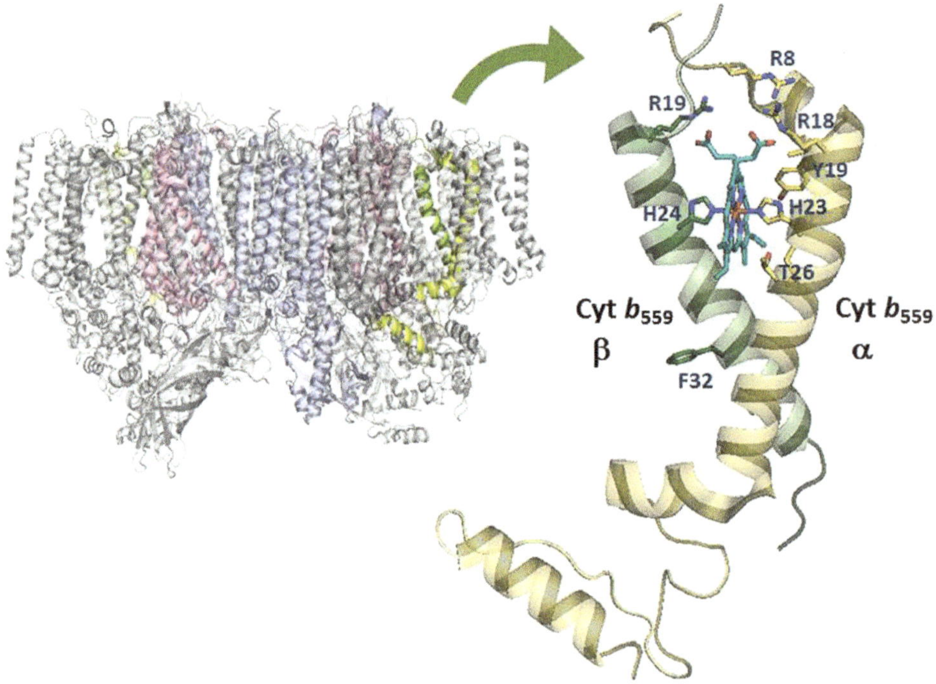

FIGURE 1 | Structural model of the cytochrome b_{559} in Photosystem II. Amino acid residues of cytochrome b_{559} targeted by recent site-directed mutagenesis work are labeled by using amino acid sequences of cytochrome b_{559} from *Thermosynechococcus elongatus*. The figure was created using PyMol and the PDB file 4UB8.

ambient conditions. In inactive or less intact PSII preparations, Cyt b_{559} is typically in the LP or IP form and mostly oxidized (presumably by molecular oxygen) under ambient conditions (Barber and De Las Rivas, 1993; Poulson et al., 1995; Pospíšil et al., 2006).

Several studies have proposed that Cyt b_{559} participates in secondary electron transfer pathways that protect PSII against photoinhibition (see **Figure 2**; Heber et al., 1979; Falkowski et al., 1986; Thompson and Brudvig, 1988; Barber and De Las Rivas, 1993; Poulson et al., 1995; Magnuson et al., 1999; Faller et al., 2001; Tracewell and Brudvig, 2008 and references therein). In these models, the HP form of Cyt b_{559} is thought to donate its electron, via a β-carotene molecule (Car_{D2}), to reduce highly oxidized chlorophyll radicals in PSII under donor-side photoinhibitory conditions (e.g., the oxygen-evolving complex is impaired or under assembly). Oxidized Cyt b_{559} may accept an electron from the acceptor side of PSII (Q_B^-, or reduced PQH_2 from the pool), thus forming a cyclic pathway of electron transfer within PSII. On the other hand, when the electron transfer on the acceptor side of PSII is inhibited (e.g., under high-light conditions), the oxidized Cyt b_{559} might accept an electron from the acceptor side of PSII to prevent the formation of damaging singlet oxygen species (Nedbal et al., 1992; Vass et al., 1992; Barber and De Las Rivas, 1993; Bondarava et al., 2010). In addition, several different enzymatic functions of Cyt b_{559} have been proposed, such as superoxide dismutase (Ananyev et al., 1994) and PQH_2 oxidase in intact PSII (Kruk and Strzałka, 1999, Kruk and Strzalka, 2001; Bondarava et al., 2003, 2010) and superoxide oxidase and reductase in tris-washed PSII (Tiwari and Pospísil, 2009; Pospíšil, 2011). Moreover, a novel quinone-binding site (Q_C) was identified close (about 15 Å) to the heme of Cyt b_{559} in the 2.9-Å PSII crystal structure from *T. elongatus* (Guskov et al., 2009). The occupancy of this Q_C site with PQ (or PQH_2) has been proposed to modulate the redox equilibration between Cyt b_{559} and the PQ pool (Kaminskaya et al., 2006, 2007; Kaminskaya and Shuvalov, 2013) or be involved in the exchange of PQ on the Q_B site from the pool (Guskov et al., 2009). However, the Q_C site was not detected in the more recent 1.9-Å PSII crystal structure (Umena et al., 2011). Despite the recent remarkable progress in understanding the structure and function of PSII, the exact function of Cyt b_{559} in PSII remains unclear.

This mini-review gives an overview of important progress in recent site-directed mutagenesis studies performed to reveal the physiological function(s) of Cyt b_{559} in PSII. More comprehensive reviews on the structure and functions of Cyt b_{559} are available (Whitmarsh and Pakrasi, 1996; Stewart and Brudvig, 1998; Faller et al., 2005; Pospíšil, 2011; Shinopoulos and Brudvig, 2012).

FUNCTION IN ASSEMBLY AND STABILITY OF PSII REACTION CENTERS

Prior mutagenesis studies with *Synechocystis* sp. PCC 6803, *Chlamydomonas reinhardtii* or *Nicotiana tabacum* showed no

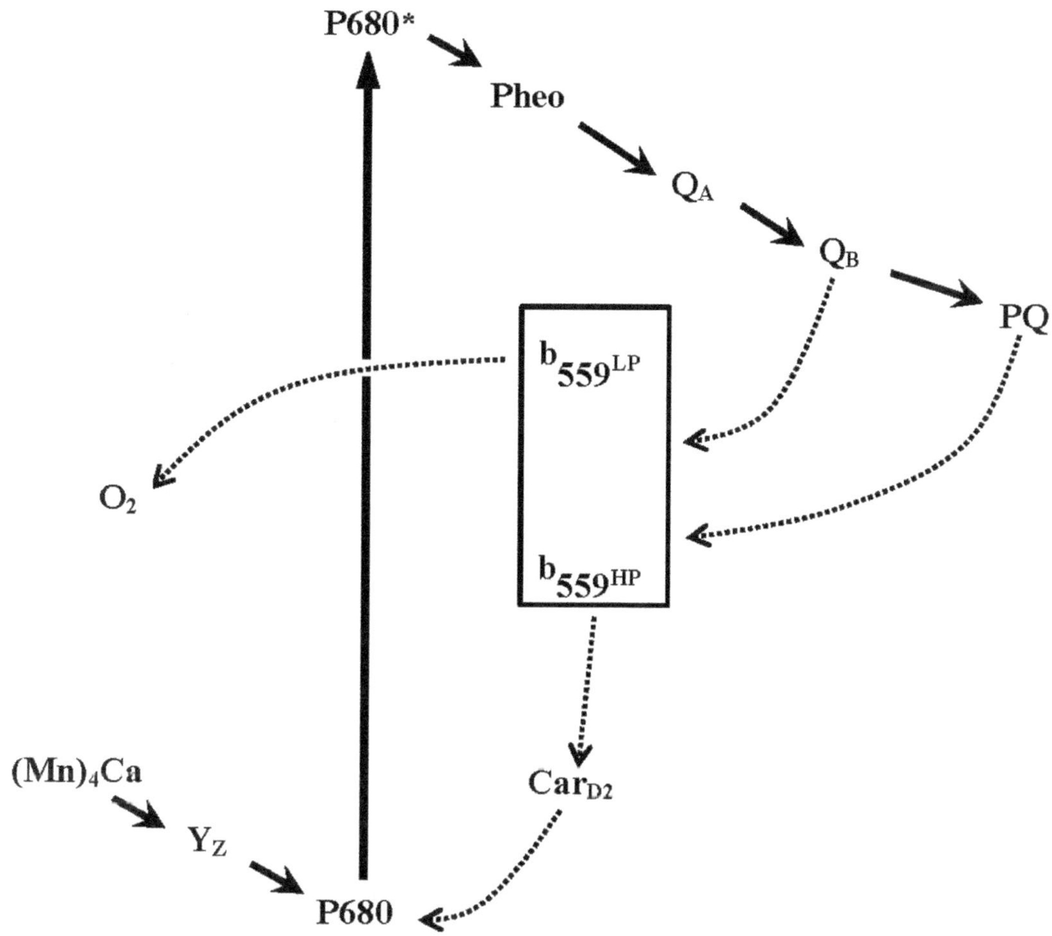

FIGURE 2 | A simplified scheme for electron transfer pathways within Photosystem II reaction centers. Primary electron transfer pathway (bold solid lines with arrows) and possible secondary electron transfer pathways involving cytochrome b_{559} (dotted line with arrows) are shown. The scheme is modified from Whitmarsh and Pakrasi (1996).

stable PSII reaction centers assembled in the absence of either Cyt b_{559} subunit (Pakrasi et al., 1988, 1989, 1990; Morais et al., 1998; Swiatek et al., 2003; Suorsa et al., 2004). Several authors proposed that Cyt b_{559} plays an important structural role, such as being a nucleating factor, during the early stage of PSII assembly (Pakrasi et al., 1988; Morais et al., 1998; Komenda et al., 2004).

In addition, mutagenesis studies with *Synechocystis* sp. PCC 6803 showed that substituting either of the heme axial ligands (His22 of the α-subunit or His22 of the β-subunit of Cyt b_{559} in *Synechocystis* sp. PCC 6803) with Leu, Met, Glu, Gln, Tyr, Lys, Arg, or Cys abolished the photoautotrophic growth and severely diminished the assembly or stability of PSII in the mutant cells, except for H22Kα mutant cells, which were able to grow photoautotrophically and accumulated stable PSII reaction centers (~81% as compared with wild-type cells; Pakrasi et al., 1991; Hung et al., 2007, 2010). Electron paramagnetic resonance results indicated the displacement of one of the two axial ligands to the heme of Cyt b_{559} in H22Kα mutant reaction centers, at least in isolated reaction centers (Hung et al., 2010). In addition, H22Kα and Y18Sα (corresponding to Y19Sα in *T. elongatus*) in mutant PSII core complexes contained predominately the LP form of Cyt b_{559}. The findings support the concept that the redox properties of Cyt b_{559} are strongly influenced by the hydrophobicity and ligation environment of the heme (Krishtalik et al., 1993; Gadjieva et al., 1999; Roncel et al., 2001; Pospíšil and Tiwari, 2010).

Spectroscopic and functional characterizations of the cyanobacterium *Synechocystis* sp. PCC 6803 with mutation of charged residues on the cytoplasmic side of Cyt b_{559} in PSII have been reported (Chiu et al., 2013). All mutant cells grew photoautotrophically and assembled stable PSII. However, R7Eα, R17Eα, and R17Lβ mutant cells grew significantly slower and were more susceptible to photoinhibition as compared with wild-type cells. In addition, the PSII core complexes from R7Eα and R17Lβ cells contained predominantly the LP form of Cyt b_{559}. Electron paramagnetic resonance results indicated the displacement of one of the two axial ligands to the heme of Cyt b_{559} in the reaction centers of the R7Eα and R17Lβ mutants. In recent PSII crystal structural models (Guskov et al., 2009; Umena et al., 2011), the side chains of these Arg residues of Cyt b_{559} (corresponding to Arg8 and Arg18 residues of the α-subunit and Arg19 residue of the β-subunit of Cyt b_{559} in *T. elongatus*) are in close contact with the heme propionates of Cyt b_{559} (see **Figure 1**). Thus, the electrostatic interactions between these Arg residues and the heme propionates of Cyt b_{559} may affect the ligation structure and redox properties of the heme in Cyt b_{559} (Chiu et al., 2013).

Furthermore, mutagenesis studies of *C. reinhardtii* showed that the H23Yα, H23Mα, and H23Cα mutant cells were unable to grow photoautotrophically, were sensitive to photoinhibition, accumulated 10–20% of the PSII (compared to wild-type cells), and contained a disrupted heme pocket while still retaining significant O_2 evolution activity (Morais et al., 2001; Hamilton et al., 2014). Thus, the heme of Cyt b_{559} was not required for photosynthetic water oxidation by PSII (Morais et al., 2001). A recent study also presented evidence to ascribe the photoinhibition phenotype of H23Cα mutant cells to a faster rate of photodamage and an impaired PSII repair cycle (Hamilton et al., 2014). Hence, Cyt b_{559} may play important roles in the assembly, repair and maintenance of the PSII complex *in vivo*.

In the other recent mutant study of *T. elongatus* that took advantage of the robustness of the PSII variant with PsbA3 as the D1 subunit, the four constructed Cyt b_{559} mutants (H23Aα, H23Mα, Y19Fα, and T26Pα) grew photoautotrophically (*T. elongatus* is an obligate photoautotroph; Sugiura et al., 2015). Although the H23Aα and H23Mα mutants assembled only an apo-Cyt b_{559}, the steady-state level of active PSII was comparable to that in the wild-type control. The results suggest that the heme has no structural role in the assembly of PSII in the presence of α- and β-subunits of Cyt b_{559}. This finding is in strong contrast to the *Synechocystis* sp. PCC 6803 mutant showing that proper coordination of the heme cofactor in Cyt b_{559} is important to the assembly or stability of PSII (Pakrasi et al., 1991; Hung et al., 2007). In addition, Cyt b_{559} mutant cells of *T. elongatus* showed no correlation between the rate of photoinhibition and the redox potential of the heme. However, the recovery of the oxygen-evolving activity of PSII after photoinhibition was significantly slower in these mutant cells. PsbA3 is the D1 isoform expressed in *T. elongatus* under high-light conditions (Nakamura et al., 2002; Kós et al., 2008). The high-light D1 isoform in cyanobacteria has a Glu instead of a Gln residue (for the low-light D1 isoform) at position 130 in the D1 protein sequence (for a review, see Mulo et al., 2009) and this Glu residue forms hydrogen-bonding interactions with pheophytinD1 (Dorlet et al., 2001; Shibuya et al., 2010). Cyanobacterial PSIIs with the high-light D1 isoform showed increased photo-tolerance and accelerated non-radiative charge recombination (Tichy et al., 2003). This phototolerant property has been attributed to a photoprotection mechanism involving the redox potential of pheophytinD1, which enhances the probability for non-radiative recombination of the singlet radical pair and prevents the formation of potentially damaging ^{3}P680 and singlet oxygen species (Vass and Cser, 2009; Sugiura et al., 2014). Further investigation could determine whether PsbA3 may compensate the photoprotective function of Cyt b_{559} in the assembly and stability of PSII in these Cyt b_{559} mutant cells.

FUNCTION IN PROTECTING PSII AGAINST PHOTOINHIBITION

Numerous site-directed mutagenesis studies have investigated the role of Cyt b_{559} in protecting PSII against photoinhibition under high light. In the photoprotective models, oxidized Cyt b_{559} may accept an electron from the acceptor side of PSII (Q_B^-, Q_C, or reduced PQH_2 from the pool). Previous mutant studies of dark-adapted leaves of the F26Sβ Cyt b_{559} tobacco mutant showed a greatly reduced PQ pool, conversion of the redox-potential form of Cyt b_{559} to the LP form, and photosynthetic activities sensitive to high light (Bondarava et al., 2003, 2010). In addition, R7Eα and R17Lβ Cyt b_{559} mutant cells of *Synechocystis* sp. PCC 6803 and R18Sα Cyt b_{559} mutant cells of *T. elongatus* showed markedly reduced PQ pools, altered redox-potential forms of Cyt b_{559}, and high susceptibility to light stress (Chiu et al., 2013; Guerrero et al., 2014). A defect in PQH_2 oxidase activity of Cyt b_{559} due to altered redox-potential forms of Cyt b_{559} in these mutant strains could explain their high susceptibility to strong light and greatly reduced PQ pools. Therefore, Cyt b_{559} may function as a PQH_2 oxidase to keep the PQ pool and the acceptor side of PSII oxidized in the dark, thereby preventing PSII from acceptor-side photoinhibition (Kruk and Strzałka, 1999, Kruk and Strzalka, 2001; Bondarava et al., 2003, 2010). However, one recent study reported no defect in PQH_2 oxidation in the dark in H23Cα mutant cells of *C. reinhardtii*, even though H23Cα mutant cells contained a disrupted heme-binding pocket of Cyt b_{559} and were sensitive to photoinhibition (Hamilton et al., 2014). Further studies are required to clarify this discrepancy. In addition, study of the 2.9-Å resolution PSII crystal structure reported the binding of Q_C at a hydrophobic cavity near Cyt b_{559} (Guskov et al., 2009). Several spectroscopic studies have provided evidence that the occupancy of the Q_C site by PQ (or PQH_2) may modulate the redox potential of Cyt b_{559} and mediate the redox equilibration between Cyt b_{559} and the PQ pool (Kaminskaya et al., 2006, 2007; Kaminskaya and Shuvalov, 2013). A recent study provided evidence of a possible one-electron oxidation of PQH_2 by Cyt b_{559} at the Q_C site involved in the formation of a superoxide anion radical (Yadav et al., 2014). The above results are consistent with Cyt b_{559} possibly accepting an electron from PQH_2 via the Q_C site in PSII. However, the Q_C site was not present in the more recent 1.9-Å PSII crystal structure (Umena et al., 2011). Further investigations are needed to solve this important issue.

Spectroscopic and functional characterization of the H22Kα and Y18Sα Cyt b_{559} mutant cells of *Synechocystis* sp. PCC 6803 showed that both mutants have functional PSII and exhibited the normal period-four oscillation in oxygen yield (Hung et al., 2010). However, both mutants were more susceptible to photoinhibition than the wild type under high-light conditions. In addition, PSII core complexes from the H22Kα and Y18Sα mutants predominantly contained the oxidized LP form of Cyt b_{559} (~79 and 86%, respectively). A defect in the photoprotective function of Cyt b_{559} in H22Kα and Y18Sα mutants could explain their high susceptibility to strong light. Furthermore, H22Kα and Y18Sα Cyt b_{559} mutants in a D1-D170A genetic background that prevented assembly of the Mn cluster showed almost completely abolished accumulation of PSII even under normal-growth-light conditions. The data support an important redox role of Cyt b_{559} in protecting PSII under donor-side photoinhibition conditions (Hung et al., 2010).

Furthermore, under low light, the H23Cα Cyt b_{559} mutant showed more rapid assembly of the Mn_4CaO_5 cluster than the wild-type control in *C. reinhardtii* (Hamilton et al., 2014). However, the photoactivation of oxygen-evolving PSII in the H23Cα mutant was inhibited under high light. The results suggest that reduction of $P680^+$ via cyclic electron flow within PSII (via Cyt b_{559} and Car_{D2}, **Figure 2**) may compete with the photoactivation process and provides important *in vivo* evidence for a photoprotective role of Cyt b_{559} in photo-assembly of the Mn_4CaO_5 cluster in PSII (Hamilton et al., 2014).

A recent mutant study involving *T. elongatus* showed that the midpoint redox potential of the HP form of Cyt b_{559} was significantly destabilized (converted to the IP form) in mutant PSII core complexes of Cyt b_{559} mutant strains (I14Aα, I14Sα, R18Sα, I27Aα, I27Tα, and F32Yβ; Guerrero et al., 2014). When the oxygen-evolving complex was inactive, the yield of dark-reduction of Cyt b_{559} was lower and the kinetics was slower in the R18Sα mutant than in wild-type cells. The results support the concept that the HP form of Cyt b_{559} may function as a PQH_2 oxidase to keep the PQ pool oxidized and also as an electron reservoir for the cyclic electron flow within PSII when the donor-side of PSII is impaired (Guerrero et al., 2014).

Moreover, a previous spectroscopic study showed that different spectral forms of Car were oxidized in PSII samples containing different redox forms of Cyt b_{559} (Tracewell and Brudvig, 2008). The authors proposed that the quenching properties of PSII may be controlled by the redox form of Cyt b_{559} by modulating the different type of oxidized Car species (radical cation or neutral radical) formed in PSII. Future study could investigate the quenching properties of PSII in the wild type versus Cyt b_{559} mutant strains of cyanobacteria with different redox forms of Cyt b_{559} to validate this proposal.

EFFECTS ON PHOTOSYNTHETIC LIGHT HARVESTING

Recent mutant studies revealed a novel role of Cyt b_{559} in modulating photosynthetic light harvesting in PSII reaction centers. A spontaneously generated mutant from *Synechocystis* sp. PCC 6803 wild-type cells grown in BG-11 agar plates containing 5 mM Glu and 10 μM DCMU carried an Arg7 to Leu mutation on the alpha-subunit of Cyt b_{559} in PSII (Chiu et al., 2009). Results of 77-K fluorescence and room-temperature chlorophyll *a* fluorescence spectra indicated that the energy transfer from phycobilisomes to PSII reaction centers was partially inhibited or uncoupled in this mutant. In addition, the cytoplasmic side of Cyt b_{559} is located within the predicted contact sites in PSII for the APC core complex of the phycobilisome (Barber et al., 2003). The Arg7 to Leu mutation of Cyt b_{559} may alter the interaction between the APC core complex and PSII reaction centers, thereby reducing energy delivery from the antenna to the reaction center and protecting mutant cells against DCMU-induced photo-oxidative stress (Rutherford and Krieger-Liszkay, 2001).

Many cyanobacteria including *Synechocystis* sp. PCC 6803 have a novel blue-green light-induced NPQ mechanism to protect PSII reaction centers against photodamage under high-light stress (Kirilovsky and Kerfeld, 2012). Under high-light conditions, a soluble orange carotenoid protein is able to absorb blue–green light and undergoes photo-conversion into the active red form, which interacts with the APC core of the phycobilisome and dissipates excess excitation energy from the phycobilisome as heat. Interestingly, several *Synechocystis* sp. PCC 6803 mutant cells (e.g., R7Lα and R17Lβ) with mutations on the cytoplasmic side of Cyt b_{559} in PSII showed significant inhibition of the effects of blue–green light-induced NPQ and apparent acceleration on its recovery (Chiu et al., 2013). These results are consistent with the proposal that the mutations on Cyt b_{559} may alter the interaction between the phycobilisome and PSII reaction centers, thereby affecting the regulation of photosynthetic light harvesting in *Synechocystis* sp. PCC 6803.

CONCLUSION AND PERSPECTIVES

Site-directed mutagenesis studies combined with spectroscopic and functional characterization have revealed multiple roles of Cyt b_{559} in the assembly and photoprotection of PSII reaction centers. The findings provide convincing evidence for the physiological role(s) of Cyt b_{559} in a photoprotective secondary electron transfer pathway within PSII reaction centers, as was suggested from earlier studies of isolated PSII complexes (reviews in Whitmarsh and Pakrasi, 1996; Stewart and Brudvig, 1998; Shinopoulos and Brudvig, 2012). In the near future, site-directed mutagenesis studies combined with advanced high-resolution protein crystallography and spectroscopic and functional analysis will provide further new insights (e.g., structure and function relationships for different redox forms of Cyt b_{559}) and possibly the final proof of the molecular mechanisms of Cyt b_{559} in PSII.

AUTHOR CONTRIBUTIONS

H-AC wrote the major part of the manuscript. Y-FC wrote the minor part of the manuscript, contributed the **Figure 1** and edited the references.

ACKNOWLEDGMENT

This work is supported by the Ministry of Science and Technology in Taiwan (104-2311-B-001-035 and 104-2627-M-001-001) and Academia Sinica to H-AC.

REFERENCES

Ananyev, G., Renger, G., Wacker, U., and Klimov, V. (1994). The photoproduction of superoxide radicals and the superoxide dismutase activity of Photosystem II. The possible involvement of cytochrome b559. *Photosynth. Res.* 41, 327–338. doi: 10.1007/BF00019410

Barber, J., and De Las Rivas, J. (1993). A functional model for the role of cytochrome b559 in the protection against donor and acceptor side photoinhibition. *Proc. Natl. Acad. Sci. U.S.A.* 90, 10942–10946. doi: 10.1073/pnas.90.23.10942

Barber, J., Morris, E. P., and da Fonseca, P. C. A. (2003). Interaction of the allophycocyanin core complex with Photosystem II. *Photochem. Photobiol. Sci.* 2, 536–541. doi: 10.1039/b300063j

Bondarava, N., De Pascalis, L., Al-Babili, S., Goussias, C., Golecki, J. R., Beyer, P., et al. (2003). Evidence that cytochrome b559 mediates the oxidation of reduced plastoquinone in the dark. *J. Biol. Chem.* 278, 13554–13560. doi: 10.1074/jbc.M212842200

Bondarava, N., Gross, C. M., Mubarakshina, M., Golecki, J. R., Johnson, G. N., and Krieger-Liszkay, A. (2010). Putative function of cytochrome b559 as a plastoquinol oxidase. *Physiol. Plant.* 138, 463–473. doi: 10.1111/j.1399-3054.2009.01312.x

Chiu, Y.-F., Chen, Y.-H., Roncel, M., Dilbeck, P. L., Huang, J.-Y., Ke, S.-C., et al. (2013). Spectroscopic and functional characterization of cyanobacterium *Synechocystis* PCC 6803 mutants on the cytoplasmic-side of cytochrome b559 in Photosystem II. *Biochim. Biophys. Acta* 1827, 507–519. doi: 10.1016/j.bbabio.2013.01.016

Chiu, Y.-F., Lin, W.-C., Wu, C.-M., Chen, Y.-H., Hung, C.-H., Ke, S.-C., et al. (2009). Identification and characterization of a cytochrome b559 *Synechocystis* 6803 mutant spontaneously generated from DCMU-inhibited photoheterotrophical growth conditions. *Biochim. Biophys. Acta* 1787, 1179–1188. doi: 10.1016/j.bbabio.2009.05.007

Dorlet, P., Xiong, L., Sayre, R. T., and Un, S. (2001). High field EPR study of the pheophytin anion radical in wild type and D1-E130 mutants of Photosystem II in *Chlamydomonas reinhardtii*. *J. Biol. Chem.* 276, 22313–22316. doi: 10.1074/jbc.M102475200

Falkowski, P. G., Fujita, Y., Ley, A., and Mauzerall, D. (1986). Evidence for cyclic electron flow around Photosystem II in *Chlorella pyrenoidosa*. *Plant Physiol.* 81, 310–312. doi: 10.1104/pp.81.1.310

Faller, P., Fufezan, C., and Rutherford, A. W. (2005). "Side-path electron donors: cytochrome b559, chlorophyll Z and β-Carotene," in *Photosystem II*, eds T. J. Wydrzynski and K. Satoh (Dordrecht: Springer), 347–365.

Faller, P., Pascal, A., and Rutherford, A. W. (2001). Beta-carotene redox reactions in Photosystem II: electron transfer pathway. *Biochemistry* 40, 6431–6440. doi: 10.1021/bi0026021

Gadjieva, R., Mamedov, F., Renger, G., and Styring, S. (1999). Interconversion of low- and high-potential forms of cytochrome b559 in tris-washed Photosystem II membranes under aerobic and anaerobic conditions. *Biochemistry* 38, 10578–10584. doi: 10.1021/bi9904656

Guerrero, F., Zurita, J. L., Roncel, M., Kirilovsky, D., and Ortega, J. M. (2014). The role of the high potential form of the cytochrome b559: study of *Thermosynechococcus elongatus* mutants. *Biochim. Biophys. Acta* 1837, 908–919. doi: 10.1016/j.bbabio.2014.02.024

Guskov, A., Kern, J., Gabdulkhakov, A., Broser, M., Zouni, A., and Saenger, W. (2009). Cyanobacterial Photosystem II at 2.9-Å resolution and the role of quinones, lipids, channels and chloride. *Nat. Struct. Mol. Biol.* 16, 334–342. doi: 10.1038/nsmb.1559

Hamilton, M. L., Franco, E., Deák, Z., Schlodder, E., Vass, I., and Nixon, P. J. (2014). Investigating the photoprotective role of cytochrome b559 in Photosystem II in a mutant with altered ligation of the haem. *Plant Cell Physiol.* 55, 1276–1285. doi: 10.1093/pcp/pcu070

Heber, U., Kirk, M. R., and Boardman, N. K. (1979). M.R. Kirk a, N.K. BoardmanPhotoreactions of cytochrome b559 and cyclic electron flow in Photosystem II of intact chloroplasts. *Biochim. Biophys. Acta* 546, 292–306. doi: 10.1016/0005-2728(79)90047-1

Hung, C.-H., Huang, J.-Y., Chiu, Y.-F., and Chu, H.-A. (2007). Site-directed mutagenesis on the heme axial-ligands of cytochrome b559 in Photosystem II by using cyanobacteria *Synechocystis* PCC 6803. *Biochim. Biophys. Acta* 1767, 686–693. doi: 10.1016/j.bbabio.2007.02.016

Hung, C.-H., Hwang, H.-J., Chen, Y.-H., Chiu, Y.-F., Ke, S.-C., Burnap, R. L., et al. (2010). Spectroscopic and functional characterizations of cyanobacterium *Synechocystis* PCC 6803 mutants on and near the heme axial ligand of cytochrome b559 in Photosystem II. *J. Biol. Chem.* 285, 5653–5663. doi: 10.1074/jbc.M109.044719

Kaminskaya, O., Shuvalov, V. A., and Renger, G. (2006). Evidence for a novel quinone-binding site in the Photosystem II (PS II) complex that regulates the redox potential of cytochrome b559. *Biochemistry* 46, 1091–1105. doi: 10.1021/bi0613022

Kaminskaya, O., Shuvalov, V. A., and Renger, G. (2007). Two reaction pathways for transformation of high potential cytochrome b559 of PS II into the intermediate potential form. *Biochim. Biophys. Acta* 1767, 550–558. doi: 10.1016/j.bbabio.2007.02.005

Kaminskaya, O. P., and Shuvalov, V. A. (2013). Biphasic reduction of cytochrome b559 by plastoquinol in Photosystem II membrane fragments: evidence for two types of cytochrome b559/plastoquinone redox equilibria. *Biochim. Biophys. Acta* 1827, 471–483. doi: 10.1016/j.bbabio.2013.01.007

Kirilovsky, D., and Kerfeld, C. A. (2012). The orange carotenoid protein in photoprotection of Photosystem II in cyanobacteria. *Biochim. Biophys. Acta* 1817, 158–166. doi: 10.1016/j.bbabio.2011.04.013

Komenda, J., Reisinger, V., Muller, B. C., Dobakova, M., Granvogl, B., and Eichacker, L. A. (2004). Accumulation of the D2 protein is a key regulatory step for assembly of the Photosystem II reaction center complex in *Synechocystis* PCC 6803. *J. Biol. Chem.* 279, 48620–48629. doi: 10.1074/jbc.M405725200

Kós, P. B., Deák, Z., Cheregi, O., and Vass, I. (2008). Differential regulation of psbA and psbD gene expression, and the role of the different D1 protein copies in the cyanobacterium *Thermosynechococcus elongatus* BP-1. *Biochim. Biophys. Acta* 1777, 74–83. doi: 10.1016/j.bbabio.2012.04.010

Krishtalik, L. I., Tae, G.-S., Cherepanov, D. A., and Cramer, W. A. (1993). The redox properties of cytochromes b imposed by the membrane electrostatic environment. *Biophys. J.* 65, 184–195. doi: 10.1016/S0006-3495(93)81050-6

Kruk, J., and Strzałka, K. (1999). Dark reoxidation of the plastoquinone-pool is mediated by the low-potential form of cytochrome b559 in spinach thylakoids. *Photosynth. Res.* 62, 273–279. doi: 10.1023/a:1006374319191

Kruk, J., and Strzalka, K. (2001). Redox changes of cytochrome b559 in the presence of plastoquinones. *J. Biol. Chem.* 276, 86–91. doi: 10.1074/jbc.M0036 02200

Magnuson, A., Rova, M., Mamedov, F., Fredriksson, P.-O., and Styring, S. (1999). The role of cytochrome b559 and tyrosineD in protection against photoinhibition during in vivo photoactivation of Photosystem II. *Biochim. Biophys. Acta* 1411, 180–191. doi: 10.1016/s0005-2728(99)00044-4

Morais, F., Barber, J., and Nixon, P. J. (1998). The chloroplast-encoded α subunit of cytochrome b559 is required for assembly of the Photosystem Two complex in both the light and the dark in *Chlamydomonas reinhardtii*. *J. Biol. Chem.* 273, 29315–29320. doi: 10.1074/jbc.273.45.29315

Morais, F., Kuhn, K., Stewart, D. H., Barber, J., Brudvig, G. W., and Nixon, P. J. (2001). Photosynthetic water oxidation in cytochrome b559 mutants containing a disrupted heme-binding pocket. *J. Biol. Chem.* 276, 31986–31993. doi: 10.1074/jbc.M103935200

Mulo, P., Sicora, C., and Aro, E.-M. (2009). Cyanobacterial psbA gene family: optimization of oxygenic photosynthesis. *Cell. Mol. Life Sci.* 66, 3697–3710. doi: 10.1007/s00018-009-0103-6

Nakamura, Y., Kaneko, T., Sato, S., Ikeuchi, M., Katoh, H., Sasamoto, S., et al. (2002). Complete genome structure of the thermophilic cyanobacterium *Thermosynechococcus elongatus* BP-1. *DNA Res.* 9, 123–130. doi: 10.1093/dnares/9.4.123

Nedbal, L., Samson, G., and Whitmarsh, J. (1992). Redox state of a one-electron component controls the rate of photoinhibition of Photosystem II. *Proc. Natl. Acad. Sci. U.S.A.* 89, 7929–7933. doi: 10.1073/pnas.89.17.7929

Pakrasi, H. B., Ciechi, P. D., and Whitmarsh, J. (1991). Site directed mutagenesis of the heme axial ligands of cytochrome b559 affects the stability of the Photosystem II complex. *EMBO J* 10, 1619–1627.

Pakrasi, H. B., Diner, B. A., Williams, J. G. K., and Arntzen, C. J. (1989). Deletion mutagenesis of the cytochrome b559 protein inactivates the reaction center of Photosystem II. *Plant Cell* 1, 591–597. doi: 10.2307/3868946

Pakrasi, H. B., Nyhus, K. J., and Granok, H. (1990). Targeted deletion mutagenesis of the beta subunit of cytochrome b559 protein destabilizes the reaction center of Photosystem II. *Z. Naturforsch. C* 45, 423–429.

Pakrasi, H. B., Williams, J. G., and Arntzen, C. J. (1988). Targeted mutagenesis of the psbE and psbF genes blocks photosynthetic electron transport: evidence for a functional role of cytochrome b559 in Photosystem II. *EMBO J.* 7, 325–332.

Pospíšil, P. (2011). Enzymatic function of cytochrome b559 in Photosystem II. *J. Photochem. Photobiol. B* 104, 341–347. doi: 10.1016/j.jphotobiol.2011.02.013

Pospíšil, P., Šnyrychová, I., Kruk, J., Strzałka, K., and Nauš, J. (2006). Evidence that cytochrome b559 is involved in superoxide production in Photosystem II: effect of synthetic short-chain plastoquinones in a cytochrome b559 tobacco mutant. *Biochem. J.* 397, 321–327. doi: 10.1042/BJ20060068

Pospíšil, P., and Tiwari, A. (2010). Differential mechanism of light-induced and oxygen-dependent restoration of the high-potential form of cytochrome b559 in Tris-treated Photosystem II membranes. *Biochim. Biophys. Acta* 1797, 451–456. doi: 10.1016/j.bbabio.2009.12.023

Poulson, M., Samson, G., and Whitmarsh, J. (1995). Evidence that cytochrome b559 protects Photosystem II against photoinhibition. *Biochemistry* 34, 10932–10938. doi: 10.1021/bi00034a027

Roncel, M., Ortega, J. M., and Losada, M. (2001). Factors determining the special redox properties of photosynthetic cytochrome b559. *Eur. J. Biochem.* 268, 4961–4968. doi: 10.1046/j.0014-2956.2001.02427.x

Rutherford, A. W., and Krieger-Liszkay, A. (2001). Herbicide-induced oxidative stress in Photosystem II. *Trends Biochem. Sci.* 26, 648–653. doi: 10.1016/S0968-0004(01)01953-3

Shibuya, Y., Takahashi, R., Okubo, T., Suzuki, H., Sugiura, M., and Noguchi, T. (2010). Hydrogen bond interaction of the pheophytin electron acceptor and its radical anion in Photosystem II as revealed by Fourier transform infrared difference spectroscopy. *Biochemistry* 49, 493–501. doi: 10.1021/bi9018829

Shinopoulos, K. E., and Brudvig, G. W. (2012). Cytochrome b559 and cyclic electron transfer within Photosystem II. *Biochim. Biophys. Acta* 1817, 66–75. doi: 10.1016/j.bbabio.2011.08.002

Stewart, D. H., and Brudvig, G. W. (1998). Cytochrome b559 of Photosystem II. *Biochim. Biophys. Acta* 1367, 63–87. doi: 10.1016/s0005-2728(98)00139 x

Sugiura, M., Azami, C., Koyama, K., Rutherford, A. W., Rappaport, F., and Boussac, A. (2014). Modification of the pheophytin redox potential in *Thermosynechococcus elongatus* Photosystem II with PsbA3 as D1. *Biochim. Biophys. Acta* 1837, 139–148. doi: 10.1016/j.bbabio.2013.09.009

Sugiura, M., Nakamura, M., Koyama, K., and Boussac, A. (2015). Assembly of oxygen-evolving Photosystem II efficiently occurs with the apo-Cyt b559 but the holo-Cyt b559 accelerates the recovery of a functional enzyme upon photoinhibition. *Biochim. Biophys. Acta* 1847, 276–285. doi: 10.1016/j.bbabio.2014.11.009

Suorsa, M., Regel, R. E., Paakkarinen, V., Battchikova, N., Herrmann, R. G., and Aro, E.-M. (2004). Protein assembly of Photosystem II and accumulation of subcomplexes in the absence of low molecular mass subunits PsbL and PsbJ. *Eur. J. Biochem.* 271, 96–107. doi: 10.1046/j.1432-1033.2003.03906.x

Swiatek, M., Regel, R. E., Meurer, J., Wanner, G., Pakrasi, H. B., Ohad, I., et al. (2003). Effects of selective inactivation of individual genes for low-molecular-mass subunits on the assembly of Photosystem II, as revealed by chloroplast transformation: the psbEFLJ operon in *Nicotiana tabacum*. *Mol. Genet. Genomics* 268, 699–710. doi: 10.1007/s00438-002-0791-1

Thompson, L. K., and Brudvig, G. W. (1988). Cytochrome b559 may function to protect Photosystem II from photoinhibition. *Biochemistry* 27, 6653–6658. doi: 10.1021/bi00418a002

Tichy, M., Lupinkova, L., Sicora, C., Vass, I., Kuvikova, S., Prasil, O., et al. (2003). *Synechocystis* 6803 mutants expressing distinct forms of the Photosystem II D1 protein from *Synechococcus* 7942: relationship between the psbA coding region and sensitivity to visible and UV-B radiation. *Biochim. Biophys. Acta* 1605, 55–66. doi: 10.1016/S0005-2728(03)00064-1

Tiwari, A., and Pospísil, P. (2009). Superoxide oxidase and reductase activity of cytochrome b559 in Photosystem II. *Biochim. Biophys. Acta* 1787, 985–994. doi: 10.1016/j.bbabio.2009.03.017

Tracewell, C. A., and Brudvig, G. W. (2008). Characterization of the secondary electron-transfer pathway intermediates of Photosystem II containing low-potential cytochrome b559. *Photosynth. Res.* 98, 189–197. doi: 10.1007/s11120-008-9360-8

Umena, Y., Kawakami, K., Shen, J. R., and Kamiya, N. (2011). Crystal structure of oxygen-evolving Photosystem II at a resolution of 1.9 Å. *Nature* 473, 55–60. doi: 10.1038/nature09913

Vass, I., and Cser, K. (2009). Janus-faced charge recombinations in Photosystem II photoinhibition. *Trends Plant Sci.* 14, 200–205. doi: 10.1016/j.tplants.2009.01.009

Vass, I., Styring, S., Hundal, T., Koivuniemi, A., Aro, E., and Andersson, B. (1992). Reversible and irreversible intermediates during photoinhibition of Photosystem II: stable reduced QA species promote chlorophyll triplet formation. *Proc. Natl. Acad. Sci. U.S.A.* 89, 1408–1412. doi: 10.1073/pnas.89.4.1408

Whitmarsh, J., and Pakrasi, H. (1996). "Form and function of cytochrome b559," in *Oxygenic Photosynthesis: The Light Reactions*, eds D. Ort and C. Yocum (Dordrecht: Springer), 249–264.

Yadav, D. K., Prasad, A., Kruk, J., and Pospíšil, P. (2014). Evidence for the involvement of loosely bound plastosemiquinones in superoxide anion radical production in Photosystem II. *PLoS ONE* 9:e115466. doi: 10.1371/journal.pone.0115466

Conflict of Interest Statement: The authors declare that the research was conducted in the absence of any commercial or financial relationships that could be construed as a potential conflict of interest.

21

Beyond the Divide: Boundaries for Patterning and Stem Cell Regulation in Plants

Shelley R. Hepworth[1] and Véronique A. Pautot[2]**

[1] Department of Biology, Institute of Biochemistry, Carleton University, Ottawa, ON, Canada, [2] Institut Jean-Pierre Bourgin, Institut National de la Recherche Agronomique, AgroParisTech, CNRS, Université Paris-Saclay, Versailles, France

The initiation of plant lateral organs from the shoot apical meristem (SAM) is closely associated with the formation of specialized domains of restricted growth known as the boundaries. These zones are required in separating the meristem from the growing primordia or adjacent organs but play a much broader role in regulating stem cell activity and shoot patterning. Studies have revealed a network of genes and hormone pathways that establish and maintain boundaries between the SAM and leaves. Recruitment of these pathways is shown to underlie a variety of processes during the reproductive phase including axillary meristems production, flower patterning, fruit development, and organ abscission. This review summarizes the role of conserved gene modules in patterning boundaries throughout the life cycle.

Keywords: meristem, lateral organ boundary, organ separation, inflorescence architecture, fruit patterning, flower patterning, abscission, dehiscence

Edited by:
Lin Xu,
Shanghai Institutes for Biological Sciences, China

Reviewed by:
Javier Palatnik,
Instituto de Biologia Molecular y Celular de Rosario, Argentina
Patricia Springer,
University of California, Riverside, USA
Long Mao,
Chinese Academy of Agricultural Sciences, China

****Correspondence:***
Shelley R. Hepworth
shelley_hepworth@carleton.ca;
Véronique A. Pautot
veronique.pautot@versailles.inra.fr

INTRODUCTION

The shoot apical meristem (SAM) plays a crucial role in plant development as a continuous source of founder cells for provision of new leaves, shoots, and internodes throughout the life cycle. The SAM is organized into a central zone composed of slowly dividing stem cells, a peripheral zone where lateral organs initiate, and a rib zone that provides cells for internodes (Aichinger et al., 2012). The maintenance of meristems depends on the balance between two antagonistic activities: propagation of stem cells at the center of the meristem and the initiation of organs at the periphery. Boundaries are domains of restricted growth that maintain this balance by separating the meristem from the growing primordia and by forming an interface between organs (Žádníková and Simon, 2014). These interfaces play a critical role by influencing cell fate in adjacent tissues. The best-characterized boundary is the domain that separates leaves from the SAM during the vegetative phase. How principles governing the activity of this boundary apply to other developmental contexts is an important question. For example, boundaries in the leaf control shape and complexity whereas boundaries in the inflorescence have specialized functions such as axillary meristem (AM) production, fruit dehiscence, and organ abscission. Thus, many aspects of plant architecture are dependent on the boundary. In this review, we first describe the genetic control of boundaries during the vegetative phase, and then focus on elaboration of these pathways for specialized functions during the reproductive phase focusing on the model plant species *Arabidopsis thaliana* (*Arabidopsis*).

SAM INITIATION, SAM MAINTENANCE AND COTYLEDON SEPARATION

The NAM-ATAF-CUC (NAC)-type CUP-SHAPED COTYLEDON1 (CUC1), CUC2, and CUC3 transcription factors confer boundary identity in higher land plants (Maugarny et al., 2015). These factors initiate the SAM and establish boundaries in conjunction with SHOOT MERISTEMLESS (STM), a three-amino acid loop extension (TALE) class I KNOTTED1-like (KNOX) homeodomain protein [(**Figure 1A**) and (Hamant and Pautot, 2010; Hay and Tsiantis, 2010)]. CUC-STM forms a conserved module in development that was first identified in embryos (Aida et al., 1999; Takada et al., 2001; Aida et al., 2002). During the globular stage of embryogenesis, *CUC1* and *CUC2* genes are activated in a narrow band between the presumptive cotyledons (Aida et al., 1999; Takada et al., 2001), where auxin is depleted (Benková et al., 2003) based on positional cues provided by WUSCHEL-RELATED HOMEOBOX (WOX2) and WOX8/STIMPY-LIKE (Lie et al., 2012). CUC1/2 factors activate *STM* at late globular stage to initiate the meristem and separate the cotyledons (Aida et al., 1999). STM in turn maintains *CUC* expression (Aida et al., 1999). More recently, STM has also been identified as a direct regulator of *CUC1* (Spinelli et al., 2011). This pattern is reinforced by two SWI/SNF chromatin remodeling ATPase complexes BRAMHA (BRM) and SPLAYED (SYD) acting independently of STM: BRM is a positive regulator of all three *CUC* genes whereas SYD is required for *CUC2* expression (Kwon et al., 2006). By late torpedo stage, *STM* marks the central region of the meristem and is slightly detected in boundaries while expression of *CUC* genes is restricted to boundaries (Long and Barton, 1998; Aida et al., 1999). Double mutant analyses show that contributions of the different *CUC* genes are partially redundant. For example, *CUC1/2* are essential for meristem initiation while *CUC3* plays a more prominent role in organ separation (Vroemen et al., 2003) and AM production (Hibara et al., 2006). *CUC1* and *CUC2* transcripts are targeted by *microRNA164* (*miR164*) to restrict their expression domain while *CUC3* from a different subclade does not contain a *miR164* binding site (Laufs et al., 2004; Mallory et al., 2004).

Other three-amino acid loop extension (TALE) homeodomain transcription factors contribute redundantly with STM in SAM initiation and maintenance. The TALE superfamily is divided into KNOX and BELL classes, whose members function as heterodimers (Hamant and Pautot, 2010; Hay and Tsiantis, 2010). Formation of KNOX-BELL heterodimers regulates nuclear localization (Cole et al., 2006; Rutjens et al., 2009; Kim et al., 2013) and influences binding site selection (Smith et al., 2002). Within the *KNOX* subclass, *BREVIPEDICELLUS* (*BP*)/*KNOTTED1-LIKE FROM A. THALIANA1* (*KNAT1*) is expressed in the peripheral and rib zones of the SAM (Lincoln et al., 1994) whereas *KNAT6* is expressed in boundaries (Belles-Boix et al., 2006). Mutation in *BP* enhances only the meristem defect of weak *stm* mutants (Byrne et al., 2002) whereas *knat6* mutation also impairs cotyledon separation showing a specific role for *KNAT6* in boundaries (Belles-Boix et al., 2006). *KNAT2*, the fourth *KNOX* class I member is expressed at the base of the meristem and in lateral organ boundaries but its inactivation does not enhance the meristem defects of weak *stm* mutants (Byrne et al., 2002; Belles-Boix et al., 2006). *KNAT2* role in the SAM remains undetermined.

At least three BELL homeodomain proteins encoded by *PENNYWISE* (*PNY*), *POUND-FOOLISH* (*PNF*), and *ARABIDOPSIS THALIANA HOMEOBOX GENE1* (*ATH1*) interact with STM to maintain the SAM (Byrne et al., 2003; Kanrar et al., 2006; Rutjens et al., 2009). *PNY* is expressed in the central zone of the SAM (Smith et al., 2004); *PNF* is expressed in the central and rib zones of the SAM (Smith et al., 2004), whereas *ATH1* is more broadly expressed in the SAM, young leaves, and boundaries and is shown to control patterning in the basal region of shoot organs (Proveniers et al., 2007; Gómez-Mena and Sablowski, 2008). PNY and ATH1 contribute redundantly with STM in SAM initiation and maintenance (Byrne et al., 2003; Kanrar et al., 2006; Rutjens et al., 2009). PNY and PNF maintain the integrity of the central zone since the expression domain of *STM* is narrower in *pny pnf* double mutants (Ung et al., 2011). Meristem termination defects in this mutant are attributed to depletion of nuclear localized BELL-STM complexes (Rutjens et al., 2009), but recent data show that PNY and PNF negatively regulate lateral organ boundary genes including *ATH1* and *KNAT6* expression to maintain SAM function (Khan et al., 2015).

TALE transcription factors repress cellular differentiation in the meristem in part by regulating the abundance of hormones including gibberellins (GA), cytokinins (CK), and brassinosteroids (BR) (**Figures 2A,B**). A WUSCHEL-CLAVATA (WUS-CLV) feedback loop functions in parallel to keep the stem-cell niche constant in size (Schoof et al., 2000). A high CK: low GA ratio promotes meristem maintenance since high CK sustains cell division and low GA inhibits cell differentiation (Aichinger et al., 2012). Accordingly, meristem activity in *stm* mutants can be restored by elevating CK biosynthesis and inhibited by elevating GA abundance or signaling or by reducing CK content (Hay et al., 2002; Jasinski et al., 2005; Yanai et al., 2005). KNOX proteins raise CK levels by activating *ISOPENTENYL TRANSFERASE7* for CK biosynthesis (Jasinski et al., 2005; Yanai et al., 2005) and lower GA levels by directly inhibiting biosynthetic genes encoding GA20-oxidases (Sakamoto et al., 2001) and activating catabolic genes encoding GA2-oxidases (Bolduc and Hake, 2009). Genes encoding GA2-oxidase are expressed at the boundary between the SAM and leaves confining GA to leaves where growth is taking place (Jasinski et al., 2005; Bolduc and Hake, 2009). WUS contributes to this network by lowering the abundance of *ARABIDOPSIS* RESPONSE REGULATOR ARR7 and ARR15 (Leibfried et al., 2005) thus increasing sensitivity to CK in the central zone and promoting its own expression (Gordon et al., 2009). BR are a class of growth-promoting hormones recently shown to play a role in meristem maintenance (Tsuda et al., 2014). New work in rice and maize show that KNOX factors in the SAM maintain indeterminacy in part via direct activation of BR catabolism genes thereby downregulating BR signaling in the meristem. Inactivation of three rice orthologs of *Arabidopsis* catabolic gene *BAS1* (*PHYTOCHROME B ACTIVATION TAGGED SUPPRESSOR1*)

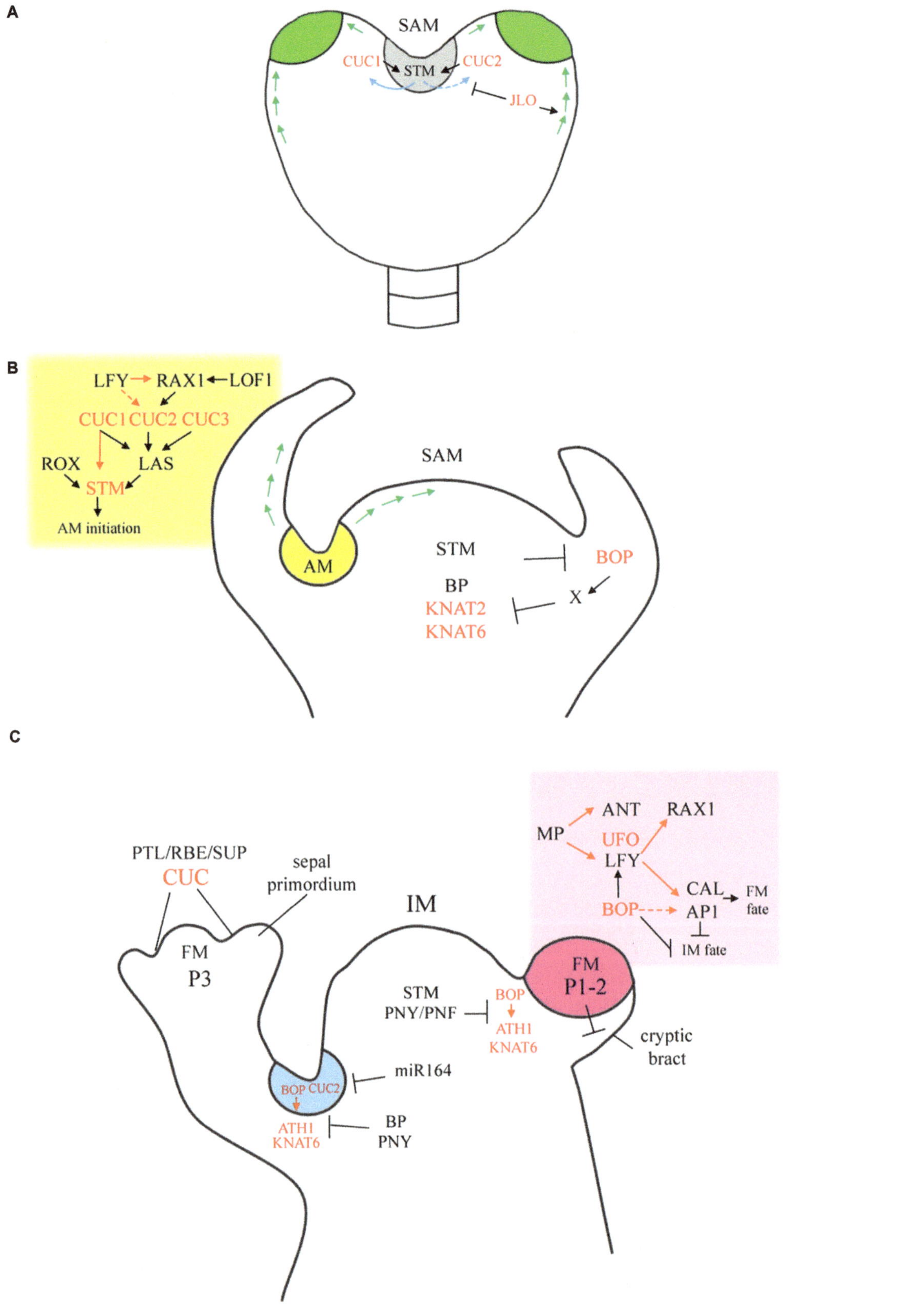

FIGURE 1 | Continued

FIGURE 1 | Continued

Homologous boundary gene networks controlling biogenesis of shoot apical meristem (SAM), axillary meristems (AMs), and floral meristems (FMs). **(A)** SAM initiation (gray). Early heart stage is shown. *CUC* genes are activated in a narrow band of auxin-depleted cells located between the presumptive cotyledons at globular stage. Activation of *CUCs* in this domain is partly dependent on chromatin remodeling ATPases and various other factors not depicted. Once activated, CUC1 and CUC2 are redundantly required for *STM* expression to form the presumptive SAM. STM in turn, directly maintains expression of *CUC1* and indirectly promotes *CUC2* and *CUC3* in establishing a feedback loop that ultimately restricts *CUC* expression to the axils of cotyledons. Reciprocally, *STM* expression is restricted to the SAM and slightly detected in boundaries. LBD family member JLO promotes *PIN1* expression required for formation of auxin maxima and represses *STM* and *KNOX* members to allow cotyledon outgrowth. Green arrows, direction of auxin flow. Green, auxin maxima at the cotyledon primordia. **(B)** SAM maintenance and AM formation (yellow). STM represses *BOP1/2* to maintain indeterminacy in the SAM. Conversely, BOP1/2 restrict *KNOX* expression in the proximal region of leaves to control patterning. Formation of an AM requires depletion of auxin from the leaf axil followed by a burst of CK. CUC1-3 are redundantly required for AM initiation functioning downstream of LFY and RAX1 to promote *LAS*. LOF1/2 contribute to *RAX1* promotion. CUC1, LAS, and ROX activities are required for sustained expression of *STM* and establishment of the AM. Green arrows, direction of auxin flow. **(C)** IM activity. PNY and PNF restrict *BOP1/2-ATH1-KNAT6* expression to boundary domains flanking the IM essential for meristem maintenance and flowering. FMs (pink). FMs are AMs with determinate fate that form in the axil of leaves whose development are repressed (cryptic bract). Auxin responsive transcription factor MP directly activates *ANT* and *LFY* to initiate FM formation. LFY directly promotes the expression of *RAX1* and *AP1* and *CAL* whose products confer floral fate. BOPs facilitate establishment of FMs via promotion of *LFY* expression, activation of *AP1*, and repression of IM identity genes. UFO is a LFY co-activator also required for formation of boundaries in the flower. Later, CUC factors are required to separate floral organs and maintain boundaries between whorls in association with numerous stage-specific factors including PTL, RBE, and SUP required for localized repression of growth. Inflorescence architecture. BP and PNY are expressed in the stem cortex where they collectively promote internode elongation, stem differentiation, phyllotaxy, and pedicel angle by restricting boundary genes *BOP1/2* and downstream effectors *ATH1* and *KNAT6* to the pedicel axil (blue). Misexpression of these genes in the BP-PNY domain restricts growth, disrupts vascular patterning, and causes ectopic lignification. *CUC2* expression is restricted by *miR164* to the pedicel axil to maintain internode patterning. FM, floral meristem; IM, inflorescence meristem. P, primordia; stages as indicated. Red lettering, SAM-leaf boundary genes. Red arrows, direct regulation. Dashed line, putative interaction.

results in the premature differentiation of meristematic cells (Tsuda et al., 2014).

BOUNDARY FORMATION

The boundary that separates the SAM from the primordia is a domain of restricted growth. This feature relies on depletion of auxin and BR from boundary cells thereby maintaining a low rate of growth relative to surrounding tissues (**Figures 2A,C**). Spatial regulation of polar auxin transporters establishes a minimum for auxin. This is accomplished in part by PIN-FORMED1 (PIN1) transporters oriented outwardly along the long axis of cells in the plane of the groove such that auxin is drained away from the boundary into the adjacent organ and meristem compartments (Heisler et al., 2010). Striking images of auxin depletion from the adaxial boundary domain of leaf primordia are observed using the auxin concentration sensor DII-Venus or auxin-responsive reporter gene *DR5:VENUS* (Wang et al., 2014a,b). ABC/PGP (ATP-binding cassette/*P*-glycoprotein) pumps are a secondary type of auxin transporter. ABC19 in this family depletes auxin from the boundary creating a low-auxin niche necessary for promotion of *CUC2* and *LOF2* expression (Zhao et al., 2013). An auxin minimum is essential in several boundary-related processes including AM formation (Tian et al., 2014; Wang et al., 2014a,b), differentiation of valve margins in the fruit (Sorefan et al., 2009), and timing of floral organ abscission (Estornell et al., 2013).

The boundary is also a minimum for BR (Bell et al., 2012; Gendron et al., 2012). The LATERAL ORGAN BOUNDARY DOMAIN (LBD) transcription factor LATERAL ORGAN BOUNDARIES (LOB) maintains low levels of BR to inhibit growth at boundaries. Ectopic expression of *LOB* reduces growth similar to BR defective mutants, while loss of *LOB* function causes overgrowth of the boundary region and organ fusion (Bell et al., 2012; Gendron et al., 2012). A feedback loop is required in establishing this pattern. Auxin-induced BR in the leaf activates *LOB1* (Chung et al., 2011) which in turn directs activation of cytochrome P450 gene *BAS1* to inhibit BR accumulation at the boundary (Bell et al., 2012). Two BTB-ankyrin transcriptional co-activators, BLADE-ON-PETIOLE1 (BOP1) and BOP2, reinforce this pattern by promoting *LOB* expression in the boundary domain (Ha et al., 2007). Fluorescent reporters show that BR-activated transcription factor BRASSINOZOLE-RESISTANT1 (BZR1) fails to accumulate in the nuclei of boundary cells thereby allowing expression of *CUC* genes which in turn repress growth at the boundary (Gendron et al., 2012).

Emerging data suggest that KNOX activity provides a positional cue in establishing the SAM-leaf boundary (Bolduc et al., 2012; Johnston et al., 2014; Tsuda et al., 2014). Several mechanisms are identified. One study shows that rice *KNOX* gene *Oryza sativa homeobox1* (*OSH1*) expressed in the meristem and base of emerging leaves facilitates SAM function and boundary formation by lowering BR abundance (Tsuda et al., 2014). Transcriptomic studies in maize focusing on the blade-sheath boundary of leaves further reveal that *CUCs*, *TALEs*, and *BOPs* are downstream targets of KNOTTED1 (KN1) under positive regulation (Bolduc et al., 2012; Johnston et al., 2014). Grasses have a blade-sheath boundary containing hinge-like auricles that control leaf angle and a fringe of epidermal tissue called the ligule whose formation is under the control of boundary genes. Barley *UNICULME4* is a *BOP* homolog required for ligule outgrowth (Tavakol et al., 2015). Maize genes required for ligule development include *LIGULELESS2 (LG2)* which encodes a TGA bZIP factor (Walsh et al., 1998), *LG3/LG4* which are closely related genes to *Arabidopsis KNAT2* and *KNAT6*; and *KNOTTED1/ROUGH SHEATH1* homologs of *STM* or *BP* (Bolduc et al., 2012; Johnston et al., 2014). Pre-ligule tissue is enriched for homologs of *CUC2*, *BOP1/2*, *KNAT6*, and two *BELL* genes, whose loci in several cases are bound directly by KN1 as identified through chromatin immunoprecipitation assays (Bolduc et al., 2012; Johnston et al., 2014). These data support

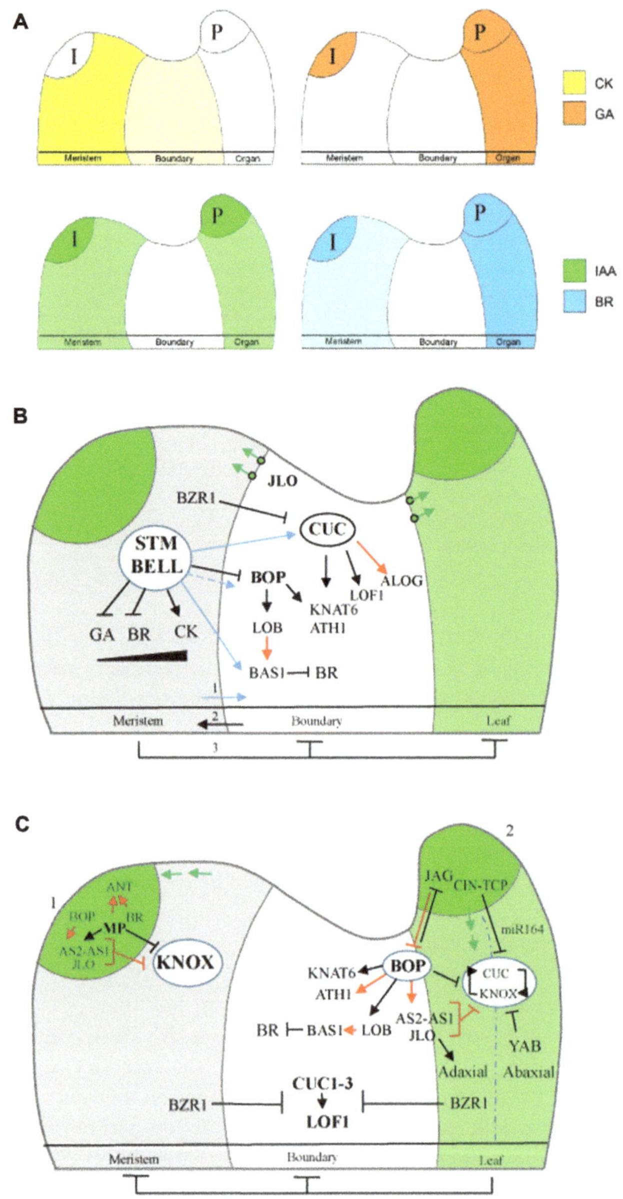

FIGURE 2 | Summary of hormone profiles and genetic interactions that maintain the leaf-SAM boundary. (A) The shoot apex contains: the meristem, the initium, boundaries, and organ primordia. The predicted distribution of key hormones are summarized (see text for details). Meristem, high CK/IAA ratio and low GA/BR ratio promotes indeterminate growth. Primordia, high IAA activates BR and represses CK biosynthesis and GA increases to promote determinate growth. Boundary, depletion for growth-promoting hormones IAA, GA, BR, and CK inhibits cell division allowing separation of meristem-organ compartments. I, Initium; CK, cytokinin; BR, brassinosteroid; GA, gibberellin; and IAA, auxin. **(B)** Summary of gene networks at the meristem-boundary interface. Meristem, STM maintains indeterminate growth by promoting CK and repressing GA/BR accumulation (black gradient indicates hormone abundance). Auxin, shown in green, marks the site of primordia initiation and distal blade of emerging leaf. (1) KNOX proteins initialize the boundary through promotion of BR catabolic genes (*BAS1*) and boundary transcription genes including *CUC* and *BOP2* (blue arrows). CUC factors confer boundary identity required for activation of other classes of boundary regulators including BOP, KNAT-BELL, LOF1, *(Continued)*

FIGURE 2 | Continued

ALOG, and LBD members that collectively restrict growth, modulate meristematic activity, and pattern the boundary. PIN1 auxin efflux carriers (green circles) are orientated facing outward such that auxin is drained away from the boundary. Green arrows indicate direction of auxin flow. (2) Boundary genes contribute to meristem maintenance (see text). (3) STM-BELL meristem factors preserve meristem integrity by restricting *BOP1/2* and *KNAT6* to boundaries. BR-activated transcription factor BZR1 represses *CUC/LOF1* in the meristem domain. **(C)** Summary of interactions at the leaf-boundary interface. (1) Polar auxin transport establishes auxin maxima in the peripheral zone where leaf initiation takes place. Auxin response factor MP initiates primordium formation by repressing *KNOX* genes, activating *ANT* members and leaf identity genes including AS1 and stimulating synthesis of BR where BZR1 binds to the *ANT* promoter as a positive regulator. Boundary genes *BOP1/2* and *JLO* expressed in the organ initial contribute to organ polarity and stable repression of *KNOX* genes. (2) Primordium outgrowth coincides with synthesis of auxin and repolarization of PIN transports toward the leaf base, which becomes a low IAA/BR domain. BOPs and JLO now restricted to the boundary reinforce this pattern in hormones via regulation of LOB and PIN1, respectively. CIN-TCPs and BZR1 in leaves maintain repression of *CUC/LOF1*. JAG in the distal blade represses *BOPs*. *BOPs* expressed in the proximal petiole domain of leaves maintain organ polarity and repress *KNOX* and *JAG* genes required for simple leaf shape indirectly in part via activation of *AS2*. YAB contributes to the repression of *KNOX* and *CUC* in the abaxial domain.

a model in which KNOX accumulation at the base of the leaf primordia and auxin accumulation in the distal portion of the primordia provide opposing positional cues in demarcating the blade-sheath boundary (Bolduc et al., 2012; Johnston et al., 2014). *Arabidopsis* studies showing that STM directly activates *CUC1* and indirectly promotes *CUC2*, *CUC3*, and *BOP2* expression (Spinelli et al., 2011) support this model.

Genetic studies show that *CUC* genes play a central role in maintaining growth repression in boundaries. Inactivation of any two *CUCs* leads to ectopic growth at cotyledon boundaries causing fusion along their margins (Aida et al., 1997; Vroemen et al., 2003). GROWTH-REGULATING FACTORS (GRFs) which act as broad regulators of cell proliferation function synergistically with CUCs in this role. Leaf fusion defects in *grf1 grf2 grf3* triple mutants are dramatically enhanced by inactivation of *GRF4* or *CUC* genes resulting in cup-shaped cotyledons and embryos that lack a functional SAM (Lee et al., 2015). Two *ALOG* (*Arabidopsis LSH1* and *Oryza G1*) family members *ORGAN BOUNDARY1/LIGHT-DEPENDENT SHORT HYPOCOTYL* (*OBO1/LSH3*) and *OBO4/LSH4* are direct targets of CUC1 and thought to repress differentiation of boundary cells (Cho and Zambryski, 2011; Takeda et al., 2011). Another regulator is the MYB transcription factor LATERAL ORGAN FUSION1 (LOF1), which promotes organ separation and meristem maintenance. Inactivation of *LOF1* enhances *stm-10* meristem termination and organ fusion defects (Lee et al., 2009).

ORGAN INITIATION

One of the earliest steps in initiation of lateral organs is down-regulation of *STM* at sites of auxin maxima in the peripheral zone of the meristem. Boundary and leaf identity genes are transiently expressed in the same compartment undergoing differentiation. Various studies show that organ initiation and boundary formation are interconnected processes (Heisler et al., 2005; Besnard et al., 2011). Auxin in the distal portion of the primordia controls the localization of boundary genes ultimately restricting their expression to the base of the emerging leaf (**Figures 2A,C**). In brief, cotyledons and leaves are initiated at auxin maxima generated by polar auxin transport. Polar auxin distribution is dependent on a family of efflux carriers including PIN1 whose membrane localization is controlled by the serine/threonine kinase PINOID. Threshold levels of auxin trigger activation of the auxin-responsive transcription factor MONOPTEROS (MP), which down-regulates *STM* and activates *AINTEGUMENTA (ANT)*, *ANT-like (AIL)*, and *ASYMMETRIC LEAVES1 (AS1)* genes to initiate leaf development (Long and Barton, 1998; Besnard et al., 2011; Yamaguchi et al., 2013). As the primordium emerges, PIN1 polarity reverses to generate new auxin peaks coinciding with a narrow band of cells marked by *CUC* expression (Heisler et al., 2005). Mutations in *PIN1*, *PID*, or *MP* that disrupt auxin transport or signaling lead to expansion of *STM* and *CUC* expression to the periphery where they suppress cotyledon outgrowth (Aida et al., 2002; Furutani et al., 2004; Schuetz et al., 2008). Proper distribution of auxin in forming this pattern is dependent on SEUSS and SEUSS-like components of the LEUNIG repressor complex although the mechanism is still unknown (Lee et al., 2014). Auxin in the leaf initial further alters the balance of hormones to favor growth and determinacy. In particular, auxin stimulates BR (Chung et al., 2011) and GA synthesis (Frigerio et al., 2006) and represses CK production (Nordstrom et al., 2004; Besnard et al., 2011). Primordium outgrowth also depends on physical changes in cell wall stiffness (Besnard et al., 2011; Peaucelle et al., 2011). Auxin stimulates the active transport of protons into the extracellular space required in activating enzymes that relax the cell wall and promotes the transcription of remodeling factors including expansions, pectin methylesterase, and hydrolases (Besnard et al., 2011; Peaucelle et al., 2011). These changes are coupled with a shift toward growth isotropy, which facilitates organ outgrowth (Sassi et al., 2014).

LEAF DIFFERENTIATION

Leaf differentiation requires the maintenance of *KNOX* repression and the restriction of *CUC2/3* expression along the leaf margin. *KNOX* repression is accomplished by an interacting network of leaf and boundary factors (**Figure 2C**). A key player in this network is the MYB transcription factor AS1, which acts in a trimeric complex with the LBD transcription factors AS2 and JAGGED LATERAL ORGANS (JLO) (Guo et al., 2008; Rast and Simon, 2012). AS1 and AS2 bind to distinct sites in the *BP* and *KNAT2* promoter where they interact through looping to induce silencing via recruitment of the histone chaperone HIRA and Polycomb-repressive complex2 (Guo et al., 2008; Lodha et al., 2013). *STM* is also a target of PRC but how this complex is recruited to the promoter is unknown (Lodha et al., 2013). *JLO* is transiently expressed at sites of organ initiation and resolves to the leaf-meristem boundary during outgrowth. Loss-of-function mutations in *JLO* impair organ outgrowth and enhance the margin patterning defects of *as2* mutants. This phenotype is caused in part by ectopic expression of *BP* and *STM* at the base of leaf primordia combined with defects in auxin distribution (Rast and Simon, 2012). JLO promotes *PIN* expression for auxin build-up at organ initiation sites and later for auxin efflux from the boundary (Bureau et al., 2010; Rast and Simon, 2012; Žádníková and Simon, 2014).

BOP1/2 activity in organ initials partially overlaps with *JLO* and likewise resolves to the boundary of emerging leaves and petiole domains during outgrowth (Ha et al., 2004; Hepworth et al., 2005; Norberg et al., 2005; Borghi et al., 2007). BOP1/2 have a dual function. They repress genes that confer meristem cell fate and induce genes that promote lateral organ fate and polarity (Ha et al., 2007). *BOP1/2* transcripts are first detected in the boundaries of torpedo stage embryos consistent with a function downstream or in parallel with CUCs (Ha et al., 2004). STM represses *BOP1/2* to maintain indeterminacy and conversely, BOP1/2 restrict *KNOX* expression to pattern the leaf (Jun et al., 2010). A prolonged phase of morphogenetic competence in *bop1 bop2* petioles coupled with *KNOX* reactivation results in initiation of ectopic leaflets reminiscent of development in a compound leaf (Ha et al., 2003, 2007; Khan et al., 2014). BOP1 binds directly to the promoter of *AS2* likely recruited by a TGA factor (Jun et al., 2010). Synergistic enhancement of meristematic activity in *bop1 bop2 as1* and *bop1 bop2 as2* petioles shows that BOP1/2 repression of *KNOX* genes is not entirely via AS1–AS2 and is likely indirect. Leaf patterning defects in *bop1 bop2* are also attributed to misexpression of abaxial/adaxial organ polarity determinants and the C2H2 zinc finger transcription factor JAGGED (JAG) which promotes cell proliferation (Norberg et al., 2005; Ha et al., 2007, 2010). JAG is normally restricted to the distal blade where it represses *BOP2* to allow extension of the leaf margin (Schiessl et al., 2014).

CINCINNATA (CIN)-like TEOSINTE BRANCHED1/CYCLOIDEA/PCF (TCP) factors are another class that contribute to negative regulation of *CUC* and *KNOX* activity in leaves to promote organ outgrowth and simple leaf shape (Koyama et al., 2007, 2010). Consistent with this view, *CIN-TCPs* are predominantly expressed in leaves and depleted from the boundary (Tian et al., 2014). Several mechanisms are identified. First, TCP3 directly promotes *miR164,* which targets *CUC1* and *CUC2* transcripts for cleavage. Second, TCP3 directly promotes *AS1* whose product represses *CUC3* and *KNOX* expression (Koyama et al., 2010). In addition, TCP3 targets auxin inducible genes that repress SAM function and cause cotyledon fusion when overexpressed (Koyama et al., 2010). Lastly, TCP4 binds to CUC2 and inhibits its activity by blocking the formation of homo-dimers and hetero-dimers with CUC3. TCP4 also impairs CUC3 transactivation ability (Rubio-Somoza et al., 2014).

LEAF SHAPE

Variations in the KNOX-PIN-CUC module play a central role in controlling leaf shape and complexity. The leaf margin, due to its meristematic feature, is particularly sensitive to alterations in this module. In simple leaf species such as *Arabidopsis, CUC2,* and *CUC3* expression is restricted to the sinus of serrations along the leaf margin while *CUC1* expression is not detected (Nikovics et al., 2006; Hasson et al., 2011). The balance between *CUC2* and *miRNA164* transcripts controls the degree of leaf serration (Nikovics et al., 2006). CUC3 contributes to leaf shape at a later stage (Hasson et al., 2011). Similar to the primordia initiation, the formation of serrations depends on auxin. To explain how serrations form on a leaf margin, Bilsborough et al. (2011) proposed a model in which two feedback loops work in concert. In the first loop, PIN1 convergence in the leaf margin generates an auxin maximum, reinforced by auxin feedback on PIN1. In the second loop, CUC2 acts non-cell autonomously to promote growth through the generation of PIN1-dependent auxin maxima and contributes to tooth outgrowth (Kawamura et al., 2010). Auxin in turn downregulates *CUC2* restricting expression to regions between serrations where growth is repressed.

The rachis of a compound leaf is likewise sensitive to alterations in *KNOX-PIN-CUC* expression. While simple leaves have a single undivided blade in which *KNOX* repression is continuous, compound leaves have a divided blade consisting of pairs of leaflets attached to a central rachis. This morphology is associated with an extended primary morphogenesis phase during which reactivation of *KNOX* genes begins the cycle by promoting auxin accumulation thereby directing leaflet initiation on the rachis (Di Giacomo et al., 2013). Down-regulation of tomato *BOPa* (one of three homologs) further enhances leaf complexity by extending the window for rachis responsiveness to auxin (Ichihashi et al., 2014). BOPa fulfills this function in part by forming a complex with LIGHT-DEPENDENT SHORT HYPOCOTYL3b (an ALOG family member) that represses tomato *KNATM* encoded by *KD1/PETROSELINUM* to modulate KNOX activity (Ichihashi et al., 2014). KNATM, a mini-KNOX lacking the homeodomain, modulates KNOX-BELL activity by competing for BELL binding partners (Kimura et al., 2008; Magnani and Hake, 2008). Analysis of the maize KN1 cistrome confirms that a majority of directly regulated genes are involved in auxin signaling, biosynthesis, and transport including PIN1 (Bolduc et al., 2012). Several legume species such as *Medicago truncatula* and pea use orthologs of *Arabidopsis LEAFY (LFY)* as an alternate source of meristem activity but remain reliant on CUC2 function for creation of auxin peaks required in leaflet initiation (Nikovics et al., 2006; Blein et al., 2008; Efroni et al., 2010). These data illustrate that variations in the KNOX-PIN-CUC module cause diversity in leaf patterning. Recruitment of this same module at later stages governs AM formation, gynoecium, and ovule development (Ishida et al., 2000; Hibara et al., 2006; Scofield et al., 2007; Raman et al., 2008).

NEW FRONTIERS

In the next part of the review, we examine the role of boundary genes during the reproductive phase. Boundaries in the inflorescence determine plant architecture through the separation of organs and the distribution of flowers on the stem but also constitute a source of AM for production of branches and flowers. Boundaries are also sites where abscission and dehiscence take place. A number of genes are recurrently expressed in these boundaries including *KNOX-BELL, BOP,* and *CUC/miR164* regulators. Similar to the role of TALE factors in the SAM, PNY and PNF preserve meristem integrity essential for flowering by excluding *BOP1/2/KNAT6-ATH1* from the meristem. Inflorescence architecture is likewise controlled by restricting *CUC2* and *BOP1/2-KNAT6/ATH1* to boundaries at the base of floral shoots. CUC-STM factors play a conserved role in formation of new meristems, including AMs that give rise to lateral branches and flowers and meristematic tissues internal to the fruit. CUC factors also play a critical role in separation of floral organs and ovules in developing flowers. TALE factors including BP and PNY preserve formation of meristematic replum tissue in fruits whereas BOP1/2-KNAT6/2, which are expressed in adjacent valve margin boundary tissues potentially contribute to dehiscence. This same network of TALE and BOP factors regulates abscission. How these conserved modules are integrated during reproductive development is now discussed.

AXILLARY MERISTEMS

The boundary located between the stem and the leaf base constitutes a source of AMs, which can remain dormant or produce secondary inflorescences and flowers (**Figure 1B**). The specification and the development of AMs involves numerous transcription factors and is modulated by hormones such as auxin, CK, BR, and strigolactones (Janssen et al., 2014). Other hormones may also be involved based on specific patterns of enrichment for abscisic acid and ethylene or depletion of jasmonic acid (JA) responsive genes in the boundary domain (Tian et al., 2014). Similar to the SAM, formation of AMs requires CUC-STM factors whose activity at the boundary is dependent on auxin and CK.

Recent studies in *Arabidopsis* and tomato show that the establishment of a stem cell niche in leaf axils requires auxin depletion followed by pulse of CK (Wang et al., 2014a,b). Manipulation of the auxin gradient using chemical inhibitors of auxin transport or mutations in auxin transport machinery including PIN1 or PID showed that disruption of auxin minima strongly inhibits AM initiation (Wang et al., 2014a,b). STM, which is a marker of AMs in the mature leaf axil (Grbic and Bleecker, 2000; Long and Barton, 2000), fails to accumulate in a strong *pid-9* mutant indicating that its expression is dependent on an auxin minimum (Wang et al., 2014a). CK perception and signaling is enhanced in leaf axils prior to AM initiation and the TCS::GFP (two-component output sensor) synthetic reporter used to visualize CK response indicates that a pulse of CK follows the auxin minimum and is required to stimulate AM production (Wang et al., 2014b). STM may contribute to this pulse based on its CK promoting activity in the SAM (Jasinski et al., 2005; Yanai et al., 2005). *Arabidopsis* mutants affected in CK perception (histidine kinase receptor mutants) or CK signaling (ARR-B type transcription factor mutants) show reduced AM production whereas overproduction of CK restores AM initiation in a *rax1* mutant (Wang et al., 2014b).

REGULATOR OF AXILLARY MERISTEM1 (RAX1) is a MYB transcription factor that specifies AMs in redundancy with RAX2 and RAX3 (Keller et al., 2006; Muller et al., 2006). One regulator of RAX1 is the MYB transcription factor LOF1 which also functions to promote AM and organ separation (Lee et al., 2009). RAX1 acts through CUC2 and is required in conjunction with CUC3 and the GRAS-domain protein LATERAL SUPPRESSOR (LAS) to maintain *STM* expression in AMs. RAX1 maintains the boundary zone through the repression of GA similarly to STM in the SAM (Keller et al., 2006; Muller et al., 2006). *LAS* and *RAX1* promote AMs via the bHLH transcription factor REGULATOR OF AM FORMATION (ROX; Yang et al., 2012). LFY, which is involved in flower specification has recently been shown to promote AM proliferation through its direct target *RAX1* and potentially others (Chahtane et al., 2013). Redundant pathways mask this role as mutations in *LFY* combined with mutations affecting various pathways including meristem, auxin signaling, floral transition and patterning, or boundary genes such *BOP1/2* show defects in AM formation (Chahtane et al., 2013). LFY further contributes to meristem emergence via CK signaling potentially through its interaction with WUS. The negative regulator of CK signaling, *ARABIDOPSIS RESPONSE REGULATOR7* (*ARR7*), was found to be up-regulated in *lfy-12* inflorescences, and LFY interacts directly with the *ARR7* promoter (Moyroud et al., 2011; Winter et al., 2011; Chahtane et al., 2013). LFY may also act through *CUC2* and the auxin signaling pathway as genes from this pathway are bound by LFY (Moyroud et al., 2011; Winter et al., 2011; Yamaguchi et al., 2013). Thus, meristem emergence results from the convergence of LFY and LAS pathways.

Genetic studies show that *CUC* genes contribute redundantly and differently to AM initiation and boundary maintenance with *CUC3* playing a prominent role (Hibara et al., 2006). Inactivation of *CUC1* and *CUC2* has no effect on AM initiation and an effect in *cuc2* mutants is seen only in the absence of *CUC3* (Hibara et al., 2006). Further studies show a contribution of CUC1/2 to AMs. Overexpression of *miR164* dramatically reduces the initiation of AMs in the *cuc3-2* mutant, and reciprocally, *miR164*-resistant versions of *CUC1* and *CUC2* form extra accessory side shoots (Raman et al., 2008). Key downstream targets of the CUC pathway include *STM* and *LAS* required in AM establishment (Raman et al., 2008), with *LAS* being a direct target of CUC2 (Tian et al., 2014). *STM* expression is missing in the adaxial boundary domain of *cuc3-2* mutants and *LAS* expression is reduced in *miR164* overexpressing lines, which diminishes *STM* expression (Greb et al., 2003; Raman et al., 2008).

Translatome analysis of *LAS*-expressing boundary and *AS1*-expressing leaf primordia cells coupled with genome-scale mapping of transcription factor binding sites reveals that CUC2 and LAS are regulatory hubs for AM initiation (Tian et al., 2014). This work identifies the auxin-induced APETALA2 domain transcription factor DORNRÖSCHEN as a direct activator of *CUC2*. This work also identifies SQUAMOSA PROMOTER BINDING PROTEIN-LIKE9 and 15 as repressors of *LAS* and *CUC2* that regulate AM initiation likely in response to environmental signals (Tian et al., 2014).

Separation of axillary shoots from subtending leaves and formation of accessory side shoots requires LOF1 and LOF2 acting downstream of *CUC* genes (Lee et al., 2009; Gendron et al., 2012). *Lof1* defects in organ separation are enhanced by mutation of the closely related *LOF2* whose expression is more widespread but dependent on *LOF1* at the boundary (Lee et al., 2009). *STM* expression in AMs is reduced in *lof1 lof2* double mutants (Lee et al., 2009). Overexpression of a cysteine-rich signaling peptide TAXIMIN1 mimics the phenotype of *lof1 lof2* double mutants (Colling et al., 2015). Interestingly, this phenotype is not due to a reduction of *LOF1/2* or other boundary transcripts suggesting an independent mechanism (Colling et al., 2015). A peptide signaling cascade has not been previously linked to formation of boundaries in plants.

Studies in other species show this hierarchy to be highly conserved (Janssen et al., 2014). Tomato *GOBLET (GOB)* encodes a NAC-domain transcription factor similar to CUC2 (Berger et al., 2009); LAS is an ortholog of *LAS* (Schumacher et al., 1999; Greb et al., 2003); and *BLIND* is an ortholog of *RAX1* (Schmitz et al., 2002). Remarkably these same genes are regulators of leaf complexity. Homologous genes have also been identified in cereals as regulators of tillering and panicle architecture. LAX PANICLE1 in rice and BARREN STALK in maize encode bHLH proteins orthologous to ROX in *Arabidopsis*. Mutants in these genes show a reduction in panicle branches and spikelets and fail to form AMs during the vegetative phase resulting in a reduction in tillers (Komatsu et al., 2003; Gallavotti et al., 2004; Oikawa and Kyozuka, 2009). These proteins sustain early proliferation of the AM by forming a boundary between the meristem and axillary bud (Oikawa and Kyozuka, 2009; Yang et al., 2012). Barley *CUL4* is a BOP homolog required for tiller formation (Tavakol et al., 2015). BOP1/2 are required for production of various determinate axillary shoots including stipules, nectaries, and flowers in dicots (Khan et al., 2014). *BOP* expression is down-regulated at an early stage of indeterminate IM formation and moves to the boundary between the meristem and AM demonstrating a transient role similar to ROX (Xu et al., 2010; Yang et al., 2012). The contribution of CUL4 in AMs production suggests a partial conservation of BOP function in monocots and dicots.

INFLORESCENCE ARCHITECTURE

The maintenance of boundaries during stem growth is critical in preserving plant architecture. Ectopic expression of boundary genes prevents the elongation and proper differentiation of stem internodes resulting in aberrant phyllotaxy. This is illustrated by clustering of flowers on the stems of plants expressing a *miR164*-resistant version of *CUC2* (Peaucelle et al., 2007). The restriction of *CUC2* expression to the floral stem axil by *miR164* in the IM is required to maintain the boundary between the pedicel and the stem (**Figure 1C**). The TALE transcription factors BP and PNY constitute another set of architecture determinants (**Figure 1C**). These factors are required to maintain internode patterning during stem growth and radial patterning in both primary and secondary phases of stem development (Smith and Hake, 2003). BP and PNY are expressed in the stem cortex and adjacent vascular tissues and form a boundary between the IM and lateral organs (Smith and Hake, 2003). Mutations in *BP* lead to short compact internodes, downward pointing siliques, and precocious outgrowth of paraclades (Douglas et al., 2002; Venglat et al., 2002). Vascular bundles in *bp* mutants are often irregular in size and/or spacing. Bundles tend to be underdeveloped with xylem elements reduced or lacking in lignin (Smith and Hake, 2003). Mutations in *PNY* cause shortened internodes and clusters of flowers on stems and partial loss of apical dominance (Byrne et al., 2003; Smith and Hake, 2003). These phenotypes are enhanced in the double mutant showing that BP-PNY have partially overlapping functions in specification of boundaries during internode growth (Smith and Hake, 2003). Genetic and transcriptome studies indicate that PNY modulates the activity of plant cell wall modifying enzymes required in loosening cell walls to allow organ initiation and internode elongation (Peaucelle et al., 2011; Etchells et al., 2012). BP regulates an overlapping set of genes and prevents premature deposition of lignin in elongating stems by direct repression of genes in the lignin biosynthetic pathway (Mele et al., 2003; Wang et al., 2006).

Genetic and expression studies show that *bp* and *pny* inflorescence defects are caused by the localized misexpression of lateral organ boundary genes *KNAT6, ATH1, BOP1/2* and to a lesser extent *KNAT2* in stems (Ragni et al., 2008; Khan et al., 2012a,b). Inactivation of *BOP1/2* or *KNAT6* or *ATH1* fully rescues *pny* defects to restore wild type inflorescence architecture. Similarly, inactivation of *BOP1/2* or *KNAT6* in combination with *KNAT2* or *ATH1* rescues *bp* defects in internode elongation and pedicel orientation. The regular pattern of vascular bundles and the pattern of lignin deposition in stems during secondary growth are reestablished in these mutants (Khan et al., 2012a,b). BOP1/2 require the functions of these downstream genes to exert changes in inflorescence architecture suggesting a linear pathway (Khan et al., 2012a,b). Further analysis of this module shows that BOP1 directly activates *ATH1* whereas activation of *KNAT6* is indirect (Khan et al., 2015). BP/STM are recently shown to promote xylem differentiation in the cambium through the repression of *BOP1* and *BOP2* (Liebsch et al., 2014). Thus, restriction of the *BOP1/2-ATH1-KNAT6* boundary module by BP-PNY is critical for plant architecture. Recent data reveal that BP directly represses *KNAT2* and *KNAT6* expression by recruiting the chromatin remodeling ATPase BRAHMA to the promoter (Zhao et al., 2015).

FLOWER INITIATION AND PATTERNING

Floral inductive signals acting on the SAM cause restructuring to form the IM. Completion of this process requires the PNY and PNF BELL members. In *pny pnf* mutants, apices support the production of leaves, but internode elongation and flower initiation are blocked (Smith et al., 2004; Kanrar et al., 2008; Lal et al., 2011). Recent data show that this block is due to misexpression of BOP1/2 and its downstream effectors KNAT6 and ATH1 which prevent accumulation of floral meristem identity genes including *LFY, CAULIFLOWER (CAL)*, and *APETALA1 (AP1)* required for flower production (Khan et al., 2015). PNY in this network directly represses *BOP1/2* to maintain its expression at boundaries. One study shows that ectopic *BOP1/2* expression reduces responsiveness to FT by lowering the abundance of its binding partner FD (Andrés et al., 2015). Transcript profiling of *BOP1* overexpressing plants further identifies promotion of JA as a potential mechanism for inhibiting accumulation of SQUAMOSA PROMOTER BINDING-LIKE PROTEINS and counteracting responsiveness to GAs (Khan et al., 2015). Thus, the setting of lateral boundaries by PNY and PNF via the restriction of *BOP1/2-ATH1-KNAT6* expression is critical for meristem integrity and specification of flowers.

The floral meristem constitutes an AM whose rapid proliferation represses outgrowth of the subtending leaf (Long and Barton, 2000). Initiation of flowers is an auxin-dependent process similar to that in leaves (**Figure 1C**). Mutations in *MP* or *PIN1* result in naked IM "pins" lacking flowers due to misexpression of meristem/organ/boundary markers including *STM, LFY, CUC*, and *ANT* throughout the peripheral zone (Vernoux et al., 2000; Hay et al., 2006; Schuetz et al., 2008). MP integrates auxin and floral signals by directly activating *ANT/AIL6* which triggers proliferation in combination with *LFY* which activates flower development (Yamaguchi et al., 2013). Interestingly, MP does not bind to the *LFY* promoter during the vegetative stage indicating that binding is stage-specific (Yamaguchi et al., 2013). LFY reinforces this loop via direct activation of genes in the auxin pathway, direct activation of *RAX1*, and direct activation of *AP1* and *CAL* whose products confer floral fate (Wagner et al., 1999; William et al., 2004; Winter et al., 2011; Yamaguchi et al., 2013).

BOP1/2 and *UNUSUAL FLORAL ORGANS (UFO)* are boundary regulators that facilitate LFY function. Genetic studies reveal that BOPs play a supporting role in the promotion of *LFY* expression (Karim et al., 2009), proliferation of the floral meristem, and determinacy in part through direct activation of *AP1* (Xu et al., 2010). Several of these functions are shared with UFO (Norberg et al., 2005; Xu et al., 2010; Risseeuw et al., 2013). Outgrowth of the floral meristem is delayed in *bop1 bop2, ufo-1*, and *lfy* mutants or absent in *bop1 bop2 lfy-1* triple mutants resulting in barren axils (Levin and Meyerowitz, 1995; Wilkinson and Haughn, 1995; Norberg et al., 2005; Xu et al., 2010). Inactivation of *BOP1/2* or *UFO CAL* greatly enhances the floral branching defect in *ap1* mutants caused by derepression of CK biosynthesis in sepal axils leading to ectopic FM initiation and loss of shoot determinacy (Levin and Meyerowitz, 1995; Xu et al., 2010; Han et al., 2014). UFO is the F-box subunit of an SCF-based E3 ubiquitin ligase complex which binds to LFY and functions as a transcriptional co-activator (Lee et al., 1997; Chae et al., 2008). Paradoxically, UFO stimulates LFY activity by directing ubiquitination of its transcriptional activation domain thus marking the protein for turnover which is required for maximal induction of target genes (Chae et al., 2008). Similar functions are shown for BOP and UFO orthologs in a variety of species (Khan et al., 2014; Vlad et al., 2014).

Arabidopsis flowers are composed of sepals, petals, stamens, and carpels arranged in four concentric whorls. LFY is responsible for this patterning by activating three sets of homeotic genes that function combinatorially according to the ABC model (Lohmann and Weigel, 2002). In *ufo* mutants, petals and stamens are reduced or absent and organs are fused or chimeric indicating disrupted boundaries in the flower (Levin and Meyerowitz, 1995; Wilkinson and Haughn, 1995). In *bop1 bop2* mutants, sepal-to-petal conversions and sepal-whorl organ fusions are localized to the abaxial side of flowers where *BOP1/2* are transcribed during late stage 2 (Hepworth et al., 2005; Xu et al., 2010). Inactivation of *BOP1/2* or *UFO* in a weak *lfy* background creates a strong *lfy* phenotype indicating closely related functions for these genes (Levin and Meyerowitz, 1995; Wilkinson and Haughn, 1995; Xu et al., 2010). *UFO* is activated in the dome of stage 2 flowers and resolves to a cup-shaped domain around *STM*-expressing cells in the central zone of stage 3 flowers (Lee et al., 1997; Samach et al., 1999) and is possibly involved in creating a boundary. Embryo expression of *UFO* is dependent on STM (Long and Barton, 1998) but the mutant has no obvious defects during this stage due to genetic redundancy.

Boundaries in the flower are maintained by various other boundary genes and stage-specific factors including the zinc finger repressors RABBIT EARS (RBE; Takeda et al., 2004), and SUPERMAN (SUP; Sakai et al., 1995; Nibau et al., 2011) and HANABA TARANU, a GATA3-type transcriptional repressor (Zhao et al., 2004; Nawy et al., 2010). *CUC* genes are expressed between organ primordia and at the edges of whorls where they repress growth required to separate the floral organs and maintain boundaries between whorls (Ishida et al., 2000; Takada et al., 2001). *ATH1* controls basal floral organ boundaries and functions downstream of *CUC* genes (Gómez-Mena and Sablowski, 2008). *LOF1* (Gomez et al., 2011), *OBO1/LSH3*, and *OBO4/LSH4* (Cho and Zambryski, 2011) functioning downstream of *CUC1* are likely to contribute based on their expression patterns or overexpression phenotypes in the flower. Floral boundary defects are also observed in *bop1 bop2* and *jlo* flowers (Hepworth et al., 2005; Rast and Simon, 2012). Analogous to leaves, AS1–AS2, and JAG repress boundary genes including *CUC1* and *CUC2* to promote sepal and petal development (Xu et al., 2008).

PETAL LOSS (PTL) encodes a stage-specific trihelix transcription factor that represses growth in inter-sepal boundaries. In *ptl* mutants, petals are often absent, or show changes in shape, polarity, and fusion with sepals (Griffith et al., 1999). Sepal fusion is increased in *ptl cuc* mutants consistent with analysis showing that CUCs suppress upward growth of inter-sepal tissue at the boundary whereas PTL limits overgrowth of the inter-sepal zone required for petal initiation (Lampugnani et al., 2012, 2013). Petal initiation is highly sensitive to perturbations in growth and auxin distribution because primordia arise from 2 to 3 founder cells located in close proximity to the inter-sepal boundary (Lampugnani et al., 2012, 2013).

FRUIT PATTERNING AND DEHISCENCE

The *Arabidopsis* fruit is derived from the gynoecium, which consists of two fused carpels representing modified leaves (Ferrándiz et al., 2010). The carpels (termed valves after fertilization) are joined to a central replum whose internal surface or carpel margin meristem provides ovules and a septum with transmitting tract. Valve margins are a specialized lateral organ boundary that forms at the valve/replum interface. They ensure the release of the seeds. Thus, three patterning elements define the transverse axis of the mature fruit: valves, valve margins, and replum. Many interactions defining the SAM-leaf boundary are conserved in ovule and fruit development (Ferrándiz et al., 2010; Reyes-Olalde et al., 2013; Arnaud and Pautot, 2014) and (**Figures 3** and **4**).

The carpel marginal meristem (CMM) forms internally at the junction between fused carpels, which is homologous to the marginal meristem of leaves (Pautot et al., 2001). The CMM produces two outgrowths fused centrally to form the septum and is flanked on both sides by the placenta that gives rise to ovules. *CUC* genes promote fusion of the carpel margins and CMM initiation via the activation of *STM* similar to their role in SAM (Kamiuchi et al., 2014). Defects in either of these genes impair placental function leading to a reduction in ovules (Endrizzi et al., 1996; Ishida et al., 2000; Scofield et al., 2007). CK also plays an important role in promoting meristematic activity. Visualization of CK using the synthetic TCS::GFP reporter shows CK in the CMM of young gynoecia. Increased levels of CK enhance replum size while decreased levels reduce replum size (Marsch-Martinez et al., 2012). Following establishment of the CMM, *CUC1/2* transcripts are detected in the placenta where together with CK they control the localization of PIN1 transporters in creating auxin maxima required for ovule initiation (Bencivenga et al., 2012; Galbiati et al., 2013; Cucinotta et al., 2014). In mutants that overexpress CK, the number of ovules increases (Bartrina et al., 2011). Conversely, where there is a reduction in CK response or defects in auxin synthesis, transport or signaling, the number of ovules decreases (Galbiati et al., 2013). This process is comparable to formation of serrations on the leaf margin (Bilsborough et al., 2011). Once an ovule is initiated, auxin and BR converge to activate *ANT* for proliferation of the ovule primordia (Galbiati et al., 2013; Huang et al., 2013). Auxin accumulation in the distal tip of the ovule ultimately restricts *CUC1/2* expression to boundaries in the ovule (Ishida et al., 2000). *CUC3* and *CUC2* are later expressed between ovule primordia overlapping in a few cells where they are redundantly required for ovule separation (Goncalves et al., 2015). *Cuc3* single mutants show rare fused ovules with defects more severe in the *cuc3 cuc2* double mutant. LOF1 may also play a role based on its expression pattern in the inner medial ridges, the placenta, and at the base of the ovules marking these domains as lateral organ boundaries (Gomez et al., 2011). MADS-box transcription factors AGAMOUS, SHATTERPROOF1/2 (SHP1/2), and SEEDSTICK (STK) confer ovule identity (Ishida et al., 2000; Pinyopich et al., 2003; Galbiati et al., 2013).

After fertilization, the ovules develop into seeds and the fruit enlarges. At maturity, the valve margins undergo secondary

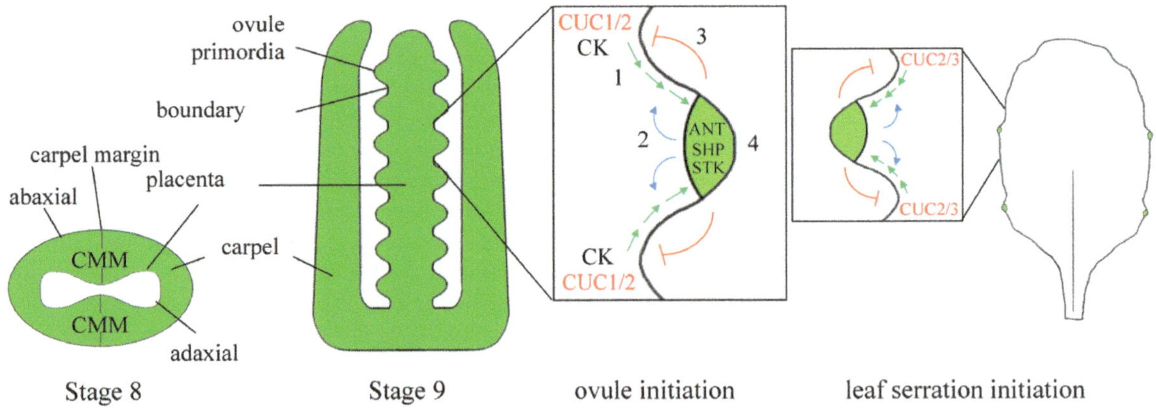

FIGURE 3 | Homologous networks for ovule initiation by the carpel marginal meristem (CMM) and patterning of leaf margin. Two fused carpels in the center of the flower form a tubular structure that constitutes the gynoecium (stages 8 and 9). The inner fused surfaces of the carpels (equivalent to an adaxial leaf surface) form a ridge of meristematic tissue called the CMM that gives rise to ovules and septum. Ovules are initiated in a process that resembles creation of serrations on the leaf margin where CUC2 is required for leaf serration and CUC3 promotes serration growth (see text). (1) CUC1/2 in the placenta together with CK promote the formation of an auxin maximum. (2) Auxin positive feedback reinforces flow of auxin to the primordia tip. (3) Auxin negative feedback restricts CUC1/2 expression to the base of the ovule. (4) Once auxin reaches threshold levels, it switches on *ANT* which promotes outgrowth of the shoot and MADS box genes *SHP1/2* and *STK* which confer ovule identity. CUC2/3 expression overlaps between ovule primordia and is required for ovule separation (not shown). Adapted from (Cucinotta et al., 2014). Red lettering, SAM-leaf boundary genes.

differentiation to form the dehiscence zone where the fruit opens. The dehiscence zone has two cell layers: a separation layer adjacent to the replum and a lignified layer. The separation layer produces enzymes that break down the middle lamella, a layer of pectin that cements cell walls together. The lignified layer is continuous with the inner layer of the fruit and is required for spring-like opening of the fruit (Ferrándiz et al., 2010).

The valve margin expresses a pair of stage-specific MADS-box transcription factors encoded by *SHP1/2* and boundary genes *KNAT2/6* and *BOP1/2* that are activated later during carpel development (Liljegren et al., 2000; Ragni et al., 2008; Khan et al., 2012b). SHP1/2 confer valve margin identity via activation of downstream bHLH transcription factors INDEHISCENT (IND) and ALCATRAZ (ALC) (Liljegren et al., 2004). IND is required for differentiation of lignified and separation layers of the valve margin whereas ALC/SPATULA (SPT) are required for differentiation of the separation layer (Rajani and Sundaresan, 2001; Liljegren et al., 2004; Girin et al., 2011; Groszmann et al., 2011). IND leads to the depletion of auxin in valve margins by relocating PIN1 transporters (Sorefan et al., 2009). IND also promotes GA production, which releases ALC and SPT proteins from DELLA repression allowing formation of a productive complex to specify the separation layer (Arnaud et al., 2010). The auxin and GA pathways seem to be independent, since the auxin minimum is maintained in GA deficient mutants. Visualization of CK in mature gynoecium shows CK in valve margins, and this localization depends on IND and SHP1/2 activity (Marsch-Martinez et al., 2012). Interestingly, CK restores valve margins in *shp1 shp2* and *ind* mutants indicating that CK functions downstream of these regulators. In contrast, a complementary pattern is observed for auxin with a synthetic DR5 reporter detected only in replum and valves. Thus, CK promotes valve margins. CK may contribute to the depletion of auxin in valve margins via the localization of PIN transporters (Marsch-Martinez et al., 2012) but it is unknown if the depletion of auxin is required for the accumulation of CK in valve boundaries or if the valve margin regulators IND and SHP1/2 activate this pathway.

The formation and relative size of external domains in the fruit: valves, valve margin, and replum are governed by antagonistic interactions analogous to those at the leaf-boundary-SAM interface (González-Reig et al., 2012). JAG and YABBY (YAB) members FILAMENTOUS (FIL)/YAB3 also found in leaves activate the MADS-box gene *FRUITFULL* (*FUL*), which is required for elongation and differentiation of the valves, and *SHP1/2*, which confer valve-margin identity (Dinneny et al., 2005). FUL in turn represses *SHP/IND/ALC* to set the valve margin boundary. Fruits in a *ful* mutant are constricted and ectopically lignified due to the misexpression of valve margin genes (Ferrándiz et al., 2000; Liljegren et al., 2004). AS1–AS2 and JAG/FIL/YAB3 reprise their roles in the leaf by restricting *BP* expression to the replum. In *as1* or *as2* mutants or in *jag fil yab3* triple mutants, valve width is reduced and the replum is expanded due to an increase in *BP* expression (Alonso-Cantabrana et al., 2007; González-Reig et al., 2012). BP, which interacts with PNY, also known as REPLUMLESS (Roeder et al., 2003), activates its expression, and contributes redundantly with PNY in maintaining the replum in part by repressing valve and valve-margin identity genes (Alonso-Cantabrana et al., 2007). Thus, inactivation of *JAG/FIL/YAB3* or *SHP/IND* genes partially rescues replum formation in a *pny* mutant (Roeder et al., 2003; Dinneny et al., 2005; Alonso-Cantabrana et al., 2007). Other factors identified are APETALA2 which prevents overgrowth of the replum and valve margin by repressing *BP/PNY* and valve margin identity genes (Ripoll et al., 2011) and the zinc finger transcription factor NO TRANSMITTING

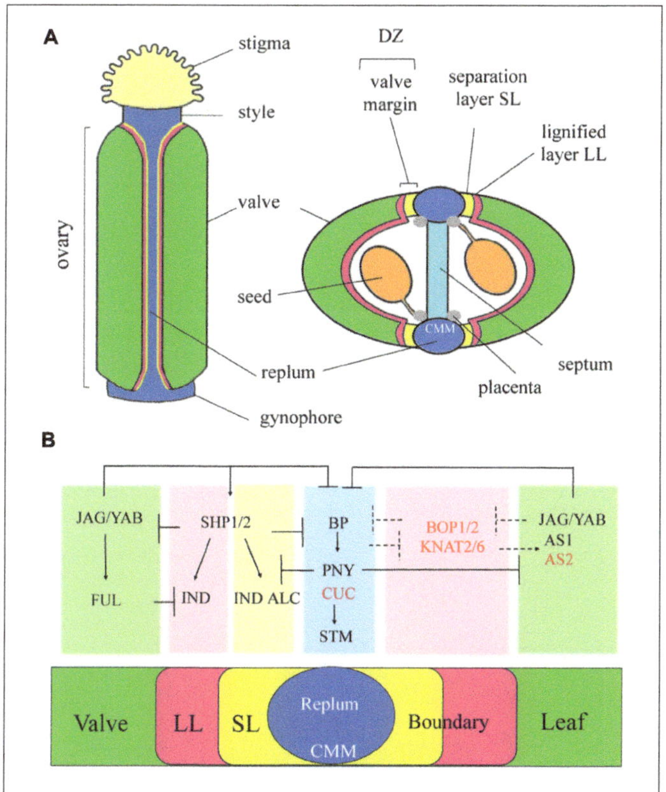

FIGURE 4 | Schematic of an *Arabidopsis* fruit and summary of networks for fruit patterning compared to the SAM-leaf boundary. **(A)** The *Arabidopsis* fruit derived from two fused carpel valves that represent modified leaves. Valve margins are a lateral organ boundary specialized for dehiscence that joins the carpels to a meristematic tissue called the replum. The internal surface of the replum or CMM provides septum and placenta that gives rise to ovules that develop into seeds at fertilization. Differentiation of the valve margins requires GA, CK, and depletion of auxin (see text). When the fruit is mature, the valve margin differentiates to form the dehiscence zone (DZ) comprising two layers: a separation layer (SL) where the fruit will open and a lignified layer (LL) continuous with the lining of the fruit that provides tension required in spring-like opening of the fruit for seed dispersal. **(B)** Similar to their role in leaves, JAG and YAB factors together with AS1–AS2 are required in restricting expression of meristematic genes *BP* and *PNY* to the central replum domain and in restricting valve margin identity genes to the boundary junction. FUL is a stage-specific factor that confers valve identity and functions similarly to AS1–AS2 and JAB/YAB to correctly position the replum and valve margin identity domains. Red lettering, SAM-leaf boundary genes: CUC1/2 activate *STM* required in formation of the CMM and *BOP1/2* and *KNAT2/6* expressed in the valve margin of fruit are predicted to antagonize BP-PNY activity in the replum. Dashed arrows, hypothetical interactions.

TRACT which promotes replum development by activating BP (Marsch-Martinez et al., 2014).

The role of boundary genes in fruit patterning and dehiscence is worth exploring. Mutations in *SHP1/2* or the different stage-specific *bHLH* genes block dehiscence (Ferrándiz et al., 2010) but this is not case for *KNAT2/6* or *BOP1/2*. Nevertheless, inactivation of these genes restores replum formation in *pny* mutants showing that the antagonistic interaction between BOP1/2- KNAT6/2 and PNY also control fruit patterning (Ragni et al., 2008; Khan et al., 2012b). Based on the role of this module in other boundary contexts, BOP1/2 and KNAT2/6 likely contribute to repression of *BP/PNY* to control replum size, specialization of cells in the separation layer, and formation of lignified cell layers (McKim et al., 2008; Khan et al., 2012a,b). Similar interactions among meristem and boundary genes regulate abscission.

ABSCISSION

Abscission zones (AZs) are typically located at lateral organ boundaries in the plant at the base of leaves, floral organs, or seeds [(Estornell et al., 2013) and **Figure 5**]. In *Arabidopsis*, AZs comprised of small, densely cytoplasmic cells form simultaneously with the boundary for detachment of floral organs and seeds (McKim et al., 2008). At anthesis, AZ cells acquire competence to respond to abscission signals and secrete cell-wall modifying and hydrolyzing enzymes that degrade the middle lamella between two adjacent cell files. Ethylene, JA, and abscisic acid are promotive signals for abscission whereas auxin, GA, and BRs are inhibitory (for reviews: Estornell et al., 2013; Niederhuth et al., 2013; Kim, 2014). Depletion of auxin from the AZ is shown to improve sensitivity to ethylene in controlling the timing of abscission (Estornell et al., 2013) reminiscent of other boundaries.

Boundary genes are required for both differentiation and separation phases of abscission. A variety of plant species lacking BOP activity fail to form an AZ (McKim et al., 2008; Wu et al., 2012; Couzigou et al., 2015). BOP1/2 quite possibly perform this function via ATH1 and KNAT2/KNAT6. Inactivation of *ATH1* has a mild abscission defect in which formation of the stamen AZs is delayed. A functional AZ eventually develops and organs detach (Gómez-Mena and Sablowski, 2008). Reprising its role in leaves, AS1 positions the medial sepal and petal AZs in *Arabidopsis* via restriction of BP activity, which in turn restricts expression of the HAESA receptor-like kinase (Gubert et al., 2014). This was discovered through isolation of a new allele, *as1-22*, which shows a delayed abscission defect (Gubert et al., 2014).

The activation of floral organ abscission involves the peptide INFLORESCENCE DEFICIENT IN ABSCISSION (IDA) and the receptor like kinase HAESA and HAESA-LIKE2 (HAE-HSL2) signaling pathway. Low levels of *IDA* and *HAESA* transcripts are also expressed in the mature dehiscence zones of the fruit (Stenvik et al., 2008). Activation of cell separation by IDA-HAE/HSL2 signaling antagonizes BP activity leading to up-regulation of *KNAT2* and *KNAT6* and accumulation of cell-wall modification and hydrolytic enzymes that mediate separation (Shi et al., 2011). Waxes, suberin, lignin, and pathogenesis-related genes are also induced in protecting exposed cells from dehydration and pathogen attack (Estornell et al., 2013; Kim, 2014). Plants overexpressing *BOP1* show significant enrichment for genes involved in lignin biosynthesis, stress and pathogen resistance (Khan et al., 2012a, 2015) suggesting a potential role in post-abscission events at the boundary.

Not all AZs form at a lateral organ boundary (Estornell et al., 2013). The pedicel AZ in tomato is a well-studied example in which a small groove in the floral pedicel leads to differentiation of a "joint" where abscission takes place. Two MADS box

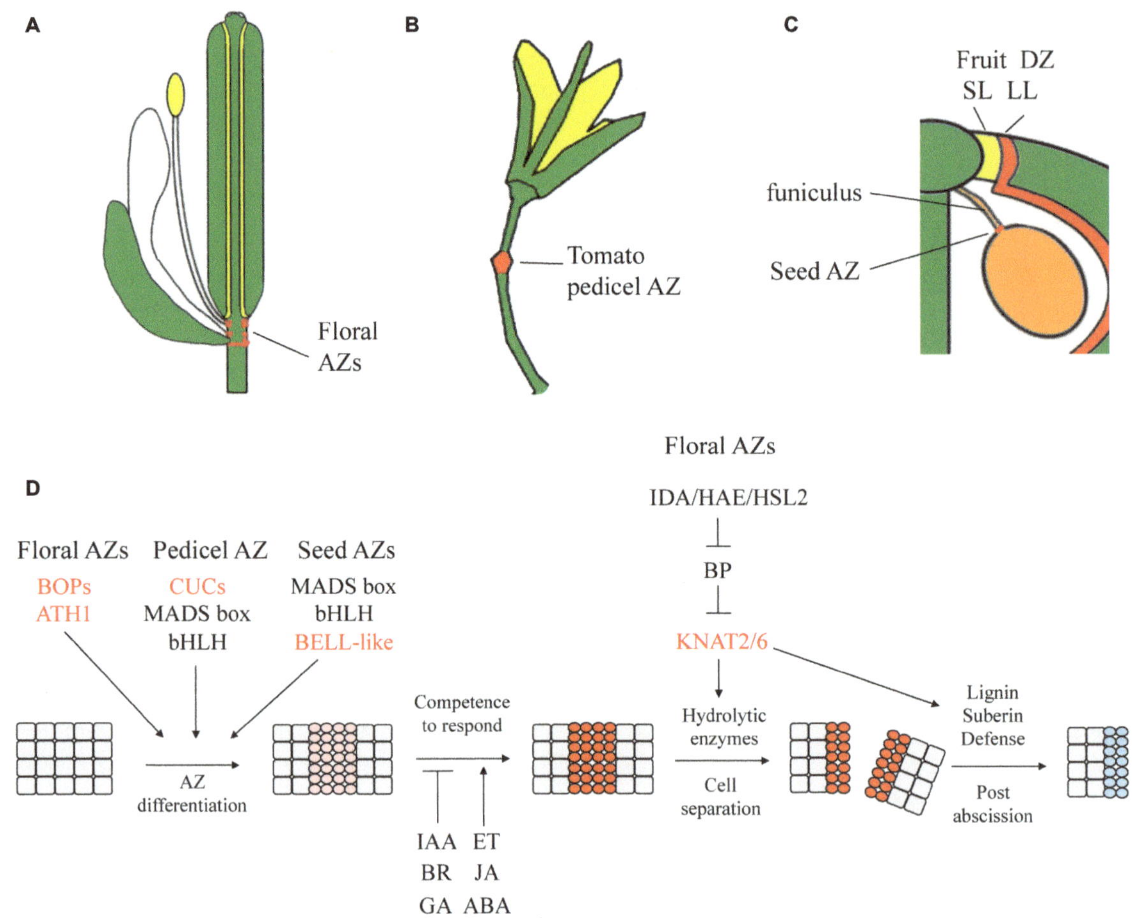

FIGURE 5 | Summary of pathways that control the development of abscission zones (AZs). Placement of AZs (in red) are shown for **(A)** *Arabidopsis* flower, **(B)** tomato pedicel, and **(C)** *Arabidopsis* seed and dehiscence zone. **(D)** The current accepted model for abscission defines four key stages: differentiation of AZ cells (pink), AZ cells acquire competence to respond to hormone inductive signals (red), activation of abscission and detachment (red), and differentiation of a protective layer over the scar (blue). Red lettering, SAM-leaf boundary genes. Dashed lines, hypothetical pathway. IAA, auxin; BR, brassinosteroid; GA, gibberellin; ET, ethylene; CK, cytokinin; JA, jasmonic acid. Adapted from (Leibfried et al., 2005; Niederhuth et al., 2013).

proteins JOINTLESS and MACROCALYX form a complex that regulates formation of the pedicel AZ together with LAS required in AM production (Schumacher et al., 1999; Nakano et al., 2012). CUC2 homolog GOB has been proposed to be involved in the regulation of the onset of abscission based on its expression in tomato pedicel AZs (Nakano et al., 2013). Tomato *KNAT6* (*TKN3*), *KNAT2* (*TKN4*), and the *KNATM* homolog *KD1* are all highly expressed in pedicel AZs similar to *Arabidopsis*. Silencing of *KD1* delays abscission by increasing auxin content and overexpression of *KD1* has the opposite effect (Ma et al., 2015). Hormonal control and transcript profiling of activated AZs in *Arabidopsis* and tomato are very similar despite these apparent differences (Estornell et al., 2013; Ito and Nakano, 2015).

An AZ at the base of the seed allows detachment from the funiculus. In *Arabidopsis*, specification of this AZ requires the MADS-box protein SEEDSTICK and the bHLH factor HECATE3 (Pinyopich et al., 2003; Ogawa et al., 2009). BOP1/2 do not seem to be expressed in the ovule nor is there an obvious requirement for the IDA-HAE/HLS2 signaling pathway (Estornell et al., 2013). This suggests a greater alignment with processes that control dehiscence. The role of PNY has yet to be investigated in seed dehiscence. Interestingly, domesticated japonica rice is selected for a promoter mutation in the *PNY/RPL* homolog *qSHATTERING1* that depletes expression from the abscission layer to inhibit seed shatter (Konishi et al., 2006). *qSH1* and a related gene *SH5* are required for development of the pedicel AZ and inhibition of these genes reduces shatter by promoting lignin biosynthesis (Yoon et al., 2014). Overexpression of wheat *TaqSH1* in *Arabidopsis* delays abscission and down regulates abscission-promoting genes suggesting that TaqSH1 can function as an upstream regulator of the *IDA-HAESA-KNAT* pathway (Zhang et al., 2013). Remarkably, the same mutation in *qSH1* was found in *Brassica rapa* which exhibit a narrow replum compared to *Arabidopsis* (Arnaud et al., 2011). Collectively, these data show remarkable overlap between separation processes involved in pod shatter, floral organ abscission, and seed dehiscence in which lateral organ boundary genes play a key role.

CONCLUDING REMARKS

In this review, we illustrate the importance of boundaries throughout development. Studies have revealed a number of genes including *CUCs*, *TALEs*, and *BOP* that play a recurring role thoughout the life cycle. During the reproductive phase, their activities are embedded within specialized networks required for inflorescence, flower, and fruit development. How these pathways are integrated is only partly understood. Many questions also remain concerning the role of these genes in the SAM. Molecular links between *CUC* genes that confer boundary identity and BOP-TALE factors are not well-established. Recent studies in monocots have shed light on the role of KNOX transcription factors in initiating boundaries but application of this model to dicots is not yet confirmed. While significant progress has been made in understanding how KNOX factors regulate hormone abundance, such links are still largely missing for CUC and BOP factors at the boundary. These factors also repress growth and cell division but few targets have been identified to date and their hierarchy is unclear. Identification of transcriptional targets is key to understanding how these factors pattern the boundary. Understanding how these networks translate to boundaries during reproductive development is still in its infancy. In particular, the influence of BR and the role of JLO is not yet explored. The function of BOP and TALE factors in fruit and in abscission is also unclear. Finally, the contribution of these networks to the activity of lateral meristems responsible for secondary growth in stems and roots is only partly understood. Transcriptome analysis and further exploration of these factors in boundaries in other species will establish the extent to which the mode of action of these factors is conserved in development.

Wolfgang von Goethe in 1790 first proposed that plants are formed from repeating units called phytomers (Bossinger, 2009). As developmental contexts are further explored this notion can be extended to the molecular level. Work in the last decade in a variety of species makes it obvious that simple patterns established with formation of the embryonic plant underlie pattern development at all stages of plant development. As such, there is much to be learned from drawing parallels and insights from different developmental contexts. Lateral organ boundaries are a particularly intriguing part of the puzzle as they regulate the balance between meristem activity and growth and represent a fundamental gatekeeper of meristem function in plants.

FUNDING

The IJPB benefits from the support of the Labex Saclay Plant Sciences-SPS (ANR-10-LABX-0040-SPS). This work was supported by a Natural Sciences and Engineering Research Council Discovery Grant (no. 327195 to SH). This work was supported by the ANR CHARMFUL program (N° ANR-11BSV2 005 02 to VP).

ACKNOWLEDGMENTS

Special thanks to Bronwyn Rowland for illustrations. We acknowledge Aude Maugarny, Owen Rowland, Adina Popescu, and Nicolas Arnaud for critical comments on the manuscript.

REFERENCES

Aichinger, E., Kornet, N., Friedrich, T., and Laux, T. (2012). Plant stem cell niches. *Annu. Rev. Plant Biol.* 63, 615–636. doi: 10.1146/annurev-arplant-042811-105555

Aida, M., Ishida, T., Fukaki, H., Fujisawa, H., and Tasaka, M. (1997). Genes involved in organ separation in *Arabidopsis*: an analysis of the cup-shaped cotyledon mutant. *Plant Cell* 9, 841–857. doi: 10.1105/tpc.9.6.841

Aida, M., Ishida, T., and Tasaka, M. (1999). Shoot apical meristem and cotyledon formation during *Arabidopsis* embryogenesis: interaction among the CUP-SHAPED COTYLEDON and SHOOT MERISTEMLESS genes. *Development* 126, 1563–1570.

Aida, M., Vernoux, T., Furutani, M., Traas, J., and Tasaka, M. (2002). Roles of PIN-FORMED1 and MONOPTEROS in pattern formation of the apical region of the *Arabidopsis* embryo. *Development* 129, 3965–3974.

Alonso-Cantabrana, H., Ripoll, J. J., Ochando, I., Vera, A., Ferrándiz, C., and Martínez-Laborda, A. (2007). Common regulatory networks in leaf and fruit patterning revealed by mutations in the *Arabidopsis* ASYMMETRIC LEAVES1 gene. *Development* 134, 2663–2671. doi: 10.1242/dev.02864

Andrés, F., Romera-Branchat, M., Martínez-Gallegos, R., Patel, V., Schneeberger, K., Jang, S., et al. (2015). Floral induction in *Arabidopsis thaliana* by FLOWERING LOCUS T requires direct repression of BLADE-ON-PETIOLE genes by homeodomain protein PENNYWISE. *Plant Physiol.* 169, 2187–2199. doi: 10.1104/pp.15.00960

Arnaud, N., Girin, T., Sorefan, K., Fuentes, S., Wood, T. A., Lawrenson, T., et al. (2010). Gibberellins control fruit patterning in *Arabidopsis thaliana*. *Genes Dev.* 24, 2127–2132. doi: 10.1101/gad.593410

Arnaud, N., Lawrenson, T., Østergaard, L., and Sablowski, R. (2011). The same regulatory point mutation changed seed-dispersal structures in evolution and domestication. *Curr. Biol.* 21, 1215–1219. doi: 10.1016/j.cub.2011.06.008

Arnaud, N., and Pautot, V. (2014). Ring the BELL and tie the KNOX: roles for TALEs in gynoecium development. *Front. Plant Sci.* 5:93. doi: 10.3389/fpls.2014.00093

Bartrina, I., Otto, E., Strnad, M., Werner, T., and Schmulling, T. (2011). Cytokinin regulates the activity of reproductive meristems, flower organ size, ovule formation, and thus seed yield in *Arabidopsis thaliana*. *Plant Cell* 23, 69–80. doi: 10.1105/tpc.110.079079

Bell, E. M., Lin, W. C., Husbands, A. Y., Yu, L., Jaganatha, V., Jablonska, B., et al. (2012). *Arabidopsis* LATERAL ORGAN BOUNDARIES negatively regulates brassinosteroid accumulation to limit growth in organ boundaries. *Proc. Natl. Acad. Sci. U.S.A.* 109, 21146–21151. doi: 10.1073/pnas.1210789109

Belles-Boix, E., Hamant, O., Witiak, S. M., Morin, H., Traas, J., and Pautot, V. (2006). KNAT6: an *Arabidopsis* homeobox gene involved in meristem activity and organ separation. *Plant Cell* 18, 1900–1907. doi: 10.1105/tpc.106.041988

Bencivenga, S., Simonini, S., Benkova, E., and Colombo, L. (2012). The transcription factors BEL1 and SPL are required for cytokinin and auxin signaling during ovule development in *Arabidopsis*. *Plant Cell* 24, 2886–2897. doi: 10.1105/tpc.112.100164

Benková, E., Michneiwicz, M., Sauer, M., Teichmann, T., Seifertová, D., Jürgens, G., et al. (2003). Local, efflux-dependent auxin gradients as a common module for plant organ formation. *Cell* 115, 591–602. doi: 10.1016/S0092-8674(03)00924-3

Berger, Y., Harpaz-Saad, S., Brand, A., Melnik, H., Sirding, N., Alvarez, J. P., et al. (2009). The NAC-domain transcription factor GOBLET specifies leaflet boundaries in compound tomato leaves. *Development* 136, 823–832. doi: 10.1242/dev.031625

Besnard, F., Vernoux, T., and Hamant, O. (2011). Organogenesis from stem cells in planta: multiple feedback loops integrating molecular and mechanical signals. *Cell. Mol. Life Sci.* 68, 2885–2906. doi: 10.1007/s00018-011-0732-4

Bilsborough, G. D., Runions, A., Barkoulas, M., Jenkins, H. W., Hasson, A., Galinha, C., et al. (2011). Model for the regulation of *Arabidopsis thaliana* leaf margin development. *Proc. Natl. Acad. Sci. U.S.A.* 108, 3424–3429. doi: 10.1073/pnas.1015162108

Blein, T., Pulido, A., Vialette-Guiraud, A., Nikovics, K., Morin, H., Hay, A., et al. (2008). A conserved molecular framework for compound leaf development. *Science* 322, 1835–1839. doi: 10.1126/science.1166168

Bolduc, N., and Hake, S. (2009). The maize transcription factor KNOTTED1 directly regulates the gibberellin catabolism gene ga2ox1. *Plant Cell* 21, 1647–1658. doi: 10.1105/tpc.109.068221

Bolduc, N., Yilmaz, A., Mejia-Guerra, M. K., Morohashi, K., O'Connor, D., Grotewold, E., et al. (2012). Unraveling the KNOTTED1 regulatory network in maize meristems. *Genes Dev.* 26, 1685–1690. doi: 10.1101/gad.193433.112

Borghi, L., Bureau, M., and Simon, R. (2007). *Arabidopsis* JAGGED LATERAL ORGANS is expressed in boundaries and coordinates KNOX and PIN activity. *Plant Cell* 19, 1795–1808. doi: 10.1105/tpc.106.047159

Bossinger, G. (2009). "Segments (Phytomers)," in *Encyclopedia of Life Sciences*, ed. A. M. Hetherington (Chichester: John Wiley and Sons). doi: 10.1002/9780470015902.a0002093.pub2

Bureau, M., Rast, M. I., Illmer, J., and Simon, R. (2010). JAGGED LATERAL ORGAN (JLO) controls auxin dependent patterning during development of the *Arabidopsis* embryo and root. *Plant Mol. Biol.* 74, 479–491. doi: 10.1007/s11103-010-9688-2

Byrne, M. E., Groover, A. T., Fontana, J. R., and Martienssen, R. A. (2003). Phyllotactic pattern and stem cell fate are determined by the *Arabidopsis* homeobox gene BELLRINGER. *Development* 130, 3941–3950. doi: 10.1242/dev.00620

Byrne, M. E., Simorowski, J., and Martienssen, R. A. (2002). ASYMMETRIC LEAVES1 reveals knox gene redundancy in *Arabidopsis*. *Development* 129, 1957–1965.

Chae, E., Tan, Q. K.-G., Hill, T. A., and Irish, V. F. (2008). An *Arabidopsis* F-box protein acts as a transcriptional co-factor to regulate floral development. *Development* 135, 1235–1245. doi: 10.1242/dev.015842

Chahtane, H., Vachon, G., Le Masson, M., Thevenon, E., Perigon, S., Mihajlovic, N., et al. (2013). A variant of LEAFY reveals its capacity to stimulate meristem development by inducing RAX1. *Plant J.* 74, 678–689. doi: 10.1111/tpj.12156

Cho, E., and Zambryski, P. C. (2011). ORGAN BOUNDARY1 defines a gene expressed at the junction between the shoot apical meristem and lateral organs. *Proc. Natl. Acad. Sci. U.S.A.* 108, 2154–2159. doi: 10.1073/pnas.1018542108

Chung, Y., Maharjan, P. M., Lee, O., Fujioka, S., Jang, S., Kim, B., et al. (2011). Auxin stimulates DWARF4 expression and brassinosteroid biosynthesis in *Arabidopsis*. *Plant J.* 66, 564–578. doi: 10.1111/j.1365-313X.2011.04513.x

Cole, M., Nolte, C., and Werr, W. (2006). Nuclear import of the transcription factor SHOOT MERISTEMLESS depends on heterodimerization with BLH proteins expressed in discrete subdomains of the shoot apical meristem of *Arabidopsis thaliana*. *Nucleic Acids Res.* 35, 1281–1292. doi: 10.1093/nar/gkl016

Colling, J., Tohge, T., De Clercq, R., Brunoud, G., Vernoux, T., Fernie, A. R., et al. (2015). Overexpression of the *Arabidopsis thaliana* signalling peptide TAXIMIN1 affects lateral organ development. *J. Exp. Bot.* 66, 5337–5349. doi: 10.1093/jxb/erv291

Couzigou, J. M., Magne, K., Mondy, S., Cosson, V., Clements, J., and Ratet, P. (2015). The legume NOOT-BOP-COCH-LIKE genes are conserved regulators of abscission, a major agronomical trait in cultivated crops. *New Phytol.* doi: 10.1111/nph.13634 [Epub ahead of print].

Cucinotta, M., Colombo, L., and Roig-Villanova, I. (2014). Ovule development, a new model for lateral organ formation. *Front. Plant Sci.* 5:117. doi: 10.3389/fpls.2014.00117

Di Giacomo, E., Lanneli, M., and Frugis, G. (2013). TALE and shape: how to make a leaf different. *Plants* 2, 317–342. doi: 10.3390/plants2020317

Dinneny, J. R., Weigel, D., and Yanofsky, M. F. (2005). A genetic framework for fruit patterning in *Arabidopsis thaliana*. *Development* 132, 4687–4696. doi: 10.1242/dev.02062

Douglas, S. J., Chuck, G., Dengler, R. E., Pelecanda, L., and Riggs, C. D. (2002). KNAT1 and ERECTA regulate inflorescence architecture in *Arabidopsis*. *Plant Cell* 14, 547–558. doi: 10.1105/tpc.010391

Efroni, I., Eshed, Y., and Lifschitz, E. (2010). Morphogenesis of simple and compound leaves: a critical review. *Plant Cell* 22, 1019–1032. doi: 10.1105/tpc.109.073601

Endrizzi, K., Moussian, B., Haecker, A., Levin, J. Z., and Laux, T. (1996). The SHOOT MERISTEMLESS gene is required for maintenance of undifferentiated cells in *Arabidopsis* shoot and floral meristems and acts at a different regulatory level than the meristem genes WUSCHEL and ZWILLE. *Plant J.* 10, 967–979. doi: 10.1046/j.1365-313X.1996.10060967.x

Estornell, L. H., Agusti, J., Merelo, P., Talon, M., and Tadeo, F. R. (2013). Elucidating mechanisms underlying organ abscission. *Plant Sci.* 19, 48–60. doi: 10.1016/j.plantsci.2012.10.008

Etchells, J. P., Moore, L., Jiang, W. Z., Prescott, H., Capper, R., Saunders, N. J., et al. (2012). A role for BELLRINGER in cell wall development is supported by loss-of-function phenotypes. *BMC Plant Biol.* 12:212. doi: 10.1186/1471-2229-12-212

Ferrándiz, C., Fourquin, C., Prunet, N., Scutt, C. P., Trehin, C., and Vialette-Giraud, A. (2010). "Carpel Development," in *Advances in Botanical Research*, eds J.-C. Kader and M. Delseny (Burlington, MA: Academic Press), 1–73.

Ferrándiz, C., Liljegren, S. J., and Yanofsky, M. F. (2000). Negative regulation of the SHATTERPROOF genes by FRUITFULL during *Arabidopsis* fruit development. *Science* 289, 436–438. doi: 10.1126/science.289.5478.436

Frigerio, M., Alabadi, D., Perez-Gomez, J., Garcia-Carcel, L., Phillips, A. L., Hedden, P., et al. (2006). Transcriptional regulation of gibberellin metabolism genes by auxin signaling in *Arabidopsis*. *Plant Physiol.* 142, 553–563. doi: 10.1104/pp.106.084871

Furutani, M., Vernoux, T., Traas, J., Kato, T., Tasaka, M., and Aida, M. (2004). PIN-FORMED1 and PINOID regulate boundary formation and cotyledon development in *Arabidopsis* embryogenesis. *Development* 131, 5021–5030. doi: 10.1242/dev.01388

Galbiati, F., Sinha Roy, D., Simonini, S., Cucinotta, M., Ceccato, L., Cuesta, C., et al. (2013). An integrative model of the control of ovule primordia formation. *Plant J.* 76, 446–455. doi: 10.1111/tpj.12309

Gallavotti, A., Zhao, Q., Kyozuka, J., Meeley, R. B., Ritter, M. K., Doebley, J. F., et al. (2004). The role of barren stalk1 in the architecture of maize. *Nature* 432, 630–635. doi: 10.1038/nature03148

Gendron, J. M., Liu, J. S., Fan, M., Bai, M. Y., Wenkel, S., Springer, P. S., et al. (2012). Brassinosteroids regulate organ boundary formation in the shoot apical meristem of *Arabidopsis*. *Proc. Natl. Acad. Sci. U.S.A.* 109, 21152–21157. doi: 10.1073/pnas.1210799110

Girin, T., Paicu, T., Stephenson, P., Fuentes, S., Korner, E., O'Brien, M., et al. (2011). INDEHISCENT and SPATULA interact to specify carpel and valve margin tissue and thus promote seed dispersal in *Arabidopsis*. *Plant Cell* 23, 3641–3653. doi: 10.1105/tpc.111.090944

Gomez, M. D., Urbez, C., Perez-Amador, M. A., and Carbonell, J. (2011). Characterization of constricted fruit (ctf) mutant uncovers a role for AtMYB117/LOF1 in ovule and fruit development in *Arabidopsis thaliana*. *PLoS ONE* 6:e18760. doi: 10.1371/journal.pone.0018760

Gómez-Mena, C., and Sablowski, R. (2008). *Arabidopsis thaliana* HOMEOBOX GENE1 establishes the basal boundaries of shoot organs and controls stem growth. *Plant Cell* 20, 2059–2072. doi: 10.1105/tpc.108.059188

Goncalves, B., Hasson, A., Belcram, K., Cortizo, M., Morin, H., Nikovics, K., et al. (2015). A conserved role for CUP-SHAPED COTYLEDON genes during ovule development. *Plant J.* 83, 732–742. doi: 10.1111/tpj.12923

González-Reig, S., Ripoll, J. J., Vera, A., Yanofsky, M. F., and Martinez-Laborda, A. (2012). Antagonistic gene activities determine the formation of pattern elements along the mediolateral axis of the *Arabidopsis* fruit. *PLoS Genet.* 8:e1003020. doi: 10.1371/journal.pgen.1003020

Gordon, S. P., Chickarmane, V. S., Ohno, C., and Meyerowitz, E. M. (2009). Multiple feedback loops through cytokinin signaling control stem cell number within the *Arabidopsis* shoot meristem. *Proc. Natl. Acad. Sci. U.S.A.* 106, 16529–16534. doi: 10.1073/pnas.0908122106

Grbic, V., and Bleecker, A. B. (2000). Axillary meristem development in *Arabidopsis thaliana*. *Plant J.* 21, 215–223. doi: 10.1046/j.1365-313x.2000.00670.x

Greb, T., Clarenz, O., Schafer, E., Muller, D., Herrero, R., Schmitz, G., et al. (2003). Molecular analysis of the LATERAL SUPPRESSOR gene in *Arabidopsis* reveals a conserved control mechanism for axillary meristem formation. *Genes Dev.* 17, 1175–1187. doi: 10.1101/gad.260703

Griffith, M. E., da Silva Conceição, A., and Smyth, D. R. (1999). PETAL LOSS gene regulates initiation and orientation of second whorl organs in the *Arabidopsis* flower. *Development* 126, 5635–5644.

Groszmann, M., Paicu, T., Alvarez, J. P., Swain, S. M., and Smyth, D. R. (2011). SPATULA and ALCATRAZ, are partially redundant, functionally diverging bHLH genes required for *Arabidopsis* gynoecium and fruit development. *Plant J.* 68, 816–829. doi: 10.1111/j.1365-313X.2011.04732.x

Gubert, C. M., Christy, M. E., Ward, D. L., Groner, W. D., and Liljegren, S. J. (2014). ASYMMETRIC LEAVES1 regulates abscission zone placement in *Arabidopsis* flowers. *BMC Plant Biol.* 14:195. doi: 10.1186/s12870-014-0195-5

Guo, M., Thomas, J., Collins, G., and Timmermans, M. C. P. (2008). Direct repression of KNOX loci by the ASYMMETRIC LEAVES1 complex of *Arabidopsis*. *Plant Cell* 20, 48–58. doi: 10.1105/tpc.107.056127

Ha, C. M., Jun, J. H., and Fletcher, J. C. (2010). Control of *Arabidopsis* leaf morphogenesis through regulation of the YABBY and KNOX families of transcription factors. *Genetics* 186, 197–206. doi: 10.1534/genetics.110.118703

Ha, C. M., Jun, J. H., Nam, H. G., and Fletcher, J. (2004). BLADE-ON-PETIOLE1 encodes a BTB/POZ domain protein required for leaf morphogenesis in *Arabidopsis thaliana*. *Plant Cell Physiol.* 45, 1361–1370. doi: 10.1093/pcp/pch201

Ha, C. M., Jun, J. H., Nam, H. G., and Fletcher, J. C. (2007). BLADE-ON-PETIOLE 1 and 2 control *Arabidopsis* lateral organ fate through regulation of LOB domain and adaxial-abaxial polarity genes. *Plant Cell* 19, 1809–1825. doi: 10.1105/tpc.107.051938

Ha, C. M., Kim, G.-T., Kim, B. C., Jun, J. H., Soh, M. S., Ueno, Y., et al. (2003). The BLADE-ON-PETIOLE1 gene controls leaf pattern formation through the modulation of meristematic activity in *Arabidopsis*. *Development* 130, 161–172. doi: 10.1242/dev.00196

Hamant, O., and Pautot, V. (2010). Plant development: a TALE story. *C. R. Biol.* 333, 371–381. doi: 10.1016/j.crvi.2010.01.015

Han, Y., Zhang, C., Yang, H., and Jiao, Y. (2014). Cytokinin pathway mediates APETALA1 function in the establishment of determinate floral meristems in *Arabidopsis*. *Proc. Natl. Acad. Sci. U.S.A.* 111, 6840–6845. doi: 10.1073/pnas.1318532111

Hasson, A., Plessis, A., Blein, T., Adroher, B., Grigg, S., Tsiantis, M., et al. (2011). Evolution and diverse roles of the CUP-SHAPED COTYLEDON genes in *Arabidopsis* leaf development. *Plant Cell* 23, 54–68. doi: 10.1105/tpc.110.081448

Hay, A., Barkoulas, M., and Tsiantis, M. (2006). ASYMMETRIC LEAVES1 and auxin activities converge to repress BREVIPEDICELLUS expression and promote leaf develoment in *Arabidopsis*. *Development* 133, 3955–3961. doi: 10.1242/dev.02545

Hay, A., Kaur, H., Phillips, A., Hedden, P., Hake, S., and Tsiantis, M. (2002). The gibberellin pathway mediates KNOTTED1-type homeobox function in plants with different body plans. *Curr. Biol.* 12, 1557–1565. doi: 10.1016/S0960-9822(02)01125-9

Hay, A., and Tsiantis, M. (2010). KNOX genes: versatile regulators of plant development and diversity. *Development* 137, 3153–3165. doi: 10.1242/dev.030049

Heisler, M. G., Hamant, O., Krupinski, P., Uyttewaal, M., Ohno, C., Jönsson, H., et al. (2010). Alignment between PIN1 polarity and microtubule orientation in the shoot apical meristem reveals a tight coupling between morphogenesis and auxin transport. *PLoS Biol.* 8:e1000516. doi: 10.1371/journal.pbio.100 0516

Heisler, M. G., Ohno, C., Das, P., Sieber, P., Reddy, G. V., Long, J. A., et al. (2005). Patterns of auxin transport and gene expression during primordium development revealed by live imaging of the *Arabidopsis* inflorescence meristem. *Curr. Biol.* 15, 1899–1911. doi: 10.1016/j.cub.2005.09.052

Hepworth, S. R., Zhang, Y., Mckim, S., Li, X., and Haughn, G. W. (2005). BLADE-ON-PETIOLE-dependent signaling controls leaf and floral patterning in *Arabidopsis*. *Plant Cell* 17, 1434–1448. doi: 10.1105/tpc.104.030536

Hibara, K., Karim, M. R., Takada, S., Taoka, K., Furutani, M., Aida, M., et al. (2006). *Arabidopsis* CUP-SHAPED COTYLEDON3 regulates postembryonic shoot meristem and organ boundary formation. *Plant Cell* 18, 2946–2957. doi: 10.1105/tpc.106.045716

Huang, H. Y., Jiang, W. B., Hu, Y. W., Wu, P., Zhu, J. Y., Liang, W. Q., et al. (2013). BR signal influences *Arabidopsis* ovule and seed number through regulating related genes expression by BZR1. *Mol. Plant* 6, 456–469. doi: 10.1093/mp/sss070

Ichihashi, Y., Aguilar-Martinez, J. A., Farhi, M., Chitwood, D. H., Kumar, R., Millon, L. V., et al. (2014). Evolutionary developmental transcriptomics reveals a gene network module regulating interspecific diversity in plant leaf shape. *Proc. Natl. Acad. Sci. U.S.A.* 111, 2616–2621. doi: 10.1073/pnas.1402835111

Ishida, T., Aida, M., Takada, S., and Tasaka, M. (2000). Involvement of CUP-SHAPED COTYLEDON genes in gynoecium and ovule development in *Arabidopsis thaliana*. *Plant Cell Physiol.* 41, 60–67. doi: 10.1093/pcp/41.1.60

Ito, Y., and Nakano, T. (2015). Development and regulation of pedicel abscission in tomato. *Front. Plant Sci.* 6:442. doi: 10.3389/fpls.2015.00442

Janssen, B. J., Drummond, R. S., and Snowden, K. C. (2014). Regulation of axillary shoot development. *Curr. Biol.* 17, 28–35.

Jasinski, S., Piazza, P., Craft, J., Hay, A., Woolley, L., Rieu, I., et al. (2005). KNOX action in *Arabidopsis* is mediated by coordinate regulation of cytokinin and gibberellin activities. *Curr. Biol.* 15, 1560–1565. doi: 10.1016/j.cub.2005.07.023

Johnston, R., Wang, M., Sun, Q., Sylvester, A. W., Hake, S., and Scanlon, M. J. (2014). Transcriptomic analyses indicate that maize ligule development recapitulates gene expression patterns that occur during lateral organ initiation. *Plant Cell* 26, 4718–4732. doi: 10.1105/tpc.114.132688

Jun, J. H., Ha, C. M., and Fletcher, J. C. (2010). BLADE-ON-PETIOLE1 coordinates organ determinacy and axial polarity in *Arabidopsis* by directly activating ASYMMETRIC LEAVES2. *Plant Cell* 22, 62–76. doi: 10.1105/tpc.109.070763

Kamiuchi, Y., Yamamoto, K., Furutani, M., Tasaka, M., and Aida, M. (2014). The CUC1 and CUC2 genes promote carpel margin meristem formation during *Arabidopsis* gynoecium development. *Front. Plant Sci.* 5:165. doi: 10.3389/fpls.2014.00165

Kanrar, S., Bhattacharya, M., Arthur, B., Courtier, J., and Smith, H. M. (2008). Regulatory networks that function to specify flower meristems require the function of homeobox genes PENNYWISE and POUND-FOOLISH in *Arabidopsis*. *Plant J.* 54, 924–937. doi: 10.1111/j.1365-313X.2008.03458.x

Kanrar, S., Onguka, O., and Smith, H. M. (2006). *Arabidopsis* inflorescence architecture requires the activities of KNOX-BELL homeodomain heterodimers. *Planta* 224, 1163–1173. doi: 10.1007/s00425-006-0298-9

Karim, M. R., Hirota, A., Kwiatkowska, D., Tasaka, M., and Aida, M. (2009). A role for *Arabidopsis* PUCHI in floral meristem identity and bract suppression. *Plant Cell* 21, 1360–1372. doi: 10.1105/tpc.109.067025

Kawamura, E., Horiguchi, G., and Tsukaya, H. (2010). Mechanisms of leaf tooth formation in *Arabidopsis*. *Plant J.* 62, 429–441. doi: 10.1111/j.1365-313X.2010.04156.x

Keller, T., Abbott, J., Moritz, T., and Doerner, P. (2006). *Arabidopsis* REGULATOR OF AXILLARY MERISTEMS1 controls a leaf axil stem cell niche and modulates vegetative development. *Plant Cell* 18, 598–611. doi: 10.1105/tpc.105.038588

Khan, M., Ragni, L., Tabb, P., Salasini, B. C., Chatfield, S., Datla, R., et al. (2015). Repression of lateral organ boundary genes by PENNYWISE and POUND-FOOLISH is essential for meristem maintenance and flowering in *Arabidopsis thaliana*. *Plant Physiol.* 169, 2166–2186. doi: 10.1104/pp.15.00915

Khan, M., Tabb, P., and Hepworth, S. R. (2012a). BLADE-ON-PETIOLE1 and 2 regulate *Arabidopsis* inflorescence architecture in conjunction with homeobox genes KNAT6 and ATH1. *Plant Signal. Behav.* 7, 788–792. doi: 10.4161/psb.20599

Khan, M., Xu, M., Murmu, J., Tabb, P., Liu, Y., Storey, K., et al. (2012b). Antagonistic interaction of BLADE-ON-PETIOLE1 and 2 with BREVIPEDICELLUS and PENNYWISE regulates *Arabidopsis* inflorescence architecture. *Plant Physiol.* 158, 946–960. doi: 10.1104/pp.111.188573

Khan, M., Xu, H., and Hepworth, S. R. (2014). BLADE-ON-PETIOLE genes: setting boundaries in development and defense. *Plant Sci.* 21, 157–171. doi: 10.1016/j.plantsci.2013.10.019

Kim, D., Cho, Y. H., Ryu, H., Kim, Y., Kim, T. H., and Hwang, I. (2013). BLH1 and KNAT3 modulate ABA responses during germination and early seedling development in *Arabidopsis*. *Plant J.* 75, 755–766. doi: 10.1111/tpj.12236

Kim, J. (2014). Four shades of detachment: regulation of floral organ abscission. *Plant Signal. Behav.* 9, e976154. doi: 10.4161/15592324.2014.976154

Kimura, S., Koenig, D., Kang, J., Yoong, F. Y., and Sinha, N. (2008). Natural variation in leaf morphology results from mutation of a novel KNOX gene. *Curr. Biol.* 18, 672–677. doi: 10.1016/j.cub.2008.04.008

Komatsu, K., Maekawa, M., Ujiie, S., Satake, Y., Furutani, I., Okamoto, H., et al. (2003). LAX and SPA: major regulators of shoot branching in rice. *Proc. Natl. Acad. Sci. U.S.A.* 100, 11765–11770. doi: 10.1073/pnas.1932414100

Konishi, S., Izawa, T., Lin, S. Y., Ebana, K., Fukuta, Y., Sasaki, T., et al. (2006). An SNP caused loss of seed shattering during rice domestication. *Science* 312, 1392–1396. doi: 10.1126/science.1126410

Koyama, T., Furutani, M., Tasaka, M., and Ohme-Takagi, M. (2007). TCP transcription factors control the morphology of shoot lateral organs via negative regulation of the expression of boundary-specific genes in *Arabidopsis*. *Plant Cell* 19, 473–484. doi: 10.1105/tpc.106.044792

Koyama, T., Mitsuda, N., Seki, M., Shinozaki, K., and Ohme-Takagi, M. (2010). TCP transcription factors regulate the activities of ASYMMETRIC LEAVES1 and miR164, as well as the auxin response, during differentiation of leaves in *Arabidopsis*. *Plant Cell* 22, 3574–3588. doi: 10.1105/tpc.110.075598

Kwon, C. S., Hibara, K., Pfluger, J., Bazhani, S., Metha, S., Metha, H., et al. (2006). A role for chromatin remodeling in regulation of CUC gene expression in the *Arabidopsis* cotyledon boundary. *Development* 133, 3223–3230. doi: 10.1242/dev.02508

Lal, S., Pacis, L. B., and Smith, H. M. (2011). Regulation of the SQUAMOSA PROMOTER-BINDING PROTEIN-LIKE genes/microRNA156 module by the homeodomain proteins PENNYWISE and POUND-FOOLISH in *Arabidopsis*. *Mol. Plant* 4, 1123–1132. doi: 10.1093/mp/ssr041

Lampugnani, E. R., Kilinc, A., and Smyth, D. R. (2012). PETAL LOSS is a boundary gene that inhibits growth between developing sepals in *Arabidopsis thaliana*. *Plant J.* 71, 724–735. doi: 10.1111/j.1365-313X.2012.05023.x

Lampugnani, E. R., Kilinc, A., and Smyth, D. R. (2013). Auxin controls petal initiation in *Arabidopsis*. *Development* 140, 185–194. doi: 10.1242/dev.084582

Laufs, P., Peaucelle, A., Morin, H., and Traas, J. (2004). MicroRNA regulation of the CUC genes is required for boundary size control in *Arabidopsis* meristems. *Development* 131, 4311–4322. doi: 10.1242/dev.01320

Lee, B. H., Jeon, J. O., Lee, M. M., and Kim, J. H. (2015). Genetic interaction between GROWTH-REGULATING FACTOR and CUP-SHAPED COTYLEDON in organ separation. *Plant Signal. Behav.* 10, e988071. doi: 10.4161/15592324.2014.988071

Lee, D. K., Geisler, M., and Springer, P. S. (2009). LATERAL ORGAN FUSION1 and LATERAL ORGAN FUSION2 function in lateral organ separation and axillary meristem formation in *Arabidopsis*. *Development* 136, 2423–2432. doi: 10.1242/dev.031971

Lee, I. H., Wolfe, D. S., Nilsson, O., and Weigel, D. (1997). A LEAFY co-regulator encoded by UNUSUAL FLORAL ORGANS. *Curr. Biol.* 7, 95–104. doi: 10.1016/S0960-9822(06)00053-4

Lee, J. E., Lampugnani, E. R., Bacic, A., and Golz, J. F. (2014). SEUSS and SEUSS-LIKE 2 coordinate auxin distribution and KNOXI activity during embryogenesis. *Plant J.* 80, 122–135. doi: 10.1111/tpj.12625

Leibfried, A., To, J. P., Busch, W., Stehling, S., Kehle, A., Demar, M., et al. (2005). WUSCHEL controls meristem function by direct regulation of cytokinin-inducible response regulators. *Nature* 438, 1172–1175. doi: 10.1038/nature04270

Levin, J. Z., and Meyerowitz, E. (1995). UFO: an *Arabidopsis* gene involved in both floral meristem and floral organ development. *Plant Cell* 7, 529–548. doi: 10.1105/tpc.7.5.529

Lie, C., Kelsom, C., and Wu, X. (2012). WOX2 and STIMPY-LIKE/WOX8 promote cotyledon boundary formation in *Arabidopsis*. *Plant J.* 72, 674–682.

Liebsch, D., Sunaryo, W., Holmlund, M., Norberg, M., Zhang, J., Hall, H. C., et al. (2014). Class I KNOX transcription factors promote differentiation of cambial derivatives into xylem fibers in the *Arabidopsis* hypocotyl. *Development* 141, 4311–4319. doi: 10.1242/dev.111369

Liljegren, S. J., Ditta, G. S., Eshed, Y., Savidge, B., Bowman, J. L., and Yanofsky, M. F. (2000). SHATTERPROOF MADS-box genes control seed dispersal in *Arabidopsis*. *Nature* 404, 766–770. doi: 10.1038/35008089

Liljegren, S. J., Roeder, A. H., Kempin, S. A., Gremski, K., Ostergaard, L., Guimil, S., et al. (2004). Control of fruit patterning in *Arabidopsis* by INDEHISCENT. *Cell* 116, 843–853. doi: 10.1016/S0092-8674(04)00217-X

Lincoln, C., Long, J., Yamaguchi, J., Serikawa, K., and Hake, S. (1994). A knotted1-like homeobox gene in *Arabidopsis* is expressed in the vegetative meristem and dramatically alters leaf morphology when overexpressed in transgenic plants. *Plant Cell* 6, 1859–1876. doi: 10.1105/tpc.6.12.1859

Lodha, M., Marco, C. F., and Timmermans, M. C. P. (2013). The ASYMMETRIC LEAVES complex maintains repression of KNOX homeobox genes via direct recruitment of Polycomb-repressive complex2. *Genes Dev.* 27, 596–601. doi: 10.1101/gad.211425.112

Lohmann, J. U., and Weigel, D. (2002). Building beauty: the genetic control of floral patterning. *Dev. Cell* 2, 135–142. doi: 10.1016/S1534-5807(02)00122-3

Long, J. A., and Barton, M. K. (1998). The development of apical embryonic pattern in *Arabidopsis*. *Development* 125, 3027–3035.

Long, J., and Barton, M. K. (2000). Initiation of axillary and floral meristems in *Arabidopsis*. *Dev. Biol.* 218, 341–353. doi: 10.1006/dbio.1999.9572

Ma, C., Meir, S., Xiao, L., Tong, J., Liu, Q., Reid, M. S., et al. (2015). A KNOTTED1-LIKE HOMEOBOX protein regulates abscission in tomato by modulating the auxin pathway. *Plant Physiol.* 167, 844–853. doi: 10.1104/pp.114.253815

Magnani, E., and Hake, S. (2008). KNOX lost the OX: the *Arabidopsis* KNATM gene defines a novel class of KNOX transcriptional regulators missing the homeodomain. *Plant Cell* 20, 875–887. doi: 10.1105/tpc.108.058495

Mallory, A. C., Dugas, D. V., Bartel, D. P., and Bartel, B. (2004). MicroRNA regulation of NAC-domain targets is required for proper formation and separation of adjacent embryonic, vegetative, and floral organs. *Curr. Biol.* 14, 1035–1046. doi: 10.1016/j.cub.2004.06.022

Marsch-Martinez, N., Ramos-Cruz, D., Irepan Reyes-Olalde, J., Lozano-Sotomayor, P., Zuniga-Mayo, V. M., and De Folter, S. (2012). The role of cytokinin during *Arabidopsis* gynoecia and fruit morphogenesis and patterning. *Plant J.* 72, 222–234. doi: 10.1111/j.1365-313X.2012.05062.x

Marsch-Martinez, N., Zuniga-Mayo, V. M., Herrera-Ubaldo, H., Ouwerkerk, P. B., Pablo-Villa, J., Lozano-Sotomayor, P., et al. (2014). The NTT transcription factor promotes replum development in *Arabidopsis* fruits. *Plant J.* 80, 69–81. doi: 10.1111/tpj.12617

Maugarny, A., Gonçalves, B., Arnaud, N., and Laufs, P. (2015). "CUC transcription factors: to the meristem and beyond," in *Plant Transcription Factors: Evolutionary, Structural and Functional Aspects*, ed. D. H. Gonzalez (Amsterdam: Elsevier).

McKim, S. M., Stenvik, G. E., Butenko, M. A., Kristiansen, W., Cho, S. K., Hepworth, S. R., et al. (2008). The BLADE-ON-PETIOLE genes are essential for abscission zone formation in *Arabidopsis*. *Development* 135, 1537–1546. doi: 10.1242/dev.012807

Mele, G., Ori, M., Sato, Y., and Hake, S. (2003). The knotted1-like homeobox gene BREVIPEDICELLUS regulates cell differentiation by modulating metabolic pathways. *Genes Dev.* 17, 2088–2093. doi: 10.1101/gad.1120003

Moyroud, E., Minguet, E. G., Ott, F., Yant, L., Pose, D., Monniaux, M., et al. (2011). Prediction of regulatory interactions from genome sequences using a biophysical model for the *Arabidopsis* LEAFY transcription factor. *Plant Cell* 23, 1293–1306. doi: 10.1105/tpc.111.083329

Muller, D., Schmitz, G., and Theres, K. (2006). BLIND homologous R2R3 Myb genes control the pattern of lateral meristem initiation in *Arabidopsis*. *Plant Cell* 18, 586–597. doi: 10.1105/tpc.105.038745

Nakano, T., Fujisawa, M., Shima, Y., and Ito, Y. (2013). Expression profiling of tomato pre-abscission pedicels provides insights into abscission zone properties including competence to respond to abscission signals. *BMC Plant Biol.* 13:40. doi: 10.1186/1471-2229-13-40

Nakano, T., Kimbara, J., Fujisawa, M., Kitagawa, M., Ihashi, N., Maeda, H., et al. (2012). MACROCALYX and JOINTLESS interact in the transcriptional regulation of tomato fruit abscission zone development. *Plant Physiol.* 158, 439–450. doi: 10.1104/pp.111.183731

Nawy, T., Bayer, M., Mravec, J., Friml, J., Birnbaum, K. D., and Lukowitz, W. (2010). The GATA factor HANABA TARANU is required to position the proembryo boundary in the early *Arabidopsis* embryo. *Dev. Cell* 19, 103–113. doi: 10.1016/j.devcel.2010.06.004

Nibau, C., Di Stilio, V. S., Wu, H. M., and Cheung, A. Y. (2011). *Arabidopsis* and Tobacco SUPERMAN regulate hormone signalling and mediate cell proliferation and differentiation. *J. Exp. Bot.* 62, 949–961. doi: 10.1093/jxb/erq325

Niederhuth, C. E., Cho, S. K., Seitz, K., and Walker, J. C. (2013). Letting go is never easy: abscission and receptor-like protein kinases. *J. Integr. Plant Biol.* 55, 1251–1263. doi: 10.1111/jipb.12116

Nikovics, K., Blein, T., Peaucelle, A., Ishida, T., Morin, H., Aida, M., et al. (2006). The balance between the MIR164A and CUC2 genes controls leaf margin serration in *Arabidopsis*. *Plant Cell* 18, 2929–2945. doi: 10.1105/tpc.106.045617

Norberg, M., Holmlund, M., and Nilsson, O. (2005). The BLADE ON PETIOLE genes act redundantly to control the growth and development of lateral organs. *Development* 132, 2203–2213. doi: 10.1242/dev.01815

Nordstrom, A., Tarkowski, P., Tarkowska, D., Norbaek, R., Astot, C., Dolezal, K., et al. (2004). Auxin regulation of cytokinin biosynthesis in *Arabidopsis thaliana*: a factor of potential importance for auxin-cytokinin-regulated development. *Proc. Natl. Acad. Sci. U.S.A.* 101, 8039–8044. doi: 10.1073/pnas.0402504101

Ogawa, M., Kay, P., Wilson, S., and Swain, S. M. (2009). *ARABIDOPSIS* DEHISCENCE ZONE POLYGALACTURONASE1 (ADPG1), ADPG2, and QUARTET2 are Polygalacturonases required for cell separation during reproductive development in *Arabidopsis*. *Plant Cell* 21, 216–233. doi: 10.1105/tpc.108.063768

Oikawa, T., and Kyozuka, J. (2009). Two-step regulation of LAX PANICLE1 protein accumulation in axillary meristem formation in rice. *Plant Cell* 21, 1095–1108. doi: 10.1105/tpc.108.065425

Pautot, V., Dockx, J., Hamant, O., Kronenberger, J., Grandjean, O., Jublot, D., et al. (2001). KNAT2: evidence for a link between KNOTTED-like genes and carpel development. *Plant Cell* 13, 1719–1734. doi: 10.1105/tpc.13.8.1719

Peaucelle, A., Braybrook, S. A., Le Guillou, L., Bron, E., Kuhlemeier, C., and Höfte, H. (2011). Pectin-induced changes in cell wall mechanics underlie organ initiation in *Arabidopsis*. *Curr. Biol.* 21, 1720–1726. doi: 10.1016/j.cub.2011.08.057

Peaucelle, A., Morin, H., Traas, J., and Laufs, P. (2007). Plants expressing a miR164-resistant CUC2 gene reveal the importance of post-meristematic maintenance of phyllotaxy in *Arabidopsis*. *Development* 134, 1045–1050. doi: 10.1242/dev.02774

Pinyopich, A., Ditta, G. S., Savidge, B., Liljegren, S. J., Baumann, E., Wisman, E., et al. (2003). Assessing the redundancy of MADS-box genes during carpel and ovule development. *Nature* 424, 85–88. doi: 10.1038/nature01741

Proveniers, M., Rutjens, B., Brand, M., and Smeekens, S. (2007). The *Arabidopsis* TALE homeobox gene ATH1 controls floral competency through positive regulation of FLC. *Plant J.* 52, 899–913. doi: 10.1111/j.1365-313X.2007.03285.x

Ragni, L., Belles-Boix, E., Gunl, M., and Pautot, V. (2008). Interaction of KNAT6 and KNAT2 with BREVIPEDICELLUS and PENNYWISE in *Arabidopsis* inflorescences. *Plant Cell* 20, 888–900. doi: 10.1105/tpc.108.058230

Rajani, S., and Sundaresan, V. (2001). The *Arabidopsis* myc/bHLH gene ALCATRAZ enables cell separation in fruit dehiscence. *Curr. Biol.* 11, 1914–1922. doi: 10.1016/S0960-9822(01)00593-0

Raman, S., Greb, T., Peaucelle, A., Blein, T., Laufs, P., and Theres, K. (2008). Interplay of miR164, CUP-SHAPED COTYLEDON genes and LATERAL SUPPRESSOR controls axillary meristem formation in *Arabidopsis thaliana*. *Plant J.* 55, 65–76. doi: 10.1111/j.1365-313X.2008.03483.x

Rast, M. I., and Simon, R. (2012). *Arabidopsis* JAGGED LATERAL ORGANS acts with ASYMMETRIC LEAVES2 to coordinate KNOX and PIN expression in shoot and root meristems. *Plant Cell* 24, 2917–2933. doi: 10.1105/tpc.112.099978

Reyes-Olalde, J. I., Zuniga-Mayo, V. M., Chavez Montes, R. A., Marsch-Martinez, N., and De Folter, S. (2013). Inside the gynoecium: at the carpel margin. *Trends Plant Sci.* 18, 644–655. doi: 10.1016/j.tplants.2013.08.002

Ripoll, J. J., Roeder, A. H., Ditta, G. S., and Yanofsky, M. F. (2011). A novel role for the floral homeotic gene APETALA2 during *Arabidopsis* fruit development. *Development* 138, 5167–5176. doi: 10.1242/dev.073031

Risseeuw, E., Venglat, P., Xiang, D., Komendant, K., Daskalchuk, T., Babic, V., et al. (2013). An activated form of UFO alters leaf development and produces ectopic floral and inflorescence meristems. *PLoS ONE* 8:e83807. doi: 10.1371/journal.pone.0083807

Roeder, A. H., Ferrándiz, C., and Yanofsky, M. F. (2003). The role of the REPLUMLESS homeodomain protein in patterning the *Arabidopsis* fruit. *Curr. Biol.* 13, 1630–1635. doi: 10.1016/j.cub.2003.08.027

Rubio-Somoza, I., Zhou, C.-M., Confraria, A., Martinho, C., Von Born, P., Baena-Gonzalez, E., et al. (2014). Temporal control of leaf complexity by miRNA-regulated licensing of protein complexes. *Curr. Biol.* 24, 2714–2719. doi: 10.1016/j.cub.2014.09.058

Rutjens, B., Bao, D., Van Eck-Stouten, E., Brand, M., Smeekens, S., and Proveniers, M. (2009). Shoot apical meristem function in *Arabidopsis* requires the combined activities of three BEL1-like homeodomain proteins. *Plant J.* 58, 641–654. doi: 10.1111/j.1365-313X.2009.03809.x

Sakai, H., Medrano, L., and Meyerowitz, E. (1995). Role of SUPERMAN in maintaining *Arabidopsis* floral whorl boundaries. *Nature* 378, 199–203. doi: 10.1038/378199a0

Sakamoto, T., Kamiya, N., Ueguchi-Tanaka, M., Iwahori, S., and Matsuoka, M. (2001). KNOX homeodomain protein directly suppresses the expression of a gibberellin biosynthetic gene in the tobacco shoot apical meristem. *Genes Dev.* 15, 581–590. doi: 10.1101/gad.867901

Samach, A., Klenz, J. E., Kohalmi, S. E., Risseeuw, E., Haughn, G. W., and Crosby, W. L. (1999). The UNUSUAL FLORAL ORGANS gene of *Arabidopsis thaliana* is an F-box protein required for normal patterning and growth in the floral meristem. *Plant J.* 20, 433–445. doi: 10.1046/j.1365-313x.1999.00617.x

Sassi, M., Ali, O., Boudon, F., Cloarec, G., Abad, U., Cellier, C., et al. (2014). An auxin-mediated shift toward growth isotropy promotes organ formation at the shoot meristem in *Arabidopsis*. *Curr. Biol.* 24, 2335–2342. doi: 10.1016/j.cub.2014.08.036

Schiessl, K., Muino, J. M., and Sablowski, R. (2014). *Arabidopsis* JAGGED links floral organ patterning to tissue growth by repressing Kip-related cell cycle inhibitors. *Proc. Natl. Acad. Sci. U.S.A.* 111, 2830–2835. doi: 10.1073/pnas.1320457111

Schmitz, G., Tillmann, E., Carriero, F., Fiore, C., Cellini, F., and Theres, K. (2002). The tomato BLIND gene encodes a MYB transcription factor that controls the formation of lateral meristems. *Proc. Natl. Acad. Sci. U.S.A.* 99, 1064–1069. doi: 10.1073/pnas.022516199

Schoof, H., Lenhard, M., Haecker, A., Mayer, K. F., Jurgens, G., and Laux, T. (2000). The stem cell population of *Arabidopsis* shoot meristems in maintained by a regulatory loop between the CLAVATA and WUSCHEL genes. *Cell* 100, 635–644. doi: 10.1016/S0092-8674(00)80700-X

Schuetz, M., Berleth, T., and Mattsson, J. (2008). Multiple MONOPTEROS-dependent pathways are involved in leaf initiation. *Plant Physiol.* 148, 870–880. doi: 10.1104/pp.108.119396

Schumacher, K., Schmitt, T., Rossberg, M., Schmitz, G., and Theres, K. (1999). The Lateral suppressor (Ls) gene of tomato encodes a new member of the VHIID protein family. *Proc. Natl. Acad. Sci. U.S.A.* 96, 290–295. doi: 10.1073/pnas.96.1.290

Scofield, S., Dewitte, W., and Murray, J. A. H. (2007). The KNOX gene SHOOT MERISTEMLESS is required for the development of reproductive meristematic tissues in *Arabidopsis*. *Plant J.* 50, 767–781.

Shi, C. L., Stenvik, G. E., Vie, A. K., Bones, A. M., Pautot, V., Proveniers, M., et al. (2011). *Arabidopsis* class I KNOTTED-like homeobox proteins act downstream in the IDA-HAE/HSL2 floral abscission signaling pathway. *Plant Cell* 23, 2553–2567. doi: 10.1105/tpc.111.084608

Smith, H. M., and Hake, S. (2003). The interaction of two homeobox genes, BREVIPEDICELLUS and PENNYWISE, regulates internode patterning in the *Arabidopsis* inflorescence. *Plant Cell* 15, 1717–1727.

Smith, H. M. S., Boschke, I., and Hake, S. (2002). Selective interaction of plant homeodomain proteins mediates high DNA-binding affinity. *Proc. Natl. Acad. Sci. U.S.A.* 99, 9579–9584. doi: 10.1073/pnas.092271599

Smith, H. M. S., Campbell, B. C., and Hake, S. (2004). Competence to respond to floral inductive signals requires the homeobox genes PENNYWISE and POUND-FOOLISH. *Curr. Biol.* 14, 812–817. doi: 10.1016/j.cub.2004.04.032

Sorefan, K., Girin, T., Liljegren, S. J., Ljung, K., Robles, P., Galván-Ampudia, C. S., et al. (2009). A regulated auxin minimum is required for seed dispersal in *Arabidopsis*. *Nature* 459, 583–586. doi: 10.1038/nature07875

Spinelli, S. V., Martin, A. P., Viola, I. L., Gonzalez, D. H., and Palatnik, J. F. (2011). A mechanistic link between STM and CUC1 during *Arabidopsis* development. *Plant Physiol.* 156, 1894–1904. doi: 10.1104/pp.111.177709

Stenvik, G. E., Tandstad, N. M., Guo, Y., Shi, C. L., Kristiansen, W., Holmgren, A., et al. (2008). The EPIP peptide of INFLORESCENCE DEFICIENT IN ABSCISSION is sufficient to induce abscission in *Arabidopsis* through the receptor-like kinases HAESA and HAESA-LIKE2. *Plant Cell* 20, 1805–1817. doi: 10.1105/tpc.108.059139

Takada, S., Hibara, K., Ishida, T., and Tasaka, M. (2001). The CUP-SHAPED COTYLEDON1 gene of *Arabidopsis* regulates shoot apical meristem formation. *Development* 128, 1127–1135.

Takeda, S., Hanano, K., Kariya, A., Shimizu, S., Zhao, L., Matsui, M., et al. (2011). CUP-SHAPED COTYLEDON1 transcription factor activates the expression of LSH4 and LSH3, two members of the ALOG gene family, in shoot organ boundary cells. *Plant J.* 66, 1066–1077. doi: 10.1111/j.1365-313X.2011.04571.x

Takeda, S., Matsumoto, N., and Okada, K. (2004). RABBIT EARS, encoding a SUPERMAN-like zinc finger protein, regulates petal development in *Arabidopsis thaliana. Development* 131, 425–434. doi: 10.1242/dev.00938

Tavakol, E., Okagaki, R., Verderio, G., Shariati, J. V., Hussien, A., Bilgic, H., et al. (2015). The barley Uniculme4 gene encodes a BLADE-ON-PETIOLE-like protein that controls tillering and leaf patterning. *Plant Physiol.* 168, 164–174. doi: 10.1104/pp.114.252882

Tian, C., Zhang, X., He, J., Yu, H., Wang, Y., Shi, B., et al. (2014). An organ boundary-enriched gene regulatory network uncovers regulatory hierarchies underlying axillary meristem initiation. *Mol. Syst. Biol.* 10, 755. doi: 10.15252/msb.20145470

Tsuda, K., Kurata, N., Ohyanagi, H., and Hake, S. (2014). Genome-wide study of KNOX regulatory network reveals brassinosteroid catabolic genes important for shoot meristem function in rice. *Plant Cell* 26, 3488–3500. doi: 10.1105/tpc.114.129122

Ung, N., Lal, S., and Smith, H. M. S. (2011). The role of PENNYWISE and POUND-FOOLISH in the maintenance of the shoot apical meristem in *Arabidopsis. Plant Physiol.* 156, 605–614. doi: 10.1104/pp.110.171462

Venglat, S. P., Dumonceaux, T., Rozwadowski, K., Parnell, L., Babic, V., Keller, W., et al. (2002). The homeobox gene BREVIPEDICELLUS is a key regulator of inflorescence architecture in *Arabidopsis. Proc. Natl. Acad. Sci. U.S.A.* 99, 4730–4735. doi: 10.1073/pnas.072626099

Vernoux, T., Kronenberger, J., Grandjean, O., Laufs, P., and Traas, J. (2000). PIN-FORMED 1 regulates cell fate at the periphery of the shoot apical meristem. *Development* 127, 5157–5165.

Vlad, D., Kierzkowski, D., Rast, M. I., Vuolo, F., Dello Ioio, R., Galinha, C., et al. (2014). Leaf shape evolution through duplication, regulatory diversification, and loss of a homeobox gene. *Science* 343, 780–783. doi: 10.1126/science.1248384

Vroemen, C. W., Mordhorst, A. P., Albrecht, C., Kwaaitaal, M. A. C. J., and De Vries, S. C. (2003). The CUP-SHAPED COTYLEDON3 gene is required for boundary and shoot meristem formation in *Arabidopsis. Plant Cell* 15, 1563–1577.

Wagner, D., Sablowski, R. W., and Meyerowitz, E. M. (1999). Transcriptional activation of APETALA1 by LEAFY. *Science* 285, 582–584. doi: 10.1126/science.285.5427.582

Walsh, J., Waters, C. A., and Freeling, M. (1998). The maize gene liguleless2 encodes a basic leucine zipper protein involved in the establishment of the leaf blade-sheath boundary. *Genes Dev.* 12, 208–218. doi: 10.1101/gad.12.2.208

Wang, Q., Kohlen, W., Rossmann, S., Vernoux, T., and Theres, K. (2014a). Auxin depletion from the leaf axil conditions competence for axillary meristem formation in *Arabidopsis* and tomato. *Plant Cell* 26, 2068–2079. doi: 10.1105/tpc.114.123059

Wang, Y., Wang, J., Shi, B., Yu, T., Qi, J., Meyerowitz, E. M., et al. (2014b). The stem cell niche in leaf axils is established by auxin and cytokinin in *Arabidopsis. Plant Cell* 26, 2055–2067. doi: 10.1105/tpc.114.123083

Wang, X., Xu, W., Ma, L., Fu, Z., Deng, X., Li, J., et al. (2006). Requirement of KNAT1/BP for the development of abscission zones in *Arabidopsis thaliana. J. Integr. Plant Biol.* 48, 15–26. doi: 10.1111/j.1744-7909.2005.00085.x-i1

Wilkinson, M. D., and Haughn, G. W. (1995). UNUSUAL FLORAL ORGANS controls meristem identity and organ primordia fate in *Arabidopsis. Plant Cell* 7, 1485–1499. doi: 10.2307/3870137

William, D. A., Su, Y., Smith, M. R., Lu, M., Baldwin, D. A., and Wagner, D. (2004). Genomic identification of direct target genes of LEAFY. *Proc. Natl. Acad. Sci. U.S.A.* 101, 1775–1780. doi: 10.1073/pnas.0307842100

Winter, C. M., Austin, R. S., Blanvillain-Baufume, S., Reback, M. A., Monniaux, M., Wu, M. F., et al. (2011). LEAFY target genes reveal floral regulatory logic, cis motifs, and a link to biotic stimulus response. *Dev. Cell* 20, 430–443. doi: 10.1016/j.devcel.2011.03.019

Wu, X. M., Yu, Y., Han, L. B., Li, C. L., Wang, H. Y., Zhong, N. Q., et al. (2012). The tobacco BLADE-ON-PETIOLE2 gene mediates differentiation of the corolla abscission zone by controlling longitudinal cell expansion. *Plant Physiol.* 159, 835–850. doi: 10.1104/pp.112.193482

Xu, B., Li, Z., Zhu, Y., Wang, H., Ma, H., Dong, A., et al. (2008). *Arabidopsis* genes AS1, AS2, and JAG negatively regulate boundary-specifying genes to promote sepal and petal development. *Plant Physiol.* 146, 566–575.

Xu, M., Hu, T., Mckim, S. M., Murmu, J., Haughn, G. W., and Hepworth, S. R. (2010). *Arabidopsis* BLADE-ON-PETIOLE1 and 2 promote floral meristem fate and determinacy in a previously undefined pathway targeting APETALA1 and AGAMOUS-LIKE24. *Plant J.* 63, 974–989. doi: 10.1111/j.1365-313X.2010.04299.x

Yamaguchi, J., Wu, M.-F., Winter, C. M., Berns, M. C., Nole-Wilson, S., Yamaguchi, A., et al. (2013). A molecular framework for auxin-mediated initiation of flower primordia. *Dev. Cell* 24, 271–282. doi: 10.1016/j.devcel.2012.12.017

Yanai, O., Shani, E., Dolezal, K., Tarkowski, P., Sablowski, R., Sandberg, G., et al. (2005). *Arabidopsis* KNOXI proteins activate cytokinin biosynthesis. *Curr. Biol.* 15, 1566–1571. doi: 10.1016/j.cub.2005.07.060

Yang, F., Wang, Q., Schmitz, G., Muller, D., and Theres, K. (2012). The bHLH protein ROX acts in concert with RAX1 and LAS to modulate axillary meristem formation in *Arabidopsis. Plant J.* 71, 61–70. doi: 10.1111/j.1365-313X.2012.04970.x

Yoon, J., Cho, L. H., Kim, S. L., Choi, H., Koh, H. J., and An, G. (2014). The BEL1-type homeobox gene SH5 induces seed shattering by enhancing abscission-zone development and inhibiting lignin biosynthesis. *Plant J.* 79, 717–728. doi: 10.1111/tpj.12581

Žádníková, P., and Simon, R.(2014). How boundaries control plant development. *Curr. Opin. Plant Biol.* 17, 116–125. doi: 10.1016/j.pbi.2013.11.013

Zhang, L., Liu, D., Wang, D., Hang, R., Geng, S., Wu, L., et al. (2013). Over expression of the wheat BEL1-like gene TaqSH1 affects floral organ abscission in *Arabidopsis thaliana. J. Plant Biol.* 56, 98–105. doi: 10.1007/s12374-012-0438-7

Zhao, H., Liu, L., Mo, H., Qian, L., Cao, Y., Cui, S., et al. (2013). The ATP-binding cassette transporter ABCB19 regulates postembryonic organ separation in *Arabidopsis. PLoS ONE* 8:e60809. doi: 10.1371/journal.pone.0060809

Zhao, M., Yang, S., Chen, C. Y., Li, C., Shan, W., Lu, W., et al. (2015). *Arabidopsis* BREVIPEDICELLUS interacts with the SWI2/SNF2 chromatin remodeling ATPase BRAHMA to regulate KNAT2 and KNAT6 expression in control of inflorescence architecture. *PLOS Genet.* 11:e1005125. doi: 10.1371/journal.pgen.1005125

Zhao, Y., Medrano, L., Ohashi, K., Fletcher, J. C., Yu, H., Sakai, H., et al. (2004). HANABA TARANU is a GATA transcription factor that regulates shoot apical meristem and flower development in *Arabidopsis. Plant Cell* 16, 2586–2600. doi: 10.1105/tpc.104.024869

Conflict of Interest Statement: The authors declare that the research was conducted in the absence of any commercial or financial relationships that could be construed as a potential conflict of interest.

22

Structure and Function of Centromeric and Pericentromeric Heterochromatin in *Arabidopsis thaliana*

*Lauriane Simon†, Maxime Voisin†, Christophe Tatout and Aline V. Probst**

CNRS UMR6293, INSERM U1103, Clermont University, GReD, Aubière, France

The centromere is a specific chromosomal region where the kinetochore assembles to ensure the faithful segregation of sister chromatids during mitosis and meiosis. Centromeres are defined by a local enrichment of the specific histone variant CenH3 mostly at repetitive satellite sequences. A larger pericentromeric region containing repetitive sequences and transposable elements surrounds the centromere that adopts a particular chromatin state characterized by specific histone variants and post-translational modifications and forms a transcriptionally repressive chromosomal environment. In the model organism *Arabidopsis thaliana* centromeric and pericentromeric domains form conspicuous heterochromatin clusters called chromocenters in interphase. Here we discuss, using *Arabidopsis* as example, recent insight into mechanisms involved in maintenance and establishment of centromeric and pericentromeric chromatin signatures as well as in chromocenter formation.

Keywords: centromere, chromocenter, histone variants, 3D nucleus, lamina, nuclear envelope

Edited by:
Inna Lermontova,
Institute of Plant Genetics and Crop Plant Research, Germany

Reviewed by:
Célia Baroux,
University of Zürich, Switzerland
Stefan Heckmann,
University of Birmingham, UK

****Correspondence:***
Aline V. Probst
aline.probst@univ-bpclermont.fr

† These authors have contributed equally to this work.

Centromeres are essential chromosomal structures that were first defined as central restrictions of the mitotic chromosomes that function in chromosome segregation during cell division. Except for *Saccharomyces cerevisiae*, centromeres are not defined genetically by a specific DNA sequence but rather epigenetically by a particular chromatin environment and the presence of the specific histone variant CenH3. The centromeric and the surrounding pericentromeric chromosomal regions form heterochromatin domains that remain condensed during interphase (Heitz, 1928) and in some species like *Arabidopsis thaliana* these are clustered into chromocenter structures (**Figure 1**; Fransz et al., 2002). Here we discuss our current knowledge concerning sequence composition, chromatin features and interphase higher-order organization of centromeric and pericentromeric regions into chromocenters, referring to the centromeric region specifically as the part of the chromosome involved in kinetochore formation, while we refer to the pericentromeric domains as the adjacent chromatin regions (according to Gent and Dawe, 2011).

SEQUENCE COMPOSITION OF *ARABIDOPSIS* CENTROMERIC AND PERICENTROMERIC REGIONS

Centromeric sequences consist in most organisms of short repetitive DNA sequences arranged in tandem and/or transposable elements (Plohl et al., 2014). New centromeres can emerge from anonymous sequences but they gradually incorporate repetitive arrays (Han et al., 2009; Plohl et al., 2014) suggesting that DNA repeats may be a preferred DNA environment for centromere formation. In *A. thaliana*, centromeric regions mainly consist of a 178 bp long sequence also called AtCon, pAL1,

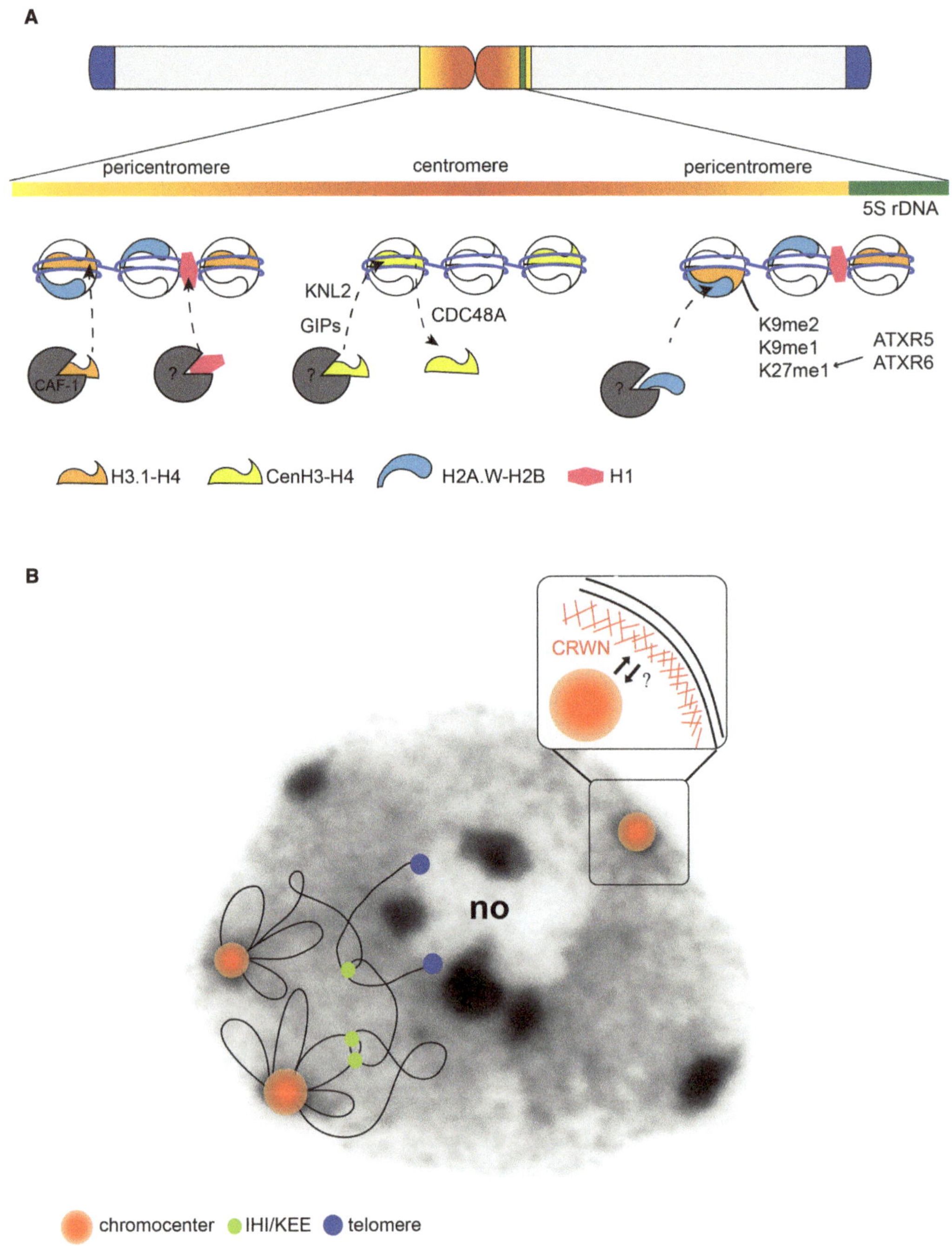

FIGURE 1 | (A) Model of an *Arabidopsis thaliana* chromosome. Enlargement of the central part of a chromosome (top) shows the sequence composition of the centromeric (red, mainly consisting of repetitive 180 bp repeats) and the pericentromeric region (orange, containing interspersed repeats, transposable elements and their derivatives) embedding a 5S rRNA gene locus (green) as well as their chromatin composition. The pericentromeric domains are enriched in the canonical histone H3.1 (orange), the H2A.W variant (blue) and the linker histone H1 (pink) as well as in repressive histone modifications such as H3K9me1/me2 and H3K27me1. The centromere is defined by enrichment in the H3 variant CenH3 (yellow). Whether CenH3 nucleosomes form large blocks or are interspersed with nucleosomes composed of canonical or specific variant histones and which histones are incorporated as placeholders during replication remains to be determined. CenH3, H1 and H2A.W are deposited by yet unknown histone chaperones. GIP1/GIP2 and KNL2 play a role in CenH3 deposition. In specific cell types, CenH3 is actively removed by the AAA-ATPase CDC48A. **(B)** Model of the organization of a chromosome in nuclear space overlaid on a DAPI stained image of an *Arabidopsis* leaf nucleus. Centromeric and pericentromeric regions of the 5 *Arabidopsis* chromosomes are tightly packed into chromocenters (red/orange). Chromocenters structure the chromosome in nuclear space by anchoring proximal euchromatic loops, while distal chromosomal regions tend to cluster with telomeres (blue) next to the nucleolus (no). Interactive heterochromatic island (IHI)/KNOT engaged element (KEE) regions, identified in Hi-C maps, form additional intra- and inter-chromosomal contacts (green). Enlargement at the top right shows the nuclear envelope coated internally by a lamina-like structure, which includes CRWN proteins. Chromocenters tend to localize at the nuclear periphery, but the physical link between nuclear envelope components or the nuclear lamina-like components remains to be identified.

or 180 bp (Kumekawa et al., 2000, 2001; Nagaki et al., 2003) that is arranged in head-to-tail tandem repeats ranging from ~0.4 to 3 Mb (**Figure 1A**). The 180 bp repeats are interrupted by a 398 bp fragment of the Athila2 LTR called 106B (Thompson et al., 1996). While highly similar sequences are found on all chromosomes, some 180 bp repeat variants are specific to one centromere (Heslop-Harrison et al., 1999). Centromeres are flanked by a pericentromeric region, which in *Arabidopsis* contains various types of repeat sequences such as Athila retrotransposons, 500 bp and 160 bp repeats (Bauwens et al., 1991), "Transcriptional Silent Information" (TSI) composed of the 3′ half of an Athila retrotransposon (Steimer et al., 2000) and on some chromosomes the 5S rDNA clusters (Fransz et al., 1998; Cloix et al., 2000). Most repetitive sequences from the pericentromeric region and transposons are kept silent, while others like the 5S rRNA gene clusters (Cloix et al., 2002) are highly transcribed. Despite the efforts in genome assembly (Schneeberger et al., 2011), the establishment of the exact reference sequence for these highly repetitive chromosomal regions remains a challenge for the future.

THE CENTRAL ROLE OF CenH3 IN CENTROMERE DEFINITION

A high frequency of DNA repeats is not sufficient to define centromeres (Han et al., 2006; Birchler et al., 2010); instead, centromeres are determined by a specific chromatin environment. The basic subunit of chromatin, the nucleosome, comprises 146 bp of DNA that wrap around an octamer of core histones H3, H4, H2A, and H2B. A specific histone variant, called CenH3, replaces the canonical H3.1 in centromeric nucleosomes. CenH3 is enriched at 180 bp repeats as shown by Chromatin Immunoprecipitation (ChIP) and Fluorescence *in situ* hybridization (FISH) experiments (Nagaki et al., 2003; Shibata and Murata, 2004). In agreement with its central role in centromere definition, homozygous *cenH3* mutants are lethal and plants expressing RNAi constructs leading to reduced CenH3 levels show meiosis defects, partial sterility, and in older plants an increased 4C:2C ratio indicating G2 arrest (Lermontova et al., 2011). CenH3 proteins evolve rapidly, as example 23 out of 178 amino acids differ between the closely related species *A. thaliana* and *A. arenosa* (Talbert et al., 2002). The N-terminal tail is substantially longer compared to the canonical H3.1 or the variant H3.3 and particularly divergent between species, revealing adaptive evolution with the species-specific centromeric repeats (Talbert et al., 2002; Maheshwari et al., 2015). An additional domain involved in adaptive evolution is the histone fold domain including the loop 1 region that makes multiple contacts with DNA (Cooper and Henikoff, 2004). This histone fold domain has been found sufficient for CenH3 loading at centromeric sequences (Lermontova et al., 2006) and a single point mutation close to the loop1 region reduces CenH3 loading substantially (Karimi-Ashtiyani et al., 2015). While plants carrying this loading deficient CenH3 are fertile when selfed, backcrossing to wild type (WT) plants leads to haploid and aneuploid progeny, retaining only the WT CenH3. Unexpectedly, even CenH3 from distant monocotyledon species can complement *A. thaliana cenh3* mutants (Maheshwari et al., 2015). These results differ from other studies using N-or C-terminal GFP tagged versions of heterologous CenH3, where only tagged CenH3 from a closely related species was properly targeted or functionally complemented a *cenH3* mutant (Ravi et al., 2010). This shows that a GFP tagged CenH3 version is not functionally equivalent, as the large GFP tag may interfere with proper CenH3 loading or the assembly of kinetochore proteins.

Together, these studies underline a central role for CenH3 in centromere definition. Fast co-evolution of CenH3 and centromeric repeats is proposed to contribute to reproductive isolation and speciation (Ma et al., 2007; Plohl et al., 2014). Understanding these key mechanisms may have major application in breeding programs when interspecific crosses between cultivated species and their WT relatives are involved.

CenH3 DEPOSITION

CenH3 needs to be deposited in a controlled manner to avoid mis-localization to ectopic sites (Lacoste et al., 2014) that might seed neo-centromeres (Shang et al., 2013). Appropriate incorporation of CenH3 is therefore controlled by specific histone chaperones. In mammals, CenH3 is deposited post-mitotically (Jansen et al., 2007), while CenH3 nucleosome assembly takes place in G2 phase in *Arabidopsis* (Lermontova et al., 2006). In contrast to mitotic nuclei, meiosis includes also a post-divisional loading step during interkinesis (Schubert et al., 2014) and is associated with a specific loading pathway or quality check that eliminates modified CenH3 proteins (Lermontova et al., 2011; Ravi et al., 2011). No functional homolog of CenH3 chaperones known in humans (HJURP, Dunleavy et al., 2009; Foltz et al., 2009), Drosophila (CAL1, Chen et al., 2014) or yeast (SCM3, Camahort et al., 2007) has yet been identified in plants. In contrast, a homolog of yeast Mis18, which is implicated in forming the correct epigenetic context for CenH3 loading (Hayashi et al., 2004), has been identified in *Arabidopsis* and termed KINETOCHORE NULL2 (KNL2; Lermontova et al., 2013). KNL2 is mainly expressed in meristem tissues similar to CenH3 and except during mitosis localizes to centromeres during the whole cell cycle including G2 phase when CenH3 is loaded. Loss of KNL2 negatively impacts CenH3 expression and deposition (Lermontova et al., 2013), but also reduces DNA methylation and affects histone methyltransferase expression, suggesting that the chromatin context of centromeric or pericentromeric sequences repeats may play a role in CenH3 loading. Furthermore, recent work suggests a role for the γ-tubulin complex protein 3-interacting proteins (GIPs) in CenH3 loading or maintenance at centromeres (Batzenschlager et al., 2015). GIP proteins are found in a complex with CenH3 and a double *gip1 gip2* mutant shows reduced intensity of CenH3 signals, centromere cohesion defects and aneuploidy, despite increased levels of KNL2 in the *gip1 gip2* mutant background (Batzenschlager et al., 2015). Given that the CenH3 deposition machinery evolved rapidly and involves distinct players in yeast, Drosophila, mammals and plants, it can be speculated that GIP proteins are part of a plant-specific pathway contributing to CenH3 assembly.

Interestingly, CenH3 is associated only with specific subsets of 180 bp repeats (Shibata and Murata, 2004) and these are hypomethylated (Zhang et al., 2008) compared to other subsets of 180 bp repeats that are hypermethylated, enriched in H3K9me2 and associated with the canonical histone H3.1 (Stroud et al., 2012; Wollmann et al., 2012; Vaquero-Sedas and Vega-Palas, 2013). To which extent CenH3 containing nucleosomes are interspersed with nucleosomes containing canonical H3.1 or its variant H3.3, and which of the CenH3 types is deposited as placeholder upon chromatin assembly during S-phase, remains to be elucidated in plants. Furthermore, the identification of the histone chaperone involved in CenH3 deposition and further characterization of the role of GIP proteins will be critical to better understand how CenH3 is specifically targeted to centromeric repeats.

While controlled CenH3 deposition is important, in some differentiated cells, CenH3 is also actively removed, such as in the vegetative pollen nucleus that does not contain visible CenH3 enrichment at centromeres compared to the sperm cell nuclei (Ingouff et al., 2009; Schoft et al., 2009). In the vegetative nucleus, CenH3 is sumoylated and removed by the AAA-ATPase molecular chaperone CDC48A to be targeted for proteolysis (Mérai et al., 2014).

THE PERICENTROMERIC REGION

The CenH3 containing centromere domain is flanked by pericentromeric heterochromatin that is highly DNA methylated, shows more regular nucleosome spacing than euchromatin (Chodavarapu et al., 2010) and is characterized by inaccessibility to DNAse I (Shu et al., 2012). Pericentromeric nucleosomes carry histone modifications repressive for transcription such as H4K20me1, H3K9me1, H3K9me2, and H3K27me1 (Tariq et al., 2003; Naumann et al., 2005; Fransz et al., 2006; Roudier et al., 2011; Shu et al., 2012; Sequeira-Mendes et al., 2014). As an example, H3K27me1 loss is associated with release of transcriptional silencing of TSI and certain transposons in the pericentromeric region (Jacob and Feng, 2009) as well as over-replication of pericentromeric sequences (Jacob et al., 2010). The histone-methyltransferases ATXR5 and ATXR6 preferentially mono-methylate lysine 27 of the canonical histone H3.1 (Jacob et al., 2014), which is highly enriched in pericentromeric regions (Vaquero-Sedas and Vega-Palas, 2013). Furthermore, plants deficient in the Chromatin Assembly Factor 1 (CAF-1) complex that deposits histone H3.1 in a replication-coupled manner in mammals (Smith and Stillman, 1989; Tagami et al., 2004) show stochastic reactivation of TSI and CACTA transposable elements (Takeda et al., 2004; Ono et al., 2006). Additional core histones also exist as specialized variants enriched in heterochromatin such as the H2A.W variants H2A.W.6, H2A.W.7 and H2A.W.12 (Yelagandula et al., 2014), which colocalize as RFP fusion proteins with H3K9me2 at pericentromeric regions by microscopy. Simultaneous loss of H3K9me2 enhances the phenotype of double h2a.w.6 h2a.w.7 mutants and leads to increased expression of certain transposons suggesting that histone/DNA methylation and H2A.W incorporation present two parallel pathways involved in heterochromatin maintenance. Therefore, an important role can be assigned to the incorporation of specific histone types in the establishment of the particular chromatin environment of the pericentromeric region. Canonical or histone variants affect chromatin organization both through their inherent physico-chemical properties and through their specific post-translational modifications that might be set in a nucleosomal context, e.g., H3K27me1 by ATXR5 and ATXR6 (Jacob et al., 2014), or already during assembly (Loyola et al., 2009) and synthesis (Rivera et al., 2015) of the respective histone as recently described in mammals. Despite the advances in the description of the pericentromeric heterochromatin signature, not much is known whether and how pericentromeric heterochromatin contributes to centromere function in plants. Chromatin or sequence features of the pericentromeric domain may play a role in loading of CenH3 at the centromere as it is the case in fission yeast (Folco et al., 2008; Catania et al., 2015) but this remains to be investigated.

ORGANIZATION OF CENTROMERIC AND PERICENTROMERIC CHROMATIN INTO CHROMOCENTERS

In interphase nuclei, FISH experiments indicated that centromeric and pericentromeric repeats cluster together in chromocenter structures (Fransz et al., 2002). Recent Hi-C analyses confirmed that repeated sequences are grouped together and revealed further intra and inter-chromosomal interactions, **Figure 1B**. Multiple reasons have been brought forward to explain the particular organization of centromeric and pericentromeric sequences into chromocenters: the clustering may compartmentalize silent chromatin away from euchromatin, help concentrate chromatin modifiers setting repressive chromatin marks or coordinate replication of this domain in time and space (Heitz, 1928; Quivy et al., 2004; Almouzni and Probst, 2011). Chromocenters are not randomly organized in nuclear space but instead preferentially localize into the most outer zone next to the nuclear periphery (Fransz et al., 2002; Fang and Spector, 2005; Andrey et al., 2010; Poulet et al., 2015). To date, there is no clear explanation for this preferential localization and several hypotheses can be proposed. First, this organization can be the result of non-specific forces acting on heterochromatin because of its elevated thickness and rigidity in respect to euchromatin (Cook and Marenduzzo, 2009; de Nooijer et al., 2009). Second, peripheral position may be advantageous to allow rapid contact between the centromere and microtubules at the beginning of cell division. Interestingly, GIP proteins have been shown to localize to both sides of the nuclear envelope and close to the chromocenters and may therefore be seen as good candidates to connect the microtubule machinery and the centromeres upon nuclear envelope breakdown (Batzenschlager et al., 2014). The identification of the structural components linking heterochromatin to the nuclear periphery is an active area of research in plants and the lamin-like structures including CRWN1-4 (CRoWded Nuclei) proteins are intriguing candidates (Dittmer et al., 2007; Fiserova et al., 2009; Goto et al., 2014). Indeed, *crwn1 crwn2* mutants show reduced nuclear volume and increased chromocenter clustering, while chromocenters are more dispersed in *crwn4* (Dittmer et al., 2007; Wang et al., 2013a; Poulet et al., 2015). Hi-C data in *crwn1* and *crwn4* mutants reveal higher chromosomal compaction and increased interactions among pericentromeric regions, which reflects the altered chromocenter organization detected by FISH (Grob et al., 2014). Altered chromocenter organization was also observed in *syn4* and *cap-d3* mutants, lacking subunits of cohesin or condensin complexes respectively (Schubert et al., 2009, 2013). Given that both CRWN proteins and condensing/cohesion complexes affect chromocenter organization and chromosome compaction it might be interesting to further investigate whether a functional relationship exists between these complexes.

CHROMOCENTER MAINTENANCE AND DYNAMICS

Mutants impaired in factors involved in setting of epigenetic marks such as DNA methyltransferases (Soppe et al., 2002; Mathieu et al., 2007; Stroud et al., 2014) and histone K27 and K9 methyltransferases (Jacob and Feng, 2009; Yelagandula et al., 2014), in chromatin remodeling (Probst et al., 2003) or in chromatin assembly (Schönrock et al., 2006) affect heterochromatin organization in chromocenters. Furthermore, recent data suggest a role for the histone variant H2A.W in heterochromatin condensation into chromocenters based on its capacity to promote chromatin fiber-to-fiber interactions through its C-terminal end *in vitro* (Yelagandula et al., 2014) and accordingly *h2a.w* double or triple mutants show chromocenter decondensation. In addition to H2A.W, the linker histone H1 facilitates folding of the nucleosome into higher-order structures (Zhou et al., 2013 and references therein). The observation that some plant cells (such as the spore mother cells) show a drastic reduction in chromocenter compaction, concomitantly to H1 depletion (She et al., 2013; She and Baroux, 2015) suggests a role for the linker histone in pericentromeric chromatin organization, but a causal relationship remains to be established. During development, the organization of centromeric and pericentromeric sequences in chromocenters is dynamic (Benoit et al., 2013). For example during germination, chromocenter organization is lost 1 to 3 days after imbibition and only small and diffuse pre-chromocenters can be detected (van Zanten et al., 2011). Chromocenter assembly then takes place in cotyledons during a short time window between 3 and 5 days after germination (Mathieu et al., 2003; Douet et al., 2008; Bourbousse et al., 2015). Decondensation of chromocenters was also observed at later developmental stages such as during floral transition, when the plant undergoes reprogramming from vegetative to reproductive state, or during protoplast formation, which trigger a partial decondensation of 5S rDNA and 180 bp repeats (Tessadori et al., 2007a,b). Furthermore, the organization of chromocenters dynamically changes upon pathogen infection (Pavet et al., 2006), or under abiotic stresses (Probst and Mittelsten Scheid, 2015). As an example, chromocenters decondense during prolonged heat stress (Pecinka et al., 2010), which requires HEAT-INTOLERANT 4 (HIT4; Wang et al., 2013b). These dynamic changes in chromocenter organization during development or stress might reflect global chromatin changes, revealing the role of chromocenters in the organization of euchromatic loops in nuclear space thereby potentially contributing to gene expression regulation.

A ROLE FOR NON-CODING RNA IN CENTROMERE FUNCTION?

Centromeric and pericentromeric regions are essential for chromosome segregation in mitosis and meiosis and help to structure chromosomes through the formation of chromocenters in interphase. These are complex functions requesting many factors including specific DNA sequences, deposition of histone variants and epigenetic marks, as well as chromatin organization in nuclear space. While centromeric and pericentromeric regions form a generally repressive chromatin environment, some of these repetitive elements are expressed at low level in specific tissues or developmental stages and processed by the RNAi pathway (May et al., 2005; Slotkin et al., 2009; Slotkin, 2010). In recent years, evidence for RNA in centromere regulation and function accumulated in different organisms (reviewed in Gent et al., 2012). Examples include a role for non-coding RNAs in heterochromatin assembly in fission yeast (Volpe et al., 2002), HP1 recruitment (Maison et al., 2002) and chromocenter organization (Probst et al., 2010) in mammals, as well as CenH3 deposition in mammals and Drosophila (Quénet and Dalal, 2014; Rošić et al., 2014). Furthermore, the passage of RNA polymerase II itself is critical for centromere function (Catania et al., 2015; Chen et al., 2015). Understanding the complex interplay between DNA sequence, transcription, non-coding RNA, chromatin and nuclear environment in centromere function in plants will be a major challenge for the future.

AUTHOR CONTRIBUTIONS

LS and MV contributed equally to the content and to the drafting of the manuscript. CT and AP edited the manuscript. All authors read and approved the manuscript.

ACKNOWLEDGMENTS

This work was supported by INSERM, CNRS and Clermont universities, by the French National Research Agency (ANR-11 JSV2 009 01, ANR-12-ISV6-0001 to AVP) and by the Region Auvergne and FEDER (Life GRID to CT). LS was supported by a PhD fellowship from the Region Auvergne.

REFERENCES

Almouzni, G., and Probst, A. V. (2011). Heterochromatin maintenance and establishment: lessons from the mouse pericentromere. *Nucleus* 2, 332–338. doi: 10.4161/nucl.2.5.17707

Andrey, P., Kiêu, K., Kress, C., Lehmann, G., Tirichine, L., Liu, Z., et al. (2010). Statistical analysis of 3D images detects regular spatial distributions of centromeres and chromocenters in animal and plant nuclei. *PLoS Comput. Biol.* 6:27. doi: 10.1371/journal.pcbi.1000853

Batzenschlager, M., Herzog, E., Houlné, G., Schmit, A.-C., and Chabouté, M.-E. (2014). GIP/MZT1 proteins orchestrate nuclear shaping. *Front. Plant Sci.* 5:29. doi: 10.3389/fpls.2014.00029

Batzenschlager, M., Lermontova, I., Schubert, V., Fuchs, J., Berr, A., Koini, M. A., et al. (2015). *Arabidopsis* MZT1 homologs GIP1 and GIP2 are essential for centromere architecture. *Proc. Natl. Acad. Sci. U.S.A.* 112, 8656–8660. doi: 10.1073/pnas.1506351112

Bauwens, S., Van Oostveldt, P., Engler, G., and Van Montagu, M. (1991). Distribution of the rDNA and three classes of highly repetitive DNA in the chromatin of interphase nuclei of *Arabidopsis thaliana*. *Chromosoma* 101, 41–48.

Benoit, M., Layat, E., Tourmente, S., and Probst, A. V. (2013). Heterochromatin dynamics during developmental transitions in *Arabidopsis*—a focus on ribosomal DNA loci. *Gene* 526, 39–45. doi: 10.1016/j.gene.2013.01.060

Birchler, J. A., Krishnaswamy, L., Gaeta, R. T., Masonbrink, R. E., and Zhao, C. (2010). Engineered minichromosomes in plants. *CRC. Crit. Rev. Plant Sci.* 29, 135–147. doi: 10.1080/07352681003709918

Bourbousse, C., Mestiri, I., Zabulon, G., Bourge, M., Formiggini, F., Koini, M. A., et al. (2015). Light signaling controls nuclear architecture reorganization during seedling establishment. *Proc. Natl. Acad. Sci. U.S.A.* 112, E2836–E2844. doi: 10.1073/pnas.1503512112

Camahort, R., Li, B., Florens, L., Swanson, S. K., Washburn, M. P., and Gerton, J. L. (2007). Scm3 is essential to recruit the histone h3 variant cse4 to centromeres and to maintain a functional kinetochore. *Mol. Cell* 26, 853–865. doi: 10.1016/j.molcel.2007.05.013

Catania, S., Pidoux, A. L., and Allshire, R. C. (2015). Sequence features and transcriptional stalling within centromere DNA promote establishment of CENP-A chromatin. *PLoS Genet.* 11:e1004986. doi: 10.1371/journal.pgen.1004986

Chen, C.-C., Bowers, S., Lipinszki, Z., Palladino, J., Trusiak, S., Bettini, E., et al. (2015). Establishment of centromeric chromatin by the CENP-A assembly factor CAL1 requires FACT-mediated transcription. *Dev. Cell* 34, 73–84. doi: 10.1016/j.devcel.2015.05.012

Chen, C. C., Dechassa, M. L., Bettini, E., Ledoux, M. B., Belisario, C., Heun, P., et al. (2014). CAL1 is the Drosophila CENP-A assembly factor. *J. Cell Biol.* 204, 313–329. doi: 10.1083/jcb.201305036

Chodavarapu, R. K., Feng, S., Bernatavichute, Y. V., Chen, P.-Y., Stroud, H., Yu, Y., et al. (2010). Relationship between nucleosome positioning and DNA methylation. *Nature* 466, 388–392. doi: 10.1038/nature09147

Cloix, C., Tutois, S., Mathieu, O., Cuvillier, C., Espagnol, M. C., Picard, G., et al. (2000). Analysis of 5S rDNA arrays in *Arabidopsis thaliana*: physical mapping and chromosome-specific polymorphisms. *Genome Res.* 10, 679–690. doi: 10.1101/gr.10.5.679

Cloix, C., Tutois, S., Yukawa, Y., Mathieu, O., Cuvillier, C., Espagnol, M.-C., et al. (2002). Analysis of the 5S RNA pool in *Arabidopsis thaliana*: RNAs Are heterogeneous and only two of the genomic 5S loci produce mature 5S RNA. *Genome Res.* 12, 132–144. doi: 10.1101/gr.181301

Cook, P. R., and Marenduzzo, D. (2009). Entropic organization of interphase chromosomes. *J. Cell Biol.* 186, 825–834. doi: 10.1083/jcb.200903083

Cooper, J. L., and Henikoff, S. (2004). Adaptive evolution of the histone fold domain in centromeric histones. *Mol. Biol. Evol.* 21, 1712–1718. doi: 10.1093/molbev/msh179

de Nooijer, S., Wellink, J., Mulder, B., and Bisseling, T. (2009). Non-specific interactions are sufficient to explain the position of heterochromatic chromocenters and nucleoli in interphase nuclei. *Nucleic Acids Res.* 37, 3558–3568. doi: 10.1093/nar/gkp219

Dittmer, T. A., Stacey, N. J., Sugimoto-Shirasu, K., and Richards, E. J. (2007). LITTLE NUCLEI genes affecting nuclear morphology in *Arabidopsis thaliana*. *Plant Cell* 19, 2793–2803. doi: 10.1105/tpc.107.053231

Douet, J., Blanchard, B., Cuvillier, C., and Tourmente, S. (2008). Interplay of RNA Pol IV and ROS1 during post-embryonic 5S rDNA chromatin remodeling. *Plant Cell Physiol.* 49, 1783–1791. doi: 10.1093/pcp/pcn152

Dunleavy, E. M., Roche, D., Tagami, H., Lacoste, N., Ray-Gallet, D., Nakamura, Y., et al. (2009). HJURP Is a cell-cycle-dependent maintenance and deposition factor of CENP-A at centromeres. *Cell* 137, 485–497. doi: 10.1016/j.cell.2009.02.040

Fang, Y., and Spector, D. L. (2005). Centromere positioning and dynamics in living *Arabidopsis* plants. *Mol. Biol. Cell* 16, 5710–5718. doi: 10.1091/mbc.E05-08-0706

Fiserova, J., Kiseleva, E., and Goldberg, M. W. (2009). Nuclear envelope and nuclear pore complex structure and organization in tobacco BY-2 cells. *Plant J.* 59, 243–255. doi: 10.1111/j.1365-313X.2009.03865.x

Folco, H. D., Pidoux, A. L., Urano, T., and Allshire, R. C. (2008). Heterochromatin and RNAi are required to establish CENP-A chromatin at centromeres. *Science* 319, 94–97. doi: 10.1126/science.1150944

Foltz, D. R., Jansen, L. E. T., Bailey, A. O., Yates, J. R., Bassett, E. A., Wood, S., et al. (2009). Centromere-specific assembly of CENP-A nucleosomes is mediated by HJURP. *Cell* 137, 472–484. doi: 10.1016/j.cell.2009.02.039

Fransz, P. F., Armstrong, S., Alonso-Blanco, C., Fischer, T. C., Torres-Ruiz, R. A., and Jones, G. (1998). Cytogenetics for the model system *Arabidopsis thaliana*. *Plant J.* 13, 867–876.

Fransz, P. F., De Jong, J. H., Lysak, M., Castiglione, M. R., and Schubert, I. (2002). Interphase chromosomes in *Arabidopsis* are organized as well defined chromocenters from which euchromatin loops emanate. *Proc. Natl. Acad. Sci. U.S.A.* 99, 14584–14589. doi: 10.1073/pnas.212325299

Fransz, P. F., ten Hoopen, R., and Tessadori, F. (2006). Composition and formation of heterochromatin in *Arabidopsis thaliana*. *Chromosom. Res.* 14, 71–82. doi: 10.1007/s10577-005-1022-5

Gent, J. I., and Dawe, R. K. (2011). RNA as a structural and regulatory component of the centromere. *Annu. Rev. Genet.* 46, 443–453. doi: 10.1146/annurev-genet-110711-155419

Gent, J. I., Dong, Y., Jiang, J., and Dawe, R. K. (2012). Strong epigenetic similarity between maize centromeric and pericentromeric regions at the level of small RNAs, DNA methylation and H3 chromatin modifications. *Nucleic Acids Res.* 40, 1550–1560. doi: 10.1093/nar/gkr862

Goto, C., Tamura, K., Fukao, Y., Shimada, T., and Hara-Nishimura, I. (2014). The novel nuclear envelope protein KAKU4 modulates nuclear morphology in *Arabidopsis*. *Plant Cell* 26, 2143–2155. doi: 10.1105/tpc.113.122168

Grob, S., Schmid, M. W., and Grossniklaus, U. (2014). Hi-C analysis in *Arabidopsis* identifies the KNOT, a structure with similarities to the flamenco locus of drosophila. *Mol. Cell* 55, 678–679. doi: 10.1016/j.molcel.2014.07.009

Han, F., Lamb, J. C., and Birchler, J. A. (2006). High frequency of centromere inactivation resulting in stable dicentric chromosomes of maize. *Proc. Natl. Acad. Sci. U.S.A.* 103, 3238–3243. doi: 10.1073/pnas.0509650103

Han, Y., Zhang, Z., Liu, C., Liu, J., Huang, S., Jiang, J., et al. (2009). Centromere repositioning in cucurbit species: implication of the genomic impact from centromere activation and inactivation. *Proc. Natl. Acad. Sci. U.S.A.* 106, 14937–14941. doi: 10.1073/pnas.0904833106

Hayashi, T., Fujita, Y., Iwasaki, O., Adachi, Y., Takahashi, K., and Yanagida, M. (2004). Mis16 and Mis18 are required for CENP-A loading and histone deacetylation at centromeres. *Cell* 118, 715–729. doi: 10.1016/j.cell.2004.09.002

Heitz, E. (1928). Das heterochromatin der Moose. *Jahrb Wiss Bot.* 69, 762–818.

Heslop-Harrison, J. S., Murata, M., Ogura, Y., Schwarzacher, T., and Motoyoshi, F. (1999). Polymorphisms and genomic organization of repetitive DNA from centromeric regions of *Arabidopsis* chromosomes. *Plant Cell* 11, 31–42. doi: 10.1105/tpc.11.1.31

Ingouff, M., Sakata, T., Li, J., Sprunck, S., Dresselhaus, T., and Berger, F. (2009). The two male gametes share equal ability to fertilize the egg cell in *Arabidopsis thaliana*. *Curr. Biol.* 19, 19–20. doi: 10.1016/j.cub.2008.11.025

Jacob, Y., Bergamin, E., Donoghue, M. T. A., Mongeon, V., LeBlanc, C., Voigt, P., et al. (2014). Selective methylation of histone H3 variant H3.1 regulates heterochromatin replication. *Science* 343, 1249–1253. doi: 10.1126/science.1248357

Jacob, Y., and Feng, S. (2009). ATXR5 and ATXR6 are H3K27 monomethyltransferases required for chromatin structure and gene silencing. *Nat. Struct.* 16, 763–768. doi: 10.1038/nsmb.1611

Jacob, Y., Stroud, H., Leblanc, C., Feng, S., Zhuo, L., Caro, E., et al. (2010). Regulation of heterochromatic DNA replication by histone H3 lysine 27 methyltransferases. *Nature* 466, 987–991. doi: 10.1038/nature09290

Jansen, L. E. T., Black, B. E., Foltz, D. R., and Cleveland, D. W. (2007). Propagation of centromeric chromatin requires exit from mitosis. *J. Cell Biol.* 176, 795–805. doi: 10.1083/jcb.200701066

Karimi-Ashtiyani, R., Ishii, T., Niessen, M., Stein, N., Heckmann, S., Gurushidze, M., et al. (2015). Point mutation impairs centromeric CENH3 loading and induces haploid plants. *Proc. Natl. Acad. Sci. U.S.A.* 201504333. doi: 10.1073/pnas.1504333112

Kumekawa, N., Hosouchi, T., Tsuruoka, H., and Kotani, H. (2000). The size and sequence organization of the centromeric region of *Arabidopsis thaliana* chromosome 5. *DNA Res.* 7, 315–321. doi: 10.1093/dnares/7.6.315

Kumekawa, N., Hosouchi, T., Tsuruoka, H., and Kotani, H. (2001). The size and sequence organization of the centromeric region of *Arabidopsis thaliana* chromosome 4. *DNA Res.* 8, 285–290. doi: 10.1093/dnares/8.6.285

Lacoste, N., Woolfe, A., Tachiwana, H., Garea, A. V., Barth, T., Cantaloube, S., et al. (2014). Mislocalization of the centromeric histone variant CenH3/CENP-A in human cells depends on the chaperone DAXX. *Mol. Cell* 53, 631–644. doi: 10.1016/j.molcel.2014.01.018

Lermontova, I., Koroleva, O., Rutten, T., Fuchs, J., Schubert, V., Moraes, I., et al. (2011). Knockdown of CENH3 in *Arabidopsis* reduces mitotic divisions and causes sterility by disturbed meiotic chromosome segregation. *Plant J.* 68, 40–50. doi: 10.1111/j.1365-313X.2011.04664.x

Lermontova, I., Kuhlmann, M., Friedel, S., Rutten, T., Heckmann, S., Sandmann, M., et al. (2013). *Arabidopsis* kinetochore null2 is an upstream component for centromeric histone H3 variant cenH3 deposition at centromeres. *Plant Cell* 25, 3389–3404. doi: 10.1105/tpc.113.114736

Lermontova, I., Schubert, V., Fuchs, J., Klatte, S., Macas, J., and Schubert, I. (2006). Loading of *Arabidopsis* centromeric histone CENH3 occurs mainly during G2 and requires the presence of the histone fold domain. *Plant Cell* 18, 2443–2451. doi: 10.1105/tpc.106.043174

Loyola, A., Tagami, H., Bonaldi, T., Roche, D., Quivy, J. P., Imhof, A., et al. (2009). The HP1alpha-CAF1-SetDB1-containing complex provides H3K9me1 for Suv39-mediated K9me3 in pericentric heterochromatin. *EMBO Rep.* 10, 769–775. doi: 10.1038/embor.2009.90

Ma, J., Wing, R. A., Bennetzen, J. L., and Jackson, S. A. (2007). Plant centromere organization: a dynamic structure with conserved functions. *Trends Genet.* 23, 134–139. doi: 10.1016/j.tig.2007.01.004

Maheshwari, S., Tan, E. H., West, A., Franklin, F. C. H., Comai, L., and Chan, S. W. L. (2015). Naturally occurring differences in CENH3 affect chromosome segregation in zygotic mitosis of hybrids. *PLoS Genet.* 11:e1004970. doi: 10.1371/journal.pgen.1004970

Maison, C., Bailly, D., Peters, A. H. F. M., Quivy, J.-P., Roche, D., Taddei, A., et al. (2002). Higher-order structure in pericentric heterochromatin involves a distinct pattern of histone modification and an RNA component. *Nat. Genet.* 30, 329–334. doi: 10.1038/ng843

Mathieu, O., Jasencakova, Z., Vaillant, I., Gendrel, A.-V., Colot, V., Schubert, I., et al. (2003). Changes in 5S rDNA chromatin organization and transcription during heterochromatin establishment in *Arabidopsis. Plant Cell* 15, 2929–2939. doi: 10.1105/tpc.017467

Mathieu, O., Reinders, J., Čaikovski, M., Smathajitt, C., and Paszkowski, J. (2007). Transgenerational stability of the *Arabidopsis* epigenome is coordinated by CG methylation. *Cell* 130, 851–862. doi: 10.1016/j.cell.2007.07.007

May, B. P., Lippman, Z. B., Fang, Y., Spector, D. L., and Martienssen, R. a. (2005). Differential regulation of strand-specific transcripts from *Arabidopsis* centromeric satellite repeats. *PLoS Genet.* 1:e79. doi: 10.1371/journal.pgen.0010079

Mérai, Z., Chumak, N., García-Aguilar, M., Hsieh, T.-F., Nishimura, T., Schoft, V. K., et al. (2014). The AAA-ATPase molecular chaperone Cdc48/p97 disassembles sumoylated centromeres, decondenses heterochromatin, and activates ribosomal RNA genes. *Proc. Natl. Acad. Sci. U.S.A.* 111, 16166–16171. doi: 10.1073/pnas.1418564111

Nagaki, K., Talbert, P. B., Zhong, C. X., Dawe, R. K., Henikoff, S., and Jiang, J. (2003). Chromatin immunoprecipitation reveals that the 180-bp satellite repeat is the key functional DNA element of *Arabidopsis thaliana* centromeres. *Genetics* 163, 1221–1225.

Naumann, K., Fischer, A., Hofmann, I., Krauss, V., Phalke, S., Irmler, K., et al. (2005). Pivotal role of AtSUVH2 in heterochromatic histone methylation and gene silencing in *Arabidopsis. EMBO J.* 24, 1418–1429. doi: 10.1038/sj.emboj.7600604

Ono, T., Kaya, H., Takeda, S., Abe, M., Ogawa, Y., Kato, M., et al. (2006). Chromatin assembly factor 1 ensures the stable maintenance of silent chromatin states in *Arabidopsis. Genes Cells* 11, 153–162. doi: 10.1111/j.1365-2443.2006.00928.x

Pavet, V., Quintero, C., Cecchini, N. M., Rosa, A. L., and Alvarez, M. E. (2006). *Arabidopsis* displays centromeric DNA hypomethylation and cytological alterations of heterochromatin upon attack by *Pseudomonas syringae. Mol. Plant Microbe Interact.* 19, 577–587. doi: 10.1094/MPMI-19-0577

Pecinka, A., Dinh, H. Q., Baubec, T., Rosa, M., Lettner, N., and Scheid, O. M. (2010). Epigenetic regulation of repetitive elements is attenuated by prolonged heat stress in *Arabidopsis. Plant Cell* 22, 3118–3129. doi: 10.1105/tpc.110.078493

Plohl, M., Meštrović, N., and Mravinac, B. (2014). Centromere identity from the DNA point of view. *Chromosoma* 123, 313–325. doi: 10.1007/s00412-014-0462-0

Poulet, A., Arganda-Carreras, I., Legland, D., Probst, A. V., Andrey, P., and Tatout, C. (2015). NucleusJ: an ImageJ plugin for quantifying 3D images of interphase nuclei. *Bioinformatics* 31, 1144–1146. doi: 10.1093/bioinformatics/btu774

Probst, A. V., Fransz, P. F., Paszkowski, J., and Mittelsten Scheid, O. (2003). Two means of transcriptional reactivation within heterochromatin. *Plant J. cell Mol. Biol.* 33, 743–749. doi: 10.1046/j.1365-313X.2003.01667.x

Probst, A. V., and Mittelsten Scheid, O. (2015). Stress-induced structural changes in plant chromatin. *Curr. Opin. Plant Biol.* 27, 8–16. doi: 10.1016/j.pbi.2015.05.011

Probst, A. V., Okamoto, I., Casanova, M., El Marjou, F., Le Baccon, P., and Almouzni, G. (2010). A strand-specific burst in transcription of pericentric satellites is required for chromocenter formation and early mouse development. *Dev. Cell* 19, 625–638. doi: 10.1016/j.devcel.2010.09.002

Quénet, D., and Dalal, Y. (2014). A long non-coding RNA is required for targeting centromeric protein A to the human centromere. *Elife* 3, e03254. doi: 10.7554/eLife.03254

Quivy, J.-P., Roche, D., Kirschner, D., Tagami, H., Nakatani, Y., and Almouzni, G. (2004). A CAF-1 dependent pool of HP1 during heterochromatin duplication. *EMBO J.* 23, 3516–3526. doi: 10.1038/sj.emboj.7600362

Ravi, M., Kwong, P. N., Menorca, R. M. G., Valencia, J. T., Ramahi, J. S., Stewart, J. L., et al. (2010). The rapidly evolving centromere-specific histone has stringent functional requirements in *Arabidopsis thaliana. Genetics* 186, 461–471. doi: 10.1534/genetics.110.120337

Ravi, M., Shibata, F., Ramahi, J. S., Nagaki, K., Chen, C., Murata, M., et al. (2011). Meiosis-specific loading of the centromere-specific histone CENH3 in *Arabidopsis thaliana. PLoS Genet.* 7:e1002121. doi: 10.1371/journal.pgen.1002121

Rivera, C., Saavedra, F., Alvarez, F., Díaz-Celis, C., Ugalde, V., Li, J., et al. (2015). Methylation of histone H3 lysine 9 occurs during translation. *Nucleic Acids Res.* 43, 9097–9106. doi: 10.1093/nar/gkv929

Rošić, S., Köhler, F., and Erhardt, S. (2014). Repetitive centromeric satellite RNA is essential for kinetochore formation and cell division. *J. Cell Biol.* 207, 335–349. doi: 10.1083/jcb.201404097

Roudier, F., Ahmed, I., Bérard, C., Sarazin, A., Mary-Huard, T., Cortijo, S., et al. (2011). Integrative epigenomic mapping defines four main chromatin states in *Arabidopsis. EMBO J.* 30, 1928–1938. doi: 10.1038/emboj.2011.103

Schneeberger, K., Ossowski, S., Ott, F., Klein, J. D., Wang, X., Lanz, C., et al. (2011). Reference-guided assembly of four diverse *Arabidopsis thaliana* genomes. *Proc. Natl. Acad. Sci. U.S.A.* 108, 10249–10254. doi: 10.1073/pnas.1107739108

Schoft, V. K., Chumak, N., Mosiolek, M., Slusarz, L., Komnenovic, V., Brownfield, L., et al. (2009). Induction of RNA-directed DNA methylation upon decondensation of constitutive heterochromatin. *EMBO Rep.* 10, 1015–1021. doi: 10.1038/embor.2009.152

Schönrock, N., Exner, V., Probst, A., Gruissem, W., and Hennig, L. (2006). Functional genomic analysis of CAF-1 mutants in *Arabidopsis thaliana. J. Biol. Chem.* 281, 9560–9568. doi: 10.1074/jbc.M513426200

Schubert, V., Lermontova, I., and Schubert, I. (2013). The *Arabidopsis* CAP-D proteins are required for correct chromatin organisation, growth and fertility. *Chromosoma* 122, 517–533. doi: 10.1007/s00412-013-0424-y

Schubert, V., Lermontova, I., and Schubert, I. (2014). Loading of the centromeric histone H3 variant during meiosis-how does it differ from mitosis? *Chromosoma* 123, 491–497. doi: 10.1007/s00412-014-0466-9

Schubert, V., Weißleder, A., Ali, H., Fuchs, J., Lermontova, I., Meister, A., et al. (2009). Cohesin gene defects may impair sister chromatid alignment and genome stability in *Arabidopsis thaliana. Chromosoma* 118, 591–605. doi: 10.1007/s00412-009-0220-x

Sequeira-Mendes, J., Aragüez, I., Peiró, R., Mendez-Giraldez, R., Zhang, X., Jacobsen, S. E., et al. (2014). The functional topography of the *Arabidopsis* genome is organized in a reduced number of linear motifs of chromatin states. *Plant Cell.* 26, 2351–2366. doi: 10.1105/tpc.114.124578

Shang, W.-H., Hori, T., Martins, N. M. C., Toyoda, A., Misu, S., Monma, N., et al. (2013). Chromosome engineering allows the efficient isolation of vertebrate neocentromeres. *Dev. Cell* 24, 635–648. doi: 10.1016/j.devcel.2013.02.009

She, W., and Baroux, C. (2015). Chromatin dynamics in pollen mother cells underpin a common scenario at the somatic-to-reproductive fate transition of both the male and female lineages in *Arabidopsis. Front. Plant Sci.* 6:294. doi: 10.3389/fpls.2015.00294

She, W., Grimanelli, D., Rutowicz, K., Whitehead, M. W. J., Puzio, M., Kotlinski, M., et al. (2013). Chromatin reprogramming during the somatic-to-reproductive cell fate transition in plants. *Development* 140, 4008–4019. doi: 10.1242/dev.095034

Shibata, F., and Murata, M. (2004). Differential localization of the centromere-specific proteins in the major centromeric satellite of *Arabidopsis thaliana*. *J. Cell Sci.* 117, 2963–2970. doi: 10.1242/jcs.01144

Shu, H., Wildhaber, T., Siretskiy, A., Gruissem, W., and Hennig, L. (2012). Distinct modes of DNA accessibility in plant chromatin. *Nat. Commun.* 3, 1281. doi: 10.1038/ncomms2259

Slotkin, R. K. (2010). The epigenetic control of the athila family of retrotransposons in *Arabidopsis*. *Epigenetics* 5, 483–490. doi: 10.4161/epi.5.6.12119

Slotkin, R. K., Vaughn, M., Borges, F., Tanurdzić, M., Becker, J. D., Feijó, J. A., et al. (2009). Epigenetic reprogramming and small RNA silencing of transposable elements in pollen. *Cell* 136, 461–472. doi: 10.1016/j.cell.2008.12.038

Smith, S., and Stillman, B. (1989). Purification and characterization of CAF-I, a human cell factor required for chromatin assembly during DNA replication *in vitro*. *Cell* 58, 15–25. doi: 10.1016/0092-8674(89)90398-X

Soppe, W. J. J., Jasencakova, Z., Houben, A., Kakutani, T., Meister, A., Huang, M. S., et al. (2002). DNA methylation controls histone H3 lysine 9 methylation and heterochromatin assembly in *Arabidopsis*. *EMBO J.* 21, 6549–6559. doi: 10.1093/emboj/cdf657

Steimer, A., Amedeo, P., Afsar, K., Fransz, P. F., Mittelsten Scheid, O., and Paszkowski, J. (2000). Endogenous targets of transcriptional gene silencing in *Arabidopsis*. *Plant Cell* 12, 1165–1178. doi: 10.1105/tpc.12.7.1165

Stroud, H., Do, T., Du, J., Zhong, X., Feng, S., Johnson, L., et al. (2014). Non-CG methylation patterns shape the epigenetic landscape in *Arabidopsis*. *Nat. Struct. Mol. Biol.* 21, 64–72. doi: 10.1038/nsmb.2735

Stroud, H., Otero, S., Desvoyes, B., Ramirez-Parra, E., Jacobsen, S. E., and Gutierrez, C. (2012). Genome-wide analysis of histone H3.1 and H3.3 variants in *Arabidopsis thaliana*. *Proc. Natl. Acad. Sci. U.S.A.* 109, 5370–5375. doi: 10.1073/pnas.1203145109

Tagami, H., Ray-Gallet, D., Almouzni, G., and Nakatani, Y. (2004). Histone H3.1 and H3.3 complexes mediate nucleosome assembly pathways dependent or independent of DNA synthesis. *Cell* 116, 51–61. doi: 10.1016/S0092-8674(03)01064-X

Takeda, S., Tadele, Z., Hofmann, I., Probst, A. V., Angelis, K. J., Kaya, H., et al. (2004). BRU1, a novel link between responses to DNA damage and epigenetic gene silencing in *Arabidopsis*. *Genes Dev.* 18, 782–793. doi: 10.1101/gad.295404

Talbert, P. B., Masuelli, R., Tyagi, A. P., Comai, L., and Henikoff, S. (2002). Centromeric localization and adaptive evolution of an *Arabidopsis* histone H3 variant. *Plant Cell* 14, 1053–1066. doi: 10.1105/tpc.010425

Tariq, M., Saze, H., Probst, A. V., Lichota, J., Habu, Y., and Paszkowski, J. (2003). Erasure of CpG methylation in *Arabidopsis* alters patterns of histone H3 methylation in heterochromatin. *Proc. Natl. Acad. Sci. U.S.A.* 100, 8823–8827. doi: 10.1073/pnas.1432939100

Tessadori, F., Chupeau, M.-C., Chupeau, Y., Knip, M., Germann, S., van Driel, R., et al. (2007a). Large-scale dissociation and sequential reassembly of pericentric heterochromatin in dedifferentiated *Arabidopsis* cells. *J. Cell Sci.* 120, 1200–1208. doi: 10.1242/jcs.000026

Tessadori, F., Schulkes, R. K., Van Driel, R., and Fransz, P. (2007b). Light-regulated large-scale reorganization of chromatin during the floral transition in *Arabidopsis*. *Plant J.* 50, 848–857. doi: 10.1111/j.1365-313X.2007.03093.x

Thompson, H. L., Schmidt, R., and Dean, C. (1996). Identification and distribution of seven classes of middle-repetitive DNA in the *Arabidopsis thaliana* genome. *Nucleic Acids Res.* 24, 3017–3022.

van Zanten, M., Koini, M. A., Geyer, R., Liu, Y., Brambilla, V., Bartels, D., et al. (2011). Seed maturation in *Arabidopsis thaliana* is characterized by nuclear size reduction and increased chromatin condensation. *Proc. Natl. Acad. Sci. U.S.A.* 108, 20219–20224. doi: 10.1073/pnas.1117726108

Vaquero-Sedas, M. I., and Vega-Palas, M. A. (2013). Differential association of *Arabidopsis* telomeres and centromeres with histone H3 variants. *Sci. Rep.* 3, 1202. doi: 10.1038/srep01202

Volpe, T. A., Kidner, C., Hall, I. M., Teng, G., Grewal, S. I. S., and Martienssen, R. A. (2002). Regulation of heterochromatic silencing and histone H3 lysine-9 methylation by RNAi. *Science* 297, 1833–1837. doi: 10.1126/science.1074973

Wang, H., Dittmer, T. A., and Richards, E. J. (2013a). *Arabidopsis* CROWDED NUCLEI (CRWN) proteins are required for nuclear size control and heterochromatin organization. *BMC Plant Biol.* 13:200. doi: 10.1186/1471-2229-13-200

Wang, L.-C., Wu, J.-R., Chang, W.-L., Yeh, C.-H., Ke, Y.-T., Lu, C.-A., et al. (2013b). *Arabidopsis* HIT4 encodes a novel chromocentre-localized protein involved in the heat reactivation of transcriptionally silent loci and is essential for heat tolerance in plants. *J. Exp. Bot.* 64, 1689–1701. doi: 10.1093/jxb/ert030

Wollmann, H., Holec, S., Alden, K., Clarke, N. D., Jacques, P.-E., and Berger, F. (2012). Dynamic deposition of histone variant H3.3 accompanies developmental remodeling of the *Arabidopsis* transcriptome. *PLoS Genet.* 8:e1002658. doi: 10.1371/journal.pgen.1002658

Yelagandula, R., Stroud, H., Holec, S., Zhou, K., Feng, S., Zhong, X., et al. (2014). The histone variant H2A.W defines heterochromatin and promotes chromatin condensation in *Arabidopsis*. *Cell* 158, 98–109. doi: 10.1016/j.cell.2014.06.006

Zhang, W., Lee, H.-R., Koo, D.-H., and Jiang, J. (2008). Epigenetic modification of centromeric chromatin: hypomethylation of DNA sequences in the CENH3-associated chromatin in *Arabidopsis thaliana* and maize. *Plant Cell* 20, 25–34. doi: 10.1105/tpc.107.057083

Zhou, B., Feng, H., and Kato, H. (2013). Structural insights into the histone H1-nucleosome complex. *Proc. Natl. Acad. Sci. U.S.A.* 111, 1222.

Conflict of Interest Statement: The authors declare that the research was conducted in the absence of any commercial or financial relationships that could be construed as a potential conflict of interest.

23

Involvement of Plant Stem Cells or Stem Cell-Like Cells in Dedifferentiation

Fangwei Jiang[1], *Zhenhua Feng*[1], *Hailiang Liu*[2*] *and Jian Zhu*[1*]

[1] Department of Molecular and Cell Biology, School of Life Science and Technology, Tongji University, Shanghai, China,
[2] Translational Center for Stem Cell Research, Tongji University School of Medicine, Tongji Hospital, Shanghai, China

Dedifferentiation is the transformation of cells from a given differentiated state to a less differentiated or stem cell-like state. Stem cell-related genes play important roles in dedifferentiation, which exhibits similar histone modification and DNA methylation features to stem cell maintenance. Hence, stem cell-related factors possibly synergistically function to provide a specific niche beneficial to dedifferentiation. During callus formation in *Arabidopsis* petioles, cells adjacent to procambium cells (stem cell-like cells) are dedifferentiated and survive more easily than other cell types. This finding indicates that stem cells or stem cell-like cells may influence the dedifferentiating niche. In this paper, we provide a brief overview of stem cell maintenance and dedifferentiation regulation. We also summarize current knowledge of genetic and epigenetic mechanisms underlying the balance between differentiation and dedifferentiation. Furthermore, we discuss the correlation of stem cells or stem cell-like cells with dedifferentiation.

Edited by:
Anja Geitmann, Université de Montréal, Canada

Reviewed by:
Venugopala R. Gonehal, University of California, Riverside, USA
Kaori M. Furuta, Nara Institute of Science and Technology, Japan

***Correspondence:**
Hailiang Liu
hailiang_1111@126.com;
Jian Zhu
zhujian1@tongji.edu.cn

Keywords: plant stem cell, dedifferentiation, epigenetic, callus, asexual reproduction

INTRODUCTION

Stem cells are a major research hotspot because of their therapeutic potential in several applications, such as human tissue replacement and drug development (Heidstra and Sabatini, 2014; Morus et al., 2014). Stem cell activity affects the remarkable longevity and strong regeneration capacity of plants (Heidstra and Sabatini, 2014). Animal stem cells reside in stem cell niches, which produce signals that balance self-renewal (division) and differentiation by generating daughter cells to form new tissues. Plant stem cells are also maintained in specific niches, called meristems, which are organized structures involved in post-embryonic development (Aichinger et al., 2012; Sozzani and Iyer-Pascuzzi, 2014). Plant stem cells can divide and further differentiate into various tissues and organs or form a new plant under specific environmental conditions (specific hormone backgrounds; Galinha et al., 2009).

Plant stem cells are maintained in primary and secondary meristems. Primary meristems possess three parts: apical meristem (shoots and roots), intercalary meristem, and floral meristem. Secondary meristems comprise lateral meristems, which include cambium and phellogen, and traumatic (callus) meristem (Morus et al., 2014). Intercalary meristem is present between mature tissues in certain plants, does not differentiate with other adjacent meristems into mature tissues, and is silently preserved. However, intercalary meristem can divide and differentiate during a specific period of development or under specific environmental conditions (Beveridge et al., 2007). Numerous studies and reviews have focused on the stem cell behavior (Aichinger et al., 2012; Lee et al., 2013; Yadav et al., 2013; Heidstra and Sabatini, 2014).

Procambium/cambium cells satisfy the criteria for stem cells because of their capacity for long-term self-renewal and ability to differentiate into one or more specialized cell types; as such, these tissues serve as niches for plant stem cells (Alison et al., 2002). Several types of cells also possess stem cell properties. For instance, differentiated pericycle cells initiate lateral root formation after leaving the root meristem (Beeckman and De Smet, 2014). Sena et al. (2009) reported that stem cell-like properties of differentiated cells can mediate complete organ regeneration when plant root meristems are dispersed. These cells exhibit properties similar to those of stem cells and are thus defined as stem cell-like cells in this review.

Callus is predominantly formed from a pre-existing population of stem cells (Sugimoto et al., 2010, 2011) and from cell dedifferentiation. This presumption means that stem cells or stem cell-like cells play important roles in callus formation. Hence, stem cells or stem cell-like cells may be involved in dedifferentiation of other cells to a certain degree. Other studies also consider that regeneration or dedifferentiation does not require a functional stem cell niche in plants (Sena et al., 2009). Thus far, the relationship between stem cells and dedifferentiation remains unclear. In this study, we reviewed the mechanisms of stem cell niches and dedifferentiation in plant growth. We also analyzed the potential correlation between stem cells and plant cell dedifferentiation. This review provides insights for understanding plant growth and behavior.

DEDIFFERENTIATION

Dedifferentiation is the transformation of cells from a given differentiated state to a less differentiated or stem cell-like state and leads to reacquisition of pluripotency; dedifferentiation is a cellular process associated with reentry into the cell cycle, trans/redifferentiation, or even cell death (Grafi, 2004). This process is commonly studied in plants and amphibians, zebrafish, and other basal life forms capable of organ regeneration (Brockes and Kumar, 2002; Poss et al., 2002; Jopling et al., 2010; Sugimoto et al., 2011; Xu and Huang, 2014). Cellular dedifferentiation mostly occurs in aged or damaged tissues for regeneration (Sanchez Alvarado and Tsonis, 2006).

Plant cellular dedifferentiation is manifested by transition of differentiated cells into protoplasts (Jamet et al., 1990; Zhao et al., 2001) and features widespread chromatin decondensation, similar to that of stem cells (Zhao et al., 2001; Tessadori et al., 2007; Ondrej et al., 2010). Auxin and cytokinin induce protoplasts to reenter the cell cycle, proliferate, and form callus (Grafi, 2004; Muraro et al., 2011). Stress conditions, including protoplasting (plant cell wall is degraded by enzymes), wounding, and exposure to dark or heat, induce somatic cells to change their fate via dedifferentiation (Malamy, 2005; Grafi et al., 2011a,b; Florentin et al., 2013; Feher, 2014). Researchers propose that the dedifferentiated state is a transient state of senescence because senescing cells share similar features, such as gene expression and chromatin modification, with dedifferentiating protoplast cells (Balazadeh et al., 2008; Damri et al., 2009; Yadav et al., 2009). This hypothesis indicates that dedifferentiation precedes cell death (Grafi, 2004). Thus, dedifferentiation may be involved in cell fate switch.

A study revealed that cell dedifferentiation in animals is triggered by signals from the stem cell niche and leads to dynamic cellular rearrangements. This process possibly involves Janus kinase–signal transducer and activator of transcription signaling pathway because its downregulation leads to inhibition of dedifferentiation (Sheng et al., 2009). In addition, epimorphosis requires preexisting stem cells or dedifferentiation-generated progenitor cells to proliferate, differentiate, and replace lost cells (Cai et al., 2007). These findings support that plant stem cells and/or stem cell niches are significantly associated with dedifferentiation.

INVOLVEMENT OF STEM CELL-RELATED GENES IN DEDIFFERENTIATION

Dedifferentiation is a complex process related to the expression and regulation of specific genes. Several studies investigated the relationship between genes and dedifferentiation (Berdasco et al., 2008). Several lines of evidence support the hypothesis that stem cell-related genes are involved in dedifferentiation (Yadav et al., 2010; Jopling et al., 2011).

Our previous studies showed that stem cells initially remain in the vascular bundle and divide, and callus is formed after petiole explants are cultured for 24–36 h (Yu et al., 2010; Li et al., 2011a). The mRNA chips were analyzed to screen related genes. The results showed that *WUSCHEL* (*WUS*) and *NO APICAL MERISTEM* (*NAM*) gene families are strongly expressed in petiole dedifferentiation (Liu et al., 2010). Other studies also reported that *WUS* is highly expressed in several callus lines (Iwase et al., 2011a), and overexpressing this gene forms callus and somatic embryos (Zuo et al., 2002). Moreover, increased *WUS* levels lead to dedifferentiation of stem cell progenitors into stem cells (Reddy and Meyerowitz, 2005; Yadav et al., 2010). *WUS* is a stem cell niche signal important to maintain stem cells in a relatively undifferentiated state (Laux et al., 1996; Mayer et al., 1998; Yadav et al., 2013; Zhou et al., 2015). Thus, *WUS*, a well-known stem cell-related gene, is involved in dedifferentiation. Similarly, teosinte branched1/Cycloidea/proliferating cell factor is a transcriptional activator that negatively regulates shoot meristem maintenance (Koyama et al., 2010) and cell dedifferentiation (Ikeda and Ohme-Takagi, 2014). Overexpressing stem cell-related genes, including master regulators in egg cell fate (*RKD1* and *RKD2*, Koszegi et al., 2011) and embryonic fate (*RKD4*, *LEC1*, *LEC2*, *AGL15*, and *BBM*, Srinivasan et al., 2007; Thakare et al., 2008; Waki et al., 2011; Guo et al., 2013), is sufficient to induce callus formation (Ikeuchi et al., 2013). Most LATERAL ORGAN BOUNDARIES DOMAIN (LBD) transcription factors are involved in a regulatory loop that maintains shoot meristem and defines the lateral organ boundary (Bell et al., 2012). For instance, LBD15 is involved in shoot apical meristem development by regulating *WUS* expression (Sun et al., 2013). As such, numerous *LBD* genes are possibly involved in dedifferentiation (Liu et al., 2010). Furthermore, *LBDs* directly form callus in *Arabidopsis* regeneration (Fan et al., 2012). In rice, *ADVENTITIOUS ROOTLESS1* (*ARL1*), an *LBD* family member, is involved in hormone-mediated pericycle cell dedifferentiation and promotes initial cell division (Liu et al., 2005). These findings show that stem cell-related genes play an important role in dedifferentiation. Hence, we presume that dedifferentiation may share a similar regulatory mechanism with the stem cell niche.

The AP2/ERF transcription factor WOUND INDUCED DEDIFFERENTIATION 1 (WIND1) and its close homologs, including WIND2 to WIND4, induce wounding and promote cell dedifferentiation in *Arabidopsis* (Iwase et al., 2011b). A similar homologous gene, namely, *ThWIND1-L*, was found in *Thellungiella halophila* (Zhou et al., 2012). However, the direct relationship of *WINDs* to stem cell niche remains inconclusive. WIND activates cytokinin signaling but not auxin signaling, whereas auxin alone, not cytokinin alone, can induce callus formation (Li et al., 2011a). As such, dedifferentiation may involve several pathways comprising stem cell-related genes. The dedifferentiation mechanism is not a precise copy of the regulatory mechanism in a stem cell niche. Therefore, numerous genes regulate one phenomenon by different pathways and coordinate with each other to maintain a specific niche. The balance in niches can decide the cell fate and facilitates plant growth, development, asexual reproduction, and pluripotency. This phenomenon is represented in a "seesaw model," which posits that the reprogramming of animal cells is affected by the balance in interactions among genes (Shu et al., 2013).

The types and levels of cell differentiation differ in explants. Specific cells, such as differentiated cells, switch fate during dedifferentiation, whereas other cells, such as stem cells, are not affected by differentiation. However, not all parenchymal cells in explants can reach a stem cell-like status because some of these cells may die. Hence, when the explants encounters a cell fate decision, a certain signal should indicate which cells should survive. This signal may be secreted by the cell itself to determine autonomous events in each cell. Moreover, signal communication may exhibit similar characteristics to the mode used by stem cells to decide their number in the microenvironment. In several cases and in organisms ranging from bacteria to humans, cells adopt a particular fate stochastically without apparent regard to the environment or history (Losick and Desplan, 2008). In the large majority of cases, cells acquire their fate by virtue of lineage and/or proximity to an inductive signal from another cell (Losick and Desplan, 2008). Signals exchanged between neighboring cells, similar to the Notch receptor in animals, can amplify and consolidate molecular differences, which eventually dictate cell fates (Artavanis-Tsakonas et al., 1999; Drevon and Jaffredo, 2014). Limited direct evidence confirms that the cell–cell communication plays an important role in dedifferentiation. However, cell-to-cell transport through plasmodesmata was detected in tree callus (Pina et al., 2009). We assume that the signal from another cell also plays an important role in callus formation and may exhibit similar characteristics to the signal used by stem cells to decide their number in the microenvironment. Communication is a fundamental mechanism for coordinating developmental and physiological events in multicellular organisms. This process is also widely shared as a key mechanism in a stem cell niche (Geiger and Van Zant, 2002; Oatley and Brinster, 2012). For instance, heterotrimeric G proteins are key molecules that transmit extracellular signals. The G protein beta-subunit1 AGB1 and RPK2, a major CLV3 peptide hormone receptor, work synergistically in stem cell homeostasis through physical interactions to facilitate meristem development (Ishida et al., 2014). Communication in dedifferentiation may be the next significant study area.

SIMILAR EPIGENETIC MODIFICATIONS SHARED IN STEM CELL MAINTENANCE AND CELLULAR DEDIFFERENTIATION

Dedifferentiation is controlled by various epigenetic mechanisms, including chromatin structural changes, as well as DNA and gene methylation. Epigenetic changes in stem cells are reversible and provide plasticity in plants; plasticity allows differentiated cells to recover totipotency under certain physiological and environmental conditions.

Open chromatin conformation characterizes dedifferentiated cells in plants and animals (Grafi, 2004; Meshorer and Misteli, 2006; Gaspar-Maia et al., 2011; Grafi et al., 2011b). This characteristic is necessary to maintain stem cell developmental capacities, including self-renewal and differentiation into multiple cell types (Melcer and Meshorer, 2010; Gaspar-Maia et al., 2011). To support transcriptional promiscuity, chromatin in stem cells is "plastic" or "open," with a decondensed heterochromatin architecture, enriched active histone modifications, and hyperdynamic association of chromatin proteins with chromatin (Zipori, 2004; Meshorer and Misteli, 2006). The overexpression of genes involved in chromatin modification pathways in stem cells confirms the existence of flexible chromatin structures in shoot apical meristems (Yadav et al., 2009).

Similar to stem cells, dedifferentiating plant cells acquire an open, decondensed chromatin architecture (Zhao et al., 2001; Meshorer and Misteli, 2006; Tessadori et al., 2007), which is essential but not sufficient to initiate gene transcription. In *Arabidopsis* protoplasts, the acquisition of pluripotency is associated with chromatin reorganization in specific domains (Jiang et al., 2013). Grafi et al. (2007) demonstrated that histone methylation is required to establish the dedifferentiated state.

In mammals, DNA methyltransferase 1 (DNMT1) is responsible for self-renewal and differentiation of stem cells, as well as self-renewal of mammalian somatic tissues (Chan et al., 2005; Sen et al., 2010). Similar to DNMT1, MET1 is a methyltransferase involved in maintaining DNA methylation in *Arabidopsis* (Chan et al., 2005). Root explants with *MET1* mutants exhibit reduced callus formation in culture with 2,4-D (analog of auxin); this finding indicates that DNMT1-like methyltransferases are related to dedifferentiation process-callus formation (Berdasco et al., 2008). Active DNA demethylation in plants is generally performed through a base excision repair pathway, which can be induced following the activity of the DME/ROS1 family of DNA glycosylases or by the coupled activities of 5-methylcytosine deaminase (Zhu, 2009; Zhang and Zhu, 2012).

In cooperation with DNA methylation, histone modifications regulate gene expression by altering chromatin structure and transcriptional activity, thereby influencing callus formation

(Desvoyes et al., 2010). KRYPTONITE (*KYP*)/*SUVH4* mutants, which encode a histone H3 lysine 9 (H3K9) methyltransferase, produces defects during stem cell proliferation and callus formation (Grafi et al., 2007). Moreover, *MET1* and *KYP* mutants alter the expression of *WUS*, a marker for plant stem-like cells (Li et al., 2011b). These findings support that histone methylation activity is required to establish/maintain the dedifferentiated state and/or re-entry into the cell cycle, as well as callus formation, in plant cells (Berdasco et al., 2008).

The promoter activity of stress-induced *ANAC2* [*abscisic acid-responsive NAC* (*NAM/ATAF1/2/CUC2*), also known as *Arabidopsis* transcription activation factor (*ATAF1*)] is associated with widespread decondensation of pericentric chromatin (Florentin et al., 2013); the *ANAC2* promoter is also meristem specific (Damri et al., 2009). The promoter of *ANAC2* activity is upregulated in dedifferentiating protoplast cells and in cells subjected to various biotic and abiotic stress conditions (Damri et al., 2009; Yadav et al., 2009; Florentin et al., 2013).

The *ANAC2* promoter is active not only in meristems but also in leaf primordia and young leaves (Florentin et al., 2013). Thus, stem cell state may be maintained in plant tissues and organs during the early development stages in the form of stem cell-like cells (Laux, 2003). In humans, stem cells may stimulate paracrine effects to orchestrate wound healing and cancer progression (Dittmer and Leyh, 2014). These stem cells may secrete growth factors or cytokines, as well as non-protein factors, such as RNAs and lipids, to communicate with other cells. Plant stem cells may also produce factors to regulate hormones in a paracrine effect in response to wound or other stresses. Thus, the widely retained stem cells or stem cell-like cells may give rise to dedifferentiation. However, whether stem cells assist neighboring cells to alter gene expression and influence dedifferentiation process require further research.

STEM CELLS OR STEM CELL-LIKE CELLS INVOLVED IN THE DEDIFFERENTIATION PROCESS

Asexual reproduction maintains the characteristics of species in a relatively stable environment and is considered dedifferentiation. *Kalanchoe daigremontiana* produces viviparous seedlings to accomplish asexual reproduction (Garces et al., 2014). Some retained initiating cells located in the nicks of mature leaf edges can divide and differentiate into viviparous seedlings (Garces et al., 2007). The young plants of *Graptopetalum paraguayense* and *Crassula portulacea* originate from a wound in the leaf. The young plant of *G. paraguayense* is only generated at the proximal petiole, in which several layers of stem cell-like cells are located; conversely, the adventitious buds of *C. portulacea* are produced by the activities of procambium cells (stem cell-like cells) in the vascular bundle in any wound of a leaf. These observations imply that the retained stem cells or stem cell-like cells may initiate asexual reproduction of plants (Murashige, 1974).

During callus formation in petiole explants, cells located near vascular tissues and are adjacent to procambium cells (stem cell-like cells), more easily survive than other cells (Bongso and Richards, 2004; Yu et al., 2010). Pre-existing stem cells or stem cell-like cells may produce signals that promote cell dedifferentiation through cell-to-cell communication. However, direct evidence of cell-to-cell communication remains insufficient. In addition, the dedifferentiation activities of cells may be determined by their cell types, age, or other features.

In *Drosophila*, airway stem cells can influence the dedifferentiation of airway epithelial cells into stable and functional basal stem cells *in vivo* and *ex vivo* (Tata et al., 2013). These findings suggest that stem cells and their progeny can reciprocally modulate one another through a precise and local control mechanism. This premise supports that plant stem cells/stem cell-like cells perform a significant function in dedifferentiation.

CONCLUSION

Significant advances have been achieved in stem cell research. Accumulating evidence suggests that a complicated regulatory network contributes to modulated plant stem cell maintenance, although the detailed mechanisms have not been completely revealed. This complicated regulatory network may include genes, hormones, and epigenetic factors, which play significant roles in dedifferentiation. These factors build and maintain a specific niche. In addition, stem cells or stem cell-like cells may influence the dedifferentiating niche. Studies documented that stem cells retained in the tissues contribute to callus formation under certain environmental conditions. Moreover, plant cells that underwent stress or senescence present a dedifferentiation patter similar to the stem cell-like state prior to entry into new cell fates. Stem cells may communicate with neighboring cells by releasing specific signals during dedifferentiation of neighboring cells. This communication process may be enhanced by adding stem cells. However, further research must be performed to confirm this assumption. Elucidating signals/factors involved in the governing dedifferentiating niche, including stem cells, can provide insights into designing sophisticated and specific molecular tools to systematically manipulate organ regeneration.

ACKNOWLEDGMENT

This paper was supported by the Natural Science Foundation of China (Grant No. 31370214).

REFERENCES

Aichinger, E., Kornet, N., Friedrich, T., and Laux, T. (2012). Plant stem cell niches. *Annu. Rev. Plant Biol.* 63, 615–636. doi: 10.1146/annurev-arplant-042811-105555

Alison, M. R., Poulsom, R., Forbes, S., and Wright, N. A. (2002). An introduction to stem cells. *J. Pathol.* 197, 419–423. doi: 10.1002/path.1187

Artavanis-Tsakonas, S., Rand, M. D., and Lake, R. J. (1999). Notch signaling: cell fate control and signal integration in development. *Science* 284, 770–776. doi: 10.1126/science.284.5415.770

Balazadeh, S., Riano-Pachon, D. M., and Mueller-Roeber, B. (2008). Transcription factors regulating leaf senescence in *Arabidopsis thaliana*. *Plant Biol. (Stuttg.)* 10(Suppl. 1), 63–75. doi: 10.1111/j.1438-8677.2008.00088.x

Beeckman, T., and De Smet, I. (2014). Pericycle. *Curr. Biol.* 24, R378–R379. doi: 10.1016/j.cub.2014.03.031

Bell, E. M., Lin, W. C., Husbands, A. Y., Yu, L., Jaganatha, V., Jablonska, B., et al. (2012). *Arabidopsis* lateral organ boundaries negatively regulates brassinosteroid accumulation to limit growth in organ boundaries. *Proc. Natl. Acad. Sci. U.S.A.* 109, 21146–21151. doi: 10.1073/pnas.1210789109

Berdasco, M., Alcazar, R., Garcia-Ortiz, M. V., Ballestar, E., Fernandez, A. F., Roldan-Arjona, T., et al. (2008). Promoter DNA hypermethylation and gene repression in undifferentiated *Arabidopsis* cells. *PLoS ONE* 3:e3306. doi: 10.1371/journal.pone.0003306

Beveridge, C. A., Mathesius, U., Rose, R. J., and Gresshoff, P. M. (2007). Common regulatory themes in meristem development and whole-plant homeostasis. *Curr. Opin. Plant Biol.* 10, 44–51. doi: 10.1016/j.pbi.2006.11.011

Bongso, A., and Richards, M. (2004). History and perspective of stem cell research. *Best Pract. Res. Clin. Obstet. Gynaecol.* 18, 827–842. doi: 10.1016/j.bpobgyn.2004.09.002

Brockes, J. P., and Kumar, A. (2002). Plasticity and reprogramming of differentiated cells in amphibian regeneration. *Nat. Rev. Mol. Cell Biol.* 3, 566–574. doi: 10.1038/nrm881

Cai, S., Fu, X., and Sheng, Z. (2007). Dedifferentiation: a new approach in stem cell research. *Bioscience* 57, 655–662. doi: 10.1641/B570805

Chan, S. W., Henderson, I. R., and Jacobsen, S. E. (2005). Gardening the genome: DNA methylation in *Arabidopsis thaliana*. *Nat. Rev. Genet.* 6, 351–360. doi: 10.1038/nrg1601

Damri, M., Granot, G., Ben-Meir, H., Avivi, Y., Plaschkes, I., Chalifa-Caspi, V., et al. (2009). Senescing cells share common features with dedifferentiating cells. *Rejuvenation Res.* 12, 435–443. doi: 10.1089/rej.2009.0887

Desvoyes, B., Sanchez, M. P., Ramirez-Parra, E., and Gutierrez, C. (2010). Impact of nucleosome dynamics and histone modifications on cell proliferation during *Arabidopsis* development. *Heredity (Edinb.)* 105, 80–91. doi: 10.1038/hdy.2010.50

Dittmer, J., and Leyh, B. (2014). Paracrine effects of stem cells in wound healing and cancer progression (Review). *Int. J. Oncol.* 44, 1789–1798. doi: 10.3892/ijo.2014.2385

Drevon, C., and Jaffredo, T. (2014). Cell interactions and cell signaling during hematopoietic development. *Exp. Cell Res.* 329, 200–206. doi: 10.1016/j.yexcr.2014.10.009

Fan, M., Xu, C., Xu, K., and Hu, Y. (2012). LATERAL ORGAN BOUNDARIES DOMAIN transcription factors direct callus formation in *Arabidopsis* regeneration. *Cell Res.* 22, 1169–1180. doi: 10.1038/cr.2012.63

Feher, A. (2014). Somatic embryogenesis—stress-induced remodeling of plant cell fate. *Biochim. Biophys Acta.* 1849, 385–402. doi: 10.1016/j.bbagrm.2014.07.005

Florentin, A., Damri, M., and Grafi, G. (2013). Stress induces plant somatic cells to acquire some features of stem cells accompanied by selective chromatin reorganization. *Dev. Dyn.* 242, 1121–1133. doi: 10.1002/dvdy.24003

Galinha, C., Bilsborough, G., and Tsiantis, M. (2009). Hormonal input in plant meristems: a balancing act. *Semin. Cell Dev. Biol.* 20, 1149–1156. doi: 10.1016/j.semcdb.2009.09.007

Garces, H. M., Champagne, C. E., Townsley, B. T., Park, S., Malho, R., Pedroso, M. C., et al. (2007). Evolution of asexual reproduction in leaves of the genus *Kalanchoe*. *Proc. Natl. Acad. Sci. U.S.A.* 104, 15578–15583. doi: 10.1073/pnas.0704105104

Garcês, H. M., Koenig, D., Townsley, B. T., Kim, M., and Sinha, N. R. (2014). Truncation of LEAFY COTYLEDON1 protein is required for asexual reproduction in *Kalanchoë daigremontiana*. *Plant Physiol.* 165, 196–206. doi: 10.1104/pp.114.237222

Gaspar-Maia, A., Alajem, A., Meshorer, E., and Ramalho-Santos, M. (2011). Open chromatin in pluripotency and reprogramming. *Nat. Rev. Mol. Cell Biol.* 12, 36–47. doi: 10.1038/nrm3036

Geiger, H., and Van Zant, G. (2002). The aging of lympho-hematopoietic stem cells. *Nat. Immunol.* 3, 329–333. doi: 10.1038/ni0402-329

Grafi, G. (2004). How cells dedifferentiate: a lesson from plants. *Dev. Biol.* 268, 1–6. doi: 10.1016/j.ydbio.2003.12.027

Grafi, G., Ben-Meir, H., Avivi, Y., Moshe, M., Dahan, Y., and Zemach, A. (2007). Histone methylation controls telomerase-independent telomere lengthening in cells undergoing dedifferentiation. *Dev. Biol.* 306, 838–846. doi: 10.1016/j.ydbio.2007.03.023

Grafi, G., Chalifa-Caspi, V., Nagar, T., Plaschkes, I., Barak, S., and Ransbotyn, V. (2011a). Plant response to stress meets dedifferentiation. *Planta* 233, 433–438. doi: 10.1007/s00425-011-1366-3

Grafi, G., Florentin, A., Ransbotyn, V., and Morgenstern, Y. (2011b). The stem cell state in plant development and in response to stress. *Front. Plant Sci.* 2:53. doi: 10.3389/fpls.2011.00053

Guo, F., Liu, C., Xia, H., Bi, Y., Zhao, C., Zhao, S., et al. (2013). Induced expression of AtLEC1 and AtLEC2 differentially promotes somatic embryogenesis in transgenic tobacco plants. *PLoS ONE* 8:e71714. doi: 10.1371/journal.pone.0071714

Heidstra, R., and Sabatini, S. (2014). Plant and animal stem cells: similar yet different. *Nat. Rev. Mol. Cell Biol.* 15, 301–312. doi: 10.1038/nrm3790

Ikeda, M., and Ohme-Takagi, M. (2014). TCPs, WUSs, and WINDs: families of transcription factors that regulate shoot meristem formation, stem cell maintenance, and somatic cell differentiation. *Front. Plant Sci.* 5:427. doi: 10.3389/fpls.2014.00427

Ikeuchi, M., Sugimoto, K., and Iwase, A. (2013). Plant callus: mechanisms of induction and repression. *Plant Cell* 25, 3159–3173. doi: 10.1105/tpc.113.116053

Ishida, T., Tabata, R., Yamada, M., Aida, M., Mitsumasu, K., Fujiwara, M., et al. (2014). Heterotrimeric G proteins control stem cell proliferation through CLAVATA signaling in *Arabidopsis*. *EMBO Rep.* 15, 1202–1209. doi: 10.15252/embr.201438660

Iwase, A., Mitsuda, N., Koyama, T., Hiratsu, K., Kojima, M., Arai, T., et al. (2011a). The AP2/ERF transcription factor WIND1 controls cell dedifferentiation in *Arabidopsis*. *Curr. Biol.* 21, 508–514. doi: 10.1016/j.cub.2011.02.020

Iwase, A., Ohme-Takagi, M., and Sugimoto, K. (2011b). WIND1: a key molecular switch for plant cell dedifferentiation. *Plant Signal. Behav.* 6, 1943–1945. doi: 10.4161/psb.6.12.18266

Jamet, E., Durr, A., Parmentier, Y., Criqui, M. C., and Fleck, J. (1990). Is ubiquitin involved in the dedifferentiation of higher plant cells? *Cell Differ. Dev.* 29, 37–46. doi: 10.1016/0922-3371(90)90022-O

Jiang, F., Zhu, J., and Liu, H. L. (2013). Protoplasts: a useful research system for plant cell biology, especially dedifferentiation. *Protoplasma* 250, 1231–1238. doi: 10.1007/s00709-013-0513-z

Jopling, C., Boue, S., and Izpisua Belmonte, J. C. (2011). Dedifferentiation, transdifferentiation and reprogramming: three routes to regeneration. *Nat. Rev. Mol. Cell Biol.* 12, 79–89. doi: 10.1038/nrm3043

Jopling, C., Sleep, E., Raya, M., Marti, M., Raya, A., and Izpisua Belmonte, J. C. (2010). Zebrafish heart regeneration occurs by cardiomyocyte dedifferentiation and proliferation. *Nature* 464, 606–609. doi: 10.1038/nature08899

Koszegi, D., Johnston, A. J., Rutten, T., Czihal, A., Altschmied, L., Kumlehn, J., et al. (2011). Members of the RKD transcription factor family induce an egg cell-like gene expression program. *Plant J.* 67, 280–291. doi: 10.1111/j.1365-313X.2011.04592.x

Koyama, T., Mitsuda, N., Seki, M., Shinozaki, K., and Ohme-Takagi, M. (2010). TCP transcription factors regulate the activities of ASYMMETRIC LEAVES1 and miR164, as well as the auxin response, during differentiation of leaves in *Arabidopsis*. *Plant Cell* 22, 3574–3588. doi: 10.1105/tpc.110.075598

Laux, T. (2003). The stem cell concept in plants: a matter of debate. *Cell* 113, 281–283. doi: 10.1016/S0092-8674(03)00312-X

Laux, T., Mayer, K. F., Berger, J., and Jurgens, G. (1996). The *WUSCHEL* gene is required for shoot and floral meristem integrity in *Arabidopsis*. *Development* 122, 87–96.

Lee, Y., Lee, W. S., and Kim, S. H. (2013). Hormonal regulation of stem cell maintenance in roots. *J. Exp. Bot.* 64, 1153–1165. doi: 10.1093/jxb/ers331

Li, F., Cui, X., Feng, Z., Du, X., and Zhu, J. (2011a). The effect of 2,4-D and kinetin on dedifferentiation of petiole cells in *Arabidopsis thaliana*. *Biol. Plant.* 56, 121–125. doi: 10.1007/s10535-011-0201-9

Li, W., Liu, H., Cheng, Z. J., Su, Y. H., Han, H. N., Zhang, Y., et al. (2011b). DNA methylation and histone modifications regulate *de novo* shoot regeneration in *Arabidopsis* by modulating *WUSCHEL* expression and auxin signaling. *PLoS Genet.* 7:e1002243. doi: 10.1371/journal.pgen.1002243

Liu, H., Wang, S., Yu, X., Yu, J., He, X., Zhang, S., et al. (2005). ARL1, a LOB-domain protein required for adventitious root formation in rice. *Plant J.* 43, 47–56. doi: 10.1111/j.1365-313X.2005.02434.x

Liu, H.-L., Wang, G.-C., Feng, Z., and Zhu, J. (2010). Screening of genes associated with dedifferentiation and effect of LBD29 on pericycle cells in *Arabidopsis thaliana*. *Plant Growth Regul.* 62, 127–136. doi: 10.1007/s10725-010-9495-4

Losick, R., and Desplan, C. (2008). Stochasticity and cell fate. *Science* 320, 65–68. doi: 10.1126/science.1147888

Malamy, J. E. (2005). Intrinsic and environmental response pathways that regulate root system architecture. *Plant Cell Environ.* 28, 67–77. doi: 10.1111/j.1365-3040.2005.01306.x

Mayer, K. F., Schoof, H., Haecker, A., Lenhard, M., Jurgens, G., and Laux, T. (1998). Role of *WUSCHEL* in regulating stem cell fate in the *Arabidopsis* shoot meristem. *Cell* 95, 805–815. doi: 10.1016/S0092-8674(00)81703-1

Melcer, S., and Meshorer, E. (2010). Chromatin plasticity in pluripotent cells. *Essays Biochem.* 48, 245–262. doi: 10.1042/bse0480245

Meshorer, E., and Misteli, T. (2006). Chromatin in pluripotent embryonic stem cells and differentiation. *Nat. Rev. Mol. Cell Biol.* 7, 540–546. doi: 10.1038/nrm1938

Morus, M., Baran, M., Rost-Roszkowska, M., and Skotnicka-Graca, U. (2014). Plant stem cells as innovation in cosmetics. *Acta Pol. Pharm.* 71, 701–707.

Muraro, D., Byrne, H., King, J., Voss, U., Kieber, J., and Bennett, M. (2011). The influence of cytokinin-auxin cross-regulation on cell-fate determination in *Arabidopsis thaliana* root development. *J. Theor. Biol.* 283, 152–167. doi: 10.1016/j.jtbi.2011.05.011

Murashige, T. (1974). Plant propagation through tissue cultures. *Plant Physiol.* 25, 135–166. doi: 10.1146/annurev.pp.25.060174.001031

Oatley, J. M., and Brinster, R. L. (2012). The germline stem cell niche unit in mammalian testes. *Physiol. Rev.* 92, 577–595. doi: 10.1152/physrev.00025.2011

Ondrej, V., Navratilova, B., Protivankova, I., Piterkova, J., Sedlarova, M., Luhova, L., et al. (2010). Recondensation level of repetitive sequences in the plant protoplast nucleus is limited by oxidative stress. *J. Exp. Bot.* 61, 2395–2401. doi: 10.1093/jxb/erq067

Pina, A., Errea, P., Schulz, A., and Martens, H. J. (2009). Cell-to-cell transport through plasmodesmata in tree callus cultures. *Tree Physiol.* 29, 809–818. doi: 10.1093/treephys/tpp025

Poss, K. D., Wilson, L. G., and Keating, M. T. (2002). Heart regeneration in zebrafish. *Science* 298, 2188–2190. doi: 10.1126/science.1077857

Reddy, G. V., and Meyerowitz, E. M. (2005). Stem-cell homeostasis and growth dynamics can be uncoupled in the *Arabidopsis* shoot apex. *Science* 310, 663–667. doi: 10.1126/science.1116261

Sanchez Alvarado, A., and Tsonis, P. A. (2006). Bridging the regeneration gap: genetic insights from diverse animal models. *Nat. Rev. Genet.* 7, 873–884. doi: 10.1038/nrg1923

Sen, G. L., Reuter, J. A., Webster, D. E., Zhu, L., and Khavari, P. A. (2010). DNMT1 maintains progenitor function in self-renewing somatic tissue. *Nature* 463, 563–567. doi: 10.1038/nature08683

Sena, G., Wang, X., Liu, H. Y., Hofhuis, H., and Birnbaum, K. D. (2009). Organ regeneration does not require a functional stem cell niche in plants. *Nature* 457, 1150–1153. doi: 10.1038/nature07597

Sheng, X. R., Brawley, C. M., and Matunis, E. L. (2009). Dedifferentiating spermatogonia outcompete somatic stem cells for niche occupancy in the *Drosophila* testis. *Cell Stem Cell* 5, 191–203. doi: 10.1016/j.stem.2009.05.024

Shu, J., Wu, C., Wu, Y., Li, Z., Shao, S., Zhao, W., et al. (2013). Induction of pluripotency in mouse somatic cells with lineage specifiers. *Cell* 153, 963–975. doi: 10.1016/j.cell.2013.05.001

Sozzani, R., and Iyer-Pascuzzi, A. (2014). Postembryonic control of root meristem growth and development. *Curr. Opin. Plant Biol.* 17, 7–12. doi: 10.1016/j.pbi.2013.10.005

Srinivasan, C., Liu, Z., Heidmann, I., Supena, E. D., Fukuoka, H., Joosen, R., et al. (2007). Heterologous expression of the BABY BOOM AP2/ERF transcription factor enhances the regeneration capacity of tobacco (*Nicotiana tabacum* L.). *Planta* 225, 341–351. doi: 10.1007/s00425-006-0358-1

Sugimoto, K., Gordon, S. P., and Meyerowitz, E. M. (2011). Regeneration in plants and animals: dedifferentiation, transdifferentiation, or just differentiation? *Trends Cell Biol.* 21, 212–218. doi: 10.1016/j.tcb.2010.12.004

Sugimoto, K., Jiao, Y., and Meyerowitz, E. M. (2010). *Arabidopsis* regeneration from multiple tissues occurs via a root development pathway. *Dev. Cell* 18, 463–471. doi: 10.1016/j.devcel.2010.02.004

Sun, X., Feng, Z., Meng, L., Zhu, J., and Geitmann, A. (2013). *Arabidopsis* ASL11/LBD15 is involved in shoot apical meristem development and regulates WUS expression. *Planta* 237, 1367–1378. doi: 10.1007/s00425-013-1844-x

Tata, P. R., Mou, H., Pardo-Saganta, A., Zhao, R., Prabhu, M., Law, B. M., et al. (2013). Dedifferentiation of committed epithelial cells into stem cells *in vivo*. *Nature* 503, 218–223. doi: 10.1038/nature12777

Tessadori, F., Chupeau, M. C., Chupeau, Y., Knip, M., Germann, S., Van Driel, R., et al. (2007). Large-scale dissociation and sequential reassembly of pericentric heterochromatin in dedifferentiated *Arabidopsis* cells. *J. Cell Sci.* 120, 1200–1208. doi: 10.1242/jcs.000026

Thakare, D., Tang, W., Hill, K., and Perry, S. E. (2008). The MADS-domain transcriptional regulator AGAMOUS-LIKE15 promotes somatic embryo development in *Arabidopsis* and soybean. *Plant Physiol.* 146, 1663–1672. doi: 10.1104/pp.108.115832

Waki, T., Hiki, T., Watanabe, R., Hashimoto, T., and Nakajima, K. (2011). The *Arabidopsis* RWP-RK protein RKD4 triggers gene expression and pattern formation in early embryogenesis. *Curr. Biol.* 21, 1277–1281. doi: 10.1016/j.cub.2011.07.001

Xu, L., and Huang, H. (2014). Genetic and epigenetic controls of plant regeneration. *Curr. Top. Dev. Biol.* 108, 1–33. doi: 10.1016/B978-0-12-391498-9.00009-7

Yadav, R. K., Girke, T., Pasala, S., Xie, M., and Reddy, G. V. (2009). Gene expression map of the *Arabidopsis* shoot apical meristem stem cell niche. *Proc. Natl. Acad. Sci. U.S.A.* 106, 4941–4946. doi: 10.1073/pnas.0900843106

Yadav, R. K., Perales, M., Gruel, J., Ohno, C., Heisler, M., Girke, T., et al. (2013). Plant stem cell maintenance involves direct transcriptional repression of differentiation program. *Mol. Syst. Biol.* 9, 654. doi: 10.1038/msb.2013.8

Yadav, R. K., Tavakkoli, M., and Reddy, G. V. (2010). *WUSCHEL* mediates stem cell homeostasis by regulating stem cell number and patterns of cell division and differentiation of stem cell progenitors. *Development* 137, 3581–3589. doi: 10.1242/dev.054973

Yu, Y., Feng, Z., Wang, G., Li, F., Du, X., and Zhu, J. (2010). Initiation of dedifferentiation and structural changes in *in vitro* cultured petiole of *Arabidopsis thaliana*. *Protoplasma* 241, 75–81. doi: 10.1007/s00709-010-0108-x

Zhang, H., and Zhu, J. K. (2012). Active DNA demethylation in plants and animals. *Cold. Spring Harb. Symp. Quant. Biol.* 77, 161–173. doi: 10.1101/sqb.2012.77.014936

Zhao, J., Morozova, N., Williams, L., Libs, L., Avivi, Y., and Grafi, G. (2001). Two phases of chromatin decondensation during dedifferentiation of plant cells: distinction between competence for cell fate switch and a commitment for S phase. *J. Biol. Chem.* 276, 22772–22778. doi: 10.1074/jbc.M101756200

Zhou, C., Guo, J., Feng, Z., Cui, X., and Zhu, J. (2012). Molecular characterization of a novel AP2 transcription factor ThWIND1-L from *Thellungiella halophila*. *Plant Cell Tissue Organ Cult.* 110, 423–433. doi: 10.1007/s11240-012-0163-4

Zhou, Y., Liu, X., Engstrom, E. M., Nimchuk, Z. L., Pruneda-Paz, J. L., Tarr, P. T., et al. (2015). Control of plant stem cell function by conserved interacting transcriptional regulators. *Nature* 517, 377–380. doi: 10.1038/nature13853

Zhu, J. K. (2009). Active DNA demethylation mediated by DNA glycosylases. *Annu. Rev. Genet.* 43, 143–166. doi: 10.1146/annurev-genet-102108-134205

Zipori, D. (2004). The nature of stem cells: state rather than entity. *Nat. Rev. Genet.* 5, 873–878. doi: 10.1038/nrg1475

Zuo, J., Niu, Q. W., Frugis, G., and Chua, N. H. (2002). The *WUSCHEL* gene promotes vegetative-to-embryonic transition in *Arabidopsis*. *Plant J.* 30, 349–359. doi: 10.1046/j.1365-313X.2002.01289.x

Conflict of Interest Statement: The authors declare that the research was conducted in the absence of any commercial or financial relationships that could be construed as a potential conflict of interest.

24

Induction of Embryogenesis in *Brassica Napus* Microspores Produces a Callosic Subintinal Layer and Abnormal Cell Walls with Altered Levels of Callose and Cellulose

Verónica Parra-Vega, Patricia Corral-Martínez, Alba Rivas-Sendra and Jose M. Seguí-Simarro*

Cell Biology Group, COMAV Institute, Universitat Politècnica de València, Valencia, Spain

The induction of microspore embryogenesis produces dramatic changes in different aspects of the cell physiology and structure. Changes at the cell wall level are among the most intriguing and poorly understood. In this work, we used high pressure freezing and freeze substitution, immunolocalization, confocal, and electron microscopy to analyze the structure and composition of the first cell walls formed during conventional *Brassica napus* microspore embryogenesis, and in cultures treated to alter the intracellular Ca^{2+} levels. Our results revealed that one of the first signs of embryogenic commitment is the formation of a callose-rich, cellulose-deficient layer beneath the intine (the subintinal layer), and of irregular, incomplete cell walls. In these events, Ca^{2+} may have a role. We propose that abnormal cell walls are due to a massive callose synthesis and deposition of excreted cytoplasmic material, and the parallel inhibition of cellulose synthesis. These features were absent in pollen-like structures and in microspore-derived embryos, few days after the end of the heat shock, where abnormal cell walls were no longer produced. Together, our results provide an explanation to a series of relevant aspects of microspore embryogenesis including the role of Ca^{2+} and the occurrence of abnormal cell walls. In addition, our discovery may be the explanation to why nuclear fusions take place during microspore embryogenesis.

Edited by:
Bo Liu,
University of California-Davis, USA

Reviewed by:
Yoshinobu Mineyuki,
University of Hyogo, Japan
Georgia Drakakaki,
University of California-Davis, USA

Correspondence:
Jose M. Seguí-Simarro
seguisim@btc.upv.es

Keywords: androgenesis, benzyl alcohol, caffeine, doubled haploids, microspore embryogenesis, rapeseed, cell wall carbohydrates

INTRODUCTION

Microspore embryogenesis is a fascinating experimental process whereby a haploid microspore is reprogrammed to become a haploid or doubled haploid (DH) embryo (Seguí-Simarro and Nuez, 2008a). This inducible pathway has a great biotechnological potential, since among many other advantages, it allows for faster and cheaper ways to obtain pure lines for hybrid seed production (Forster et al., 2007). In addition, it offers the possibility to be used as an *in vitro* model to study different basic and induced processes. Indeed, the androgenic switch is induced by the application of different types of abiotic stresses, including heat shock, cold, and starvation, among others (Shariatpanahi et al., 2006). Once induced, the cellular responses to abiotic stresses coexist with a developmental switch toward embryogenesis, and with the cessation of the old gametophytic program (Malik et al., 2007; Seguí-Simarro and Nuez, 2008a). Conceivably, all these changes must imply a profound remodeling at the genetic and molecular levels, and also in cell architecture. Among all the changes undergone by the embryogenic microspore, one of the aspects that attracted the attention of the first cell biologists that studied this process was how induced cells are divided (Zaki and Dickinson, 1991; Simmonds and Keller, 1999).

In somatic-type plant cells, the first structural marker of cell division is the microtubular pre-prophase band (PPB), which defines the future division plane (Pickett-Heaps and Northcote, 1966). By late anaphase phragmoplast initials are formed, and at early telophase, a tubulo-vesicular network (TVN) cell plate is assembled in the middle of a solid phragmoplast (Seguí-Simarro et al., 2004; Austin et al., 2005). At mid telophase, a ring-shaped transitional phragmoplast marks the transformation of the central region of the cell plate into a wide tubular network and then into a maturing, planar fenestrated sheet, while the actively growing peripheral zone expands centrifugally and eventually fuses with the mother cell wall. Finally, at late telophase the peripheral zone matures too, and the cell plate is transformed into a new cell wall (reviewed in Seguí-Simarro et al., 2008). These orchestrated changes in cell plate structure are accompanied by the deposition of different polysaccharides in a timely manner (reviewed in Worden et al., 2012; Drakakaki, 2015). The first polysaccharides present in the nascent cell plate would be pectins and hemicelluloses. Then, the synthesis of copious amounts of callose in the cell plate lumen is responsible for the transformation of the TVN cell plate into a maturing tubular network, and for the widening of these tubules into fenestrated sheets (Samuels et al., 1995). The final transformation of the planar fenestrated sheet-type cell plate into a new primary cell wall involves the progressive replacement of callose deposits by cellulose fibrils (Kakimoto and Shibaoka, 1992; Samuels et al., 1995; Otegui and Staehelin, 2000). Proper cellulose deposition appears essential for cell plate stabilization, as revealed by the aborted cell plates present in cellulose-deficient mutants (Zuo et al., 2000; Beeckman et al., 2002). Finally, cellulose combines with the already secreted hemicellulose molecules into a cellulose-hemicellulose network, while pectic polysaccharides reorganize to form the pectin-rich middle lamella (Carpita and McCann, 2000). In the final, somatic-type primary cell wall, callose is absent with the exception of the region around plasmodesmata, where it is supposed to play a regulatory role in cell-to-cell movement of molecules (Levy et al., 2007).

Based on this canonical pattern, some specialized cell types have developed alternative division mechanisms adapted to their function. This is the case, for example, of microspores. The first pollen mitosis (PMI) that transforms a microspore into a young pollen grain is characterized by the absence of a previous PPB (Van Lammeren et al., 1985), and by the building of an asymmetric phragmoplast (Brown and Lemmon, 1991), giving rise to the large, vegetative cell and the small, generative cell of the pollen grain. The cell wall formed around the generative cell is also special, since it is hemispherical and transiently rich in callose (Park and Twell, 2001). However, soon it was found that microspores committed to *in vitro* embryogenesis significantly diverge from these cell division patterns. First, embryogenic microspores usually divide symmetrically, as opposed to *in vivo* microspores, in a mechanism more similar to somatic-type cytokinesis than to PMI cytokinesis (Zaki and Dickinson, 1991). Accordingly, it was found that embryogenic microspores show the typical PPB of somatic-type dividing cells (Simmonds and Keller, 1999). The cell walls of induced microspores and microspore-derived embryos (MDEs) present other unique features. For example, it was recently shown that embryogenic microspores exhibit abundant deposits of excreted cytoplasmic material (Corral-Martínez et al., 2013). These walls are also characterized by altered levels of certain cell wall components, such as xyloglucans, pectins (Barany et al., 2010), and arabinogalactan proteins (El-Tantawy et al., 2013). The major cell wall polymer, cellulose, has also been investigated, but unfortunately, the results are somehow confusing. For example, Schulze and Pauls (2002) assumed that the first cell walls produced in *Brassica napus* embryogenic microspores are somatic-type and therefore, cellulose-rich. In olive, Solis et al. (2008) described the presence of cellulose in some embryogenic microspores, at both the inner cell walls and the *"thick peripheral wall localized below the exine."* However, Dubas et al. (2013) recently demonstrated by calcofluor white staining that cellulose was absent from the first cell walls formed in few-celled *B. napus* embryogenic microspores, whereas older MDEs presented abundant cellulose signal in their walls.

The other major cell plate-forming polymer, callose, is a β-1,3-glucan essential not only for somatic-type cell plate formation, but also for the cellular response to biotic and abiotic stresses (reviewed in Stone and Clarke, 1992; Chen and Kim, 2009). During microspore development, the presence of callose as the main component of the tetrad cell walls appears critical for a proper development of the microspore, since rice *GSL5* knock-out mutants, defective in callose deposition during post-meiotic cytokinesis, exhibit a severe male-sterile phenotype (Shi et al., 2014). Later on, the deposition of a thin callose layer below the exine was proposed to have a role in exine patterning (Dong et al., 2005). In immature pollen grains, callose is found in the cell walls of the generative cell (Gorska-Brylass, 1967; Heslop-Harrison, 1968). Upon germination, callose progressively accumulates beneath the intine and is a major component of the pollen tube (Meikle et al., 1991; Ferguson et al., 1998). In the particular case of the first embryogenic divisions of the microspore, Telmer et al. (1995) found callose at the connections of the new cell wall with the mother cell wall, attributing it to alterations in the plasma membrane or the cell wall physiology. Other than that, little is known about the role of callose in the development of the MDE.

In this work we present a detailed study of the cell walls present in *B. napus* microspores during the initial stages of their induction toward embryogenesis. We analyzed a number of confocal three-dimensional series obtained from cultured microspores subjected to different experimental treatments, and a collection of transmission electron microscopy (TEM) images obtained from samples processed by high pressure freezing and freeze substitution (HPF-FS). HPF-FS is known as the best procedure to preserve even the most labile and/or transient ultrastructural elements of *in vivo* physiological processes (Gilkey and Staehelin, 1986), as well as during the androgenic switch (Corral-Martínez et al., 2013). Our samples covered all the stages of *B. napus* microspore embryogenesis, the different cell types formed, and the subcellular changes undergone as a consequence of embryogenesis induction. Together, our results shed light on the dynamics of callose and cellulose deposition during the initials of the androgenic switch, and its significance for further MDE development.

MATERIALS AND METHODS

Plant Materials

Brassica napus L. donor plants of the highly embryogenic cv. Topas were grown as previously described (Seguí-Simarro et al., 2003). Plants were grown in the greenhouses of the COMAV Institute (Universitat Politècnica de València, Spain), the University of Colorado (Boulder, CO, USA), and the Plant Research International (Wageningen, The Netherlands).

Brassica napus Microspore Cultures

Flower buds containing mostly vacuolated microspores were selected and processed as previously described (Custers, 2003). Briefly, buds were surface sterilized with 2% sodium hypochlorite for 10 min, and washed three times in sterile distilled water. To release the microspores, buds were gently crushed in filter sterilized NLN-13 medium with the back of the plunger of a disposable 50 ml syringe. NLN-13 medium consists of NLN medium (Lichter, 1982) + 13% sucrose. Then, the slurry was filtered through 30 μm nylon cloths. The filtrate was transferred to 15 ml conical tubes and centrifuged at 800 rpm for 4 min. After discarding the supernatant, the pellet of microspores was resuspended in 10 ml of fresh NLN-13 medium. This procedure was additionally repeated twice for a total of three centrifugations and resuspensions. Before the last centrifugation step, microspore concentration was calculated using a hemacytometer. The required volume of NLN-13 medium was added to adjust suspension to a concentration of 4×10^4 microspores per ml. For caffeine and benzyl alcohol (BA) experiments, NLN-13 medium supplemented with 1 mM caffeine and 100 μM BA were used, respectively. Adjusted microspore suspension was distributed in sterile culture dishes. Dishes were incubated in darkness for 24 h at 32°C to induce embryogenesis, and then continuously at 25°C for MDE progression up to the time of sample collection and processing.

Callose and Cellulose Staining for Confocal Laser Scanning Microscopy

Samples of *B. napus* microspore cultures were collected at day 1, 2, 3, 4, and 6 after isolation, and fixed with 4% paraformaldehyde in phosphate buffered saline (PBS) pH 7.4. For callose staining, samples were placed in glass slides with 0.9% agarose, and stained with 10 μg/ml propidium iodide (PI) in PBS for 10 min. PI was used in order to have a reference of the subcellular staining pattern of aniline blue. PI is a general stain for nucleic acids, binding both DNA and RNA. Thus, when a previous RNAse treatment is omitted, the cytoplasm is also stained with PI (Suzuki et al., 1997). After three washes with PBS, samples were stained with 0.1% aniline blue (Evans et al., 1984) dissolved in PBS for 20 min, and then washed with the same buffer.

For cellulose staining, we used two different fluorescent stains, Calcofluor White ST and Pontamine Fast Scarlet (S4B). Calcofluor White was used diluted to 0.05% in water (Dubas et al., 2013). PI was also used better identification of the developmental stage of the microspore as described above. Both stains were added directly to the fixed microspores and slides were mounted with Vectashield and kept in darkness for at least 15 min before microscopic analysis. For S4B staining, samples were stained with 0.01% S4B in PBS for 30 min (Anderson et al., 2010), then washed thrice with PBS, and finally mounted with a mix of Vectashield and DAPI (1.25 μg/ml according to Custers et al., 1994) in a 1:4 proportion. Double staining with aniline blue and S4B was also performed as described above but combining both stains at 0.05 and 0.01%, respectively. Finally, preparations were observed with a Leica CTR 5500 and a Zeiss LSM 780 confocal laser scanning microscopes. Digital 3-D series of images were processed with Leica Application Suite Advanced Fluorescence (LAS AF) and FIJI software.

Processing of *B. napus* Anthers and Microspore Cultures for TEM

Anthers carrying microspores and pollen grains at different stages of microsporogenesis and microgametogenesis were excised, transferred to aluminum sample holders, cryoprotected with 150 mM sucrose, frozen in a Baltec HPM 010 high-pressure freezer (Technotrade, Manchester, NH, USA) and a Leica EM HPM-100 high-pressure freezer (Leica Microsystems, Vienna), and then transferred to LN_2. Cultured microspores and small MDEs were recovered from culture dishes by gently spinning culture media. Larger MDEs were manually picked up from cultures. These samples were transferred to aluminum sample holders, cryoprotected with their same glucose-rich culture medium and high-pressure frozen. All samples were freeze substituted in a Leica AFS2 system (Leica Microsystems, Vienna) with 2% OsO_4 in anhydrous acetone at –80°C for 7 days, followed by slow warming to room temperature over a period of 2 days. After rinsing in several acetone washes, they were removed from the holders, incubated in propylene oxide for 30 min, rinsed again in acetone, and infiltrated with increasing concentrations of Epon resin (Ted Pella, Redding, CA, USA) in acetone according to the following schedule: 4 h in 5% resin, 4 h in 10% resin, 12 h in 25% resin, and 24 h in 50, 75, and 100% resin, respectively. Polymerization was performed at 60°C for 2 days. Using a Leica UC6 ultramicrotome, thin sections (1 μm) were obtained for light microscopy observation, and ~80 nm sections were obtained for electron microscopy. Sections were mounted on formvar-coated, 200 mesh copper grids, stained with uranyl acetate and lead citrate, and observed in a Philips CM10 electron microscope.

Immunogold Labeling

For the immunodetection of callose we used an anti-callose, mouse IgG monoclonal antibody (Biosupplies, Australia) crossreacting with linear β-1,3-oligosaccharide segments in β-1,3-glucans (Meikle et al., 1991). Immunogold labeling was performed in HPF-fixed, OsO_4-treated, epoxy-embedded samples. This type of processing is not usually employed for immunogold labeling because it may preclude the immunolocalization of protein epitopes. Notwithstanding this, we used these samples for immunolocalization because the epitope studied (callose) is a carbohydrate, insensitive to the effects of OsO_4 and the heat used to polymerize the resin. This way, we were able to combine specific immunolabeling with the excellent ultrastructural preservation provided by the use of these methods. 80–100 nm sections were deposited on Formvar and carbon-coated, 200-mesh nickel grids. Sections were hydrated with distilled water for 1 min, 1x PBS for 1 min, and blocked with 5%BSA in PBS for 5 min. Then, sections were incubated for 1 h at 25°C with anti-callose antibody, diluted 1:5,000 in 1% BSA. Next, sections were subjected to three 4-min washes with PBS and incubated for 45 min at 25°C with a goat anti-mouse secondary antibody conjugated to 10 nm colloidal gold (BBI Solutions, UK), diluted 1:25 in 1% BSA. Then, sections were subjected to three 4-min washes with PBS and one with distilled water. Finally, sections were counterstained with 0.5% uranyl acetate in 70% methanol and lead citrate, 10 min each.

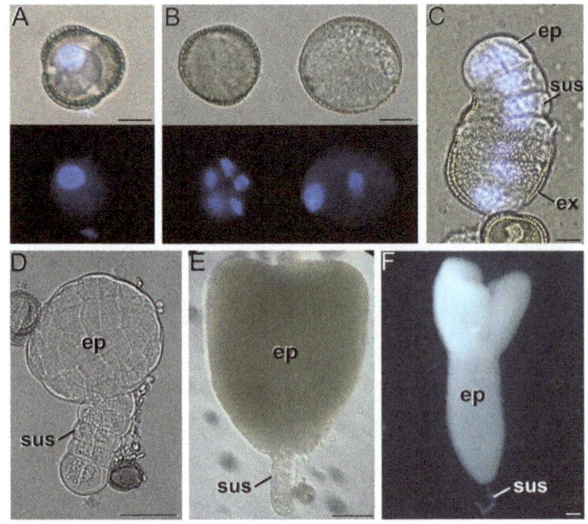

FIGURE 1 | The different stages of *Brassica napus* microspore cultures. (A) Isolated microspore, prior to induction. **(B)** Induced, embryogenic microspore (left) and pollen-like structure (right). Top and bottom images in **(A,B)** were taken under phase contrast and fluorescence with DAPI-stained nuclei, respectively. Note the presence of up to five nuclei in the embryogenic microspore, and the presence of a brighter (generative) and a faintly fluorescing (vegetative) nucleus in the pollen-like structure. **(C)** Quadrant-like embryo emerging from the exine (ex), where the suspensor (sus) and the embryo proper (ep) domains can be clearly identified. Superimposed phase contrast + fluorescence image where the DAPI-stained nuclei are shown in blue. **(D)** Globular embryo. **(E)** Heart-shaped embryo. **(F)** Cotyledonar embryo. Bars in A–C: 10 μm; D: 50 μm; E, F: 100 μm.

RESULTS

In *B. napus* vacuolate microspores (**Figure 1A**), embryogenesis starts with a set of cell divisions still within the microspore exine (**Figure 1B**, left). In parallel, other microspores insensitive to induction may develop as pollen-like structures (**Figure 1B**, right) or just arrest and die. Successive divisions in embryogenic microspores give rise to exine rupture and to the emergence of young MDEs where the embryo proper and suspensor domains can soon be distinguished (**Figure 1C**). MDEs then follow a typical embryogenic pattern proceeding through the globular (**Figure 1D**), heart-shaped (**Figure 1E**), torpedo, and cotyledonar stages (**Figure 1F**). Samples of all the stages shown in **Figure 1** were collected at different culture stages and studied as follows.

Embryogenic Microspores Occasionally Present Discontinuous, Incomplete Cell Plates, and Nuclear Fusion Profiles

Some embryogenic microspores presented continuous cell walls morphologically similar to conventional walls with the only exception of the irregularities and deposits of excreted material (arrowheads in **Figure 2A**) previously described (Corral-Martínez et al., 2013). However, we also observed often dividing cells with abnormal cell plates and cell walls. Abnormal cell plates were characterized by an irregular architecture, with numerous tubular profiles and abundant openings (fenestrae) that permitted the contact between the cytoplasm of daughter cells (white arrows in **Figures 2A,B**). In conventional cell plates, planar fenestrated profiles typically arise at the mid stage of cytokinesis. This stage is defined by the presence of a transitional, ring-shaped phragmoplast (Austin et al., 2005) and the targeted delivery of Golgi-derived vesicles to the growing edges of the cell plate, but not to the maturing central part (Seguí-Simarro et al., 2004). However, in the unusual cell plates of embryogenic microspores such a ring phragmoplast was absent, as revealed by the close proximity of cell organelles and dense vacuoles (Corral-Martínez et al., 2013) to both central and peripheral domains of the cell plate (asterisks in **Figures 2A,B**). In addition, the few Golgi-derived vesicles identified were observed randomly dispersed throughout the tubular cell plate (**Figure 2B**). These data would be indicating that in these cells, the final stage of cytokinesis (the formation of a planar fenestrated sheet) is disturbed. In parallel, we identified embryogenic microspores with binucleated cells (**Figure 2C**) and cells containing nuclei larger than usual, occasionally with a peanut-like morphology (**Figure 2A**), indicative of the recent occurrence of nuclear fusion events (Seguí-Simarro and Nuez, 2007, 2008b; Corral-Martínez et al., 2011).

Thus, we analyzed embryogenic microspores at previous stages of cell division in order to identify the initials of the cytokinesis defects. A detailed observation of mitotic cells did not reveal any structural abnormality with respect to the

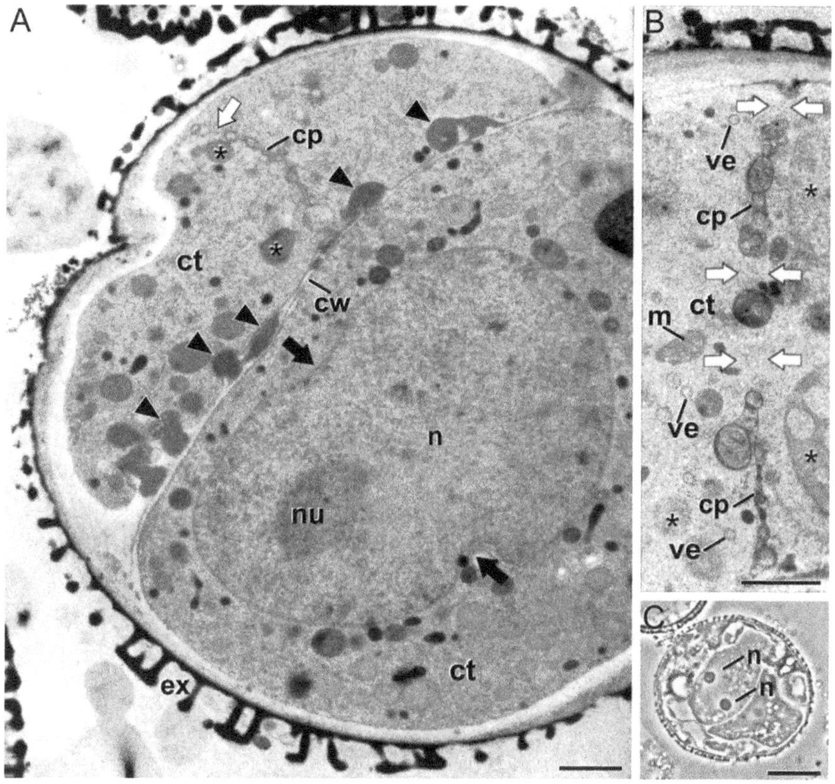

FIGURE 2 | Embryogenic *B. napus* microspores with abnormal cell plates and signs of nuclear fusion. (A) Induced microspore with a complete cell wall (cw) with numerous deposits of excreted material (arrowheads), a gapped cell plate (cp), and a peanut-shaped nucleus (n). The black arrows point to nuclear constrictions indicative of a recent nuclear fusion. **(B)** Abnormal, incomplete cell plate with abundant gaps (white arrows) that connect the cytoplasms of the daughter cells, and organelles such as dense vacuolar compartments (asterisks) close to the cell plate. **(C)** Embryogenic microspore with two nuclei coexisting in the same cytoplasm. Ct, cytoplasm; ex, exine; m, mitochondrion; nu, nucleolus; ve, Golgi-derived vesicle. Bars in A: 10 μm; B: 1 μm; C: 200 nm.

normal architecture of the mitotic machinery. As seen in **Figure 3A**, metaphasic cells presented the typical condensed chromosomes and an organelle-free region where the abundant parallel microtubules of the mitotic spindle could be identified. Images of cells showing the microtubular scaffold of the mitotic spindle and the solid phragmoplast appeared normal, as well as the cell plate initials that appeared correctly assembled (data not shown). Later on, at early telophase, the TVN cell plate typical from the solid phragmoplast stage was clearly identified together with abundant vesicles and microtubules near the cell plate (**Figure 3B**). Together, these observations pointed to the existence of problems during late stages of cytokinesis, but not before. It seemed that at the ring phragmoplast stage, where cell plate flattening and maturation starts, cytokinesis was somehow blocked and abnormal cell walls were finally formed.

The Presence of a Thick, Distinct Layer Beneath the Intine is a Differential Feature of Embryogenic Microspores

Soon after induction, we observed that cultured microspores presented an extra layer deposited just between the intine and the plasma membrane (**Figure 4A**). From now on, we will refer to this layer as the *subintinal layer*. The subintinal layer is defined as a continuous layer of irregular thickness, alternating thick and thin domains, with an electron light appearance in osmium-treated, epoxy-embedded samples. Such a layer, including all the mentioned features, was exclusively observed in embryogenic microspores (**Figure 4B**). Pollen-like structures also presented a thickened inner layer (**Figure 4C**). However, this layer was homogeneous in thickness and very similar to the intine in terms of electron density. Indeed, at low magnification the coat of pollen-like structures appeared to be composed by the exine and a thickened intine, in contrast to the three layers of embryogenic microspores (compare **Figures 4B,C**). This suggested an independent nature for the subintinal layer. As expected, this layer was absent from young microspores and pollen grains not subjected to isolation and induction treatments (**Figures 6A,B**, respectively). Therefore, it seemed that the subintinal layer is a structure assembled exclusively in embryogenic microspores.

Embryogenic Microspores Present Callose-rich Subintinal Layers and Cell Walls

The subintinal layer of embryogenic microspores had in general a texture resembling the callose-rich wall of meiocytes

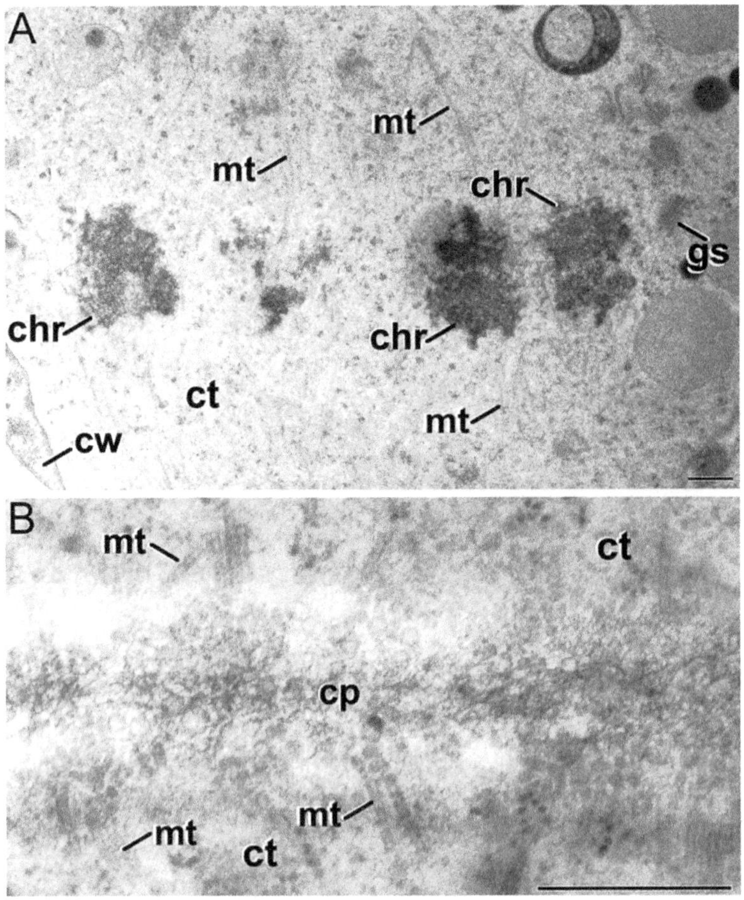

FIGURE 3 | Dividing cells of *B. napus* embryogenic microspores. (A) Metaphasic chromosomes (chr) at the cell equator of a dividing cell. **(B)** Tubulo-vesicular network (TVN) cell plate (cp) of a dividing cell at early telophase. ct, cytoplasm; cw, cell wall; gs, Golgi stack; mt, microtubules. Bars: 500 nm.

(Supplementary Figures S1A,B), as it can be deduced by comparing **Figure 4B** and Supplementary Figure S1C. For this reason, we investigated the putative callosic nature of this layer by two parallel approaches: staining with aniline blue, a callose-binding fluorescent dye, and immunogold labeling with anti-callose antibodies. Just after the induction treatment, abundant aniline blue fluorescence was found at specific regions throughout the embryogenic microspore. **Figure 5** shows a series of confocal slices from a representative embryogenic microspore, covering different planes from pole to pole. In this series, it is observed that callose accumulated below the exine in a non-uniform manner, alternating wide regions with bright fluorescence and thin regions with faint fluorescence (**Figure 5**). Interestingly, we also found aniline blue staining in the newly formed cell walls separating the daughter cells. Fluorescence was not uniform, indicating a differential callose accumulation at different cell wall domains. In some cell walls, cytoplasmic bridges were found connecting daughter cells through the regions devoid of aniline blue staining (Supplementary Figure S2). This finding, consistent with our observations in TEM micrographs (**Figure 2B**), confirmed the existence of incomplete, callose-rich cell walls. On the contrary, no evident aniline blue fluorescent signal was observed at the thickened coat of pollen-like structures or in the cell walls of small globular MDEs (Supplementary Figures S3A,B). Therefore, we confirmed the unusual and transient accumulation of callose in embryogenic microspores.

Then, we performed a more comprehensive study at different stages before induction (during *in vivo* microspore development), just after induction, and 10 days after induction (in MDEs) by means of immunogold labeling with anti-callose antibodies. Positive controls with tomato meiocytes (Supplementary Figure S1D) showed an intense labeling of the callosic walls. Negative controls excluding the secondary antibody provided no labeling at all (Supplementary Figures S1E,F). Vacuolate microspores, during their *in vivo* development within the anther, showed almost no callose at their coat (**Figure 6A**). The exine was completely devoid of labeling and only clusters of 2–3 gold particles could be rarely observed at the intine, usually close to the plasma membrane. This was consistent with the thin callose layer described as necessary for exine formation and pollen viability (Dong et al., 2005). Pollen grains maturing within the anther showed no labeling at any layer of the pollen coat (**Figure 6B**) except for mature pollen, where anti-callose labeling was found below the pollen apertures (data

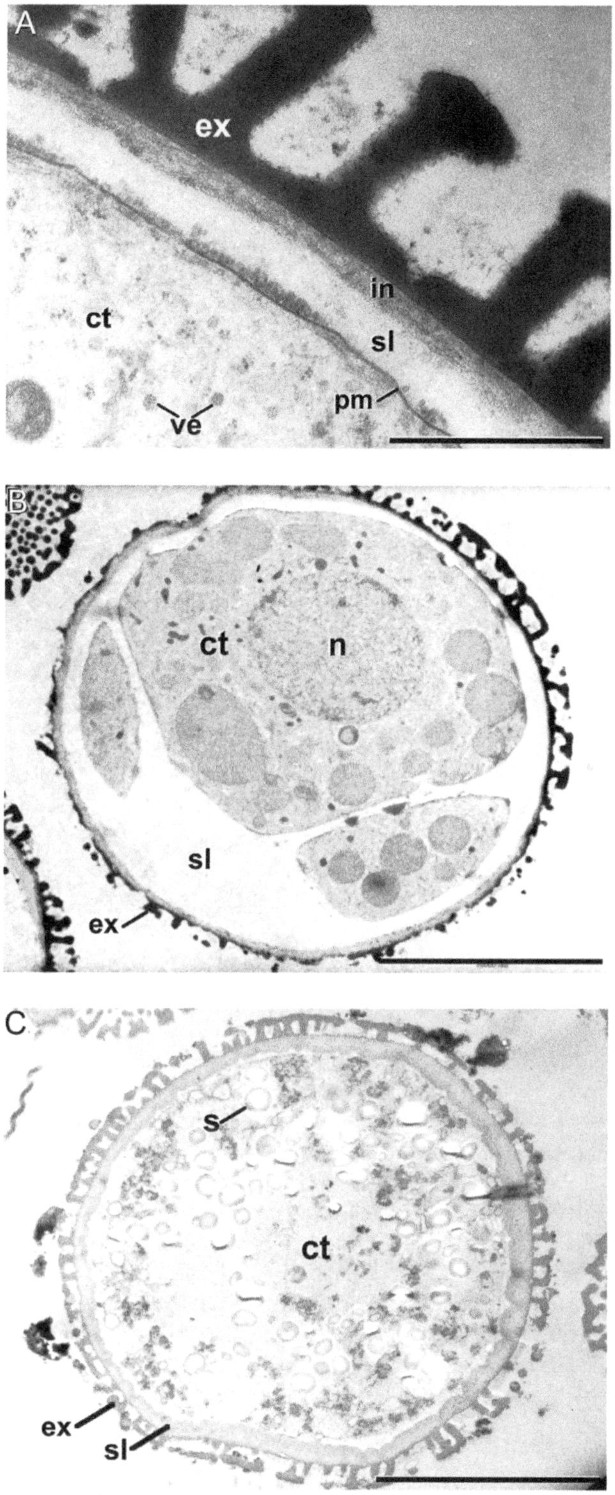

FIGURE 4 | The subintinal layer of *B. napus* embryogenic microspores. **(A)** Electron light subintinal layer (sl) deposited beneath the intine (in) of a embryogenic microspore. **(B)** Overview of an embryogenic microspore showing a continuous subintinal layer. **(C)** Pollen-like structure showing a thickened, uniform and electron dense intine. Ct, cytoplasm; ex, exine; n, nucleus; pm, plasma membrane; s, starch granule; ve, vesicle. Bars in A: 200 nm, B, C: 10 μm.

not shown). After induction, the few cells enclosed within the embryogenic microspores presented a specific anti-callose labeling restricted to the subintinal layer (**Figure 6C**). In contrast, the thickened inner layer of pollen-like structures showed no labeling (**Figure 6D**). This different labeling pattern between embryogenic and pollen-like structures was consistent with their differences at the ultrastructural level. In addition to their different thickness and electron density, these layers also differed in callose contents.

Anti-callose antibodies also decorated the new cell walls created after the first embryogenic divisions. Young, developing cell plates (**Figure 6E**) but also older, mature cell walls (**Figure 6F**) showed immunogold labeling dispersed throughout the wall. Callose is a common component of developing somatic-type cell plates, but is progressively replaced by cellulose during the maturation stages (Samuels et al., 1995). With the exception of plasmodesmata, callose is absent from mature cell walls. Therefore, its persistence should be considered an abnormal feature of embryogenic microspores. In walls with deposits of excreted material (Corral-Martínez et al., 2013), callose was always absent from these deposits (**Figure 6G**). This suggested that callose deposition and secretion of cytoplasmic material were independent processes. In older stages of microspore embryogenesis, the cell walls of suspensor-bearing MDEs did not show decoration with anti-callose antibodies (**Figure 6H**), as expected for mature cell walls derived from somatic-type cytokinesis. In other words, the unusual presence of callose in mature cell walls of few-celled embryogenic microspores appeared to be a transient phenomenon, associated to the first stages of this developmental switch, and absent from MDEs with a clearly established embryogenic pattern.

Cellulose is Absent in the Callose-Rich Cell Walls of Just Induced Embryogenic Microspores

Due to the abnormal pattern of callose deposition observed in embryogenic cells, we next studied whether this might have an influence on its subsequent replacement by cellulose. For this, we stained microspore cultures with two cellulose-specific stains: calcofluor white, conventionally used to stain cellulose, and Pontamine Fast Scarlet 4B (S4B), described as more specific for cellulose than calcofluor white (Thomas et al., 2013). Calcofluor white staining revealed a noticeable and continuous cellulose signal at the intine of pollen-like structures, being more intense at the regions just below the apertures (Supplementary Figure S4A). In contrast, the induced microspores and MDEs exhibited variable patterns of cellulose deposition (Supplementary Figures S4B,C). To further investigate this unusual observation and the possible interaction between cellulose and callose, we performed a double staining of just induced microspore culture samples with aniline blue and S4B. We analyzed 289 cultured microspores from different randomly chosen microscope fields. From this microspore population, 217 (75.1%) showed no staining, only exine autofluorescence. Most likely, these microspores would be either dead or arrested in development, as usual in microspore

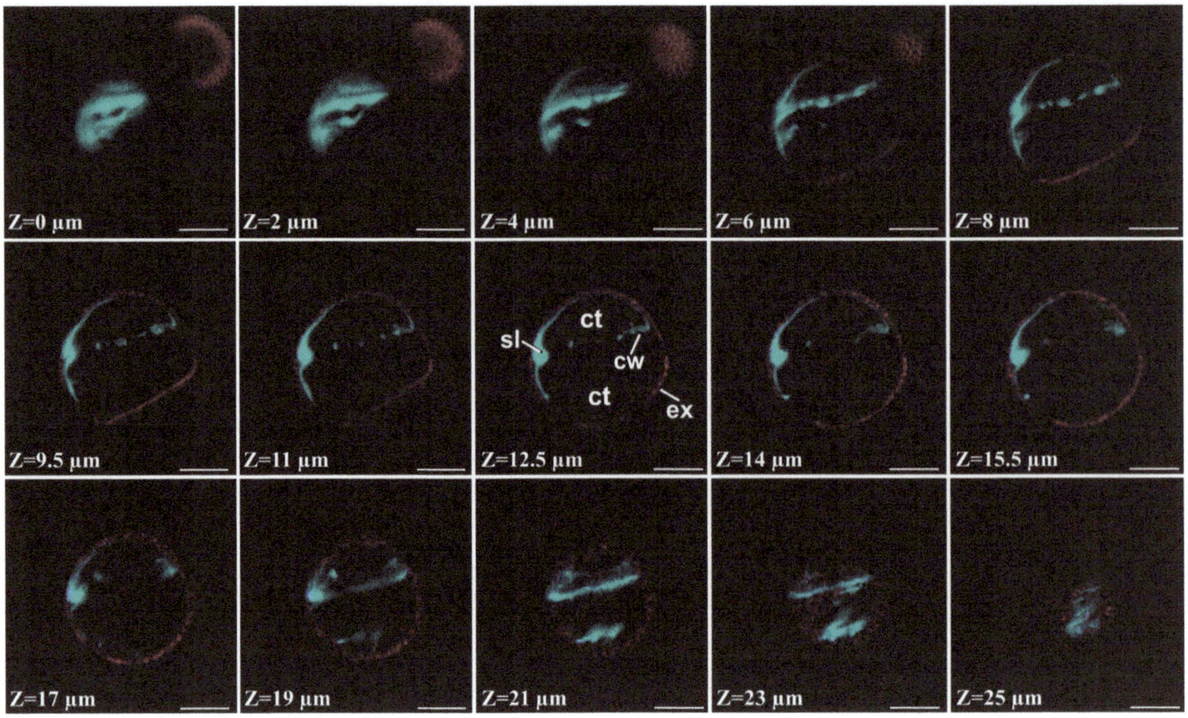

FIGURE 5 | Aniline blue staining of a 3 day-old embryogenic microspore. Confocal Z-series covering an entire micropore where the blue signal corresponds to aniline blue staining and the red signal corresponds to exine (ex) autofluorescence. ct, cytoplasm; cw, cell wall; sl, subintinal layer. Bar: 10 μm.

cultures of all inducible species (Seguí-Simarro and Nuez, 2008a). The remaining 72 microspores (24.9%) showed aniline blue staining, S4B staining, or both. We carefully analyzed these 72 microspores and categorized them into four groups according to their morphology, internal architecture and double staining pattern.

- *Group 1*: 12.5% of the microspores presented a slightly lobulated shape and a size similar to vacuolate microspores (**Figures 7A–D**). No embryogenic divisions were identified in their cytoplasm. No S4B staining was observed in any part of any of these cells. However, intense aniline blue fluorescence was consistently found at the region below the apertures (**Figure 7A**), in some cases extended beyond these regions (**Figure 7B**), and even along the entire subintinal layer (**Figure 7C**). Some of them also presented small stubs penetrating into the cytoplasm (arrows in **Figure 7D**), although a clear cell wall could not be observed in phase contrast images. Together, these microspores with increasing levels of aniline blue staining at their periphery were suggestive of different stages in the formation of the subintinal layer, prior to the first embryogenic division. Such formation would start at the region of the apertures, and would extend centrifugally to eventually cover the whole microspore.

- *Group 2*: 40.3% of the microspores presented the typical features of embryogenic microspores. In addition to the morphological features mentioned above for the first group, we clearly observed in all of them continuous (but variable in thickness) subintinal layers and cell walls, positive for aniline blue staining but negative for S4B staining (**Figure 7E**).

- *Group 3*: 43.1% of the microspores clearly showed pollen-like features, such as enlarged size, oval shape, dense cytoplasm, absence of symmetric divisions, and abundance of starch granules. When observing the S4B staining, these cells revealed another common feature: all of them but one presented a bright and continuous staining at the thickened intine layer (**Figure 7F**). This trait was remarkably different from the embryogenic microspores, devoid of cellulose at any cell wall region. Some of the cellulose-containing, pollen-like structures (15.3% of the total) were also positive for aniline blue staining at the regions below the apertures (data not shown), and only one (1.4% of the total) was positive only for aniline blue. In some double stained pollen-like structures, starch granules seemed clearly delineated by aniline blue staining (**Figure 7F**). To verify this, we isolated starch rich, mature pollen grains from anthers and stained them with aniline blue, confirming that pollen starch granules were also stained by aniline blue (Supplementary Figure S5).

- *Group 4*: 4.2% of the microspores presented abnormal patterns of development, combining features of pollen-like structures such as starch grains and large sizes with features of embryogenic microspores, such as cell divisions and new cell walls (**Figure 7G**). In addition, these cells presented high levels of cytoplasmic vacuolation, exine rupture at different regions of their surface, and asymmetric division patterns. These structures have been previously described as non-embryogenic, arresting, and dying after one or few division rounds (Corral-Martínez et al., 2013). Two microspores of this group (2.8% of the total) showed only cellulose staining, which delineated both the inner

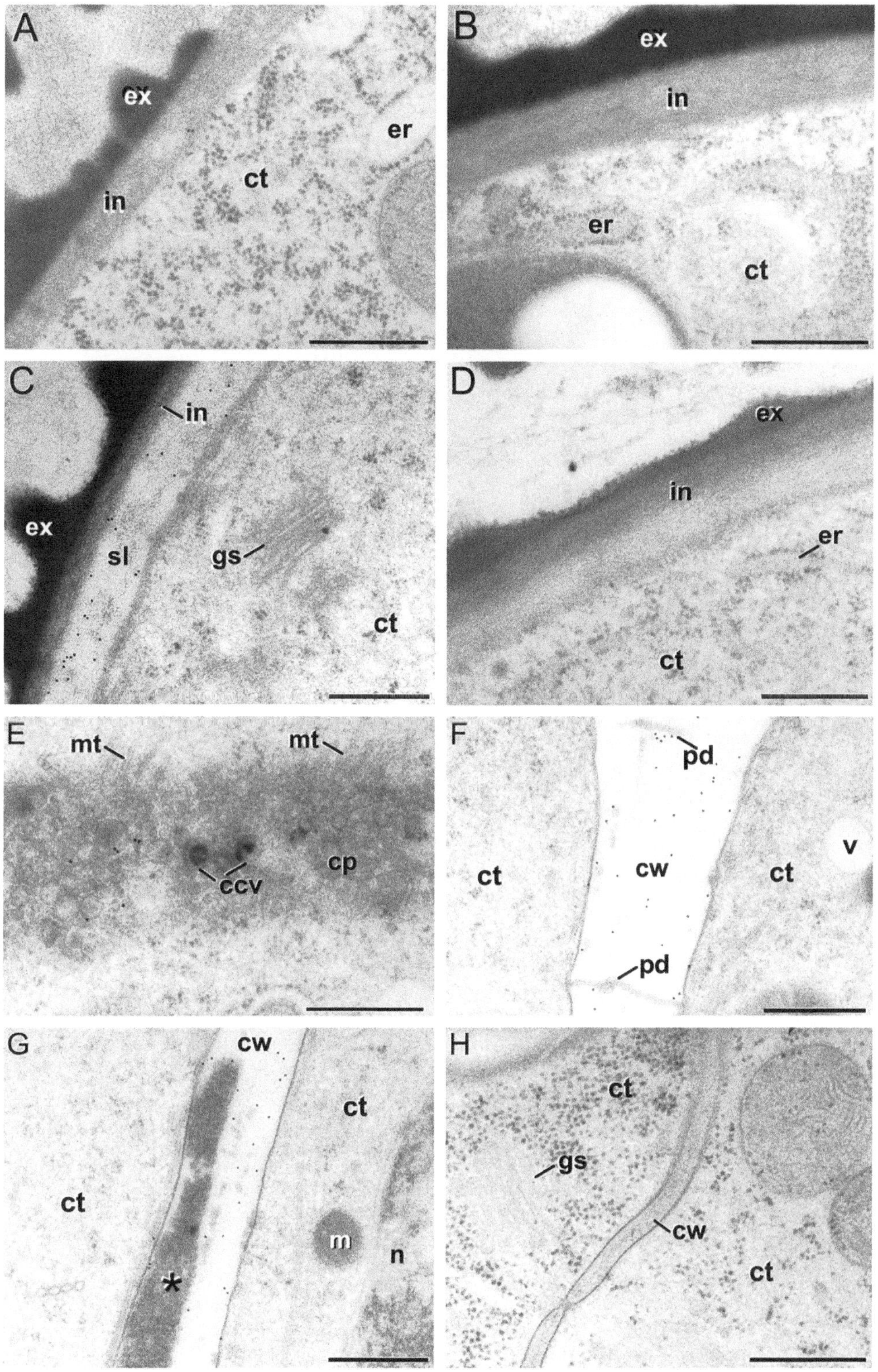

FIGURE 6 | Anti-callose immunogold labeling. (A,B) *In vivo* vacuolate microspore **(A)** and pollen grain **(B)**. Note the absence of a subintinal layer in these cell types. **(C)** Embryogenic microspore with abundant gold particles decorating the subintinal layer (sl). **(D)** Pollen-like structure with a thickened intine without anti-callose gold particles. **(E–G)** Developing cell plate **(E)** and mature cell wall **(F,G)** of embryogenic microspore cells, decorated with anti-callose gold particles. Note in **(G)** that callose labeling is absent from the deposits of excreted material (asterisk). **(H)** MDE cell without anti-callose gold particles in the cell wall. Ccv, clathrin-coated vesicle; ct, cytoplasm; cp, cell plate; cw, cell wall; er, endoplasmic reticulum; ex, exine; gs, golgi stack; in, intine; m, mitochondria; mt, microtubules; n, nucleus; pd, plasmodesma; v, vacuole. Bars: 500 nm.

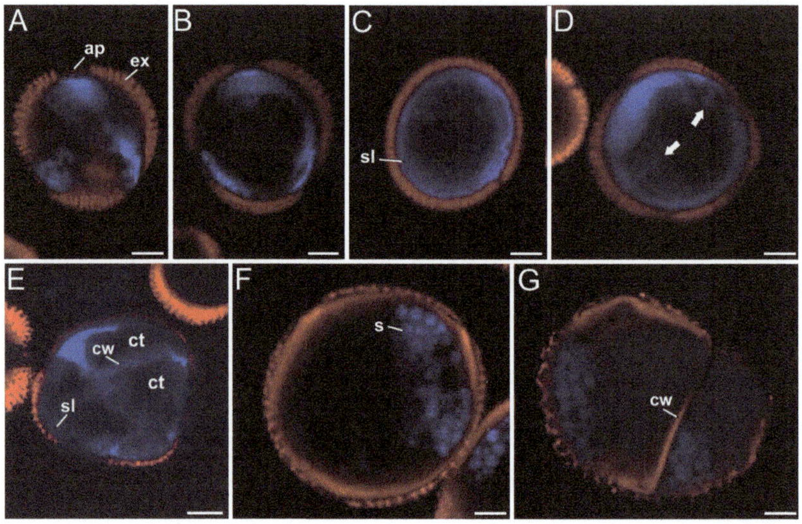

FIGURE 7 | Callose (blue) and cellulose (red) double staining with aniline blue and S4B. (A–D) Unicellular (not yet divided) microspores developing a callose-rich subintinal layer (group 1). **(E)** Three-celled microspore (group 2) with callose-rich subintinal layer (sl) and inner cell walls (cw). Note the absence of red cellulose signal. **(F)** Pollen-like structure (group 3) showing cellulose signal (red) below the exine, but not blue callose signal. **(G)** Microspore with unusual patterns of development (group 4) showing cellulose signal (red) below the exine and in the inner cell wall. The blue spots in **(F–G)** correspond to unspecific staining of starch granules by aniline blue. The dark red signal corresponds to exine (ex) autofluorescence. Bars: 5 μm.

cell walls and the thickened intine (**Figure 7G**), and only one (1.4% of the total) presented both, cellulose and callose signal, mostly at the regions below the apertures.

As for MDEs, in young 4–8 MDEs, cellulose deposition was very scarcely found in the outer walls, being absent from the inner walls (data not shown). Equivalent (quadrant and octant) suspensor-bearing MDEs presented a similar cellulose pattern at the embryo proper domain, whereas suspensor cell walls were clearly delineated by cellulose staining (Supplementary Figure S4B). This pattern, however, changed from globular MDEs onward (Supplementary Figure S4C), where staining consistently decorated all the cell walls of the embryo proper in a pattern similar to a conventional mature cell wall. Together, these results indicated that the first embryogenic divisions of the induced microspore are defined by abnormal patterns of callose and cellulose deposition, establishing a clear correlation between the morphological features that define a microspore as embryogenic, the presence of callose at both the subintinal layer and the cell walls, and the total absence of cellulose. This pattern, however, changes in MDEs, where cellulose is progressively more present in the cell walls of developing MDEs, being abundant at the stages of globular and beyond, as expected for normal somatic-type cell walls.

Altered Ca^{2+} Levels Modify the Patterns of Callose and Cellulose Deposition in the New Cell Walls of Embryogenic Microspores

Once demonstrated that the abundance of callose in the newly formed cell walls is a feature inherent to the developmental switch toward embryogenesis, we wanted to find the cause of such abundance. Due to the role assigned to Ca^{2+} in the regulation of callose deposition and in the inhibition of cellulose synthesis at the cell plate (Hayashi et al., 1987; Verma and Hong, 2001), we applied caffeine and BA to microspore cultures, and observed their effects in embryogenic microspores. Caffeine is known to redistribute the intracellular Ca^{2+} by altering transmembrane calcium gradients, thereby reducing the levels of Ca^{2+} available for callose synthases at the cell plate (Samuels and Staehelin, 1996). As a consequence, Ca^{2+}-deprived callose synthases cannot synthesize callose at normal levels (Kakimoto and Shibaoka, 1992). In turn, BA is a membrane fluidizer that induces major but transient elevations of cytosolic Ca^{2+} (Saidi et al., 2009).

When 1 mM caffeine was added to microspore cultures, callose-containing cell walls were observed between daughter cells, but they were thinner and aniline blue staining was less intense and extense (**Figure 8A**) than in control cells. The subintinal layer was also thinner, discontinuous and irregularly stained by aniline blue. Interestingly, S4B staining revealed the continuous presence of cellulose along the cell walls and subintinal layer of ~50% of the cells (**Figure 8A'**). As opposed to control cultures, 21 day-old caffeine-treated cultures did not produce MDEs. In contrast, 100 μM BA-mediated increase of intracellular Ca^{2+} levels produced a increase in callose deposition, principally in the subintinal layer as revealed by the abundant aniline blue staining (**Figure 8B**). S4B staining, however, evidenced the absence of cellulose in these cells (**Figure 8B'**). Interestingly, 21 days after induction, BA-treated cultures produced ~600 MDEs/ml (Supplementary Figure S6), which represented 240% of the yield of control cultures (~250 MDEs/ml). Together, these results indicated that alteration of Ca^{2+} levels by caffeine and BA affected the amount of callose deposited in the cell walls of embryogenic

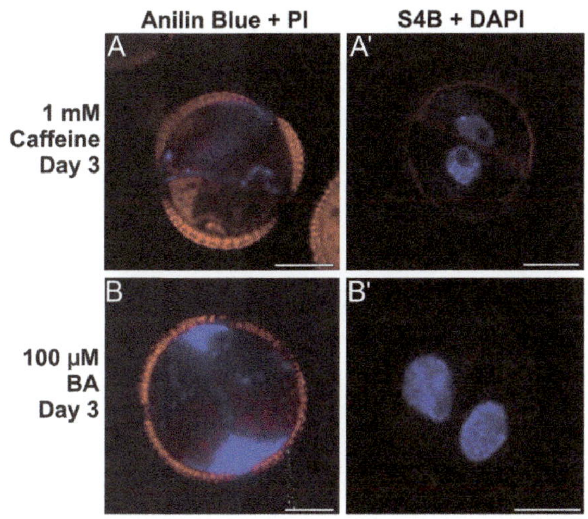

FIGURE 8 | Effect of caffeine and benzyl alcohol (BA) on callose and cellulose deposition. (A,A') Three day-old embryogenic microspores cultured with 1 mM caffeine and stained with aniline blue + PI **(A)** and S4B + DAPI **(A')**. **(B,B')** Three day-old embryogenic microspores cultured with 100 μM BA and stained with aniline blue + PI **(B)** and S4B + DAPI **(B')**. Bars: 10 μm.

microspores. Caffeine-mediated Ca^{2+} redistribution reduced callose deposition and allowed for the synthesis of cellulosic (conventional) walls in some microspores, which suggested that the lack of cellulose was due to a Ca^{2+}-mediated massive deposition of callose. On the other hand, BA-mediated increase of intracellular Ca^{2+} produced more callose in the subintinal layer and, as expected, cellulose-free cell walls. In addition to this, it seemed that the addition of BA was markedly beneficial for microspore embryogenesis, since 2.4x more MDEs were produced.

DISCUSSION

It is clear from this and other works (Solis et al., 2008; Barany et al., 2010; Dubas et al., 2013) that the cell wall is a highly dynamic structure whose architecture, arrangement, and composition changes dramatically as the microspore enters embryogenesis and then transforms into a MDE. We have demonstrated that a direct consequence of the exposure of microspores to the inductive treatment is the production of unusual walls with a unique composition. The microspore intine, described to have a pectocellulosic nature (Sitte, 1953), retained a cellulose-rich composition in pollen-like structures, as previously described for pollen (Heslop-Harrison, 1979). In contrast, the subintinal layer presented a dramatically different composition, indicating that it is not an extension of the intine, but a different structure originated *de novo* upon induction. Interestingly, the two cell walls created *de novo* in embryogenic microspores (inner cell walls and the subintinal layer) had the same appearance and composition, which was different from previously existing walls (the intine), but also from the walls of MDEs. It seems that during the first stages of the switch, the mechanisms of synthesis and/or deposition of cell wall components are altered, giving rise to the transient formation of atypical walls. The origins, implications and consequences of the presence of these walls are discussed next.

The Subintinal Layer is an Early Marker of the Androgenic Switch

The presence of additional layers beneath the coat of embryogenic microspores in different species has been previously reported in the literature, but only occasionally (Rashid et al., 1982; Zaki and Dickinson, 1990; González-Melendi et al., 1995; Solis et al., 2008; Dubas et al., 2013). Perhaps, the reason why this layer has not been studied more in detail relays on the fact that TEM images of chemically fixed embryogenic microspores frequently showed cell retraction and waviness of plasma membrane (Telmer et al., 1993; González-Melendi et al., 1995; Maraschin et al., 2005; Seguí-Simarro et al., 2006), making it difficult to figure out whether the electron light spaces observed are real structures or just retraction artifacts. In addition, it is known that aldehyde fixation may induce callose deposition (Hughes and Gunning, 1980), which may raise doubts about the presence of callose in chemically fixed embryogenic microspores. We demonstrated that the callose-rich subintinal layer of embryogenic microspores is a real entity, not induced by chemical fixation. Thus, it is pertinent to ask whether this layer is a cause or a consequence of the developmental switch.

In plant development, the synthesis of a callose-rich layer is generally related to the developmental fate of the cells that synthesize it. The self-aggregating properties of the helical callose molecules, and the dense, insoluble nature of the highly hydrated but semipermeable callose gels make callose deposits ideal for creating physical and chemical barriers (Albersheim et al., 2011). Indeed, during meiosis callosic walls isolate the developing microspore and megaspore mother cells, acting as a physical barrier against premature swelling and bursting (Zhang et al., 2002), and also as a 'molecular filter' to allow for the expression of their specific developmental programs without the interference of the surrounding diploid tissues (Heslop-Harrison and Mackenzie, 1967; Abramova et al., 2003). Aside of natural processes, callose has been observed in internodal cells of *Humulus lupulus* committed to organogenesis (Fortes et al., 2002), and in cells of *Trifolium* (Maheswaran and Williams, 1985) and *Cichorium* (Dubois et al., 1991) reprogrammed to somatic embryogenesis. The deposition of a callose layer covering *in vitro*-induced young globular embryos was reported during somatic embryogenesis in *Camellia japonica* leaves (Pedroso and Pais, 1992). Most importantly, it was found that when zygotic embryos of *Eleutherococcus senticosus* are exposed to osmotic stressors such as mannitol or sucrose, single epidermal cells develop a surrounding callose layer between the plasma membrane and the cell walls, similar to the subintinal layer we hereby describe, and then enter somatic embryogenesis (You et al., 2006). These authors stated that plasmolysis-induced callose would interrupt cell-to-cell communication, which in turn would stimulate the embryogenic reprogramming. In line with these findings, we

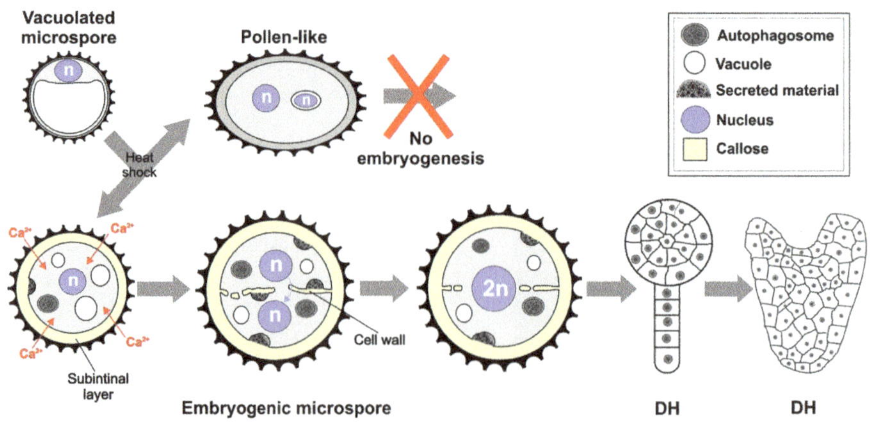

FIGURE 9 | Formation of the callose-rich subintinal layer and abnormal cell walls upon heat shock induction of embryogenesis in *B. napus* microspores. See text for further details.

observed that all the embryogenic microspores developed a callosic subintinal layer, whereas those non-embryogenic, did not develop it. Since both structures were exposed to the inductive heat shock and only the embryogenic developed an additional subintinal layer, we deduce that this layer would not be a direct consequence of the heat shock. It could also be argued that this layer could be related to the callose layer that appears beneath the intine in germinating pollen grains (Ferguson et al., 1998). However, we never found such a layer in the non-embryogenic, pollen-like structures, which accounts against this possibility. Instead, we propose that the subintinal layer is related to the change in developmental fate toward embryogenesis. We even found some microspores that developed the callosic layer even before the first clear evidence of inner cell walls was found in them (**Figures 7A–D**). Thus, we propose that the callosic subintinal layer would be formed in microspores prior to the first embryogenic division, and this would isolate them from the outer environment (the liquid culture media or the cultured anther tissue), creating a suitable environment to start a different developmental program, the embryogenic program. It can also be concluded that the deposition of a callose-rich layer, as a step previous to *in vitro* embryo development, would be a common marker associated to reprogrammed cells during different types of *in vitro* morphogenesis, including somatic and microspore embryogenesis.

The Abnormal Inner Walls of Embryogenic Microspores are Due to Disrupted Cytokinesis

We showed that the cell walls of embryogenic microspores are abnormal in terms of both structure and composition. For us, the most reasonable cause for this is that cytokinesis proceeds normally up to the fenestrated sheet-transitional phragmoplast stage of cytokinesis, and then it is disrupted. Our results are consistent with this hypothesis, since we did not observe any structural alteration up to this stage. Once the cell plate is established at the cell equator, callose synthesis is the driving force that transforms the architecture of the developing cell plate. Callose is an amorphous polymer proposed to provide the fluidity needed to rapidly develop an initial membranous scaffold (Samuels et al., 1995) and to respond to the guidance mechanisms that insert the cell plate at specific sites of the parental wall (Xie et al., 2012). Newly synthesized callose widens the cell plate membranous tubules, as it spreads over the membrane inner surface (Samuels et al., 1995). At the end of this spreading phase, somatic-type cell plates resemble irregular sheets with numerous openings (Samuels et al., 1995; Seguí-Simarro et al., 2004, 2008). This is how the cell walls look like in our micrographs of abnormally walled *B. napus* embryogenic microspores (**Figures 2A,B**). Once the spreading phase is accomplished in somatic-type cytokinesis, the transition from a fluid and wrinkled immature cell plate to a stiff and straight cell wall is associated to deposition of cellulose and pectins and the parallel removal of callose (Kakimoto and Shibaoka, 1992; Samuels et al., 1995; Drakakaki, 2015). However, we showed that callose is not replaced by cellulose in the gapped and incomplete walls of embryogenic microspores. Callose does not form crystalline associations with adjacent strands, as cellulose does (Samuels et al., 1995). Thus, a callose-rich cell wall would in principle be more fluid than a conventional, cellulose-rich cell wall with no callose. In addition, β-1,3-glucans such as callose tend to gelate. It was suggested that the formation of hydrogels by callose would allow insertion or co-gelation with other polymers in the cell plate (Brown and Lemmon, 2009). It is also known that callose gelation can be increased upon heating (Stone and Clarke, 1992). Conceivably, the combination of a mild heat stress, callose accumulation and reduced cellulose contents would confer the newly formed cell walls of embryogenic microspores new properties, different from those of conventional walls. These walls would be more difficult to flatten and close through the conventional mechanisms of cell plate maturation, giving rise to irregular final walls, wrinkled and incomplete, as those shown in **Figures 2A,B**.

All this considered, we propose that the very first division rounds in embryogenic microspores are impaired at the stage of callose replacement by cellulose and as a consequence, the cell walls produced are defective. In support of this, similar phenotypes of abnormal cell plates linked to callose abundance and/or scarce or absent cellulose have also been reported in other experimental systems. For example, in the *Arabidopsis* cytokinesis-defective *cyt1* mutant, the persistence of callose affected cell plate maturation, preventing cellulose synthesis and producing incomplete cell walls (Nickle and Meinke, 1998). Onion root cells treated with the herbicide dichlobenil, that prevents cellulose synthesis, showed reticulate and wavy cell plates with dramatically increased levels of callose (Vaughn et al., 1996). Furthermore, the *Arabidopsis rsw1* null mutant, where the activity of the RSW1/CeSA1 cellulose synthase gene is knocked down, produced thin, highly undulated, and frequently interrupted primary cell walls (Beeckman et al., 2002).

Increased Ca^{2+} Levels may be Responsible for Prolonged Callose Synthase Activity and Cellulose Inhibition

According to our results, it is pertinent to ask why callose persists and cellulose is not deposited in the first cell walls formed in embryogenic microspores. One candidate to mediate in this process is Ca^{2+} signaling, a highly conserved mechanism for temperature sensing among land plants (Horvath et al., 2012). Heat stress, among other treatments, is known to alter the properties of plant cell plasma membranes (reviewed in Horvath et al., 2012). Heat shock tends to fluidize plasma membranes, becoming leaky to ions and thus favoring uncontrolled influxes of Ca^{2+} and rapid, transient and proportional elevations of cytosolic Ca^{2+} in moss cells (Saidi et al., 2009), higher plants (Gong et al., 1998), and in *B. napus* embryogenic microspores (Pauls et al., 2006). In addition, high levels of intracellular Ca^{2+} are required for the activity of cell plate-specific, Ca^{2+}-dependent callose synthases (reviewed in Verma, 2001; Drakakaki, 2015). Related to this, our experiments with caffeine and BA suggested a possible relationship between the intracellular Ca^{2+} levels and the amount of callose deposited, which shows that the synthases involved in the development of the new cell wall and the subintinal layer are Ca^{2+}-sensitive. In somatic-type cell walls, high Ca^{2+} levels are known to cause the shift toward callose production, whereas Ca^{2+} levels must be reduced for cellulose synthesis to initiate during cell plate maturation. Indeed, prolonged high Ca^{2+} levels may extend callose synthesis in time and inhibit the shift from callose to cellulose synthesis (Verma, 2001). Our hypothesis is that a similar scenario might be occurring during the formation of the first cell walls in *B. napus* embryogenic microspores. Our experiments showed that caffeine decreased the presence of callose in both the subintinal layer and the inner cell walls, probably due to a reduction of the Ca^{2+} available for callose synthases. As a consequence, cellulose could be produced. In turn, the use of BA produced even more callose in the subintinal layer, probably due to an increase in the Ca^{2+} levels. In summary, we postulate that heat shock would alter the plasma membrane properties to allow for a Ca^{2+} influx that, among other effects (alteration of phragmoplast integrity, for example), would keep callose synthases active and, at the same time, would inhibit cellulose synthases. In this context, it is interesting to note that the effect of BA added to the effect of heat shock not only in terms of Ca^{2+} influx and callose synthesis, but also of MDE yield. Indeed, several lines of evidence point to the redundancies of the cellular responses triggered by temperature-induced Ca^{2+} signaling and BA-mediated Ca^{2+} influx (Sangwan et al., 2001; Saidi et al., 2009; Horvath et al., 2012). Thus, it is tempting to speculate that BA may enhance or replace the effect of heat shock in cell wall remodeling and perhaps, in the activation of the whole embryogenic program in *B. napus* microspores.

Callose-rich and Cellulose-deficient Walls may Promote Genome Doubling

Cytokinetic defects typically induce the eventual formation of polyploid cells. Incomplete cell walls with large holes allow for the contact between the nuclei of daughter cells. Even those with smaller gaps may end up producing larger holes, since they are prone to break and collapse into pieces. These abnormal walls are known to facilitate the coexistence of nuclei of different cells in the same cytoplasm (Sunderland and Dunwell, 1974) and their eventual fusion, as demonstrated to occur in many other *in vitro* culture systems and cytokinesis-defective mutants (Mayer et al., 1999; Kasha et al., 2001; Muller et al., 2002; Strompen et al., 2002; Testillano et al., 2004; González-Melendi et al., 2005; Seguí-Simarro and Nuez, 2007; Corral-Martínez et al., 2011; De Storme et al., 2013). Our results showed that the instability and fluidity of callose-rich and cellulose-deficient cell walls, together with the severe cell wall deformations produced by the accumulation of excreted cytoplasmic material (Corral-Martínez et al., 2013), are behind the occurrence of incomplete, broken, or fenestrated cell walls. Hence in some embryogenic microspores, impairment of cytokinesis mediated by excessive callose deposition would generate the structural context necessary to facilitate nuclear fusion and therefore genome duplication, a necessary step widely accepted to produce DH cells, embryos and plants. On the other hand, cells where callose deposition is not so persistent would develop normal walls, which, according to this hypothesis, would preclude nuclear fusion during these initial stages, giving rise to haploid embryos. In turn, it is reasonable to assume that there will also be few embryogenic microspores whose initial phase of defective cytokinesis would last longer, being their cells more prone to undergo more than one round of nuclear fusions. It may also be possible that in a single embryogenic microspore, normally dividing cells coexist with others undergoing defective cytokinesis, giving rise to mixoploid and/or polyploid MDEs, known to occur in *B. napus* (Abdollahi et al., 2012) as well as in other species (Sato et al., 2005).

CONCLUSION

We showed in this work that one of the first signs of embryogenic commitment is the formation of callose-rich, cellulose-deficient subintinal layers and irregular, gapped, and incomplete cell walls (**Figure 9**), both of them absent in pollen-like structures. This is due to the combination of three factors: the formation of large deposits of excreted cytoplasmic material described in Corral-Martínez et al. (2013), the massive deposition of callose, and the inhibition of cellulose synthesis. Callose would be produced by Ca^{2+}-dependent callose synthases activated by a raise in intracellular Ca^{2+} levels, which is a direct consequence of the increased permeability of the plasma membrane as a result of the inductive heat shock. Few days after the heat shock, abnormal cell walls are no longer produced, and MDEs present conventional, somatic-type cell walls.

These findings provide new insights that contribute to understand this intriguing, yet largely unknown developmental switch. In turn, our results pose new, open questions. One of them is why pollen-like structures, exposed to the same physico-chemical environment, do not produce callose-rich cell walls. A possible explanation is that these structures may have irreversibly entered the gametophytic-like pathway before the application of the inductive treatment, and this may somehow make them insensitive to the effects of heat shock exposure in terms of induction of embryogenesis. Another question is how and when the cell wall-associated phenomena we described cease, and conventional cell walls are formed in MDEs. It is possible that once the cells return to normal conditions, newly formed conventional walls coexist with the few preexisting callose-rich walls, as we have observed in some globular MDEs. After 1000s of conventional cytokinesis, the relevance of callosic walls would become negligible. However, we cannot rule out the possibility of callose being eliminated after the heat shock by specific 1,3-β glucanases. In favor of this hypothesis, Borderies et al. (2004) correlated the development of maize MDEs with the secretion of several proteins to the culture medium, one of them being a 1,3-β glucanase. It would be interesting to know whether *B. napus* MDEs produce 1,3-β glucanases as well. Finally, it would also be interesting to know whether the process we hereby describe or similar phenomena are triggered in other species where microspore embryogenesis is induced by other stresses different from temperature changes.

AUTHOR CONTRIBUTIONS

JMSS designed the research; VPV, PCM, ARS, and JMSS obtained and processed the samples. VPV, ARS, and PCM performed the experiments. JMSS wrote the manuscript.

ACKNOWLEDGMENTS

We want to express our thanks to the staff of the Electron Microscopy Service of Universitat Politècnica de València. Thanks are also due to Dr. Kim Boutilier (PRI Wageningen, The Netherlands) for her help during the stays of VPV and ARS at her lab, to Dr. Samantha Vernhettes (INRA Versailles, France) for her kind gift of S4B staining, and especially to Prof. L. A. Staehelin for his help and friendship during the stay of JMSS at UC Boulder. This work was supported by grants BEST/2008/154 from Generalitat Valenciana and AGL2014-55177-R from Spanish MINECO to JMSS.

REFERENCES

Abdollahi, M. R., Ghazanfari, P., Corral-Martínez, P., Moieni, A., and Seguí-Simarro, J. M. (2012). Enhancing secondary embryogenesis in *Brassica napus* by selecting hypocotyl-derived embryos and using plant-derived smoke extract in culture medium. *Plant Cell Tissue Organ Cult.* 110, 307–315. doi: 10.1007/s11240-012-0152-7

Abramova, L. I., Avalkina, N. A., Golubeva, E. A., Pyzhenkova, Z. S., and Golubovskaya, I. N. (2003). Synthesis and deposition of callose in anthers and ovules of meiotic mutants of maize (*Zea mays*). *Russ. J. Plant Physiol.* 50, 324–329. doi: 10.1023/a:1023866019102

Albersheim, P., Darvill, A., Roberts, K., Sederoff, R., and Staehelin, A. (2011). *Plant Cell Walls.* New York, NY: Garland Science.

Anderson, C. T., Carroll, A., Akhmetova, L., and Somerville, C. (2010). Real-time imaging of cellulose reorientation during cell wall expansion in *Arabidopsis* roots. *Plant Physiol.* 152, 787–796. doi: 10.1104/pp.109.150128

Austin, J. R., Seguí-Simarro, J. M., and Staehelin, L. A. (2005). Quantitative analysis of changes in spatial distribution and plus-end geometry of microtubules involved in plant-cell cytokinesis. *J. Cell Sci.* 118, 3895–3903. doi: 10.1242/jcs.02512

Barany, I., Fadon, B., Risueno, M. C., and Testillano, P. S. (2010). Cell wall components and pectin esterification levels as markers of proliferation and differentiation events during pollen development and pollen embryogenesis in *Capsicum annuum* L. *J. Exp. Bot.* 61, 1159–1175. doi: 10.1093/jxb/erp392

Beeckman, T., Przemeck, G. K. H., Stamatiou, G., Lau, R., Terryn, N., De Rycke, R., et al. (2002). Genetic complexity of cellulose synthase a gene function in *Arabidopsis* embryogenesis. *Plant Physiol.* 130, 1883–1893. doi: 10.1104/pp.102.010603

Borderies, G., Le Bechec, M., Rossignol, M., Lafitte, C., Le Deunff, E., Beckert, M., et al. (2004). Characterization of proteins secreted during maize microspore culture: arabinogalactan proteins (AGPs) stimulate embryo development. *Eur. J. Cell Biol.* 83, 205–212. doi: 10.1078/0171-9335-00378

Brown, R. C., and Lemmon, B. E. (1991). Pollen development in orchids. 3. A novel generative pole microtubule system predicts unequal pollen mitosis. *J. Cell Sci.* 99, 273–281.

Brown, R. C., and Lemmon, B. E. (2009). "Callose in cell division," in *Chemistry, Biochemistry, and Biology of (1-3)-b-Glucans and Related Polysaccharides*, eds A. Bacic, G. B. Fincher, and B. A. Stone (Amsterdam: Elsevier Science), 425–438.

Carpita, N. C., and McCann, M. C. (2000). "Chapter 2: the cell wall," in *Biochemistry and Molecular Biology of Plants*, eds B. B. Buchanan, W. Gruissem, and R. L. Jones (Rockville, MD: American Society of Plant Physiologists).

Chen, X.-Y., and Kim, J.-Y. (2009). Callose synthesis in higher plants. *Plant Signal. Behav.* 4, 1–4.

Corral-Martínez, P., Nuez, F., and Seguí-Simarro, J. M. (2011). Genetic, quantitative and microscopic evidence for fusion of haploid nuclei and growth of somatic calli in cultured ms1035 tomato anthers. *Euphytica* 178, 215–228. doi: 10.1007/s10681-010-0303-z

Corral-Martínez, P., Parra-Vega, V., and Seguí-Simarro, J. M. (2013). Novel features of *Brassica napus* embryogenic microspores revealed by high pressure freezing and freeze substitution: evidence for massive autophagy and excretion-based cytoplasmic cleaning. *J. Exp. Bot.* 64, 3061–3075. doi: 10.1093/jxb/ert151

Custers, J. (2003). "Microspore culture in rapeseed (*Brassica napus* L.)," in *Doubled Haploid Production in Crop Plants*, eds M. Maluszynski, K. J. Kasha, B. P. Forster, and I. Szarejko (Dordrecht: Kluwer Academic Publishers), 185–193.

Custers, J. B. M., Cordewener, J. H. G., Nöllen, Y., Dons, J. J., and Van Lookeren-Campagne, M. M. (1994). Temperature controls both gametophytic and sporophytic development in microspore cultures of *Brassica napus*. *Plant Cell Rep.* 13, 267–271. doi: 10.1007/BF00233317

De Storme, N., De Schrijver, J., Van Criekinge, W., Wewer, V., Dörmann, P., and Geelen, D. (2013). GLUCAN SYNTHASE-LIKE8 and STEROL METHYLTRANSFERASE2 are required for ploidy consistency of the sexual reproduction system in *Arabidopsis*. *Plant Cell* 25, 387–403. doi: 10.1105/tpc.112.106278

Dong, X., Hong, Z., Sivaramakrishnan, M., Mahfouz, M., and Verma, D. P. S. (2005). Callose synthase (CalS5) is required for exine formation during microgametogenesis and for pollen viability in *Arabidopsis*. *Plant J.* 42, 315–328. doi: 10.1111/j.1365-313X.2005.02379.x

Drakakaki, G. (2015). Polysaccharide deposition during cytokinesis: challenges and future perspectives. *Plant Sci.* 236, 177–184. doi: 10.1016/j.plantsci.2015.03.018

Dubas, E., Custers, J., Kieft, H., Wedzony, M., and Van Lammeren, A. A. (2013). Characterization of polarity development through 2- and 3-D imaging during the initial phase of microspore embryogenesis in *Brassica napus* L. *Protoplasma* 251, 103–113. doi: 10.1007/s00709-013-0530-y

Dubois, T., Guedira, M., Dubois, J., and Vasseur, J. (1991). Direct somatic embryogenesis in leaves of *Cichorium*: a histological and SEM study of early stages. *Protoplasma* 162, 120–127. doi: 10.1007/BF02562555

El-Tantawy, A.-A., Solís, M.-T., Costa, M. L., Coimbra, S., Risueño, M.-C., and Testillano, P. S. (2013). Arabinogalactan protein profiles and distribution patterns during microspore embryogenesis and pollen development in *Brassica napus*. *Plant Reprod.* 26, 231–243. doi: 10.1007/s00497-013-0217-8

Evans, N. A., Hoyne, P. A., and Stone, B. A. (1984). Characteristics and specificity of the interaction of a fluorochrome from aniline blue (Sirofluor) with polysaccharides. *Carbohydr. Polym.* 4, 215–230. doi: 10.1016/0144-8617(84)90012-2

Ferguson, C., Teeri, T. T., Siika-Aho, M., Read, S. M., and Bacic, A. (1998). Location of cellulose and callose in pollen tubes and grains of *Nicotiana tabacum*. *Planta* 206, 452–460. doi: 10.1007/s004250050421

Forster, B. P., Heberle-Bors, E., Kasha, K. J., and Touraev, A. (2007). The resurgence of haploids in higher plants. *Trends Plant Sci.* 12, 368–375. doi: 10.1016/j.tplants.2007.06.007

Fortes, A. M., Testillano, P. S., Risueno, M. D., and Pais, M. S. (2002). Studies on callose and cutin during the expression of competence and determination for organogenic nodule formation from internodes of *Humulus lupulus* var. Nugget. *Physiol. Plant.* 116, 113–120. doi: 10.1034/j.1399-3054.2002.1160114.x

Gilkey, J. C., and Staehelin, L. A. (1986). Advances in ultrarapid freezing for the preservation of cellular ultrastructure. *J. Electron Microsc. Tech.* 3, 177–210. doi: 10.1002/jemt.1060030206

Gong, M., Van Der Luit, A. H., Knight, M. R., and Trewavas, A. J. (1998). Heat-shock-induced changes in intracellular Ca2+ level in tobacco seedlings in relation to thermotolerance. *Plant Physiol.* 116, 429–437. doi: 10.1104/pp.116.1.429

González-Melendi, P., Ramírez, C., Testillano, P. S., Kumlehn, J., and Risueño, M. C. (2005). Three dimensional confocal and electron microscopy imaging define the dynamics and mechanisms of diploidisation at early stages of barley microspore-derived embryogenesis. *Planta* 222, 47–57. doi: 10.1007/s00425-005-1515-7

González-Melendi, P., Testillano, P. S., Ahmadian, P., Fadón, B., Vicente, O., and Risueño, M. C. (1995). In situ characterization of the late vacuolate microspore as a convenient stage to induce embryogenesis in *Capsicum*. *Protoplasma* 187, 60–71. doi: 10.1007/BF01280233

Gorska-Brylass, A. (1967). Transitory callose envelope surrounding the generative cell in pollen grains. *Acta Soc. Bot. Pol.* 36, 419–422. doi: 10.5586/asbp.1967.038

Hayashi, T., Read, S. M., Bussell, J., Thelen, M., Lin, F. C., Brown, R. M., et al. (1987). UDP-glucose: (1-- > 3)-beta-glucan synthases from mung bean and cotton: differential effects of Ca and Mg on enzyme properties and on macromolecular structure of the glucan product. *Plant Physiol.* 83, 1054–1062. doi: 10.1104/pp.83.4.1054

Heslop-Harrison, J. (1968). Synchronous pollen mitosis and the formation of the generative cell in massulate orchids. *J. Cell Sci.* 3, 457–466.

Heslop-Harrison, J. (1979). Aspects of the structure, cytochemistry and germination of the pollen of rye (*Secale cereale* L.). *Ann. Bot.* 44(Suppl. 1), 1–47.

Heslop-Harrison, J., and Mackenzie, A. (1967). Autoradiography of soluble [2-I4C] thymidine derivatives during meiosis and microsporogenesis in *Lilium* anthers. *J. Cell Sci.* 2, 387–400.

Horvath, I., Glatz, A., Nakamoto, H., Mishkind, M. L., Munnik, T., Saidi, Y., et al. (2012). Heat shock response in photosynthetic organisms: membrane and lipid connections. *Prog. Lipid Res.* 51, 208–220. doi: 10.1016/j.plipres.2012.02.002

Hughes, J. E., and Gunning, B. E. S. (1980). Glutaraldehyde-induced deposition of callose. *Can. J. Bot.* 58, 250–258. doi: 10.1139/b80-025

Kakimoto, T., and Shibaoka, H. (1992). Synthesis of polysaccharides in phragmoplasts isolated from tobacco BY-2 cells. *Plant Cell Physiol.* 33, 353–361.

Kasha, K. J., Hu, T. C., Oro, R., Simion, E., and Shim, Y. S. (2001). Nuclear fusion leads to chromosome doubling during mannitol pretreatment of barley (*Hordeum vulgare* L.) microspores. *J. Exp. Bot.* 52, 1227–1238. doi: 10.1093/jexbot/52.359.1227

Levy, A., Erlanger, M., Rosenthal, M., and Epel, B. L. (2007). A plasmodesmata-associated beta-1,3-glucanase in *Arabidopsis*. *Plant J.* 49, 669–682. doi: 10.1111/j.1365-313X.2006.02986.x

Lichter, R. (1982). Induction of haploid plants from isolated pollen of *Brassica napus*. *Z. Pflanzenphysiol.* 105, 427–434. doi: 10.1016/S0044-328X(82)80040-8

Maheswaran, G., and Williams, E. G. (1985). Origin and development of somatic embryoids formed directly on immature embryos of *Trifolium repens* in vitro. *Ann. Bot.* 56, 619–630.

Malik, M. R., Wang, F., Dirpaul, J. M., Zhou, N., Polowick, P. L., Ferrie, A. M. R., et al. (2007). Transcript profiling and identification of molecular markers for early microspore embryogenesis in *Brassica napus*. *Plant Physiol.* 144, 134–154. doi: 10.1104/pp.106.092932

Maraschin, S. D., Vennik, M., Lamers, G. E. M., Spaink, H. P., and Wang, M. (2005). Time-lapse tracking of barley androgenesis reveals position-determined cell death within pro-embryos. *Planta* 220, 531–540. doi: 10.1007/s00425-004-1371-x

Mayer, U., Herzog, U., Berger, F., Inze, D., and Jürgens, G. (1999). Mutations in the pilz group genes disrupt the microtubule cytoskeleton and uncouple cell cycle progression from cell division in *Arabidopsis* embryo and endosperm. *Eur. J. Cell Biol.* 78, 100–108. doi: 10.1016/S0171-9335(99)80011-9

Meikle, P. J., Bonig, I., Hoogenraad, N. J., Clarke, A. E., and Stone, B. A. (1991). The location of (1→3)-b-glucans in the walls of pollen tubes of *Nicotiana alata* using a (1→3)-b-glucan-specific monoclonal antibody. *Planta* 185, 1–8. doi: 10.1007/BF00194507

Muller, S., Fuchs, E., Ovecka, M., Wysocka-Diller, J., Benfey, P. N., and Hauser, M. T. (2002). Two new loci, PLEIADE and HYADE, implicate organ-specific regulation of cytokinesis in *Arabidopsis*. *Plant Physiol.* 130, 312–324. doi: 10.1104/pp.004416

Nickle, T. C., and Meinke, D. W. (1998). A cytokinesis-defective mutant of *Arabidopsis* (cyt1) characterized by embryonic lethality, incomplete cell walls, and excessive callose accumulation. *Plant J.* 15, 321–332. doi: 10.1046/j.1365-313X.1998.00212.x

Otegui, M. S., and Staehelin, L. A. (2000). Cytokinesis in flowering plants: more than one way to divide a cell. *Curr. Opin. Plant Biol.* 3, 493–502. doi: 10.1016/S1369-5266(00)00119-9

Park, S. K., and Twell, D. (2001). Novel patterns of ectopic cell plate growth and lipid body distribution in the *Arabidopsis* gemini pollen1 mutant. *Plant Physiol.* 126, 899–909. doi: 10.1104/pp.126.2.899

Pauls, K. P., Chan, J., Woronuk, G., Schulze, D., and Brazolot, J. (2006). When microspores decide to become embryos – cellular and molecular changes. *Can. J. Bot.* 84, 668–678.

Pedroso, M. C., and Pais, M. S. (1992). A scanning electron-microscopy and X-ray-microanalysis study during induction of morphogenesis in *Camellia japonica* L. *Plant Sci.* 87, 99–108. doi: 10.1016/0168-9452(92)90197-t

Pickett-Heaps, J. D., and Northcote, D. H. (1966). Organization of microtubules and endoplasmic reticulum during mitosis and cytokinesis in wheat meristems. *J. Cell Sci.* 1, 109–120.

Rashid, A., Siddiqui, A. W., and Reinert, J. (1982). Subcellular aspects of origin and structure of pollen embryos of *Nicotiana*. *Protoplasma* 113, 202–208. doi: 10.1007/bf01280908

Saidi, Y., Finka, A., Muriset, M., Bromberg, Z., Weiss, Y. G., Maathuis, F. J. M., et al. (2009). The heat shock response in moss plants is regulated by specific calcium-permeable channels in the plasma membrane. *Plant Cell* 21, 2829–2843. doi: 10.1105/tpc.108.065318

Samuels, A. L., Giddings, T. H. Jr., and Staehelin, L. A. (1995). Cytokinesis in tobacco BY-2 and root tip cells: a new model of cell plate formation in higher plants. *J. Cell Biol.* 130, 1345–1357. doi: 10.1083/jcb.130.6.1345

Samuels, A. L., and Staehelin, L. A. (1996). Caffeine inhibits cell plate formation by disrupting membrane reorganization just after the vesicle fusion step. *Protoplasma* 195, 144–155. doi: 10.1007/BF01279193

Sangwan, V., Foulds, I., Singh, J., and Dhindsa, R. S. (2001). Cold-activation of *Brassica napus* BN115 promoter is mediated by structural changes in membranes and cytoskeleton, and requires Ca2+ influx. *Plant J.* 27, 1–12. doi: 10.1046/j.1365-313x.2001.01052.x

Sato, S., Katoh, N., Iwai, S., and Hagimori, M. (2005). Frequency of spontaneous polyploidization of embryos regenerated from cultured anthers or microspores of *Brassica rapa* var. *pekinensis* L. and *B. oleracea* var. *capitata* L. *Breed. Sci.* 55, 99–102. doi: 10.1270/jsbbs.55.99

Schulze, D., and Pauls, K. P. (2002). Flow cytometric analysis of cellulose tracks development of embryogenic *Brassica* cells in microspore cultures. *New Phytol.* 154, 249–254. doi: 10.1046/j.1469-8137.2002.00350.x

Seguí-Simarro, J. M., Austin, J. R., White, E. A., and Staehelin, L. A. (2004). Electron tomographic analysis of somatic cell plate formation in meristematic cells of *Arabidopsis* preserved by high-pressure freezing. *Plant Cell* 16, 836–856. doi: 10.1105/tpc.017749

Seguí-Simarro, J. M., Bárány, I., Suárez, R., Fadón, B., Testillano, P. S., and Risueño, M. C. (2006). Nuclear bodies domain changes with microspore reprogramming to embryogenesis. *Eur. J. Histochem.* 50, 35–44.

Seguí-Simarro, J. M., and Nuez, F. (2007). Embryogenesis induction, callogenesis, and plant regeneration by in vitro culture of tomato isolated microspores and whole anthers. *J. Exp. Bot.* 58, 1119–1132. doi: 10.1093/jxb/erl271

Seguí-Simarro, J. M., and Nuez, F. (2008a). How microspores transform into haploid embryos: changes associated with embryogenesis induction and microspore-derived embryogenesis. *Physiol. Plant.* 134, 1–12. doi: 10.1111/j.1399-3054.2008.01113.x

Seguí-Simarro, J. M., and Nuez, F. (2008b). Pathways to doubled haploidy: chromosome doubling during androgenesis. *Cytogenet. Genome Res.* 120, 358–369. doi: 10.1159/000121085

Seguí-Simarro, J. M., Otegui, M. S., Austin, J. R., and Staehelin, L. A. (2008). "Plant cytokinesis – insights gained from electron tomography studies," in *Cell Division Control in Plants*, eds D. P. S. Verma and Z. Hong (Berlin: Springer), 251–287.

Seguí-Simarro, J. M., Testillano, P. S., and Risueño, M. C. (2003). Hsp70 and Hsp90 change their expression and subcellular localization after microspore embryogenesis induction in *Brassica napus* L. cv Topas. *J. Struct. Biol.* 142, 379–391. doi: 10.1016/S1047-8477(03)00067-4

Shariatpanahi, M. E., Bal, U., Heberle-Bors, E., and Touraev, A. (2006). Stresses applied for the re-programming of plant microspores towards in vitro embryogenesis. *Physiol. Plant.* 127, 519–534. doi: 10.1111/j.1399-3054.2006.00675.x

Shi, X., Sun, X., Zhang, Z., Feng, D., Zhang, Q., Han, L., et al. (2014). GLUCAN SYNTHASE-LIKE 5 (GSL5) plays an essential role in male fertility by regulating callose metabolism during microsporogenesis in rice. *Plant Cell Physiol.* 56, 497–509. doi: 10.1093/pcp/pcu193

Simmonds, D. H., and Keller, W. A. (1999). Significance of preprophase bands of microtubules in the induction of microspore embryogenesis of *Brassica napus*. *Planta* 208, 383–391. doi: 10.1007/s004250050573

Sitte, P. (1953). Untersuchungen zur submikroskopischen morphologie der pollen und sporoenmembranen. *Mikroskopie* 8, 290–299.

Solis, M.-T., Pintos, B., Prado, M.-J., Bueno, M.-A., Raska, I., Risueno, M.-C., et al. (2008). Early markers of in vitro microspore reprogramming to embryogenesis in olive (*Olea europaea* L.). *Plant Sci.* 174, 597–605. doi: 10.1016/j.plantsci.2008.03.014

Stone, B. A., and Clarke, A. E. (1992). "Chemistry and physiology of higher plant 1,3-β-glucans (callose)," in *Chemistry and Biology of (1-3)-β-Glucans*, eds B. A. Stone and A. E. Clarke (Bundoora, VIC: La Trobe University Press), 365–429.

Strompen, G., El Kasmi, F., Richter, S., Lukowitz, W., Assaad, F. F., Jürgens, G., et al. (2002). The *Arabidopsis* HINKEL gene encodes a kinesin-related protein involved in cytokinesis and is expressed in a cell cycle-dependent manner. *Curr. Biol.* 12, 153–158. doi: 10.1016/S0960-9822(01)00655-8

Sunderland, N., and Dunwell, J. M. (1974). "Anther and pollen culture," in *Plant Tissue and Cell Culture*, ed. H. E. Street (Oxford: Blackwell Scientific Publications), 223–265.

Suzuki, T., Fujikura, K., Higashiyama, T., and Takata, K. (1997). DNA staining for fluorescence and laser confocal microscopy. *J. Histochem. Cytochem.* 45, 49–53. doi: 10.1177/002215549704500107

Telmer, C. A., Newcomb, W., and Simmonds, D. H. (1993). Microspore development in *Brassica napus* and the effect of high temperature on division in vivo and in vitro. *Protoplasma* 172, 154–165. doi: 10.1007/BF01379373

Telmer, C. A., Newcomb, W., and Simmonds, D. H. (1995). Cellular changes during heat shock induction and embryo development of cultured microspores of *Brassica napus* cv. Topas. *Protoplasma* 185, 106–112. doi: 10.1007/BF01272758

Testillano, P. S., Georgiev, S., Mogensen, H. L., Coronado, M. J., Dumas, C., Risueno, M. C., et al. (2004). Spontaneous chromosome doubling results from nuclear fusion during in vitro maize induced microspore embryogenesis. *Chromosoma* 112, 342–349. doi: 10.1007/s00412-004-0279-3

Thomas, J., Ingerfeld, M., Nair, H., Chauhan, S. S., and Collings, D. A. (2013). Pontamine fast scarlet 4B: a new fluorescent dye for visualising cell wall organisation in radiata pine tracheids. *Wood Sci. Technol.* 47, 59–75. doi: 10.1007/s00226-012-0483-x

Van Lammeren, A. A. M., Keijzer, C. J., Willemse, M. T., and Kieft, H. (1985). Structure and function of the microtubular cytoskeleton during pollen development in *Gasteria verrucosa* (Mill) H. Duval. *Planta* 165, 1–11. doi: 10.1007/BF00392205

Vaughn, K. C., Hoffman, J. C., Hahn, M. G., and Staehelin, L. A. (1996). The herbicide dichlobenil disrupts cell plate formation: immunogold characterization. *Protoplasma* 194, 117–132. doi: 10.1007/bf01882020

Verma, D. P. S. (2001). Cytokinesis and building of the cell plate in plants. *Annu. Rev. Plant Physiol. Plant Mol. Biol.* 52, 751–784. doi: 10.1146/annurev.arplant.52.1.751

Verma, D. P. S., and Hong, Z. (2001). Plant callose synthase complexes. *Plant Mol. Biol.* 47, 693–701. doi: 10.1023/A:1013679111111

Worden, N., Park, E., and Drakakaki, G. (2012). Trans-Golgi network: an intersection of trafficking cell wall components. *J. Integr. Plant Biol.* 54, 875–886. doi: 10.1111/j.1744-7909.2012.01179.x

Xie, B., Deng, Y., Kanaoka, M. M., Okada, K., and Hong, Z. (2012). Expression of *Arabidopsis* callose synthase 5 results in callose accumulation and cell wall permeability alteration. *Plant Sci.* 183, 1–8. doi: 10.1016/j.plantsci.2011.10.015

You, X. L., Yi, J. S., and Choi, Y. E. (2006). Cellular change and callose accumulation in zygotic embryos of *Eleutherococcus senticosus* caused by plasmolyzing pretreatment result in high frequency of single-cell-derived somatic embryogenesis. *Protoplasma* 227, 105–112. doi: 10.1007/s00709-006-0149-3

Zaki, M. A., and Dickinson, H. G. (1990). Structural changes during the first divisions of embryos resulting from anther and free microspore culture in *Brassica napus*. *Protoplasma* 156, 149–162. doi: 10.1007/BF01560653

Zaki, M. A. M., and Dickinson, H. G. (1991). Microspore-derived embryos in *Brassica*: the significance of division symmetry in pollen mitosis I to embryogenic development. *Sex. Plant Reprod.* 4, 48–55. doi: 10.1007/BF00194572

Zhang, C., Guinel, F. C., and Moffatt, B. A. (2002). A comparative ultrastructural study of pollen development in *Arabidopsis thaliana* ecotype Columbia and male-sterile mutant apt1-3. *Protoplasma* 219, 59–71. doi: 10.1007/s007090200006

Zuo, J. R., Niu, Q. W., Nishizawa, N., Wu, Y., Kost, B., and Chua, N. H. (2000). KORRIGAN, an *Arabidopsis* endo-1,4-beta-glucanase, localizes to the cell plate by polarized targeting and is essential for cytokinesis. *Plant Cell* 12, 1137–1152. doi: 10.2307/3871261

Conflict of Interest Statement: The authors declare that the research was conducted in the absence of any commercial or financial relationships that could be construed as a potential conflict of interest.

Permissions

All chapters in this book were first published in FPLS, by Frontiers; hereby published with permission under the Creative Commons Attribution License or equivalent. Every chapter published in this book has been scrutinized by our experts. Their significance has been extensively debated. The topics covered herein carry significant findings which will fuel the growth of the discipline. They may even be implemented as practical applications or may be referred to as a beginning point for another development.

The contributors of this book come from diverse backgrounds, making this book a truly international effort. This book will bring forth new frontiers with its revolutionizing research information and detailed analysis of the nascent developments around the world.

We would like to thank all the contributing authors for lending their expertise to make the book truly unique. They have played a crucial role in the development of this book. Without their invaluable contributions this book wouldn't have been possible. They have made vital efforts to compile up to date information on the varied aspects of this subject to make this book a valuable addition to the collection of many professionals and students.

This book was conceptualized with the vision of imparting up-to-date information and advanced data in this field. To ensure the same, a matchless editorial board was set up. Every individual on the board went through rigorous rounds of assessment to prove their worth. After which they invested a large part of their time researching and compiling the most relevant data for our readers.

The editorial board has been involved in producing this book since its inception. They have spent rigorous hours researching and exploring the diverse topics which have resulted in the successful publishing of this book. They have passed on their knowledge of decades through this book. To expedite this challenging task, the publisher supported the team at every step. A small team of assistant editors was also appointed to further simplify the editing procedure and attain best results for the readers.

Apart from the editorial board, the designing team has also invested a significant amount of their time in understanding the subject and creating the most relevant covers. They scrutinized every image to scout for the most suitable representation of the subject and create an appropriate cover for the book.

The publishing team has been an ardent support to the editorial, designing and production team. Their endless efforts to recruit the best for this project, has resulted in the accomplishment of this book. They are a veteran in the field of academics and their pool of knowledge is as vast as their experience in printing. Their expertise and guidance has proved useful at every step. Their uncompromising quality standards have made this book an exceptional effort. Their encouragement from time to time has been an inspiration for everyone.

The publisher and the editorial board hope that this book will prove to be a valuable piece of knowledge for researchers, students, practitioners and scholars across the globe.

List of Contributors

Chiung-Chih Chu and Hsou-min Li
Institute of Molecular Biology, Academia Sinica, Taipei, Taiwan

Sergey Mursalimov, Natalya Permyakova and Elena Deineko
Institute of Cytology and Genetics, Siberian Branch, Russian Academy of Sciences, Novosibirsk, Russia

Andreas Houben and Dmitri Demidov
Leibniz Institute of Plant Genetics and Crop Plant Research, Stadt Seeland, Germany

Kiyotaka Nagaki, Naoki Yamaji and Minoru Murata
Applied Genomics Unit, Institute of Plant Science and Resources, Okayama University, Kurashiki, Japan

Keisuke Tanaka and Hisato Kobayashi
NODAI Genome Research Center, Tokyo University of Agriculture, Setagaya, Japan

Yan Cheng, Han Liu, Yongpeng Li, Yuanyuan Zhang, Wei Jiang and Yongming Zhou
National Key Laboratory of Crop Genetic Improvement, College of Plant Science and Technology, Huazhong Agricultural University, Wuhan, China

Ling Cao
National Key Laboratory of Crop Genetic Improvement, College of Plant Science and Technology, Huazhong Agricultural University, Wuhan, China
Department of Biochemistry, University of Saskatchewan, Saskatoon, SK, Canada

Sheng Wang and Hong Wang
Department of Biochemistry, University of Saskatchewan, Saskatoon, SK, Canada

Ivan Radin and Gerhard Rödel
Institute for Genetics, Department of Biology, Technische Universität Dresden, Dresden, Germany

Natanael Mansilla
Instituto de Agrobiotecnología del Litoral-Consejo Nacional de Investigaciones Científicas y Técnicas, Universidad Nacional del Litoral, Santa Fe, Argentina

Iris Steinebrunner
Department of Biology, Technische Universität Dresden, Dresden, Germany

Yasunori Ichihashi
RIKEN Center for Sustainable Resource Science, Yokohama, Japan

HirokazuTsukaya
Department of Biological Sciences, Graduate School of Science, The University of Tokyo, Tokyo, Japan Bio-Next Project, Okazaki Institute for Integrative Bioscience, National Institutes of Natural Sciences, Okazaki, Japan

Mingkun Huang and, Yilong Hu
Key Laboratory of South China Agricultural Plant Molecular Analysis and Genetic Improvement, South China Botanical Garden, Chinese Academy of Sciences, Guangzhou, China
University of the Chinese Academy of Sciences, Beijing, China

Xu Liu, Yuge Li and Xingliang Hou
Key Laboratory of South China Agricultural Plant Molecular Analysis and Genetic Improvement, South China Botanical Garden, Chinese Academy of Sciences, Guangzhou, China

Maria Cuacos, F. Chris H. Franklin and Stefan Heckmann
School of Biosciences, University of Birmingham, Birmingham, UK

Patrycja Haniewicz
Laboratory of Structural Biology, Department of Molecular Biology, International Institute of Molecular and Cell Biology, Warsaw, Poland

Davide Floris, Domenica Farci and Maria C. Loi
Laboratory of Photosynthesis and Photobiology, Department of Life and Environmental Sciences, University of Cagliari, Cagliari, Italy

Joanna Kirkpatrick
Proteomics Core Facility, European Molecular Biology Laboratory, Heidelberg, Germany

Claudia Büchel
Laboratory of Plant Cell Physiology, Institute of Molecular Biosciences, Goethe-University Frankfurt, Frankfurt am Main, Germany

Matthias Bochtler
Laboratory of Structural Biology, Department of Molecular Biology, International Institute of Molecular and Cell Biology, Warsaw, Poland
Department of Bioinformatics, Institute of Biochemistry and Biophysics, Warsaw, Poland

Dario Piano
Laboratory of Structural Biology, Department of Molecular Biology, International Institute of Molecular and Cell Biology, Warsaw, Poland
Laboratory of Photosynthesis and Photobiology, Department of Life and Environmental Sciences, University of Cagliari, Cagliari, Italy

TaniaIslas-Flores and Marco A. Villanueva
Unidad Académica de Sistemas Arrecifales, Instituto de Ciencias del Mary Limnología, Universidad Nacional Autónomade México, Puerto Morelos, México

Ahasanur Rahman and Hemayet Ullah
Department of Biology, Howard University, Washington, DC, USA

Yalin Liu, Handong Su and Yang Liu
State Key Laboratory of Plant Cell and Chromosome Engineering, Institute of Genetics and Developmental Biology, Chinese Academy of Sciences, Beijing, China
University of Chinese Academy of Sciences, Beijing, China

Jing Zhang and Fangpu Han
State Key Laboratory of Plant Cell and Chromosome Engineering, Institute of Genetics and Developmental Biology, Chinese Academy of Sciences, Beijing, China

James A. Birchler
Division of Biological Sciences, University of Missouriat Columbia, Columbia, MO, USA

Thilo Rühle
Plant Molecular Biology, Department of Biology, Ludwig-Maximilians-University Munich, Munich, Germany

Dario Leister
Plant Molecular Biology, Department of Biology, Ludwig-Maximilians-University Munich, Munich, Germany
Department of Plant and Environmental Sciences, Copenhagen Plant Science Centre, University of Copenhagen, Copenhagen, Denmark

Gea Guerriero and Jean-Francois Hausman
Environmental Research and Innovation, Luxembourg Institute of Science and Technology, Esch-sur-Alzette, Luxembourg

Inés Ezcurra
School of Biotechnology, Division of Industrial Biotechnology, KTH Royal Institute of Technology, Alba Nova University Center, Stockholm, Sweden

Jimi C. Miller
Department of Molecular Biophysics and Biochemistry, Yale University, New Haven, CT, USA

William R. Chezem and Nicole K. Clay
Department of Molecular, Cellular and Developmental Biology, Yale University, New Haven, CT, USA

Sophia Henry, Fanchon Divol, Mathilde Bettembourg, Charlotte Bureau, Emmanuel Guiderdoni, Christophe Périn and AnneDiévart
CIRAD, UMR AGAP, Montpellier, France

Erica-Ashley Jaipargas, Kiah A. Barton, Neeta Mathur and Jaideep Mathur
Laboratory of Plant Development and Interactions, Department of Molecular and Cellular Biology, University of Guelph, ON, Canada

Jian-Ping Huang
College of Life Science, Northeast Agricultural University, Harbin, China
Department of Biology, University of North Carolina, Chapel Hill, NC, USA

Meral Tunc-Ozdemir
Department of Biology, University of North Carolina, Chapel Hill, NC, USA

Ying Chang
College of Life Science, Northeast Agricultural University, Harbin, China

Alan M. Jones
Department of Biology, University of North Carolina, Chapel Hill, NC, USA
Department of Pharmacology, University of North Carolina, Chapel Hill, NC, USA

Natalia Nikonorova
Department of Plant Systems Biology, Flanders Institute for Biotechnology (VIB), Ghent, Belgium
Department of Plant Biotechnology and Bioinformatics, Ghent University, Ghent, Belgium

Lam D. Vu
Department of Plant Systems Biology, Flanders Institute for Biotechnology (VIB), Ghent, Belgium
Department of Plant Biotechnology and Bioinformatics, Ghent University, Ghent, Belgium
Department of Medical Protein Research, Flanders Institute for Biotechnology (VIB), Ghent University, Ghent, Belgium
Department of Biochemistry, Ghent University, Ghent, Belgium

Nathan Czyzewicz
Division of Plant and Crop Sciences, School of Biosciences, University of Nottingham, Lough borough, UK

Kris Gevaert
Department of Medical Protein Research, Flanders Institute for Biotechnology (VIB), Ghent University, Ghent, Belgium
Department of Biochemistry, Ghent University, Ghent, Belgium

Ive De Smet
Department of Plant Systems Biology, Flanders Institute for Biotechnology (VIB), Ghent, Belgium
Department of Plant Biotechnology and Bioinformatics, Ghent University, Ghent, Belgium
Division of Plant and Crop Sciences, School of Biosciences, University of Nottingham, Lough borough, UK
Center for Plant Integrative Biology, University of Nottingham, Lough borough, UK

Ladislav Dokládal and Eva Sýkorová
Mendel Centre for Plant Genomics and Proteomics, Central European Institute of Technology and Faculty of Science, Masaryk University, Brno, Czech Republic
Institute of Biophysics-Academy of Sciences of the Czech Republic v.v.i., Brno, Czech Republic

David Honys and Rajiv Rana
Institute of Experimental Botany-Academy of Sciences of the Czech Republic v.v.i., Prague, Czech Republic

Lan-Ying Lee and Stanton B. Gelvin
Department of Biological Sciences, Purdue University, West Lafayette, IN, USA

Hsiu-An Chu and Yi-Fang Chiu
Institute of Plant and Microbial Biology-Academia Sinica, Taipei, Taiwan

Shelley R. Hepworth
Department of Biology, Institute of Biochemistry, Carleton University, Ottawa, ON, Canada

Véronique A. Pautot
Institut Jean-Pierre Bourgin, Institut National dela Recherche Agronomique, Agro Paris Tech, CNRS, Université Paris-Saclay, Versailles, France

Lauriane Simon, Maxime Voisin, Christophe Tatout and Aline V. Probst
CNRS UMR6293, INSERM U1103, Clermont University, GReD, Aubière, France

Fangwei Jiang, Zhenhua Feng and Jian Zhu
Department of Molecular and Cell Biology, School of Life Science and Technology, Tongji University, Shanghai, China

Hailiang Liu
Translational Center for Stem Cell Research, Tongji University School of Medicine, Tongji Hospital, Shanghai, China

Verónica Parra-Vega, Patricia Corral-Martínez, Alba Rivas-Sendraand and Jose M. Seguí-Simarro
Cell Biology Group, COMAV Institute, Universitat Politècnica de Valéncia, Valencia, Spain